T0211749

Communications in Computer and Information Science 1090

More information about this series at http://www.springer.com/series/7899

Igor Bykadorov · Vitaly Strusevich · Tatiana Tchemisova (Eds.)

Mathematical Optimization Theory and Operations Research

18th International Conference, MOTOR 2019 Ekaterinburg, Russia, July 8–12, 2019 Revised Selected Papers

Editors
Igor Bykadorov
Sobolev Institute of Mathematics
Novosibirsk, Russia

Vitaly Strusevich
University of Greenwich
London, UK

Tatiana Tchemisova
University of Aveiro
Aveiro, Portugal

ISSN 1865-0929 ISSN 1865-0937 (electronic)
Communications in Computer and Information Science
ISBN 978-3-030-33393-5 ISBN 978-3-030-33394-2 (eBook)
https://doi.org/10.1007/978-3-030-33394-2

This Springer imprint is published by the registered company Springer Nature Switzerland AG
The registered company address is: Gewerbestrasse 11, 6330 Cham, Switzerland

- Prof. Oleg Khamisov (Melentiev Energy Systems Institute SB RAS, Russia): "The Fundamental Role of Concave Programming in Continuous Global Optimization"
- Prof. Alexander Kononov (Sobolev Institute of Mathematics, Russia): "Primal-dual Method and Online Problems"
- Prof. Nenad Mladenovic (Mathematical Institute SANU, Serbia): "Solving Non-linear System of Equations as an Optimization Problem"
- Prof. Evgeni A. Nurminski (Far Eastern Federal University, Russia): "Projection Problems and Problems with Projection"
- Prof. Alexander Strekalovsky (Matrosov Institute for System Dynamics and Control Theory SB RAS, Russia): "Modern Methods of Non-convex Optimization"

MOTOR 2019 was a successor of the following well-known series of international and Russian conferences, which were previously organized in Ural, Siberia, and the Far East regions of the Russian Federation:

- Baikal International Triennial School Seminar on Methods of Optimization and Their Applications (BITSS MOPT) was established in 1969 by academician N. N. Moiseev; the 17th event in this series was held in 2017, in Buryatia (http://isem.irk.ru/conferences/mopt2017/en/index.html)
- All-Russian Conference on Mathematical Programming and Applications, (MPA) was established in 1972 by academician I. I. Eremin; the 15th conference in this series was held in 2015, near Ekaterinburg (http://mpa.imm.uran.ru/96/en)
- International Conference on Discrete Optimization and Operations Research, (DOOR) was organized nine times from 1996, and the most recent event was held in 2016 in Vladivostok (http://www.math.nsc.ru/conference/door/2016/)
- International Conference on Optimization Problems and Their Applications, (OPTA) has been organized regularly in Omsk since 1997, the 7th event in this series was held in 2018 (http://opta18.oscsbras.ru/en/)

Starting from different origins, today these conference series have grown very close to each other, having much in common in their research topics, scientific community, and organizers. Therefore, this year the united Program Committee (PC) decided to organize a joint meeting inheriting the long history of all the events and to call it the 18th International Conference on Mathematical Optimization Theory and Operations Research (MOTOR). This name will be given to the subsequent conferences of the chain, and the 19th MOTOR conference will take place in July 2020 at some beautiful place near Novosibirsk, Russia.

Following the tradition, the main conference scope includes but is not limited to mathematical programming, bi-level and global optimization, integer programming and combinatorial optimization, approximation algorithms with theoretical performance guarantees and approximation schemes, heuristics and meta-heuristics, optimal control and game theory, optimization problems in function approximation, optimization in machine learning and data analysis, and valuable practical applications to operations research and economics.

In response to the call for papers, MOTOR 2019 received 232 submissions. Out of 170 full papers considered for reviewing (62 abstracts and short communications were excluded because of formal reasons), 48 papers were selected by the PC for publication

Preface

This volume contains the refereed and selected papers presented at the 18th International Conference on Mathematical Optimization Theory and Operations Research (MOTOR 2019) (http://motor2019.uran.ru) held during July 8–12, 2019, in the picturesque Obukhovky resort near Ekaterinburg, Russia. This scientific forum brought together a wide research community in the fields of mathematical programming and global optimization, discrete optimization, complexity theory and combinatorial algorithms, optimal control and games, and their applications to relevant practical issues that require the use of operations research, mathematical economics, and data analysis.

The conference featured ten invited lectures:

- Prof. Olga Battaïa (ISAE-SUPAERO, France): "Decision Under Ignorance: A Comparison of Existing Criteria in a Context of Linear Programming"
- Prof. Oleg Burdakov (Linköping University, Sweden): "Node Partitioning and Cycles Creation Problem"
- Prof. Christoph Dürr (Sorbonne Université, France): "Bijective Analysis of Online Algorithms"
- Prof. Alexander Grigoriev (Maastricht University, The Netherlands): "A Survey on Possible and Impossible Attempts to Solve the Treewidth Problem via ILPs"
- Prof. Mikhail Kovalyov (United Institute of Informatics Problems NASB, Belarus): "No-Idle Scheduling of Unit-Time Jobs with Release Dates and Deadlines on Parallel Machines"
- Prof. Vadim Levit (Ariel University, Israel): "Critical and Maximum Independent Sets Revisited"
- Prof. Bertrand M. T. Lin (National Chiao Tung University, Taiwan): "An Overview of the Relocation Problem"
- Prof. Natalia Shakhlevich (University of Leeds, UK): "On a New Approach to Optimization Under Uncertainty"
- Prof. Angelo Sifaleras (University of Macedonia, Greece): "Exterior Point Simplex-Type Algorithms for Linear and Network Optimization Problems"
- Prof. Vitaly Strusevich (University of Greenwich, UK): "Design of Fully-Polynomial Time Approximation Schemes for Non-linear Boolean Programming Problems"

The following seven tutorials were given by the outstanding scientists:

- Prof. Tatjana Davidović (Mathematical Institute of the Serbian Academy of Sciences and Arts, Serbia): "Distributed Memory-Based Parallelization of Metaheuristic Methods"
- Prof. Stephan Dempe (TU Bergakademie Freiberg, Germany): "Bilevel optimization: The Model and its Transformations"

in the first volume of proceedings (published in Springer LNCS, Vol. 11548). Out of the remaining submissions the PC selected 44 revised papers for publication in this volume. Thus, the acceptance rates for the two volumes are about 28% and 36% respectively. Each submission was reviewed by at least three PC members or invited reviewers, experts in their fields, in order to supply detailed and helpful comments.

We would like to thank all the authors for their submissions, as well as all members of the PC and external reviewers for their efforts in providing exhaustive reviews. We thank our sponsors, the Russian Foundation for Basic Research, Higher School of Economics (Campus Nizhny Novgorod), Ural Federal University, and Novosibirsk State University. In addition, we are grateful to Alfred Hofmann, Aliaksandr Birukou, Anna Kramer, and their colleagues from Springer LNCS and CCIS editorial board for their kind and helpful support.

September 2019

Igor Bykadorov
Vitaly Strusevich
Tatiana Tchemisova

Organization

Program Committee Chairs

Michael Khachay	Krasovsky Institute of Mathematics and Mechanics, Russia
Yury Kochetov	Sobolev Institute of Mathematics, Russia
Panos M. Pardalos	University of Florida, USA

Program Committee

Alexander Afanasiev	IITP RAS, Russia
Edilkhan Amirgaliev	Suleyman Demirel University, Kazakhstan
Anatoly Antipin	Dorodnicyn Computing Centre FRC CSC RAS, Russia
Adil Bagirov	Federation University Australia, Australia
Evripidis Bampis	Sorbonne Université, France
Olga Battaïa	ISAE-SUPAERO, France
Vitaly I. Berdyshev	Krasovsky Institute of Mathematics and Mechanics, Russia
Vladimir Beresnev	Sobolev Institute of Mathematics, Russia
René van Bevern	Novosibirsk State University, Russia
Givi Bolotashvili	Georgian Technical University, Georgia
Oleg Burdakov	Linköping University, Sweden
Sergiy Butenko	Texas A & M University, USA
Igor Bykadorov	Sobolev Institute of Mathematics, Russia
Tatjana Davidović	Mathematical Institute SANU, Serbia
Vladimir Deineko	Warwick University, UK
Stephan Dempe	Freiberg University, Germany
Alexander Dolgui	IMT Atlantique, France
Anton Eremeev	Dostoevsky Omsk State University, Russia
Adil Erzin	Novosibirsk State University, Russia
Yuri G. Evtushenko	Dorodnicyn Computing Centre FRC CSC RAS, Russia
Fedor Fomin	University of Bergen, Norway
Edward Gimadi	Sobolev Institute of Mathematics, Russia
Alexander Gornov	Matrosov Institute of System Dynamics and Control Theory, Russia
Alexander Grigoriev	Maastricht University, The Netherlands
Mikhail Gusev	Krasovsky Institute of Mathematics and Mechanics, Russia
Milojica Jacimović	University of Montenegro, Montenegro
Vyacheslav Kalashnikov	Instituto Tecnológico y de Estudios Superiores de Monterrey, Mexico
Valeriy Kalyagin	Higher School of Economics, Russia

Vladimir N. Ushakov	Krasovsky Institute of Mathematics and Mechanics, Russia
Vladimir V. Vasin	Krasovsky Institute of Mathematics and Mechanics, Russia

Additional Reviewers

Alexander Adelshin
Vera Afreixo
Alexander L. Ageev
Natalia Aizenberg
Elena Akimova
Ekaterina Alekseeva
Ricardo Almeida
Zhazira Amirgaliyeva
Boris Ananyev
Aram V. Arutyunov
Sergey Astrakov
Pasqualy Avella
Yuri Averboukh
Konstantin Avrachenkov
Artem Baklanov
Nuno Bastos
Arnab Basu
Pavel Borisovsky
Endre Boros
Edmund Burke
Valentina Cacchiani
Gruia Calinescu
S. P. Chakrabarty
Pavel Chebotarev
Ilya Chernykh
A. Yu. Chirkov
Dimitrije D. Ĉvokić
Alexey Danilin
Ivan Davydov
Maxim Demenkov
Vitali Demidenko
Yury Dor
Vladimir Fedorov
Alexander Filatov
Tatiana Filippova
Matteo Fischetti
Sergey Foss
Stefania Funari
Anindita Ganguly
Alexander Gasnikov
Mikhail Gomojunov
Danijel Grahovac
Tatiana Gruzdeva
Mikhail Gudyma
Jae Jin Hwang
Alexei Ignatov
Victor Il'ev
Evgeny Ivanko
Sergey Ivanov
Viktor Izhutkin
Igor' Izmest'ev
Slobodan Jelić
Uwe Kahler
Yuri Kan
Igor Kandoba
Daniel Karapetyan
Margarita Karaseva
Lev Kazakovtsev
Farid Khan
Vladimir Khandeev
Dmitry Khlopin
Elena Khoroshilova
Anatoly Kleimenov
Xenia Klimentova
Konstantin Kobylkin
Sergey Kokovin
Polina Kononova
Elena Kostousova
Julia Kovalenko
Igor Kozin
Denis Krotov
Konstantin Kudryavtsev
Ivana Kuzmanović Ivičić
Sergey Lavlinskii
Pavel Lebedev
Seokjun Lee
Anna Lempert
Tatyana Levanova
Snježana Majstorović
Vittorio Maniezzo
Natalia Martins
Igor Masich
Domagoj Matijević
Oxana Matviychuk
Andrey Melnikov
Elena Musatova
Andrey Naumov
Katherine Neznakhina
Nataliia Obrosova
Timm Oertel
Yuri Ogorodnikov
Andrei Orlov
Idowu Ademola Osinuga
Anna Panasenko
Ilya Panfilov
Artem Panin
Sophie Parragh
Valerii Patsko
Jiming Peng
Alexander Petunin
Alexander Plakhov
Roman Plotnikov
Alexander Plyasunov
Nick Pogodaev
Dmitry Pokrovsky
Bernhard Primas
Franz Rendl
Anna Rettieva
Aleksandr Rogozin
Evgeny Rudoy
Pavel Ruzankin

Ivan Ryzhikov
Yaroslav Salii
Marina Sandomirskaia
Alexander Semenov
Daehee Seo
Dmitry Serkov
Vladimir Servakh
Alexander Sesekin
Alexander Shapiro
Jhilakshi Sharma
Vladimir Shenmaier
Alexander Sidorov
Denis Sidorov
Konstantin Siemenikhin
Cristiana Silva
Ruslan Simanchev
Gaurav Singh
Evgeny Skvortsov
Andrei Sleptchenko
Gueorgui Smirnov
Olga Sokolova
Evgenii Sopov
Stepan Sorokin
Vladimir A. Srochko
Vasile Staicu
Predrag Stanimirovic
Vladimir Stanovov
Maxim Staritsyn
Dmitry Stashkov
Fedor Stonyakin
Alena Stupina
Alexander Tarasyev
Galina Timofeeva
Paul Tochilin
Zoran Tomljanović
Ya-Chih Tsai
Olga Tsekhan
Oxana Tsidulko
Yury Tsoy
Inna Urazova
Dragan Urošević
Igor Vasilyev
Stefan Voß
Gyung Soo Woo
Sergei Yakovlev
Elena Yanovskaya
Alexander Yurin
Gennady Zabudsky
Vyacheslav Zalyubovsky
Lidia Zaozerskaya
Vannel Zeufack
Nikolai Zolotykh
Alexander Zyryanov

Industry Section Chair

Alexander Kurochkin — Sobolev Institute of Mathematics, Russia

Organizing Committee

Michael Khachay — Krasovsky Institute of Mathematics and Mechanics, Russia
Konstantin Kobylkin — Krasovsky Institute of Mathematics and Mechanics, Russia
Nina Kochetova — Sobolev Institute of Mathematics, Russia
Polina Kononova — Sobolev Institute of Mathematics, Russia
Galina F. Kornilova — Krasovsky Institute of Mathematics and Mechanics, Russia
Maria A. Kostina — Krasovsky Institute of Mathematics and Mechanics, Russia
Timur Medvedev — Higher School of Economics, Russia
Katherine Neznakhina — Krasovsky Institute of Mathematics and Mechanics, Russia
Yuri Ogorodnikov — Krasovsky Institute of Mathematics and Mechanics, Russia
Maxim Pasynkov — Krasovsky Institute of Mathematics and Mechanics, Russia
Maria Poberiy — Krasovsky Institute of Mathematics and Mechanics, Russia

Organizers

Krasovsky Institute of Mathematics and Mechanics, Russia
Sobolev Institute of Mathematics, Russia
Melentiev Energy Systems Institute SB RAS, Russia

Sponsors

Russian Foundation for Basic Research, grant no. 19-07-20007
Higher School of Economics (Campus Nizhny Novgorod)
Ural Federal University
Novosibirsk State University

Contents

Mathematical Programming

Operations Research

Optimal Control and Applications

Invited Paper

Committees: History and Applications in Machine Learning

Vladimir D. Mazurov[1,2(✉)] and Ekaterina Yu. Polyakova[2]

[1] Krasovsky Institute of Mathematics and Mechanics, Ekaterinburg, Russia
vldmazurov@gmail.com
[2] Ural Federal University, Ekaterinburg, Russia
ekaterina.y.polyakova@gmail.com

Abstract. The article outlines a brief history and applications of the committee theory. The use of committees in the problems of recognition and optimization is discussed. The application of the committee structures, ambiguous interpretation of non-formalized and contradictory data are given. The ways of rational regard on environmental factors in the context of a lack of resources are considered. The question of the numerical finding of committee structures is discussed, and these results are directly related to the theory of voting. The class of non-classical logics also contains MK-logic (Mazurov, Khachay). This section of non-classical logic includes the works by N. A. Vasiliev, L. Wittgenstein, J. Lukashevich, and Latin American mathematicians having a wrong term in their titles parainconsistent logic. One of the important results achieved by M. Yu. Khachay: For arbitrary positive integers q and k, $k < q$, the minimum estimate of the subsystem power is given that is resolvable by a committee of k-elements for the inconsistent system having a committee of q-elements. Further the history of this field will be mentioned.

Keywords: Committee · Existence · Linear inequalities · Affine case

1 Introduction

In 1965, S.B. Stechkin and I.I. Eremin set a problem of substantiating the necessary and sufficient condition for the existence of a linear inequalities system committee. We solved this problem in 1966, and further ways were opened to continue the research and applications. First, we obtained the results on the conditions for the existence of various modifications of the committees and their applications in economics, engineering, medicine and biology (see. e.g. [17–19]). Further, some of them were extended by Khachay [11,13], Rybin [9,20,24], Kobylkin [15], Gainanov [5] who obtained a number of valuable results in the field of the algorithmic analysis of the committee constructions.

1.1 Some Applications

As the examples of applications the following ones can be referred to:

I. Bykadorov et al. (Eds.): MOTOR 2019, CCIS 1090, pp. 3–16, 2019.
https://doi.org/10.1007/978-3-030-33394-2_1

- Issues of substantiation and application of numerous specific problems of mathematical programming and recognition solved by us with post-optimal solutions analysis;
- tasks of researching operations with the Ural theme (geology, mining, metallurgy);
- algebraic factor analysis;
- factors and their names;
- design issues;
- mathematical structuralism;
- morphology and structural analysis;
- applied structuralism.

1.2 Non-classical Logics

MK-logic (logic of Mazurov-Khachay);
Vasiliev logic;
Lukashevich logic;
parainconsistent logic;
post-optimization analysis and MK-logic.

By its very nature, recognition is associated with mathematical epistemology, mathematical theory of neural networks. Within these disciplines, we voluntarily or involuntarily approach to the questions of the essence of human intelligence, human mentality, formal or informal logic. What is absolutely certain is the fact that we at least imitate human mentality as much as we can. In this case, we use the following scheme of operation of neural networks:

$$? \to S \to A \to R,$$

where "?" denotes the reality that is unknown for us. This reality affects the S-layer—the network sensor unit, and we obtain the pixel array x. It arrives at the input of the block A—the block of associative neurons. The result of the work of these neurons is received on the R-block – the block of resulting neurons. If the experimental data diverges from the ones constructed according to the $S \to A \to R \to x$ scheme, then a network correction is done. A. Novikov investigated the linear correction method, having proved that if there is a solution x, then the result is obtained through a finite number of corrections. Researchers of neural networks unintentionally seek to interpret the work of the network as a learning process of artificial intelligence. It is characteristic that Rosenblatt entitled his book Principles of Neurodynamics [23], Nilson—Learning Machines [21], and Vapnik in his book "Statistical Learning Theory" [25] entitled the philosophical section "Some General Remarks".

Vapnik poses a question why the pattern recognition problem arouses such a great interest among scientists of various specialties. It seems that the answer to this question was obtained both in the works of Zhuravlev [27,28] and in the seminal works of Vapnik (see, e.g. [26]).

The third direction, close to the theory of Yu.I. Zhuravlev, is the analysis of collective decisions—the method of committees.

2 Committee Solutions: Basic Concepts

Let we be given by a ground set X and a family of its non-empty subsets $D_1, D_2, \ldots, D_m$. Let us consider the system of abstract inclusions

$$x \in D_j \quad (j \in [m]), \tag{1}$$

where $[m]$ is the integer segment $\{1, 2, \ldots, m\}$. If $\bigcap_{j=1}^{m} D_j = \varnothing$, system (1) is known as *infeasible*. Any non-empty subset $L \subset [m]$ induces a *subsystem* of system (1). Without loss of generality, we do not distinguish a subset L and the appropriate subsystem. If $D(L) = \bigcap_{j \in L} D_j \neq \varnothing$, the subsystem is known as feasible. Any subsystem maximal by inclusion of system (1) is known as its *maximal feasible subsystem* or m.f.s.

Definition 1. *A finite sequence* $Q = (x^1, \ldots, x^q)$, $x^i \in X$, *such that, for any* $j \in [m]$,

$$|\{i\colon x^i \in D_j\}| > q/2$$

is known as a committee generalized solution of system (1).

The number q is called *a length* of the committee Q, and we state that system (1) has a committee solution (or is solvable by a committee) of length q. A committee of minimal length q (for a given system (1)) is known as its *minimum* committee solution or just a minimum committee.

We can easily represent the set of all committee solutions of the system (1) as follows. Introduce the vector-function

$$\varphi\colon X \to \{-1, 1\}^m, \quad \text{where } \varphi_j(x) = \begin{cases} 1, \text{if } x \in D_j, \\ -1, \text{otherwise.} \end{cases}$$

By its construction, the image $\varphi(X)$ is a finite set. Let $\varphi(X) = \{\varphi^1, \ldots, \varphi^s\}$. Without loss of generality, we can suppose that, the vectors φ^i are incomparable, i.e., for any i_1 and i_2, the equation $\varphi^{i_1} \geq \varphi^{i_2}$ implies $i_1 = i_2$. This implies that any φ^i is a characteristic vector of some m.f.s. of system (1).

According to Definition 1, a finite sequence Q is a committee solution of (1) if and only if, by some permutation, Q can be represented in the form

$$(\underbrace{y^{1,1}, \ldots, y^{1,z_1}}_{z_1}, \ldots, \underbrace{y^{s,1}, \ldots, y^{s,z_s}}_{z_s}), \tag{2}$$

where $\varphi(y^{i,l}) = \varphi^i$ and $z_1, \ldots, z_s$ are nonnegative integers, such that

$$\sum_{i=1}^{s} z_i \varphi^i \geq e = [1, 1, \ldots, 1]^T.$$

It can be easily verified that a sequence Q is a minimum committee of system (1) if and only if the vector z is an optimal solution in the following integer linear program

$$\min \left\{ \sum_{i=1}^{s} z_i\colon \sum_{i=1}^{s} z_i \varphi^i \geq e,\ z \in \mathbb{Z}_+^s \right\}.$$

Evidently, if a system (1) is feasible, its minimum committees are of length 1 and coincide with regular solutions. In other circumstances, any minimum committee has more than 1 entry.

Existence conditions for committee solutions of system (1) can be easily represented in terms of graph and hypergraph of its m.f.s.

Definition 2. *A finite graph $G = (V, E)$ is known as the m.f.s. graph of system* (1) *if its nodeset V consists of index sets $J_1, J_2, \ldots, J_s$ of the system, and*

$$\{J_i, J_j\} \in E \iff J_i \cup J_j = [m].$$

Assertion 1. *Let s be a natural number, $J_1, \ldots, J_{2s-1}$ be a cycle in the m.f.s. graph of system* (1)*, and $x^i \in D(J_i)$. Then, the sequence $Q = (x^1, \ldots, x^{2s-1})$ is a committee solution of system* (1).

To obtain necessary and sufficient conditions we need to introduce a more general notion (see, e.g. [10]).

Definition 3. *A finite hypergraph $G = (V, E)$, whose nodeset V coincides with the family $\{J_1, \ldots, J_s\}$ of index sets of the m.f.s. of system* (1) *such that*

$$\{J_{i_1}, \ldots, J_{i_t}\} \in E \iff \bigcup_{k=1}^{t} = [m],$$

is known as a m.f.s. hypergraph of system (1).

Theorem 1. *Let $\Gamma = (V\Gamma, E\Gamma)$ be a non-empty*[1] *finite hypergraph without multiple edges. Γ is isomorphic to a m.f.c. hypergraph $G = (V, E)$ of an inclusions system*

$$x \in D_j(\Gamma) \quad (j \in [m]) \tag{3}$$

for some $m = m(\Gamma)$ if and only if Γ satisfies the following conditions

$$\text{if } |V\Gamma| > 1, \text{ then } E\Gamma \text{ has no loops} \tag{4}$$

$$(u \in E\Gamma, u \subset w) \Rightarrow w \in E\Gamma. \tag{5}$$

Definition 4. *For a m.f.s. hypergraph $\Gamma = (V\Gamma, E\Gamma)$ and natural numbers σ and τ, a finite sequence of nodes $S = (v_{i_1}, \ldots, v_{i_{\sigma+1}})$ is known as (σ, τ)-simplex in the hypergraph Γ, if the following inclusion*

$$\{v_{i_j} : j \in L\} \in E\Gamma$$

is valid for any $L \subset [\sigma + 1]$, $|L| = \tau + 1$.

The concept of (σ, τ)-simplex takes its origin from the geometric reasonings. It can be easily verified that $(2, 1)$-simplex induces a triangle (a cycle of length 3) in the hypergraph Γ.

Theorem 2. *System* (1) *has a committee generalized solution of length q if and only if its m.f.s. hypergraph has a (σ, τ)-simplex for $\sigma = q-1$ and $\tau = \lfloor (q-1)/2 \rfloor$.*

[1] i.e. $E\Gamma \neq \varnothing$.

3 Game Theoretic Conditions of Committee Existence

In this section, we set forth a series of necessary existence conditions for committee generalized solutions in terms of optimal strategies for the appropriate antagonistic games between *the Nature* and *the Researcher.* For any natural numbers q, $k < q$ and any system of abstract inclusions which has a generalized committee solution of length q, we set forth an attainable relative lower cardinality bound for the maximum subsystem having a committee solution of length k. To obtain the result, we calculate an upper value of the corresponding zero-sum two-player game.

For any setting of the game under discussion, the first player, we will call him or her *Researcher*, tries to choose a finite sequence of length k attempting to enlarge the subsystem resolved by the sequence as a generalized committee solution. The second player, *Nature*, tries to stop these attempts proposing infeasible systems, which can hardly be solved with committees of length less than q.

We demonstrate that, almost always, this game does not have any value when being set in pure strategies, but has a value in mixed ones. Further, we research the asymptotic behavior of our bound in the case of $k = q - h$ for any fixed h and $q \to \infty$. Here, we mainly follow the papers [11,12].

3.1 Problem Statement

By some non-empty ground let we be given a set X and a family of its subsets $D_1, D_2, \ldots, D_m$ defining the following abstract system of inclusions

$$x \in D_j \quad (j \in [m]), \tag{6}$$

and natural numbers q and $k < q$. We suppose that system (6) has a generalized committee solution of length q and is meant to provide a lower cardinality bound for a maximum subsystem (of system (6)) satisfied by a committee of length k.

To obtain the bound, we consider an arbitrary committee solution $Q = (x^1, \ldots, x^q)$ of system (6). In the same manner as in Sect. 2, we assign to the committee Q an incidence $m \times q$-matrix $A = A(Q)$ with entries

$$a_{ji} = \begin{cases} 1, \text{ if } x^i \in D_j \\ -1, \text{otherwise.} \end{cases}$$

Denote by a_j the j-th row of the matrix A. By construction, for any a_j,

$$\sum_{i=1}^{q} a_{ji} \geq 1. \tag{7}$$

Subsequently, we denote the set of $m \times q$-matrices A satisfying equation (7) by $M(q)$.

To any subset $I \subset [q]$ of cardinality k, we assign

- the characteristic vector $\tau = \tau(I) = \sum_{e \in I} e_i^q$, where e_i^q is the i-th orth of the q-dimensional Euclidean space E_q;
- the subcommittee $Q(I) = (x^i : i \in I)$ of length k that is defined by the subset I;
- maximal subsystem of system (6) that is resolved by the subcommittee $Q(ISuffi)$; without loss of generality, subsequently, we do not distinguish by construction this subsystem and its index set $J(I)$, which satisfies the equation $J(I) = \{j : (a_j, \tau(I)) \geq 1\}$;
- the relative cardinality of the subsystem $J(I)$

$$\delta_{q,k}(I, A) = \frac{|J(I)|}{m}.$$

To any matrix $A \in M(q)$, we assign the number

$$\delta_{q,k}(A) = \max\{\delta_{q,k}(I, A) : I \subset [q], |I| = k\}. \tag{8}$$

Let us consider the following zero-sum two-player game $\Gamma = (X, Y, K)$, where the strategy sets of the first and second players are

$$X = \{I \subset [q] : |I| = k\} \text{ and } Y = M(q)$$

respectively and the payoff function is $K(I, A) = \delta_{q.k}(I, A)$. To answer the principal question of this Section, we need to calculate an upper value

$$\delta_{q,k} = \min_{A \in M(q)} \delta_{q,k}(A) = \min_{A \in M(q)} \max_{J \subset [q], |I| = k} \delta_{q,k}(I, A)$$

of this game. For the ensuing constructions, we need the following standard notation:

$\binom{n}{i} = \dfrac{n!}{i!(n-i)!}$	the binomial coefficient
$b(i; n, p) = \binom{n}{i} p^i (1-p)^{n-i}$	the binomial distribution mass function, i.e. probability of i successes in n trials
$\varphi(x) = \dfrac{1}{\sqrt{2\pi}} e^{-x^2/2}$	the standard Gaussian density
$\lfloor x \rfloor$ and $\lceil x \rceil$	the floor and ceiling functions of real argument.

Also, we call real-valued sequences $\{\xi_n\}$ and η_n *asymptotically equivalent* and use the notation $\xi_n \sim \eta_n$, if

$$\lim_{n \to \infty} \frac{\xi_n}{\eta_n} = 1.$$

Let $s = \lceil \frac{q+1}{2} \rceil$ and $t = \lceil \frac{k+1}{2} \rceil$. Similarly to $\tau(I)$, to any subset $S \subset [q]$ of size s, we assign the characteristic vector $\sigma(S) = \sum_{i \in S} e_i^q$. Without loss of generality, we suppose that the vector sets $\Sigma = \{\sigma(S)\}$ and $\Theta = \{\tau(I)\}$ are ordered lexicographically by descending and their elements σ^i and τ^j are labelled by natural numbers $i = 1, 2, \ldots, \binom{q}{s}$ and $j = 1, 2, \ldots, \binom{q}{k}$, respectively.

3.2 Upper Value

In this subsection, we prove the exact formulas for calculating $\delta_{q,k}$. Then, we find a mixed strategy equilibrium for the game in question.

Theorem 3. *For any natural numbers $k < q$, the following equations*

$$\delta_{q,k} = \frac{s}{q} \sum_{l=t-1}^{k-1} \frac{\binom{l}{t-1}\binom{(q-1)-l}{(s-1)-(t-1)}}{\binom{q-1}{s-1}} = \frac{k}{q} \sum_{l=t-1}^{s-1} \frac{\binom{l}{t-1}\binom{(q-1)-l}{(k-1)-(t-1)}}{\binom{q-1}{k-1}} \qquad (9)$$

are valid.

Theorem 3 generalizes several known results. For example, when substituting in (9) $k = t = 1$, we easily obtain the following.

Corollary 1. *Any system* (6) *that has a committee solution of length q contains a feasible subsystem of relative size at least $\lceil \frac{q+1}{2} \rceil / q$.*

3.3 Mixed Strategy Equilibrium

To prove the existence of a mixed strategy equilibrium of the game Γ, we demonstrate that its mixed extension coincides with a mixed extension of a corresponding matrix game.

By construction the pure strategy set of *Researcher* is finite. His or her set $\bar{X}$ of mixed strategies is

$$\bar{X} = \left\{ x \in \mathbb{R}^{\binom{q}{k}} : \sum_{j=1}^{\binom{q}{k}} x_j = 1, x \geq 0 \right\}.$$

Let us consider the pure strategy set $M(q)$ of *the Nature*. We begin with exclusion from $M(q)$ the matrices containing a row with more than $\lceil \frac{q+1}{2} \rceil$ ones, since these matrices are dominated by some other pure strategies of the second player. Then, because of the evident invariance of the payoff function to any permutation of rows in a matrix and simultaneous cloning of them, we proceed with exclusion of one matrix from any couple of equivalent strategies.

3.4 Asymptotic Bounds

In this section we introduce approximate formulas to calculate $\delta_{q,k}$ for large values of q assuming that $k = q - n$ for some fixed integer n.

Theorem 4. *Let λ be an arbitrary natural number.*

(i) If $n = 2\lambda$, then

$$\lim_{q\to\infty} \delta_{q,q-n} = \frac{1}{2}\left(1 + b(\lambda; 2\lambda - 1, 0.5)\right). \tag{10}$$

(ii) If $n = 2\lambda - 1$, then the limit $\lim_{q\to\infty} \delta_{q,q-n}$ does not exist, since

$$\lim_{s\to\infty} \delta_{2s,2(s-\lambda)+1} = \frac{1}{2} + b(\lambda; 2\lambda - 1, 0.5) \tag{11}$$

$$\lim_{s\to\infty} \delta_{2s-1,2(s-\lambda)} = \frac{1}{2}. \tag{12}$$

Corollary 2. *For any fixed natural k and $q > k$ the following equations*

$$\delta_{q,k} \geq 1/2 \quad \textit{and} \quad \lim_{q\to\infty} \delta_{q,k} = \frac{1}{2}$$

are valid.

Corollary 3. *Limit Eqs.* (10) *and* (11) *depend asymptotically on λ as follows*

$$\lim_{q\to\infty} \delta_{q,q-2\lambda} \sim \frac{1}{2} + \frac{1}{\sqrt{2\lambda-1}}\,\varphi\left(\frac{1}{\sqrt{2\lambda-1}}\right)$$

and

$$\lim_{s\to\infty} \delta_{2s,2(s-\lambda)+1} \sim \frac{1}{2} + \frac{2}{\sqrt{2\lambda-1}}\,\varphi\left(\frac{1}{\sqrt{2\lambda-1}}\right).$$

Proof. The given asymptotic equations follow straight-forwardly from the famous De Moivre – Laplace theorem. According to it, the equation

$$b(i; n, p) \sim \frac{1}{\sqrt{np(1-p)}}\,\varphi\left(\frac{i - np}{\sqrt{np(1-p)}}\right).$$

holds uniformly by i. In our case,

$$n = 2\lambda - 1,\ i = \lambda,\ p = \frac{1}{2},$$

which completes the proof.

It is remarkable that the Gaussian approximation for the biggest value of the binomial distribution probability mass function performs well even for rather small values of the parameter λ. In particular, this is fortified by the following numeric data (Tables 1 and 2).

Table 1. Numeric evaluation, part 1

λ	1	2	3	4	5	6	7	8	9	10
$\lim_{q\to\infty} \delta_{q,q-2\lambda}$	0.750	0.688	0.656	0.637	0.623	0.613	0.605	0.598	0.593	0.588
Gaussian appr.	0.742	0.695	0.661	0.640	0.626	0.615	0.606	0.600	0.594	0.589
Relative error	0.011	0.010	0.008	0.005	0.005	0.003	0.003	0.003	0.002	0.002

Table 2. Numeric evaluation, part 2

λ	1	2	3	4	5	6	7	8	9	10
$\lim_{s\to\infty} \delta_{2s,2(s-\lambda)+1}$	1.000	0.875	0.812	0.773	0.746	0.726	0.709	0.696	0.685	0.676
Gaussian appr.	0.984	0.891	0.823	0.781	0.752	0.730	0.713	0.699	0.688	0.678
Relative error	0.016	0.018	0.014	0.010	0.008	0.006	0.006	0.004	0.004	0.003

4 Affine Separating Committees and Ensembles of Linear Classifiers

In this section we study the properties of committee solutions of infeasible system of linear inequalities, which is a special kind of an abstract system of constraints (1). In this case, committee solutions are closely connected with the special type of learning algorithms known in literature as *ensemble learning techniques.*

We begin with the common setting of the two-pattern classification problem (see, e.g. [3]). Suppose, we are given a probabilistic triple $(X \times Y, \mathcal{A}, P)$. Here the feature space X and the set $Y = \{-1, 1\}$ of class labels. In many cases, we can suppose that X is a subset of the n-dimensional Euclidean space E_n. It is required, in the preliminary given family of classifiers $\mathcal{H} \subset [X \to Y]$, to find "the most accurate" $\bar{h}$. Numerous formalizations are admitted due to an accuracy criterion. For simplicity, we focus on the following one

$$\bar{h} = \arg\min\{P(f(x) \neq y) \colon h \in \mathcal{H}\},$$

i.e. on finding a classifier that minimizes the misclassification probability.

If the probabilistic measure is known, this problem has the well-known closed form solution—the Bayes classifier. In the general case studied in this section, when all information about the unknown measure P is exhausted by the finite i.i.d. *training* sample

$$(x_1, y_1), \ldots, (x_m, y_m), \tag{13}$$

the goal is to propose an efficient *learning* algorithm that could find a good approximation to the desired optimal classifier. Within the famous Vapnik-Chervonenkis structural risk minimization learning approach, it is important to design learning algorithms minimizing frequency of misclassification on sample (13) regularized by a capacity of the family $\mathcal{H}$ in terms of its VC-dimension.

We consider the setting of such a learning problem, where it is needed to fit a piecewise linear classifier

$$h(x) = \text{sign} \sum_{j=1}^{k} \alpha_j \, \text{sign}(c_j^T x - d_j) \tag{14}$$

for some non-negative weights α_j, which without loss of generality can be assumed as integers, vectors c_j and real biases d_j. In literature (see, e.g. [13], classifier (14) is called an *affine separating committee*. The motivation to study such classifiers arises from the following points:

(i) for any non-contradictory[2] sample (13), there exists a *perfect* affine committee classifier that makes no classification errors on this sample [17]

(ii) the family of affine committees (14) defined over the n-dimensional feature space E_n and sharing the property $\sum_{j=1}^{k} \alpha_j = q$ has bounded VC-dimension [14].

We continue with the following notation. Let subsets A and B be defined (by sample (13)) as follows

$$A = \{x_i \colon y_i = 1\}, \quad B = \{x_i \colon y_i = -1\}. \tag{15}$$

Any classifier h determined by Eq. (14) can be equivalently represented by the following finite sequence $K = K(h) = (f_1, \ldots, f_q)$, such that $q = \sum_{j=1}^{k} \alpha_j$ and

$$\begin{aligned} f_1(x) \equiv & \ldots \equiv f_{\alpha_1}(x) \equiv c_1^T x - d_1, \\ f_{\alpha_1+1}(x) \equiv & \ldots \equiv f_{\alpha_1+\alpha_2}(x) \equiv c_2^T x - d_2, \\ & \ldots \\ f_{q-\alpha_k+1}(x) \equiv & \ldots \equiv f_q(x) \equiv c_k^T x - d_k. \end{aligned}$$

It can be easily seen that an affine separating committee is a natural generalization of the concept of a separating hyperplane in Euclidean spaces. By means of the famous Hyperplane Separation Theorem (see, e.g. [4]), for any finite sets A and B, the equation

$$\text{conv}(A) \cap \text{conv}(B) = \varnothing$$

presumes the existence of a linear function $f(x) = c^T x - d$ such that the hyperplane $H = \{x \in E_n \colon c^T x - d = 0\}$ separates these sets, i.e., $f(a) > 0$ and $f(b) < 0$ for any $a \in A$ and $b \in B$, respectively. Therefore, if the sets A and B are separable in the regular case, then there exists an affine committee of length 1 that separates them. For the general case, the following criterion is valid.

Theorem 5 ([17]). *Finite subsets $A, B \subset E_n$ can be separated by an affine committee if an only if $A \cap B = \varnothing$.*

[2] For which the condition $x_{i_1} = x_{i_2}$ implies $y_{i_1} = y_{i_2}$.

Problem 1 (Minimum Affine Separating Committee (MASC)). For the given sets $A, B \subset E_n$ it is necessary to find an affine separating committee of the minimum length.

In the conclusion of this section, we give a brief outline of the recent results concerning algorithmic analysis of the MASC problem following the paper [13].

Theorem 6 ([8]). *The Minimum Affine Separating Committee problem is strongly NP-hard and remains intractable even in the case, when*

$$A \cup B \subset \{x \in \{0,1,2\}^n : \|x\|_2 \leq 2\}.$$

The MASC problem does not refer to the APX approximability class, unless $P \neq NP$.

According to Theorem 6, the MASC problem is hard to solve not only in the class of exact algorithms but even with any constant approximation ratio. The following theorem extends this result to the spaces of any fixed dimension.

Theorem 7 ([14]). *The MASC problem is polynomially solvable in the real line and strongly NP-hard in n-dimensional Euclidean space for any fixed dimension $n > 1$.*

It is noteworthy that the claim of Theorem 7 remains valid even in the case, when the set $A \cup B$ is in the *general position*. Usually, a finite set $D \subset E_n$ of size $|D| > n$ is said to be *in general position*, if, for any $D' \subset D$, $|D'| = n + 1$, dimension of the affine hull $\text{aff}(D')$ is equal to n. The special setting of the MASC problem given in the n-dimensional Euclidean space with additional condition on general position of $A \cup B$ is known as MASC-GP(n).

Nearly all known results in the scope of efficient algorithm construction for the MASC problem are based on the following theorem, which can be considered as a specification of Theorem 1 to the case, when training sets are in general position.

Theorem 8 ([9]). *For any finite subsets $A, B \subset E_n$ being in general position, for which $A \cap B = \varnothing$ and $|A \cup B| = m$, there exists an affine separation committee of length*

$$q \leq 2 \left\lceil \frac{\lfloor (m-n) \rfloor}{n} \right\rceil + 1. \tag{16}$$

Two subsequent geometric properties of finite dimensional Euclidean spaces lead to the proof of Theorem 8 mainly.

Property 1. Let Z be a finite subset of E_n and $\varnothing \neq Z' \subset Z$ such that $|Z'| \leq n$ and be in general position. Then, there exist open half-spaces $L_1 = \{x : c_1^T - d_1 < 0\}$ and $L_2 = \{x : c_2^T - d_2 < 0\}$ such that $Z \subset L_1 \cup L_2$ and $Z' \subset L_1 \cap L_2$.

Property 2. Let A and B be non-empty finite subsets of E_n, where $A \cup B$ is in general position and of size $m > n$. Then, for any subsets $A' \subset A$ and $B' \subset B$ of common size $A' \cup B' = n$ there exist $A'' \supseteq A'$ and $B'' \supseteq B'$ and a function $f(x) = c^T x - d$, such that $f(a) > 0$ and $f(b) < 0$ for any $a \in A''$ and $b \in B''$ respectively, and $|A'' \cup B''| \geq \lceil (m+n)/2 \rceil$.

Indeed, the proof of Theorem 8 extends the proof of Theorem 5 and propose a polynomial time approximation algorithm for the problem MASC-GP(n) with time complexity bound $O(m/n \times T_n)$ and the approximation ratio $O(m/n)$. Here, T_n signifies the difficulty of solving a Kramer system of linear equations over n variables.

Remark 1. Bound (16) is tight. In particular, it is attained on sets mentioned in [6] and called *uniformly distributed sets*. In [7], the MASC problem is shown to be polynomially solvable over such sets. The formal definition is as follows

Definition 5. *A finite set $Z = A \cup B \subset E_n$ is known as uniformly distributed (by Gale), if $A \cap B = \varnothing$, $|A \cup B| = n + 2k$ for some natural k and, for any non-trivial hyperplane $H = \{x \in E_n : f(x) \equiv c^T x - d = 0\}$, there exist $A' \subset A$ and $B' \subset B$, $|A' \cup B'| \geq k$, such that $f(a) > 0$ and $f(b) < 0$ for any $a \in A'$ and $b \in B'$, respectively.*

It is known that, for any natural numbers n and k, there exists a uniformly distributed subset $Z = A \cup B \subset E_n$ of size $2k + n$. Thus, in terms of machine learning, it can be stated that, any time, when a training sample is defined by a uniformly distributed subset, the algorithm proposed in the proof of Theorem 5, in the family of the smallest VC dimension, in linear time with respect to the sample length, will obtain a committee classifier (14) without making any classification errors.

Nowadays, the advanced approximation algorithms for the MASC problem are based on the synthesis of the mentioned above approach and the famous Multiple Weights Update technique (see, e.g. [2]). Characteristics of the *BoostedGreedyCommittee* algorithm [13] which has the best known approximation factor are shown in the following theorem.

Theorem 9. *BoostedGreedyCommittee finds an $O(((m \ln m)/n)^{1/2})$-approximate solution for the MASC problem in time $m^{O(n)}$. If, for the given sets A and B, there is a minimum committee $(f_0, f_1, \ldots, f_{q-1})$ such that, for any $t = 1, \ldots, (q-1)/2$ and any $a \in A$, $b \in B$ the following equation*

$$(f_{2t-1}(a) > 0 \vee f_{2t}(a) > 0) \wedge (f_{2t-1}(b) < 0 \vee f_{2t}(b) < 0)$$

is valid, then the approximation factor of this algorithm is $O(\ln m)$.

5 Conclusion

This survey does not pretend to be called exhaustive. We intentionally restrict ourselves to some theoretic results concerning the committees, leaving without considering their applications to numerous practical decision making problem in economy, industry, and medicine forwarding the interested reader to recent papers presenting interesting results in procatice, e.g. [1,5,16,22].

References

1. Akberdina, V., Chernavin, N., Chernavin, F.: Application of the committee machine method to forecast the movement of exchange rates and oil prices. Digest Finan. **23**(1), 108–120 (2018). https://doi.org/10.24891/df.23.1.108
2. Arora, S., Hazan, E., Kale, S.: The multiplicative weights update method: a meta-algorithm and applications. Theory Comput. **8**(1), 121–164 (2012)
3. Bishop, C.M.: Pattern Recognition and Machine Learning. Information Science and Statistics. Springer, New York (2007)
4. Boyd, S., Vandenberghe, L.: Convex Optimization. Cambridge University Press, Cambridge (2009)
5. Gainanov, D., Berenov, D.: Algorithm for predicting the quality of the product of metallurgical production. In: Evtushenko, Y., Khachay, M., Khamisov, O., Kochetov, Y., Malkova, V., Posypkin, M. (eds.) Proceedings of the VIII International Conference on Optimization and Applications (OPTIMA-2017), Petrovac, Montenegro, 2–7 October 2017, pp. 194–200. No. 1987 in CEUR Workshop Proceedings, Aachen (2017). http://ceur-ws.org/Vol-1987/paper29.pdf
6. Gale, D.: Neighboring vertices on a convex polyhedron. Linear Inequalities Relat. Syst. **38**, 255–263 (1956)
7. Khachai, M.: Computational complexity of the minimum committee problem and related problems. Dokl. Math. **73**, 138–141 (2006). https://doi.org/10.1134/S1064562406010376
8. Khachai, M.: Computational and approximational complexity of combinatorial problems related to the committee polyhedral separability of finite sets. Pattern Recogn. Image Anal. **18**(2), 236–242 (2008)
9. Khachai, M., Rybin, A.: A new estimate of the number of members in a minimum committee of a system of linear inequalities. Pattern Recogn. Image Anal. **8**, 491–496 (1998)
10. Khachai, M.: On the existence of majority committee. Discrete Math. Appl. **7**(4), 383–397 (1997). https://doi.org/10.1515/dma.1997.7.4.383
11. Khachai, M.: A relation connected with a decision making procedure based on mojority vote. Dokl. Math. **64**(3), 456–459 (2001)
12. Khachai, M.: A game against nature related to majority vote decision making. Comput. Math. Math. Phys. **42**(10), 1547–1555 (2002)
13. Khachay, M.: Committee polyhedral separability: complexity and polynomialapproximation. Mach. Learn. **101**(1–3), 231–251 (2015). https://doi.org/10.1007/s10994-015-5505-0
14. Khachay, M., Poberii, M.: Complexity and approximability of committee polyhedral separability of sets in general position. Informatica **20**(2), 217–234 (2009)
15. Kobylkin, K.: Constraint elimination method for the committee problem. Autom. Remote Control **73**, 355–368 (2012). https://doi.org/10.1134/s0005117912020130
16. Lebedeva, E., Kobzeva, N., Gilev, D., Kislyak, N., Olesen, J.: Psychosocial factors associated with migraine and tension-type headache in medical students. Cephalalgia **37**(13), 1264–1271 (2017). https://doi.org/10.1177/0333102416678389

17. Mazurov, V.: Committees of inequalities systems and the pattern recognition problem. Kibernetika **3**, 140–146 (1971)
18. Mazurov, V.: Committee Method in Problems of Optimization and Classification. Nauka, Moscow (1990)
19. Mazurov, V., Khachai, M.: Committees of systems of linear inequalities. Autom. Remote Control **65**, 193–203 (2004). https://doi.org/10.1023/b:aurc.0000014716.77510.61
20. Mazurov, V., Khachai, M., Rybin, A.: Committee constructions for solving problems of selection, diagnostics, and prediction. In: Proceedings of the Steklov Institute of Mathematics (suppl. 1), pp. S67–S101 (2002)
21. Nilsson, N.: Learning Machines: Foundations of Trainable Pattern Classifying Systems. McGraw-Hill, New York (1965)
22. Pandey, T.N., Jagadev, A.K., Dehuri, S., Cho, S.B.: A novel committee machineand reviews of neural network and statistical models for currency exchangerate prediction: an experimental analysis. J. King Saud Univ. Comput. Inf. Sci. (2018). https://doi.org/10.1016/j.jksuci.2018.02.010. http://www.sciencedirect.com/science/article/pii/S1319157817303816
23. Rosenblatt, F.: Principles of neurodynamics: perceptrons and the theory of brain mechanisms (1986). https://doi.org/10.1007/978-3-642-70911-1_20
24. Rybin, A.: On some sufficient conditions of existence of a majority committee. Pattern Recorgn. Image Anal. **10**(3), 297–302 (2000)
25. Vapnik, V.N.: Statistical Learning Theory. Adaptive and Learning Systems for Signal Processing, Communications, and Control. Wiley (1998)
26. Vapnik, V.N.: The Nature of Statistical Learning Theory. Statistics for Engineering and Information Science, 2nd edn. Springer, New York (2000). https://doi.org/10.1007/978-1-4757-3264-1
27. Zhuravlev, Y.I.: Correct algebras over sets of incorrect (heuristic) algorithms. I. Cybernetics **13**(4), 489–497 (1977). https://doi.org/10.1007/BF01069539
28. Zhuravlev, Y.I.: Correct algebras over sets of incorrect (heuristic) algorithms. II. Cybernetics **13**(6), 814–821 (1977). https://doi.org/10.1007/BF01068848

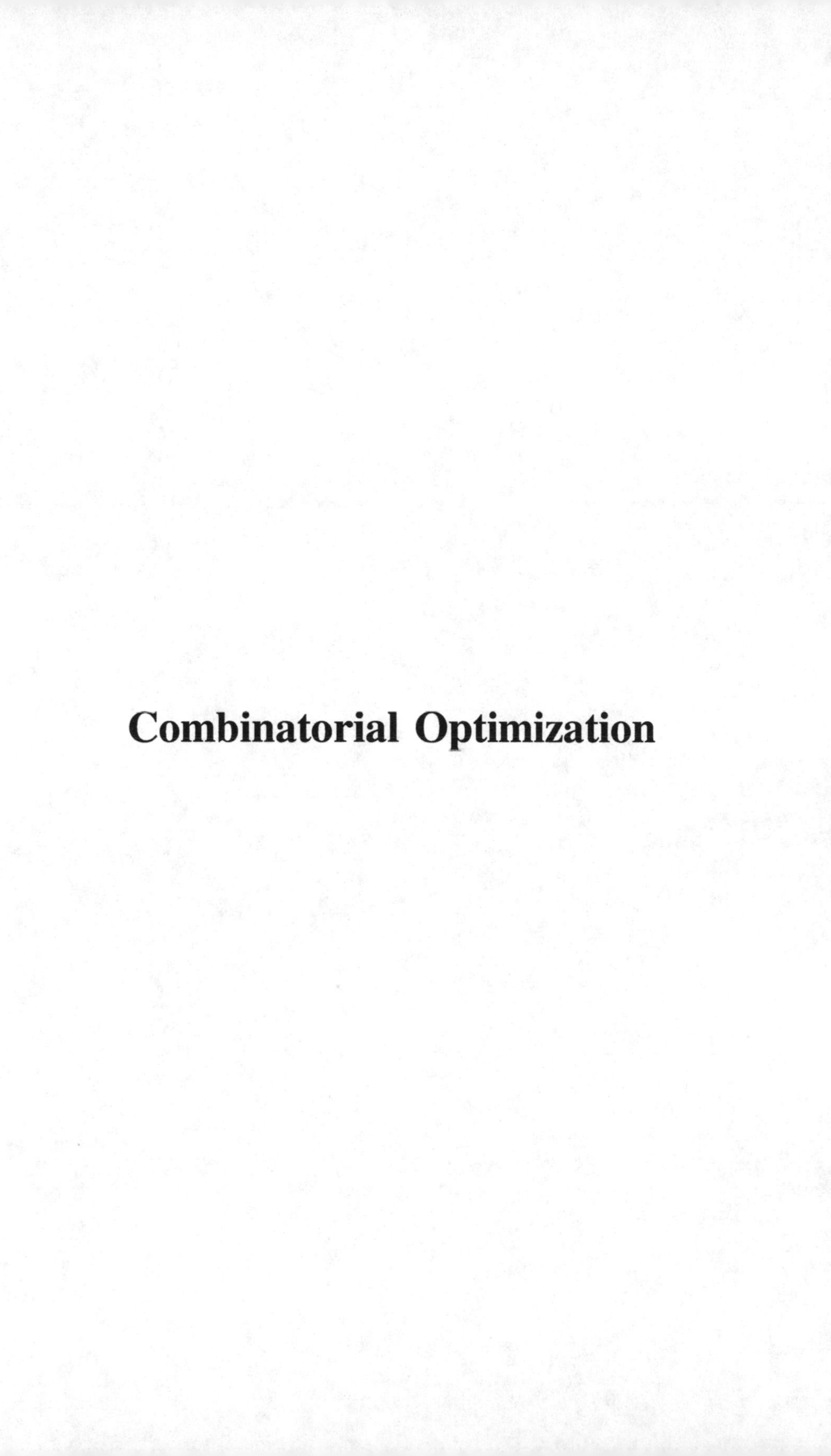

Combinatorial Optimization

An Evolutionary Based Approach for the Traffic Lights Optimization Problem

Ivan Davydov[1,2(✉)] and Daniil Tolstykh[2]

[1] Sobolev Institute of Mathematics, Novosibirsk, Russia
vann.davydov@gmail.com
[2] Department of Mechanics and Mathematics, Novosibirsk State University, Novosibirsk, Russia
daniil.tolstykh.1996@gmail.com
https://www.researchgate.net/profile/Ivan_Davydov

Abstract. We consider the traffic lights optimization problem which arises in city management due to continuously growing traffic. Given a road network and predictions (or statistical data) about the traffic flows through the arcs of this network the problem is to define the offsets and phase length for each traffic light in order to improve the overall quality of the service. The latter can be defined through a number of criteria, such as average speed, average trip duration, total waiting time etc. For this problem, we present an evolutionary based heuristic approach. We use a simulation model on the basis of the SUMO modeling system to evaluate the quality of obtained solutions. The results of numerical experiments on real data confirm the efficiency of the proposed approach.

Keywords: Simulation modeling · Evolutionary algorithm · SUMO · Traffics lights sheduling

1 Introduction

Due to the continuous growth of traffic in urban area a number of various problems arise. In order to avoid serious congestions on the roads only a few solutions may be applied. Among them we can mark three most effective. The first one is to restrict the possibility of personal car owners to enter the urban area of the city. It can be done in different ways: reduced parking space, paid parking in the central area, paid entrance to the central area on working days, etc. Although this is a very efficient way to reduce the traffic jams and the pollution level, this is usually quite unpopular step for the citizens. It can only be considered in case of a perfectly organized public transportation system, as it should provide the carrying service almost equal to a private one. The second approach consists in the development of the road network. New crossings, bridges, roads, additional

Supported by RFBR according to the research project 19-01-00562.

I. Bykadorov et al. (Eds.): MOTOR 2019, CCIS 1090, pp. 19–29, 2019.
https://doi.org/10.1007/978-3-030-33394-2_2

lanes can significantly reduce the congestion. On the cons this approach is quite expensive. Also it can hardly be applied in the historical centers of big cities, with its narrow streets and buildings standing close to each other. The last but not the least way to improve the quality of the city road network is to perform a fine tuning of all of its existing components. This includes a possible redefining of the routes and lane connections on the intersections, assigning special lanes, prohibiting parking in rush hours etc. One out of the most effective steps in this area is the precise tuning of the traffic lights (TL for short). While this procedure is almost free of charge, if made in an optimal way, it provides quite efficient results. Depending on the initial setup and current load of the network in some cases total travel time might be reduced by 20% and even more.

A number of studies have been done on this topic. Majority of the papers can be roughly divided into three sets. The first ones consider the problem of optimal schedule for isolated crossing. In [13] authors propose a dual step approach for fine tuning of TL on the intersection. Using a number of mobility patterns on intersection an off-line scheme is applied first. The resulting optimal schedules are used then in the on-line settings. In [4] it is supposed that flow is unstable and may be different from hour to hour. To overcome this problem, the authors intend to estimate the quality of a traffic light schedule according to the worst case traffic scenario. It is assumed that for each lane the maximal and the minimal flow values are known, and the total deviation from the median values is bounded from above. A proposed dynamic programming approach allows finding an optimal schedule, although the computational time can exceed 10 hours on an average PC. Solution evaluation is made according to the model, described in Highway Capacity Manual [8]. Synchronization of such isolated crossings is sometimes considered as a problem itself. In [11] authors propose a differential evolution approach and investigate the benefits of parallelism for this complex problem. In [12] authors propose two models to tackle traffic signal coordination problems for long arterials and grid networks.

The second direction of research deals with more complicated systems which consist of several intersection. In [3] a cell-type road network is considered (although the approach can be generalized). It is assumed that the traffic loads are known and fixed. Green lights and offsets for each traffic light are under optimization together with the cycle length, unique for all TL objects. To tackle the problem the authors propose a heuristic approach, based on a Bee Colony algorithm. Although the authors claim the effectiveness of the approach, the tests were implemented only on artificially generated data, so it is impossible to compare the solution provided by the approach with the real life behaviour. The common point in all such works is a formulation of the problem in terms of mathematical programming. It means that the quality of the road network is somehow measured via explicit functions, while the possibility itself to create a model, truthful enough, to simulate the real traffic is quite doubtful.

In order to overcome this difficulty, one can use a simulation model. In general a simulation model is a kind of "black box" for a goal function calculation. Given an input, it performs a number of calculations, and provides clear and understandable output. A concept of simulation modelling is widely applied nowadays in operations research, when the problem under consideration deals with complex systems with many agents. The reason is that inside a sophisticated simulation model a lot of different relationships between the agents of the problem might be incorporated. Being applied to the traffic problems, this approach allows simulating every vehicle, pedestrian, lane of the road, traffic light independently from the other objects, assign unique properties to this object and provide a detailed specification of its interaction with other objects. The result is a microscopic simulation, which can easily be tuned in order to provide simulation as close to reality as needed. Among other simulation tools, SUMO - Simulation of Urban Mobility [10] appears to be both quite accessible and popular among researchers. In [7] the authors propose an evolution approach, based on a Particle Swarm heuristic to tackle the traffic light optimization problem. The criterion to be optimized is the function, which depends on a number of flow measurements, like the number of cars that have reached their destination within the simulation scenario, total travel time, total delay, etc. The components of this criterion were obtained during the SUMO simulation run. Test results on the road networks of Malaga and Sevilla showed that a reasonable tuning of the traffic lights may bring up to 15% raise in the efficiency of the road network.

In this paper we continue the efforts aimed at producing a good union of evolution approaches and SUMO. In order to rise the outcome of the simulation model we have incorporated a set of detectors into the simulated road network. These detectors provide additional data on the traffic flows. This information is then used to make a local improvement of the solution during the search process.

The paper is organized as follows. The second section provides a detailed formulation of the problem. The description of the proposed PSO-based approach is given in Sect. 3. Section 4 contains the results of the computational experiment. Section 5 concludes the research and provides the directions for future investigations.

2 Problem Formulation

We consider the traffic light optimization problem as follows. We are given a road network and an information about the traffic flows. The road network is described via set of arcs and edges. The road network is presented in a OpenStreetMaps format, and contains all the information, including the number of lanes, road marking, crosswalks, etc. There is also a set of traffic light objects given. It is assumed that the number of phases, and their order are predescribed and fixed for every TL. The values of the flows are presented as a statistical data and contain the information on the type, speed and the number of vehicles, that are passing through the measuring sensor. We distinguish between two types ov vehicles: long heavy trucks and private cars. The instance of the problem (scenario) consists of a road network with specified intersection, traffic lights, lane connections, crosswalks and road signs. The second term is a set of vehicles. Each vehicle has its own predefined route and a specified departure time. A traffic lights schedule defines the number of the vehicles that cross the intersections and their order. The aim is to define a schedule for the traffic lights, specifying the length of the green light for each phase of each traffic light, cycle length and phase shift time (offset) in order to minimize the cost function.

We assume that each intersection is controlled by one traffic light, although physically there might be more than one device per crossing. Also, it is assumed that the number of different states of TL and its order is fixed. Thus, the setting of an isolated TL can be encoded as an integer vector. The number of components of this vector corresponds to the number of phases. Each phase itself is also a k-vector, where k is the number of connections, controlled by a particular TL. Each connection represents a possible direction of movement. An example, provided in Fig. 1 demonstrates the concept. Here the number of connections is $k = 12$, and they are ordered in a clockwise direction. During the first phase eight of them are "green" while the other four are "red". Thus each phase of the TL can be denoted by a string of k characters. "GGggrrrGGGg" denotes the phase in which 4 connections have a green light, next 4 are red, and the other 4 are green once again. The difference between the small "g" and the capital "G" denotes that, although both connections are allowed and "green", vehicles moving along "G" connections have higher priority. Phase durations, written in a specified order, like $60, 6, 31, 6, 30, 6$ define the regime of the whole TL object. Here 6 phases follow one after another in a cycle. Thus, the length of a whole cycle is defined as a sum of its phases (139 s for the case). While the cycle length can be excluded from the set of variables, it is not the case for the offset. The latter defines the shift between the beginning of cycles of different TL objects. This option is highly likely to be used, while managing big systems, since it introduces another level of interaction between the TL objects. The set of variables related to one intersection consist of the lengths of the phases and the offset value. Solution of the whole problem can be encoded as a vector which contains an ordered list of all TL's phases and offsets, one for each TL. We assume that both the phase length and the offset can take only integer values. The minimal value of the greenred phase is limited by 8 s due to the safety reasons. The maximal value of the cycle is also bounded, and that induces the boundaries on the length of each phase. The offset value is between zero and the total cycle time. While the choice of the offsets is usually considered as a stand-alone problem, it might cause an inappropriate interrelation between the crossings, so in this model finding the offset values is incorporated into an optimization process.

The second part of the problem which should be defined as well is the optimization criterion. A number of different parameters can be used to estimate the efficiency of the TL schedule. Among them are the number of cars that have reached their destination within a scenario, total travel time and total delay. In [7] the authors propose the following fitness function as a measure of quality:

$$fitness = \frac{TT + SW + (NV \cdot ST)}{V^2 + P}.$$

Here TT denotes total travel time of the vehicles, ST stands for simulation time. SW represents the amount of time that vehicles had to spend waiting on

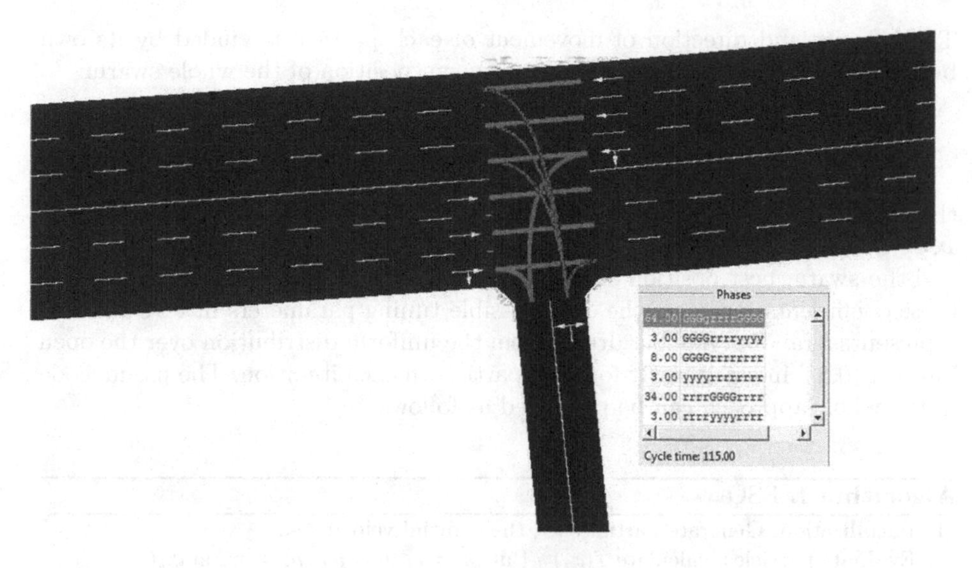

Fig. 1. Phase distribution

the red light. NV is the number of cars that have not reached their destination during the simulation. The denominator is the sum of squared number V of vehicles that have reached their destination within a simulation run and an additional parameter P, standing for phase balance. This parameter is defined as follows

$$P = \sum_{k=0}^{tl} \sum_{j=0}^{ph} s_{k,j} \frac{G_{k,j}}{r_{k,j}},$$

where $G_{k,j}$ is the number of traffic lights in green and $r_{k,j}$ is the number of traffic lights in red in the phase state j of duration $s_{k,j}$ on the intersection k.

3 PSO Based Heuristic

Particle Swarm Heuristic firmly took its place in the list of the most simple and at the same time effective evolutionary based approaches. It simultaneously combines the advantages of the trajectory based approaches together with pluses of the evolutionary methods. It was first proposed in [9] as a concept for optimization of nonlinear functions. The idea of the approach was inspired by the behaviour of the organisms in a bird flock or fish school. Being initially designed for continuous optimization problems nowadays PSO is efficiently applied to the discrete problems as well. It is a population-based iterative approach. In each step of the algorithm a number of particles (represent solutions) form a population, a swarm. Each particle corresponds to an encoded solution x_i. In each iteration k each particle i updates the corresponding solution x_i according to its velocity of movement v_i:

$$x_i := x_i + v_i.$$

The velocity and direction of movement of each particle is guided by its own best known position as well as the best known position of the whole swarm:

$$v_i := \omega v_i + \phi_p U(0,1)(p_i - x_i) + \phi_b U(0,1)(b - x_i).$$

Here p_i denotes the best known position of particle i during the history of the search, b denotes the best known position of the whole swarm. Coefficients ω, ϕ_p and ϕ_b corresponds to the inertia force, and an impact of the personal and the swarm best position on the direction of movement. As a matter of fact these coefficients represent the only possible tuning parameters in PSO. $U(0,1)$ represents a random number drawn from the uniform distribution over the open interval $(0,1)$, independently for each particle in each iteration. The pseudocode of the whole approach can be presented as follows:

Algorithm 1. PSO

1: Initialization. Generate particles x_i, their initial velocity v_i.
2: Evaluate particles, calculate $f(x_i)$. Put $p_i = x_i$, $b = p_m, m = \arg\max_i(p_i)$
3: **while** a termination criterion is not met **do**:
4:     **for** each particle $i = 1, ..., S$ **do**
5:         Pick random numbers: r_p, r_b from $U(0,1)$
6:         Update the particle's velocity:$v_i := \omega v_i + \phi_p r_p (p_i - x_i) + \phi_b r_b (b - x_i)$
7:         Update the particle's position: $x_i := x_i + v_i$
8:         Evaluate particles, calculate $f(x_i)$
9:         **if** $f(x_i) < f(p_i)$ **then**
10:             update the particle's best known position $p_i := x_i$
11:         **end if**
12:         **if** $f(x_i) < f(b)$ **then**
13:             update the swarm's best known position $b := x_i$
14:         **end if**
15:     **end for**
16: **end while**
17: Return b as best solution found

Algorithm 1 describes the pseudo-code of PSO. The algorithm starts by initializing the swarm. Each component of each particle is generated at random using a uniform distribution over a predefined interval. The same is then done with the velocities. Then all particles are evaluated via a simulation run. Then the main cycle is started. During a predefined number of iterations the following cycle is processed: particles velocities are updated and each particles position is updated according to its velocity. There are three forces that drive each particle - its own inertia weight which draws the particle in the same direction, traction for the best position of this particle and for the best position found by all particles in the swarm. The values of the coefficients ω, ϕ_p and ϕ_b should be chosen accordingly. Higher values of ω correspond to the exploration search, while lower values lead to exploitation of the promising region. The balance between ϕ_p and ϕ_b defines the advantage of individual solution over the general one. Optimal values of the parameters can be found during a meta-optimization process, or optimized during the search process. After the movement each particle is evaluated during the simulation run. Then the best known positions of particles and the whole swarm are updated and the process repeats. As no convergence is guaranteed the algorithm terminates after a predefined number of steps. The best particle found so far is taken as an answer.

The idea of a union between heuristics and simulation modelling is not new. But, when dealing with complex systems the simulation run becomes an expensive procedure which requires a lot of computational time. While simulation itself is unavoidable being the only way to estimate the solution quality, it can be used more efficiently. In this work we propose the following local search procedure aimed at improving the outcome from a simulation. Among other options, SUMO allows installing virtual detecting loops on selected sections of road network. These loops serves as detectors and accumulate traffic data on the segment of the road. The quantity of such loops does not affect the simulation runtime, so one is able to use this feature during every run of SUMO. Being placed before and after the intersection these detectors can provide a precise information on the number of cars, that have passed the crossing on green or stopped on red. This information, collected from all the directions of the intersection is then used to improve the schedule of a particular TL. If we observe that one direction is overcrowded and the queue on the stop line is only growing, we redistribute the green time in favor of this direction in a predefined proportion. The schema of this improvement for Energetikov roundabout can be presented as follows.

Algorithm 2. Local improvement

1: Initialization. Collect data on the vehicles flows from all directions Q_1, Q_2, Q_3.
2: **if** $Q_1 \geq Q_2 + Q_3$ **then** add 2s to the 1_{st} phase
3:     **for** $i \leq 2$ **do**
4:         subtract 1sec from the 4_{th} phase with probability $p = Q_2/(Q_2 + Q_3)$,
        subtract 1sec from the 3_{rd} phase otherwise
6:     **end for**
7: **end if**
8: **if** $Q_3 \geq Q_1 + Q_2$ **then** subtract 2s from the 1_{st} phase
9:     **for** $i \leq 2$ **do**
10:         add 1sec to the 3_{rd} phase with probability $p = Q_1/(Q_1 + Q_2)$,
        add 1sec to the 4_{th} phase otherwise
12:     **end for**
13: **end if**
14: **if** $Q_2 \geq Q_1 + Q_3$ **then** subtract 2s from the 4_{th} phase
15:     **for** $i \leq 2$ **do**
16:         add 1sec to the 1_{st} phase with probability $p = Q_1/(Q_1 + Q_3)$,
        add 1sec to the 3_{rd} phase otherwise
18:     **end for**
19: **end if**

4 Numerical Experiments

The proposed approach was implemented in Python environment and tested on a real data instances. We considered the Stancionnaya street and Energeticov roundabout, city of Novosibirsk, Russia, for our setting. The Softline company provided us with the measurements of the real traffic flow on this road network, which is known to be one of the most congestioned part of the city. Together with the traffic flow we have obtained a real-life schedule for all of the TL objects on the segment under consideration. On the basis of this data a set of instances was created, representing different times of the day - morning rush hour, evening jams and mid-day traffic. We considered a 30-min simulation settings. The values of the total traffic flow for a usual Friday on the considered segment are respectively 3820, 3660 and 1672 vehicles. Initial simulation runs showed that while in small to normal traffic conditions the real static schedule performs mostly satisfactory it is not the case for the morning and evening rush hours. Also, we noted that the most overloaded part of the considered network is the Energetikov roundabout. Despite its size, this TL regulated junction is unable to carry its functions during rush hours under the current TL schedule. Figure 2 represents the morning jams on this roundabout.

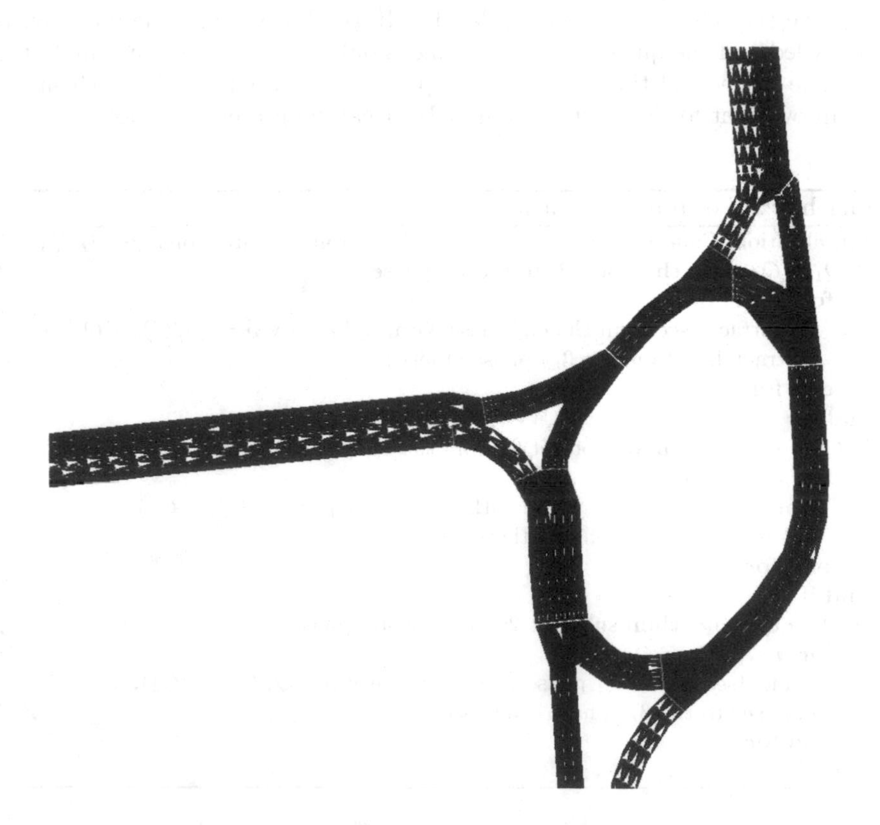

Fig. 2. Morning congestion on the Energetikov roundabout.

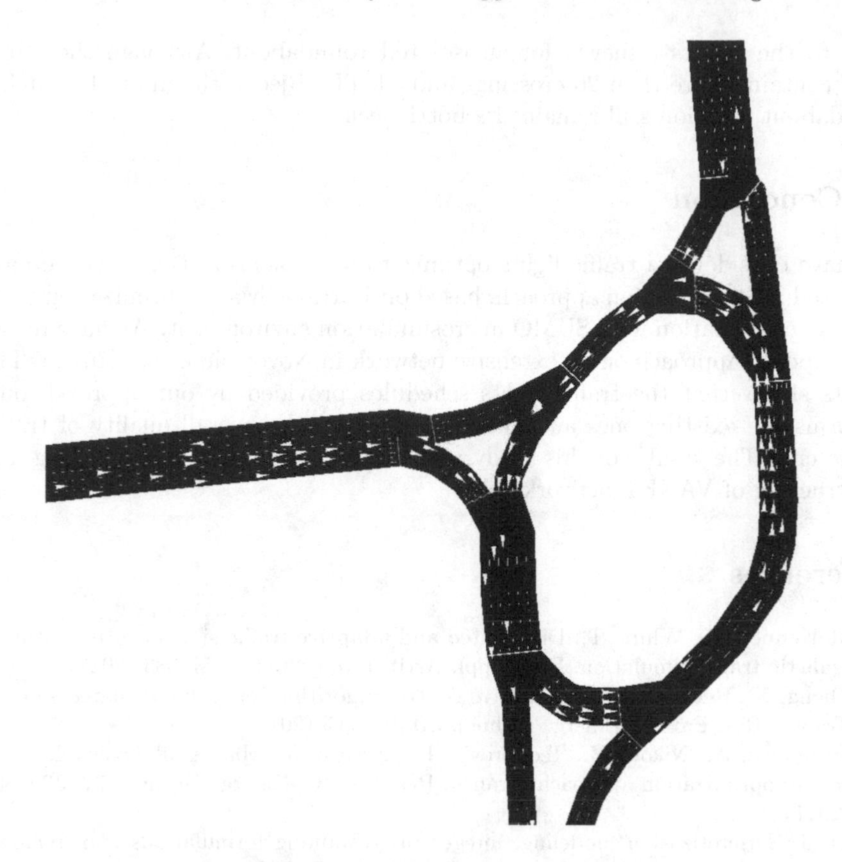

Fig. 3. Optimized scheduling reduces the congestion.

During the first computational experiment we considered only the roundabout itself in a setting of the morning rush hour. For this setting the value of the fitness function with real TL schedule equals 1328 with 2166 vehicles being able to finish their route during the simulation run. We performed 10 runs of our approach, each starting from a randomly generated solution (schedule) and obtained the following results. 9 out of 10 runs converged to the same solution with the value of the fitness function of 964, which gives a 27% improvement. The number of vehicles that arrived to their destination also increased by 13% and reached 2452 vehicles. Similar improvement was also achieved for other instances. The fitness function for the evening hours gained 23%, mid-day case gained 11% rise. Figure 3 demonstrates the resulting improvement. The screenshot is taken in the same moment of the simulation as in Fig. 2. We observe no queue from the south direction and a reduced queue from the west direction.

In the second computational experiment we have considered the whole Stancionnaya str. together with Energeticov roundabout. The results showed that a total improvement of the fitness function value, reached on this segment is close to the results achieved for an isolated roundabout. Although the street itself contains more than 20 crossings and 11 TL objects, the most challenging roundabout junction still remains its bottleneck.

5 Conclusion

We have considered a traffic lights optimization problem. In this work we have proposed an optimization approach, based on Particle Swarm optimization technique in combination with SUMO microsimulation environment. We have tested the proposed approach on an extensive network in Novosibirsk city, Russia. The results shows that the traffic lights schedules provided by our approach outperforms the existing ones and allows to improve the overall quality of traffic in the city. The results of this study may further be used in the planning and constructing of VANET networks.

References

1. McKenney, D., White, T.: Distributed and adaptive traffic signal control within a realistic traffic simulation. Eng. Appl. Artif. Intell. **26**(1), 574–583 (2013)
2. Zheng, X., Recker, W.: An adaptive control algorithm for traffic-actuated signals. Transp. Res. Part C: Emerg. Technol. **30**, 93–115 (2013)
3. Jovanovic, A., Nikoli, M., Teodorovic, D.: Area-wide urban traffic control: a bee colony optimization approach. Transp. Res. Part C: Emerg. Technol. **77**, 329–350 (2017)
4. Li, J.: Discretization modeling, integer programming formulations and dynamic programming algorithms for robust traffic signal timing. Transp. Res. Part C: Emerg. Technol. **19**(4), 708–719 (2011)
5. Coogan, S., Kim, E., Gomes., G., Arcak, M., Varaiya, P.: Offset optimization in signalized traffic networks via semidefinite relaxation. Transp. Res. Part B: Methodol. **100**, 82–92 (2017)
6. Gao, K., Zhang, Y., Sadollah, A., Su, R.: Optimizing urban traffic light scheduling problem using harmony search with ensemble of local search. Appl. Soft Comput. **48**, 359–372 (2016)
7. Garcia-Nieto, J., Alba, E., Carolina Olivera, A.: Swarm intelligence for traffic light scheduling: application to real urban areas. Eng. Appl. Artif. Intell. **25**(2), 274–283 (2013)
8. Transportation Research Record: Highway capacity manual. Technical report, Transportation Research Record (2000)
9. Kennedy, J., Eberhart, R.: Particle swarm optimization. In: Proceedings of IEEE International Conference on Neural Networks, vol. 4, pp. 1942–1948 (1995)
10. Krajzewicz, D., Bonert, M., Wagner, P.: The open source traffic simulation package SUMO. In: RoboCup 2006 Infrastructure Simulation Competition (2006)
11. Souravlias, D., Luquey, G., Albay, E., Parsopoulos, K.E.: Smart traffic lights: a first parallel computing approach. In: 2016 International Conference on Intelligent Networking and Collaborative Systems (INCoS), Ostrawva, pp. 229–236 (2016)

12. Zhang, L., Song, Z., Tang, X., Wang, D.: Signal coordination models for long arterials and grid networks. Transp. Res. Part C: Emerg. Technol. **71**, 215–230 (2016)
13. Angulo, E., Romero, F.P., Garcia, R., Serrano-Guerrero, J., Olivas, J.A.: An adaptive approach to enhanced traffic signal optimization by using soft-computing techniques. Expert Syst. Appl. **38**(3), 2235–2247 (2011)

On Given Diameter MST Problem on Random Input Data

Edward Kh. Gimadi[1,2](✉) and Ekaterina Yu. Shin[2]

[1] Sobolev Institute of Mathematics, 4 Acad. Koptyug Avenue, 630090 Novosibirsk, Russia
gimadi@math.nsc.ru
[2] Novosibirsk State University, 2 Pirogova Str., 630090 Novosibirsk, Russia
katherine15963@gmail.com

Abstract. We give an approximation deterministic algorithm for solving the Random MST with given diameter of directed graph. The problem is NP-hard. Algorithm has a quadratic time complexity. A probabilistic analysis was performed under conditions that edge weights of given graph are identically independent uniformly distributed random variables on an interval with positive ends. Sufficient conditions of asymptotic optimality are presented.

Keywords: Graph · Minimum spanning tree · Asymptotically optimal algorithm · Probabilistic analysis · Performance guarantees · Random inputs · Uniform

1 Introduction

The Minimum Spanning Tree (MST) problem is a one of the classic discrete optimization problems. Given weighted graph $G = (V, E)$, MST is to find a spanning tree of a minimal weight. The polynomial solvability of the problem was shown in the classic algorithms by Boruvka (1926), Kruskal (1956) and Prim (1957). These algorithms have complexity $\mathcal{O}(n^2)$ and $\mathcal{O}(M \log n)$, where $M = |E|$ and $n = |V|$.

It is interesting to note that the mathematical expectation of weight MST on a random graph can be unexpectedly small. So for example on a complete graph with weights of edges from class $UNI(0;1)$, the weight of a MST w.h.p. (with high probability) is close to the constant 2,02... [3]. Similar results were obtained in [1,2].

A generalization of the problem is a diameter-bounded MST problem. The diameter of a tree is the maximum number of edges within the tree connecting a pair of vertices. For this problem, given a graph and a number $d = d_n$, the goal is to find in the graph a spanning tree T_n of minimal total weight having its diameter bounded above to given number d (this problem we call d-BAMST), or from below to given number d (this problem we call d-BBMST).

Both problems are NP-hard in general.

I. Bykadorov et al. (Eds.): MOTOR 2019, CCIS 1090, pp. 30–38, 2019.
https://doi.org/10.1007/978-3-030-33394-2_3

The problem d-BAMST is polynomially solvable for diameters two or three, and NP-hard for any diameter between 4 and $(n-1)$, even for the edge weights equal to 1 or 2 (see in [4], p. 206).

The problem d-BBMST is NP-hard, because its particular case for $d = n-1$ is the problem "Hamiltonian Path" [4]. In the papers [7–9] it was studied a problem d-BBMST on undirected complete graphs. In the earliest of these articles, the asymptotically optimal approach was presented by an algorithm $\widetilde{A}$ with time-complexity $O(n^2)$ for graphs which belong to UNI$(a_n; b_n)$-class. It is a class of complete n-vertex graphs where edge weights are independent identically distributed random variables with uniform distribution on a segment $(a_n; b_n)$.

Next, we recall two important concepts: an algorithm with estimates and an asymptotically optimal algorithm.

By $F_A(I)$ and $OPT(I)$ we denote respectively the approximate (obtained by some approximation algorithm A) and the optimum value of the objective function of the problem on the input I. An algorithm A is said to have *performance guarantees* $(\varepsilon_A(n), \delta_A(n))$ on the set of all random inputs I for the n-sized problem, if

$$\mathbf{Pr}\Big\{F_A(I) > \big(1+\varepsilon_A(n)\big)OPT(I)\Big\} \leq \delta_A(n), \tag{1}$$

where $\varepsilon_A(n)$ is an estimation of *the relative error* of the solution obtained by algorithm A, $\delta_A(n)$ is an estimation of *the failure probability* of the algorithm, which is equal to the proportion of cases when the algorithm does not hold the relative error $\varepsilon_A(n)$ or does not produce any answer at all.

Following [6], we say that an algorithm A is called *asymptotically optimal* on the n-sized problem, if there are exist such performance guarantees that $\varepsilon_A(n) \to 0$ and $\delta_A(n) \to 0$ as $n \to \infty$. Apparently, judging by the review article [11], the first examples of asymptotically optimal algorithms were presented in the works [5,10] for the traveling salesman problem on random input data.

In current paper we study a given-diameter minimum spanning tree problem (d-MST) on the directed complete graph G_n. We introduce a polynomial-time algorithm to solve this problem and provide sufficient conditions for this algorithm to be asymptotically optimal. A probabilistic analysis was performed under conditions that edge weights of given graph are identically independent distributed random variables.

Next, let's proceed to the description of the Algorithm $\mathcal{A}$ for solving the MST problem with given diameter (d-MST).

2 An Algorithm $\mathcal{A}$ for Finding d-MST

Let $d = d_n$ be a parameter of the tree diameter.

Stage 1. On the first stage we build a path $P(d) = (v_0, v_1, \ldots, v_d)$ of d edges. For v_0 start with an arbitrary vertex of a graph and let $P(0) = (v_0)$. On the $(k+1)$-th step, having path $P(k) = (v_0, v_1, \ldots, v_k)$, a vertex v_{k+1} not laying in $P(k)$ is taken, which is closest to the v_k.

Let $P = P(d)$, $V_1 = V \setminus \{v_0, v_1, \ldots, v_d\}$, $V_2 = V \setminus V_1$.

Stage 2. Every vertex $u \in V_2$ is connected by the shortest possible edge with a vertex $v(u) \in P \setminus \{v_0, v_d\}$. By S we denote the set of edges $(u, v(u))$, $u \in V_2$.

As a result we obtain n-vertex spanning tree $T_{\mathcal{A}}$ with a diameter which equal to $d = d_n$ which is an approximate solution of the problem.

Statement 1. The diameter of spanning tree $T_{\mathcal{A}}$ is equal to d, since when connecting any vertex from V_2 to path $P \setminus \{v_0, v_d\}$ during the Stage 2, the diameter does not change.

Further, we denote by $W(G')$ the weight of the subgraph G' of the given graph G.

The weight $W_{\mathcal{A}} = W(T_{\mathcal{A}})$ of the resulting spanning tree is equal to

$$W_{\mathcal{A}} = W(P) + W(S),$$

where $W(P) = \sum_{k=1}^{d} c(v_{k-1}, v_k)$, $W(S) = \sum_{u \in V_2} c(u, v(u))$.

3 Analysis of Algorithm $\mathcal{A}$

The algorithm has polynomial complexity $O(n^2)$, since the construction of the path P in Stage 1 is done by greedy algorithm in time $O((n-d)^2)$, and in the Stage 2 it takes about $d(n-d)$ comparison operations.

A probabilistic analysis we perform under conditions that weights of graph edges are random variables η from the class $\mathrm{UNI}(a_n, b_n)$, namely, are identically independent distributed random variables with uniform distribution on a set (a_n, b_n), $0 < a_n \leq b_n < \infty$. Obviously, normalized variables $\xi = \frac{\eta - a_n}{b_n - a_n} \in \{0; 1\}$ belong to the class $\mathrm{UNI}(0; 1)$.

Further we suppose that the parameter d is defined on the set of values d in the range $\ln n \leq d < n$.

Put $\psi = \frac{1}{e} - 1 \approx 0,63$.

By η_k (correspondingly, ξ_k) we denote random variable equal to minimum over k variables from the class $\mathrm{UNI}(a_n, b_n)$ (correspondingly, from $\mathrm{UNI}(0; 1)$).

According to the description of the algorithm $\mathcal{A}$, the weight $W_{\mathcal{A}}$ of the constructed spanning tree $T_{\mathcal{A}}$ is a random value equal to

$$\begin{aligned} W_{\mathcal{A}} = W(P) + W(S) &= \sum_{k=n-d}^{n-1} \eta_k + \sum_{v \in V_2} \eta_{d-1} \\ &= \sum_{k=n-d}^{n-1} \eta_k + (n-d-1)\eta_{d-1} = (n-1)a_n + (b_n - a_n) W'_{\mathcal{A}}, \end{aligned}$$

where

$$W'_{\mathcal{A}} = \sum_{k=n-d}^{n-1} \xi_k + (n-d-1)\xi_{d-1}.$$

Statement 2.

$$\mathbf{E}W'_{\mathcal{A}} \leq \widetilde{\mathbf{E}W}'_{\mathcal{A}} = \ln\frac{n}{n-d} + \frac{n-d-1}{d}.$$

Since

$$\sum_{k=n-d}^{n-1} \xi_k = \sum_{k=1}^{n-1} \xi_k - \sum_{k=1}^{n-d-1} \xi_k,$$

and $\mathbf{E}\xi_k = 1/(k+1)$, $0 \leq k < n$, we have

$$\mathbf{E}W'_{\mathcal{A}} = \sum_{k=1}^{n}\frac{1}{k} - \sum_{k=1}^{n-d}\frac{1}{k} + \frac{n-d-1}{d} \leq \widetilde{\mathbf{E}W}'_{\mathcal{A}} = \ln\frac{n}{n-d} + \frac{n-d-1}{d}.$$

Statement 3. In the case $d < n\psi$

$$\ln\frac{n-1}{n-d} < 1.$$

Proof. It follows from $d < \psi n$ and $\psi = \frac{1}{e} - 1$:

$$\ln\frac{n}{n-d} < \ln\frac{n}{n-\psi n} = \ln e = 1.$$

From Statements 2 and 3 we have

Statement 4. In the case 1 ($d < n\psi$) the following inequality holds:

$$\mathbf{E}W'_{\mathcal{A}} \leq \widetilde{\mathbf{E}W}'_{\mathcal{A}} = \frac{n-1}{d}.$$

Statement 5. In the case 2 ($\psi n \leq d < n$) the following estimate is correct:

$$\mathbf{E}W'_{\mathcal{A}} \leq \widetilde{\mathbf{E}W}'_{\mathcal{A}} = \ln n.$$

Proof. Show that for all $\psi n \leq d < n$ the following inequality holds:

$$\ln(n-d) \geq \frac{n-d-1}{d}. \tag{2}$$

Indeed:

For $d = n-1$ we have the equality $\ln(n-d) = \frac{n-d-1}{d} = 0$.

For $d = n-2$ the inequality is true, because we have $\ln 2 - \frac{1}{n-2} > 0$,

For $\psi n \leq d \leq n-3$ the inequality is true, since on the one hand $\ln(n-d) \geq \ln 3 > 1$, and on the other hand $\frac{n-d-1}{d} < 1$ because

$$\frac{n-d-1}{d} \leq \frac{n-n\psi-1}{n\psi} = \frac{1}{e-1} - \frac{1}{n\psi} < 1.$$

Taking into account Statement 2 we obtain

$$\mathbf{E}W'_{\mathcal{A}} \leq \ln\frac{n}{n-d} + \frac{n-d-1}{d} = \ln n - \Big(\ln(n-d) - \frac{n-d-1}{d}\Big) \leq \widetilde{\mathbf{E}W}'_{\mathcal{A}} = \ln n.$$

Statement 5 proved.

Lemma 1. *The Algorithm $\mathcal{A}$ For solving the d-MST on entries UNI($a_n; b_n$) has the following estimates of the relative error ε_n and the fault probability δ_n:*

$$\varepsilon_n = (1+\lambda_n)\frac{(b_n - a_n)}{(n-1)a_n}\widetilde{\mathbf{E}W}'_{\mathcal{A}}, \tag{3}$$

where $\lambda_n > 0$:

$$\delta_n = \lambda_n \widetilde{\mathbf{E}W}'_{\mathcal{A}}. \tag{4}$$

Proof.

$$\mathbf{P}\Big\{W_{\mathcal{A}} > (1+\varepsilon_n)OPT\Big\} \le \mathbf{P}\Big\{W_{\mathcal{A}} > (1+\varepsilon_n)(n-1)a_n\Big\}$$
$$= \mathbf{P}\Big\{(n-1)a_n + (b_n - a_n)W'_{\mathcal{A}} > (1+\varepsilon_n)(n-1)a_n\Big\}$$
$$= \mathbf{P}\Big\{W'_{\mathcal{A}} - \mathbf{E}W'_{\mathcal{A}} > \frac{\varepsilon_n(n-1)a_n}{(b_n - a_n)} - \mathbf{E}W'_{\mathcal{A}}\Big\} = \mathbf{P}\Big\{\widetilde{W}'_{\mathcal{A}} > \frac{\varepsilon_n(n-1)a_n}{(b_n - a_n)} - \mathbf{E}W'_{\mathcal{A}}\Big\}$$
$$\le \mathbf{P}\Big\{\widetilde{W}'_{\mathcal{A}} > \frac{\varepsilon_n(n-1)a_n}{(b_n - a_n)} - \widetilde{\mathbf{E}W}'_{\mathcal{A}}\Big\} = \lambda_n \widetilde{\mathbf{E}W}'_{\mathcal{A}} = \delta_n.$$

Further for the probabilistic analysis of Algorithm $\mathcal{A}$ we need the following probabilistic statement

Petrov's Theorem. [12] *Consider independent random variables $X_1, \dots, X_n$. Let there be constants T and $h_1, \dots, h_n > 0$ such that for all $k = 1, \dots, n$ and $0 \le t \le T$*

$$\mathbf{E}e^{tX_k} \le \exp\Big\{\frac{h_k t^2}{2}\Big\}. \tag{5}$$

Define $S = \sum_{k=1}^n X_k$ and $H = \sum_{k=1}^n h_k$. Then

$$\mathbf{Pr}\{S > x\} \le \begin{cases} \exp\{-\frac{x^2}{2H}\}, & \text{for } 0 \le x \le HT, \\ \exp\{-\frac{Tx}{2}\}, & \text{if } x \ge HT. \end{cases}$$

Theorem 1. *Let the parameter $d = d_n$ be defined so that*

$$\ln n \le d < n, \tag{6}$$

Then Algorithm $\mathcal{A}$ solves the problem d-MST on entries UNI($a_n; b_n$) asymptotically optimal w.h.p., if

$$\frac{b_n}{a_n} = \begin{cases} o(d), & \text{for } \ln n \le d < n\psi, \\ o(\frac{n}{\ln n}), & \text{if } n\psi \le d < n. \end{cases} \tag{7}$$

Proof. We will carry a proof for two cases of possible values of the parameter d: $\ln n \leq d < n\psi$ and $n\psi \leq d < n$.

Case 1: $\ln n \leq d < n\psi$.

According to the formula (3) and Statement 4 for the relative error we have

$$\begin{aligned} \varepsilon_n &= (1+\lambda_n)\frac{(b_n - a_n)}{(n-1)a_n}\widetilde{\mathbf{E}W}'_{\mathcal{A}} \\ &= (1+\lambda_n)\frac{(b_n - a_n)}{(n-1)a_n}\frac{(n-1)}{d} \leq (1+\lambda_n)\frac{b_n/a_n}{d}. \end{aligned}$$

Within considered Case 1, we set $\lambda_n = \sqrt{\frac{2\ln n}{n-d-1}}$. Since $\ln n \leq d < n\psi$, it is true: $\lambda_n < 1$, and we see that $\varepsilon_n \to 0$ under condition

$$\frac{b_n}{a_n} = o(d_n).$$

Now using Petrov's Theorem and Statement 4, estimate the fault probability

$$\delta_n = \Pr\left\{W'_{\mathcal{A}} > \lambda_n \widetilde{\mathbf{E}W}'_{\mathcal{A}}\right\} = \Pr\left\{W'_{\mathcal{A}} > \lambda_n \frac{(n-1)}{d}\right\}.$$

Set constants

$$h_k = \begin{cases} \frac{1}{k^2}, & \text{if } n-d \leq k < n, \\ \frac{1}{d^2}, & \text{if } 1 \leq k < n-d. \end{cases}$$

Summing the constants h_k, $k = 1, \ldots, n-1$, we obtain :

$$H = \sum_{k=1}^{n-d-1} h_k + \sum_{k=n-d}^{n-1} h_k = \sum_{k=1}^{n-d-1} \frac{1}{d^2} + \sum_{k=n-d}^{n-1} \frac{1}{k^2} = \frac{n-d-1}{d^2} + \sum_{k=n-d}^{n-1} \frac{1}{k^2}.$$

We see that

$$H \geq \frac{n-d-1}{d^2}.$$

Set $T = d$; $x = \lambda_n \frac{(n-1)}{d}$.

Taking into account the values λ_n, T, H, and x, the following inequality is satisfied:

$$TH \geq \frac{n-d-1}{d} \geq \lambda_n \frac{n-1}{d} = x$$

According to Petrov's Theorem, we have an estimate for the failure probability of the algorithm $\mathcal{A}$:

$$\delta_n = \Pr\{\widetilde{W}'_{\mathcal{A}} > x\} \leq \exp\Big\{-\frac{x^2}{2H}\Big\}.$$

Now show that

$$\frac{x^2}{2H} \geq \ln n.$$

Indeed, since $n - d \geq n(1 - \psi)$, according to the inequality (6), we get

$$\frac{x^2}{2H} \geq \frac{(\lambda_n \frac{(n-1)}{d})^2}{2\frac{(n-d-1)}{d^2}} \geq \frac{\left(\lambda_n \frac{(n-d-1)}{d}\right)^2}{2\frac{(n-d-1)}{d^2}}$$
$$= \frac{(n-d-1)}{2}\lambda_n^2 = \frac{(n-d-1)}{2}\frac{2\ln n}{(n-d-1)} = \ln n.$$

From this it follows that

$$\delta_n = \Pr\{W'_{\mathcal{A}} > x\} \leq \exp\left\{-\frac{x^2}{2H}\right\} \leq \exp(-\ln n) = \frac{1}{n} \to 0,$$

as $n \to \infty$. So in the Case 1 Algorithm $\mathcal{A}$ solves the problem d-MST on entries UNI$(a_n; b_n)$ asymptotically optimal.

Case 2 : $n\psi \leq d < n$.

According to the formula (3) and Statement 5 for the relativ error we have

$$\varepsilon_n = (1+\lambda_n)\frac{(b_n - a_n)}{(n-1)a_n}\widetilde{\mathbf{E}W}'_{\mathcal{A}} = (1+\lambda_n)\frac{(b_n - a_n)}{(n-1)a_n}\ln n \leq (1+\lambda_n)\frac{(b_n/a_n)\ln n}{(n-1)}.$$

Within considered Case 2, we set $\lambda_n = \frac{2}{d}$. From this $\lambda_n < 1$, and we see that $\varepsilon_n \to 0$ under condition

$$\frac{b_n}{a_n} = o\Big(\frac{n}{\ln n}\Big).$$

Now using Petrov's Theorem and Statement 5, estimate the fault probability

$$\delta_n = \Pr\left\{W'_{\mathcal{A}} > \lambda_n \widetilde{\mathbf{E}W}'_{\mathcal{A}}\right\} = \Pr\left\{W'_{\mathcal{A}} > \lambda_n \ln n\right\},$$

Set constants

$$h_k = \begin{cases} \frac{1}{k^2}, \text{ if } n-d \leq k < n, \\ \frac{1}{d^2}, \text{ if } \ 1 \leq k < n-d. \end{cases}$$

Summing the constants h_k, $k = 1, \ldots, n-1$, we obtain:

$$H = \sum_{k=1}^{n-d-1} h_k + \sum_{k=n-d}^{n-1} h_k = \sum_{k=1}^{n-d-1}\frac{1}{d^2} + \sum_{k=n-d}^{n-1}\frac{1}{k^2} = \frac{n-d-1}{d^2} + \sum_{k=n-d}^{n-1}\frac{1}{k^2}.$$

We see that

$$H < \frac{n-d-1}{d^2} + 1,$$

because

$$\sum_{k=n-d}^{n-1}\frac{1}{k^2}\leq\sum_{k=n-d}^{n-1}\frac{1}{k(k-1)}\leq\sum_{k=2}^{n-1}\frac{1}{k(k-1)}=\sum_{k=2}^{n-1}\Big(\frac{1}{k-1}-\frac{1}{k}\Big)$$
$$=\Big(1-\frac{1}{2}\Big)+\Big(\frac{1}{2}-\frac{1}{3}\Big)+\ldots+\Big(\frac{1}{n-2}-\frac{1}{n-1}\Big)<1.$$

Set $T = d$; $x = \lambda_n \ln n$.

Taking into account the values λ_n, T, H, and x, the following inequality is satisfied:

$$TH \leq \frac{n-d-1}{d}+1 \leq \lambda_n \ln n = x.$$

According to Petrov's Theorem, we have an estimate for the failure probability of the algorithm $\mathcal{A}$:

$$\delta_n = \Pr\{\widetilde{W}'_{\mathcal{A}} > x\} \leq \exp\Big\{-\frac{Tx}{2}\Big\}.$$

Since $T = d$, $\lambda_n = \frac{2}{d}$, $x = \lambda_n \ln n$, we have

$$\frac{Tx}{2} = \ln n.$$

From this it follows that

$$\delta_n = \Pr\{W'_{\mathcal{A}} > x\} \leq \exp\Big\{-\frac{Tx}{2}\Big\} = \exp(-\ln n) = \frac{1}{n} \to 0,$$

as $n \to \infty$. So in the Case 2 Algorithm $\mathcal{A}$ also solves the problem d-MST on entries $\mathrm{UNI}(a_n; b_n)$ asymptotically optimal.

We conclude that within the values of the parameter d for both cases, under conditions (7) we have estimates of the relative error $\varepsilon_n \to 0$ and the fault probability $\delta_n \to 0$ as $n \to \infty$.

Theorem 1 is completely proved.

Conclusion. It would be interesting to investigate

(a) the Random d-MST problem on input data with infinite support like exponential or trunketed-normal distribution;
(b) the Random d-MST problem on input data on undirected graph;
(c) the problem of finding several edge-disjoined spanning trees with a diameter which is given or bounded above.

Acknowledgments. The authors are supported by the program of fundamental scientific researches of the SB RAS, project 0314-2019-0014 and by the Ministry of Science and Higher Education of the Russian Federation under the 5–100 Excellence Programme.

References

1. Angel, O., Flaxman, A.D., Wilson, D.B.: A sharp threshold for minimum bounded-depth and bounded-diameter spanning trees and Steiner trees in random networks. arXiv:0810.4908v2 [math.PR], 5 May 2011
2. Cooper, C., Frieze, A., Ince, N., Janson, S., Spencer, J.: On the length of a random minimum spanning tree. Comb. Probab. Comput. **25**(1), 89–107 (2016)
3. Frieze, A.: On the value of a random MST problem. Discrete Appl. Math. **10**, 47–56 (1985)
4. Garey, M.R., Johnson, D.S.: Computers and Intractability. Freeman, San Francisco (1979)
5. Gimadi, E.Kh., Perepelitsa, V.A.: Asymptotic approach for solving the traveling salesman problem. Upravlyaemye Sistemy **12**, 35–45 (1974). (in Russian)
6. Gimadi, E.Kh., Glebov, N.I., Perepelitsa, V.A.: Algorithms with estimates for discrete optimization problems. Problemy Kibernetiki **31**, 35–42 (1974). (in Russian)
7. Gimadi, E.K., Serdyukov, A.I.: A probabilistic analysis of an approximation algorithm for the minimum weight spanning tree problem with bounded from below diameter. In: Inderfurth, K., Schwödiauer, G., Domschke, W., Juhnke, F., Kleinschmidt, P., Wäscher, G. (eds.) Operations Research Proceedings 1999. ORP, vol. 1999, pp. 63–68. Springer, Heidelberg (2000). https://doi.org/10.1007/978-3-642-58300-1_12
8. Gimadi, E.K., Shin, E.Y.: Probabilistic analysis of an algorithm for the minimum spanning tree problem with diameter bounded below. J. Appl. Ind. Math. **9**(4), 480–488 (2015)
9. Gimadi, E.Kh., Istomin, A., Shin, E.: On algorithm for the minimum spanning tree problem bounded below. In: Proceedings DOOR 2016, Vladivostok, Russia, 19–23 September 2016. CEUR-WS, vol. 1623, pp. 11–17 (2016)
10. Perepelitsa, V.A., Gimadi, E.Kh.: Problem of finding the minimum hamiltonian cycle in a weighted graph. Discrete Anal. **15**, 57–65 (1969). (Inst. Math., Novosibirsk) (in Russian)
11. Slominski, L.: Probabilistic analysis of combinatorial algorithms: a bibliography with selected annotations. Computing **28**, 257–267 (1982)
12. Petrov V.V.: Limit Theorems of Probability Theory. Sequences of Independent Random Variables, p. 304. Clarendon Press, Oxford (1995)

Variable Neighborhood Search for the Resource Constrained Project Scheduling Problem

Evgenii N. Goncharov[1,2(✉)]

[1] Sobolev Institute of Mathematics, prosp. Akad. Koptyuga, 4, Novosibirsk, Russia
[2] Novosibirsk State University, str. Pirogova, 1, Novosibirsk, Russia
gon@math.nsc.ru
http://www.math.nsc.ru/

Abstract. We consider the resource-constrained project scheduling problem (RCPSP) with respect to the makespan minimization criterion. The problem accounts for technological constraints of activities precedence together with resource constraints. Activities preemptions are not allowed. The problem with renewable resources is NP-hard in the strong sense. We propose a variable neighborhood search algorithm with two neighborhoods. Numerical experiments based on standard RCPSP test dataset j120 from the PCPLIB library demonstrated that the proposed algorithm produces better results than existing algorithms in the literature for large-sized instances. For some instances from the dataset j120 the best known heuristic solutions were improved.

Keywords: Project management · Resource-constrained project scheduling problem · Renewable resources · Variable neighborhood search

1 Introduction

We consider the resource constrained project scheduling problem (RCPSP) with precedence and resource constraints. The RCPSP can be defined as a combinatorial optimization problem. A partial order on the set of activities is defined with a directed acyclic graph. We know duration, the set and amounts of consumed resources, for every activity. We assume that at every unit interval of the planning horizon $\hat{T}$ the same number of resources is allotted, and the resources are assumed to be unbounded outside the project horizon $\hat{T}$. All resources are renewable. Activities preemptions are not allowed. The objective is to schedule the activities of a project so as to minimize the project makespan. According to the classification scheme proposed in [19] this problem is denoted as $m, 1|cpm|C_{\max}$. According to the classification proposed in [3], this problem is

The work was supported by the program of fundamental scientific researches of the SB RAS, project No. 0314-2019-0014.

I. Bykadorov et al. (Eds.): MOTOR 2019, CCIS 1090, pp. 39–50, 2019.
https://doi.org/10.1007/978-3-030-33394-2_4

denoted as $PS \mid prec \mid C_{\max}$. As a generalization of the job-shop scheduling problem the RCPSP is NP-hard in the strong sense [1] and is actually one of the most intractable classical problems in practice. Worth noting that introducing cumulative resources into the same problem makes the problem solvable with polynomial complexity [9].

It may be conceivable to use optimal methods only for projects of small size. An exact-solution approaches have been developed in [2,8,27,31]. For larger problems, one needs heuristics to get the best solution within a convenient response time, and heuristics remain the best way to solve these problems efficiently. The construction methods are based on a scheduling scheme and an activity selection mechanism made by one or more priority rules or sampling techniques. Papers by Brucker [3], Kolisch and Hartmann [23], Herroelen et al. [18], Kolisch and Padman [21], Kolisch and Hartmann [22], Hartmann and Briskorn [15] survey the RCPSP, its numerous variants, and the solution techniques.

Many local search methods have been proposed to solve the RCPSP. These methods provide, in most cases, solutions better than construction methods as they proceed with a starting feasible schedule generated by one or many construction methods. As the local search methods are more effective than construction methods for large problems, this paper introduces a new hybrid approach combining concepts of tabu search [10] and variable neighborhood search [14] algorithm that uses the activity sequence encoding as a basis to search a neighborhood. This allows fast computation strategies in order to provide very good schedules in a real-time environment. The neighborhood search (NS) algorithm is used to improve one feasible solution by fixing the start times of some activities and rescheduling other activities. Palpant et al. [29] proposed five selection methods. Their numerical experiments showed that the "Block" selection method clearly outperformed other methods. Therefore, we use only the Block selection method to form the sub-problem in all NS operators. In this paper we propose a variable neighborhood search [14]. We use two alternative versions of neighborhoods. In fact, one of them is a modification of that proposed in [29], and the other is a modification of that proposed in [20]. The quality of the proposed algorithm has been examined for dataset j120 from the electronic library PSPLIB [25]. We provide results of the numerical experiments. For the dataset j120 (50000 and 500000 iterations) we have obtained one of the best average deviations from the critical path value. For 8 instances from the dataset j120 we have found the best (previously unknown) heuristic solutions.

2 Problem Setting

The RCPSP problem can be defined as follows. A project is taken as a directed acyclic graph $G = (N, A)$. We denote by $N = \{1, ..., n\} \cup \{0, n+1\}$ the set of activities in the project where activities 0 and $n+1$ are dummy. The latter activities define the start and the completion of the project, respectively. The precedence relation on the set N is defined with a set of pairs $A = \{(i, j) \mid i -$

$-precedes\ j\}$. If $(i, j) \in A$, then activity j cannot start before activity i has been completed. The set A contains all pairs $(0, j)$ and $(j, n+1)$, $j = 1, ..., n$.

We have a set of renewable resources K, for each resource type $k \in K$ there is a constant availability $R_k \in Z^+$ throughout the project horizon $\hat{T}$. Activity j has deterministic duration $p_j \in Z^+$. The profile of resource consumption is assumed to be constant for every activity. So, activity j requires $r_{jk} \geq 0$ units of resource of type k, $k \in K$ at every time instant when it is processed. We assume that $r_{jk} \leq R_k$, $j \in N$, $k \in K$.

Now, we introduce the problem variables. We denote by $s_j \geq 0$ the starting time of activity $j \in N$. Since activities are executed without preemptions, the completion time of activity j is equal to $c_j = s_j + p_j$. We define a schedule S as an $(n+2)$-vector $(s_0, ..., s_{n+1})$. The completion time $T(S)$ of the project corresponds to the moment when the last activity $n+1$ is completed, i.e., $T(S) = c_{n+1}$. We denote by $J(t) = \{j \in N \mid s_j < t \leq c_j\}$ the set of activities which are executed in the unit time interval $[t-1, t)$ under schedule S. The problem is to find a feasible schedule $S = \{s_j\}$ respecting the resource and precedence constraints so that the completion time of the project is minimized. It can be formalized as follows: minimize the makespan of the project

$$T(S) = \max_{j \in N}(s_j + p_j) \tag{1}$$

under constraints

$$s_i + p_i \leq s_j, \quad \forall (i, j) \in A; \tag{2}$$

$$\sum_{j \in J(t)} r_{jk} \leq R_k,\ k \in K,\ t = 1, ..., \hat{T}; \tag{3}$$

$$s_j \in \mathbf{Z}^+, \quad j \in N. \tag{4}$$

Inequalities (2) define activities precedence constraints. Relation (3) corresponds to the resource constraints. Finally, (4) defines the variables in question.

3 Variable Neighborhood Search

3.1 Solution Representation

We represent a feasible solution as an activity list [23]. Feasible solution is encoded by the list of activities $L = (j_0, ..., j_{n+1})$. All lists under consideration are assumed to be compatible with the precedence relations. For an arbitrary list L, the serial decoding procedure (S-SGS) calculates the active schedule $S(L)$ [23]. It is known that there is an optimal schedule among the active schedules. A schedule is called active if the starting times of the activities are such that no activity can be started earlier of its starting time without violating either a precedence relation or a resource constraint. The parallel decoder (P-SGS) sequentially considers increasing moments of time, and schedules a subset of the eligible activities to start at this moment.

3.2 Resource Weights

We use the following heuristic rule to operate with a solution being evaluated. In the preliminary stage, before VNS algorithm has started, we find the degree of scarcity for each resource and rank them, assigning them with a weight. We denote w_k the weight of a resource of type k, $k = 1, ..., K$. If we have resources weights, we can compare them, giving priority to those where the higher priority (scarce) resources are used rationally, i.e., give less surplus of unused resources. We denote the weight v_j of activity j as

$$v_j = \sum_{k \in K} r_{jk} w_k / R_k.$$

We determine the degree of relative scarcity for the resources by solving a relaxed problem. For this purpose, we weaken the renewability condition for the resources and consider a problem with cumulative resources.

$$T(S) = \max_{j \in N}(s_j + p_j) \tag{5}$$

under constraints

$$s_i + p_i \le s_j, \quad \forall (i,j) \in A; \tag{6}$$

$$\sum_{t'=1}^{t} \sum_{j \in J(t')} r_{jk} \le \sum_{t'=1}^{t} R_k, \quad k \in K, \ t = 1, ..., \hat{T}. \tag{7}$$

$$s_j \in \mathbf{Z}^+, \quad j \in N. \tag{8}$$

The fast approximated algorithm to solve problem (5)–(8) is known, it's computational complexity depends on the number n of activities as a function of order n^2. In the case of real-valued activity durations, the algorithm is asymptotically exact with absolute error that tends to zero as the problem dimension grows [7]. In addition, for integer-valued activity durations the exact algorithm is developed [8,9]. In this work we consider a problem with integer-valued activity durations, so we can use any one of these two algorithms to solve the relaxed problem (5)–(8). We choose the first one. By applying this algorithm, in addition to the solution of the relaxed problem, we also get the residue for each cumulative resource that allows us to define the degree of scarcity for all resources: the less is a resource's surplus the scarcer it is. As a final step we apply the resulting resource ranking rule obtained in the relaxed problem to the original problem (1)–(4).

3.3 Block of Activities

For a given feasible schedule $S = (s_0, ..., s_{n+1})$ and a core activity $j = 1, ..., n$, the NS operator reschedules a set of activities, A_j^s, while keeping the start times of all other activities. Let P be a predetermined number of activities that will be rescheduled. The value of P influences the computational time to obtain a neighborhood solution by rescheduling. Smaller value of P usually means fewer activities to be rescheduled and less time to obtain the new schedule. The following Block selection method is used to create A_j^s [29].

$CreateBlock(j, S) \rightarrow A_j^s$.

1. $A_j^s = j$; $b = 0$; create a random order for all activities in $A/\{j\}$. Let i is the first activity in the order.
2. If $s_j - p_i - b \leq s_i \leq s_j + p_j + b$, $A_j^s = A_j^s \bigcup \{i\}$.
3. If $|A_j^s| = P$, go to Step 6.
4. If i is the last activity among the ones not belonging to A_j^s based on the order defined in Step 1, $b = b + 1$.
5. Let i be the next activity among the ones not belonging to A_j^s based on the order defined in Step 1. Go to Step 2.
6. END.

The Block selection method basically selects a set of P activities that are overlapped or close to activity j in a given feasible schedule.

3.4 The Initial Solution

We can use any available method to generate a good initial solution rapidly. The choice of the algorithm for the initial solution is not critical for the local search methods. Gagnon et al. [6] noted that there is some dilemma concerning the choice of the initial solution used by a NS method adaptation. Starting with a very good solution doesn't let enough space to find a significant improvement. On the other side, it may take a long computation time to improve a bad starting solution. We can use, for example, the S-SGS and P-SGS schemes, stochastic methods with the forward-backward improvement procedure (FBI) [32], greedy algorithms. In this paper we use the stochastic greedy algorithm [13] to build up a starting schedule solution.

3.5 Tabu List Management

The NS method exploits the knowledge gained from the solutions considered previously. This knowledge is maintained in a tabu list used as a memory in order to avoid cyclicality, i.e. repeating of recent transformations applied to obtain the solutions under evaluation. Recency tabu tenure is recorded by keeping at each entry of the tabu list, of attributes on the last visited solutions. We use tabu list of the constant length. As a tabu status of an arbitrary solution L we consider the sum of the starting times:

$$TS(L) = \sum_{j=1}^{n} s_j$$

for the schedule $S(L)$.

3.6 Neighborhood A

The first neighborhood $N_A(S)$ is the modification of the scheme proposed in [30]. For a given feasible schedule $S = \{s_0, s_1, ..., s_n, s_{n+1}\}$ and a core activity $j \in A$, we determine the block of activities A_j^s. The NS operator reschedules a set of activities A_j^s, while keeping the start times of other activities. The rescheduling sub-problem is formed by the following steps. We fix the start times of all the activities not belonging to the set A_j^s and release resources used by all the activities from A_j^s in each time period t. The available amount of resource k for activities from A_j^s in period t is R_k minus the resource used by all the activities not belonging to the set A_j^s in period t. Then we derive an earliest start time (EST) and a latest finish time (LFT) for each activity $i \in A_j^s$ as $EST_i = max\{s_l + p_l, \forall\ l \notin A_j^s\ and\ (l,i) \in A\}$ and $LFT_i = max\{s_l, \forall\ l \notin A_j^s\ and\ (l,i) \in A\}$.

(EST_i, LFT_i) defines a time window for activity i that could be rescheduled in order to guarantee that the new schedule is still feasible for all activities that will not be rescheduled. The rescheduling problem is to reschedule all the activities from A_j^s to minimize their makespan while meeting the resource restriction of each period and time window constraints defined by (EST_i, LFT_i).

Palpant et al. [29] used a commercial integer linear programming solver to obtain the optimal solution for the rescheduling problem. Proon and Jin [30] adopts the forward or backward serial scheduling generation schemes (S-SGS) [23] to solve the rescheduling sub-problem. In this paper we propose a modification of this algorithm [30]. In each iteration, a new random vector is produced for the activities $i \in A_j^s$ as a priority list. We order the activities in A_j^s by decreasing their weights v_j. The vector is created iteratively by randomly picking the next activity from the ordered list among all unselected activities whose precedent activities in A_j^s have been selected. Following the priority list, one activity by one is moved to the earliest (latest) start time that is precedence- and resource-feasible and satisfies the time window (EST_i, LFT_i). Once all the activities $i \in A_j^s$ are rescheduled, the activities that do not belong to A_j^s are added to form a complete feasible solution. A global left shift is then performed on all the activities in A to possibly reduce the makespan. The resulting new schedule is compared with the previous solution before applying the NS operator. If the makespan is improved, the resulting schedule replaces the previous schedule and the NS operator stops. If there is no improvement, as long as the number of iterations has not reached a predefined limit, λ, the S-SGS is applied on the schedule with a new random priority list as the next iteration.

3.7 Neighborhood B

As the neighborhood $N_B(S)$ we use the modification of the scheme proposed in [20]. For a given list of activities L (and a correspondent active schedule $S = \{s_0, ..., s_{n+1}\}$) and a core activity $j \in A$, we determine the block of activity A_j^s. If the block contains at least one predecessor of the activity j then we put

A_j^s to be empty. We represent the list L in the form of three consecutive lists $L = A^1, A_j^s, A^2$.

The element L' of the neighborhood $N_B(S)$ is constructed for each activity $j \in A$ by using the non-empty block A_j^s. The list L' is obtained from the list L by the following steps. We fix the start times of all the activities from the set A^1 and release resources used by all the activities from set A^1 in each time period t. We calculate a partial schedule for the activities belonging to set A^1 via the serial decoding procedure. Then we extend the partial schedule by scheduling activities belonging to set A_j^s via the parallel decoding procedure. According to the procedure for each schedule time t we have the corresponding eligible set E_t, i.e. a set of activities which could be started at t without violation of any constraints. There are exponentially many possibilities to select a subset of activities from the eligible set to include into the schedule. We solve the multi-dimensional knapsack problem with objective function maximizing the weighted resource utilization ratio [35]

$$max \sum_{j \in E_t} x_j \sum_{k \in K} \frac{w_k r_{jk}}{R_k}, \tag{9}$$

$$\sum_{j \in E_t} r_{jk} x_j \le R_k - \sum_{j \in J(t)} r_{jk} \quad k \in K, \tag{10}$$

$$x_j \in \{0, 1\}, \quad j \in N. \tag{11}$$

The right-hand side of the restriction is a remaining capacity of the resource type k at the time t. We use Greedy Randomized Adaptive Search Procedures (GRASP) to solve the problem. Note that in the process of solving the problem (9)–(11), the advantage may be gained not by the activities that make the best use of resources, but by those that make the best use of more scarced resources.

Finally, we construct the list L' as follows. We put the activities $\in A_j^s$ into the list L' in non-decreasing order of its starting times in the partial schedule. Remaining activities are listed in the list L' at the same order as in the list L. The schedule $S(L')$ is called the neighbor sample for the schedule S. The set that contains all neighbor samples is called a neighborhood of the schedule S and is denoted by $N_B(S)$.

3.8 Algorithm Outline

Step 1. Generate the initial schedule S, and set $T^* := T(S)$, $S^* := S$. Tabu list TL is set empty.

Step 2. Until the stopping criterion is satisfied, the following is done.
Step 2.1. Choose a neighborhood equally probable.
Step 2.2. Find the neighbor sample S', not prohibited by the tabu list TL.
Step 2.3. If $T(S') < T^*$, then we assume $T^* := T(S')$, $S^* := S'$.
Step 2.4. Update the tabu list TL and set $S := S'$.

As a stopping criteria we consider reaching the maximum number of sequences evaluated, denoted as λ. The value of T^* is the result of the algorithm. If the value of T^* does not change for a certain (predefined) iterations limit, then we change the block size (parameter P). We make this change of parameter P a predefined number of times. Finally, if the value of T^* does not change a predefined number of times, we generate a new initial schedule.

4 Numerical Experiments

The VNS algorithm was coded in C++ in the Visual Studio system and run on a 3.4 GHz CPU and 8 Gb RAM computer under the operating system Windows 7. In order to evaluate the performance of the proposed VNS algorithm, we use the standard set presented in Kolisch and Sprecher [25] referred as j120. These instances are available in the project scheduling library PSPLIB along with their the best-known values. The dataset j120 contains 60 series of instances, 10 instances in each series, 600 instances in total. Each instance considers four types of resources. Three parameters: network complexity (NC), resource factor (RF) and resource strength (RS) are combined together to define the full factorial experimental design. The NC defines the average number of precedence relations per activity. The RF sets the average percent of various resource type demand by activities. The RS measures scarcity of the resources. Zero value of the RS factor corresponds to the minimum need for each resource type to execute all activities while the RS value of one corresponds to the required amount of each resource type obtained from the early start time schedule. The parameter values used to built up these instances for the set j120 are: $NC \in \{1.5, 1.8, 2.1\}$, $RF \in \{0.25, 0.5, 0.75, 1\}$ and $RS \in \{0.1, 0.2, 0.3, 0.4, 0.5\}$. It is known [37] that values of the parameters $RF = 4$, $RS = 0.2$ match hard enough series. Identifiers j12016, j12036, j12056, j12011, j12031, j12051 with $n = 120$ correspond to the series with the largest gap between the best solutions found and the length of the critical path. Each triplet of such identifications matches values $NC = \{1.5; 1.8; 2.1\}$, respectively.

See Kolisch et al. [24] for the process of how the instances were created. The instances can be found in Kolisch and Sprecher [25] and are downloadable at http://www.om-db.wi.tum.de/psplib/.

The measure of the solution quality is the average percent deviation (APD) from the lower bounds obtained by the critical path algorithm [24] for instances in the dataset j120 which optimal solutions are unknown. It is customary to compare the heuristic efficiency by restricting to the same number of schedules evaluated. In Table 1 one can find comparison the VNS algorithm performance with the previous results of experimental evaluation of competitive heuristics for the dataset j120. Limit of schedules λ we set at 50000 and 500000. The scrutiny of the presented results clearly shows the good performance of the proposed VNS algorithm: for $\lambda = 50000$ it was the third, and for $\lambda = 500000$ it shows the second result.

Furthermore, for 8 instances from the dataset j120 we obtained the best (previously unknown) heuristic solutions, they are currently presented in the

Table 1. Average deviations from the critical path for dataset j120.

Algorithm	Reference	APD, %	
		$\lambda = 50000$	$\lambda = 500000$
GA	Goncharov and Leonov [12]	30,50	**29,74**
VNS	This paper	30,56	29,88
Biased random-key GA	Goncalves [11]	32,76	30,08
GANS	Proon and Jin [30]	**30,45**	30,78
ACOSS	Chen et al. [36]	30,56	–
DBGA	Debels and Vanhoucke [5]	30,69	–
GA	Debels and Vanhoucke [5]	30,82	–
GA - Hybrid, FBI	Valls et al. [34]	31,24	30,95
Enhanced SS	Mobini et al. [28]	31,37	–
Scatter search - FBI	Debels et al. [4]	31,57	30,48
GAPS	Mendes et al. [26]	31,44	31,20
GA, FBI	Valls et al. [33]	31,58	–
GA, TS-Path re-linking	Kochetov and Stolyar [20]	32,06	–
GA-Self adapting	Hartmann [17]	33,21	–
GA-Activity list	Hartmann [16]	34,04	–
Sampling-LFT, FBI	Tormos and Lova [32]	35,01	–
SGE-Priority rule, FBI	Goncharov [13]	35,08	–
GA-Priority rule	Hartmann [16]	36,51	–

PSPLIB library. We provide the list of the mentioned instances in Table 2.As one can see from Table 1, the VNS algorithm conceded to the genetic algorithm on the whole dataset j120. But at the same time, as one can see from Table 2, the VNS algorithm has found the previously unknown best heuristic solutions exclusively on the series with the largest gap between the best solutions found and the length of the critical path. Therefore, we can conclude that on such "hard series" of instances, the VNS algorithm shows better results in comparison with the GA algorithm.

Average processing time is 16 s for $\lambda = 50000$ and 150 s for $\lambda = 500000$.

Table 2. List of instances for which new heuristic solutions are obtained.

Dataset	Series	Instances
j120	11	3
j120	16	8
j120	31	4
j120	36	3
j120	51	3, 4
j120	56	7, 8

5 Conclusion

Authors have proposed a variable neighborhood search algorithm for the resource-constrained project scheduling problem with respect to the makespan minimization criterion. We have developed two versions of the neighborhoods. The algorithm uses a heuristic that takes into account the degree of criticality (scarcity) of the resources, which is derived from the solution of the relaxed problem with a constraint on the cumulative resources. We have conducted numerical experiments on sets of instances from the PSPLIB electronic library. The results of the computational experiments suggest that the proposed VNS algorithm is a very competitive heuristic and yields better results than several heuristics presented in the literature. For some instances from the dataset j120 the best known heuristic solutions were improved.

Further studies will be focused on constructing hybrid algorithms for the RCPSP problem.

References

1. Blażewicz, J., Lenstra, J.K., Rinnoy Kan, A.H.G.: Scheduling subject to resource constraints: classification and complexity. Discrete Appl. Math. **5**(1), 11–24 (1983)
2. Brucker, P., Knust, S., Schoo, A., Thiele, O.: A branch and bound algorithm for the resource-constrained project scheduling problem. Eur. J. Oper. Res. **107**, 272–288 (1998)
3. Brucker, P., Drexl, A., Möhring, R., et al.: Resource-constrained project scheduling: notation, classification, models, and methods. Eur. J. Oper. Res. **112**(1), 3–41 (1999)
4. Debels, D., De Reyck Leus, B.R., Vanhoucke, M.: A hybrid scatter search electromagnetism meta-heuristic for project scheduling. Eur. J. Oper. Res. **169**, 638–653 (2006)
5. Debels, D., Vanhoucke, M.: Decomposition-based genetic algorithm for the resource-consrtained project scheduling problem. Oper. Res. **55**, 457–469 (2007)
6. Gagnon, M., Boctor, F.F., d'Avignon, G.: A Tabu Search Algorithm for the Resource-constrained Project Scheduling Problem. ASAC (2004)
7. Gimadi, E.Kh.: On some mathematical models and methods for planning large-scale projects. models and optimization methods. In: Proceedings AN USSR Sib. Branch, Math. Inst., Novosibirsk. Nauka, vol. 10, pp. 89–115 (1988)
8. Gimadi, E.Kh., Goncharov, E.N., Mishin, D.V.: On some implementations of solving the resource-constrained project scheduling problem. Yugoslav J. Oper. Res. **29**(1), 31–42 (2019)
9. Gimadi, E.Kh., Zalyubovskii, V.V., Sevast'yanov, S.V.: Polynomial solvability of scheduling problems with storable resources and deadlines. Diskretnyi Analiz i Issledovanie Operazii, Ser. 2 **7**(1), 9–34 (2000)
10. Glover, F., Laguna, M.: Tabu Search. Kluwer Academic Publishers, Boston (1997)
11. Goncalves, J., Resende, M.G.C., Mendes, J.: A biased random key genetic algorithm with forward-backward improvement for resource-constrained project scheduling problem. J. Heuristics **17**, 467–486 (2011)
12. Goncharov, E.N., Leonov, V.V.: Genetic algorithm for the resource-constrained project scheduling problem. Autom. Remote Control **78**(6), 1101–1114 (2017)

13. Goncharov, E.N.: Stochastic greedy algorithm for the resource-constrained project scheduling problem. Diskret. Anal. Issled. Oper. **21**(3), 10–23 (2014)
14. Hansen, P., Mladenovic, N.: Developments of variable neighborhood search. In: Ribeiro, C., Hansen, P. (eds.) Essays and Surveys of Metaheuristics, pp. 415–440. Kluwer Academic Publishers, Boston (2002)
15. Hartmann, S., Briskorn, D.: A survey of variants and extentions of the resource-constrained project scheduling problem. Eur. J. Oper. Res. **207**, 1–14 (2010)
16. Hartmann, S.: A competitive genetic algorithm for the resource-constrained project scheduling. Naval Res. Logistics. **45**, 733–750 (1998)
17. Hartmann, S.: A self-adaptive genetic algorithm for project scheduling under resource constraints. Naval Res. Logistics. **49**, 433–448 (2002)
18. Herroelen, W., De Reyck, B., Demeulemeester, E.: Resource-constrained project scheduling: a survey of recent developments. Comput. Oper. Res. **25**(4), 279–302 (1998)
19. Herroelen, W., Demeulemeester, E., De Reyck, B.: A classification scheme for project scheduling. In: Weglarz, J. (Ed.) Project Scheduling-Recent Models, Algorithms and Applications, International Series in Operations Research and Management Science, vol. 14(1), pp. 77–106. Kluwer Academic Publishers, Dordrecht (1998)
20. Kochetov, Yu., Stolyar, A.: Evolutionary local search with variable neighborhood for the resource-constrained project scheduling problem. In: Proceedings of 3rd International Workshop of Computer Science and Information Technologies. Russia, pp. 96–99 (2003)
21. Kolisch, R., Padman, R.: An integrated survey of deterministic project scheduling. Omega **49**(3), 249–272 (2001)
22. Kolisch, R., Hartmann, S.: Experimental investigation of heuristics for resource-constrained project scheduling: an update. Eur. J. Oper. Res. **174**, 23–37 (2006)
23. Kolisch, R., Hartmann, S.: Heuristic algorithms for solving the resource-constrained project scheduling problem: classification and computational analysis. In: Weglarz, J. (ed.) Project Scheduling: Recent Models, Algorithms and Applications, pp. 147–178. Kluwer Academic Publishers (1999)
24. Kolisch, R., Sprecher, A., Drexl, A.: Characterization and generation of a general class of resource-constrained project scheduling problems. Manage. Sci. **41**, 1693–1703 (1995)
25. Kolisch, R., Sprecher, A.: PSPLIB - a project scheduling problem library. Eur. J. Oper. Res. **96**, 205–216 (1996). http://www.om-db.wi.tum.de/psplib/
26. Mendes, J.J.M., Goncalves, J.F., Resende, M.G.C.: A random key based genetic algorithm for the resource constrained project scheduling problem. Comput. Oper. Res. **36**, 92–109 (2009)
27. Mingozzi, A., Maniezzo, V., Ricciardelli, S., Bianco, L.: An exact algorithm for the resource-constrained project scheduling problem based on a new mathematical formulation. Manage. Sci. **44**, 715–729 (1998)
28. Mobini, M.D.M., Rabbani, M., Amalnik, M.S., et al.: Using an enhanced scatter search algorithm for a resource-constrained project scheduling problem. Soft Comput. **13**, 597–610 (2009)
29. Palpant, M., Artigues, C., Michelon, P.: LSSPER: solving the resource-constrained project scheduling problem with large neighborhood search. Ann. Oper. Res. **131**, 237–257 (2004)
30. Proon, S., Jin, M.: A genetic algorithm with neighborhood search for the resource-consrtained project scheduling problem. Naval Res. Logist. **58**, 73–82 (2011)

31. Sprecher, A.: Scheduling resource-constrained projects competitively at modest resource requirements. Manage. Sci. **46**, 710–723 (2000)
32. Tormos, P., Lova, A.: A competitive heuristic solution techniques for resource-consrtained project scheduling. Ann. Oper. Res. **102**, 65–81 (2001)
33. Valls, V., Ballestin, F., Quintanilla, M.S.: Justification and RCPSP: a technique that pays. Eur. J. Oper. Res. **165**, 375–386 (2005)
34. Valls, V., Ballestin, F., Quintanilla, S.: A hybrid genetic algorithm for the resource-consrtained project scheduling problem. Eur. J. Oper. Res. **185**(2), 495–508 (2008)
35. Valls, V., Ballestin, F., Quintanilla, S.: A population-based approach to the resource-constrained project scheduling problem. Ann. Oper. Res. **131**, 305–324 (2004)
36. Chen, W., Shi, Y.J., Teng, H.F., et al.: An efficient hybrid algorithm for resource-constrained project scheduling. Inf. Sci. **180**(6), 1031–1039 (2010)
37. Weglarz, J.: Project Scheduling. Recent Models, Algorithms and Applications. Kluwer Academic Publishers, Boston (1999)

The VNS Approach for a Consistent Capacitated Vehicle Routing Problem Under the Shift Length Constraints

Igor Kulachenko[1] and Polina Kononova[1,2](✉)

[1] Novosibirsk State University, Novosibirsk, Russia
soge.ink@gmail.com

[2] Sobolev Institute of Mathematics SB RAS, Novosibirsk, Russia
pkononova@math.nsc.ru

Abstract. We consider a new real-world application of vehicle routing planning in a finite time horizon. A company has a set of capacitated vehicles in some depots and must serve a set of clients. There is a frequency for each client stating how often this client must be visited. Time intervals between two consecutive visits must be the same but the visiting schedule is flexible. To get some competitive advantage, the company tries to increase its service quality. To this end, each client should be visited by one driver only. The goal is to minimize the total length of vehicles' paths over the planning horizon under the frequency constraints and driver shift length constraints. We present an integer linear programming model for this new consistent capacitated vehicle routing problem. To find near-optimal solutions, we design the Variable Neighborhood Search metaheuristic with eleven neighborhood structures. The driver shift length and capacity constraints are penalized and included into the objective function. Empirical results for real test instances from Orenburg region in Russia with up to 900 clients and four weeks in the planning horizon are discussed.

Keywords: Operations research · Mathematical models · Optimization problems · Time scheduling · Search methods · Routing algorithms · Computer experiments

1 Introduction

The literature on vehicle routing problems has become very rich and covers nowadays a variety of applications, modeling approaches, and solution methods [16]. Due to their huge importance in practice, these problems have attracted attention of many researchers and motivate a large number of collaborations between

The reported study was funded by RFBR and Novosibirsk region according to the research project N 19-47-540005.

I. Bykadorov et al. (Eds.): MOTOR 2019, CCIS 1090, pp. 51–67, 2019.
https://doi.org/10.1007/978-3-030-33394-2_5

companies and academia. In addition, vehicle routing problems lead to challenging formulations that require the development of sophisticated solution strategies and motivates the design of clever heuristics and meta-heuristics.

Earlier [25] we considered the uncapacitated variant of our problem, and now we consider a more general case. We have a capacitated heterogeneous fleet of vehicles and a finite set of clients with their demand. Our goal is to find a set of routes for the vehicles to service all clients with minimal total distance. This optimization problem and its variants have been extensively studied for nearly 60 years (see the early work of [6]). It is the consistent vehicle routing problem (ConVRP) where the companies focus on client satisfaction to get some competitive advantage [23,24]. Over a given time horizon we need to construct a set of routes for the vehicles such that to service all clients. The consistency is modeled as follows:

- a client can be visited by one driver only, and split deliveries are not allowed;
- a client should be visited at about the same time of a specific day selected by the client in advance.

Thus, a company can increase client satisfaction by providing consistent service [12,23]. In this paper, we consider a new ConVRP assuming that each client is served by the same vehicle, and a frequency is given for each client indicating how often this client should be visited. Each client is visited on the same day of the week one, two or four times a month. These consistency requirements were suggested by a Russian logistics company interested in results. Each vehicle has a maximum capacity that limits the number of clients it can visit before returning to the depot. All vehicles start from and return to their depot in the given working interval. Our goal is to find a visiting schedule for each client and a set of routes for each vehicle that jointly service all clients under the frequency constraints and driver shift length constraints. The objective is to minimize the total traveling distance for all vehicles over the planning horizon.

In [5], a similar periodic VRP was studied without consistency requirements. In [12,24], the ConVRP was studied with fixed visiting scheduling for clients and unlimited fleet. In our problem, we consider ConVRP with a flexible schedule and limited fleet. To solve this real-world routing problem, we design the Variable Neighborhood Search heuristic (VNS) [26,27]. We use eleven neighborhood structures for local search including four large neighborhoods of Kernighan–Lin [17,22]. To enlarge the search space, we relax the shift length and capacity constraints and include them into the objective function with non-negative penalties that are modified during the search [5,10]. Intensification and diversification strategies are applied in the VNS framework as well.

The rest of this paper is structured as follows. We first introduce the mathematical model in Sect. 2. Neighborhood structures are presented in Sect. 3. The framework of the VNS heuristic is described in Sect. 4. Computational results for real-world instances are discussed in Sect. 5. The last Sect. 6 concludes the paper.

2 Mathematical Model

Let us consider a complete directed graph $G = (V, A)$ with the set of nodes V and the set of arcs A. The set V is the union of the set of depots M and the set of clients I. Each depot $m \in M$ has a heterogeneous fleet of vehicles. The set K defines the total vehicle park. For each vehicle $k \in K$, we know its depot $m(k)$ and its capacity v_k. For each arc $(i, j) \in A$, we have two parameters: the length of arc d_{ij} and traveling time t_{ij}. We denote the length of a driver's shift by T. Each client $i \in I$ has a given frequency of visits μ_i in the planning horizon D. Time intervals between two consecutive visits of client i should be the same and equal to $\tau_i = \lfloor |D|/\mu_i \rfloor$. A demand q_i for each client i is given. By s_i we denote the service time which is positive for each client and 0 for each depot.

We introduce the following binary decision variables:

$$x_{ijkd} = \begin{cases} 1, & \text{if vehicle } k \text{ on day } d \text{ traverses arc } (i,j), \\ 0, & \text{otherwise,} \end{cases}$$

$$y_{ikd} = \begin{cases} 1, & \text{if vehicle } k \text{ on day } d \text{ visits client } i, \\ 0, & \text{otherwise,} \end{cases}$$

$$w_{id} = \begin{cases} 1, & \text{if client } i \text{ is visited on day } d, \\ 0, & \text{otherwise.} \end{cases}$$

The auxiliary non-negative variables u_{ikd} will be used for subtour elimination.

Now we can present the consistent capacitated vehicle routing problem under the shift length constraints as the mixed integer linear program:

$$\min \sum_{d \in D} \sum_{k \in K} \sum_{i \in V} \sum_{j \in V} d_{ij} x_{ijkd} \tag{1}$$

subject to

$$\sum_{i \in I} q_i y_{ikd} \le v_k, \quad k \in K, d \in D, \tag{2}$$

$$y_{mkd} = \begin{cases} 1, m = m(k), \\ 0, m \ne m(k), \end{cases} \quad m \in M, k \in K, d \in D, \tag{3}$$

$$\sum_{k \in K} y_{ikd} = w_{id}, \quad i \in I, d \in D, \tag{4}$$

$$\sum_{d \in D} w_{id} = \mu_i, \quad i \in I, \tag{5}$$

$$\sum_{t=0}^{\tau_i - 1} w_{i(d+t)} = 1, \quad i \in I, d \in \{0, \dots, (\mu_i - 1)\tau_i\}, \tag{6}$$

$$w_{i\alpha} + w_{i\beta} - 2 \le y_{ik\alpha} - y_{ik\beta}, \quad i \in I, k \in K, \alpha, \beta \in D, \alpha \ne \beta, \tag{7}$$

$$\sum_{i \in V} x_{ijkd} = \sum_{i \in V} x_{jikd} = y_{jkd}, \quad j \in V, k \in K, d \in D, \tag{8}$$

$$u_{ikd} - u_{jkd} + n x_{ijkd} \le n - 1, \quad i, j \in I, \ k \in K, \ d \in D, \tag{9}$$

$$\sum_{i \in V} \sum_{j \in V} x_{ijkd}(t_{ij} + s_j) \le T, \qquad k \in K, d \in D, \tag{10}$$

$$u_{ikd} \ge 0, \;\; i \in I, \;\; k \in K, \; d \in D, \tag{11}$$

$$w_{id}, x_{ijkd}, y_{ikd} \in \{0, 1\}, \;\; i, j \in V, \;\; k \in K, d \in D. \tag{12}$$

The objective function (1) minimizes the total traveling distance for all vehicles and all days of the planning horizon. In constraint (2), the total load of vehicle k should not exceed its capacity. Equalities (3) show the distribution of vehicles by depots. Equations (4) and (5) ensure that each client is visited according to its frequency. Constraints (6) guarantee that time intervals between two consecutive visits of each client are the same. Driver consistency is guaranteed in (7). Constraints (8) make sure that each client has exactly one predecessor and one successor and each vehicle returns to its own depot. Inequalities (9) prevent subtours on the set of clients, $n = |I|$. The completion of the routes within the driver shift is enforced by inequalities (10). The last two constraints define the types of variables.

It is easy to see that variables u_{ikd} can be replaced by new variables u_{id} without loss of generality and dimension of the program can be reduced. Note that the problem (1)–(12) can be infeasible because of the limited fleet of vehicles in each depot and the driver shift constraints. To overcome this, we relax the constraints (2) and (10) and include them into the objective function with penalties $\gamma_{kd} \ge 0, \lambda_{kd} \ge 0$, $k \in K, d \in D$. As a result, we have got a relaxation of the original problem (1)–(12) as follows:

$$L(x, \gamma, \lambda) = \min \sum_{d \in D} \sum_{k \in K} \sum_{i \in V} \sum_{j \in V} d_{ij} x_{ijkd} + \sum_{d \in D} \sum_{k \in K} (\gamma_{kd} \kappa_{kd} + \lambda_{kd} \varepsilon_{kd}) \tag{13}$$

subject to (3)–(9), (11), (12) and additional constraints for new variables $\kappa_{kd}, \varepsilon_{kd} \ge 0$ which indicate the excess capacity in kilograms and the over-hours in minutes for each pair (k, d):

$$\kappa_{kd} \ge \sum_{i \in V} q_i y_{ikd} - v_k, \;\; k \in K, d \in D, \tag{14}$$

$$\varepsilon_{kd} \ge \sum_{i \in V} \sum_{j \in V} x_{ijkd}(t_{ij} + s_j) - T, \;\; k \in K, d \in D. \tag{15}$$

Now the relaxed problem (3)–(9), (11)–(15) is feasible even if there is just one vehicle at any depot, and we can solve it by local search metaheuristics [28]. The penalties $\gamma_{kd}, \lambda_{kd}$ will be modified during the search in order to get a feasible solution.

3 Neighborhoods

In the past four decades, local search has grown from a simple heuristic idea into a mature field of research in combinatorial optimization [1]. Local search is often used to solve NP-complete problems since it provides a reliable approach for obtaining high-quality solutions for realistic-size problems in a reasonable time. For partition and permutation problems, many small and large neighborhoods are introduced and studied from a theoretical and an empirical point of views [2, 3,11,13,14,16,21]. Below we present eleven neighborhoods for the problem which is a special case of partition and permutation problems. We already considered all these neighborhoods in [25].

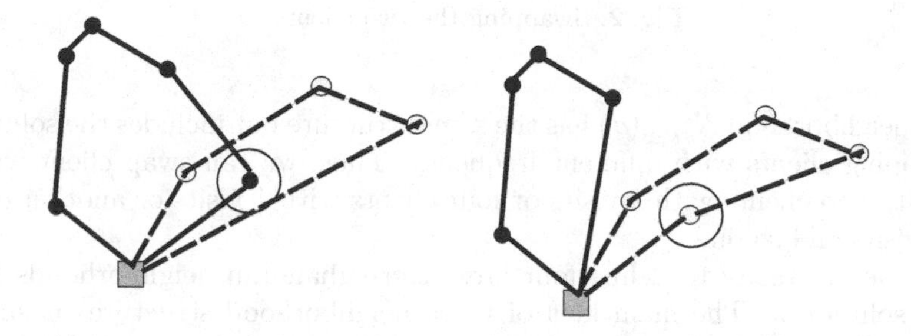

Fig. 1. Moving the client to another route

Let us denote by σ a feasible solution to the problem. For each vehicle $k \in K$ and each day $d \in D$ we have a route (the order of clients). We say that a driver of vehicle k is *happy* on day d if $\kappa_{kd} = \varepsilon_{kd} = 0$ and *unhappy* if these constraints are violated. We want to move clients from unhappy pairs (k, d) to happy ones.

Now we define the following neighborhood structures for solution σ.

The move neighborhood $N_{\text{move}}(\sigma)$ consists of some feasible solutions resulting from σ by *moving* a client to another vehicle or the same vehicle but another day (Fig. 1). If the client must be visited several times, we move all his visits respectively. Moreover, we move an unhappy pair to a happy one only. In order to find the best permutation for new schedules, we select the best positions of new visits in previous schedules. The cardinality of this neighborhood is $O(|I||D||K|)$. It is a large set. Thus, we will use a randomized neighborhood $N^q_{\text{move}}(\sigma)$, $0 < q < 1$, which is a random part of the neighborhood $N_{\text{move}}(\sigma)$. Each element of the set $N_{\text{move}}(\sigma)$ is included in the set $N^q_{\text{move}}(\sigma)$ with probability q independently of other elements.

The neighborhood $\tilde{N}^q_{\text{move}}(\sigma)$ has the same structure but includes the solutions for all moves, except those from happy pairs to unhappy.

The swap neighborhood $N_{\text{swap}}(\sigma)$ consists of some feasible solutions resulting from σ by *swapping* two clients with the same frequency for the same or different vehicles (Fig. 2). We consider only the clients which are close enough to each other, the mutual distance between them is at most R, where R is a parameter of the neighborhood. The cardinality of the neighborhood is $O(|I|^2)$. Thus, we apply the same randomization trick and use $N^q_{\text{swap}}(\sigma)$ neighborhood instead of the deterministic case.

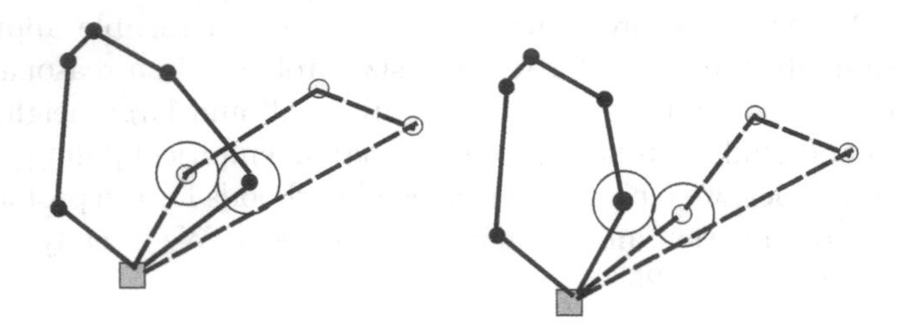

Fig. 2. Swapping the two clients

The neighborhood $\tilde{N}^q_{\text{swap}}(\sigma)$ has the same structure but includes the solutions for swapping clients with different frequency. Thus, we can swap client with 4 visits with two clients with 2 visits or four clients with 1 visits or another client with 4 visits and so on.

Now we are ready to define four large Kernighan–Lin neighborhoods for a feasible solution σ. The main idea of these neighborhood structures is similar to the truncated Tabu Search method by a small neighborhood, say $N(\sigma)$. The neighborhood $KL(\sigma)$ consists of l solutions resulting from σ by the following rule [17,19]:

1. Find the best feasible solution σ' in the neighborhood $N(\sigma)$.
2. Set $\sigma := \sigma'$, even if σ' is worse than σ.
3. Repeat steps 1 and 2 l times, if a move or swap is used at step 1 or 2 of previous iterations, it can not be used anymore.

The sequence of $\sigma_1, \ldots, \sigma_l$ defines l neighbors of the solution σ. We say that σ_b is a local minimum with respect to the KL–neighborhood if σ_b is the best solution of $\sigma_1, \ldots, \sigma_l$.

Using the six basic neighborhoods $N_{\text{move}}(\sigma)$, $N^q_{\text{move}}(\sigma)$, $\tilde{N}^q_{\text{move}}(\sigma)$, $N_{\text{swap}}(\sigma)$, $N^q_{\text{swap}}(\sigma)$, and $\tilde{N}^q_{\text{swap}}(\sigma)$ instead of the neighborhood $N(\sigma)$, we may have four Kernighan–Lin neighborhoods $KL_{\text{move}}(\sigma)$, $\tilde{KL}_{\text{move}}(\sigma)$, $KL_{\text{swap}}(\sigma)$, and $\tilde{KL}_{\text{swap}}(\sigma)$ respectively. We illustrate the idea of the Kernighan–Lin neighborhoods in Fig. 3.

As we have mentioned above, the position of a new client in scheduling is selected without reordering other clients for the same pair (k, d). For improving the final scheduling, we apply local descent algorithm by the well-known *2-opt* neighborhood for each pair (k, d). The idea of this neighborhood is to choose two non-adjacent arcs and replace them by two other arcs for creating a new tour. The main goal is removing intersections of arcs (see Fig. 4). In fact, we divide the problem into $|K||D|$ the traveling salesman subproblems, and a local optimum by the *2-opt* neighborhood is obtained for each subproblem independently [15].

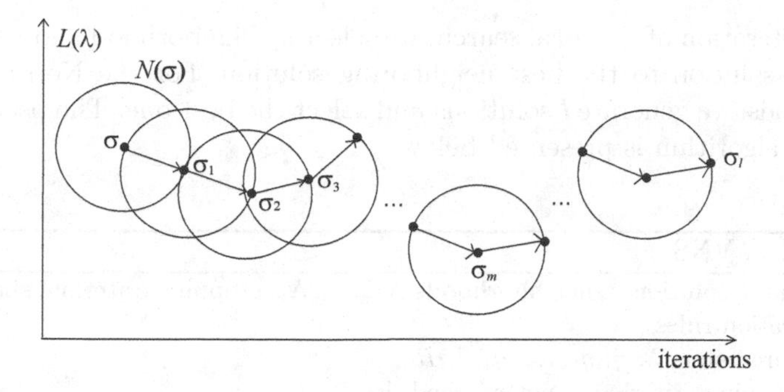

Fig. 3. The Kernighan–Lin neighborhood

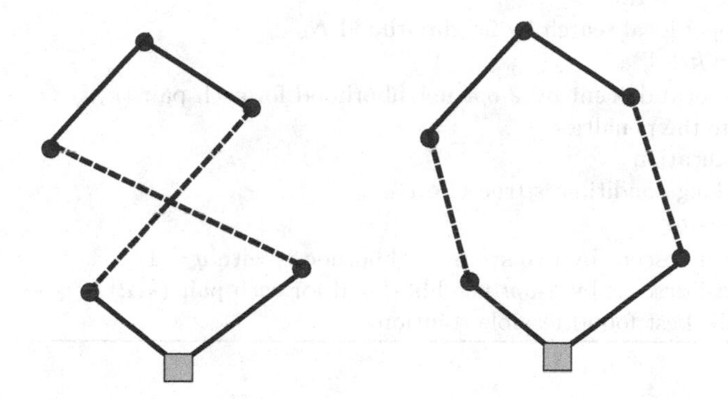

Fig. 4. Neighborhood *2-opt*. Removing the intersection

4 Optimization Method

Variable Neighborhood Search is an efficient framework of local search invented about 20 years ago by Pierre Hansen and Nenad Mladenovich [26]. It is based upon a simple, but a strong principle: a systematic change of a neighborhood within the search. Its development has been rapid and successful in many real-world applications [27], including hard routing problems [18,20] and games [7–9]. The main idea is to focus on local optima and change the landscape of search assuming that the local optimum for one neighborhood may not be the local optimum for another neighborhood. Below we apply this method to the relaxed problem. We used this method for a simpler problem in [25].

To start the method, we need to create an initial solution σ and define the penalties. The VNS method can start from an arbitrary solution, but we use a greedy solution to get a uniform distribution of clients through all pairs (k, d), $k \in K, d \in D$. We start from clients with high frequency and wish to minimize the maximal number of clients per day and per vehicle [4]. We put identical initial values of all penalties, $\lambda_{kd} = 2.5$, $\gamma_{kd} = 3$ in such a way to get approximately the same values of the items in the objective function $L(x, \gamma, \lambda)$.

In each iteration of the local search, we select a neighborhood and move from the current solution to the best neighboring solution. For the Kernighan–Lin neighborhoods, we generate l solutions and select the best one. The pseudo-code of the VNS algorithm is presented below.

Algorithm 1. VNS

Require: initial solution σ, neighborhoods $N_1, \ldots, N_8$, stopping criterion, shaking and intensification rules
1: Define parameters λ, γ, $q_1, \ldots, q_4$, l, R
2: **while** stopping criterion is not reached **do**
3: $\quad k = 1$
4: $\quad$ **while** $k \leq 8$ **do**
5: $\quad\quad$ Apply local search by neighborhood N_k
6: $\quad\quad$ $k = k + 1$
7: $\quad$ Apply local descent by *2-opt* neighborhood for each pair (k, d)
8: $\quad$ Update the penalties
9: $\quad$ Intensification
10: $\quad$ **if** shaking condition is true **then**
11: $\quad\quad$ Shaking
12: Apply local descent by two swap neighborhoods with $q = 1$
13: Apply local descent by *2-opt* neighborhood for each pair (k, d)
14: **return** the best found feasible solution

At the initialization step, we generate an initial solution by a greedy algorithm and define the parameters of the method (line 1). Note that a randomization of the first four neighborhoods may be different and $q_i \neq q_j$, $1 \leq i \neq j \leq 4$. We define these values in such a way that the cardinality of each randomized neighborhood is the same and equal to 200 on average. Thus, we accelerate the search, reduce the running time per iteration, and add a diversification aspect into the search process. The shaking procedure (line 11) is an additional diversification rule. We use some random steps by the swap or move neighborhoods in this procedure if the best found solution does not change for a long time.

The stopping criterion (line 2) is the total number of iterations which depends on the number of clients and their frequency. We use up to $O(n_1^2)$ iterations in our experiments, where $n_1 = \sum_{i \in I} \mu_i$.

Local search (lines 5, 6) is applied by the move and swap neighborhoods and then by the Kernighan–Lin neighborhoods. In the latter case, we use local descents only and terminate the process in a local optimum. For the four basic neighborhoods, we terminate the process after a prescribed number of iterations. Further (line 7), we get a local optimum by *2-opt* neighborhood for each pair (k, d) of vehicle and day. As a rule, we discover a new best solution at this stage. If we find a solution with $\kappa_{kd} > 0$ for some pair (k, d) then we increase the penalties $\gamma_{kd} := 1.05\gamma_{kd}$. If a new solution has $\varepsilon_{kd} = 0$ for all pairs (k, d), then we decrease the penalties λ. Otherwise, we increase them. In general case, we modify the penalties by the following rule (line 8):

$$\gamma_{kd} = \begin{cases} 1.05\gamma_{kd}, \text{ if } \kappa_{kd} > 0, \\ \gamma_{kd}, \text{ if } \kappa_{kd} = 0, \end{cases}$$

$$\lambda_{kd} = \begin{cases} 1.03\lambda_{kd}, \text{ if } \varepsilon_{kd} > 0, \\ 0.97\lambda_{kd}, \text{ if } \varepsilon_{kd} = 0. \end{cases}$$

In the intensification procedure (line 9), we return to the best found solution and increase all randomization parameters $q_1, \ldots, q_4$ and the value of penalties λ_{kd} again to check the most promising area more carefully. If we discover a new best solution, we return to the previous values of these parameters. In the shaking procedure (line 11), we do the same for $q_1, \ldots, q_4$ and λ_{kd} to start the search in a new area of the feasible domain. Finally, we apply deterministic local descent (lines 12, 13) by N_{swap} and $\tilde{N}_{\text{swap}}$ neighborhoods ($q_3 = q_4 = 1$) and get local optima by *2-opt* neighborhood for each pair (k, d).

5 Computational Results

The described VNS algorithm was implemented in C++ with MSVC++ 14.16 compiler using standard release options. All experiments were conducted on a computer with an AMD Ryzen 5 2600 3.4 GHz processor and 16 GB of RAM running under Microsoft Windows 10 (64-bit).

The data set used to test the algorithm is proposed by a Russian logistics company with 892 clients from Orenburg region. Among them, one third are clients of frequency 1, and slightly more than half are clients of frequency 2. There are three depots located at a distance of 250 km from each other. We randomly select a part of the large instance to get small ones. For this purpose, we varied certain parameters used during client selection. These parameters include a number necessary for localization of the depots (radius), selection probabilities different for different depots, and a number specifying the random shift of the vehicles relative to their initial position. For the client selection and further in the algorithm, a 32-bit Mersenne Twister pseudo-random number generator was used. As a result, we generated 10 various instances with 600–700 clients. This range of the number of clients allows obtaining diverse large instances with the same number of vehicles in each depot. Besides, this range is close to the actual number of clients served by the company in one region. We assigned two vehicles in each depot for these test instances.

Also, to compare the algorithm with an optimization solver, data set with 672 clients from Orenburg region was used. Slightly less than three-quarters of these clients have a frequency of 1, while the numbers of clients of frequency 2 and 4 are approximately equal. There are the same three depots for this set. Using the same method as for the larger data set, we generated instances with 320–350 and 130–150 clients. We assigned one vehicle in each depot for the former instances and one vehicle in a single depot for the latter ones.

Client attributes include name, GPS coordinates (latitude and longitude), the frequency of visits, service time, and demand. The shift length is 8 h, including 40 min for a break. The time for a break is not fixed in drivers' schedules, and they can spend it at any free from client service time of the working day. The problem we are investigating does not include time windows. Hence, we can just adjust the shift length to $T = 7\,\text{h}\ 20\,\text{min}$.

Vehicles can leave the depot starting at 8:30, but must arrive at the first client no earlier than 9:00. The last client must be serviced before 17:00, but the vehicle must return to the depot no later than 18:00. To include these additional requirements into the model, we modify the matrix (t_{ij}) by the following rule:

$$t_{ij} = \begin{cases} t_{ij}, \text{ if } i, j \in I, \\ \max\{0, t_{ij} - 30'\}, \text{ if } i \in M, \quad j \in I, \\ \max\{0, t_{ij} - 60'\}, \text{ if } i \in I, \quad j \in M. \end{cases}$$

The planning horizon is 20 days. The speed of each vehicle is 50 km/h. For Kernighan–Lin neighborhoods we generate $l = 25$ neighboring solutions. The threshold R for the swap neighborhoods is defined as 20 km. Local search by the all basic neighborhoods of the VNS algorithm (line 5) is set as 1300 iterations. Results are obtained by running the VNS in 10 min 10 times per instance.

Figure 5 illustrates the typical behavior of the method. The initial value of the objective function $L(x, \gamma, \lambda)$ is huge with excess capacity $\kappa = \sum_{kd} \kappa_{kd}$ and total over-hours $\varepsilon = \sum_{kd} \varepsilon_{kd}$. And after 2000 iterations we found a solution with $\kappa = \sum_{kd} \kappa_{kd} = 0$. Note that the total over-hours and overload decrease as iterations grow. We denoted by $\times$ new record values of the objective function for a feasible solution. We see that the value decreases for a feasible solution too.

In Tables 1, 2, 3 and 4, we show computational results for these 10 small instances. The purpose is to study the impact of the capacity constraint on the solution. Each instance was run 10 times and the minimal, average, and

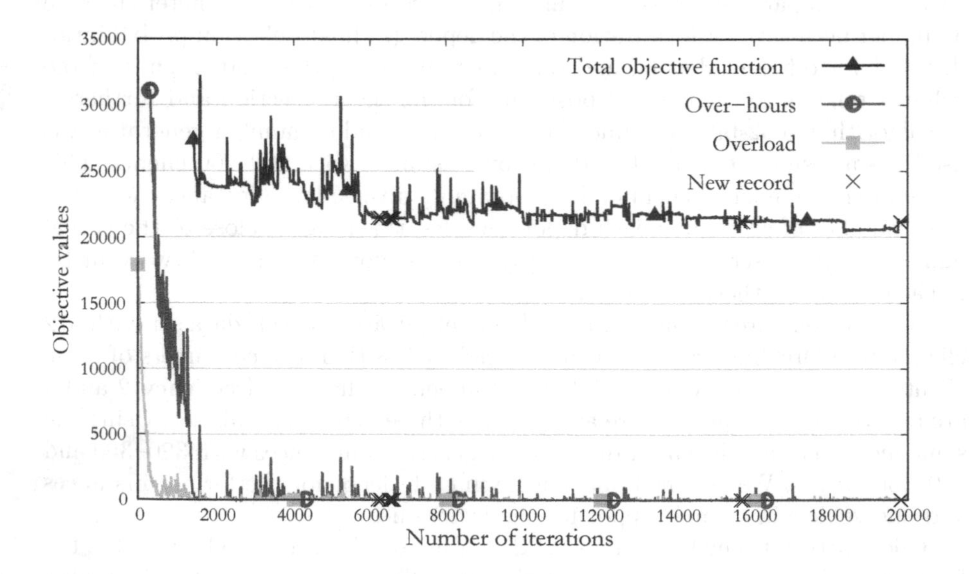

Fig. 5. Easy instance with three vehicles in each depot

maximum values for these runs are presented in the tables. To compare the obtained solutions the objective function (13) with penalties $\gamma_{kd} = 50, \lambda_{kd} = 40$ was used.

Table 1. Capacity $v_k \in [800, 1000]$

Instance	Min		Avg		Max	
	ε	Distance	ε	Distance	ε	Distance
1	0	15360	0	15867.6	0	16195
2	0	16060	0	16298.4	0	16586
3	4	16282	9	16357.6	28	16260
4	0	16961	0	17203	0	17370
5	0	15434	3.5	15803.1	0	16796
6	0	14334	0	14639.4	0	15204
7	0	15521	0	16118.5	0	16704
8	0	16324	0	16695.3	0	16993
9	0	16381	5	16886.3	10	17519
10	4	17158	6.4	17572.5	0	18443

At first (Tables 1, 2 and 3), we used the model which does not allow violation of the capacity constraint at any step of the algorithm (including shakes). Table 1 show results for the case when the vehicles have enough capacity for all clients they may need to serve. By doing this, we were able to get an average daily load for the vehicles from different depots. For most of the instances, this number turned out to be 400–500 kg. From Tables 2 and 3 we see that the results become worse with the tightening of the capacity. Also, Table 3 shows that the search space can become so narrow that for some vehicles there will be no other options but to serve clients intended for another depot (instances 2–4).

Next, the relaxed formulation of the problem with capacity penalties was used (Table 4). This allowed us to find much better solutions in most of the instances. It confirms the need for the penalties.

It should be taken into account that despite the presence of penalties in some solutions, in most of the instances they are small enough to be neglected.

In order to study the efficiency of the algorithm, we compared it with the results obtained by metaheuristic solver LocalSolver. We chose it instead of such classical MILP solvers as CPLEX and Gurobi, since the latter cannot find the exact solution in a reasonable time for even quite small instances of the problem. The results are presented in Tables 5, 6, 7 and 8.

Table 2. Capacity $v_k \in [500, 600]$

Instance	Min		Avg		Max	
	ε	Distance	ε	Distance	ε	Distance
1	0	15467	0	16125.4	0	16686
2	0	17354	17.3	17446.5	20	18271
3	13	16319	29.9	17295.2	75	18631
4	0	17308	0	17546.5	0	17905
5	0	16207	4.8	16489	0	17241
6	0	14599	0	14986.9	0	15324
7	0	15896	0.6	16513.6	0	16900
8	0	16798	6.5	17347.1	48	18106
9	33	16180	45	16362	101	16201
10	1	17667	1.9	18149.1	1	18464

Table 3. Capacity $v_k \in [450, 500]$

Instance	Min		Avg		Max	
	ε	Distance	ε	Distance	ε	Distance
1	0	16083	0	16373.6	0	16700
2	326	17968	1236.7	19365.9	1693	19335
3	1023	18604	1651.5	18985.9	1992	18722
4	389	18886	1400.2	20205.4	2309	20979
5	0	16704	2.1	17420.2	1	18091
6	0	15893	22.8	16584.8	162	16623
7	18	16529	14.5	17306.4	51	18197
8	2	17822	28	17718.9	126	18523
9	104	16073	111.7	16314.3	101	17389
10	36	19520	87.8	19101.7	111	19416

Since LocalSolver did not allow us to obtain satisfactory solutions to the problem with all the hard constraints, we decided to use the minimized objective function (16) for it. The main challenges for the solver were caused by the constraints (5)–(7) of the problem. We include constraints (6)–(7) into the objective function (16) with penalty ψ, while constraints (5) remained hard. Variable ζ in (16) denotes the total number of constraints (6)–(7) violations. We used penalties $\gamma_{kd} = 100, \lambda_{kd} = 200, \psi = 10^5$ for this objective function. The constant ϕ necessary for determining the variable χ was set to $3 \cdot 10^5$ for Table 6

Table 4. Capacity $v_k \in [450, 500]$ with the penalties

Instance	Min			Avg			Max		
	ε	κ	Distance	ε	κ	Distance	ε	κ	Distance
1	4	0	15388	6.1	0	15896.2	36	0	16411
2	178	0	17449	270.5	11	18472.9	468	0	18909
3	280	0	17978	276.5	88	18606.8	595	0	19204
4	0	0	19710	24.7	0	19920.5	34	0	20324
5	0	0	16335	9.3	0	16560.7	31	0	16563
6	22	0	14638	26.5	0	15131.6	79	0	14767
7	7	0	16092	2.5	0	16700	0	0	17212
8	0	0	16883	0.3	0	17200.3	1	0	17930
9	114	0	16208	115.7	0	16338.7	130	0	16084
10	11	20	18837	21.5	18	19037.7	18	20	19503

and $8 \cdot 10^4$ for Table 8. To provide non-deterministic results for LocalSolver runs, we added to the model one excessive constraint repeating already contained one.

$$\chi = \begin{cases} 1000, L(x, \gamma, \lambda) > \phi, \\ 1, \text{ otherwise.} \end{cases}$$

$$L_s(x, \gamma, \lambda, \psi) = \min \left(\sum_{d \in D} \sum_{k \in K} \sum_{i \subset V} \sum_{j \subset V} d_{ij} x_{ijkd} \right.$$
$$\left. + \sum_{d \in D} \sum_{k \in K} (\gamma_{kd} \kappa_{kd} + \lambda_{kd} \varepsilon_{kd}) \right) \chi + \psi \zeta \tag{16}$$

In Tables 5, 6, 7 and 8, we show computational results obtained by our VNS algorithm and by LocalSolver. There were 10 runs for each instance. For Tables 5 and 8, computational time was set to 5 min, and for Tables 6 and 7, it was set to 10 and 2 min, respectively. To compare the obtained solutions, penalties $\lambda_{kd} = 40$ for Tables 5 and 7 and $\gamma_{kd} = 25, \lambda_{kd} = 20, \psi = 300$ for Tables 6 and 8 were used. All the solutions obtained for the former tables have $\kappa = 0$.

It is clearly visible that the results obtained by our algorithm are better than those of LocalSolver in all instances. It is also worth noting the decrease in the difference between the maximum and minimum objective values of the results with a decrease in the dimension of the problem.

Table 5. The results for instances with 320–350 clients for VNS algorithm

Instance	Min		Avg		Max	
	ε	Distance	ε	Distance	ε	Distance
1	0	10473	0.1	10486	1	10483
2	0	10543	2.1	10643	0	10874
3	0	10821	0	10853.3	0	10892
4	2	9950	3	10018.5	3	10141
5	0	8834	0.7	9042.2	0	9399
6	0	8830	0	8913.4	0	9301
7	0	9748	1.4	9867.3	11	9870
8	0	9204	0.6	9266.5	0	9383
9	0	10904	0.8	11014.9	6	11134
10	0	10217	0	10243.7	0	10277

Table 6. The results for instances with 320–350 clients for LocalSolver

Instance	Min				Avg				Max			
	ε	κ	ζ	Distance	ε	κ	ζ	Distance	ε	κ	ζ	Distance
1	48	0	0	10673	307.2	2	16.9	11713.1	312	0	73	12336
2	316	0	0	11094	404.7	3	46.7	11490.6	236	0	243	11874
3	120	0	0	10659	304.3	1	67.3	11525.5	377	0	219	11586
4	163	0	0	10790	326	0	17	11708.9	294	0	96	12603
5	12	0	0	11309	223.2	2	2.1	10722.3	474	20	3	10730
6	189	0	0	9980	329.9	10	3.7	10120	567	0	14	10240
7	145	0	0	11016	381	4	4.2	11163.1	585	20	11	10929
8	296	0	0	9827	482.1	7	29.5	10531.7	502	0	78	10452
9	374	0	0	11202	405.1	1	47.6	11993	240	10	124	12520
10	67	0	0	10974	216.2	0	4.3	11258.5	361	0	14	11891

The large instance with 892 clients can be effectively solved using developed VNS algorithm as well. However, to achieve an acceptable value of the variance for the results of applying the algorithm for an instance with such a number of clients, more computational time is required. Although smaller instances are usually examined in the literature, large-scale ones are also of interest to study, as they are often applicable in the real world.

Table 7. The results for instances with 130–150 clients for VNS algorithm

Instance	Min		Avg		Max	
	ε	Distance	ε	Distance	ε	Distance
1	0	3199	0	3205.5	0	3217
2	0	3388	0	3394.8	0	3405
3	0	2747	0	2759.4	0	2769
4	0	1657	0	1666.7	0	1687
5	0	3040	0.2	3053.6	2	2992
6	0	2110	0	2115.5	0	2126
7	0	2818	1.1	2819.3	11	2818
8	0	2960	0	2972.8	0	2987
9	0	2879	0	2883.5	0	2894
10	0	3061	3	2994.1	5	2958

Table 8. The results for instances with 130–150 clients for LocalSolver

Instance	Min				Avg				Max			
	ε	κ	ζ	distance	ε	κ	ζ	Distance	ε	κ	ζ	Distance
1	0	0	0	3430	20.4	4	0.4	3626.4	98	40	0	3650
2	65	0	0	3382	123.9	3	1	3683	241	0	6	3668
3	0	0	0	3130	40	0	0.7	3309.9	236	0	7	3357
4	0	0	0	1784	1.6	0	0.7	2481.8	16	0	3	3074
5	2	0	0	3207	129.4	8	1.3	3440.3	305	0	9	3502
6	0	0	0	2273	2.3	0	0.9	2637	23	0	4	2281
7	0	0	0	3212	89.6	1	0	3238.8	277	0	0	3223
8	0	0	0	3219	87.5	2	3.2	3258.5	321	0	32	3475
9	0	0	0	3128	17.9	0	0.5	3413.8	62	0	5	3451
10	61	0	0	3373	150.5	17	2	3286.5	265	0	13	3515

6 Conclusion

In this paper, we have studied a new consistent capacitated vehicle routing problem and designed the VNS algorithm for real-world instances. This algorithm is able to solve large-scale instances and reduce the total traveling distance.

Companies today are increasingly focused on customer satisfaction to achieve a competitive advantage. One of the components of these customer-first strategies is service consistency [12,23]. Thus, it is important to study problems with consistency requirements addressing real-world challenges. In our version of ConVRP, it is required that the same driver visit the same clients on the same day of the week according to their frequency of visits. We presented our program to the logistics company, and they were satisfied with it.

One of the new research directions is the control of time for client visits. As we have mentioned before, a logistics company can get an additional competitive advantage if each client is visited at about the same time. Such type of constraints can be incorporated into the model to improve the service of clients. Sure, new constraints will increase the total traveling distance and may require additional vehicles. The optimal balance here is an important line for research as well.

References

1. Aarts, E., Lenstra, J.: Local Search in Combinatorial Optimization. Wiley, New York (1997)
2. Ahuja, R.K., Ergun, Ö., Orlin, J.B., Punnen, A.P.: A survey of very large-scale neighborhood search techniques. Discrete Appl. Math. **123**(1–3), 75–102 (2002)
3. Alekseeva, E., Kochetov, Y., Plyasunov, A.: Complexity of local search for the p-median problem. Eur. J. Oper. Res. **191**(3), 736–752 (2008)
4. Coene, S., Arnout, A., Spieksma, F.: On a periodic vehicle routing problem. J. Oper. Res. Soc. **61**(12), 1719–1728 (2010)
5. Cordeau, J.F., Gendreau, M., Laporte, G.: A tabu search heuristic for periodic and multi-depot vehicle routing problems. Networks **30**(2), 105–119 (1997)
6. Dantzig, G.B., Ramser, J.H.: The truck dispatching problem. Manage. Sci. **6**(1), 80–91 (1959)
7. Davydov, I., Kochetov, Y., Carrizosa, E.: VNS heuristic for the $(r|p)$-centroid problem on the plane. Electron. Notes Discrete Math. **39**, 5–12 (2012)
8. Davydov, I., Kochetov, Y., Carrizosa, E.: A local search heuristic for the $(r|p)$-centroid problem in the plane. Comput. OR **52**, 334–340 (2014)
9. Diakova, Z., Kochetov, Y.: A double VNS heuristic for the facility location and pricing problem. Electron. Notes Discrete Math. **39**, 29–34 (2012)
10. Gendreau, M., Hertz, A., Laporte, G.: A tabu search heuristic for the vehicle routing problem. Manage. Sci. **40**(10), 1276–1290 (1994)
11. Golden, B.L., Raghavan, S., Wasil, E.A.: The Vehicle Routing Problem: Latest Advances and New Challenges. Springer, US (2008). https://doi.org/10.1007/978-0-387-77778-8
12. Groër, C., Golden, B., Wasil, E.: The consistent vehicle routing problem. Manuf. Serv. Oper. Manag. **11**(4), 630–643 (2009)
13. Grover, L.K.: Local search and the local structure of NP-complete problems. Oper. Res. Lett. **12**(4), 235–243 (1992)
14. Gutin, G.Z., Yeo, A.: Small diameter neighbourhood graphs for the traveling salesman problem: at most four moves from tour to tour. Comput. Oper. Res. **26**(4), 321–327 (1999)
15. Hemmelmayr, V.C., Doerner, K.F., Hartl, R.F.: A variable neighborhood search heuristic for periodic routing problems. Eur. J. Oper. Res. **195**(3), 791–802 (2009)
16. Irnich S., Toth, P., Vigo, D.: The family of vehicle routing problems.In: Vehicle Routing: Problems, Methods, and Applications, pp. 1–33. Society for Industrial and Applied Mathematics, Philadelphia (2014)
17. Kernighan, B., Lin, S.: An efficient heuristic procedure for partitioning graphs. Bell Syst. Tech. J. **49**(2), 291–307 (1970)
18. Khmelev, A., Kochetov, Y.: A hybrid VND method for the split delivery vehicle routing problem. Electron. Notes Discrete Math. **47**, 5–12 (2015)

19. Kochetov, Y., Kononova, P., Paschenko, M.: Formulation space search approach for the teacher/class timetabling problem. Yugoslav J. Oper. Res. **18**(1), 1–11 (2008)
20. Kochetov, Y., Khmelev, A.: A hybrid algorithm of local search for the heterogeneous fixed fleet vehicle routing problem. J. Appl. Ind. Math. **9**(4), 503–518 (2015)
21. Kochetov, Y.: Computational bounds for local search in combinatorial optimization. Comput. Math. Math. Phys. **48**(5), 747–763 (2008)
22. Kononova, P., Kochetov, Y.: The variable neighborhood search for the two machine flow shop problem with a passive prefetch. J. Appl. Ind. Math. **7**(1), 54–67 (2013)
23. Kovacs, A.A., Golden, B.L., Hartl, R.F., Parragh, S.N.: Vehicle routing problems in which consistency considerations are important: a survey. Networks **63**(3), 192–213 (2014)
24. Kovacs, A.A., Parragh, S.N., Hartl, R.F.: A template-based adaptive large neighborhood search for the consistent vehicle routing problem. Networks **63**(1), 60–81 (2014)
25. Kulachenko, I., Kononova, P., Kochetov, Y., Kurochkin, A.: The variable neighborhood search for a consistent vehicle routing problem under the shift length constraints. In: Proceedings of the 9-th IFAC Conference Manufacturing Modelling, Management and Control MIM 2019 (2019, Accepted)
26. Mladenovic, N., Hansen, P.: Variable neighborhood search. Comput. Oper. Res. **24**, 1097–1100 (1997)
27. Mladenovic, N., Hansen, P.: Developments of variable neighborhood search. In: Ribeiro, C., Hansen, P. (eds.) Essays and Surveys in Metaheuristics, vol. 2, 1st edn, pp. 415–439. Springer, Boston (2002). https://doi.org/10.1007/978-1-4615-1507-4_19
28. Talbi, E.G.: Metaheuristics: From Design to Implementation. Wiley, USA (2009)

Development of Ant Colony Optimization Algorithm for Competitive p-Median Facility Location Problem with Elastic Demand

Tatyana Levanova[1,2(✉)] and Alexander Gnusarev[1]

[1] Sobolev Institute of Mathematics, Omsk Branch, Omsk, Russia
levanova@ofim.oscsbras.ru, alexander.gnussarev@gmail.com
[2] Dostoevsky Omsk State University, Omsk, Russia

Abstract. In this paper, we consider the competitive p-median facility location and design problem with elastic demand that we have earlier formulated based on the problem with elastic demand and the classical p-median problem. The situation that arises in a new company planning to enter the existing market of goods and services is considered. The firm wants to locate its businesses in p points, capturing as much of the profits from competitors as possible. The problem has a mathematical model with a non-linear objective function. Searching the optimal solution to the constructed problem is difficult. The CPU-time of commercial software is significant even for not too large dimension. For the new model, we have previously proposed variants of local search algorithms, and created a series of test instances based on real data. In this paper, an ant colony algorithm is developed, and an artificial ant algorithm is proposed. The algorithm's parameters are adjusted taking into account the specifics of the problem. Experimental studies and comparison of the ant colony optimization algorithm with the simulated annealing are carried out.

Keywords: Discrete optimization · Location problem · p-Median · Elastic demand · Nature-inspired algorithm · Ant colony

1 Introduction

At the present time it is often necessary to solve the applied problems with large data volume. Their size makes the usage of the commercial software more complicated. Moreover, the theoretical difficulty of such problems testifies the creation of the multi-purpose effective algorithm to be impossible. That is why the creation of a problem-specific method for approximate calculation has become more important. During the last twenty years the nature inspired methods (based on the natural processes and events) have been actively developed. Those algorithms comprise one of the artificial intelligence branches. Its subject is modelling multi-agent systems, evolutionary processes, physical phenomena and others from the

I. Bykadorov et al. (Eds.): MOTOR 2019, CCIS 1090, pp. 68–78, 2019.
https://doi.org/10.1007/978-3-030-33394-2_6

problem solving point of view. Collective intelligence methods take special place among the artificial intelligence methods. Swarm intelligence systems usually consist of a set of agents interacting with each other and with the environment. They are represented by some self-organizing systems, the agents of which are relatively simple but, at the same time, able to make collective decisions. The Ant Colony algorithm, the Firefly algorithm, The Bee Colony optimization, the Bat algorithm, Intellectual Water Drops algorithm and others are among them [2,10]. During some years the development of different approximation algorithms for the discrete Location problems has been held in Sobolev Institute of Mathematics SB RAS [3,5,6,8]. Special attention should be paid to the development of the Ant Colony algorithm, which helped to gain the quality solutions rather fast. In this paper we continue that line of research and offer the Ant Colony Optimization Algorithm for competitive p-median facility location and design problem with elastic demand. The special aspects of the algorithm development are described, the results of research are included, and the obtained results are discussed.

2 Competitive p-Median Problem with Elastic Demand

Let us consider the situation in which some new company is to locate a fixed number of its facilities in the market according to the amount of finances. It can define the locations and the types of the facilities, for instance, a supermarket, a market, etc. It has to rival for the demand with the competitors existing in the market. The utility of facilities depends on the distance, their attractiveness and other factors. The serviced share of the demand changes flexibly depending on the new company's decisions and the choice of the customers. The company's aim is to define the locations and types of its facilities in order to attract the largest share of total demand. Customers choose the facilities depending on the distance, attractiveness and other factors. That is why the share of the serviced demand changes elastically depending on the companys and customers decisions. The problem was formulated in this form by Aboolian, Berman, Krass [1].

In this paper, we consider the p-median competitive facility location and design problem (CPFLDP) that we have formulated in [9], combining the conditions of the problem with elastic demand and the classical p-median problem. Let us construct a mathematical CPFLDP model. Introduce the following notation, considering [1]: N is a discrete set of demand points, where a facility can be located; $C \subset N$ are the competitor's points; $S = N \setminus C$ are the points of possible location of the new company; R is set of facility types; w_i is the demand weight at point $i \in N$; c_{jr} is the opening cost of a facility in case $r \in R$ at point $j \in S$; p is a number of facilities to be opened; $p \leq |S|$. The problem variables take the value $x_{jr} = 1$, if a facility of the type $r \in R$ is located in point $i \in N$, $x_{jr} = 0$ otherwise. The utility $u_{ij} = \sum_{r=1}^{R} k_{ijr} x_{jr}$, where the supplementary coefficients $k_{ijr} = a_{jr}(d_{ij} + 1)^{-\beta}$ take into account the distance d_{ij} between the points $i, j \in N$, the customers' sensitivity to it β and the attractiveness a_{jr} of the facility type r. The total utility $U_i(S)$ from the facilities opened by the company in points $i \in N$ is determined by the formula $U_i(C) = \sum_{j \in C} u_{ij}$. The notation for the competitors are similar. The demand function is nonlinear: $g(U_i) = 1 - \exp\bigg(- \lambda_i U_i \bigg)$, where U_i is the total utility from all the company's and competitor's facilities for a customer at $i \in N$, λ_i is the characteristic of the elastic demand in point i, $\lambda_i > 0$. The company's total share of facility $i \in N$ is measured by:

$$MS_i = \frac{U_i(S)}{U_i(S) + U_i(C)} = \frac{\sum_{j \in S} \sum_{r=1}^{R} k_{ijr} x_{jr}}{\sum_{j \in S} \sum_{r=1}^{R} k_{ijr} x_{jr} + \sum_{j \in C} u_{ij}}.$$

According to the notations above, the mathematical model can be written as:

$$\max \sum_{i \in N} w_i \cdot \Bigg(1 - \exp\Bigg(- \lambda_i \Bigg(\sum_{j \in S} \sum_{r \in R} k_{ijr} x_{jr} + \sum_{j \in C} \sum_{r \in R} k_{ijr} x_{jr} \Bigg)\Bigg)\Bigg) \cdot MS_i \quad (1)$$

$$\sum_{j \in S} \sum_{r \in R} c_{jr} x_{jr} \leq B, \quad (2)$$

$$\sum_{r \in R} x_{jr} \leq 1, j \in S, \quad (3)$$

$$\sum_{j \in S} \sum_{r \in R} x_{jr} = p, \quad (4)$$

$$x_{jr} \in \{0, 1\}, \quad j \in S, r \in R. \quad (5)$$

The objective function (1) reproduces the company's goal to maximize its share of the demand. The inequality (2) allows the facilities' allocation based on the available budget. The conditions (3) show the possibility of opening only one type of the facility in each point $j \in S$. The equation (4) sets the condition for the number of facilities to be opened.

3 Ant Colony Algorithm

Ant Colony algorithm (AC) appeared due to the investigation of life ants behavior while searching for the shortest route between a source of food and the anthill [4]. It was discovered that while moving an ant produces a substance called pheromone which is left behind as a track. The pheromone track is used by the other members of the colony in searching for the food source. Moreover, the probability of choosing the way increases with the pheromone concentration on it. Such ants' behavior can be interpreted as an optimization process. The pheromone contains the information about the quality of the route and is used as a data transfer method. The Ant Colony algorithm uses an artificial ant to represent the probability greedy algorithm which constructs solutions. During the construction of those solutions the useful data is accumulated and processed in the Ant Colony algorithm. It influences the further search and can be interpreted as analogue of life ants' pheromone. Thus, the behavior of some ants is modeled in the Ant Colony algorithm. The stopping criterion can be expressed by a number of iterations, computational time and other. During last few decades the Ant Colony algorithms have been continually developed and used in discrete optimization problems [4]. In this paper we present the development of the Ant Colony algorithm for the p-median competitive facility location and design problem. The specifics of the problem have been taken into account while developing the algorithm. We have constructed the artificial ant algorithm using an original greedy algorithm; considered the special types of neighborhoods; and offered the rules for the attractiveness evaluation, the pheromone value, the probability of the facility opening and etc.

The Ant Colony algorithm is based on the Artificial Ant algorithm (AA). Let us introduce the AA algorithm first in order to describe the AC algorithm. The AA is represented as an original greedy algorithm. Its main idea can be described as follows: at first, possible facility locations with the highest objective function value are defined. After that the search of the best type for the open facilities is carried out. Let us introduce the following notations: t is the iteration number, b^t is the available budget at the current iteration t. L^t is a set of facilities which can be opened in iteration t. It consists of pairs (j, r), where $j \in S$ is a possible facility location and $r \in R$ is the type of facility at this location. Let D^t be the current set of located facilities. It consists of pairs (j, r), where $j \in S$ is the location of the opened facility and $r \in R$ is the type of facility at the given location;

$f(D^t)$ be the objective function value at set D^t;

$\rho_{jr}(D^t) = f(D^t \cup (j, r)) - f(D^t)$ be the objective function improvement if the pair $(j, r) \in L^t$ is added into the current location D^t;

Vector $\Delta_{jr} = c_{j(r+1)} - c_{j(r)}, r > 0$, shows how many budget units have to be spent if a facility of type r is substituted by a facility of type $r + 1$.

Assume that the number of types R is equal for all the facilities. It is assumed that for any subset $Z \subset S$ of cardinality p the budget size $B \geq \sum_{j \in Z} c_{j1}$, otherwise the problem is unsolvable. The greedy algorithm constructs the feasible solution that defines the location and the type of a facility to be opened with the limited budget B. Scanning of all the possible locations set S takes considerable time, that is why the subset $Z_1 \subset S$ will be browsed. The cardinal number Z_1 is equal to $2p$ ($|Z_1| = 2p$). The proposed algorithm consists of two stages. In the first stage the facility locations are defined. For this purpose p facilities of the first type are opened. Thereby the number of budget units $\sum_{j \in Z_1} c_{j1}$ is spent. In the second stage, a subset $Z_2 \subset Z_1$ of cardinality p ($|Z_2| = p$) is formed. During the second stage the left budget units $B - \sum_{j \in Z_1} c_{j1}$ are assigned to improve the types of the facilities opened in the first stage.

Let us describe a scheme of the artificial ant algorithm for competitive p-median problem with elastic demand (AACP).

First Stage
0. Identify the initial set of possible locations $L^0 := \{(j,1) | j \in Z_1\}$. Elements of set S are included into set Z_1 with the probability φ (see below). Define the set of located facilities $D^0 = \varnothing$, the auxiliary subset $Z_2^0 = \varnothing$, the initial budget volume $b^0 = B$, the number p of the facilities to be opened, and the initial value of the first stage iteration counter $t := 0$.

Iteration t
Execute the following steps until $t \leq p$.

1. Search for a pair $(j, 1)$ out of set L^t where the objective function gains its best improvement, specifically

$$(\hat{\jmath}, 1) = argmax_{(j,1) \in L^{t-1}} \{\rho_{j1}(D^t)\}.$$

2. Move the chosen pair away from the set of possible locations L^t and add it to the subset of opened facilities D^t, i.e. $L^{t+1} = L^t \backslash (\hat{\jmath}, 1)$, $D^{t+1} = D^t \cup (\hat{\jmath}, 1)$. If the improvement of the facility type in the location $\hat{\jmath}$ is possible, i.e. $|R| \geqslant 2$, then add the new pair into the subset $Z_2^{t+1} = Z_2^t \cup (\hat{\jmath}, 2)$.
3. Reduce the current budget by the spent amount $b^{t+1} := b^t - c_{\hat{\jmath},1}$, $t := t+1$. If $t \leq p$ then move to the next iteration of the first stage, otherwise $B_1 = b^{t+1}$ move to the second stage.

Second Stage
0. If $Z_2 = \varnothing$ or $B_1 = 0$, then the greedy algorithm completes its work. That means it is impossible to improve the functional variant of the opened facilities in the second stage or all the budget has been spent. Otherwise the initial sets $L^0 := Z_2$, $D^0 = D$ are to be formed, the initial budget volume $b^0 := B_1$ and value of the second stage iteration counter $t := 0$ are to be defined.

Iteration t
Implement the following steps until $b^t > 0$.

1. Search for a pair (j, r) out of set L^t where the objective function gains its best improvement, specifically

$$(\hat{\jmath}, \hat{r}) = argmax_{(j,r) \in L^{t-1}} \{\rho_{jr}(D^t)\}.$$

2. Move the chosen pair away from the set of possible locations L^t and add it to the subset of opened facilities D^t, i.e. $L^{t+1} = L^t \backslash (\hat{\jmath}, \hat{r})$, $D^{t+1} = D^t \cup (\hat{\jmath}, \hat{r})$. If a facility has already been placed in the location $\hat{\jmath}$, then remove the previous location from D^t:

$$D^{t+1} = D^t \backslash (\hat{\jmath}, \hat{r} - 1).$$

If the improvement of the facility type in the location $\hat{\jmath}$ is possible, i.e. $\hat{r}+1 \leqslant |R|$, then add the new pair into the subset $L^{t+1} = L^t \cup (\hat{\jmath}, \hat{r} + 1)$.

3. Reduce the size of the available budget by the spent amount $b^{t+1} := b^t - \Delta_{\hat{j},\hat{r}}$. If $b^{t+1} = 0$, then all the budget has been spent and the permissible solution has been found. The algorithm finishes its work. Otherwise go to the next iteration of the second stage $t := t + 1$.

The following parameters are used in the Ant Colony algorithm:
G is the quantity of artificial ants;
g is the quantity of better solutions found by the artificial ants;
$\alpha^t = (\alpha_j^t)$ is the numeric vector which contains some information about the solution search at the iteration t. This vector is an analogue of the real ants pheromone and is traditionally called 'pheromone vector'. Its components $\alpha_j^t, j \in S$ contain the values of this parameter (the pheromone level) and are calculated using the formula

$$\alpha_j^t = \frac{\alpha_{min} + q^{1-\gamma_j^t}(\alpha_j^{t-1} - \alpha_{min})}{\rho}, j \in S, \tag{6}$$

where γ_j^t is the frequency of the facility j appearance in g best solutions; $\rho \in (0, 1)$ is the decay coefficient (pheromone evaporation); $q \in (0, 1)$ is the algorithm parameter; α_{min} is the minimal pheromone level value.

The possibility for the facility j to be included into set Z_1 on iteration t is calculated the following way:

$$\varphi_j^t = \frac{\alpha_j^t(\Delta f_j^t + \varepsilon)}{\sum_{k \in S} \alpha_k^t(\Delta f_j^t + \varepsilon)}, j \in S, \tag{7}$$

where Δf_j^t is the objective function change as a result of the functional improvement in the location $j \in S$ at iteration t; $\varepsilon > 0$ excludes 0 division.

Let us introduce the Ant Colony algorithm scheme for the problem under consideration.

Ant Colony Algorithm
0. Set the initial values $\rho, \alpha_{min}, \alpha_j, \varphi_j^1, \gamma_j^1, j \in S$; the best objective function value $F = 0$, iteration counter value $t := 1$.

Repeat the following steps until the stopping criterion is achieved.

Iteration t

1. Build G possible solutions by the artificial ant algorithm AACP, namely find $D_1^t, \ldots, D_G^t$ and $f(D_1^t), \ldots, f(D_G^t)$.
2. Arrange the solutions in non-increasing order of the objective function f values and choose g best of them; define f^* as the best objective function value at the current iteration t.
3. Renew α_j^t using g best solutions according to the formula (6).
4. Recalculate φ_j^t based on α_j^t according to the formula (7).
5. If $f^* > F$, then $F := f^*$.

Move to the next iteration $t := t + 1$.

In order to check the algorithm's quality, a computational experiment has been carried out. It consisted of the two stages: (1) setting up the algorithms and parameters; (2) an experimental evaluation of the algorithms defined in (1). The results are described further in Sects. 4 and 5.

4 Algorithms and Parameters Turning

The specifics of the problem under consideration should be taken into account in the process of developing the ant colony algorithm. It is necessary to build an original artificial ant algorithm and adjust the set of parameters of it.

All the experimental studies of tuning and determining the quality of the proposed algorithm were carried out for two series of test cases constructed using real data from [1]. In the first series (Series 1) the distances among the points were generated by the uniform distribution of distances in the interval [0; 30]; this corresponds to the case when the distance between the points is measured by the road network. In the second series (Series 2), the distances are measured on a straight line and therefore satisfy the triangle inequality. Each series consists of the set of 16 instances with the dimension $|N| = 60, 80, 100, 150, 200, 300$ with three possible projects and budget limits of 3, 5, 7 and 9 units. The demand is elastic, its parameter $\lambda = 1$. The customers are sensitive to the distance, $\beta = 2$. The number p of the facilities to be opened was selected in such a way, that the set of possible solutions was not empty. That is why the following combinations of the budget values with the numbers of facilities to be opened were used in the experimental design: $(B, p) = \{(3, 2), (5, 3), (7, 4), (9, 6)\}$.

In many cases, it is possible to compare the obtained solution with the optimal solution or with the best known solution for other tasks. Anyway, it is often impossible to find even a feasible solution for our problem. Therefore, we have to use the upper bounds constructed in [7] and the results of the Simulated Annealing (SA) algorithm proposed by us in the article [9]. SA is currently known for its successful usage for a wide range of optimization problems [8]. Before starting the SA algorithm, the initial values of the parameters are set. At each value of temperature a certain number of iterations is carried out. On each iteration a new solution is selected from neighborhood of the current solution randomly. This solution is accepted as a new current one according to some probabilistic law and the value of the temperature parameter decreases. The process continues until the system reaches the frozen state or until other stopping criteria are met. On the basis of the competitive p-median facility location and design problem specificity the special kinds of neighborhoods have been built. Best parameter values for the simulated annealing algorithm have been found [9].

In the process of tuning the algorithms, it is necessary to choose the parameter values for the best behaviour of the algorithm in most instances. Based on our previous experience of the ant colony algorithm developing, as well as the current numerical experiments, the following parameter values were chosen: the pheromone evaporation coefficient $\rho = 0.95$; the initial pheromone level $\alpha_0 = 0.5$, the minimal pheromone level $\alpha_{min} = 0.3$; the number of ants at each iteration is

$L = 20$; the maximal number of iterations is equal to $T_{max} = 10$. The probability of the facility choice used in the AACP is changed by the formula (7).

The tuning of the Ant Colony algorithm were carried out. Two Artificial Ant algorithm modifications were proposed. In variation AACP1 the points are sorted by not increasing of the values φ_i. After that a number of $2 \cdot p$ best points are chosen with the given probability and included in set Z_1. However, for such organization of the artificial ant the maximal deviations of the AC for $N = 300$ is 7% for Series 1 (27%, Series 2) (Tables 1), the calculation time for the both Series took about 700 s. That is why the 2nd modification of the Ant Colony algorithm was constructed. In the second variation AACP2 points are chosen with a given probability alternately: first, out of the front part of the arranged list; second, from its rare part its end; and then included in the set Z_1. The iterations are repeated until there are $2 \cdot p$ elements in the set Z_1. The second modification turned out to be better than the first one. That is why AACP2 was used in the main experiment. More detailed information about the maximal (max), average (aver) and minimal (min) deviations and the calculation time is given in Tables.

Table 1. Ant Colony with AACP1, deviations from upper bounds, %

	Series 1			Series 2		
N	Min	Aver	Max	Min	Aver	Max
60	0.000	0.630	5.713	19.274	29.987	49.883
80	0.000	0.545	7.306	18.802	27.285	37.001
100	0.000	0.779	7.057	10.027	22.180	33.692
150	0.000	1.194	6.992	18.648	26.692	35.501
200	0.000	0.991	6.756	10.271	17.049	25.030
300	0.000	1.133	7.313	9.330	17.108	27.483

5 Main Stage of Computational Experiments

In the main stage of the experimental studies, the numerical experiments were conducted using the algorithms with the selected settings. The algorithms were implemented, the computational experiments were carried out on a computer with CPU Intel® Xeon® X5675 @ 3.07 GHz, 32 GB RAM. In connection with the probabilistic nature of the ant colony algorithm, 100 runs of the algorithm were performed on each test instance. Table 2 contains information about the deviations of the ant colony algorithm for the series of the test cases with maximum dimension of 300 points. It can be seen that the average deviation of the ant colony algorithm for the first series is 0.758%, the maximal deviation does not exceed 4.959%. For the simulated annealing algorithm the average deviation

Table 2. Deviation from upper bounds, %

	Uniform distribution distances (Series 1)						Euclidean distances (Series 2)					
	AC(AACP2)			SA			AC(AACP2)			SA		
N	Min	Aver	Max	Min	Aver	Max	Min	Aver	Max	Min	Aver	Max
60	0.000	0.480	3.699	0.000	0.483	3.326	13.222	29.891	49.883	19.274	29.695	49.883
80	0.000	0.169	1.128	0.000	0.101	0.825	19.607	26.876	33.252	18.802	26.495	33.252
100	0.000	0.569	4.954	0.000	0.450	2.421	14.927	22.135	31.747	10.027	21.577	33.029
150	0.000	0.660	5.241	0.000	0.263	2.212	19.244	26.483	32.986	18.349	25.666	32.189
200	0.000	0.640	3.338	0.000	0.190	3.011	10.884	16.430	22.535	10.178	16.123	22.535
300	0.000	0.758	4.959	0.000	0.384	3.162	9.330	16.443	20.247	9.330	15.775	18.912

for the first series instances of 300 locations is 0.384%, and the maximal deviation not exceeds 3.162%. The minimal deviation is less then 0.001% for both algorithms. This result is very good for the problems of such dimension. As for the second series, we can see a different situation. Even the minimal deviation from the upper bound is 9.330% for AC. Here the average algorithm deviation is 16.443%, the maximal deviation is 20.247%. The average deviations for the simulated annealing algorithm in this case for the same dimension is 15.775%, the maximal deviation is 18.912%. Anyway, that is not a failure and can be explained by different reasons. We assume that the estimate is sufficiently inaccurate for this case. This is partially confirmed by our research for other test data. For Series 1, the 95% confidence interval for the probability of obtaining the deviations less than 0.001% for the ant colony algorithm is between [0.441; 0.461]. For Series 2, the 95% confidence interval for the probability of obtaining the deviations less than 20% for colony algorithm is [0.341; 0.360].

Table 3. Comparison of CPU time, sec

	AC(AACP1)			AC(AACP2)			SA		
N	Min	Aver	Max	Min	Aver	Max	Min	Aver	Max
60	5.441	17.377	39.255	3.336	9.376	20.724	5.238	12.174	24.052
80	8.337	29.085	69.263	4.686	14.831	31.555	9.126	18.855	33.777
100	11.224	43.385	102.504	7.578	20.597	49.377	12.972	25.287	48.124
150	32.589	94.605	187.371	16.582	52.049	120.010	28.346	51.234	106.724
200	48.695	148.818	287.950	29.767	81.436	177.184	43.980	75.961	146.520
300	127.028	362.436	704.494	72.948	201.838	412.701	94.818	166.465	312.111

The CPU time for both series was mostly the same, that is why there is the computation time for the first series only in Table 3. On average, the Ant Colony algorithm is faster than other algorithms for various difficult problems (for example, [5,6]). Ant colony algorithm faster then simulated annealing algorithm on dimensions $N = 60, 80, 100$. Note that well-known commercial software is rather time-consuming. Solver CoinBonmin of system GAMS was used for both series. For instance, for one of the instances of 60 locations the CPU time of CoinBonmin was 63 h and the objective function deviation from the upper bounds was 12.919%.

6 Conclusion

The paper is devoted to the study of one of the decision-making situations in the competitive environment. A new p-median competitive facility location and design problem with elastic demand is considered. The problem-oriented version of the ant colony algorithm is proposed. The adjustment of its parameters is carried out with the help of a special numerical experiment. In the main stage of the experimental studies the numerical experiments were conducted using the algorithms with selected settings, interesting data were obtained. In particular, the minimum deviation from the upper bounds did not exceed 0.001% for the series of test cases with uniform distribution of distances. However, this value was not less than 9% for the other series with Euclidean distances. The maximum computational time for the largest problem of 300 facilities did not exceed 400 s, which is comparable to the time of Simulated Annealing algorithm and it is significantly less than the time of the CoinBonmin (GAMS). It would be interesting to continue the research in this direction. At the present moment, we are investigating the multi-thread implementation of the algorithm. This should improve the running time of the algorithm. In general, the research results, as well as the relative simplicity of the proposed algorithm implementation, indicate applicability of these methods to the class of problems under consideration.

Acknowledgement. This research was supported by the Russian Foundation for Basic Research, grant 18-07-00599.

References

1. Aboolian, R., Berman, O., Krass, D.: Competitive facility location and design problem. Eur. J. Oper. Res. **182**(1), 40–62 (2007)
2. Davidović, T., Teodorović, D., Šelmić, M.: Bee colony optimization - part I: the algorithm overview. Yug. J. Oper. Res. **25**(1), 33–56 (2015)
3. Davydov, I.A., Kochetov, Y.A., Mladenovic, N., Urosevic, D.: Fast metaheuristics for the discrete $r|p$-centroid problem. Autom. Remote Control **75**(4), 677–687 (2014)
4. Dorigo, M., Stützle, T.: Ant Colony Optimization. MIT Press, Cambridge (2004)
5. Kochetov, Y., Alekseeva, E., Levanova, T., Loresh, M.: Large neighborhood local search for the p-median problem. Yug. J. Oper. Res. **15**(1), 53–63 (2005)
6. Levanova, T.V., Loresh, M.A.: Algorithms of ant system and simulated annealing for the p-median problem. Autom. Remote Control. **65**(3), 431–438 (2004)
7. Levanova, T., Gnusarev, A.: Variable neighborhood search approach for the location and design problem. In: Kochetov, Y., Khachay, M., Beresnev, V., Nurminski, E., Pardalos, P. (eds.) DOOR 2016. LNCS, vol. 9869, pp. 570–577. Springer, Cham (2016). https://doi.org/10.1007/978-3-319-44914-2_45

8. Levanova, T., Gnusarev, A.: Development of threshold algorithms for a location problem with elastic demand. In: Lirkov, I., Margenov, S. (eds.) LSSC 2017. LNCS, vol. 10665, pp. 382–389. Springer, Cham (2018). https://doi.org/10.1007/978-3-319-73441-5_41
9. Levanova, T., Gnusarev, A.: Simulated annealing for competitive p-median facility location problem. J. Phys: Conf. Ser. **1050**, 1 (2018)
10. Yang, X.-S.: Nature-Inspired Metaheuristic Algorithms. Luniver Press (2010)

Bounds for Non-IRUP Instances of Cutting Stock Problem with Minimal Capacity

Artem V. Ripatti(✉) and Vadim M. Kartak

Ufa State Aviation Technical University, Karl Marx str. 12, 450008 Ufa, Russia
ripatti@inbox.ru, kvmail@mail.ru

Abstract. We consider the well-known cutting stock problem (CSP). An instance of CSP possesses IRUP (the integer round up property) if difference (the gap) between its optimal function value and optimal value of its continuous relaxation is less than 1. If the gap is 1 or greater, then an instance is non-IRUP. Constructing non-IRUP instances is very hard and a question about how large the gap can be is an open theoretical problem. Aim of our research is to find non-IRUP instances with minimal capacity. We have found a non-IRUP instance with integer sizes of items having capacity $L = 16$, while a previously known instance of such kind had capacity $L = 18$. We prove that all instances with capacity $L \leq 10$ have IRUP.

Keywords: Cutting stock problem · Integer round up property · Capacity · Linear programming

1 Introduction

In classic formulation, the cutting stock problem (CSP) is stated as follows: there are $m \in \mathbb{N}$ groups of items of different lengths $l_1, \cdots, l_m$ and availabilities $b_1, \cdots, b_m$. The goal is to pack all items into the minimal number of containers of the same capacity L (the total length of all items inside any container should not exceed L).

The cutting stock problem is one of the earliest problems that have been studied through methods of operational research [10]. This problem has many real-world applications, especially in industries where high-value material is being cut [6] (steel industry, paper industry). No exact algorithm is known that solves practical problem instances optimality, so there are lots of heuristic approaches. Each year the number of publications about this problem increases, so we refer the reader to bibliography [20] and the most resent survey [4].

Throughout this paper we abbreviate an instance of CSP as $E := (L, l, b)$. The total number of items is $n = \sum_{i=1}^{m} b_i$. Without loss of generality, we assume

Supported by RFBR, project 19-07-00895.

I. Bykadorov et al. (Eds.): MOTOR 2019, CCIS 1090, pp. 79–85, 2019.
https://doi.org/10.1007/978-3-030-33394-2_7

that all numbers in the input data are positive integers and $0 < l_1 < \cdots < l_m \leq L$.

The classical approach for solving CSP is based on the formulation of Gilmore and Gomory [9]. Any subset of items (called a *pattern*) is formalized as a vector $a = (a_1, \cdots, a_m)^\top \in \mathbb{Z}_+^m$ where $a_i \in \mathbb{Z}_+$ denotes the number of items i in the pattern a. A pattern a of E is *feasible* if $a^\top l \leq L$. So, we can define the set of all feasible patterns $P^f(E) = \{a \in \mathbb{Z}_+^m \mid a^\top l \leq L\}$. For a given set of patterns $P = \{a^1, \cdots, a^r\}$, let $A(P)$ be the $(n \times r)$-matrix whose columns are given by the patterns a^i. Then the CSP can be formulated as follows:

$$z(E) := \sum_{i=1}^{r} x_i \to \min \text{ subject to } A(P^f(E))x = b, x \in \mathbb{Z}_+^r.$$

The common approximate solution approach involves considering *the continuous relaxation* of CSP

$$z_C(E) := \sum_{i=1}^{r} x_i^C \to \min \text{ subject to } A(P^f(E))x^C = b, x^C \in \mathbb{R}_+^r.$$

Here x and x^C are called *the optimal solutions* for the integer and continuous problems respectively, and $z(E)$ and $z_C(E)$ are called *the optimal function values*.

The difference $\Delta(E) = z(E) - z_C(E)$ is called *the gap* of instance E. Practical experience and numerous computations have shown that for most instances the gap is very small. An instance E has *the integer round up property* (IRUP) if $\Delta(E) < 1$. Otherwise, E is called a non-IRUP instance. This notation was introduced by Baum and Trotter [1]. Subsequently, the largest known gap was increased [7,8,14,15,17,19]. Currently, the largest gap known is $\frac{6}{5}$, and there is no example of a gap of at least 2.

The first known constructions of non-IRUP instances were rather huge. The example of Marcotte [14] having $L = 3\,397\,386\,355$ was decreased to $L = 1\,111\,139$ by Chan [3]. The authors of [2,5] gave an example with $L = 100$. In [13] an example with $L = 18$ has been found using so-called *equivalence of instances* (see also [11,12]). In this paper we focus on improving bounds for minimal possible L.

The paper has the following structure. In Sect. 2 we describe some theory related to our model which is presented in Sect. 3. In Sect. 4 we present computational results and, finally, we draw a conclusion in Sect. 5.

2 Preliminaries

In this section we describe some theory we use throughout the paper. We believe that results presented in this section are well-known, but we present their proofs for the sake of completeness. Anyway, the reader may be referred to [16,18] where some results are mentioned.

Given an instance $E = (L, l, b)$, let x^C be an optimal continuous solution of E. Then $\overline{x}$ is the integer part of x^C and x^* is the fractional part of x^C, so

$x^C = \overline{x} + x^*$ and $0 \le x_i^* < 1$ for all $1 \le i \le r$. Also $z_C(E) = e^\top \overline{x} + e^\top x^*$ where $e = (1, \cdots, 1) \in \mathbb{R}_+^m$. Replacing b by $\overline{b} = b - A(P^f(E))\overline{x}$ yields the *residual* instance $\overline{E} = (L, l, \overline{b})$. An instance E is called *reducible* if there exists an optimal continuous solution of E with non-zero integer part.

Lemma 1. *x^* is an optimal continuous solution of $\overline{E}$.*

Proof. On one hand, $z_C(\overline{E}) \le e^\top x^*$ because x^* is a feasible solution for $A(P^f(E))x = \overline{b}$. On the other hand $z_C(E) \le e^\top \overline{x} + z_C(\overline{E})$ which is equivalent to $z_C(\overline{E}) \ge e^\top x^*$. So, $z_C(\overline{E}) = e^\top x^*$.

Lemma 2. $\Delta(E) \le \Delta(\overline{E})$.

Proof. $z(E) - z_C(E) \le (e^\top \overline{x} + z(\overline{E})) - (e^\top \overline{x} + e^\top x^*) = z(\overline{E}) - z_C(\overline{E})$.

Consider a set of instances $\mathbb{E}(L, l) = \{E(L, l, b) \mid b \in \mathbb{Z}_+^m\}$ where L and l are fixed. Now we are interested in maximal possible the gap $\Delta(E)$ over all instances $E \in \mathbb{E}(L, l)$.

Lemma 3. *The maximal gap $\Delta(E)$ occurs over instances $E \in \mathbb{E}(L, l)$ with $z_C(E) < m$.*

Proof. Consider an instance $E \in \mathbb{E}(L, l)$ with $z_C(E) \ge m$. There exists an optimal continuous solution x^C of E which has at most m non-zero components because all feasible patterns form a m-dimensional vector space. Then by the pigeonhole principle, $x_i^C \ge 1$ for some $1 \le i \le r$. Therefore E is reducible.

Using Lemma 3 we already can find the maximal possible gap over $\mathbb{E}(L, l)$ in finite time iterating over all instances $E \in \mathbb{E}(L, l)$ with $z_C(E) < m$. However, we are going to improve this result.

Lemma 4. *The maximal gap $\Delta(E)$ occurs over instances $E \in \mathbb{E}(L, l)$ with $z(E) \le m$.*

Proof. Consider the m-dimensional vector space S induced by all feasible patterns. Let us build a convex hull H over all the feasible patterns and let $F = \{(f_1, \cdots, f_m)^\top\}$ be a set of facets of H, where $f_i \in P^f(E)$.

Consider some facet $f = (f_1, \cdots, f_m)^\top \in F$. The linear combination $x^\top f$ where $x \in \mathbb{R}_+^m$ covers some subspace S' of S. Every integer point $b \in S'$ corresponds to an instance $E = (L, l, b)$, and a vector x such that $b = x^\top f$ can always be transformed into an optimal continuous solution x^C of E by inserting zero elements. Now consider a m-dimensional parallelepiped $S_1' \subset S' \subseteq S$ formed by linear combination ${x^1}^\top f$ where $x^1 \in \mathbb{R}_+^m$ and $0 \le x_i^1 \le 1$ for all $1 \le i \le m$. All integer points b inside S_1' correspond to instances E with $z(E) \le m$. And all integer points b from S' outside S_1' correspond to reducible instances E.

We remark that Lemma 3 does not imply Lemma 4 directly because there exist instances E with $z(E) \ge m + 1$ and $z_C(E) < m$ (and with $\Delta(E) > 1$).

3 Model

Consider an instance $E = (L, l, b)$ which possesses all possible item lengths: $l = (1, \cdots, L)$. When L is fixed, l is fixed too. The matrix of patterns $A(P^f(E))$ depends on L and l only, so it is also fixed. Availabilities of the item types b we consider as variables.

Now we build the following ILP model:

$$\begin{aligned} k - \sum x_i &\to \max, \\ A(P^f(E))x &= b, \quad (1) \\ A(P^f(E))y &= b, \quad (2) \\ \sum y_i &= k, \\ x &\in \mathbb{R}^r_+, \\ y &\in \mathbb{Z}^r_+, \\ b &\in \mathbb{Z}^m_+, \end{aligned}$$

where k is the fixed value of $z(E)$ for an optimal integer solution, x is the optimal continuous solution, and y is the optimal integer solution.

Now we have to ensure that k is indeed the optimal integer function value of E, i.e. a solution of the system where $\sum y_i < k$ is impossible. To this end we add special constraints to bound the vector b:

$$\begin{aligned} &b_i \geq u + 1 - w_i^u(u+1) \quad \forall 0 \leq u \leq (k-1)L, 1 \leq i \leq m, \\ &w \in \mathbb{B}^{(k-1)L+1} \times \mathbb{B}^m. \end{aligned}$$

Here, $b_i \leq u$ implies $w_i^u = 1$. Now consider all integer solutions of size $k-1$ $C_{k-1}(E) := \{A(P^f(E))y \mid y \in \mathbb{Z}^r_+ \wedge \sum y_i = k-1\}$. To ensure that all integer solutions of the system are not less than k, we add the following constraints:

$$\sum_{i=1}^{m} w_i^{c_i} \leq m - 1, \quad \forall c \in C_{k-1}(E).$$

The latter constraint works as follows: for fixed $c \in C_{k-1}(E)$, if $b_i \leq c_i$ for all $1 \leq i \leq m$, then $\sum w_i^{c_i} = m$ and we have an integer function value less than k.

For small values of L and k the model is small enough to be solved in reasonable time. However, the model can be reduced by the following observation: l can be $(1, \cdots, L-1)$ because any number of items of size L does not change the value of $\Delta(E)$. Also, the model can be further reduced by replacing the set of feasible patterns $P^f(E)$ with a set of *inextensible* patterns $P^f_*(E) = \{a \in \mathbb{Z}^m_+ \mid a^\top l \leq L \wedge a^\top l + l_1 > L\}$. To this end we also have to transform equations (1) and (2) into inequalities with the "$\geq$" sign. The final model is the following:

$$
\begin{aligned}
& k - \sum x_i \to \max, \\
& A(P^f_*(E))x \geq b, \\
& A(P^f_*(E))y \geq b, \\
& \sum y_i = k, \\
& b_i \geq u + 1 - w^u_i(u+1) \quad \forall 0 \leq u \leq (k-1)L, 1 \leq i \leq m, \\
& \sum_{i=1}^{m} w^{c_i}_i \leq m - 1, \quad \forall c \in C_{k-1}(E), \\
& w \in \mathbb{B}^{(k-1)L+1} \times \mathbb{B}^m, \quad x \in \mathbb{R}^r_+, \quad y \in \mathbb{Z}^r_+, \quad b \in \mathbb{Z}^m_+.
\end{aligned}
$$

Now, using this model we can find instances with the maximal gap for fixed L and k, and using Lemma 4 we can build the lower bound for minimal possible L of non-IRUP instance by solving the model for all $k < L$ for some fixed L.

4 Results

We implemented our model as C++ program using CPLEX 12.7. The program was run on machine Intel Core i7-5820K 4.2 GHz with 6 cores 32 Gb RAM. Results are presented in Table 1.

Table 1. Computational results

$L \backslash k$	1	2	3	4	5	6	7	8	9
2	0.50000								
3	0.66667	0.66667							
4	0.75000	0.75000	0.75000						
5	0.80000	0.80000	0.80000	0.80000					
6	0.83333	0.83333	0.83333	0.83333	0.83333				
7	0.85714	0.85714	0.85714	0.85714	0.85714	0.85714			
8	0.87500	0.87500	0.87500	0.87500	0.87500	0.87500	0.87500		
9	0.88889	0.88889	0.88889	0.88889	0.88889	0.88889	0.88889	0.88889	
10	0.90000	0.90000	0.90000	0.90000	0.90000	0.90000	0.90000	0.90000	0.90000
11	0.90909	0.90909	0.91667	0.91667	0.91667				
12	0.91667	0.91667	0.93750	0.93750	0.93750				
13	0.92308	0.92308	0.93333	0.93333	0.93333				
14	0.92857	0.92857	0.94444	0.94444	0.94444				
15	0.93333	0.93333	0.96667	0.96667	0.96667				
16	0.93750	0.93750	1.00000	1.00000					
17	0.94118	0.94118	0.96667	0.97222					

For $L = 10$ and $k = 9$ the ILP program had $\approx 600\,000$ constraints and the running time was about 52 h. The following non-IRUP instances were found: $(16, (2, 3, 7, 8, 10)^\top, (2, 1, 1, 1, 1)^\top)$ and $(16, (2, 3, 7, 8, 10)^\top, (2, 1, 1, 3, 1)^\top)$.

It is possible to prove a more general result; namely, that a series of instances $E_t = (16, (2, 3, 7, 8, 10)^\top, (2, 1, 1, 2t + 1, 1)^\top)$ is non-IRUP for every t. $z(E_t) = t + 3$, because it is easy to pack all the items into $t + 3$ containers, but it is impossible to pack them into $t + 2$ ones. Indeed, suppose it is possible, then all containers should be fully filled. By a parity argument, items of sizes 3 and 7 should be in a single container, but then there is no way to fill this container completely. $z_C(E_t) \leq t + 2$ because there is a feasible solution $\frac{1}{2}(0, 2, 0, 0, 1)^\top + \frac{1}{2}(3, 0, 0, 0, 1)^\top + \frac{2t+1}{2}(0, 0, 0, 2, 0)^\top + \frac{1}{2}(1, 0, 2, 0, 0)^\top$. So, $\Delta(E_t) \geq 1$.

5 Conclusion

We have suggested a model to calculate the maximal possible gap for the fixed capacity L and the optimal integer function value k and have run it for small values of L and k. We have improved the best known bound for minimal capacity of non-IRUP instance from $L = 18$ to $L = 16$. Also we have computationally proved that all instances with $L \leq 10$ have IRUP.

We conjecture that $L = 16$ is the sharp bound for the minimal possible L and we plan to improve our model to prove this conjecture.

Acknowledgements. The authors would like to thank the anonymous referees for valuable remarks.

References

1. Baum, S., Trotter Jr., L.: Integer rounding for polymatroid and branching optimization problems. SIAM J. Algebraic Discrete Methods **2**(4), 416–425 (1981)
2. Caprara, A., Dell'Amico, M., Díaz Díaz, J., Iori, M., Rizzi, R.: Friendly bin packing instances without integer round-up property. Math. Program. Series A and B (2014). https://doi.org/10.1007/s10107-014-0791-z
3. Chan, L., Simchi-Levi, D., Bramel, J.: Worst-case analyses, linear programming and the bin-packing problem. Math. Program. **83**(1–3), 213–227 (1998)
4. Delorme, M., Iori, M., Martello, S.: Bin packing and cutting stock problems: mathematical models and exact algorithms. Eur. J. Oper. Res. **255**(1), 1–20 (2016)
5. Díaz Díaz, J.: Exact and heuristic solutions for combinatorial optimization problems. Ph.D. thesis, Università Degli Studi di Modena e Reggio Emilia (2012)
6. Dyckhoff, H., Kruse, H.J., Abel, D., Gal, T.: Trim loss and related problems. Omega **13**(1), 59–72 (1985)
7. Fieldhouse, M.: The duality gap in trim problems. SICUP Bull. **5**(4) (1990)
8. Gau, T.: Counter-examples to the IRU property. SICUP Bull. **12**(3) (1994)
9. Gilmore, P., Gomory, R.: A linear programming approach to the cutting-stock problem. Oper. Res. **9**(6), 849–859 (1961)
10. Kantorovich, L.V.: Mathematical methods of organizing and planning production. Manage. Sci. **6**(4), 366–422 (1960)

11. Kartak, V.M., Ripatti, A.V.: Large proper gaps in bin packing and dual bin packing problems. J. Global Optim. (2018). https://doi.org/10.1007/s10898-018-0696-0
12. Kartak, V.M., Ripatti, A.V.: The minimum raster set problem and its application to the d-dimensional orthogonal packing problem. Eur. J. Oper. Res. **271**(1), 33–39 (2018). https://doi.org/10.1016/j.ejor.2018.04.046
13. Kartak, V.M., Ripatti, A.V., Scheithauer, G., Kurz, S.: Minimal proper non-IRUP instances of the one-dimensional cutting stock problem. Discrete Appl. Math. **187**(Complete), 120–129 (2015)
14. Marcotte, O.: An instance of the cutting stock problem for which the rounding property does not hold. Oper. Res. Lett. **4**(5), 239–243 (1986)
15. Rietz, J., Dempe, S.: Large gaps in one-dimensional cutting stock problems. Discrete Appl. Math. **156**(10), 1929–1935 (2008)
16. Rietz, J., Scheithauer, G., Terno, J.: Tighter bounds for the gap and non-IRUP constructions in the one-dimensional cutting stock problem. Optimization **51**(6), 927–963 (2002)
17. Scheithauer, G., Terno, J.: About the gap between the optimal values of the integer and continuous relaxation one-dimensional cutting stock problem. In: Gaul, W., Bachem, A., Habenicht, W., Runge, W., Stahl, W.W. (eds.) Operations Research Proceedings, vol. 1991, pp. 439–444. Springer, Heidelberg (1992). https://doi.org/10.1007/978-3-642-46773-8_111
18. Scheithauer, G., Terno, J.: The modified integer round-up property of the one-dimensional cutting stock problem. Eur. J. Oper. Res. **84**(3), 562–571 (1995)
19. Scheithauer, G., Terno, J.: Theoretical investigations on the modified integer round-up property for the one-dimensional cutting stock problem. Oper. Res. Lett. **20**(2), 93–100 (1997)
20. Sweeney, P.E., Paternoster, E.R.: Cutting and packing problems: a categorized, application-orientated research bibliography. J. Oper. Res. Soc. **43**(7), 691–706 (1992)

Merging Variables: One Technique of Search in Pseudo-Boolean Optimization

Alexander A. Semenov(✉)

ISDCT SB RAS, Irkutsk, Russia
biclop.rambler@yandex.ru

Abstract. In the present paper we describe new heuristic technique, which can be applied to the optimization of pseudo-Boolean functions including Black-Box functions. This technique is based on a simple procedure which consists in transition from the optimization problem over Boolean hypercube to the optimization problem of auxiliary function in a specially constructed metric space. It is shown that there is a natural connection between the points of the original Boolean hypercube and points from new metric space. For a Boolean hypercube with fixed dimension it is possible to construct a number of such metric spaces. The proposed technique can be considered as a special case of Variable Neighborhood Search, which is focused on pseudo-Boolean optimization. Preliminary computational results show high efficiency of the proposed technique on some reasonably hard problems. Also it is shown that the described technique in combination with the well-known (1+1)-Evolutionary Algorithm allows to decrease the upper bound on the runtime of this algorithm for arbitrary pseudo-Boolean functions.

Keywords: Pseudo-Boolean optimization · Local search · Variable Neighborhood Search · (1+1)-Evolutionary Algorithm · Boolean satisfiability problem

1 Basic Notions and Methods

Let $\{0,1\}^n$ be a set of all possible binary vectors (strings) of length n. The set $\{0,1\}^n$ is sometimes called a *Boolean hypercube*. Let us associate with $\{0,1\}^n$ a set consisting of n symbols: $X = \{x_1, \ldots, x_n\}$. The elements of X will be referred to as *Boolean variables*. Further we will consider $\{0,1\}^n$ as a set of all possible assignments of variables from X. For an arbitrary $X' \subseteq X$ by $\{0,1\}^{|X'|}$ we will denote a set of all possible assignments of variables from X'.

A pseudo-Boolean function (PBF) [1] is an arbitrary total function of the kind

$$f : \{0,1\}^n \to \mathbb{R}. \tag{1}$$

Example 1. Consider an arbitrary Conjunctive Normal Form (CNF) C, where $X = \{x_1, \ldots, x_n\}$ is a set of Boolean variables from this CNF. Let us associate with an arbitrary $\alpha \in \{0,1\}^n$ the number of clauses that take the value of 1 when

I. Bykadorov et al. (Eds.): MOTOR 2019, CCIS 1090, pp. 86–102, 2019.
https://doi.org/10.1007/978-3-030-33394-2_8

their variables take the values from α. Denote the resulting function by f_C. It is easy to see that f_C is a function of the kind $f_C : \{0,1\}^n \rightarrow \mathbb{N}_0$ ($\mathbb{N}_0 = \{0,1,2,\ldots\}$) and $\max_{\{0,1\}^n} f_C \leq m$, where m is the number of clauses in C. Then CNF C is satisfiable if and only if $\max_{\{0,1\}^n} f_C = m$. The problem $f_C \rightarrow \max_{\{0,1\}^n}$ represents the optimization formulation of the Boolean Satisfiability problem (SAT) and is often referred to as MaxSAT [2]. This problem is NP-hard, so there is a huge class of combinatorial problems, which can be effectively reduced to it.

The main result of the present paper is a technique applicable in the context of several common metaheuristic schemes. Before proceeding to its description, let us briefly describe the basic metaheuristics used below.

First, we will consider the simplest computational scheme, which belongs to the class of the local search methods. The concept of a *neighborhood* in a search space is at the core of the algorithms from this class. With each point of a search space the *neighborhood function* [3] associates a set of neighboring points. This set is called the neighborhood of the considered point. For an n-dimensional Boolean hypercube the neighborhood function is of the following kind:

$$\aleph : \{0,1\}^n \rightarrow 2^{\{0,1\}^n}. \tag{2}$$

A simple way to define function (2) is to associate an arbitrary $\alpha \in \{0,1\}^n$ with all points from $\{0,1\}^n$ for which the Hamming distance [4] from α is not greater than certain d. The number d is referred to as a *radius* of Hamming neighborhood. Hereinafter by $\aleph_d(\alpha)$ we denote a neighborhood of radius d of an arbitrary point α of a search space. By $\langle\{0,1\}^n, \aleph_1\rangle$ we denote a space $\{0,1\}^n$ in which a neighborhood of an arbitrary point α is $\aleph_1(\alpha)$.

Below we give a simple example of the local search algorithm which is sometimes referred to as Hill Climbing (HC). We can use this algorithm to maximize the functions of the kind (1). One iteration of the HC algorithm consists of the following steps.

Input: an arbitrary point $\alpha \in \{0,1\}^n$, a value $f(\alpha)$;
1. α – current point;
2. traverse the points from $\aleph_1(\alpha) \setminus \{\alpha\}$, computing for each point α' from this set a value $f(\alpha')$. If there is such a point α', that $f(\alpha') > f(\alpha)$ then go to step 3, otherwise, go to step 4;
3. $\alpha \leftarrow \alpha'$, $f(\alpha) \leftarrow f(\alpha')$, go to step 1;
4. $\alpha^* \leftarrow \alpha$; $(\alpha^*, f(\alpha^*))$ is a local extremum of f on $\{0,1\}^n$;

Output: $(\alpha^*, f(\alpha^*))$.

By itself, Hill Climbing is a basic heuristic and, generally speaking, it does not guarantee that the global extremum of the considered function will be achieved (except for some specific cases). Usually, during the optimization of an arbitrary function (1) one attempts to go through a number of local extrema. As a result, a point with the best value of the *objective function* (1) is considered to be an output. The best value of this function found at the current moment is called *Best Known Value* (BKV).

Without any exaggeration it can be said that over the past half century a huge number of papers have been devoted to describing ways of escaping local extrema. Listing the key papers in this direction would take up too much space. A good review of the relevant results can be found in [3,5].

In some sense, one can view the evolutionary algorithms [5] as the alternative to local search methods. This class of algorithms can be described as "a variation on a theme of random walk". The simplest example of such algorithms is the (1+1)-Evolutionary Algorithm shortly denoted as (1+1)-EA [6]. Below we present the description of one iteration of this algorithm, which will be referred to as *(1+1)-random mutation.*

Input: an arbitrary point $\alpha \in \{0,1\}^n$, a value $f(\alpha)$;

- make (1+1)-random mutations of α: by going through α in fixed order, change every bit to the opposite with probability p; let α' be a result of a random mutation of α;
- if for a point α' it holds that $f(\alpha') \geq f(\alpha)$ (assuming that the maximization problem for function (1) is considered), then the next (1+1)-random mutation is applied to α', otherwise, (1+1)-random mutation is applied to α (this situation is called *stagnation*);

Output: $(\alpha', f(\alpha'))$, where α' is the result of several random mutations.

The probability p is usually determined as $p = 1/n$. It should be noted, that for any function of the kind (1) and points $\alpha, \alpha' \in \{0,1\}^n$ the probability of transition $\alpha \to \alpha'$ is non-zero. Let $\alpha^{\#}$ be the point of the global extremum of function (1). According to [7], the expected running time of the (1+1)-EA, denoted further as $E_{(1+1)-EA}$, is defined as the mean of the (1+1)-random mutations needed to achieve $\alpha^{\#}$ from an arbitrary initial point $\alpha \in \{0,1\}^n$.

The value $E_{(1+1)-EA}$ can be considered as a measure of efficiency for (1+1)-EA. If the value of function (1) is given by the oracle, the nature of which is not taken into account, then it could be shown (see [7]), that $E_{(1+1)-EA} \leq n^n$. It is important that this bound is reached (with minor reservations) for explicitly specified functions [7]. On the other hand, for an equiprobable choice of points from a hypercube $\{0,1\}^n$ the expected value for the number of checked points before achieving $\alpha^{\#}$ is not greater than 2^n. Thus, in the worst case scenario, (1+1)-EA is extremely inefficient. However, when applied to many practical tasks (1+1)-EA can be surprisingly productive.

2 Merging Variables Principle (MVP)

In this section we describe a simple technique which can be applied to the problems of optimization of arbitrary functions of the kind (1), including Black-Box functions.

Consider an arbitrary function (1) and the problem $f \underset{\{0,1\}^n}{\rightarrow} \max$ (or $f \underset{\{0,1\}^n}{\rightarrow} \min$). Let us associate with $\{0,1\}^n$ a set of Boolean variables $X = \{x_1, \dots, x_n\}$ (considering $\{0,1\}^n$ as a set of all possible assignments of variables from X).

Let us fix an arbitrary positive integer $r : 1 \leq r < n$ and define a new set of variables $Y = \{y_1, \dots, y_r\}$. Consider an arbitrary surjection $\mu : X \rightarrow Y$. With an arbitrary $y_j \in Y$, $j \in \{1, \dots, r\}$ we associate a set X_j of preimages of y_j in the context of mapping μ. Let us link with y_j a set D_j, which consists of $2^{|X_j|}$ different symbols of some alphabet: $D_j = \{\beta_1^j, \dots, \beta_{2^{|X_j|}}^j\}$, and fix an arbitrary bijection $\omega_j : D_j \rightarrow \{0,1\}^{|X_j|}$. Consider a set

$$D^\mu = D_1 \times \ldots \times D_r.$$

Definition 1. *The elements of D_j are called the values of variable y_j, $j \in \{1, \dots, r\}$ and D_j is called the domain of this variable. An arbitrary string $\beta \in D^\mu$ is called an assignment of variables from Y. Implying all notions which were introduced above we will say that merging mapping μ is defined. The elements of Y are referred to as merged variables.*

Regarding the set D^μ we note that the Hamming metric is naturally defined on D^μ and thus D^μ is a metric space.

Lemma 1. *An arbitrary merging mapping $\mu : X \rightarrow Y$ defines a bijective mapping*

$$\tau_\mu : D^\mu \rightarrow \{0,1\}^n.$$

Proof. Assume that for a set of Boolean variables X, $|X| = n$, a merging mapping μ, $\mu : X \rightarrow Y$, $|Y| = r$, $1 \leq r < n$ is given. The fact that μ is surjection means that sets X_j, $j \in \{1, \dots, r\}$ do not intersect, and any variable from X turns out to be in some set of the kind X_j. Consider an arbitrary assignment $\beta \in D^\mu$. Let β^j be a symbol, located in the coordinate with the number $j, j \in \{1, \dots, r\}$ of β. Consider set X_j. Let α^j be a binary string associated with an element β^j by bijection ω_j. Let us view α^j as an assignment of variables from X_j. Thus, bijections ω_j, $j \in \{1, \dots, r\}$ associate all coordinates of β with binary strings thereby setting the values of all variables from X. Consequently, an arbitrary string $\beta \in D^\mu$ is associated with some string $\alpha \in \{0,1\}^n$. Denote the resulting function by $\tau_\mu : D^\mu \rightarrow \{0,1\}^n$. Note that *Range* $\tau_\mu = \{0,1\}^n$. If we assume that there is a vector $\alpha \in \{0,1\}^n$, which does not have a preimage in D^μ for a given τ_μ, then it contradicts with the properties of bijections ω_j, $j \in \{1, \dots, r\}$. Thus, τ_μ is a surjection. Also it is easy to see, that two arbitrary different elements from D^μ have different images for a given τ_μ (injection). Consequently, τ_μ is bijection. The Lemma 1 is proved.

Definition 2. *Function τ_μ, defined in the proof of Lemma 1, is called a bijection induced by a merging mapping μ.*

Example 2. Assume that $X = \{x_1, x_2, x_3, x_4, x_5\}$. Let us define the mapping $\mu : X \to Y$, $Y = \{y_1, y_2, y_3\}$ as follows:

$$X_1 = \{x_1, x_4\}, X_2 = \{x_2\}, X_3 = \{x_3, x_5\}.$$

The domains of variables y_1, y_2, y_3 are the following: $D_1 = \{\beta_1^1, \beta_2^1, \beta_3^1, \beta_4^1\}$, $D_2 = \{\beta_1^2, \beta_2^2\}$, $D_3 = \{\beta_1^3, \beta_2^3, \beta_3^3, \beta_4^3, \}$. Bijections ω_j, $j \in \{1, 2, 3\}$ are defined as it is shown in Fig. 1. Thus, the mapping $\tau_\mu : D^\mu \to \{0, 1\}^5$ is defined. By Lemma 1 it is a bijection. For example, $\tau_\mu(\beta_3^1, \beta_2^2, \beta_4^3) = (11101)$.

ω_1

β_1^1	00
β_2^1	01
β_3^1	10
β_4^1	11

ω_2

β_1^2	0
β_2^2	1

ω_3

β_1^3	00
β_2^3	01
β_3^3	10
β_4^3	11

Fig. 1. Bijections ω_j, $j \in \{1, 2, 3\}$ which define the mapping $\tau_\mu : D^\mu \to \{0, 1\}^5$

The main idea of the technique presented below consists in transitioning from the optimization problem of the original function (1) on $\{0, 1\}^n$ to the optimization problem of specially constructed function on D^μ (for a given merging mapping $\mu : X \to Y$).

Definition 3. *Consider an optimization problem for an arbitrary function* (1). *Let $\mu : X \to Y$ be an arbitrary merging mapping. Consider the function*

$$F_{f,\mu} : D^\mu \to \mathbb{R},$$

defined in the following way: $F_{f,\mu}(\beta) = f(\tau_\mu(\beta))$, in which τ_μ is a bijection induced by μ. Function $F_{f,\mu}$ is called μ-conjugated with f.

Lemma 2.

$$\text{extr}_{\{0,1\}^n} f = \text{extr}_{D^\mu} F_{f,\mu}.$$

(here extr can be understood as min or max).

Proof. In the context of Lemma 1 this equality is in fact evident. Indeed, there is a bijection τ_μ between $\{0, 1\}^n$ and D^μ. The value of function $F_{f,\mu}$ in an arbitrary point $\beta \in D^\mu$ is equal to the value of f in point $\alpha = \tau_\mu(\beta)$. Thus, the smallest (largest) value of $F_{f,\mu}$ on D^μ is equal to the smallest (largest) value of f on $\{0, 1\}^n$. The Lemma 2 is proved.

The following property gives us the exact value of the number of different merging mappings for the set X of power n.

Lemma 3. *Let f be an arbitrary function of the kind* (1). *Then, the number of different merging mappings of the kind $\mu : X \to Y$ is $\sum_{r=1}^{n-1} r! \cdot S(n, r)$, where $S(\cdot, \cdot)$ – is a Stirling number of the second kind.*

Proof. Assume that $X = \{x_1, \ldots, x_n\}$. For an arbitrary merging mapping $\mu : X \rightarrow Y$ a set Y can contain $1, 2, \ldots, n-1$ variables. An arbitrary merging mapping is constructed in two steps. The first step is to divide X into r parts (the order of the elements in each part does not matter). As a result there is a composition of sets $X_1, \ldots, X_r$. At the second step each set X_j, $j \in \{1, \ldots, r\}$ is associated with a variable from $Y = \{y_1, \ldots, y_r\}$. The number of unordered partitionings of n-element set into r parts is $S(n, r)$ (see, for example, [8]). Each unordered partitioning of X into r parts can be mapped to Y ($|Y| = r$) in $r!$ ways. The Lemma 3 is proved.

Let us summarize the contents of the present section. The *Merging Variables Principle* (MVP) consists in the transition from the optimization of an arbitrary function f of the kind (1) over a Boolean hypercube to the optimization problem of a function which is μ-conjugated with f over metric space D^μ. The main goal of the further sections is to demonstrate the benefits of MVP.

3 Combining MVP with Local Search

For an arbitrary function f of the kind (1) consider a problem $f \underset{\{0,1\}^n}{\longrightarrow} \max$. Assume, that $\{0,1\}^n$ is a set of all possible assignments of variables from set $X = \{x_1, \ldots, x_n\}$. Consider a merging mapping $\mu : X \rightarrow Y$, $Y = \{y_1, \ldots, y_r\}$, $1 \leq r < n$ and a metric space (with Hamming metric) $D^\mu = D_1 \times \ldots \times D_r$. Let $\tau_\mu : D^\mu \rightarrow \{0,1\}^n$ be a bijection induced by μ. We solve the maximization problem of function $F_{f,\mu}$ on D^μ. Let us define the neighborhood function over D^μ in the following way. For an arbitrary $\beta \in D^\mu$ assume that

$$\aleph_1^\mu(\beta) = \{\gamma \in D^\mu : d_H(\beta, \gamma) \leq 1\}.$$

In other words, the neighborhood of an arbitrary point β contains all points from D^μ, for which the Hamming distance d_H between them and β is at most 1. Let us denote a metric space D^μ with the neighborhood structure $\aleph_1^\mu$ by $\langle D^\mu, \aleph_1^\mu \rangle$.

Below we will use a term "random merging mapping", which refers to any construction of mapping $\mu : X \rightarrow Y$ by means of a random experiment. The most natural is a scheme of random arrangements of particles in boxes [9]. Specifically, for a fixed $r, 1 \leq r < n$ assume that an arbitrary variable y_j, $j \in \{1, \ldots, r\}$ is associated with a box which can accommodate n particles. A set X is considered as a set containing n particles which are randomly scattered in r boxes according to the sampling without replacement.

Below we present a variant of Hill Climbing algorithm, which uses MVP (Merging Variable Hill Climbing algorithm, MVHC).

Input: an arbitrary point $\alpha \in \{0,1\}^n$, $f(\alpha)$;

1. define a random merging mapping $\mu : X \rightarrow Y$, $Y = \{y_1, \ldots, y_r\}$, $1 \leq r < n$;
2. construct a point $\beta = \tau_\mu^{-1}(\alpha)$ in $\langle D^\mu, \aleph_1^\mu \rangle$, $D^\mu = D_1 \times \ldots \times D_r$, where D_j, $j \in \{1, \ldots, r\}$ are domains of y_j;

3. run HC in $\langle D^\mu, \aleph_1^\mu \rangle$ starting from point β for an objective function $F_{f,\mu}$; let β^* be a local maximum, achieved in one iteration of HC;
4. construct a point $\alpha^* = \tau_\mu(\beta^*)$ ($\alpha^* \in \{0,1\}^n$);

Output: $(\alpha^*, f(\alpha^*))$.

Theorem 1. *In the context of the MVHC scheme described above let* $\beta = \tau_\mu^{-1}(\alpha)$ *be a point in* $\langle D^\mu, \aleph_1^\mu \rangle$ *which is not a local maximum. Then* $f(\alpha^*) > f(\alpha)$, *where* $\alpha^* = \tau_\mu(\beta^*)$ *and* β^* *is a local maximum, achieved by HC in* $\langle D^\mu, \aleph_1^\mu \rangle$ *in one iteration, starting from point* β.

Proof. Let $\mu, \tau_\mu, D^\mu, \alpha, \alpha^*, \beta, \beta^*$ be the objects from the description of the MVHC algorithm and the theorem formulation. Since β is not a local maximum in the space $\langle D^\mu, \aleph_1^\mu \rangle$, then $F_{f,\mu}(\beta^*) > F_{f,\mu}(\beta)$. Thus, (by the definition of function $F_{f,\mu}$) it follows that $f(\tau_\mu(\beta^*)) > f(\tau_\mu(\beta))$. Therefore, $f(\alpha^*) > f(\alpha)$. The Theorem is proved.

The MVHC algorithm can be used to construct an iterative computational scheme in which the random merging mapping is launched multiple times: in particular, the output α^* of an arbitrary iteration can be used as an input for the following iteration.

Below we would like to comment on a number of features of the proposed algorithm and show the techniques that can improve the practical effectiveness of MVHC. The proofs for the properties described below are not shown due to their simplicity and limitations on the volume of the paper.

a. Note that point α can be a local maximum of function (1) in the space $\langle \{0,1\}^n, \aleph_1 \rangle$, while point $\beta = \tau_\mu^{-1}(\alpha)$ is simultaneously not a local maximum of function μ-conjugated with (1) in $\langle D^\mu, \aleph_1^\mu \rangle$. This fact makes it possible to view MVHC as a special case of Variable Neighborhood Search (VNS) metaheuristic strategy [10–12]. Indeed, let α be an arbitrary point in $\{0,1\}^n$, $\mu : X \to Y$ be an arbitrary merging mapping and $\tau_\mu : D^\mu \to \{0,1\}^n$ be a bijection induced by μ. Define the neighborhood of α in $\{0,1\}^n$ as follows:

$$\tilde{\aleph}(\alpha) = \{\tau_\mu(\gamma) | \gamma \in \aleph_1^\mu(\tau_\mu^{-1}(\alpha))\}, \tag{3}$$

where $\aleph_1^\mu(\beta)$ is the Hamming neighborhood of radius 1 for the point β in D^μ. Note that (3) defines the neighborhood function over $\{0,1\}^n$. The different merging mappings will yield different neighborhood structures in the context of (3). From this point of view, the Theorem 1 is the variant of the main VNS principle saying that the local extremum of a function with regard to one neighborhood structure may not be a local extremum of this function with regard to a different neighborhood structure. The Lemma 3 says that in the context of MVHC there exist numerous ways to construct neighborhood structures even for small n and r (say, $n = 100$ and $r = 10$).

b. Let $\mu : X \to Y$, $|X| = n$, $|Y| = r$ be an arbitrary random mapping. Let $X_1, \ldots, X_r$ be the sets of preimages of variables from Y with respect to μ, and $|X_1| = l_1, \ldots, |X_r| = l_r$; $l_1 + \ldots + l_r = n$. Then for an arbitrary point $\beta \in D^{\mu}$ the following holds:

$$|\aleph_1^{\mu}(\beta)| = \sum_{j=1}^{r} 2^{l_j} + (1-r). \tag{4}$$

This fact means that for domains of relatively large size the traversal of points from the neighborhood $\aleph_1^{\mu}(\beta)$ can be naturally performed in parallel: each domain should be processed by an individual computing process. In more detail, assume that we have t independent computing processes. Consider an arbitrary $\beta \in D^{\mu}$ and let β^1 be an arbitrary point from D^{μ}, which differs from β in coordinate number 1 while coinciding with β in the remaining coordinates. It is clear that in total there are $2^{l_1} - 1$ points of this kind. Let us traverse such points and compute the corresponding values of function $F_{f,\mu}$ using a computing process number 1. We can treat the points which differ from β only in the second coordinate in the similar fashion, etc. For $t < r$ once the computing process finished the current task it can take any domains which have not yet been processed. One process should perform the supervisor function and track whether the current Best Known Value have been improved.

c. Let μ be an arbitrary merging mapping and β^* be a local extremum of $F_{f,\mu}$ in $\langle D^{\mu}, \aleph_1^{\mu}\rangle$. It is easy to show that in this case $\alpha^* = \tau_{\mu}(\beta^*)$ is a local extremum of f in $\langle\{0,1\}^n, \aleph_1\rangle$. Assume that μ_k, $k \in \{1, \ldots, K\}$ are random merging mappings and $\alpha^* \in \{0,1\}^n$ is such a local extremum that points $\beta_k^* = \tau_{\mu_k}^{-1}(\alpha^*)$ are local extrema in the spaces D^{μ_k}, $k \in \{1, \ldots, K\}$ for a large enough K. Then let us call the point α^* *strong local extremum*.
d. Consider an arbitrary merging mapping $\mu : X \to Y$. Let α be an arbitrary point in $\{0,1\}^n$ and $\tilde{\aleph}(\alpha)$ be the neighborhood of α defined (with respect to fixed μ) in accordance with (3). Assume that $l^* = \max\{l_1, \ldots, l_r\}$. It is easy to show that for $r \geq 2$ it holds that $\tilde{\aleph}(\alpha) \subset \aleph_{l^*}(\alpha)$. The power $\tilde{\aleph}(\alpha)$ (it is expressed by the number in the right part of (4)) can be significantly smaller than the power of $\aleph_{l^*}(\alpha)$. For example, if $n = 100$, $r = 10$ then $l_1 = \ldots = l_{10} = 10$, $|\tilde{\aleph}(\alpha)| = 10 \times 2^{10} - 9 = 10231$, while $|\aleph_{10}(\alpha)| > 1,5 \times 10^{13}$.

The property **d** essentially means that the merging mapping technique may be useless if the algorithm reached such a local extremum α^*, that the closest point (Hamming distance-wise) from $\{0,1\}^n$ with the better objective function value is at a distance $> l^*$ from α. On the first glance it might seem that this fact significantly limits the applicability of the proposed method. However, it is possible to describe the supplementary technique for MVHC which is based on the idea to store strong local extrema and use them to direct the search process. In this context we will use the tabu lists concept which serves as a basis of the tabu search strategy [13].

So, a strong local extremum is such a local extremum in $\{0,1\}^n$, for which it was not possible to improve BKV even after a significant number of different merging mappings μ_k, $k \in \{1, \ldots, K\}$. Let us denote such a point as α_1^*. The goal is to move from α_1^* to a point with the better BKV. Since we do not employ any knowledge about function f, it means that such transitions should rely on heuristic arguments. The first of the arguments is to escape the neighborhood of the kind $\aleph_{l_1^*}(\alpha_1^*)$ in $\{0,1\}^n$, where l_1^* is a "critical" domain size that is known from the search history. On the other hand, due to various reasons appealing to the "locality principle" it is undesirable to move "too far" from α_1^*. It is especially relevant if during the transition to α_1^* the BKV have been improved multiple times. Thus, the simplest step is to move to an arbitrary point situated at a distance of $l_1^* + 1$ from α_1^*. Let α_2 be such a point. Assume that we launch MVHC from this point and α_2^* is the resulting strong local extremum of f, which is different from α_1^*. Similar to l_1^* we can define critical domain size l_2^* used during the transition from α_2 to α_2^*, critical domain size l_3^* and etc.

As a result, assume that we have strong local extrema $\alpha_1^*, \ldots, \alpha_R^*$ and our goal is to construct a point $\alpha_{R+1} \in \{0,1\}^n$ to launch the $R+1$-th iteration of MVHC from it. Taking into account the above, we have a problem of choosing next current point α_{R+1} as a point which satisfies a system of constraints of the following kind:

$$(d_H(\alpha_{R+1}, \alpha_1^*) = L_1) \wedge \ldots \wedge (d_H(\alpha_{R+1}, \alpha_R^*) = L_R). \tag{5}$$

The numbers $L_1, \ldots, L_R$ can be chosen according to different criteria. Let us describe the simplest one. Consider the following system of constraints:

$$(d_H(\alpha_{R+1}, \alpha_1^*) = l_1^* + 1) \wedge \ldots \wedge (d_H(\alpha_{R+1}, \alpha_R^*) = l_R^* + 1). \tag{6}$$

If there exists a point α_{R+1} that satisfies (6) then it is chosen as a starting point for the next MVHC iteration. If such a point does not exist, then we call (6) incompatible. In this case it is possible to relax some of the constraints of the kind $d_H(\alpha_{R+1}, \alpha_q^*) = l_q^* + 1$ by replacing them with constraints of the kind $d_H(\alpha_{R+1}, \alpha_q^*) = L_q$, where $L_q \geq l_q^* + 2, q \in \{1, \ldots, R\}$. The resulting system of constraints of the kind (5) is again to be tested for compatibility.

Let us consider the problem of testing the compatibility of an arbitrary system of the kind (5). Consider an arbitrary constraint of the kind $d_H(\alpha_{R+1}, \alpha^*) = L$, where $\alpha^* = (\alpha^1, \ldots, \alpha^n)$ is a known Boolean vector and L is a known natural number. Let us represent the unknown components of vector α_{R+1} using Boolean variables $z_1, \ldots, z_n$. Now consider the expression

$$(z_1 \oplus \alpha^1) + \ldots + (z_n \oplus \alpha^n) = L, \tag{7}$$

where $\oplus$ is the sum mod2, and $+$ is an integer sum.

We can consider (7) as an equation for unknown variables $z_1, \ldots, z_n$. It is easy to see that a set of vectors α_{R+1}, which satisfy the constraint $d_H(\alpha_{R+1}, \alpha^*) = L$, coincides with the set of solutions of the Eq. (7). To solve the systems of equations of the kind (7) or to prove the inconsistent of such systems we can use any complete algorithm for solving SAT. The corresponding reduction to SAT is performed effectively using the procedures described, for example, in [14].

Thus, to choose new current points in the context of MVHC we can employ a strategy in which SAT oracles are combined with the tabu lists containing strong local extrema.

4 Combining MVP with Evolutionary Computations

Now let us consider how MVP can be combined with evolutionary algorithms. In particular, let us study the MV-variant of (1+1)-EA. As it was stated above, for an arbitrary function of the kind (1) in [7] there was obtained the following upper bound: $E_{(1+1)-EA} \leq n^n$. Also in [7] there was given an example of a function (the TRAP function) for which this bound is asymptotically achieved (in terms of [7]).

In the description of the MV-variant of (1+1)-EA (we denote the corresponding algorithm as (1+1)-MVEA) we want to preserve the following property of the original algorithm: that the expected value of the number of bits in which the Boolean vector is different from its (1+1)-random mutation should be 1.

Assume that there is an arbitrary merging mapping $\mu : X \to Y$, $|X| = n$, $|Y| = r$, $1 \leq r < n$. For an arbitrary point $\alpha \in \{0,1\}^n$ perform the following steps.

Input: arbitrary point $\alpha \in \{0,1\}^n, f(\alpha)$;

1. construct a point $\beta = \tau_{\mu}^{-1}(\alpha)$; perform r Bernoulli trials with probability of success $p = 1/r$; let $\{i_1, \ldots, i_q\} \subseteq \{1, \ldots, r\}$ be the numbers of successful trials; for each $j \in \{i_1, \ldots, i_q\}$ consider the domain D_j of a variable y_j, let X_j be the set of preimages of y_j for the mapping μ, $\omega_j : D_j \to \{0,1\}^{|X_j|}$ is a fixed bijection, β_j is the value of y_j in β;
2. consider the Boolean vector $\alpha_j = \omega_j(\beta_j)$ of size $l_j = |X_j|$; perform (1+1)-random mutation on α_j with probability of success equal to $\frac{1}{l_j}$, let α'_j be the result of the mutation, $\beta'_j = \omega_j^{-1}(\alpha'_j)$;
3. construct a point β' in D^{μ}: in the coordinate with number $j \in \{i_1, \ldots, i_q\}$ the point β' has β'_j; in the remaining coordinates with numbers from the set $\{1, \ldots, r\} \setminus \{i_1, \ldots, i_q\}$ the point β' coincides with β;
4. construct a point $\alpha' = \tau_{\mu}(\beta')$ ($\alpha' \in \{0,1\}^n$);

Output: $(\alpha', f(\alpha'))$.

Definition 4. *To the described sequence of actions the result of which is the transition $\alpha \to \alpha'$ we will refer as (1+1)-merging variable random mutation.*

Lemma 4. *For an arbitrary merging mapping μ the expected value of the number of bits in which the points α and α' differ is 1.*

Proof. Indeed, for an arbitrary merging mapping μ the expected value of the number of successful trials among r Bernoulli trials with success probability $1/r$ is equal to 1. For an arbitrary Boolean vector of the kind $\alpha_j = \omega_j(\beta_j)$ of size l_j the expected value of the number of flipped bits as a result of (1+1)-random mutation, where the probability of flipping is $\frac{1}{l_j}$, is equal to 1. The considered random variables are obviously independent, thus in the context of a single (1+1)-merging variable random mutation the expected value of the random variable $d_H(\alpha, \alpha')$ is equal to 1. The Lemma 4 is proved.

Definition 5. *For a fixed merging mapping μ, the (1+1)-merging variable evolutionary algorithm ((1+1)-MVEA) is a sequence of (1+1)-merging variable random mutations. In the context of maximization problem of an arbitrary function* (1)*: the next mutation is applied to α' if $f(\alpha') \geq f(\alpha)$. Otherwise, the next mutation is applied to α (stagnation).*

The following definition is a variant of the Definition 5 from [7] with relation to (1+1)-MVEA.

Definition 6. *Let f be an arbitrary function of the kind* (1) *and $\alpha^{\#}$ be a global extremum of function f on $\{0,1\}^n$. Let μ be an arbitrary merging mapping. We will define the expected running time of (1+1)-MVEA as the mean of the number of (1+1)-merging variable random mutations that have to be applied to an arbitrary point $\alpha \in \{0,1\}^n$ until it transforms into $\alpha^{\#}$. Denote this value by $E^{\mu}_{(1+1)-MVEA}$.*

Theorem 2. *Assume that f is an arbitrary function of the kind* (1), *$\mu : X \to Y$ is an arbitrary merging mapping: $X = \{x_1, \ldots, x_n\}$, $Y = \{y_1, \ldots, y_r\}$, $1 \leq r < n$, $l_j = |X_j| \geq 2$ for all $j \in \{1, \ldots, r\}$ and $l = \max\{l_1, \ldots, l_r\}$. Then the following estimation holds:*

$$E^{\mu}_{(1+1)-MVEA} \leq r^r \cdot l^n. \tag{8}$$

Proof. Let μ be an arbitrary merging mapping for which all the conditions of the theorem are satisfied. Now let us reason in a way similar to the proof of the Theorem 6 in [7]. Let $\alpha \in \{0,1\}^n$ be an arbitrary point and $\alpha^{\#}$ be a global extremum of the function (1) on $\{0,1\}^n$. Denote by $P_{\alpha \to \alpha^{\#}}$ the probability that α will transition into $\alpha^{\#}$ as a result of one iteration of the (1+1)-MVEA-algorithm. Consider the points $\beta = \tau_{\mu}^{-1}(\alpha)$, $\beta^{\#} = \tau_{\mu}^{-1}(\alpha^{\#})$ from the space D^{μ}. In this context, for an arbitrary $j \in \{1, \ldots, r\}$ with the coordinates β_j, $\beta_j^{\#}$ there will be associated the binary strings α_j, $\alpha_j^{\#}$.

Now let us construct the lower bound for the probability of an event that as a result of one (1+1)-MVEA iteration there will take place a transition from α to $\alpha^{\#}$. It is clear that this may happen if and only if there takes place the transition from β to $\beta^{\#}$. Let $q = d_H(\beta, \beta^{\#})$ be the Hamming distance between β and $\beta^{\#}$ in the space D^{μ}. Assume that the set $J = \{i_1, \ldots, i_q\} \subseteq \{1, \ldots, r\}$ contains the numbers of coordinates in β, in which this point differs from $\beta^{\#}$, and $U = \{1, \ldots, r\} \setminus J$. Let us denote by $\sigma = (\sigma_1, \ldots, \sigma_r)$, $\sigma_i \in \{0,1\}$, $i \in \{1, \ldots, r\}$ the set of results of a sequence of r Bernoulli trials with success probability $1/r$ (as usually, we assume that $\sigma_1 = 1$ corresponds to success).

The transition $\beta \to \beta^{\#}$ takes place if and only if within one (1+1)-merging variable random mutation the following two events denoted by A_j and B_u happen simultaneously:

a. for an arbitrary $j \in J$ the event A_j takes place if and only if $\beta_j \to \beta_j^{\#}$;
b. for an arbitrary $u \in U$ the event B_u takes place if and only if $\beta_u \to \beta_u$.

It is easy to see that all the events of the kind $A_j, B_u, j \in J, u \in U$ are independent, thus

$$P_{\alpha \to \alpha^{\#}} = \left(\prod_{j \in J} \Pr\{A_j\} \right) \cdot \left(\prod_{u \in U} \Pr\{B_u\} \right).$$

For an arbitrary $k \in \{1, \ldots, r\}$ let us denote by p_k the probability that the result of the random (1+1)-mutation with probability of success $\frac{1}{l_k}$ of the string $\alpha_k = \omega_k(\beta_k)$ is the string $\alpha_k^{\#}$. Then for any $j \in J$ it holds that $\Pr\{A_j\} = \frac{1}{r} \cdot p_j$.

For an arbitrary $u \in U$ the event B_u can happen in one of the two cases: first if $\sigma_u = 0$, and, second, if $\sigma_u = 1$, but the result of the (1+1)-random mutation with the probability of success $\frac{1}{l_u}$ of the string $\alpha_u = \omega_u(\beta_u)$ is the string α_u. In the first case, $\Pr\{B_u\} = (1 - \frac{1}{r})$. In the second case, $\Pr\{B_u\} = \frac{1}{r} \cdot (1 - \frac{1}{l_u})^{l_u}$. Thus, in any case when $r \geq 2$, $l_u \geq 2$ it holds that $\Pr\{B_u\} \geq \frac{1}{r} \cdot \frac{1}{l_u^{l_u}}$. Taking this fact into account the following bound holds:

$$P_{\alpha \to \alpha^{\#}} \geq \left(\frac{1}{r^q} \cdot \prod_{j \in J} p_j \right) \cdot \left(\frac{1}{r^{r-q}} \cdot \prod_{u \in U} \frac{1}{l_u^{l_u}} \right). \quad (9)$$

In accordance with [7] for an arbitrary $k \in \{1, \ldots, r\}$, such that $l_k \geq 2$, the following holds: $p_k \geq \frac{1}{l_k^{l_k}}$. Together with (9) this fact gives us the next bound:

$$P_{\alpha \to \alpha^{\#}} \geq \frac{1}{r^r} \cdot \prod_{k \in \{1, \ldots, r\}} \frac{1}{l_k^{l_k}}. \quad (10)$$

Let us emphasize that (10) holds for an arbitrary $\alpha \in \{0,1\}^n$. Assume that $l = \max\{l_1, \ldots, l_r\}$. Then, taking into account that $\sum_{k=1}^{r} l_k = n$, it follows from (10):

$$P_{\alpha \to \alpha^{\#}} \geq \frac{1}{r^r} \cdot \frac{1}{l^n}.$$

The bound (8) follows from the latter inequality. The Theorem 2 is thus proved.

The bound (8) looks a little surprising since it is actually easy to determine the merging mappings with such parameters r and l that the corresponding variant of the bound (8) becomes significantly better than the similar bound for (1+1)-EA shown in [7].

Definition 7. *Assume that* $|X| = n$, $|Y| = r$, $1 \leq r < n$ *and* $n = \lfloor \frac{n}{r} \rfloor \cdot r + b$, *where* $b, b \in \{0, \ldots, r-1\}$ *is the remainder from the division of* n *by* r. *Let* $\mu : X \to Y$ *be an arbitrary merging mapping, such that for* b *sets of the kind* X_j, $j \in \{1, \ldots, r\}$ *it holds that* $|X_j| = \lfloor \frac{n}{r} \rfloor + 1$, *and for the remaining* $r - b$ *sets of such kind* $|X_j| = \lfloor \frac{n}{r} \rfloor$. *Let us refer to such* μ *as uniform merging mapping.*

Corollary 1. *Let* $\mu : X \to Y$ *be an arbitrary uniform merging mapping such that* $l_j \geq 2$ *for all* $j \in \{1, \ldots, r\}$. *Then there exists such a function* $\delta(n) : 1 < \delta(n) \leq n$, *that the following evaluation holds:*

$$E^{\mu}_{(1+1)-MVEA} \leq n^{n \cdot \left(\frac{1}{\delta(n)} - \frac{\log_n \delta(n)}{\delta(n)} + \log_n(\delta(n)+1) \right)}. \quad (11)$$

Proof. Let $\mu : X \to Y$ be an arbitrary uniform merging mapping. By definition it means that $l \leq \frac{n}{r} + 1$. Since $l_j \geq 2$ for all $j \in \{1, \ldots, r\}$, let us use the evaluation (9):

$$E^{\mu}_{(1+1)-MVEA} \leq r^r \cdot \left(\frac{n}{r} + 1\right)^n. \tag{12}$$

Now introduce $\delta(n) : \delta(n) = n/r$. Then $1 < \delta(n) \leq n$. Taking this into account we can transform (12) as follows:

$$E^{\mu}_{(1+1)-MVEA} \leq \left(\frac{n}{\delta(n)}\right)^{\frac{n}{\delta(n)}} \cdot (\delta(n)+1)^n = n^{\frac{n}{\delta(n)}} \cdot (\delta(n))^{-\frac{n}{\delta(n)}} \cdot (\delta(n)+1)^n =$$

$$= n^{\frac{n}{\delta(n)}} \cdot n^{-\frac{n}{\delta(n)} \cdot \log_n \delta(n)} \cdot n^{n \cdot \log_n(\delta(n)+1)} = n^{n \cdot \left(\frac{1}{\delta(n)} - \frac{\log_n \delta(n)}{\delta(n)} + \log_n(\delta(n)+1)\right)}.$$

Thus the Corollary 1 is proved.

Based on (11) it is possible to give a number of examples of uniform merging mappings, that provide better worst-case-estimations of (1+1)-MVEA for an arbitrary function of the kind (1) compared to the similar estimation for (1+1)-EA from [7]. Indeed, for example for $\delta(n) \sim \sqrt[3]{n}$ and for any $n \geq 27$ it follows from (11) that $E^{\mu}_{(1+1)-MVEA} \lesssim n^{n \cdot (\frac{1}{\sqrt[3]{n}} - \frac{1}{3\sqrt[3]{n}} + \frac{1}{2})}$ (here it is taken into account that for $n \geq 27$ it holds that $\log_n(\sqrt[3]{n}+1) < \frac{1}{2}$). Thus in this case the following holds $E^{\mu}_{(1+1)-MVEA} \lesssim n^{(\frac{n}{2} + \frac{2}{3} n^{2/3})}$.

5 Preliminary Computational Results

The MVHC was implemented in the form of a multi-threaded C++ application. It employs the parallel variant of the procedure for traversing the neighborhoods in the search space (see Sect. 3).

In the role of test instances we considered the problems of finding preimages of some cryptographic functions reduced to the Boolean Satisfiability problem (SAT). Such instances are justified to be hard, thus they can be viewed as a good test suite to compare the effectiveness of combinatorial algorithms. At the current stage we considered the problems of finding preimages of a well-known MD4 cryptographic hash function [15] with additional constraints on the hash value. In particular, the goal was to find such 512-bit inputs that yield MD4 hash values with leading zeros. This problem can be reduced to SAT effectively. For this purpose we employed the Transalg software system [16].

Let $\{0,1\}^{512} \to \{0,1\}^{128}$ be a function which is defined by the MD4 algorithm. Let C be a CNF which encodes this algorithm. In the set of variables from C let us select two sets. First set is X^{in}, which consists of 512 Boolean variables encoding an input of MD4. Second one is X^{out} – a set of 128 Boolean variables encoding the output of MD4. In the set X^{out} select k variables encoding the leading bits of the hash value, and assign these variables with value 0. Denote the resulting CNF as C_k. This CNF is satisfiable and from any satisfying assignment one can effectively extract such $\alpha \in \{0,1\}^{512}$ for which the leading k bits of corresponding MD4 hash value are equal to zero.

To find the satisfying assignment for C_k we used two approaches. First we applied to C_k the multithreaded solvers, based on the CDCL algorithm [17], that won the yearly SAT competitions in recent years. In the second approach we used the MVHC algorithm described in the Sect. 3 of the present paper. Consider, a set of variables X^{in}, $|X^{in}| = 512$ in CNF C_k. Associate an arbitrary vector $\alpha \in \{0,1\}^{512}$ with a set of literals over variables from X^{in}. Recall, that a literal is either the variable itself or its negation. If a component of vector α corresponding to a variable x_i, $i \in \{1, \ldots, 512\}$ takes value 1, then the corresponding literal is x_i. Otherwise, the literal is $\neg x_i$. All such literals are conjunctively added to CNF C_k and the resulting CNF is denoted by $C_k(\alpha)$. It is well known that set X^{in} is a Strong Unit Propagation Backdoor Set (SUPBS) for CNF C_k [18]. This means that the satisfiability of CNF $C_k(\alpha)$ can be checked in time linear on the size of this CNF using a simple procedure of Boolean constraints propagation called Unit Propagation Rule [17]. Thus, we consider function of the kind (1) which associates with an arbitrary $\alpha \in \{0,1\}^{512}$ a number of clauses in $C_k(\alpha)$ that take the value of 1 as a result of application of Unit Propagation rule to CNF $C_k(\alpha)$. If the value of this function is equal to the number of clauses that are satisfied in $C_k(\alpha)$, then α is a MD4 preimage of a hash value with k leading zero bits. For this function the problem of maximization on $\{0,1\}^{512}$ was solved using MVHC algorithm, in which uniform merging mapping was employed.

All tested algorithms were run on a personal computer (Intel Core i7, 16 GB RAM) in 8 threads. Since these algorithms are randomized, the result of each test is an average time of three independent launches for each algorithm. The obtained results are presented in Table 1.

Table 1. An average time (in seconds) of finding a MD4 preimage for hash value with k leading zero bits. For MVHC algorithm an uniform merging mapping was used

Solver	$k = 18$	$k = 20$	$k = 22$
MVHC ($l = 4$)	244.8	1028	2126
MVHC ($l = 8$)	490.1	1044.8	2003.1
MVHC ($l = 12$)	30	105.9	1882.8
CRYPTOMINISAT [19]	429.1	1197.9	3197.5
PLINGELING [20]	2175.1	1840.3	4218.4

6 Related Work (Briefly)

As it was mentioned above, there is a large set of metaheuristics and corresponding discussion contained in the monograph [5] by S. Luke. One of the first papers in which some complexity estimations of the simplest evolutionary algorithm (1+1)-EA were presented was G. Rudolf's dissertation [6].

Variable Neighborhood Search method (VNS) was first proposed in [10] and developed in subsequent papers: [11,12] and a number of others. Also we would like to note that the ideas underlying the MVP are similar in nature to those previously used in papers dedicated to the application of Large Scale Neighborhood Search [21,22].

A number of results on the complexity estimation of evolutionary algorithms originates in [7]. These studies are actively conducted to the present day. From the latest results in this area one should note [23].

We emphasize that MaxSAT is not the main object of study of the present paper. The special case of MaxSAT, related to the preimage finding problem of cryptographic functions, was considered only as an example of the maximization problem of pseudo-Boolean function. Listing the key papers devoted to SAT and MaxSAT would take up too much space. In this context, we refer only to the well-known handbook [2] and, in particular, to its chapter on MaxSAT [24]. It should be noted that in a number of papers various metaheurists were used to solve MaxSAT, employing both local search (see [25,26], etc.) and the concept of evolutionary computations (see, for example, [26,27]).

7 Conclusion and Acknowledgements

In the present paper we described a metaheuristic technique focused on the problem of pseudo-Boolean optimization. Arguments were given for using this technique both in combination with local search methods and in conjunction with evolutionary algorithms. The proposed technique when applied to local search methods can be considered as a special case of Variable Neighborhood Search. The first program implementation of the technique turned out to be quite effective in application to some reasonably hard problems of pseudo-Boolean optimization.

The author expresses deep gratitude to Ilya Otpuschennikov for the program implementation of MVHC algorithm. The author also thanks Maxim Buzdalov for productive discussion and useful advice.

The research was funded by Russian Science Foundation (project No. 16-11-10046).

References

1. Boros, E., Hammer, P.L.: Pseudo-boolean optimization. Discrete Appl. Math. **123**(1–3), 155–225 (2002)
2. Biere, A., Heule, M., van Maaren, H., Walsh, T. (eds.): Handbook of Satisfiability, vol. 185. IOS Press, Amsterdam (2009)
3. Burke, E., Kendall, G. (eds.): Search Methodologies, 2nd edn. Springer, New York (2014). https://doi.org/10.1007/978-1-4614-6940-7
4. McWilliams, F., Sloan, N.: The Theory of Error-Correcting Codes. North Holland, Amsterdam (1983)
5. Luke, S.: Essentials of Metaheuristics, 2nd edn. George Mason University, Fairfax (2015)

6. Rudolph, G.: Convergence properties of evolutionary algorithms. Ph.D. thesis, Hamburg (1997)
7. Droste, S., Jansen, T., Wegener, I.: On the analysis of the (1+1) evolutionary algorithm. Theor. Comput. Sci. **276**(1–2), 51–81 (2002)
8. Stanley, R.: Enumerative Combinatorics. Cambridge University Press, Cambridge (2011)
9. Feller, W.: An Introduction to Probability Theory and its Applications, 3rd edn. Wiley, Hoboken (1970)
10. Mladenović, N., Hansen, P.: Variable neighborhood search. Comput. Oper. Res. **24**(11), 1097–1100 (1997)
11. Hansen, P., Mladenović, N.: Variable neighborhood search: principles and applications. Eur. J. Oper. Res. **130**(3), 449–467 (2001)
12. Hansen, P., Mladenović, N., Todosijević, R., Hanafi, S.: Variable neighborhood search: basics and variants. EURO J. Comput. Optim. **5**(3), 423–454 (2016)
13. Glover, F., Laguna, M.: Tabu Search. Kluwer Academic Publishers, Dordrecht (1997)
14. Eén, N., Sörensson, N.: Translating pseudo-boolean constraints into SAT. JSAT **2**(1–4), 1–26 (2006)
15. Rivest, R.L.: The MD4 message digest algorithm. In: Menezes, A.J., Vanstone, S.A. (eds.) CRYPTO 1990. LNCS, vol. 537, pp. 303–311. Springer, Heidelberg (1991). https://doi.org/10.1007/3-540-38424-3_22
16. Otpuschennikov, I., Semenov, A., Gribanova, I., Zaikin, O., Kochemazov, S.: Encoding cryptographic functions to SAT using TRANSALG system. In: The 22nd European Conference on Artificial Intelligence (ECAI 2016). Frontiers in Artificial Intelligence and Applications, vol. 285, pp. 1594–1595. IOS Press (2016)
17. Marques-Silva, J.P., Lynce, I., Malik, S.: Conflict-driven clause learning SAT solvers. In: Biere et al. [2], pp. 131–153
18. Williams, R., Gomes, C.P., Selman, B.: Backdoors to typical case complexity. In: The 18th International Joint Conference on Artificial Intelligence (IJCAI 2003), pp. 1173–1178 (2003)
19. Soos, M., Nohl, K., Castelluccia, C.: Extending SAT solvers to cryptographic problems. In: Kullmann, O. (ed.) SAT 2009. LNCS, vol. 5584, pp. 244–257. Springer, Heidelberg (2009). https://doi.org/10.1007/978-3-642-02777-2_24
20. Biere, A.: CaDiCaL, Lingeling, Plingeling, Treengeling, YalSAT entering the SAT competition 2017. In: Balyo, T., Heule, M.J.H., Järvisalo, M. (eds.) SAT Competition 2017, vol. B-2017-1, pp. 14–15 (2017)
21. Ahuja, R.K., Ergun, O., Orlin, J.B., Punnen, A.P.: A survey of very large-scale neighborhood search techniques. Discrete Appl. Math. **123**(1–3), 75–102 (2002)
22. Avella, P., D'Auria, B., Salerno, S., Vasil'ev, I.: A computational study of local search algorithms for italian high-school timetabling. J. Heuristics **13**(6), 543–556 (2007)
23. Doerr, B.: Analyzing randomized search heuristics via stochastic domination. Theor. Comput. Sci. **773**, 115–137 (2019)
24. Li, C., Manya, F.: MaxSAT. In: Biere et al. [2], pp. 613–632
25. Ansótegui, C., Heymann, B., Pon, J., Sellmann, M., Tierney, K.: Hyper-reactive tabu search for MaxSAT. In: Battiti, R., Brunato, M., Kotsireas, I., Pardalos, P.M. (eds.) LION 12 2018. LNCS, vol. 11353, pp. 309–325. Springer, Cham (2019). https://doi.org/10.1007/978-3-030-05348-2_27

26. Bouhmala, N., Øvergård, K.I.: Combining genetic algorithm with variable neighborhood search for MAX-SAT. In: Zelinka, I., Vasant, P., Duy, V.H., Dao, T.T. (eds.) Innovative Computing, Optimization and Its Applications. SCI, vol. 741, pp. 73–92. Springer, Cham (2018). https://doi.org/10.1007/978-3-319-66984-7_5
27. Buzdalov, M., Doerr, B.: Runtime analysis of the $(1 + (\lambda, \lambda))$ genetic algorithm on random satisfiable 3-CNF formulas. In: Proceedings of the Genetic and Evolutionary Computation Conference (GECCO 2017), pp. 1343–1350. ACM Press (2017)

Using Sat Solvers for Synchronization Issues in Partial Deterministic Automata

Hanan Shabana[1,2] and Mikhail V. Volkov[2(✉)]

[1] Faculty of Electronic Engineering, Menoufia University, Menouf, Egypt
hananshabana22@gmail.com

[2] Institute of Natural Sciences and Mathematics, Ural Federal University, Ekaterinburg, Russia
m.v.volkov@urfu.ru

Abstract. We approach the task of computing a carefully synchronizing word of minimum length for a given partial deterministic automaton, encoding the problem as an instance of SAT and invoking a SAT solver. Our experimental results demonstrate that this approach gives satisfactory results for automata with up to 100 states even if very modest computational resources are used.

Keywords: Nondeterministic automaton · Deterministic automaton · Partial deterministic automaton · Careful synchronization · Carefully synchronizing word · SAT · SAT solver

1 Introduction

A *nondeterministic finite automaton* (NFA) is a triple $\langle Q, \Sigma, \delta \rangle$, where Q and Σ are finite non-empty sets called the *state set* and the *input alphabet* respectively, and δ is a subset of $Q \times \Sigma \times Q$. The elements of Q and Σ are called *states* and *letters*, respectively, and δ is referred to as the *transition relation*[1]. For each pair $(q, a) \in Q \times \Sigma$, we denote by $\delta(q, a)$ the subset $\{q' \mid (q, a, q') \in \delta\}$ of Q; this way δ can be viewed as a function $Q \times \Sigma \to \mathcal{P}(Q)$, where $\mathcal{P}(Q)$ is the power set of Q. When we treat δ as a function, we refer to it as the *transition function*.

Let Σ^* stand for the collection of all finite words over the alphabet Σ, including the empty word ε. The transition function extends to a function $\mathcal{P}(Q) \times \Sigma^* \to \mathcal{P}(Q)$, still denoted δ, in the following inductive way: for every subset $S \subseteq Q$ and every word $w \in \Sigma^*$, we set

$$\delta(S, w) := \begin{cases} S & \text{if } w = \varepsilon, \\ \bigcup_{q \in \delta(S,v)} \delta(q, a) & \text{if } w = va \text{ with } v \in \Sigma^* \text{ and } a \in \Sigma. \end{cases}$$

[1] The conventional concept of an NFA includes distinguishing two non-empty subsets of Q consisting of *initial* and *final* states. As these play no role in our considerations, the above simplified definition well suffices for the purpose of this paper.

Supported by the Ministry of Science and Higher Education of the Russian Federation, projects no. 1.580.2016 and 1.3253.2017, and the Competitiveness Enhancement Program of Ural Federal University.

I. Bykadorov et al. (Eds.): MOTOR 2019, CCIS 1090, pp. 103–118, 2019.
https://doi.org/10.1007/978-3-030-33394-2_9

(Here the set $\delta(S, v)$ is defined by the induction assumption since v is shorter than w.) We say that a word $w \in \Sigma^*$ is *undefined at a state* $q \in Q$ if the set $\delta(q, w)$ is empty; otherwise w is said to be *defined at* q.

When we deal with a fixed NFA, we suppress the sign of the transition relation, introducing the NFA as the pair $\langle Q, \Sigma \rangle$ rather than the triple $\langle Q, \Sigma, \delta \rangle$ and writing $q.w$ for $\delta(q, w)$ and $S.w$ for $\delta(S, w)$.

A *partial* (respectively, *complete*) *deterministic* automaton is an NFA $\langle Q, \Sigma \rangle$ such that $|q.a| \leq 1$ (respectively, $|q.a| = 1$) for all $(q, a) \in Q \times \Sigma$. We use the acronyms PFA and CFA for the expressions 'partial deterministic automaton' and 'complete deterministic automaton', respectively.

A CFA $\mathscr{A} = \langle Q, \Sigma \rangle$ is called *synchronizing* if there exists a word $w \in \Sigma^*$ whose action leaves the automaton in one particular state no matter at which state in Q it is applied: $q.w = q'.w$ for all $q, q' \in Q$. Any w with this property is said to be a *synchronizing* word for the automaton.

Synchronizing automata serve as simple yet adequate models of error-resistant systems in many applied areas (system and protocol testing, information coding, robotics). At the same time, synchronizing automata surprisingly arise in some parts of pure mathematics and theoretical computer science (symbolic dynamics, theory of substitution systems, formal language theory). We refer to the survey [39] and the chapter [20] of the forthcoming 'Handbook of Automata Theory' for a discussion of synchronizing automata as well as their diverse connections and applications. From both applied and theoretical viewpoints, the key question is to find the optimal, i.e., shortest reset word for a given synchronizing automaton. Under standard assumptions of complexity theory, this optimization question is known to be computationally hard; see [20, Section 2] for a summary of various hardness results in the area. As it is quite common for hard problems of applied importance, there have been many attempts to develop practical approaches to the question. These approaches have been based on certain heuristics [1,17,18] and/or popular techniques, including (but not limiting to) binary decision diagrams [29], genetic and evolutionary algorithms [19,32], satisfiability solvers [38], answer set programming [12], hierarchical classifiers [30], and machine learning [31].

The present authors [36,37] have initiated an extension to the realm of NFAs of the approach of [38]. Here we consider a more restricted class, namely, that of PFAs, where studying synchronization issues appears to be much better motivated. While we follow the general strategy of and re-use some technical tricks from [36,37], our present constructions heavily depend on the specifics of partial automata and have not been obtained via specializing the constructions of those earlier papers.

The rest of the paper is structured as follows. In Sect. 2 we describe and motivate the version of PFA synchronization that we have studied. In Sect. 3 we first outline the approach based on satisfiability solvers and then explain in detail how we encode PFAs and their synchronization problems as instances of the Boolean satisfiability problem. In Sect. 4 we provide samples of our experimental results and conclude in Sect. 5 with a brief discussion of the future work.

We have tried to make the paper, to a reasonable extent, self-contained, except for a few discussions that involve some basic concepts of computational complexity theory. These concepts can be found, e.g., in the early chapters of the textbook [28].

2 Synchronization of NFAs and PFAs

The concept of synchronization of CFAs as defined in Sect. 1 was extended to NFAs in several non-equivalent ways. The following three nowadays popular versions were suggested and analyzed in [13] in 1999 (although, in an implicit form, some of them appeared in the literature much earlier, see, e.g., [5,11]). For $i \in \{1,2,3\}$, an NFA $\mathscr{A} = \langle Q, \Sigma \rangle$ is called *D_i-synchronizing* if there exists a word $w \in \Sigma^*$ that satisfies the condition (D_i) from the list below:

(D_1): $|q.w| = |Q.w| = 1$ for all $q \in Q$;
(D_2): $q.w = Q.w$ for all $q \in Q$;
(D_3): $\bigcap_{q \in Q} q.w \neq \varnothing$.

Any word satisfying (D_i) is called *D_i-synchronizing* for $\mathscr{A}$. The definition readily yields the following properties of D_i-synchronizing words:

Lemma 1. *(a) A D_1- or D_3-synchronizing word is defined at each state.*
(b) A D_2-synchronizing word is either defined at each state or undefined at each state.
(c) Every D_1-synchronizing word is both D_2- and D_3-synchronizing; every D_2-synchronizing word defined at each state is D_3-synchronizing.

In [37] we adapted the approach based on satisfiability solvers to finding D_3-synchronizing words of minimum length for NFAs. The first-named author used a similar method in the cases of D_1- and D_2-synchronization; results related to D_2-synchronization were reported in [36].

Yet another version of synchronization for NFAs was introduced in [15] and systematically studied in [23–27], which terminology we adopt. An NFA $\mathscr{A} = \langle Q, \Sigma \rangle$ is called *carefully synchronizing* if there is a word $w = a_1 \cdots a_\ell$, with $a_1, \dots, a_\ell \in \Sigma$, that satisfies the condition (C), being the conjunction of $(C1)$–$(C3)$ below:

$(C1)$: the letter a_1 is defined at every state in Q;
$(C2)$: the letter a_t with $1 < t \leq \ell$ is defined at every state in $Q.a_1 \cdots a_{t-1}$,
$(C3)$: $|Q.w| = 1$.

Any w satisfying (C) is called a *carefully synchronizing word* (c.s.w., for short) for $\mathscr{A}$. Thus, when a c.s.w. is applied at any state in Q, no undefined transition occurs during the course of application. Every carefully synchronizing word is clearly D_1-synchronizing but the converse is not true in general; moreover, a D_1-synchronizing NFA may admit no c.s.w.

In this paper we focus on carefully synchronizing words for PFAs. There are several theoretical and practical reasons for this.

On the theoretical side, it is easy to see that each of the conditions (C), (D_1), (D_3) leads to the same notion when restricted to PFAs. As for D_2-synchronization, if a word w is D_2-synchronizing for a PFA $\mathscr{A}$, then w carefully synchronizes $\mathscr{A}$ whenever w is defined at each state. Otherwise w is nowhere defined by Lemma 1b, and such 'annihilating' words are nothing but usual synchronizing words for the CFA obtained from $\mathscr{A}$ by adding a new sink state and making all transitions undefined in $\mathscr{A}$ lead to this sink state. Synchronization of CFAs with a sink state is relatively well understood (see [35]), and therefore, we may conclude that D_2-synchronization also reduces to careful synchronization in the realm of PFAs. On the other hand, there exists a simple transformation that converts every NFA $\mathscr{A} = \langle Q, \Sigma \rangle$ into a PFA $\mathscr{B} = \langle Q, \Sigma' \rangle$ such that $\mathscr{A}$ is D_3-synchronizing if and only if so is $\mathscr{B}$ and the minimum lengths of D_3-synchronizing words for $\mathscr{A}$ and $\mathscr{B}$ are equal; see [14, Lemma 8.3.8] and [16, Lemma 3]. These observations demonstrate that from the viewpoint of optimal synchronization, studying carefully synchronizing words for PFAs may provide both lower and upper bounds applicable to arbitrary NFAs and all aforementioned kinds of synchronization.

Probably even more important is the fact that careful synchronization of PFAs is relevant in numerous applications. Due to the page limit, we mention only two examples here.

In industrial robotics, synchronizing automata are widely used to design feeders, sorters, and orienters that work with flows of certain objects carried by a conveyer. The goal is achieved by making the flow encounter passive obstacles placed appropriately along the conveyer belt; see [21,22] for the origin of this automata approach and [2] for an illustrative example. Now imagine that the objects to be oriented or sorted have a fragile side that could be damaged if hitting an obstacle. In order to prevent any damage, we have to forbid 'dangerous' transitions in the automaton modelling the orienter/sorter so that the automaton becomes partial and the obstacle sequences must correspond to carefully synchronizing words. (Actually, the term 'careful synchronization' has been selected with this application in mind.)

Our second example relates to so-called synchronized codes[2]. Recall that a *prefix code* over a finite alphabet Σ is a set X of words in Σ^* such that no word of X is a prefix of another word of X. Decoding of a finite prefix code X over Σ can be implemented by a finite deterministic automaton $\mathscr{A}_X$ whose state Q is the set of all proper prefixes of the words in X (including the empty word ε) and whose transitions are defined as follows: for $q \in Q$ and $a \in \Sigma$,

$$q.a = \begin{cases} qa & \text{if } qa \text{ is a proper prefix of a word of } X, \\ \varepsilon & \text{if } qa \in X, \\ \text{undefined} & \text{otherwise.} \end{cases}$$

In general, $\mathscr{A}_X$ is a PFA (it is complete if and only if the code X is not contained in another prefix code over Σ). It can be shown that if $\mathscr{A}_X$ is carefully synchronizing,

[2] We refer the reader to [4, Chapters 3 and 10] for a detailed account of profound connections between codes and automata.

the code X has a very useful property: whenever a loss of synchronization between the decoder and the coder occurs (because of a channel error), it suffices to transmit a c.s.w. w of $\mathscr{A}_X$ such that w sends all states in Q to the state ε to ensure that the following symbols will be decoded correctly.

We may conclude that the problems of determining whether or not a given PFA is carefully synchronizing and of finding its shortest carefully synchronizing words are both natural and important. The bad news is that these problems turn out to be quite difficult: it is known that the first problem is PSPACE-complete and that the minimum length of carefully synchronizing words for carefully synchronizing PFAs can be exponential as a function of the number of states. (These results were found in [33,34] and later rediscovered and strengthened in [25].) Thus, one has to use some tools that have proved to be efficient for dealing with computationally hard problems. As mentioned in Sect. 1, in this paper we make an attempt to employ a satisfiability solver as such a tool.

3 Encoding

For completeness, recall the formulation of the Boolean satisfiability problem (SAT). An instance of SAT is a pair (V, C), where V is a set of Boolean variables and C is a collection of clauses over V. (A *clause* over V is a disjunction of literals and a *literal* is either a variable in V or the negation of a variable in V.) Any *truth assignment* on V, i.e., any map $\varphi\colon V \to \{0,1\}$, extends to a map $C \to \{0,1\}$ (still denoted by φ) via the usual rules of propositional calculus: $\varphi(\neg x) = 1 - \varphi(x)$, $\varphi(x \vee y) = \max\{\varphi(x), \varphi(y)\}$. A truth assignment φ *satisfies* C if $\varphi(c) = 1$ for all $c \in C$. The answer to an instance (V, C) is YES if (V, C) has a *satisfying assignment* (i.e., a truth assignment on V that satisfies C) and NO otherwise.

By Cook's classic theorem (see, e.g., [28, Theorem 8.2]), SAT is NP-complete, and by the very definition of NP-completeness, every problem in NP reduces to SAT. On the other hand, over the last score or so, many efficient *SAT solvers*, i.e., specialized programs designed to solve instances of SAT have been developed. Modern SAT solvers can solve instances with hundreds of thousands of variables and millions of clauses within a few minutes. Due to this progress, the following approach to computationally hard problems has become quite popular nowadays: one encodes instances of such problems into instances of SAT that are then fed to a SAT solver[3]. It is exactly the strategy that we want to apply.

We start with the following problem:

CSW (the existence of a c.s.w. of a given length):
INPUT: a PFA $\mathscr{A}$ and a positive integer ℓ (given in unary);
OUTPUT: YES if $\mathscr{A}$ has a c.s.w. of length ℓ;
NO otherwise.

[3] We refer the reader to the survey [10] or to the handbook [6] for a detailed discussion of the approach and examples of its successful applications in various areas.

We have to assume that the integer ℓ is given in unary because with ℓ given in binary, a polynomial time reduction from CSW to SAT is hardly possible. (Indeed, it easily follows from [25] that the version of CSW in which the integer parameter is given in binary is PSPACE-hard, and the existence of a polynomial reduction from a PSPACE-hard problem to SAT would imply that the polynomial hierarchy collapses at level 1.) In contrast, the version of CSW with the unary integer parameter is easily seen to belong to NP: given an instance $(\mathscr{A} = \langle Q, \Sigma\rangle, \ell)$ of CSW in this setting, guessing a word $w \in \Sigma^*$ of length ℓ is legitimate. Then one just checks whether or not w is carefully synchronizing for $\mathscr{A}$, and time spent for this check is clearly polynomial in the size of $(\mathscr{A}, \ell)$.

Now, given an arbitrary instance $(\mathscr{A}, \ell)$ of CSW, we construct an instance (V, C) of SAT such that the answer to $(\mathscr{A}, \ell)$ is YES if and only if so is the answer to (V, C). Our encoding follows general patterns presented in [6, Chapters 2 and 16] but has some specific features so that we describe it in full detail and provide a rigorous proof of its adequacy. In the following presentation of the encoding, precise definitions and statements are interwoven with less formal comments explaining the 'physical' meaning of variables and clauses.

So, take a PFA $\mathscr{A} = \langle Q, \Sigma\rangle$ and an integer $\ell > 0$. Denote the sizes of Q and Σ by n and m respectively, and fix some numbering of these sets so that $Q = \{q_1, \ldots, q_n\}$ and $\Sigma = \{a_1, \ldots, a_m\}$.

We start with introducing the variables used in the instance (V, C) of SAT that encodes $(\mathscr{A}, \ell)$. The set V consists of two sorts of variables: $m\ell$ *letter variables* $x_{i,t}$ with $1 \le i \le m$, $1 \le t \le \ell$, and $n(\ell + 1)$ *state variables* $y_{j,t}$ with $1 \le j \le n$, $0 \le t \le \ell$. We use the letter variables to encode the letters of a hypothetical c.s.w. w of length ℓ: namely, we want the value of the variable $x_{i,t}$ to be 1 if and only if the t-th letter of w is a_i. The intended meaning of the state variables is as follows: we want the value of the variable $y_{j,t}$ to be 1 whenever the state q_j belongs to the image of Q under the action of the prefix of w of length t, in which situation we say that *q_j is active after t steps*. We see that the total number of variables in V is $m\ell + n(\ell + 1) = (m + n)\ell + n$.

Now we turn to constructing the set of clauses C. It consists of four groups. The group I of *initial clauses* contains n one-literal clauses $y_{j,0}$, $1 \le j \le n$, and expresses the fact that all states are active after 0 steps.

For each $t = 1, \ldots, \ell$, the group L of *letter clauses* includes the clauses

$$x_{1,t} \vee \cdots \vee x_{m,t}, \quad \neg x_{r,t} \vee \neg x_{s,t}, \quad \text{where } 1 \le r < s \le m. \tag{1}$$

Clearly, the clauses (1) express the fact that the t-th position of our hypothetical c.s.w. w is occupied by exactly one letter in Σ. Altogether, L contains $\ell\left(\frac{m(m-1)}{2} + 1\right)$ clauses.

For each $t = 1, \ldots, \ell$ and each triple (q_j, a_i, q_k) in the transition relation of $\mathscr{A}$, the group T of *transition clauses* includes the clause

$$\neg y_{j,t-1} \vee \neg x_{i,t} \vee y_{k,t}. \tag{2}$$

Invoking the basic laws of propositional logic, one sees that the clause (2) is equivalent to the implication $y_{j,t-1} \,\&\, x_{i,t} \to y_{k,t}$, that is, (2) expresses the fact that if the state q_j has been active after $t-1$ steps and a_i is the t-th letter of w, then the state $q_k = q_j.a_i$ becomes active after t steps. Further, for each $t = 1, \ldots, \ell$ and each pair (q_j, a_i) such that a_i is undefined at q_j in $\mathscr{A}$, we add to T the clause

$$\neg y_{j,t-1} \vee \neg x_{i,t}. \tag{3}$$

The clause is equivalent to the implication $y_{j,t-1} \to \neg x_{i,t}$, and thus, it expresses the requirement that the letter a_i should not be occur in the t-th position of w if q_j has been active after $t-1$ steps. Obviously, this corresponds to the conditions $(C1)$ (for $t = 0$) and $(C2)$ (for $t > 0$) in the definition of careful synchronization. For each $t = 1, \ldots, \ell$ and each pair $(q_j, a_i) \in Q \times \Sigma$, exactly one of the clauses (2) or (3) occurs in T, whence T consists of ℓmn clauses.

The final group S of *synchronization clauses* includes the clauses

$$\neg y_{r,\ell} \vee \neg y_{s,\ell}, \quad \text{where } 1 \leq r < s \leq n. \tag{4}$$

The clauses (4) express the requirement that at most one state remains active when the action of the word w is completed, which corresponds to the condition $(C3)$ from the definition of careful synchronization. The group S contains $\frac{n(n-1)}{2}$ clauses.

Summing up, the number of clauses in $C := I \cup L \cup T \cup S$ is

$$n + \ell\left(\tfrac{m(m-1)}{2} + 1\right) + \ell mn + \tfrac{n(n-1)}{2} = \ell\left(\tfrac{m(m-1)}{2} + mn + 1\right) + \tfrac{n(n+1)}{2}. \tag{5}$$

In comparison with encodings used in our earlier papers [36,37], the encoding suggested here produces much smaller SAT instances. Since in the applications the size of the input alphabet is a (usually small) constant, the leading term in (5) is $\Theta(\ell n)$ while the restriction to PFAs of the encodings from [36,37] has $\Theta(\ell n^2)$ clauses.

Theorem 2. *A PFA $\mathscr{A}$ has a c.s.w. of length ℓ if and only if the instance (V, C) of SAT constructed above is satisfiable. Moreover, the carefully synchronizing words of length ℓ for $\mathscr{A}$ are in a 1-1 correspondence with the restrictions of satisfying assignments of (V, C) to the letter variables.*

Proof. Suppose that $\mathscr{A}$ has a c.s.w. of length ℓ. We fix such a word w and denote by w_t its prefix of length $t = 1, \ldots, \ell$. Define a truth assignment $\varphi\colon V \to \{0, 1\}$ as follows: for $1 \leq i \leq m$, $0 \leq j \leq n$, $1 \leq t \leq \ell$, let

$$\varphi(x_{i,t}) := \begin{cases} 1 & \text{if the } t\text{-th letter of } w \text{ is } a_i, \\ 0 & \text{otherwise;} \end{cases} \tag{6}$$

$$\varphi(y_{j,0}) := 1; \tag{7}$$

$$\varphi(y_{j,t}) := \begin{cases} 1 & \text{if the state } q_j \text{ lies in } Q.w_t, \\ 0 & \text{otherwise.} \end{cases} \tag{8}$$

In view of (6) and (7), φ satisfies all clauses in L and respectively I. As $w_\ell = w$ and $|Q.w| = 1$, we see that (8) ensures that φ satisfies all clauses in S. It remains to analyze the clauses in T. For each fixed $t = 1, \dots, \ell$, these clauses are in a 1-1 correspondence with the pairs in $Q \times \Sigma$. We fix such a pair (q_j, a_i), denote the clause corresponding to (q_j, a_i) by c and consider three cases.

Case 1: the letter a_i is not the t-th letter of w. In this case $\varphi(x_{i,t}) = 0$ by (6), and hence, $\varphi(c) = 1$ since the literal $\neg x_{i,t}$ occurs in c, independently of c having the form (2) or (3).

Case 2: the letter a_i is the t-th letter of w but it is undefined at q_j. In this case the clause c must be of the form (3). Observe that $t > 1$ in this case since the first letter of the c.s.w. w must be defined at each state in Q. Moreover, the state q_j cannot belong to the set $Q.w_{t-1}$ because a_i must be defined at each state in this state. Hence $\varphi(y_{j,t-1}) = 0$ by (8), and $\varphi(c) = 1$ since the literal $\neg y_{j,t-1}$ occurs in c.

Case 3: the letter a_i is the t-th letter of w and it is defined at q_j. In this case the clause c must be of the form (2), in which the literal $y_{k,t}$ corresponds to the state $q_k = q_j.a_i$. If the state q_j does not belong to the set $Q.w_{t-1}$, then as in the previous case, we have $\varphi(y_{j,t-1}) = 0$ and $\varphi(c) = 1$. If q_j belongs to $Q.w_{t-1}$, then the state q_k belongs to the set $(Q.w_{t-1}).a_i = Q.w_t$, whence $\varphi(y_{k,t}) = 1$ by (8). We conclude that $\varphi(c) = 1$ since the literal $y_{k,t}$ occurs in c.

Conversely, suppose that $\varphi\colon V \to \{0, 1\}$ is a satisfying assignment for (V, C). Since φ satisfies the clauses in L, for each $t = 1, \dots, \ell$, there exists a unique $i \in \{1, \dots, m\}$ such that $\varphi(x_{i,t}) = 1$. This defines a map $\chi\colon \{1, \dots, \ell\} \to \{1, \dots, m\}$. Let $w := a_{\chi(1)} \cdots a_{\chi(\ell)}$. We aim to show that w is a c.s.w. for $\mathscr{A}$, i.e., to verify that w fulfils the conditions ($C1$)–($C3$) from the definition of a c.s.w. For this, we first prove two auxiliary claims. Recall that a state is said to be active after t steps if it lies in $Q.w_t$, where, as above, w_t is the length t prefix of the word w. (By the length 0 prefix we understand the empty word ε).

Claim 1. *For each $t = 0, 1, \dots, \ell$, there are states active after t steps.*

Claim 2. *If a state q_k is active after t steps, then $\varphi(y_{k,t}) = 1$.*

We prove both claims simultaneously by induction on t. The induction basis $t = 0$ is guaranteed by the fact that all states are active after 0 steps and φ satisfies the clauses in I. Now suppose that $t > 0$ and there are states active after $t - 1$ steps. Let q_r be such a state. Then $\varphi(y_{r,t-1}) = 1$ by the induction assumption. Let $i := \chi(t)$, that is, a_i is the t-th letter of the word w. Then $\varphi(x_{i,t}) = 1$, whence φ cannot satisfy the clause of the form (3) with $j = r$. Hence this clause cannot appear in T as φ satisfies the clauses in T. This means that the letter a_i is defined at q_r in $\mathscr{A}$, and the state $q_s := q_r.a_i$ is active after t steps. Claim 1 is proved.

Now let q_k be an arbitrary state that is active after $t > 0$ steps. Since a_i is the t-th letter of w, we have $Q.w_t = (Q.w_{t-1}).a_i$, whence $q_k = q_j.a_i$ for same $q_j \in Q.w_{t-1}$. Therefore the clause (2) occurs in T, and thus, it is satisfied by φ. Since q_j is active after $t - 1$ steps, $\varphi(y_{j,t-1}) = 1$ by the induction assumption; besides that, $\varphi(x_{i,t}) = 1$. We conclude that in order to satisfy (2), the assignment φ must fulfil $\varphi(y_{k,t}) = 1$. This completes the proof of Claim 2.

We turn to prove that the word w fulfils ($C1$) and ($C2$). This amounts to verifying that for each $t = 1, \dots, \ell$, the t-th letter of the word w is defined at every state q_j that is active after $t-1$ steps. Let, as above, a_i stand for the t-th letter of w. If a_i were undefined at q_j, then by the definition of the set T of transition clauses, this set would include the corresponding clause (3). However, $\varphi(x_{i,t}) = 1$ by the construction of w and $\varphi(y_{j,t-1}) = 1$ by Claim 2. Hence φ does not satisfy this clause while the clauses from T are satisfied by φ, a contradiction.

Finally, consider ($C3$). By Claim 1, some state is active after ℓ steps. On the other hand, the assignment φ satisfies the clauses in S, which means that $\varphi(y_{j,\ell}) = 1$ for at most one index $j \in \{1, \dots, n\}$. By Claim 2 this implies that at most one state is active after ℓ steps. We conclude that exactly one state is active after ℓ steps, that is, $|Q.w| = 1$. □

4 Experimental Results

We have successfully applied the encoding constructed in Sect. 3 to solve CSW instances with the help of a SAT solver. As in [12,36–38], we have used MiniSat 2.2.0 [8,9]. In order to find a c.s.w. of minimum length for a given PFA $\mathscr{A}$, we have considered CSW instances $(\mathscr{A}, \ell)$ with fixed $\mathscr{A}$ and performed binary search on ℓ. Even though our encoding is different from those we used in [36,37], it shares with them the following useful feature: when presented in DIMACS CNF format, the 'primary' SAT instance that encodes the CSW instance $(\mathscr{A}, 1)$ can be easily scaled to the SAT instances that encode the CSW instances $(\mathscr{A}, \ell)$ with any value of ℓ. Due to this feature, one radically reduces time needed to prepare the input data for the SAT solver; we refer the reader to [36, Sect. 3] for a detailed explanation of the trick and an illustrative example. Thus, we encode $(\mathscr{A}, 1)$, write the corresponding SAT instance in DIMACS CNF format, and scale the instance to the instances encoding $(\mathscr{A}, \ell)$ with $\ell = 2, 4, 8, \dots$ until we reach an instance on which the SAT solver returns YES[4]. The corresponding value of ℓ serves as the right border $\ell_{\max}$ of the binary search while the previous value of ℓ serves as the left border $\ell_{\min}$. Then we test the SAT instance corresponding to $(\mathscr{A}, \frac{\ell_{\max}+\ell_{\min}}{2})$, etc.

We implemented the algorithm outlined above in C++ and compiled with GCC 4.9.2. In our experiments we used a personal computer with an Intel(R) Core(TM) i5-2520M processor with 2.5 GHz CPU and 4 GB of RAM. The code can be found at https://github.com/hananshabana/SynchronizationChecker.

[4] In principle, it may happen that we never reach such an instance (which indicates that either $\mathscr{A}$ is not carefully synchronizing or the minimum length of carefully synchronizing words for $\mathscr{A}$ is too big so that MiniSat cannot handle the resulting SAT instance) but we have not observed such "bad" cases in our experiments with randomly generated PFAs.

As a sample of our experimental findings, we present here our results on synchronization of PFAs with a unique undefined transition. Observe that the problem of deciding whether or not a given PFA is carefully synchronizing remains PSPACE-complete even if restricted to this rather special case [25]. We considered random PFAs with $n \leq 100$ states and two input letters. The condition $(C1)$ in the definition of a carefully synchronizing PFA implies that such a PFA must have an everywhere defined letter. We denoted this letter by a and the other letter, called b, was chosen to be undefined at a unique state. Further, it is easy to see that for a PFA $\langle Q, \{a,b\}\rangle$ with a, b so chosen to be carefully synchronizing, it is necessary that $|Q.a| < |Q|$. Therefore, we fixed a state $q_a \in Q$ and then selected a uniformly at random from all n^{n-1} maps $Q \to Q \setminus \{q_a\}$. Similarly, to ensure there is a unique undefined transition with b, we fixed a state $q_b \in Q$ (not necessarily different from q_a) and then selected b uniformly at random from all $(n-1)^n$ maps $Q \setminus \{q_b\} \to Q$. For each fixed n, we generated up to 1000 random PFAs this way and calculated the average length $\ell(n)$ of their shortest carefully synchronizing words. We used the least squares method to find a function that best reflects how $\ell(n)$ depends on n, and it turned out that our results are reasonably well approximated by the following expression:

$$\ell(n) \approx 3.92 + 0.49n - 0.005n^2 + 0.000024n^3.$$

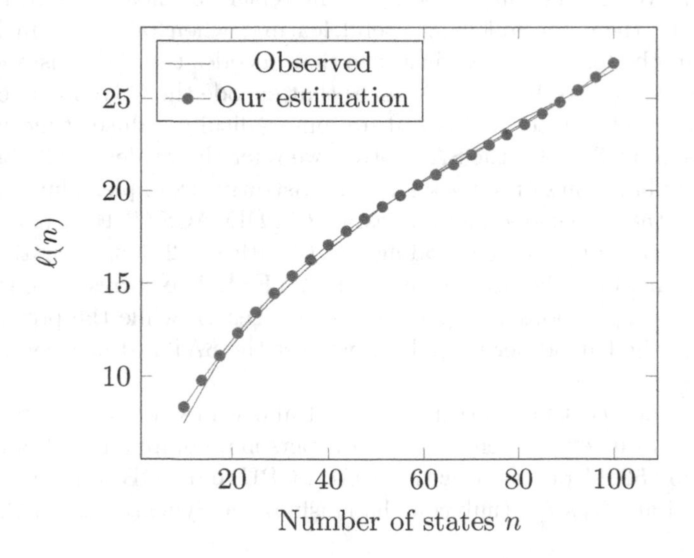

The next graph shows the relation between the relative standard deviation of our datasets and the number of states. We see that the relative standard deviation gradually decreases as the number of states grows.

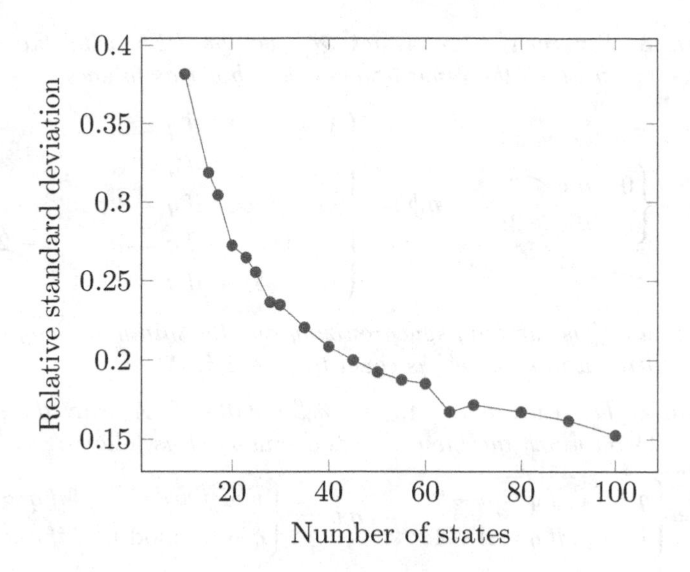

We performed similar experiments with random PFAs that have two or three undefined transition. We also tested our algorithm on PFAs from the series $\mathscr{P}_n$ suggested in [7]. The state set of $\mathscr{P}_n$ is $\{1, 2, \ldots, n\}$, $n \geq 3$, on which the input letters a and b act as follows:

$$q.a := \begin{cases} q+1 & \text{if } q = 1, 2, \\ q & \text{if } q = 3, \ldots, n; \end{cases} \qquad q.b := \begin{cases} \text{undefined} & \text{if } q = 1, \\ q+1 & \text{if } q = 2, \ldots, n-1, \\ 1 & \text{if } q = n. \end{cases}$$

We examined all automata $\mathscr{P}_n$ with $n = 4, 5, \ldots, 11$, and for each of them, our result matched the theoretical value predicted by [7, Theorem 3]. The time consumed ranges from 0.301 s for $n = 4$ to 4303 s for $n = 11$. Observe that in the latter case the shortest c.s.w. has length 116 so that the "honest" binary search started with $(\mathscr{P}_{11}, 1)$ required 14 calls of MiniSat, namely, for the encodings of $(\mathscr{P}_{11}, \ell)$ with $\ell = 1, 2, 4, 8, 16, 32, 64, 128, 96, 112, 120, 116, 114, 115$. Of course, if one just wants to confirm (or to disprove) a theoretical prediction ℓ for the minimum length of carefully synchronizing words for a given PFA $\mathscr{A}$, two calls of a SAT solver—on the encodings of $(\mathscr{A}, \ell)$ and $(\mathscr{A}, \ell - 1)$—suffice.

In our experiments, we kept track of "slowly synchronizing" PFAs, that is, PFAs with the minimum length of carefully synchronizing words close to the square of the number of states. Whenever we encountered such examples, we made an attempt to generalize them in order to get infinite series of provably "slowly synchronizing" PFAs. The following statements present two of the results we found this way.

Proposition 3. *For each $n > 4$, let $\mathscr{H}'_n$ be the PFA with the state set $\{0, 1, \dots, n-1\}$ on which the input letters a and b act as follows:*

$$q.a := \begin{cases} 0 & \text{if } q \leq 2, \\ q & \text{if } q \geq 3; \end{cases} \qquad q.b := \begin{cases} 3 & \text{if } q = 0, \\ 0 & \text{if } q = 1, \\ \text{undefined} & \text{if } q = 2, \\ q+1 & \text{if } q = 3, \dots, n-2, \\ 1 & \text{if } q = n-1. \end{cases}$$

The automaton $\mathscr{H}'_n$ is carefully synchronizing and the minimum length of carefully synchronizing words for $\mathscr{H}'_n$ is equal to $(n-2)^2$.

Proposition 4. *For each $n > 4$, let $\mathscr{H}''_n$ be the PFA with the state set $\{0, 1, \dots, n-1\}$ on which the input letters a and b act as follows:*

$$q.a := \begin{cases} q+1 & \text{if } q \leq n-2, \\ 1 & \text{if } q = n-1; \end{cases} \qquad q.b := \begin{cases} \text{undefined} & \text{if } q = 0, \\ q+1 \pmod{n} & \text{if } q \geq 1. \end{cases}$$

The automaton $\mathscr{H}''_n$ is carefully synchronizing and the minimum length of carefully synchronizing words for $\mathscr{H}'_n$ is equal to $n^2 - 3n + 3$.

We omit the proofs of Propositions 3 and 4 due to space constraints. The proofs (which are not difficult) can be obtained by a suitable adaptation of the approach developed for the case of CFAs in [3, Section 4].

From the viewpoint of our studies, the series $\mathscr{H}'_n$ and $\mathscr{H}''_n$ are of interest as they exhibit two extremes with respect to amenability of careful synchronization to the SAT solver approach. The series $\mathscr{H}'_n$ has turned to be a hard nut to crack for our algorithm: the maximum n for which the algorithm was able to find a c.s.w. of minimum length is 13, and computing this word (of length 121) took almost 4 h. In contrast, automata in the series $\mathscr{H}''_n$ turn out to be quite amenable: for instance, the algorithm found a c.s.w. of length 343 for $\mathscr{H}''_{20}$ in 13.38 s. At present, we have no explanation for what causes such a strong contrast: is this an intrinsic structure of the PFAs under consideration, or the nature of the algorithm built in MiniSat, or just a peculiarity of our implementation?

We made also a comparison with the only approach to computing carefully synchronizing words of minimum length that we had found in the literature, namely, the approach based on partial power automata; see [27, p. 295]. Given a PFA $\mathscr{A} = \langle Q, \Sigma \rangle$, its *partial power automaton* $\mathcal{P}(\mathscr{A})$ has the subsets of Q as the states, the same input alphabet Σ, and the transition function defined as follows: for each $a \in \Sigma$ and each $P \subseteq Q$,

$$P.a := \begin{cases} \{q.a \mid q \in P\} & \text{provided } q.a \text{ is defined for all } q \in P, \\ \text{undefined} & \text{otherwise.} \end{cases}$$

It is easy to see that $w \in \Sigma^*$ is a c.s.w. of minimum length for $\mathscr{A}$ if and only if w labels a minimum length path in $\mathcal{P}(\mathscr{A})$ starting at Q and ending at a singleton. Such a path can be found by breadth-first search in the underlying digraph of $\mathcal{P}(\mathscr{A})$.

The result of the comparison is presented in the picture on the next page. In this experiment we had to restrict to PFAs with at most 16 states since beyond this number of states, our implementation of the method based on partial power automata could not complete the computation due to memory restrictions (recall that we used rather modest computational resources). However, we think that the exhibited data suffice to demonstrate that the approach based on SAT solvers shows a by far better performance.

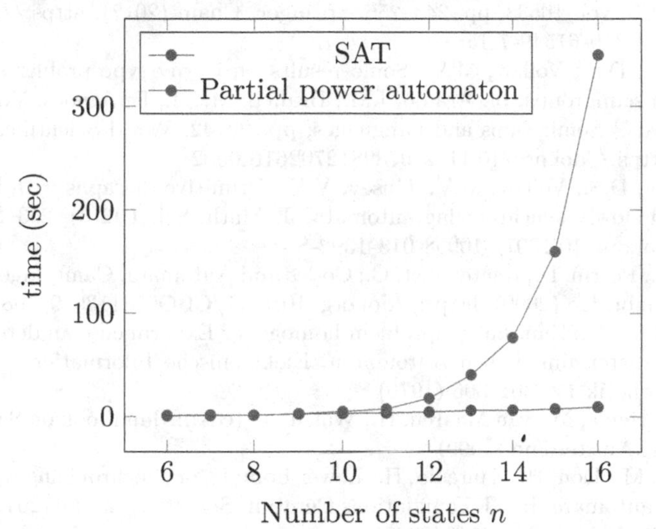

5 Conclusion and Future Work

We have presented an attempt to approach the problem of computing a c.s.w. of minimum length for a given PFA via the SAT solver method. For this, we have developed a new encoding, which, in comparison with encodings used in our earlier papers [36,37], requires a more sophisticated proof but leads to more economic SAT instances. In our future experiments, we plan to employ more advanced SAT solvers. Using more powerful computers constitutes other obvious direction for improvements. Clearly, the approach is amenable to parallelization since calculations needed for different automata are completely independent so that one can process in parallel as many automata as many processors are available.

Now we are designing new experiments. For instance, it appears to be interesting to compare the minimum lengths of a synchronizing word for a synchronizing CFA and of carefully synchronizing words for PFAs that can be obtained from the CFA by removing one or more of its transitions. We also plan to extend the SAT solver approach to so-called *exact synchronization* of PFAs which is of interest for certain applications.

Acknowledgements. The authors are very much indebted to the referees for their valuable remarks.

References

1. Altun, Ö.F., Atam, K.T., Karahoda, S., Kaya, K.: Synchronizing heuristics: speeding up the slowest. In: Yevtushenko, N., Cavalli, A.R., Yenigün, H. (eds.) ICTSS 2017. LNCS, vol. 10533, pp. 243–256. Springer, Cham (2017). https://doi.org/10.1007/978-3-319-67549-7_15
2. Ananichev, D.S., Volkov, M.V.: Some results on Černý type problems for transformation semigroups. In: Araújo, I.M., Branco, M.J.J., Fernandes, V.H., Gomes, G.M.S. (eds.) Semigroups and Languages, pp. 23–42. World Scientific, Singapore (2004). https://doi.org/10.1142/9789812702616_0002
3. Ananichev, D.S., Volkov, M.V., Gusev, V.V.: Primitive digraphs with large exponents and slowly synchronizing automata. J. Math. Sci. **192**(3), 263–278 (2013). https://doi.org/10.1007/s10958-013-1392-8
4. Berstel, J., Perrin, D., Reutenauer, C.: Codes and Automata. Cambridge University Press, Cambridge (2009). https://doi.org/10.1017/CBO9781139195768
5. Burkhard, H.-D.: Zum Längenproblem homogener Experimente an determinierten und nicht-deterministischen Automaten. Elektronische Informationsverarbeitung und Kybernetik **12**, 301–306 (1976)
6. Biere, A., Heule, M., van Maaren, H., Walsh, T. (eds.): Handbook on Satisfiability. IOS Press, Amsterdam (2009)
7. de Bondt, M., Don, H., Zantema, H.: Lower bounds for synchronizing word lengths in partial automata. Int. J. Foundations Comput. Sci. **30**(1), 29–60 (2019). https://doi.org/10.1142/S0129054119400021
8. Eén, N., Sörensson, N.: An extensible SAT-solver. In: Giunchiglia, E., Tacchella, A. (eds.) SAT 2003. LNCS, vol. 2919, pp. 502–518. Springer, Heidelberg (2004). https://doi.org/10.1007/978-3-540-24605-3_37
9. Eén, N., Sörensson, N.: The MiniSat Page. http://minisat.se
10. Gomes, C.P., Kautz, H., Sabharwal, A., Selman, B.: Satisfiability solvers. In: van Harmelen, F., Lifschitz, V., Porter, B. (eds.) Handbook of Knowledge Representation, vol. I, pp. 89–134. Elsevier, Amsterdam (2008). https://doi.org/10.1016/S1574-6526(07)03002-7
11. Goralčík, P., Hedrlín, Z., Koubek, V., Ryšlinková, J.: A game of composing binary relations. RAIRO Inform. Théor. **16**(4), 365–369 (1982)
12. Güniçen, C., Erdem, E., Yenigün, H.: Generating shortest synchronizing sequences using answer set programming. In: Fink, M., Lierler, Y. (eds.) Answer Set Programming and Other Computing Paradigms, 6th International Workshop, ASPOCP 2013, pp. 117–127 (2013). https://arxiv.org/abs/1312.6146
13. Imreh, B., Steinby, M.: Directable nondeterministic automata. Acta Cybernetica **14**, 105–115 (1999)
14. Ito, M.: Algebraic Theory of Automata and Languages. World Scientific, Singapore (2004). https://doi.org/10.1142/4791
15. Ito, M., Shikishima-Tsuji, K.: Some results on directable automata. In: Karhumäki, J., Maurer, H., Păun, G., Rozenberg, G. (eds.) Theory Is Forever. LNCS, vol. 3113, pp. 125–133. Springer, Heidelberg (2004). https://doi.org/10.1007/978-3-540-27812-2_12
16. Ito, M., Shikishima-Tsuji, K.: Shortest directing words of nondeterministic directable automata. Discrete Math. **308**(21), 4900–4905 (2008). https://doi.org/10.1016/j.disc.2007.09.010

17. Karahoda, S., Erenay, O.T., Kaya, K., Türker, U.C., Yenigün, H.: Parallelizing heuristics for generating synchronizing sequences. In: Wotawa, F., Nica, M., Kushik, N. (eds.) ICTSS 2016. LNCS, vol. 9976, pp. 106–122. Springer, Cham (2016). https://doi.org/10.1007/978-3-319-47443-4_7
18. Karahoda, S., Kaya, K., Yenigün, H.: Synchronizing heuristics: speeding up the fastest. Expert Syst. Appl. **94**, 265–275 (2018). https://doi.org/10.1016/j.eswa.2017.10.054
19. Kowalski, J., Roman, A.: A new evolutionary algorithm for synchronization. In: Squillero, G., Sim, K. (eds.) EvoApplications 2017. LNCS, vol. 10199, pp. 620–635. Springer, Cham (2017). https://doi.org/10.1007/978-3-319-55849-3_40
20. Kari, J., Volkov, M.V.: Černý's conjecture and the road coloring problem. In: Pin, J.-É (ed.) Handbook of Automata Theory, vol. I. EMS Publishing House (in print)
21. Natarajan, B.K.: An algorithmic approach to the automated design of parts orienters. In: Proceedings 27th Annual Symposium Foundations Computer Science, pp. 132–142. IEEE Press (1986). https://doi.org/10.1109/SFCS.1986.5
22. Natarajan, B.K.: Some paradigms for the automated design of parts feeders. Int. J. Robot. Res. **8**(6), 89–109 (1989). https://doi.org/10.1177/027836498900800607
23. Martyugin, P.V.: Lower bounds for the length of the shortest carefully synchronizing words for two- and three-letter partial automata. Diskretn. Anal. Issled. Oper. **15**(4), 44–56 (2008)
24. Martyugin, P.V.: A lower bound for the length of the shortest carefully synchronizing words. Russian Math. **54**(1), 46–54 (2010). https://doi.org/10.3103/S1066369X10010056. (Iz. VUZ)
25. Martyugin, P.V.: Synchronization of automata with one undefined or ambiguous transition. In: Moreira, N., Reis, R. (eds.) CIAA 2012. LNCS, vol. 7381, pp. 278–288. Springer, Heidelberg (2012). https://doi.org/10.1007/978-3-642-31606-7_24
26. Martyugin, P.V.: Careful synchronization of partial automata with restricted alphabets. In: Bulatov, A.A., Shur, A.M. (eds.) CSR 2013. LNCS, vol. 7913, pp. 76–87. Springer, Heidelberg (2013). https://doi.org/10.1007/978-3-642-38536-0_7
27. Martyugin, P.V.: Complexity of problems concerning carefully synchronizing words for PFA and directing words for NFA. Theor. Comput. Syst. **54**(2), 293–304 (2014). https://doi.org/10.1007/s00224-013-9516-6
28. Papadimitriou, C.H.: Computational Complexity. Addison-Wesley, Boston (1994)
29. Pixley, C., Jeong, S.-W., Hachtel, G.D: Exact calculation of synchronization sequences based on binary decision diagrams. In: Proceedings 29th Design Automation Conference, pp. 620–623. IEEE Press (1992)
30. Podolak, I.T., Roman, A., Jędrzjczyk, D.: Application of hierarchical classifier to minimal synchronizing word problem. In: Rutkowski, L., Korytkowski, M., Scherer, R., Tadeusiewicz, R., Zadeh, L.A., Zurada, J.M. (eds.) ICAISC 2012. LNCS, vol. 7267, pp. 421–429. Springer, Heidelberg (2012). https://doi.org/10.1007/978-3-642-29347-4_49
31. Podolak, I.T., Roman, A., Szykuła, M., Zieliński, B.: A machine learning approach to synchronization of automata. Expert Syst. Appl. **97**, 357–371 (2018). https://doi.org/10.1016/j.eswa.2017.12.043
32. Roman, A.: Genetic algorithm for synchronization. In: Dediu, A.H., Ionescu, A.M., Martín-Vide, C. (eds.) LATA 2009. LNCS, vol. 5457, pp. 684–695. Springer, Heidelberg (2009). https://doi.org/10.1007/978-3-642-00982-2_58
33. Rystsov, I.K.: Asymptotic estimate of the length of a diagnostic word for a finite automaton. Cybernetics **16**(1), 194–198 (1980). https://doi.org/10.1007/bf01069104

34. Rystsov, I.K.: Polynomial complete problems in automata theory. Inf. Process. Lett. **16**(3), 147–151 (1983). https://doi.org/10.1016/0020-0190(83)90067-4
35. Rystsov, I.K.: Reset words for commutative and solvable automata. Theor. Comput. Sci. **172**(1), 273–279 (1997). https://doi.org/10.1016/s0304-3975(96)00136-3
36. Shabana, H.: D_2-synchronization in nondeterministic automata. Ural Math. J. **4**(2), 99–110 (2018). https://doi.org/10.15826/umj.2018.2.011
37. Shabana, H., Volkov, M.V.: Using Sat solvers for synchronization issues in nondeterministic automata. Siberian Electron. Math. Rep. **15**, 1426–1442 (2018). https://doi.org/10.17377/semi.2018.15.117
38. Skvortsov, E., Tipikin, E.: Experimental study of the shortest reset word of random automata. In: Bouchou-Markhoff, B., Caron, P., Champarnaud, J.-M., Maurel, D. (eds.) CIAA 2011. LNCS, vol. 6807, pp. 290–298. Springer, Heidelberg (2011). https://doi.org/10.1007/978-3-642-22256-6_27
39. Volkov, M.V.: Synchronizing automata and the Černý Conjecture. In: Martín-Vide, C., Otto, F., Fernau, H. (eds.) LATA 2008. LNCS, vol. 5196, pp. 11–27. Springer, Heidelberg (2008). https://doi.org/10.1007/978-3-540-88282-4_4

A Computational Comparison of Parallel and Distributed K-median Clustering Algorithms on Large-Scale Image Data

Anton V. Ushakov(✉) and Igor Vasilyev

Matrosov Institute for System Dynamics and Control Theory of SB RAS,
Lermontov str. 134, 664033 Irkutsk, Russia
{aushakov,vil}@icc.ru
http://iv.icc.ru

Abstract. Most commonly used clustering algorithms are those aimed at solving the well-known k-median problem. Their main advantage is that they are simple to implement and use, and they are flexible in choosing dissimilarity measures (not necessarily metrics). K-median algorithms are also known to be more robust to noise and outliers in comparison with k-means algorithms. In spite of that, they have been of limited use for large-scale clustering problems due to their high computational and space complexity. This work aims at computational comparison of k-median clustering algorithms in a specific large-scale setting—clustering large image collections. We implement distributed versions of the most common k-median clustering algorithms and compare them with our parallel heuristic for solving large-scale k-median problem instances. We analyze clustering results with respect to external evaluation measures and run time.

Keywords: k-median problem · Clustering · Facility location · k-medoids clustering · Parallel computing · Exemplar-based clustering

1 Introduction

Clustering is one of the main unsupervised machine learning task widely used in numerous applications (from bioinformatics to the analysis of blockchains). Among many clustering approaches, one of the most common techniques are the so-called partition-based clustering algorithms relying upon solving some optimization problems, e.g. k-means and k-median (p-median). Given a set $I = \{1, \ldots, m\}$ of data items that are supposed to be clustered into the prespecified number of disjoint groups and $d(\cdot, \cdot)$ measuring dissimilarity between data items. Note that $d(\cdot, \cdot)$ may be a distance-metric on I if data items are given in the form of feature vectors. The k-median objective is to find at most k data items (medians) from I, such that the total sum of dissimilarities between each data item and their closest median is minimized.

I. Bykadorov et al. (Eds.): MOTOR 2019, CCIS 1090, pp. 119–130, 2019.
https://doi.org/10.1007/978-3-030-33394-2_10

The k-median problem is known to be NP-hard on network and plane [19,23]. Despite that, there are several popular clustering algorithms based on finding solutions of the k-median problem, e.g. k-medoids, PAM, CLARA, CLARANS, and affinity propagation. The main advantage of most k-median clustering algorithms are their simplicity, the robustness to noise and outliers in comparison to k-means-based algorithms, and their flexibility in choosing dissimilarity measures. Thus, dissimilarities may be asymmetric and need not satisfy the triangle inequality. On the other hand, the main disadvantage of such algorithms is their high time complexity and memory load. For example, a serious bottleneck is to calculate and store the whole pairwise dissimilarity matrix. For example, to store the dissimilarities between one million data items, we actually need 4 TB of memory, which is not available on a general-purpose computer. Moreover, most of k-median clustering algorithms are heuristics that in general converge only to local optimal solutions. Some algorithms (like CLARANS and k-medoids) are heavily dependent on the choice of initial solutions, thus they may require multiple reruns to find good clusterings. Many studies are focused on mitigating time and memory complexity of k-median clustering algorithms, e.g. via sampling procedures (CLARA) and/or distributed computing. However, this usually results in a much lower clustering accuracy in comparison with the base k-median clustering algorithms.

The k-median problem can be expressed as an integer program. Let $G(I, A)$ be a weighted complete simple directed graph with the arc set $A = \{(i, j) : i, j \in I;\ i \neq j\}$. Each arc $(i, j) \in A$ is assigned the weight d_{ij} corresponding to the measure of dissimilarity between data items i and j. Let us also introduce the binary variables y_i which is 1 if node $i \in I$ is a median, 0 otherwise; and the variables x_{ij} which is equal to 1 if node i is the closest median to node j. We also introduce two sets of nodes $\delta^-(j) = \{i \in I|\ (i, j) \in A\}$ and $\delta^+(i) = \{j \in I|\ (i, j) \in A\}$. Then, the k-median clustering problem is written as follows

$$\min_{(x,y)} \sum_{(i,j)\in A} d_{ij} x_{ij}, \tag{1}$$

$$\sum_{i\in\delta^-(j)} x_{ij} + y_j = 1, \qquad j \in I, \tag{2}$$

$$x_{ij} \leqslant y_i, \qquad i \in I, j \in \delta^+(i), \tag{3}$$

$$\sum_{i\in I} y_i = k, \tag{4}$$

$$y_i,\ x_{ij} \in \{0, 1\}, \qquad i \in I,\ (i, j) \in A. \tag{5}$$

The objective function (1) is to minimize the overall sum of dissimilarities between nodes and their closest medians. Constraints (2) ensure that either a node j is a median (cluster representative) or it is assigned to a median. Constraints (3) impose that each node can only be assigned to medians. Constraint (4) enforces that the number of medians must be equal to k.

Apart from the k-median clustering algorithms commonly used in diverse applications, there are exact, approximation, and heuristic methods for finding near-optimal solutions of the k-median problem (for a comprehensive reviews see [2,11,24]). For example, for more than ten years, the best approximation algorithm for the metric problem with factor $3+\varepsilon$ was the local search heuristic based on multiple swaps. Lately, this approximation ratio was improved to $1+\sqrt{3}+\varepsilon$ in [21] and finally fine tuned, using a dependent-rounding approach, to $2.675+\varepsilon$ [7], which is the best-known approximation ratio for the k-median problem to date. Among the most efficient exact algorithms are the branch-and-cut-and-price method [6,22] and the column-and-row generation algorithm [14]. The most effective heuristics are the Lagrangian heuristic with aggregation [4] and the primal-dual variable neighborhood search [18]. These approaches are capable of finding sub-optimal solutions of problem instances with hundred thousand data items. Moreover, they provide a dual bound for the objective value, which allows ascertaining the optimality of a particular solution found. Nevertheless, the high complexity and memory usage still remains the main disadvantage of these approaches preventing them from the application to large-scale clustering problems. Some researchers have attempted to reduce computational burden of some metaheuristics by leveraging parallel and distributed computing. Thus, parallel implementations of the variable neighborhood search, tabu search, and the scatter search employing multistart and domain decomposition strategies are developed in [3,10,15,16], respectively. However, low accuracy or high memory load of these algorithms remain their main bottlenecks.

In this paper, we develop a distributed Lagrangian relaxation-based approach based on [27] and distributed implementations of several most commonly used k-median clustering algorithms. Though these algorithms are very popular in pattern analysis and computer vision (see e.g. [13]), our main motivation for this paper was to assess their accuracy, speed, and effectiveness in a specific setting—clustering large-scale imageries. In particular, we study the ability of common k-median clustering algorithms to reveal clusters in large collections of face images from the large number of identities, when faces are represented using deep features. The remainder of the paper is structured as follows. In Sect. 2, we describe the most popular k-median clustering algorithms and the details of our particular distributed implementations. In Sect. 3, we briefly discuss our distributed Lagrangian relaxation-based heuristic for the k-median problem. Finally, in Sect. 4, we report computational results and a comparison of the developed clustering algorithms on the VGGFace benchmark.

2 Parallel K-median Clustering Algorithms

In this section, we review most widely-used k-median clustering algorithms and present our particular distributed implementations. Thus, we consider the following algorithms:

PAM is probably the best known k-median (or k-medoids) clustering algorithm, first proposed in [20]. The algorithm is a simple two-step heuristic, where, in the first step, the greedy search is applied to find an initial set of medians. In the second step, PAM tries to improve the greedy solution with the local search heuristic (also known as Teitz and Bart's heuristic and vertex substitution) combined with the fast interchange technique. PAM is notorious for its high time and space requirements. Indeed, it relies on searching the whole 1-swap neighborhood of each incumbent, which may be extremely challenging in the case of large-scale problems. The dissimilarity matrix may be precomputed but it obviously results in a large memory footprint. On the other hand, the use of the fast interchange technique can substantially reduce the memory load of PAM (as only two closest medians are retained for each data item), but may require a lot of extra computation for updating medians when the numbers of both clusters and features are large.

We use PAM as a subroutine, hence we have implemented its shared-memory version, using OpenMP interface. In our implementation, we improved the first step of PAM by embedding the fast greedy approach from [29]. We keep the whole dissimilarity matrix in RAM to speed up computations and median updates. The matrix can easily be computed by parallel threads that find dissimilarities only between their specific disjoint chunks of data items and the whole dataset. The first step of PAM can then be parallelized by assigning separate chunks of data items to different threads and finding closest medians for each data item in parallel. In step 2 of PAM, each thread processes only a subset of columns of the dissimilarity matrix to compute the objective value and update medians.

CLARA. The time and space complexity of PAM motivated the same authors to develop CLARA, which is a variant of PAM applicable to large-scale datasets. CLARA is rested upon drawing multiple relatively small samples from a dataset, which are then clustered with PAM. The final clustering is selected as a set of medians providing the minimal objective value over the whole dataset. Obviously, the quality of the final clustering is mostly dependent on the size of the samples. Despite its advantages, CLARA is known to be less effective when the number of clusters is relatively large.

Nowadays, most of distributed k-median clustering algorithms rely on space-partitioning techniques or sampling approaches similar to one embedded in CLARA (e.g. see [26] and references therein). However, such techniques usually results in developing fast distributed algorithms at the expense of loosing solution quality in comparison to the base algorithms (PAM and CLARA). Since CLARA relies on clustering multiple samples, we developed a distributed version of CLARA leveraging a simple multi-search strategy. Note that such a strategy will help us to assess the performance of the basic CLARA algorithm. We use the MPI-OpenMP hybrid programming model to implement a distributed CLARA algorithm. In our implementation, each MPI process independently draws a random sample from the dataset and simultaneously runs the parallel PAM algorithm to find a clustering of the sample. The final solution is then selected among all solutions found by MPI processes.

CLARANS is another common clustering algorithm based on top of PAM. It simply performs a multistart hill-climbing over the 1-swap neighborhood, each time starting from a random solution. Its two main parameters are the number of restarts and the number of consecutive iterations without a sufficient improvement of the objective value. The main drawback of CLARANS is the same as of PAM: In order to calculate the value of the objective function, it has to either keep the whole dissimilarity matrix in the memory or compute pairwise dissimilarities every time they are needed. Thus, for large-scale datasets, it is either too memory consuming or may require too much computation (especially, when the number of features is large). Another disadvantage is that the 1-swap neighborhood may become too large for the large number of data items. To parallelize CLARANS, we employ the same multi-search strategy as for CLARA. Thus, each MPI process runs an independent hill climbing search starting from a random solution. We employ OpenMP to parallelize hill climbing by sharing the work among parallel threads. The only computationally intensive operation of the hill climbing is the calculation of the objective value. We parallelize it by assigning distinct chunks of data items to parallel threads. They independently find closest median for each item from their chunk and compute the corresponding part of the objective value.

k-medoids is a neighborhood improvement heuristic for the k-median problem, similar to k-means. It starts with a random solution and assigns all data items to their closest medians. In contrast to k-means, it solves the 1-median problem within each cluster, i.e. it finds a data item that minimizes the overall dissimilarity between data items in the cluster. If the current medians are changed, the k-medoids algorithm reassigns data items and solves the 1-median problem for each cluster again. It stops when any further improvements are not available. Despite the relocation scheme similar to k-means, the k-medoids algorithm requires much more computation to update cluster representatives. Indeed, the computation of cluster means is performed in linear time in the cluster size, while solving the 1-median problem requires quadratic time. We use the same MPI-OpenMP hybrid model. OpenMP is employed to concurrently compute the value of the objective function for a current clustering. Multiple 1-median problems can also be solved in parallel by assigning each cluster to a separate thread.

3 Distributed Lagrangian Relaxation-Based Heuristic

Here, we recall the main steps of the Lagrangian heuristic for the k-median problem proposed in [4] and then describe how it can efficiently be implemented in distributed environments.

3.1 Sequentional Algorithm

We use the most common Lagrangian relaxation of the k-median problem obtained by relaxing the constraints (2) and adding them to the objective function together with Lagrange multipliers $\lambda \in \mathbb{R}^m$:

$$\mathcal{L}(\lambda) = \min_{(x,y)} \left\{ \sum_{ij \in A} d_{ij} x_{ij} - \sum_{j \in I} \lambda_j \left(\sum_{i \in \delta^-(j)} x_{ij} + y_j - 1 \right) : \text{ subject to (3)-(5)} \right\},$$

The value of the Lagrangian dual function $\mathcal{L}(\lambda)$ is a dual (lower) bound for the k-median objective value for any set of Lagrange multipliers λ.

We denote $\mu_{ij}(\lambda) = d_{ij} - \lambda_j$ and $\rho_i(\lambda) = \sum_{j\in\delta^+(i)} \mu_{ij}(\lambda)^- - \lambda_i$ to be the reduced costs of the variables x_{ij} and y_j, respectively. Let $\rho_i(\lambda)$ be ordered increasingly, i.e. there is a permutation $i_1, \ldots, i_m$ such that $\rho_{i_1}(\lambda) \leq \cdots \leq \rho_{i_m}(\lambda)$.

Thus, the value of the Lagrangean dual function is obtained by summing-up the smallest reduced costs of the variables $y_i(\lambda^s)$ plus the sum of all the multipliers λ_j [12], i.e.

$$\mathcal{L}(\lambda) = \sum_{k=1}^{p} \rho_{i_k}(\lambda) + \sum_{i\in I} \lambda_i,$$

The best dual bound is then sought by finding the optimal solution of the following Lagrangian dual problem $\max_{\lambda} \mathcal{L}(\lambda)$. This problem is convex and non-differentiable, hence we adapt a subgradient optimization algorithm with a heuristic step-size rule, which is especially suitable in the case of large-scale problem instances. Indeed, we compute the step-size of the subgradient algorithm at iteration s as

$$\alpha_s = \frac{\beta_s(UB - \mathcal{L}(\lambda^s))}{\|g(\lambda^s)\|_2^2},$$

where β_s is a gradually decreasing parameter and UB is an primal bound for the k-median objective. Note that we also use the value of the parameter β_s as a stopping criterion. We find the initial upper bound by assigning each data item to the closest medoid $i \in I$: $y(\lambda^0) = 1$.

To increase the efficiency of the subgradient algorithm, we use a delayed column generation approach and the Lagrange multipliers stabilization technique that make the sequence of Lagrangian dual function values monotonic [4,18].

After the subgradient algorithm halts, it returns a set of Lagrange multipliers $\bar{\lambda}$ and the corresponding dual bound. To improve the initial primal bound and find the corresponding good feasible clustering solution, we leverage the so-called core heuristic [4,9,28]. It consists in finding a set of promising (core) decision variables y_i and x_{ij} of the k-median problem and fixing all remained variables to zero. The k-median problem is then solved only over the core variables (the core problem), and its optimal solution provides a primal bound for the original k-median problem. The core variables are selected with respect to their reduced costs, i.e. we set thresholds Δ and γ such that only the variables y_i and x_{ij} satisfying

$$i \in I(\bar{\lambda}, \Delta) \triangleq \{i \in I : \ \rho_i(\bar{\lambda}) \leq \Delta\},$$
$$(i,j) \in W(\bar{\lambda}, \gamma) \triangleq \{(i,j) : \ i \in I(\bar{\lambda}, \Delta), j \in \delta^+(i), \mu_{ij}(\bar{\lambda}) \leq \gamma\}.$$

are not fixed to zero. The core problem is much smaller than the original problem, thus one usually uses exact methods or commercial solvers to find its optimal solution [4,5]. However, in the case of large-scale problem instances, the core problem may become too difficult to be solved exactly. On the other hand, we can solve the core problem with a fast heuristic that finds a good feasible solution and a primal bound. Following [27], we use a simple simulated annealing algorithm over 1-swap neighborhood. We adapted a simple cooling rule $T(t+1) = qT(t)$, where $q = (t_0/t_{\min})^{1/(M_{out}-1)}$, M_{out}—a fixed number of temperature reductions, t_0 and $t_{\min}$ are initial and final temperature, respectively.

3.2 Implementation in Distributed Environments

To implement a distributed version of the Lagrangian relaxation-based heuristic for the k-median problem, we use a hybrid MPI-OpenMP programming model. Our hybrid version of the subgradient algorithm is similar to the purely MPI implementation from [17] that, however, was not effective for large numbers of data items and processes due to communication overheads. We suppose that we are given a distributed environment encompassing a number of computing nodes connected by a network. Thus, MPI is used to perform inter-node communications, while OpenMP is employed to parallelize intra-node computations. Our implementation is rested upon data parallelism paradigm, i.e. each MPI process handles only a part of the data needed to fulfill an iteration of the subgradient algorithm. At first, we distributedly compute the dissimilarity matrix, each MPI process calculates only a subset of matrix columns and keeps it in local memory. Storing the dissimilarity matrix distributedly helps considerably reduce the memory usage of the algorithm. All processes then deal with parts of reduced costs $\rho_i(\lambda^s)$, Lagrange multipliers λ^s, and subgradients $s(\lambda^s)$ that "correspond" to the subset of columns stored in their local memory. Communications between processes occur to compute and sort the reduced costs, calculate the overall norm of a subgradient and distribute the stepsize α_s.

In order to find the core variables, each MPI process traverses its subset of columns of the dissimilarity matrix and identify variables with reduced costs less than the given thresholds. The core problem is then stored distributedly. Since the computation of the k-median objective is the most computationally intensive operation of the simulated annealing, we leverage a low-level parallelization strategy. The selection of neighbors of a current incumbent and the test for their acceptance are performed sequentially. The only task executed in parallel is the computation of the core problem objective.

4 Experiments

We implemented all the clustering algorithms using C++ programming language and tested them on the HPC-cluster "Akademik V.M. Matrosov" [1]. The cluster includes 60 nodes of dual 18-core Intel Xeon E5-2695 v4 processors and 120 Gb of DDR4-2400 memory. The nodes connected via QDR Infiniband network. Since a node consists of two processors, we run our distributed algorithms using two MPI processes per node and 18 OpenMP threads per MPI process.

Fig. 1. Example face images from the VGGFAce dataset. Each row contains pictures labeled in the dataset as corresponding to the same person.

One application of unsupervised machine learning algorithms that assumes handling large number of data items is clustering and analysis of large collections of social media data (e.g. images). Indeed, in recent years, we have witnessed a fast growing popularity of social media and, hence, a fast grow of data volumes. For example, in 2013, Facebook reported about more than 350 million images uploaded every day. Large collections of face images are also subject of analysis in forensic investigations and surveillance applications, where one may expect large number of both images and subjects.

In this section, we aim at the performance analysis of k-median clustering algorithms in this specific setting, where, apart from large numbers of both face images and identities, we may expect unbalanced clusters and the large number of near-duplicates, which is challenging for partition-based clustering algorithms.

As a benchmark, we utilize the VGGFace dataset [25]. It is a semi-automatically collected face dataset commonly used for training deep networks. The dataset contains URL links to 2.6 million images from 2622 subjects (celebrities and public figures). A final manually filtered collection obtained by involving human annotators consists of 982,803 images (375 images per person in average). Unfortunately, we could not download all images but only 880,501 out of 982,803 due to a lot of web links being broken. Moreover, some of the images turned out to be placeholders that replace the original image which is no longer available. After excluding placeholders, we have a collection of 690,761 images of 2510 identities from the original filtered VGGFace dataset. Even after manual filtering, VGGFace contains a relatively large number of duplicates and mis-labeled face images that can bias the accuracy analysis. Example face images from the VGGFace dataset are presented in Fig. 1.

To extract features from the VGGFace dataset, we leverage a ResNet-50 network from [8] pre-trained on MS-Celeb-1M dataset and fine-tuned on VGGFace2. We excluded overlap between VGGface and VGGFace2 during feature extraction. We detect faces using MTCNN [30] and apply the same cropping and alignment procedure to keep consistency between training and evaluation. Each image is passed as an input to the ResNet-50 network, then 2048-dimensional vector is extracted from the layer adjacent to the classifier layer. Each feature vector is then normalized into unit Euclidean norm.

Since we have to cluster the VGGFace dataset into groups of face images corresponding to the same person, we can estimate the accuracy of the k-median clustering algorithms with external evaluation measures. In our experiments, we employ pairwise precision, recall, and F-measure:

$$Precision = \frac{TP}{TP + FP},$$
$$Recall = \frac{TP}{TP + FN},$$

where TP are true positive pairs of data items (correctly clustered), FP—false positive pairs, and FN—false negative pairs. The F-measure is then computed as a harmonic mean of precision and recall:

$$F(i,j) = \frac{2 \cdot Precision \cdot Recall}{Precision + Recall},$$

We also evaluate our clustering results according to Purity, which calculated as

$$Purity = \sum_{i=1}^{k} \frac{1}{m} \max_{j} n_i^j,$$

where n_i^j is the number of face images of the subject j that are assigned to cluster i.

The results of computational experiments are presented in Table 1. Note that we report the best solution found by each algorithm over 20 runs. We use the Euclidean distance as the dissimilarity measure. The column "Objective" presents the value of the k-median objective function, the column "Time" reports the run time measured when running the competing algorithms on 5 computing nodes of "Akademik Matrosov". Recall that the accuracy of CLARA is strongly dependent of the sample size. In the literature, the most commonly used sample size is $2k + 40$, where k is the number of clusters. We also tested CLARA over larger samples with $4k$ data items. For CLARANS, each MPI process performed only one run of the parallel hill climbing. Thus, as we run CLARANS on 5 nodes, there are 10 parallel hill climbing searches run simultaneously. The parallel hill climbing halts when 100 gradually examined neighbors of a current incumbent give an objective improvement less than 1. The parallel k-medoids is terminated when the

Fig. 2. Example clusters. In each row, the first images are medians, while the last two are wrongly clustered

value of the k-median objective is improved for less than 0.01 in some iteration. In our computational experiments we set the number of clusters to be equal to the number of identities.

Table 1. Experiment results for the VGGFace dataset. Each row contains the results for the best clustering found by the corresponding algorithm

	Objective	Prec.	Recall	F-measure	Purity	Time (sec.)
Our algorithm	**491394.90**	**0.82**	**0.80**	**0.81**	**0.90**	17094
CLARA (2k)	564172.90	0.52	0.58	0.55	0.68	4572
CLARA (4k)	545375.02	0.62	0.65	0.63	0.75	32593
CLARANS	525851.07	0.69	0.71	0.69	0.82	27868
k-medoids	531334.46	0.51	0.57	0.53	0.66	7698

We can see that the distributed Lagrangian relaxation-based algorithm provides a clustering solution which has the best objective value and considerably outperforms all the k-median clustering algorithms. The relative difference between the found dual and primal bounds is only 0.03%. We can see that the k-median solution found by our algorithm also has the best accuracy according to all the external evaluation metrics. CLARA (2k) finds one of the worst solutions. This is consistent with the observation about its low effectiveness for datasets with large number of clusters. Nevertheless, we observe that CLARA (2k) has the best run time, which can be viewed as a trade of quality for speed. CLARANS provides a better solution than both CLARA implementations and the k-medoids algorithm. Moreover, it runs faster than CLARA (4k), which makes CLARANS more preferable for large multidimensional datasets with large number of clusters. Providing a much better clustering, our approach demonstrates the competitive run time, which is superior to CLARANS and CLARA (4k).

Figure 2 presents an example clusters generated by our approach. Each row represents separate impure clusters. The first face image in each row is the corresponding cluster representative (median), while the last two images are incorrectly clustered, i.e. they are not of the same class as that the corresponding median belongs to.

Acknowledgement. This work is supported by the Russian Science Foundation under grant 17-71-10176.

References

1. Irkutsk supercomputer center of SB RAS. http://hpc.icc.ru. Accessed 15 Feb 2019
2. An, H.-C., Svensson, O.: Recent developments in approximation algorithms for facility location and clustering problems. In: Fukunaga, T., Kawarabayashi, K. (eds.) Combinatorial Optimization and Graph Algorithms, pp. 1–19. Springer, Singapore (2017). https://doi.org/10.1007/978-981-10-6147-9_1
3. Arbelaez, A., Quesada, L.: Parallelising the k-medoids clustering problem using space-partitioning. In: Helmert, M., Röger, G. (eds.) Proceedings the 6th Annual Symposium on Combinatorial Search, SoCS 2013, pp. 20–28. AAAI (2013)
4. Avella, P., Boccia, M., Salerno, S., Vasilyev, I.: An aggregation heuristic for large scale p-median problem. Comput. Oper. Res. **39**(7), 1625–1632 (2012)
5. Avella, P., Boccia, M., Sforza, A., Vasilyev, I.: An effective heuristic for large-scale capacitated facility location problems. J. Heuristics **15**(6), 597–615 (2008)
6. Avella, P., Sassano, A., Vasilyev, I.: Computational study of large-scale p-median problems. Math. Program. **109**(1), 89–114 (2007)
7. Byrka, J., Pensyl, T., Rybicki, B., Srinivasan, A., Trinh, K.: An improved approximation for k-median and positive correlation in budgeted optimization. ACM Trans. Algorithms **13**(2), 23:1–23:31 (2017). https://doi.org/10.1145/2981561
8. Cao, Q., Shen, L., Xie, W., Parkhi, O.M., Zisserman, A.: VGGFace2: a dataset for recognising faces across pose and age. In: Proceedings 13th IEEE International Conference on Automatic Face & Gesture Recognition, FG 2018, pp. 67–74. IEEE (2018). https://doi.org/10.1109/FG.2018.00020
9. Carrizosa, E., Ushakov, A., Vasilyev, I.: A computational study of a nonlinear minsum facility location problem. Comput. Oper. Res. **39**(11), 2625–2633 (2012)
10. Crainic, T.G., Gendreau, M., Hansen, P., Mladenović, N.: Cooperative parallel variable neighborhood search for the p-median. J. Heuristics **10**(3), 293–314 (2004)
11. Daskin, M.S., Maass, K.L.: The *p*-median problem. In: Laporte, G., Nickel, S., da Gama, F.S. (eds.) Location Science, pp. 21–45. Springer, Cham (2015). https://doi.org/10.1007/978-3-319-13111-5_2
12. Fisher, M.L.: The lagrangian relaxation method for solving integer programming problems. Manage. Sci. **27**(1), 1–18 (1981)
13. Frahm, J.-M., et al.: Building Rome on a cloudless day. In: Daniilidis, K., Maragos, P., Paragios, N. (eds.) ECCV 2010. LNCS, vol. 6314, pp. 368–381. Springer, Heidelberg (2010). https://doi.org/10.1007/978-3-642-15561-1_27

14. García, S., Labbé, M., Marín, A.: Solving large p-median problems with a radius formulation. INFORMS J. Comput. **23**(4), 546–556 (2011)
15. Garcia-López, F., Melián-Batista, B., Moreno-Pérez, J.A., Moreno-Vega, J.M.: The parallel variable neighborhood search for the p-median problem. J. Heuristics **8**(3), 375–388 (2002)
16. Garcia-López, F., Melián-Batista, B., Moreno-Pérez, J.A., Moreno-Vega, J.M.: Parallelization of the scatter search for the p-median problem. Parallel Comput. **29**(5), 575–589 (2003). Parallel computing in logistics
17. Hanafi, S., Sterle, C., Ushakov, A., Vasilyev, I.: A parallel subgradient algorithm for lagrangean dual function of the *p*-median problem. Studia Informatica Universalis **9**(3), 105–124 (2011)
18. Hansen, P., Brimberg, J., Urosević, D., Mladenović, N.: Solving large p-median clustering problems by primal-dual variable neighborhood search. Data Min. Knowl. Discov. **19**(3), 351–375 (2009)
19. Kariv, O., Hakimi, S.: An algorithmic approach to network location problems. II: The p-medians. SIAM J. Appl. Math. **37**(3), 539–560 (1979)
20. Kaufman, L., Rousseeuw, P.J.: Clustering by means of medoids. In: Dodge, Y. (ed.) Statistical Data Analysis Based on the L_1-Norm and Related Methods, pp. 405–416. North-Holland (1987)
21. Li, S., Svensson, O.: Approximating k-median via pseudo-approximation. SIAM J. Comput. **45**(2), 530–547 (2016). https://doi.org/10.1137/130938645
22. Mancini, E.P., Marcarelli, S., Vasilyev, I., Villano, U.: A grid-aware MIP solver: implementation and case studies. Futur. Gener. Comp. Syst. **24**(2), 133–141 (2008)
23. Megiddo, N., Supowit, K.J.: On the complexity of some common geometric location problems. SIAM J. Comput. **13**(1), 182–196 (1984)
24. Mladenović, N., Brimberg, J., Hansen, P., Moreno-Pérez, J.: The p-median problem: a survey of metaheuristic approaches. Eur. J. Oper. Res. **179**(3), 927–939 (2007)
25. Parkhi, O.M., Vedaldi, A., Zisserman, A.: Deep face recognition. In: Xie, X., Jones, M.W., Tam, G.K.L. (eds.) Proceedings the British Machine Vision Conference (BMVC), pp. 41.1–41.12. BMVA Press (2015). https://doi.org/10.5244/C.29.41
26. Song, H., Lee, J.G., Han, W.S.: PAMAE: parallel k-medoids clustering with high accuracy and efficiency. In: Proceedings 23rd ACM SIGKDD International Conference on Knowledge Discovery and Data Mining, KDD 2017, pp. 1087–1096. ACM, New York (2017). https://doi.org/10.1145/3097983.3098098
27. Vasilyev, I., Ushakov, A.: A shared memory parallel heuristic algorithm for the large-scale p-median problem. In: Sforza, A., Sterle, C. (eds.) Optimization and Decision Science: Methodologies and Applications, ODS 2017. Mathematics & Statistics, vol. 217, pp. 295–302. Springer, Cham (2017). https://doi.org/10.1007/978-3-319-67308-0_30
28. Vasilyev, I., Ushakov, A.V., Maltugueva, N., Sforza, A.: An effective heuristic for large-scale fault-tolerant k-median problem. Soft Comput. (2018). https://doi.org/10.1007/s00500-018-3562-6
29. Whitaker, R.A.: A fast algorithm for the greedy interchange for large-scale clustering and median location problems. Can. J. Oper. Res. Inf. Process. **21**, 95–108 (1983)
30. Zhang, K., Zhang, Z., Li, Z., Qiao, Y.: Joint face detection and alignment using multitask cascaded convolutional networks. IEEE Signal Process. Lett. **23**(10), 1499–1503 (2016). https://doi.org/10.1109/LSP.2016.2603342

On the One–Dimensional Space Allocation Problem with Partial Order and Forbidden Zones

Gennady G. Zabudsky[1(✉)] and Natalia S. Veremchuk[2(✉)]

[1] Sobolev Institute of Mathematics, Novosibirsk, Russia
zabudsky@ofim.oscsbras.ru
[2] Siberian State Automobile and Highway University,
Omsk, Russia
n-veremchuk@rambler.ru

Abstract. In this paper we consider a generalization of the One–Dimensional Space Allocation Problem (ODSAP). It is a well–known optimization problem. The classical formulation of the problem is as follows. It is required to place rectangular connected objects (linear segments) on a line with the minimal total cost of connections between them. The generalization of the problem is that there are fixed objects (forbidden zones) on the line and between the objects a partial order of their placement on the line is given. It is impossible to place the objects in the forbidden zones. The area in which the placement is allowed consists of disjoint segments (blocks). Centers of the placed objects are connected among themselves and with centers of the zones. The structure of connections between the objects is defined using a graph. A review of the formulations and methods for solving the classical ODSAP is given. We propose a polynomial–time algorithm for finding a local optimum for a fixed partition of the objects into the blocks when the graph of connections between the objects is a composition of rooted trees and parallel–serial graphs.

Keywords: Optimal placement · Connected rectangles · Partial order · Forbidden zones

1 Introduction

Optimization problems have many practical applications and they are interesting from the mathematical point of view. Many of these problems are NP–hard. Therefore, the promising areas of research of such problems are the development of algorithms for local optimization [5], heuristics [2], approximation schemes [6]. This paper is devoted to a search for a local optimum to one of the problems of optimal placement of objects.

G.G. Zabudsky—The work was supported by the program of fundamental scientific research of the SB RAS No. I.5.1., project No. 0314-2019-0019.

I. Bykadorov et al. (Eds.): MOTOR 2019, CCIS 1090, pp. 131–143, 2019.
https://doi.org/10.1007/978-3-030-33394-2_11

To date, the problems of optimal placement of the objects with different sizes are intensively studied both theoretically and in connection with practical applications. In particular, for such problems, there are two main directions of applications: placement of elements of electronic devices and placement of units of technological equipment. In the first direction, not only the problem of placing elements is solved, but also the tracing of connections between them is performed, see, e.g., the technology of a very–large–scale integration [3].

In the design of electronic devices, the problem consists in the determination of the optimal spatial arrangement of elements on a given surface (switching field). Criteria and constraints in the problem can be divided into metric and topological ones. The metric criteria take into account the size of the elements and the distance between them, the size of the switching zero, the distance between the terminals of the elements, the allowable length of the connections. The topological ones deal with the number of spatial intersections of connections and interlayer transitions.

The problem of tracing is to determine the geometry of connections of structural elements of electronic devices. The optimization criterion for the optimal solution of the trace problem can be, e.g., the total length of connections, or the number of layers of installation.

In the second direction of applications of optimal placement problems, the focus is usually on the placement of the equipment and the valuation of the links between them. Such problems should be solved not only in the design of technological equipment placement but also at the stage of preliminary design (planning) in other industries. As an example, we can mention here the placement of shops of an enterprise, elements of a hydraulic system of a machine, facilities for laying oil and gas pipelines and so on. At the stage of planning, a preliminary valuation of the connections between the placed elements is usually performed. In particular, this applies to the stage of installation of production, as a result of which the problem of equipment placement is solved. Here, unlike electronic devices, the tracing stage is not so important, as everything is done in three–dimensional space. The placement choice is influenced by several factors. Often, the creation of direct driveways, ease of operation and maintenance of equipment may require the "regularity" of placement [17], e.g., placement along so–called "red" lines.

In these areas of practical application of the problems of placement, it is often necessary to take into account the size of objects. Taking into account this factor in an automated solution allows one to choose the best option for the equipment's placement, which more adequately reflects the real situation. The objects involved in the placement process are usually approximated by simple geometric shapes, for example, rectangles. This reduces geometric complexity when solving problems, for example, when checking conditions of mutual non–intersected of the objects.

Different approaches are developed for solving problems of optimal placement of rectangular objects [3,10,12,14,17]. The placement of rectangular parts on rectangular blanks, so–called cutting and packing problems is widely applicable in practice and well–developed from the mathematical point of view. To solve such problems, exact methods of linear programming [9], as well as heuristic methods, such as evolutionary methods, are usually used.

One of the directions of the rectangles' placing is "regular placement", for example, placing them on a line. To construct a set of Pareto–optimal solutions to the problem of the rectangles' placing on parallel lines, aimed to minimize the length and width of the area occupied by the equipment, the apparatus of integer optimization and dynamic programming were used [17]. A dynamic programming algorithm was proposed for placing interconnected rectangular objects on the line with the criterion of the minimal total cost of connections [14]. One of a well–known problems of placing the connected dimensional objects on the line is the One–Dimensional Space Allocation Problem (ODSAP).

In recent years, the described above direction of research is gaining popularity for the problems of placing objects in the presence of the forbidden zones and barriers [13,19,20]. It is not allowed to place objects in the forbidden zones. In the case of reconstruction of a plant, forbidden zones may occur, for example, because of existing premises and technological equipment. Barriers can be considered as some regions described above where the placement of objects is not allowed. It is also forbidden to make connections inside a barrier.

This paper is devoted to the problem of placement of connected objects on a line with forbidden zones (for short "zones") and with partial order of object placement on the line. The placed objects and the forbidden zones have a form of rectangles. Note that in one–dimensional space the rectangular objects are linear segments. The area in which the placement of the objects is allowed consists of the linear segments (blocks). Centers of the objects are connected between themselves and with centers of the zones. The structure of connections between the objects and the partial order of their placement on the line are given in the form of a directed graph. It is required to place the objects outside the forbidden zones so that the partial order was satisfied and the total cost of the connections was minimal. It is known that the original continuous problem for the arbitrary undirected graph can be reduced to discrete subproblems of smaller dimension [19]. In our case, the subproblem was reduced to the problem of placement of vertices of a directed graph on a line so that the given partial order was satisfied and the weighted sum of lengths of all arcs was minimal. A review of the research of the classical ODSAP is provided. A polynomial–time algorithm for finding a local optimum for a fixed partition of the objects into blocks when the graph of connections between objects is a composition of rooted trees and parallel–serial graphs is proposed.

2 Statement of the ODSAP and Review of Research

Let n be a number of objects which should be placed, and let $I = \{1, \ldots, n\}$ be a set of indices of these objects. Let m be the number of forbidden zones with numbers from the set $J = \{1, \ldots, m\}$. Each object and each zone is a rectangle with dimensions $l_i \times h_i$ and $p_j \times d_j$, where $i \in I$ and $j \in J$, respectively.

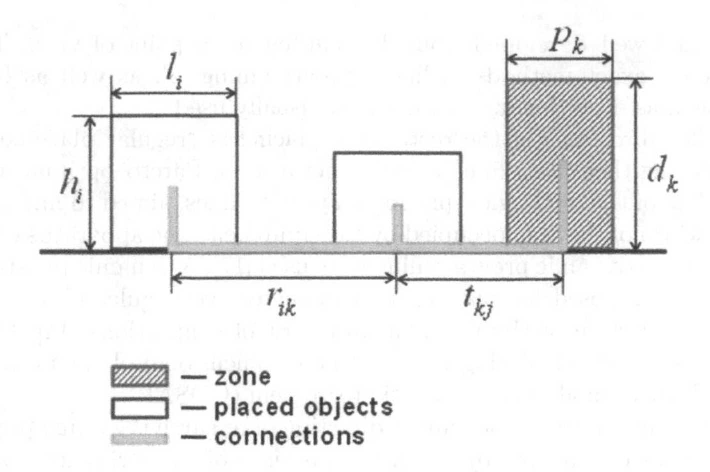

Fig. 1. The scheme of placing of objects and passing connections between objects.

The centers of the objects are connected between themselves and with the centers of the zones and the connections pass as it is shown in Fig. 1. Note that the lengths of the vertical components of connections between object i and object k, and between object i and zone j are equal to $h_i/2 + h_k/2$ and $h_i/2 + d_j/2$ respectively, and they do not depend on the placement of objects. The minimal feasible distances $l_i/2 + l_k/2$ and $l_i/2 + p_j/2$ are set between nearest points of the objects and zones. We can include the dimensions of objects and zones in minimal feasible distances and assume that there are restrictions between the projections of the centers of the objects and the zones. The problem is reduced to the placement of the point objects, i.e. projections of geometric centers of rectangles on the line. Let r_{ik}, $(r_{ii} = 0)$, $i, k \in I$, and t_{ij}, $(t_{ii} = 0)$, $i \in I$, $j \in J$, denote the minimal feasible distances between object i and object k and between object i and zone j, respectively. We denote by $R = (r_{ik})$, $T = (t_{ij})$ the symmetric matrices of the minimal feasible distances between the objects and between the objects and zones, respectively.

The structure of connections between the objects is defined using a graph $G = (V, E)$, where $V = \{1, \ldots, n\}$ and E is a edge set. The edge $(i, k) \in E$ exists if there is a connection between object i and object k. If an order of the placement for the connected objects on the line is given, for example, object i lies to the left of the object k, then (i, k) is an arc. In this case, the directed graph G also sets the partial order of the objects' placement on the line. Let $u_{ik} \geq 0$ $(u_{ik} = u_{ki})$, $w_{ij} \geq 0$ $(w_{ij} = w_{ji})$ be the specific costs of connections between object i and object k, and between object i and zone j for $i, k \in I$, $j \in J$, and $i < k$, respectively. If G is an undirected graph, then it is necessary to place the objects on the line so that the restrictions on the minimal feasible distances between the objects and between the objects and zones are satisfied and the total cost of connections between the objects and between the objects and zones is minimal [19]. If the graph G is directed, then it is additionally necessary to respect restrictions on the partial order of the objects' placement on the line.

Consider a straight–line segment of length LS containing some fixed rectilinear forbidden zones with centers at b_j, $j \in J$. Without loss of generality, we can assume that the left border of the segment of length LS is the origin. Denote by x_i the coordinate of center of object i, $i \in I$; let $x = (x_1, \ldots, x_n)$ be the placement of objects. It is needed to minimize the function:

$$F(x) = \sum_{i=1}^{n} \sum_{j=1}^{m} w_{ij}|x_i - b_j| + \sum_{i=1}^{n-1} \sum_{k=i+1}^{n} u_{ik}|x_i - x_k| \rightarrow \min, \tag{1}$$

under constraints

$$|x_i - b_j| \geq t_{ij}, \quad i \in I, j \in J, \tag{2}$$

$$|x_i - x_k| \geq r_{ik}, \quad i, k \in I, i < k, \tag{3}$$

$$\frac{l_i}{2} \leq x_i \leq LS - \frac{l_i}{2}, \quad i \in I. \tag{4}$$

Note that if (i, k) is an arc, one can omit the absolute value sign in the expression $|x_i - x_k|$.

Introduce matrixes $R = (r_{ik})$, $i, k \in I$, and $T = (t_{ij})$, $i \in I$, $j \in J$, respectively. The following conditions for the elements of the matrix R are considered:

(a) $r_{ik} = \frac{l_i + l_k}{2}$, $i, k \in I$, $i \neq k$ (non–intersected conditions);
(b) $r_{ij} + r_{jk} \geq r_{ik}$, $i, j, k \in I$, $i \neq j \neq k$ (in the case of metric problem);
(c) r_{ik} is arbitrary, $i, k \in I$, (in the case of non–metric problem).

Similar conditions can be considered for the elements of the matrix T.

Problem (1), (3) without forbidden zones ($J = \emptyset$) with conditions (a) and undirected graph G is the classical ODSAP. The classical ODSAP is NP–hard for the case when G is an arbitrary undirected graph of connections between objects [4,16].

The classical ODSAP problem (1), (3) without forbidden zones is formulated as follows in terms of permutations. Let $\pi = (\pi(1), \ldots, \pi(n))$ denote a permutation of the objects; i.e. the first (leftmost) object is $\pi(1)$, the second object is $\pi(2)$ and so on. Denote by π^{-1} the inverse of this permutation: $\pi^{-1}(i)$ is the position of object i in the permutation π. Consider the permutation π and two objects i and j. The distance between i and j with respect to this permutation, is assumed to be equal to the distance between their centers. It is equal to the half–length of object i, plus the lengths of all objects which are situated between objects i and j in π, plus the half–length of object j:

$$d(i, j, \pi) = l_i/2 + l_j/2 + \sum_{k \in B(i,j,\pi)} l_k, \tag{5}$$

where $B(i, j, \pi)$ is a set of the objects between objects i and j in π.

The ODSAP is the problem of finding a permutation π which minimizes the weighted sum of the distances between all pairs of objects, i.e.

$$Z(\pi) = \sum_{i \in I} \sum_{j \in I, j \neq i} u_{ij} d(i, j, \pi). \tag{6}$$

Defining

$$d'(i, j, \pi) = \sum_{k \in B(i,j,\pi)} l_k, \tag{7}$$

it turns out that

$$Z'(\pi) = \sum_{i \in I} \sum_{j \in I, j \neq i} u_{ij} d'(i, j, \pi) + K, \tag{8}$$

where

$$K = \sum_{i \in I} \sum_{j \in I, j \neq i} u_{ij} \frac{l_i + l_j}{2} \tag{9}$$

is a constant, independent of π.

Note that the following property is true for the classical ODSAP. There is a symmetry of the solutions to the problem. Let $\pi^{'}$ be the symmetric to π, i.e.

$$\pi^{'}(t) = \pi(n - t + 1) \quad \text{forall} \quad t = 1, \ldots, n. \tag{10}$$

Then clearly $Z(\pi') = Z(\pi)$. In other words, we can exchange the right and the left–hand sides of the line in our definition. Hence, we could somewhat simplify the problem by only considering, for instance, the permutations in which $\pi^{-1}(1) \leq n/2$, i.e. object 1 is situated on the left half of the line. This remark will be applied to reduce the computational requirements of the solution method.

To solve the ODSAP, polynomial–time algorithms were proposed for the cases when G is a rooted tree [1] and when G is a parallel–serial graph [16]. If the minimal feasible distances satisfy the constraints (b), then, for the specified graphs, the problem becomes NP–hard [15].

In [7], the classical ODSAP was formulated in terms of Integer Programming. Such formulation for ODSAP has $n(n-1)/2$ binary variables and $3[n(n-1)/2]$ constrains. This Integer Programming approach gives a possibility to obtain optimal solutions to small problems. It is not efficient in general since does not take into account the structure of the connections between objects.

In [11], the ODSAP was formulated as a generalization of the Linear and Quadratic Assignment Problems (QAP). An interesting placement problem closely related to the ODSAP is one–dimensional version of the QAP of Koopmans–Beckman type: given n points (locations) $P_1, \ldots, P_n$ with coordinates $a_1, \ldots, a_n$ on a line, and given n objects and a matrix of connections between objects, find a one–to–one assignment of the objects to the locations in order to minimize the total weighted distance. This problem is called the Generalized Linear Ordering Problem (GLOP). The generalization consists in the fact that conditions $|a_i - a_j| \neq 1$ are satisfied in this problem. Dynamic programming was used to solve the problem.

Amongst all possible formulations for the component location problem, the chosen one creates an instance of the extensively studied QAP. On the design one needs to place electronic components to established locations in a printed circuit card, building the complete electronic board. It is necessary to minimize the distance among the components that have greater levels of either interactivity and energy or data flow, to avoid excessive signal delays. This location problem can be modeled as an instance of the QAP [8,18].

The NP–hard problem (1)–(4) with conditions (a) of non–intersected of objects with each other and with forbidden zones but without partial order between objects, that is for the case when G is an undirected graph, was considered in [19,20]. A heuristic for one–line variant was described in [19]. A review of the properties and a branch and bound method for solving the problem was proposed in [20]. The results of computational experiments on comparison of the branch and bound method and a heuristic were reported. In the experiments, an integer programming model and IBM ILOG CPLEX package were used.

In this paper, problem (1)–(4) with conditions (a) and with forbidden zones for the case when G is a directed acyclic graph is considered. A polynomial–time algorithm for finding a local optimum of the problem for the case when G is a composition of rooted trees and parallel–serial graphs is proposed.

3 Local Search Algorithm

3.1 Subproblems for Rooted Trees and Parallel–Serial Graphs

Problem (1)–(4) with conditions (a) of non–intersection of objects among themselves and with forbidden zones is considered. The structure of connections between objects is defined by the directed acyclic graph $G = (V, E)$. Fix a partition of the objects into blocks that satisfies the partial order between objects.

Denote a range of feasible solutions of problem (1)–(4) by B. Range B is disconnected and it consists of r separate blocks B_k of length L_k that must contain the placed objects, $B = \bigcup_{k=\overline{1,r}} B_k$. Then a feasible solution to the problem (1)–(4) corresponds to some partition of objects into blocks.

Suppose that $x = (x_1, \ldots, x_n)$ is a feasible solution to problem (1)–(4); $I_k(x)$ is a set of object numbers in the block B_k; n_k is a cardinality of the set $I_k(x)$. We note that x can be represented as $x = (x^1, \ldots, x^r)$, where x^k are the coordinates of objects allocated in B_k with numbers from the set $I_k(x)$. If G is an arbitrary undirected graph, then, as it was shown in [19], the original continuous problem reduces to r discrete subproblems. The subproblem is the problem in the block B_k on the set of variables $I_k(x)$.

We will call a feasible solution x to the problem (1)–(4) a *local minimum* of the problem if $F(x) \leq F(x')$ for every $x' : I_k(x) = I_k(x')$, $k = 1, \ldots, r$ and the given partial order between objects is satisfied. Since describing algorithms we need only a permutation of objects, then instead of $I_k(x)$ we will write I_k.

Since $I_k \bigcap I_l = \emptyset$ for every $k, l = 1, \ldots, r$, to find a local optimum of the problem for some fixed partition of objects into blocks, it is sufficient to solve

r independent subproblems. So, in each block B_k it is possible to consider the subproblem of placement of $n_k + 2$ objects. In B_k, the subproblem contains n_k placed objects with numbers from I_k and two imaginary objects s and t which correspond to the left and the right borders of B_k with coordinates LB_k and RB_k respectively. Denote by $I_L(B_k)$ and $I_R(B_k)$ the sets of objects on the left and on the right of the block B_k; $J_L(B_k)$ and $J_R(B_k)$ being the sets of zones on the left and on the right of B_k respectively.

Introduce the necessary definitions.

Denote by $D(s,t)$ a directed graph that consists of two or more chains going from the vertex s to the vertex t and having no other common vertices except s and t.

The class of *parallel–serial graphs* is defined inductively as follows:

(a) a directed chain is the parallel–serial graph;
(b) a graph that is obtained from the parallel–serial graph by replacing any arc (s,t) with the graph $D(s,t)$ is parallel–serial.

If in a rooted tree some arbitrary arcs are replaced by the parallel–serial graph, then a resulting graph we will call *a composition of rooted trees and parallel–serial directed graphs.*

An example of the composition of rooted trees and parallel–serial directed graphs is shown in Fig. 2.

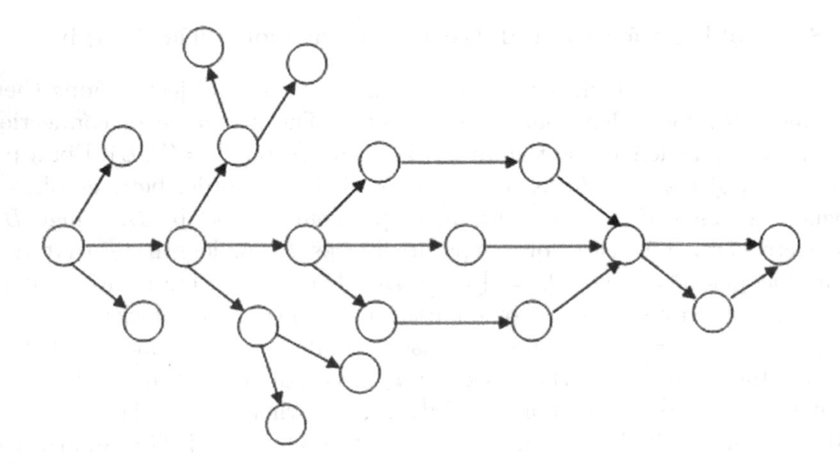

Fig. 2. The composition of rooted trees and parallel–serial directed graphs.

Consider the graph $D(s,t)$ with an arbitrary number of chains. We will call the graph $D(s,t)$ *a bipolar directed graph (BDG)*. An example of the BDG is shown in Fig. 3.

Let G be the BDG. In the case when the area of placement on the line is bounded on the left and the right by zones, we consider an arbitrary block B_k. Denote by $G' = (V', E')$ the weighted graph of connections between objects in

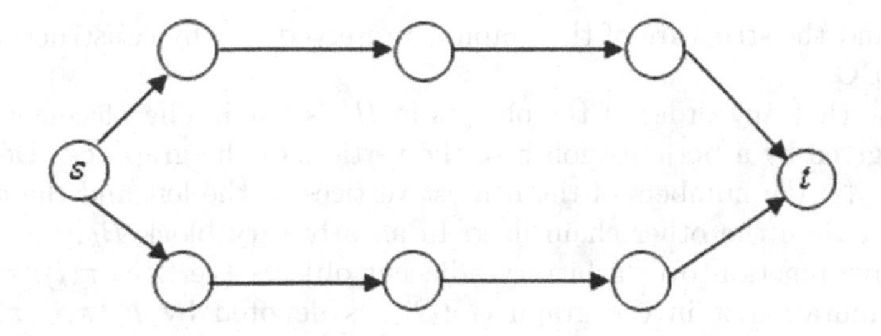

Fig. 3. Example of the BDG.

B_k, where V' is the set of vertices of G', E' is the set of its arcs. Note that $V' = I_k \bigcup\{s,t\}$.

Since the graph G has the BDG structure, then set I_k can be represented as a union of non–intersection chains. Let G' have a chains, and c_i denote the number of vertices of chain i in block B_k, $i = 1, \cdots, a$. For simplicity of notation, let set I_k include the following numbers of vertices of the chains: $I_k = \{(1, \cdots, c_1); (c_1 + 1, \cdots, c_1 + c_2); \cdots; (c_1 + c_2 + \cdots + c_{a-1} + 1, \cdots, c_1 + c_2 + \cdots + c_a)\}$.

We describe the following algorithm for constructing the arcs of the graph G'.

1. If the vertices $i, j \in I_k$ and the arc $(i,j) \in E$, then $(i,j) \in E'$.
2. Draw the arcs $(s,i) \in E'$, $i = \{1, c_1 + 1, \cdots, \sum_{y=1}^{a-1} c_y + 1\}$, from vertex s to the initial vertices of the chains of G'.
3. Draw the arcs $(j,t) \in E'$, $j = \{c_1, c_1 + c_2, \cdots, \sum_{y=1}^{a} c_y\}$, from the final vertices of the chains to vertex t of G'.

Denote by u'_{ij} the weight of the arc $(i,j) \in E'$. Denote by $l(i)$ and $r(i)$ the vertices of the graph G such that the arcs $(l(i), i)$ and $(i, r(i))$ are adjacent. Let S_y be the number of the last vertex in the chain whose number is y. Then $S_y = \sum_{i=1}^{y} c_i$, and we assume that $c_0 = 0$. Define the weights of the arcs in the graph G' as follows.

1. $u'_{ij} = u_{ij} + \sum_{p=j}^{c_b} \sum_{q \in J_L(B_k)} w_{pq} + \sum_{p=1}^{i} \sum_{q \in J_R(B_k)} w_{pq}$,

   for $\forall i, j \in I_k$, $(i,j) \in E$, $\quad b = 1, \cdots, a$;
2. $u'_{si} = u_{i\ l(i)} + \sum_{p=S_{b-1}+1}^{S_b} \sum_{q \in J_L(B_k)} w_{pq}$,

   where $b = 1, \cdots, a$, $\quad i \in \{1, c_1 + 1, \cdots, S_{a-1} + 1\}$;
3. $u'_{jt} = u_{j\ r(j)} + \sum_{p=S_{b-1}+1}^{S_b} \sum_{q \in J_R(B_k)} w_{pq}$,

   where $b = 1, \cdots, a$, $\quad j \in \{c_1, c_1 + c_2, \cdots, S_a\}$.

Thus, we have constructed the graph G' in the block B_k.

Proposition 1. *If a graph G is the BDG then the graph G' also is the BDG.*

By construction, G' differs from G in block B_k in arcs (s,j), (i,t) which are added to the set of arcs E for some $i, j \in I_k$. Other arcs are not added to the graph G' and the structure of the graph is preserved. So, by construction, G' is also the BDG.

Suppose that any order of the objects in B_k is fixed. The placement of the objects is given by a permutation π of the vertices of the graph G'. Denote by $l_\pi(i)$ and $r_\pi(i)$ the numbers of the nearest vertices to the left and the right for the vertex i from the other chain in π. In an arbitrary block B_k, the value of the objective function over arbitrary adjacent objects (vertices $\pi(i)$, $\pi(i+1)$) in the permutation π in the graph G (G'), is denoted by $F_k(\pi(i), \pi(i+1))$ ($F'_k(\pi(i), \pi(i+1))$). For simplicity, assume that the objects are located at a unit distance from each other. Then the value of the objective function in the block F'_k for G' can be written as

$$F'_k = \sum_{i=1}^{n_k-1} F'_k(\pi(i), \pi(i+1)) + F'_k(s, \pi(1)) + F'_k(\pi(n_k), t). \tag{11}$$

Denote by $P_k(G')$ and $P_k(G)$ the subproblems in the block B_k for the graphs G' and G respectively.

Proposition 2. *For a fixed π, the value of the objective function of the problem $P_k(G)$ over an arbitrary pair $(\pi(i), \pi(i+1))$ is equal to the value of the objective function of the problem $P_k(G')$ over $(\pi(i), \pi(i+1))$, i.e. $F_k(\pi(i), \pi(i+1)) = F'_k(\pi(i), \pi(i+1))$.*

Proof. Let G' be the BDG graph with two chains. For simplicity of notation, let $I_k = I_1 \cup I_2 = \{(1, \cdots, c_1); (c_1+1, \cdots, c_1+c_2)\} = \{(1, \cdots, c_1); (1', \cdots, c'_1)\}$. Given the structure of the graph G', over any pair $(\pi(i), \pi(i+1))$=(i, j), there will always be two arcs connecting either the vertices of the same chain or the vertices of different chains. Then, for a fixed π in G'

$$F'_k(\pi(i), \pi(i+1)) \equiv F'_k(i, j) = \begin{cases} u'_{ij} + u'_{l_\pi(i) r_\pi(i)}, & if \quad u'_{ij} \neq 0, \\ u'_{l_\pi(i) r_\pi(i)} + u'_{l_\pi(j) r_\pi(j)}, & if \quad u'_{ij} = 0. \end{cases}$$

We describe the value of the objective function over the pair (i, j) for each of these cases separately.

In the case when the arc (i, j) connects the vertices of the same chain ($i \in I_1$, $j = i+1 \in I_1$ and $u'_{ij} \neq 0$), the value of the objective function over the pair (i, j) is:

$$F'_k(i, j) = u'_{ij} + u'_{l_\pi(i) r_\pi(i)}.$$

Using formulas u'_{ij} for the BDG graph, we obtain

$$F'_k(i, j) = u_{ii+1} + \sum_{p=i+1}^{c_1} \sum_{q \in J_L(B_k)} w_{pq} + \sum_{p=1}^{i} \sum_{q \in J_R(B_k)} w_{pq}$$

$$+ u_{j'j'+1} + \sum_{p=j'+1}^{c'_1} \sum_{q \in J_L(B_k)} w_{pq} + \sum_{p=1'}^{j'} \sum_{q \in J_R(B_k)} w_{pq}.$$

It is easy to check that for $F_k(\pi(i), \pi(i+1))$, the formula will be the same, i.e. $F'_k(\pi(i), \pi(i+1)) = F_k(\pi(i), \pi(i+1))$.

In the case when the vertices $\pi(i)$ and $\pi(i+1)$ belong to different chains, the proof is similar.

Let G be a rooted tree. Denote a subtree with root in vertex i by $G(i)$, and the set of its vertices by $V(G(i))$. In the block B_k, we construct the graph G' by analogy with the BDG graph. Define the weights of the arcs in G'. To do this, the set of vertices included in the path from vertex s to vertex k is denoted by $V'(k)$. Then

1. $u'_{ij} = u_{ij} + \sum_{p \in V(G'(j))} \sum_{q \in J_L(B_k)} w_{pq} + \sum_{p \in V'(i)} \sum_{q \in J_R(B_k)} w_{pq}$;
2. $u'_{si} = u_{il(i)} + \sum_{p \in V(G'(i))} \sum_{q \in J_L(B_k)} w_{pq}$;
3. $u'_{jt} = u_{jr(j)} + \sum_{p \in V'(j)} \sum_{q \in J_R(B_k)} w_{pq}$.

Proposition 3. *If a graph G is a rooted tree then the graph G' is a collection of BDG graphs.*

The proof is similar to Proposition 1.

As a result, we conclude that if the graph G is the rooted tree or the BDG graph, then the graph G' is the set of BDG graphs. We consistently replace BDG on chains according to the algorithm described below. We start with specified graphs, which are the terminal subrooted trees of the graph G in the block.

3.2 Solution of Subproblems for BDG Graph

An algorithm for finding an optimal placement of vertices of BDG graph on a line consists of the following steps [16].

Step 1. In each of the chains connecting s and t, we find the arc with minimal weight. If several arcs are having this property, choose an arbitrary one.

Step 2. Divide the graph G' into two rooted trees LT and RT with the help of the arcs which found in step 1. The first tree LT has the root in the vertex s, the second tree RT has the root in the vertex t (previously changed to the opposite orientation in the tree RT), respectively.

Step 3. The optimal placement of each of the trees is found taking into account the weight of the arcs used for the cut [1].

Step 4. The resulting placement of the vertices of G' is combined as follows: first, the vertices of the tree LT are placed, and then the vertices of the tree RT are placed in the reverse order.

3.3 Search of Local Optimum

Suppose that the objects are divided into blocks satisfying a given partial order between the objects. In the case, when G is the composition of rooted trees and parallel–serial directed graphs, the algorithm for search of a local optimum is proposed. This algorithm consists of the following steps.

Step 1. Construct the sequence of the graphs $G'_1, \ldots, G'_r$ using the algorithm described in 3.1.

Step 2. In each block, find the optimal placement of objects using the algorithm from 3.2.

The complexity of the algorithm of the optimal placement of the graph's vertices on the line in each block is $O(n \cdot log\ n)$ [16]. So, the complexity of the algorithm for search of local optimum does not exceed $O(r \cdot n \cdot log\ n)$, where r is the number of blocks. Since $r \leq m$, then any local optimum of the problem can be found by $O(m \cdot n \cdot log\ n)$ operations.

4 Conclusion

In this paper, we consider a generalization of the One–Dimensional Space Allocation Problem (ODSAP) with forbidden zones. The structure of the connections between the objects is defined using a directed acyclic graph. A review of the formulations and methods for solving the classical ODSAP is given. A polynomial–time algorithm for finding a local optimum for a fixed partition of objects into blocks when the graph of connections between the objects is a composition of rooted trees and parallel–serial graphs, is proposed.

As the challenges for further research of this problem, we can specify the following.

1. Development of an algorithm for constructing of feasible partitions of the objects into blocks, taking into account a given partial order between objects. Here it is possible to modify an algorithm for constructing of feasible partitions of the objects into blocks from the work [19].
2. Search for different structures of the graph of connections between objects, for which it is possible to construct polynomial–time algorithms to find a local optimum of the problem.
3. Development of exact methods (for example, of Dynamic Programming) and heuristics for arbitrary directed graphs of connections between objects.

References

1. Adolphson, D., Hu, T.C.: Optimal linear ordering. SIAM J. Appl. Math. **25**(3), 403–423 (1973)
2. Eremin, I.I., Gimadi, E.K., Kel'manov, A.V., Pyatkin, A.V., Khachai, M.Y.: 2-Approximation algorithm for finding a clique with minimum weight of vertices and edges. Proc. Steklov Inst. Math. **284**(suppl. 1), 87–95 (2014). https://doi.org/10.1134/S0081543814020084
3. Erzin, A.I., Cho, J.D.: Concurrent placement and routing in the design of integrated circuits. Automat. Remote Control **64**(12), 1988–1999 (2003). https://doi.org/10.1023/B:AURC.0000008436.55858.41
4. Garey, M.R., Johnson, D.S.: Computers and Intractability: A Guide to the Theory of NP-Completeness. Freeman, San Francisco (1979). Mir, Moscow (1982)

5. Kochetov, Y.A., Panin, A.A., Plyasunov, A.V.: Genetic local search and hardness of approximation for the server load balancing problem. Automat. Remote Control **78**(3), 425–434 (2017). https://doi.org/10.1134/S0005117917030043
6. Khachai, M.Y., Neznakhina, E.D.: Approximation schemes for the generalized traveling salesman problem. Proc. Steklov Inst. Math. **299**(suppl. 1), 97–105 (2017). https://doi.org/10.1134/S0081543817090127
7. Love, R.F., Wong, J.Y.: On solving a one-dimensional space allocation problem with integer programming. INFOR **14**(2), 139–143 (1976)
8. Miranda, G., Luna, H.P.L., Mateus, G.R.R., Ferreira, P.M.: A performance quarantee heuristic for electronic components placement problems including thermal effects. Comput. Oper. Res. **32**, 2937–2957 (2005)
9. Mukhacheva, E.A., Zalgaller, V.A.: Linear programming cutting problems. Int. J. Softw. Eng. Knowl. Eng. **3**(4), 463–476 (1993)
10. Panyukov, A.V.: The problem of locating rectangular plants with minimal cost for the connecting network. Diskret. Anal. Issled. Oper. Ser. 2 **8**(1), 70–87 (2001)
11. Picard, J.C., Queyranne, M.: On the one-dimensional space allocation problem. Oper. Res. **29**(2), 371–391 (1981)
12. Rudnev, A.S.: Probabilistic tabu search algorithm for the packing circles and rectangles into the strip. Diskret. Anal. Issled. Oper. **16**(4), 61–86 (2009)
13. Sarkar, A., Batta, R., Nagi, R.: Placing a finite size facility with a center objective on a rectangular plane with barriers. Eur. J. Oper. Res. **179**(3), 1160–1176 (2006)
14. Simmons, D.M.: One-dimensional space allocation: an ordering algorithm. Oper. Res. **17**(5), 812–826 (1969)
15. Zabudsky, G.G.: On the complexity of the problem of placement on a line with restrictions on minimum distanses. Russ. Math. (Iz. VUZ) **49**(12), 9–12 (2005)
16. Zabudsky, G.G.: On the problem of the linear ordering of vertices of parallel-sequential graphs. Diskret. Anal. Issled. Oper. **7**(1), 61–64 (2000)
17. Zabudskii, G.G., Amzin, I.V.: Algorithms of compact location for technological equipment on parallel lines. Sib. Zh. Ind. Mat. **16**(3), 86–94 (2013). (in Russian)
18. Zabudskii, G.G., Lagzdin, A.Y.: Polynomial algorithms for solving the quadratic assignment problem on networks. Comput. Math. Math. Phys. **50**(11), 1948–1955 (2010). https://doi.org/10.1134/S0965542510110175
19. Zabudskii, G.G., Veremchuk, N.S.: An algorithm for finding an approximate solution to the Weber problem on a line with forbidden gaps. J. Appl. Ind. Math. **10**(1), 136–144 (2016). https://doi.org/10.1134/S1990478916010154
20. Zabudsky, G.G., Veremchuk, N.S.: Branch and bound method for the Weber problem with rectangular facilities on lines in the presence of forbidden gaps. In: Eremeev, A., Khachay, M., Kochetov, Y., Pardalos, P. (eds.) OPTA 2018. CCIS, vol. 871, pp. 29–41. Springer, Cham (2018). https://doi.org/10.1007/978-3-319-93800-4_3

Game Theory and Mathematical Economics

The Interaction of Consumers and Load Serving Entity to Manage Electricity Consumption

Natalia Aizenberg(✉) and Nikolai Voropai

Melentiev Energy Systems Institute SB RAS, Lermontov st., 130 Irkutsk, Russia
ayzenberg@gmail.com, voropai@isem.irk.ru

Abstract. This paper addresses the coordination of interaction between various types of consumers and a load serving entity to manage electricity consumption by using several models: the Nash equilibrium pricing, and the adverse selection model based on the contract theory. We propose a method to form rate options for load curve optimization for different types of consumers and a load serving entity for different market configurations. The utility functions describe the real situation sufficiently well and allow the implementation of a system of incentives for load curve optimization (load shifting from a peak time of the day). The rates providing a separating equilibrium are determined. We compare the effectiveness of different retail market models. We use the pricing scheme that implies the change in electricity prices depending on the electricity consumption by all users during every hour so that all users are financially motivated.

Keywords: Load-controlled consumer · Load serving entity · Coordination of interaction · Adverse selection model

1 Introduction

This paper considers interactions in the retail market in view of encouraging consumers to optimize their load. This issue has remained relevant since the 1970s. At present, the solutions have been found for the wholesale market [1], as well as for the retail market [2–7], and the wholesale market mechanisms are simpler than those of the retail market. This is stipulated by the greater behavior predictability of large industrial consumers or load serving entities (LSEs) that operate in the wholesale market. Currently, online pricing methods associated with the use of the smart grid have been actively developed. In the retail market, we deal with the interaction between several parties that have their own, sometimes opposite, interests. This can be described by game-theoretic problem

Supported by the Siberian Branch of the Russian Academy of Sciences project No. AAAA-A17-117030310449-7, grants N 019-010-00183 and N 018-010-00728 from Russian Foundation for Basic Research.

I. Bykadorov et al. (Eds.): MOTOR 2019, CCIS 1090, pp. 147–162, 2019.
https://doi.org/10.1007/978-3-030-33394-2_12

statements. The state-of-the-art models that encourage consumers to optimize their load curve can be found in [8,9]. In [8] the authors propose an on-line pricetaking model that targets consumers with different characteristics of the utility function known to the LSE. They describe a game where consumers change consumption in response to the price offered. In turn, the LSE adjust prices to gain the biggest profit. The mechanism for finding equilibrium was proposed and tested for several consumers. The key features of the model are: the retailer has no competitors in the market and operates taking into account the costs of total consumption, the price function is defined explicitly via the cost function, the price is the same for all consumers.

Models focused on finding prices to achieve the maximum welfare are given in [6], where the solution is found on the basis of bi-level programming. There are also works that propose uniform pricing for all consumers, detailing a variety of electrical appliances and their possible load shift during the off-peak time [10]. Development of modern means of communications between users and the LSE assumes different pricing for different consumers [2,3]. In this situation, we can consider pricetaking and equilibrium models. The work [9] describes pricing mechanisms based on the mechanism design (Vickrey auction) that form welfare optimum equilibrium. All works listed above deal with situations that lack arbitration: the possibility for consumers to switch to the rates offered by the LSE to other users. This cannot fully describe the behavior strategies of users. In this paper, we show a situation when all consumers are interested in choosing one rate, reducing market efficiency. We propose a problem statement with such pricing that each consumer chooses their own rate, despite the switch option. Herewith the needs of users have met at the greatest extent and the LSE profit increases.

2 Equilibrium Pricing Model

We consider the interaction between several consumers of electricity and the LSE (retailer). Consumers are divided by types, which is reflected in their utility function. The assumptions that describe economic entities in the retail electricity market are standard.

Consumer. We have N users. The consumer n, $n = \overline{1,N}$ has a utility function $u_n(x, \theta)$ for each time zone $t \in \mathbf{T} = \{1, ..., T\}$ – $u_n(x_n^t, \theta_n)$, where x_n^t is the power consumption level of user n in the time zone t, θ_n is a parameter that represents the valuation of electricity for each n-th consumer. Each consumer utility function $u_n(x, \theta)$ satisfies the following standard conditions. The utility function $u_n(x, \theta)$ is three times differentiable. Moreover, w.r.t. first variable x it is increasing and concave[1]

$$\frac{\partial u_n(x_n^t, \theta_n)}{\partial x} \geq 0, \; n = \overline{1,N}, t \in \mathbf{T}. \tag{1}$$

[1] It means that a marginal utility of users is a nonincreasing function.

and

$$\frac{\partial^2 u_n\left(x_n^t, \theta_n\right)}{\left(\partial x\right)^2} \leq 0, \; n = \overline{1, N}, t \in \mathbf{T}. \tag{2}$$

Besides, $u_n\left(x, \theta\right)$ satisfies the boundary conditions $u_n(0, \theta_n) = 0, \forall \theta_n, n = \overline{1, N}$.

Additionally, the Spence-Mirrlees condition [11] is satisfied. This is the condition for the increase of the total and marginal utility with respect to the parameter θ, as well as the condition for single-crossing of the utility functions or for the non-intersection of the consumer demands. The parameter θ provides some homogeneity for the set of utility functions with respect to θ. Usually, these conditions are required for the solvability of models with imperfect information from the contract theory.

$$\frac{\partial u_n\left(x_n^t, \theta_n\right)}{\partial \theta} > 0, \quad \frac{\partial^2 u_n\left(x_n^t, \theta_n\right)}{\partial x \partial \theta} \geq 0, \; n = \overline{1, N}, t \in \mathbf{T}. \tag{3}$$

Retailer (Load Serving Entity). The cost functions are $C_t\left(Q^t\right)$, $t \in \mathbf{T}$, where Q^t are the units of electricity offered by the LSE in each time zone t. The cost functions $C_t\left(Q^t\right)$ satisfy the following standard conditions. The cost functions are differentiable, increasing with respect to the total energy capacity Q^t

$$C_t\left(\widetilde{Q}^t\right) < C_t\left(\widehat{Q}^t\right), \; \widetilde{Q}^t < \widehat{Q}^t, \; t \in \mathbf{T},$$

and strictly convex: for $\widetilde{Q}^t$ and $\widehat{Q}^t$ for each $t \in \mathbf{T}$ and $0 < \rho < 1$ we have

$$C_t\left(\rho \cdot \widetilde{Q}^t + (1-\rho) \cdot \widehat{Q}^t\right) < \rho \cdot C_t\left(\widetilde{Q}^t\right) + (1-\rho) \cdot C_t\left(\widehat{Q}^t\right), \; t \in \mathbf{T}.$$

A quadratic cost functions is often used in electricity market models. We consider quadratic functions in the following form:

$$C_t\left(Q^t\right) = a_t + b_t \cdot Q^t + \frac{c_t}{2} \cdot \left(Q^t\right)^2, \quad t \in \mathbf{T},$$

where $a_t, b_t, c_t \geq 0$ for each $t \in \mathbf{T}$.

Consumer's Problem. Consider the profit that a consumer receives by changing their load curve, shifting part of the peak load to off-peak time. We can determine the win/loss of utility depending on changes of the curve. The initial load curve is the most convenient for the consumer and the utility of this load is the largest. Then any shift from the peak time decreases the total utility. This can be explained by the increasing of the utility function: a decrease by a certain amount of electricity consumed at peak times reduces utility, and this reduction is bigger than the increase of utility resulting from the addition of the same amount of electricity at off-peak times. Therefore, the gain in price should cover this loss in utility.

The problem is formulated under the assumption that (i) each type will have its own amount and price (rate) (x, P) and (ii) there can be several consumers of this type. For the LSE, these consumers are aggregated into a single user and then the rates are offered. Thus, the problem of the consumer of type n is to maximize the surplus $V_n(x_n, P_n)$ s.t. $x_n^t \geq 0$, $x_n = (x_n^1, ..., x_n^T)$ with the prices $P_n = (P_n^1, ..., P_n^T)$:

$$V_n(x_n, P_n) = \sum_{t=1}^{T} u_n(x_n^t, \theta_n) - \sum_{t=1}^{T} x_n^t \cdot P_n^t \to max_{x_n},\ n = \overline{1, N}, \tag{4}$$

$$\sum_{t=1}^{T} x_n^t \geq X_n, \tag{5}$$

where X_n is a minimum total load during the day for the consumer n.

According to (2) $u_n(x_n^t, \theta_n)$ is concave w.r.t. first variable, the second part of the function (4) is linear w.r.t. x_n^t; therefore $V_n(x_n, P_n)$ is concave and there exists a unique maximum. Write down the FOC conditions (the utility maximum is reached if the marginal utility in the time zone t for the consumer θ_n is equal to the price)[2]:

$$P_n^t = \frac{\partial u_n(x_n^t, \theta_n)}{\partial x}. \tag{6}$$

Equilibrium Pricing. Assume that the retailer forms prices taking into account that consumers will adjust their consumption at certain hours in accordance to their profit. The consumption vector x_n in the period t is x_n, the offer vector in t is $Q^t = \sum_{n=1}^{N} q_n^t$. Then the profit function takes the form

$$\pi_{LSE}(x) = \sum_{n=1}^{N} \sum_{t=1}^{T} P_n^t \cdot x_n^t - \sum_{t=1}^{T} C_t\left(\sum_{n=1}^{N} q_n^t\right) \to max_{q,x,P}, \tag{7}$$

under the conditions (4)–(5) and

$$\sum_{n=1}^{N} x_n^t = \sum_{n=1}^{N} q_n^t, \quad \forall t \in \mathbf{T}. \tag{8}$$

In what follows, we will search for the solution only in the variables x. Herein, $x_n^t \geq \overline{x}_n^t$, $\forall n, t$ means that consumers consume a minimum amount each time. Since electricity cannot be stocked (here we do not consider the use of batteries), optimization in each price zone can be considered as a separate problem. Then, substituting (6) into (7) for each time zone $\forall t \in \mathbf{T}$, we have

$$\pi_{LSE}^t(x) = \sum_{n=1}^{N} \frac{\partial u_n(x_n^t, \theta_n)}{\partial x^t} \cdot x_n^t - C_t\left(\sum_{n=1}^{N} x_n^t\right) \to max_{x_n}. \tag{9}$$

[2] Here the condition (5) is taken into account as equality, which makes sense for the problem under consideration due to the following reasons. We consider some variant of distribution of the same load over different time intervals. Therefore, the assumption that the user, by shifting part of the load from the peak time, will increase its consumption, does not look realistic. Therefore, we can ignore it by reducing the number of variables in the main problem (4)

For the existence of the maximum, it is necessary and sufficient that the utility function, in addition to the conditions (3), satisfies the following properties:

$$r_{u_n}(z) < 1, \;\; r_{u_n'}(z) < 2, \;\; \forall z \geq 0, \tag{10}$$

and

$$\lim_{z \to 0}\left[\frac{\partial u_n(z,\theta)}{\partial z}\right] > 0 \qquad \lim_{z \to \infty}\left[\frac{\partial u_n(z,\theta)}{\partial z} + z\frac{\partial^2 u_n(z,\theta)}{(\partial z)^2}\right] \leq 0,$$

where

$$r_{u_n}(z) \equiv r_{u_n}(z,\theta) = -\frac{\partial^2 u_n(z,\theta)}{\partial z^2} \cdot \frac{z}{\frac{\partial u_n(z,\theta)}{\partial z}} > 0, \tag{11}$$

$$r_{u_n'}(z) \equiv -\frac{\partial^3 u_n(z,\theta)}{\partial z^3} \cdot \frac{z}{\frac{\partial^2 u_n(z,\theta)}{\partial z^2}}. \tag{12}$$

Here $r_{u_n}(z)$ is the Arrow-Pratt measure [12] of concavity of the elementary utility and concavity of $\frac{\partial u_n(z,\theta)}{\partial z}$ ('of risk aversion'). The popular utility are quadratic function (we use it in this paper) and $CARA$ in [13]: $u(z,\theta_n) = 1 - \exp(-\theta_n \cdot z)$ which entails $r_{u_n}(z) = \theta_n \cdot z$. The boundary conditions indicate positive marginal utility at zero and saturable demand at infinity, for the equilibria existence. The condition imposed on r_{u_n} means that we need a greater than 1 demand elasticity for monopolistic pricing. The condition imposed on $r_{u_n'}$ means that the profit is concave, i.e., the second-order condition $2\frac{\partial^2 u_n(z,\theta)}{\partial z^2} + z \cdot \frac{\partial^3 u_n(z,\theta)}{\partial z^3} \leq 0$. Further we assume that these conditions are satisfied. See Appendix 1 for details.

We have connected problems of maximizing the concave functions (9) on a set of time intervals t under the condition (6). As a result, we obtain the Nash equilibrium: the prices $P_n = \left(P_n^1, ..., P_n^T\right)$ and the amounts of consumption $x_n = \left(x_n^1, ..., x_n^T\right)$ in each time zone $t = \overline{1,T}$ in the rate (x_n, P_n), offered to each consumer $n = \overline{1,N}$. Users optimize the load and maximize utility, the company maximizes utility. Further, in Sect. 4, we calculate the Nash equilibrium for consumers with different characteristics of the utility function θ_n, and propose rates that realize this equilibrium under the conditions of complete awareness.

3 Equilibrium Model with the Individual Rationality Condition

A weaker equilibrium variant than the one we have discussed, where consumers do not get the maximum function (4), but have some level of satisfaction. In our problem, we consider the same for everyone users n level of the received utility U. It can be called an alternative level of utility. $U = \sum_{t=1}^{T} U^t$, where U^t is level of alternative utility in time zone t. There are several reasons to discuss this approach. First, by considering the condition

$$V_n(x_n, P_n) \geq U \;\forall n = \overline{1,N}, \tag{13}$$

we provide an individual rationality (IR) constraint, when the consumer changes their consumption to increase their profit. Second, if there is competition between *load serving entities* at the market, then the alternative utility is the utility provided by another company (retailer).

$$\pi_{LSE}(x) = \sum_{n=1}^{N}\sum_{t=1}^{T} P_n^t \cdot x_n^t - \sum_{t=1}^{T} C_t \left(\sum_{n=1}^{N} x_n^t\right) \rightarrow max_{x,P}, \tag{14}$$

$$\sum_{t=1}^{T} u_n\left(x_n^t, \theta_n\right) - \sum_{t=1}^{T} x_n^t \cdot P_n^t \geq U,\ \forall n = \overline{1, N}, \tag{15}$$

$$\sum_{t=1}^{T} x_n^t \geq X_n, \forall n = \overline{1, N}, \tag{16}$$

$$x_n^t \geq 0,\ \forall n = \overline{1, N}, t = \overline{1, T}. \tag{17}$$

The constraints (13) are fulfilled as equality due to the interest of the LSE to set the highest possible prices. Only for the low prices we have $V_n(x_n, P_n) > U$ and the LSE loses extra profit. Then the objective function is a sum of concave functions with respect to the variable x_n^t and there exists a unique maximum. With certain utility functions and costs, we can estimate the welfare loss during the transition from the general market equilibrium to the equilibrium with the individual rationality condition. There exists another type of loss, not related to the imperfection of market relations. This is the imperfection of information.

4 Arbitrage Opportunity Model

The model described in the previous section could be realized in the market only if we did not have to deal with an arbitrage opportunity It might happen that one of the users is offered prices which are systematically lower than the prices offered to another user. Then this second user is interested in concealing their own real utility to get another offer or an option to choose a rate offered to the first user. Such situations should be avoided.

To avoid the situation when everyone chooses the same rate, the problem (7) should be reformulated. Now we do not simply look for equilibrium prices for each type in accordance with marginal utilities and marginal costs. It is necessary that the user, who would rather prefer to select an alternative someone else's rate, chooses prices that provide the utility of the alternative rate. Then the LSE profit will be bigger than in the situation when all users go for the same rate, but smaller than in the equilibrium described in Sect. 2. As has been done before, we simultaneously solve the demand side management problem.

The utility of the user n when they choose the rate formed for them with the prices $P_n = \left(P_n^1, ..., P_n^T\right)$ has the form:

$$V_n\left(x_n, P_n\right) = \sum_{t=1}^{T} u_n\left(x_n^t, \theta_n\right) - \sum_{t=1}^{T} x_n^t \cdot P_n^t. \tag{18}$$

The utility of the user n when they choose the rate formed for another user m with the prices $P_m = \left(P_m^1, ..., P_m^T\right)$ has the form:

$$V_n\left(x_m, P_m\right) = \sum_{t=1}^{T} u_n\left(x_m^t, \theta_n\right) - \sum_{t=1}^{T} x_m^t \cdot P_m^t. \tag{19}$$

When solving the problem outlined in the previous section, we assume that any consumer will be satisfied at some level, which is the same for everyone. This is a legitimate assumption if we consider all users to be in approximately the same position with respect to each other, while having different utility functions. Then for $\forall n$, m the individual rationality constraint (13) is satisfied.

Two types of constraints are developed to avoid the situation of adverse selection. They set an interval that meets the criteria of separating equilibrium. First, neither consumer type refuses to consume electricity at the offered rates (individual rationality (IR) constraint). Second, both types choose different rates (incentive compatibility (IC) condition).

The IR constraint for each consumer type is based on the utility U which is obtained by a consumer in the equilibrium case (4), (5), (8). The IR constraint for each user n has the form:

$$V_n\left(x_n, P_n\right) \geq U, \ \forall n = \overline{1, N}. \qquad (IR) \tag{20}$$

The IC condition for each user n is

$$V_n\left(x_n, P_n\right) \geq V_n\left(x_m, P_m\right), \ \forall n, m = \overline{1, N}. \qquad (IC) \tag{21}$$

The problem of the load serving entity is (7) under the conditions (8), (5), (20), (21). Assume that all users can be divided into two groups. The first group has a higher utility estimate with respect to the unit of electricity (this group will be included into the set H). The second group has a lower utility estimate (group L) in comparison to the first group H. In view of the importance of each user to choose "their own" rate, we replace the utility maximization problem with a system of constraints.

The efficiency of the user H choosing the rate offered to the user L.

Find out what happens if consumers can choose from a range of rates formed in the equilibrium from the problem (14)–(17). If the consumer of type L chooses the rate of the consumer of type H, then

$$V_L\left(x_H, P_H\right) = \sum_{t=1}^{T} u_L\left(x_H^t, \theta_L\right) - \sum_{t=1}^{T} x_H^t \cdot P_H^t. \tag{22}$$

Using (3), we have that $u_L\left(x_H^t, \theta_L\right) < u_H\left(x_H^t, \theta_H\right)$. Then

$$\sum_{t=1}^{T} u_L\left(x_H^t, \theta_L\right) - \sum_{t=1}^{T} x_H^t \cdot P_H^t < \sum_{t=1}^{T} u_H\left(x_H^t, \theta_H\right) - \sum_{t=1}^{T} x_H^t \cdot P_H^t = U. \quad (23)$$

Then $V_L\left(x_H, P_H\right) < U$ and consumers of type L have no incentives to choose the rate with prices P_H offered by the retailer in the equilibrium from the problems (4), (5), (7).

If the consumer of type H chose the equilibrium rate of the consumer L, we have:

$$\sum_{t=1}^{T} u_H\left(x_L^t, \theta_H\right) - \sum_{t=1}^{T} x_L^t \cdot P_L^t > \sum_{t=1}^{T} u_L\left(x_L^t, \theta_L\right) - \sum_{t=1}^{T} x_L^t \cdot P_L^t = U. \quad (24)$$

Here we employed the fact that $u_H\left(x_L^t, \theta_H\right) > u_L\left(x_L^t, \theta_L\right)$ in accordance with the property of the utility function (3).

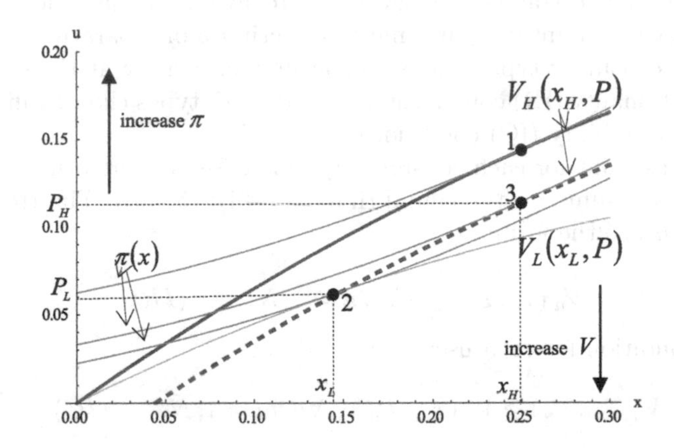

Fig. 1. Equilibria obtained in problems of interaction between consumers and retailers according to models with individual rationality constraint and with possible arbitration. (Color figure online)

We see that it is profitable for the consumer H to choose someone else's rate. This indicates an imbalance of equilibrium in the general case and reduces the retailer's profit and the total economic benefit. This situation is illustrated by Fig. 1 that shows two indifference curves of the utility functions $u_n\left(x_n^t, \theta_n\right)$ of two consumers: the green and blue solid lines. The utility level of these consumers is the same (at the figure it is equal to 0). The red lines are the indifference curves of the cost function of the LSE of the different level $C_t\left(Q^t\right)$.[3] The points 1 and

[3] Here, the level of customer satisfaction increases with the shift of the lines of the utility function level down, and the profit level increases with the shift of indifference curves of costs upwards.

2 determine the equilibrium values found when solving the problem (14)–(17). The blue dotted line indicates the utility level that the user H will reach by choosing the rate L at the point 1. This level is higher than that reached by the user at the point 2. We obtain a new equilibrium once we have solved the new problem of maximizing the profit under the conditions that the utility obtained by the user H is not less than the utility obtained when choosing the point 1 rate.

Individual rationality constraint and incentive compatibility conditions.

Proposition 1. (i) For the type of consumer L the IR-condition is satisfied as equality, but (ii) for the type of consumer H the IR-condition is satisfied as an inequality. (iii) For the type of consumer H the IC-condition is satisfied as equality, but (iv) for the type of consumer L the IC-condition is satisfied as an inequality.

Proof.

i. Any reduction of the price P_L^t by a small amount ε will increase the utility of the users of type L in the period t. On the other hand, the price reduction by the amount of ε reduces the retailer's profit $\pi_{LSE}(x)$. Since users L agree to choose the rate (x_L, P_L) while already having P_L, the retailer will offer exactly this rate.

$$\sum_{t=1}^{T} u_L\left(x_L^t, \theta_L\right) - \sum_{t=1}^{T} x_L^t \cdot P_L^t = U. \tag{25}$$

ii. We have already checked that in (24).

iii. The reasoning behind this is the same as in the entry (i). If this constraint is satisfied as a strict inequality

$$\sum_{t=1}^{T} u_H\left(x_H^t, \theta_H\right) - \sum_{t=1}^{T} x_H^t \cdot P_H^t > \sum_{t=1}^{T} u_H\left(x_L^t, \theta_H\right) - \sum_{t=1}^{T} x_L^t \cdot P_L^t. \tag{26}$$

However, in this case *the load serving entity* can raise the rate to the situation of equality and receive an extra benefit, while the consumer will still remain interested to choose their own rate. Therefore,

$$\sum_{t=1}^{T} u_H\left(x_H^t, \theta_H\right) - \sum_{t=1}^{T} x_H^t \cdot P_H^t = \sum_{t=1}^{T} u_H\left(x_L^t, \theta_H\right) - \sum_{t=1}^{T} x_L^t \cdot P_L^t. \tag{27}$$

iv.

$$\sum_{t=1}^{T} u_L\left(x_L^t, \theta_L\right) - \sum_{t=1}^{T} x_L^t \cdot P_L^t > \sum_{t=1}^{T} u_L\left(x_H^t, \theta_L\right) - \sum_{t=1}^{T} x_H^t \cdot P_H^t. \tag{28}$$

The proof of this statement is given in Appendix 2.

When the IC conditions for the user L and the IR conditions for the user H turn into inequality, it means that they are generally ineffective for the market. It follows from (24) that the consumer H receives a portion of additional utility that a retailer would receive in the case Model with the individual rationality condition. The consumer would receive a negative utility from (28) and (23). It means that the LSE can neglect these two conditions when solving the conditional profit maximization problem and use only the effective conditions: the IR condition for L (25) and the IC conditions for H (27). If the entire system comprises only two users, the profit maximization problem for the LSE has the form

$$\pi_{LSE}(x) = \sum_{t=1}^{T} P_L^t \cdot x_L^t + \sum_{t=1}^{T} P_H^t \cdot x_H^t - \sum_{t=1}^{T} C_t\left(x_L^t + x_H^t\right) \rightarrow max_{x,P}, \quad (29)$$

subject to (25), (27), (16) and (17).

In this case, the profit function is the sum of two concave functions with respect to x in accordance with the properties listed above. Therefore, we have maximization of a concave function with linear constraints.

5 Efficiency Evaluation

We carried out computations for all three models and compared the results by the welfare criterion. Compare prices of the models: the Pricing Equilibrium Model (Model I) (4), (5), (8), the Equilibrium with the individual rationality condition Model (14)–(17) (Model II) and the Arbitrage opportunity Model (Model III).

1. The model I – (4), (5), (8), is the Nash equilibrium model, therefore, it does not have incentives for all participants to change the situation. At the same time, it does not coincide with welfare maximum $W = \sum_{n=1}^{N}\sum_{t=1}^{T} u_n\left(x_n^t, \theta_n\right) - \sum_{t=1}^{T} C_t\left(\sum_{n=1}^{N} x_n^t\right)$.
2. The solution to model II is defined as the "first best": marginal utility equals marginal cost.

$$\sum_{t=1}^{T} u_n'\left(x_n^t, \theta_n\right) = \sum_{t=1}^{T} C_t'\left(\sum_{n=1}^{N} x_n^t\right), \quad n = 1, ..., N.$$

We will get a similar solution when solving the problem of maximizing social welfare.

Model I solution is determined

$$\sum_{t=1}^{T} u_n'\left(x_n^t, \theta_n\right)\left(1 - r_{u_n}\left(x_n^t\right)\right) = \sum_{t=1}^{T} C_t'\left(\sum_{n=1}^{N} x_n^t\right), \quad n = 1, ..., N.$$

If we consider the natural utility function when $r'_{u_n}(\cdot) \geq 0$, $(1 - r_{u_n}(x_n^t))$ is a decreasing function. The product of two decreasing functions $u'_n(x_n^t, \theta_n)$ and $(1 - r_{u_n}(x_n^t))$ is the decreasing function, besides on FOC we have $1 - r_{u_n}(x_n^t) < 1$. Then for $\forall x_n^t$ is true that $u'_n(x_n^t, \theta_n)(1 - r_{u_n}(x_n^t)) \leq u'_n(x_n^t, \theta_n)$. Thus, the consumption in the model I at peak time is less than in model II and prices are higher. The resulting equilibrium in model I differs from the socially optimal.

When we determine the equilibrium in model III, we obtain that for a high type of consumer H, the equilibrium is determined by (5), and for a low type of consumer, it is obtained from the following relationship and is ineffective:

$$2\sum_{t=1}^{T} u'_L(x_L^t, \theta_L) - \sum_{t=1}^{T} u'_L(x_H^t, \theta_L) = \sum_{t=1}^{T} C'_t(x_L^t + x_H^t).$$

It is not difficult to show that, based on the Spence-Merlis property, we have $2\sum_{t=1}^{T} u'_L(x_L^t, \theta_L) - \sum_{t=1}^{T} u'_L(x_H^t, \theta_L) < \sum_{t=1}^{T} u'_L(x_L^t, \theta_L)$. Accordingly, the equilibrium volumes of model III are lower and prices are higher than in model II.

The design of the retail market can be structured so that the interaction is carried out either by model I or by model III (the equilibrium in model II is not stable). But these outcomes are not optimal and the maximum welfare is not achieved. Below, in the example, we calculate the described equilibria and compare the outcomes.

Therefore, the model II is welfare better than the model I. However, the consumer is stimulated to leave this equilibrium. In particular, a possible change of behavior can be seen in the arbitrage situation from the model III.

3. The arbitrage model equilibrium is different from the welfare optimal one, but it offers less incentives to the consumer to change their behavior.

As an example, we consider an electricity supply system in the campus of Irkutsk Technical University. Three equivalent consumers $N = 3$ represent all the consumers making up the load curve of the campus. The first equivalent consumer includes polyclinics and loads of sanatorium. The second consumer includes food production facilities and hostels. The third consumer is a cafe. They receive electricity from a municipal load serving entity. The day of the winter peak load (December 28, 2016) that was observed at the Technical University campus is assumed as a calculation day. The initial data include a time-of-use rate. We have two time zone: peak and night $T = 2$.

In this paper, we consider quadratic utility functions [9] corresponding to the linear inverse demand function. These functions fully satisfy the conditions (1)–(2) and are easily interpretable. For several parameters θ_n (where example, $\theta_1 < \theta_2 < \theta_3$) they satisfy the condition (3). Electricity demand is traditionally described as linear functions with low elasticity.

$$p_n^t = \theta_n - \gamma \cdot x_n^t, \; n = \overline{1, N}, t \in \mathbf{T}, \tag{30}$$

This function satisfies the properties (10): $r_{u_n}(z) = r_{u'_n}(z) = 0$. More precisely

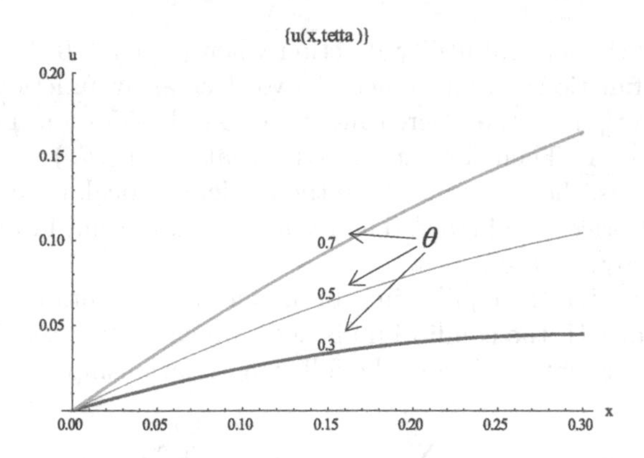

Fig. 2. The utility function of different types of consumers.

$$u_n\left(x_n^t,\theta_n\right)=\begin{cases}\theta_n\cdot x_n^t-\frac{\gamma}{2}\cdot\left(x_n^t\right)^2 & if\ 0\le x_n^t<\theta_n/\gamma,\\ \theta_n^2/2\gamma & if\ x_n^t\ge\theta_n/\gamma,\end{cases}\quad t\in\mathbf{T}.\tag{31}$$

The several examples of utility functions from this class (31) are shown in Fig. 2. The point where the utility function gets saturated and does not change corresponds to the maximum power requirement of the user.

The utility functions of users (31) with coefficients θ during nighttime {0.2, 0.3, 0.3} and {0.4, 0.6, 0.7} during daytime. The study of load of these users helped determine the necessary load. We set the parameter a of the cost function equal to 0.02 and 0.5 for the night and on-peak hours, respectively.

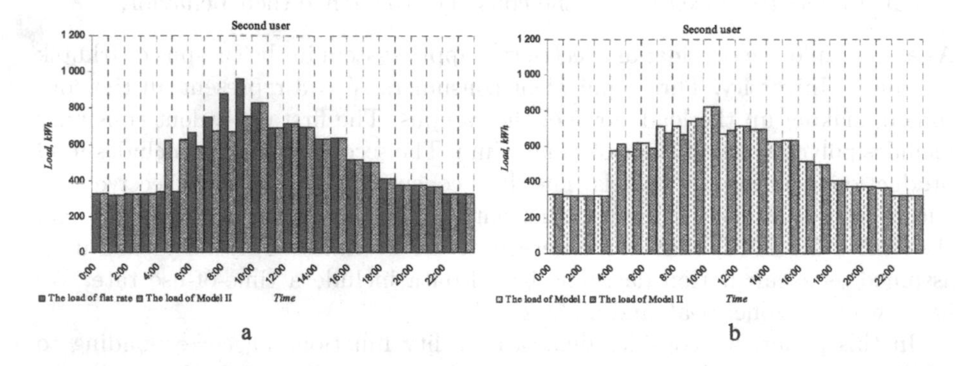

Fig. 3. a. The load of the second consumer during the day optimized at a flat rate and a rate of model II. b. The load of the second consumer during the day optimized at a rate of model II and a rate of model III.

Figure 3a shows the load of the second consumer who belongs to the group that has a higher valuation of the electricity unit and better solvency, than the first user. The dark blue color indicates the flat rate load during the daytime, the light blue color shows the load corresponding to the solution of the maximum welfare problem (individual rationality constraint). Figure 3b shows the comparison between the loads for the model II and the arbitrage model III. Figure 4 is the comparison of the total load for various techniques of modeling. It can be seen that the smallest values during peak time correspond to the individual rationality constraint model II. In this realization, the pricing equilibrium model has the worst characteristics.

Therefore, the individual rationality constraint model appears to be most effective and the arbitrage opportunity model comes second best, while the general equilibrium model reduces the efficiency even more. However, our testing showed that the efficiency loss is small, which is partly caused by the data.

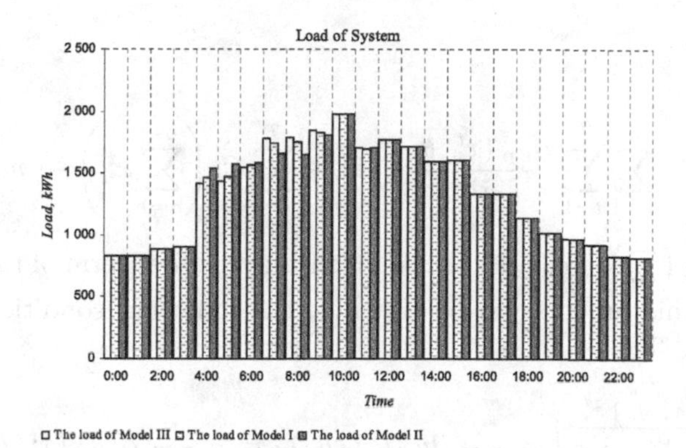

Fig. 4. The load of the electricity system (three users) during the day optimized by the rates of the three models (I, II, III).

6 Conclusions

In this paper, we have considered the infrastructure of intellectual power systems, where the load serving entity and its users are offered to optimize their load curves in accordance with their strategies. One of the goals is the peak load reduction with respect to the average value during the daytime. We use the pricing scheme that implies the change in electricity prices depending on the electricity consumption by all users during every hour so that all users are financially motivated. We considered three possible models describing the retail market and compare their efficiency. Each model has its reasons to be used. The pricing equilibrium model (I) realizes an effective scheme for all participants who do not have incentives to change their behavior (stable equilibrium). However, this model has high demands for the quality of information. The model with the individual rationality constraint (II) is less demanding for consumers and every consumer receives the same profit. This model realizes the welfare maximum but does not provide a stable equilibrium. The last model (III) describes the situation of a possible arbitrage between users. We consider a situation when some users switch to the rate of other users, thereby increasing their utility and reducing the retailer's profit. We propose a model that takes this effect into consideration, partly compensating for the loss of the LSE. In practice, the pricing equilibrium model (I) and a model of a possible arbitrage (III) can be realized, because they have a stable equilibrium. At the same time, interaction according to the model of pricing equilibrium (I) does not require a regulation unlike the model of a possible arbitrage (III). In the regulation of rates in the case of model III, the resulting equilibrium is close to the maximum social welfare. The models have been tested on the data of several users from the student campus.

Appendix 1

$$\pi_{LSE}(x) = \sum_{n=1}^{N}\sum_{t=1}^{T} \frac{\partial u_n\left(x_n^t, \theta_n\right)}{\partial x} \cdot x_n^t - \sum_{t=1}^{T} C_t\left(\sum_{n=1}^{N} x_n^t\right) \to max_{x,P}.$$

The function $C_t\left(\sum_{n=1}^{N} x_n^t\right)$ is concave. Consider the first term of the expression $\pi_{LSE}(x)$. For this function to be concave, the following condition should be satisfied for $\forall t \in \mathbf{T}$:

$$\frac{\partial\left(\sum_{n=1}^{N} \frac{\partial u_n\left(x_n^t, \theta_n\right) \cdot x_n^t}{\partial x}\right)}{\partial x} \equiv \sum_{n=1}^{N} \frac{\partial u_n\left(x_n^t, \theta_n\right) \cdot x_n^t}{\partial x} + \sum_{n=1}^{N} \frac{\partial^2 u_n\left(x_n^t, \theta_n\right)}{\partial x^2} \cdot x_n^t \geq 0 \tag{32}$$

Or

$$r_u(z) < 1, \;\; r_{u'}(z) < 2, \;\; \forall z \geq 0, \tag{33}$$

where

$$r_{u_n}(z) \equiv r_{u_n}(z, \theta) = -\frac{\partial^2 u_n(z,\theta)}{\partial z^2} \cdot \frac{z}{\frac{\partial u_n(z,\theta)}{\partial z}} > 0, r_{u_n'}(z) \equiv -\frac{\partial^3 u_n(z,\theta)}{\partial z^3} \cdot \frac{z}{\frac{\partial^2 u_n(z,\theta)}{\partial z^2}}. \tag{34}$$

Appendix 2

Assume that the condition IC is satisfied as a strict inequality for the consumer H:

$$\sum_{t=1}^{T} u_L\left(x_L^t, \theta_L\right) - \sum_{t=1}^{T} x_L^t \cdot P_L^t > \sum_{t=1}^{T} u_L\left(x_H^t, \theta_L\right) - \sum_{t=1}^{T} x_H^t \cdot P_H^t.$$

Then

$$\sum_{t=1}^{T} x_H^t \cdot P_H^t - \sum_{t=1}^{T} x_L^t \cdot P_L^t > \sum_{t=1}^{T} u_L\left(x_H^t, \theta_L\right) - \sum_{t=1}^{T} u_L\left(x_L^t, \theta_L\right).$$

It follows from (26) that

$$\sum_{t=1}^{T} x_H^t \cdot P_H^t - \sum_{t=1}^{T} x_L^t \cdot P_L^t = \sum_{t=1}^{T} u_H\left(x_H^t, \theta_H\right) - \sum_{t=1}^{T} u_H\left(x_L^t, \theta_H\right).$$

Then

$$\sum_{t=1}^{T} u_H\left(x_H^t, \theta_H\right) - \sum_{t=1}^{T} u_H\left(x_L^t, \theta_H\right) > \sum_{t=1}^{T} u_L\left(x_H^t, \theta_L\right) - \sum_{t=1}^{T} u_L\left(x_L^t, \theta_L\right),$$

or

$$\sum_{t=1}^{T} u_H\left(x_H^t, \theta_H\right) - \sum_{t=1}^{T} u_L\left(x_H^t, \theta_L\right) > \sum_{t=1}^{T} u_H\left(x_L^t, \theta_H\right) - \sum_{t=1}^{T} u_L\left(x_L^t, \theta_L\right).$$

The last equality is indeed always valid in a strict form in accordance with the condition imposed on the utility function (3). Consequently, the assumption (27) is true.

References

1. Philpott, A., Pettersen, E.: Optimizing demand-side bids in day-ahead electricity markets. IEEE Trans. Power Syst. **21**(2), 488–498 (2006)
2. Gellings, C.W.: The concept of demand-side management for electric utilities. Proc. IEEE **73**(10), 1468–1570 (1985)
3. Kim, J.H., Seo, I., Jung, H., Kim, H.-I.: Economic assessment for demand response under generation expansion planning. In: Proceedings ICEE, Hong Kong, China, 5–9 July 2015
4. Chai, B., Chen, J., Yang, Z., Zhang, Y.: Demand response management with multiple utility companies: a two-level game approach. IEEE Trans. Smart Grid **5**(2), 1340–1350 (2014)
5. Deng, R., Yang, Z., Chen, J., Rahbari-Asr, N., Chow, M.-Y.: Residential energy consumption scheduling: a coupled-constraint game approach. IEEE Trans. Smart Grid **5**(3), 722–731 (2014)
6. Fadlullah, Z.M., Quan, D.M., Kato, N., Stojmenovic, I.: GTES: an optimized game-theoretic demand-side management scheme for smart grid. IEEE Syst. J. **8**(2), 588–597 (2014)
7. Aizenberg, N., Voropai, N.: Price setting in the retail electricity market under the Bertrand competition. Procedia Comput. Sci. **122**, 649–656 (2014)
8. Mohsenian-Rad, A.H., Wong, V.W., Jatskevich, J., Schober, R., Leon-Garcia, A.: Autonomous demand-side management based on game-theoretic energy consumption scheduling for the future smart grid. IEEE Trans. Smart Grid **1**(3), 320–331 (2010)

9. Samadi, P., Mohsenian-Rad, H., Schober, R., Wong, V.W.: Advanced demand side management for the future smart grid using mechanism design. IEEE Trans. Smart Grid **3**(3), 1170–1180 (2012)
10. Hartway, R., Price, S., Woo, C.K.: Smart meter, customer choice and profitable time-of-use rate option. Energy **24**(10), 895–903 (1999)
11. Bolton, P., Dewatripont, M.: Contract Theory, p. 717. MIT Press, Cambridge (2005)
12. Pratt, J.W.: Risk aversion in the small and in the large. Econometrica **32**(12), 122–136 (1964)
13. Behrens, K., Murata, Y.: Trade, competition, and efficiency. J. Int. Econ. **87**, 1–17 (2012)

Social Optimality in International Trade Under Monopolistic Competition

Igor Bykadorov[1,2,3](✉)

[1] Sobolev Institute of Mathematics SB RAS,
4 Koptyug Ave., 630090 Novosibirsk, Russia
bykadorov.igor@mail.ru
[2] Novosibirsk State University, 1 Pirogova St., 630090 Novosibirsk, Russia
[3] Novosibirsk State University of Economics and Management,
56 Kamenskaja St., 630099 Novosibirsk, Russia

Abstract. We study the homogeneous model of international trade under the monopolistic competition of producers. The utility function assumes additive separable. The transport costs are of "iceberg types". It is known that, in the situation of market equilibrium, under linear production costs, the social welfare, as a function of transport costs, decreases near free trade while (counter-intuitively!) increases near total autarky. Instead, we study the situation of social optimality. We show that total welfare decreases. We restrict our study by the case of two countries, "big" and "small". Moreover, we study two important "limited" situations: near free trade and near total autarky. We show that near free trade, the welfare in the small country decreases; as to the big country, we find examples when (1) the welfare decreases and (2) the welfare (counter-intuitively!) increases. Besides, in the autarky case, we describe the situations of decreasing/increasing of welfare in each country.

Keywords: Monopolistic competition · International trade · Social optimality · Comparative Statics

1 Introduction

The concept of monopolistic competition, introduced by Chamberlin [17,18], widely develops now, starting with the famous paper by Dixit and Stiglitz [23] for the case of a closed economy, by Krugman [26,27] for the international trade and Melitz [29] for the heterogeneous case.

It seems that now a paper in this area can recognize as an interesting one only if it contains counter-intuitive (unexpected) results.

Usually, monopolistic competition models study the *market equilibrium*, see, e.g. [5,6,14,21,24,28,36]. One of the most interesting topic in these studies is the influence of the models' parameters (market size, transport costs, etc.) on the social welfare, see, e.g. [3,4,13,15,30,31,33,34]. In particular, in [3,4,31] the authors get that the gain from trade "is not so much". Besides, in [15] the case

I. Bykadorov et al. (Eds.): MOTOR 2019, CCIS 1090, pp. 163–177, 2019.
https://doi.org/10.1007/978-3-030-33394-2_13

of free trade and the case autarky study carefully; one of the (counterintuitive!) result is that the social welfare, as function of transport costs, increases near total autarky.

Instead, in this paper we study *social optimality* case, cf. [1,2,7,8,22]. It can be interpreted as the problem of "social planer" who optimize a scalarization of multi-criteria to find Pareto-optimal solution.[1] We show that total welfare decreases. We restrict our study by the case of two countries. Moreover, we study two important "limited" situations: near free trade and near total autarky, cf. [15]. We show that near free trade, the welfare in the small country decreases; as to the big country, we find examples when (1) the welfare decreases and (2) the welfare (counter-intuitively!) increases. As to the total autarky, we restrict our study by the case of linear production costs and show that the total welfare achieves the minimum, increasing in one country and decreasing in another country.

The paper is organized as follows. Section 2 lays out the model and contains some preliminary considerations. Here we describe the consumers and producers (Sect. 2.1), Social welfare and Social optimality (Sect. 2.2). Moreover, in Sect. 2.3 we consider the symmetric case of social optimality: get First and Second Order Conditions, and also the structure the Hesse matrix for general production costs (Proposition 1) and for linear production costs (Corollary 1). Section 3 provides the Comparative Statics w.r.t. transport costs. First, we find that the total welfare function decreases w.r.t. transport costs (Proposition 2). Further, we represent the derivative of gradient of total utility w.r.t. transport costs in terms of elasticities and Arrow-Pratt measure for general (Proposition 3) and for linear (Corollary 2) production costs. These allow to get in Sect. 3.1 the elasticities of consumptions, of firm sizes and of the mass of firms (Proposition 4); moreover, we get the derivatives (Proposition 5) and elasticities (Corollary 3) of welfare in each countries and total welfare, and an example showing that in big country the monotonicity of welfare can not be guaranteed. Section 3.2 contains the autarky study (Proposition 6). Section 4 concludes.

2 Problem

We study a homogeneous monopolistic competition model of trade between two asymmetric countries.

As it is usual in monopolistic competition, we assume that (cf. [7,18,23])

- consumers are identical, each endowed with one unit of labor;
- labor is the only production factor; consumption, output, etc. are measured in labor;
- firms are identical (have the same cost function), but produce "varieties" ("almost identical") of good;

[1] In welfare economics [25,32], a social planner is a decision-maker who attempts to achieve the best result for all parties involved. Usually this means or the maximization of a social welfare function (in neo-classical welfare economics), or Pareto optimality (in modern welfare economics).

- each variety is produced by one firm that produces a single variety;
- each demand function results from additive utility function;
- number (mass) of firms is big enough to ignore the firm's influence on the whole industry/economy;
- labor supply/demand in each country is balanced.

Let be two countries, H ("big") and F ("small").

2.1 Consumers and Producers

Let

- L be the number of consumers in country H,
- l be the number of consumers in country F.

As usual, we assume that $L \geq l$. Analogously, let

- N be the number (mass) of firms in country H,
- n be the number (mass) of firms in country F.

Note that L and l are parameters (the known constants) while N and n are the variables determined endogenously. Moreover, let us recall that, in monopolistic competition models, number of firms is big enough. Therefore, instead of standard "number of firms is N (or n)" we consider the intervals $[0, N]$ and $[0, n]$ with uniformly distributed firms.[2]

Now we introduce four kinds of the *individual* consumption. Let[3]

- X_i be the amount of the variety produced in country H by firm $i \in [0, N]$ and consumed by a consumer in country H,
- Z_i be the amount of the variety produced in country H by firm $i \in [0, N]$ and consumed by a consumer in country F,
- x_i be the amount of the variety produced in country F by firm $i \in [0, n]$ and consumed by a consumer in country F,
- z_i be the amount of the variety produced in country F by firm $i \in [0, n]$ and consumed by a consumer in country H.

To introduce the production amount of the firms (the "size" of the firm), let us introduce the parameter $\tau \geq 1$ as transport costs of "iceberg type".[4] Each firm in each country produces for consumers in each country. This way

$$Q_i = L \cdot X_i + \tau \cdot l \cdot Z_i, i \in [0, N], \tag{1}$$

[2] A popular interpretation is as follows: gas stations are equally spaced on the "long" road; we are not interested in the number of these stations, but the length of the road. In this case, N and n are called not "the number of firms", but "the mass of firms". This mass is determined endogenously and does not have to be an integer at all.

[3] Hereinafter, due to the tradition of monopolistic competition, we use the notation X_i for the function $X(i)$, etc.

[4] To sell in another country y units of the goods, the firm must produce $\tau \cdot y$ units. "During transportation, the product melts like an iceberg ...".

is the size of firm $i \in [0, N]$ in country H, while

$$q_i = l \cdot x_i + \tau \cdot L \cdot z_i, i \in [0, n], \tag{2}$$

is the size of firm $i \in [0, n]$ in country F.

Besides, let the production costs be determined for each firm in each country by the *increasing* twice differentiable function V.

Thus, the labor balances in countries H and F are, correspondingly,

$$\int_0^N V(Q_i)\, di = L, \tag{3}$$

$$\int_0^n V(q_i)\, di = l. \tag{4}$$

2.2 Social Welfare and Social Optimality

The total utility (*social welfare* function) in country H is

$$SW^H = L \cdot \left(\int_0^N u(X_i)\, di + \int_0^n u(z_i)\, di \right) \tag{5}$$

while the social welfare function in country F is

$$SW^F = l \cdot \left(\int_0^n u(x_i)\, di + \int_0^N u(Z_i)\, di \right). \tag{6}$$

Here $u(\cdot)$ is sub-utility function. As usual, we assume that $u(\cdot)$ is twice differentiable (at least near social optimality) and satisfies the conditions

$$u(0) = 0,\ u'(\xi) > 0,\ u''(\xi) < 0, \tag{7}$$

i.e., it is strictly increasing[5] and strictly concave.

[5] Usually function $u(\cdot)$ is assumed to be increasing (not necessarily strictly increasing). For example, in the case of quadratic sub-utility

$$u(\xi) = \begin{cases} A\xi - \frac{B}{2} \cdot \xi^2, & \xi \in \left[0, \frac{A}{B}\right]; \\ \frac{A^2}{2B}, & \xi \geq \frac{A}{B}; \end{cases}$$

with

$$u'(\xi) = \begin{cases} A - B \cdot \xi, & \xi \in \left[0, \frac{A}{B}\right]; \\ 0, & \xi \geq \frac{A}{B}. \end{cases}$$

We assume strictly increase only for convenient.

Now let us formulate the *social optimality* problem:

- maximize the total social welfare function $SW^H + SW^F$ (see (5) and (6)) with respect to $\{X_i\}_{i\in[0,N]}, \{Z_i\}_{i\in[0,N]}, \{x_i\}_{i\in[0,n]}, \{z_i\}_{i\in[0,n]}$ subject to labor balances (3) and (4).

2.3 Social Optimality: Symmetric Case

Let us recall that the consumers are assume identical, the producer are assumed identical. Thus, as usual, we consider the symmetric case. More precisely, we omit index i in consumption: $X_i = X, Z_i = Z, x_i = x, z_i = z$ for any i. This way (1)–(6) are

$$Q = L \cdot X + \tau \cdot l \cdot z, \tag{8}$$

$$q = l \cdot x + \tau \cdot L \cdot Z, \tag{9}$$

$$N \cdot V(Q) = L, \tag{10}$$

$$n \cdot V(q) = l, \tag{11}$$

$$SW^H = L \cdot (N \cdot u(X) + n \cdot u(z)), \tag{12}$$

$$SW^F = l \cdot (n \cdot u(x) + N \cdot u(Z)), \tag{13}$$

By substituting (10) and (11) in (12) and (13), the *symmetric social welfare* in countries H and F are, correspondingly,

$$W = L \cdot \left(L \cdot \frac{u(X)}{V(Q)} + l \cdot \frac{u(z)}{V(q)} \right), \tag{14}$$

$$w = l \cdot \left(l \cdot \frac{u(x)}{V(q)} + L \cdot \frac{u(Z)}{V(Q)} \right), \tag{15}$$

while the *total symmetric social welfare* is

$$U = W + w = L \cdot \frac{L \cdot u(X) + l \cdot u(Z)}{V(Q)} + l \cdot \frac{l \cdot u(x) + L \cdot u(z)}{V(q)}. \tag{16}$$

Thus, the *symmetric social optimality* problem is[6]

$$U \to \max_{X,Z,x,z}. \tag{17}$$

[6] It may seem more appropriate to maximize W and w separately. But here comes the problem of reconciling the resulting solutions, because of

$$\frac{\partial W}{\partial Z} = 0 \Longleftrightarrow X = 0, \quad \frac{\partial W}{\partial x} = 0 \Longleftrightarrow z = 0,$$

$$\frac{\partial w}{\partial X} = 0 \Longleftrightarrow Z = 0, \quad \frac{\partial w}{\partial z} = 0 \Longleftrightarrow x = 0.$$

Hence, we maximize the sum of welfares as a special case of scalarization in multiple-objective optimization. Of course, another linear combinations of the items can be considered and lead to the similar results.

The First Order Conditions (FOC) are

$$\frac{\partial U}{\partial X} \equiv \frac{L^2}{V(Q)} \cdot \left(u'(X) - \frac{V'(Q)}{V(Q)} \cdot (L \cdot u(X) + l \cdot u(Z)) \right) = 0, \tag{18}$$

$$\frac{\partial U}{\partial Z} \equiv \frac{L \cdot l}{V(Q)} \cdot \left(u'(Z) - \tau \cdot \frac{V'(Q)}{V(Q)} \cdot (L \cdot u(X) + l \cdot u(Z)) \right) = 0, \tag{19}$$

$$\frac{\partial U}{\partial x} \equiv \frac{l^2}{V(q)} \cdot \left(u'(x) - \frac{V'(q)}{V(q)} \cdot (l \cdot u(x) + L \cdot u(z)) \right) = 0, \tag{20}$$

$$\frac{\partial U}{\partial z} \equiv \frac{L \cdot l}{V(q)} \cdot \left(u'(z) - \frac{\tau \cdot V'(q)}{V(q)} \cdot (l \cdot u(x) + L \cdot u(z)) \right) = 0, \tag{21}$$

while the Hesse matrix of function U is block-diagonal

$$U'' = \begin{pmatrix} \frac{\partial^2 U}{\partial X^2} & \frac{\partial^2 U}{\partial X \partial Z} & 0 & 0 \\ \frac{\partial^2 U}{\partial X \partial Z} & \frac{\partial^2 U}{\partial Z^2} & 0 & 0 \\ 0 & 0 & \frac{\partial^2 U}{\partial x^2} & \frac{\partial^2 U}{\partial x \partial z} \\ 0 & 0 & \frac{\partial^2 U}{\partial x \partial z} & \frac{\partial^2 U}{\partial z^2} \end{pmatrix}. \tag{22}$$

Thus, the Second Order Conditions (SOC) are

$$\frac{\partial^2 U}{\partial X^2} < 0, \; \frac{\partial^2 U}{\partial X^2} \cdot \frac{\partial^2 U}{\partial Z^2} - \left(\frac{\partial^2 U}{\partial X \partial Z} \right)^2 > 0, \tag{23}$$

$$\frac{\partial^2 U}{\partial x^2} < 0, \; \frac{\partial^2 U}{\partial x^2} \cdot \frac{\partial^2 U}{\partial z^2} - \left(\frac{\partial^2 U}{\partial x \partial z} \right)^2 > 0. \tag{24}$$

Under FOC (18)–(21), we can calculate the elements of matrix (22).[7]

Proposition 1. *Under FOC (18)–(21),*

$$\frac{\partial^2 U}{\partial X^2} = \frac{L^2}{V(Q)} \cdot \left(u''(X) - L \cdot u'(X) \cdot \frac{V''(Q)}{V'(Q)} \right), \tag{25}$$

$$\frac{\partial^2 U}{\partial X \partial Z} = -\tau \cdot L^2 \cdot l \cdot \frac{V''(Q)}{V'(Q) \cdot V(Q)} \cdot u'(X) = -L^2 \cdot l \cdot \frac{V''(Q)}{V'(Q) \cdot V(Q)} \cdot u'(z), \tag{26}$$

$$\frac{\partial^2 U}{\partial Z^2} = \frac{L \cdot l}{V(Q)} \cdot \left(u''(Z) - \tau \cdot l \cdot u'(Z) \cdot \frac{V''(Q)}{V'(Q)} \right), \tag{27}$$

$$\frac{\partial^2 U}{\partial x^2} = \frac{l^2}{V(q)} \cdot \left(u''(x) - l \cdot u'(x) \cdot \frac{V''(q)}{V'(q)} \right), \tag{28}$$

[7] Usually in this paper we omit the proofs, they are rather technical.

$$\frac{\partial^2 U}{\partial x \partial z} = -\tau \cdot L \cdot l^2 \cdot \frac{V''(q)}{V(q) \cdot V'(q)} \cdot u'(x) = -L \cdot l^2 \cdot \frac{V''(q)}{V(q) \cdot V'(q)} \cdot u'(z), \quad (29)$$

$$\frac{\partial^2 U}{\partial z^2} = \frac{L \cdot l}{V(q)} \cdot \left(u''(z) - \tau \cdot L \cdot \frac{V''(q)}{V'(q)} \cdot u'(z) \right). \quad (30)$$

Moreover, in the case of linear production costs, i.e., if

$$V''(Q) = V''(q) = 0, \quad (31)$$

matrix U'' is diagonal. More precisely,

Corollary 1. *Under FOC (18)–(21) and linear production costs (31), SOC hold and, moreover,*

$$\frac{\partial^2 U}{\partial X^2} = \frac{L^2}{V(Q)} \cdot u''(X) < 0, \quad (32)$$

$$\frac{\partial^2 U}{\partial X \partial Z} = 0, \quad (33)$$

$$\frac{\partial^2 U}{\partial Z^2} = \frac{L \cdot l}{V(Q)} \cdot u''(Z), \quad (34)$$

$$\frac{\partial^2 U}{\partial x^2} = \frac{l^2}{V(q)} \cdot u''(x), \quad (35)$$

$$\frac{\partial^2 U}{\partial x \partial z} = 0, \quad (36)$$

$$\frac{\partial^2 U}{\partial z^2} = \frac{L \cdot l}{V(q)} \cdot u''(z). \quad (37)$$

Due to (32)–(37), SOC (23), (24) hold *automatically.*

3 Comparative Statics w.r.t. Transport Costs

Total differentiation of the system (18)–(21), i.e., the system

$$U' = \begin{pmatrix} 0 \\ 0 \\ 0 \\ 0 \end{pmatrix}$$

w.r.t. τ gives us

$$\frac{dU'}{d\tau} = \begin{pmatrix} 0 \\ 0 \\ 0 \\ 0 \end{pmatrix},$$

i.e.,

$$U'' \cdot \begin{pmatrix} \frac{dX}{d\tau} \\ \frac{dZ}{d\tau} \\ \frac{dx}{d\tau} \\ \frac{dz}{d\tau} \end{pmatrix} = - \begin{pmatrix} \frac{\partial^2 U}{\partial X \partial \tau} \\ \frac{\partial^2 U}{\partial Z \partial \tau} \\ \frac{\partial^2 U}{\partial x \partial \tau} \\ \frac{\partial^2 U}{\partial z \partial \tau} \end{pmatrix}. \tag{38}$$

First of all, let us show a simple fact.

Proposition 2. *Under FOC (18)–(21),*

$$\frac{dU}{d\tau} = -\frac{L \cdot l}{\tau} \cdot \left(\frac{u'(Z) \cdot Z}{V(Q)} + \frac{u'(z) \cdot z}{V(q)} \right) \leq 0. \tag{39}$$

Proof. Due to (18)–(21), (8), (9),

$$\begin{aligned} \frac{dU}{d\tau} &= \frac{\partial U}{\partial X} \cdot \frac{dX}{d\tau} + \frac{\partial U}{\partial Z} \cdot \frac{dZ}{d\tau} + \frac{\partial U}{\partial x} \cdot \frac{dx}{d\tau} + \frac{\partial U}{\partial z} \cdot \frac{dz}{d\tau} + \frac{\partial U}{\partial \tau} = \frac{\partial U}{\partial \tau} \\ &= -L \cdot \frac{L \cdot u(X) + l \cdot u(Z)}{(V(Q))^2} \cdot V'(Q) \cdot \frac{\partial Q}{\partial \tau} - l \cdot \frac{l \cdot u(x) + L \cdot u(z)}{(V(q))^2} \cdot V'(q) \cdot \frac{\partial q}{\partial \tau} \\ &= -L \cdot l \cdot \left(\frac{L \cdot u(X) + l \cdot u(Z)}{(V(Q))^2} \cdot V'(Q) \cdot Z + \frac{l \cdot u(x) + L \cdot u(z)}{(V(q))^2} \cdot V'(q) \cdot z \right) \\ &= -\frac{L \cdot l}{\tau} \cdot \left(\frac{u'(Z) \cdot Z}{V(Q)} + \frac{u'(z) \cdot z}{V(q)} \right) \leq 0. \end{aligned}$$

Hence (39) holds.

Hence, now we are interesting of only the behavior of the welfare in *each* country.

For function $g(\cdot)$, let us introduce the elasticity and Arrow-Pratt measure:

$$\mathcal{E}_g(\eta) \equiv \frac{g'(\eta)}{g(\eta)} \cdot \eta, \quad r_g(\eta) \equiv -\frac{g''(\eta)}{g'(\eta)} \cdot \eta \equiv -\mathcal{E}_{g'}(\eta). \tag{40}$$

Proposition 3. *Under FOC (18)–(21),*

$$\frac{\partial^2 U}{\partial X \partial \tau} = \frac{L^2 \cdot l \cdot Z}{V(Q) \cdot Q} \cdot (\mathcal{E}_V(Q) + r_V(Q)) \cdot u'(X), \tag{41}$$

$$\frac{\partial^2 U}{\partial Z \partial \tau} = \frac{L \cdot l^2 \cdot Z}{V(Q) \cdot Q} \cdot \left(-\frac{Q}{\tau \cdot l \cdot Z} + \mathcal{E}_V(Q) + r_V(Q) \right) \cdot u'(Z), \tag{42}$$

$$\frac{\partial^2 U}{\partial x \partial \tau} = \frac{L \cdot l^2 \cdot z}{V(q) \cdot q} \cdot (\mathcal{E}_V(q) + r_V(q)) \cdot u'(x), \tag{43}$$

$$\frac{\partial^2 U}{\partial z \partial \tau} = \frac{L^2 \cdot l \cdot z}{V(q) \cdot q} \cdot \left(-\frac{q}{\tau \cdot L \cdot z} + \mathcal{E}_V(q) + r_V(q) \right) \cdot u'(z). \tag{44}$$

Corollary 2. *Under FOC (18)–(21) and linear production costs (31),*

$$\frac{\partial^2 U}{\partial X \partial \tau} = \frac{L^2 \cdot l \cdot Z}{V(Q) \cdot Q} \cdot \mathcal{E}_V(Q) \cdot u'(X), \tag{45}$$

$$\frac{\partial^2 U}{\partial Z \partial \tau} = \frac{L \cdot l^2 \cdot Z}{V(Q) \cdot Q} \cdot \left(-\frac{Q}{\tau \cdot l \cdot Z} + \mathcal{E}_V(Q) \right) \cdot u'(Z), \tag{46}$$

$$\frac{\partial^2 U}{\partial x \partial \tau} = \frac{L \cdot l^2 \cdot z}{V(q) \cdot q} \cdot \mathcal{E}_V(q) \cdot u'(x), \tag{47}$$

$$\frac{\partial^2 U}{\partial z \partial \tau} = \frac{L^2 \cdot l \cdot z}{V(q) \cdot q} \cdot \left(-\frac{q}{\tau \cdot L \cdot z} + \mathcal{E}_V(q) \right) \cdot u'(z). \tag{48}$$

3.1 Free Trade

Let us consider the situation in free trade point, i.e., when $\tau = 1$. Then

$$X = Z = x = z,$$

$$Q = q = \Gamma \cdot X,$$

where $\Gamma = L + l$.

Let us denote $u = u(X), V = V(Q)$. We get from (18)–(21)

$$\frac{\partial U}{\partial X} = \frac{L^2}{V} \cdot \left(u' - \frac{V'}{V} \cdot \Gamma \cdot u \right) = \frac{l}{L} \cdot \frac{\partial U}{\partial Z} = \frac{l^2}{L^2} \cdot \frac{\partial U}{\partial x} = \frac{l}{L} \cdot \frac{\partial U}{\partial z}.$$

So FOC is

$$\mathcal{E}_u = \mathcal{E}_V. \tag{49}$$

As to SOC, we have from Proposition 1, see (25)–(30),

$$\frac{\partial^2 U}{\partial X^2} = L^2 \cdot \frac{u'}{V \cdot Q} \cdot (L \cdot r_V - \Gamma \cdot r_u) = \frac{L}{l} \cdot \frac{\partial^2 U}{\partial z^2},$$

$$\frac{\partial^2 U}{\partial Z^2} = L \cdot l \cdot \frac{u'}{V \cdot Q} \cdot (l \cdot r_V - \Gamma \cdot r_u) = \frac{L}{l} \cdot \frac{\partial^2 U}{\partial x^2},$$

$$\frac{\partial^2 U}{\partial X \partial Z} = L^2 \cdot l \cdot \frac{u'}{V \cdot Q} \cdot r_V = \frac{L}{l} \cdot \frac{\partial^2 U}{\partial x \partial z},$$

$$\frac{\partial^2 U}{\partial X^2} \cdot \frac{\partial^2 U}{\partial Z^2} - \left(\frac{\partial^2 U}{\partial X \partial Z} \right)^2 = L^3 \cdot l \cdot r_u \cdot \left(\frac{u'}{V \cdot X} \right)^2 \cdot (r_u - r_V),$$

$$\frac{\partial^2 U}{\partial x^2} \cdot \frac{\partial^2 U}{\partial z^2} - \left(\frac{\partial^2 U}{\partial x \partial z} \right)^2 = L \cdot l^3 \cdot r_u \cdot \left(\frac{u'}{V \cdot X} \right)^2 \cdot (r_u - r_V).$$

So SOC is

$$r_u > r_V. \tag{50}$$

Besides, we have from Proposition 3, see (41)–(44),

$$\frac{\partial^2 U}{\partial X \partial \tau} = \frac{L^2 \cdot l}{V \cdot \Gamma} \cdot \mathcal{E}_u \cdot u' = \frac{L}{l} \cdot \frac{\partial^2 U}{\partial x \partial \tau},$$

$$\frac{\partial^2 U}{\partial Z \partial \tau} = \frac{L \cdot l^2}{V \cdot \Gamma} \cdot \left(-\frac{\Gamma}{l} + \mathcal{E}_u\right) \cdot u',$$

$$\frac{\partial^2 U}{\partial z \partial \tau} = \frac{L^2 \cdot l}{V \cdot \Gamma} \cdot \left(-\frac{\Gamma}{L} + \mathcal{E}_u\right) \cdot u'.$$

Let us introduce the elasticity of a variable η w.r.t. τ (cf. (40))

$$E_\eta \equiv E_{\eta/\tau} = \frac{d\eta}{d\tau} \cdot \frac{\tau}{\eta}. \tag{51}$$

In terms of (51), due to (38) and SOC (50), we get

Proposition 4. *In Free Trade, the elasticities of social optimal individual consumptions are*

$$E_X = \frac{l}{\Gamma} \cdot \frac{\mathcal{E}_u \cdot r_u - r_V}{r_u \cdot (r_u - r_V)} = \frac{l}{L} \cdot E_x, \tag{52}$$

$$E_Z = \frac{1}{\Gamma \cdot r_u} \cdot \left(l \cdot \frac{(\mathcal{E}_u - 1) \cdot r_u}{r_u - r_V} - L\right) < \frac{l}{\Gamma} \cdot \frac{\mathcal{E}_u - 1}{r_u - r_V} < 0, \tag{53}$$

$$E_z = \frac{1}{\Gamma \cdot r_u} \cdot \left(L \cdot \frac{(\mathcal{E}_u - 1) \cdot r_u}{r_u - r_V} - l\right) < \frac{L}{\Gamma} \cdot \frac{\mathcal{E}_u - 1}{r_u - r_V} < 0. \tag{54}$$

Moreover, the elasticities of social optimal sizes and masses of firms are

$$E_Q = -\frac{l}{\Gamma} \cdot \frac{1 - \mathcal{E}_u - r_u + r_V}{r_u - r_V} = \frac{l}{L} \cdot E_q, \tag{55}$$

$$E_N = \mathcal{E}_u \cdot \frac{l}{\Gamma} \cdot \frac{1 - \mathcal{E}_u - r_u + r_V}{r_u - r_V} = \frac{l}{L} \cdot E_n. \tag{56}$$

Note that, in the case of linear production costs (31), SOC is $r_u > 0$ (i.e., holds automatically), while we get from (52)–(56)[8]

$$E_X = \frac{l}{\Gamma} \cdot \frac{\mathcal{E}_u}{r_u} = \frac{l}{L} \cdot E_x > 0,$$

[8] Note that concavity of sub-utility u restricts its elasticity as $\mathcal{E}_u(\xi) < 1 \ \forall \xi > 0$. Indeed, for every $\forall \xi > 0$,

$$\mathcal{E}_u(\xi) < 1 \Longleftrightarrow u'(\xi) \cdot \xi - u(\xi) < 0 \quad \forall \xi > 0.$$

Consider the function $g(\xi) = u'(\xi) \cdot \xi - u(\xi)$. One has $g'(\xi) \equiv u''(\xi) \cdot \xi < 0 \ \forall \xi > 0$ due to strictly concavity of u. But $g(0) = u(0) = 0$. Hence $g(\xi) < 0 \ \forall \xi > 0$, i.e., $u'(\xi) \cdot \xi - u(\xi) < 0 \ \forall \xi > 0$.

$$E_Q = -\frac{l}{\Gamma} \cdot \frac{\mathcal{E}_{\mathcal{E}_u}}{r_u} = \frac{l}{L} \cdot E_q = -\frac{E_N}{\mathcal{E}_u} = -\frac{l}{L} \cdot \frac{E_n}{\mathcal{E}_u}.$$

Thus, in the case of linear production costs, we now not only the signs of E_Z, E_z but also the signs of E_X, E_x; moreover, signs of E_Q, E_q, E_N, E_n are uniquely determined by the sign of $\mathcal{E}_u'$.

Now, we can calculate the derivatives w.r.t. τ of social welfare (14) in country H, social welfare (15) in country F and total social welfare (16).

Proposition 5. *In Free trade, the derivatives w.r.t. τ of social welfare in each country are*

$$\frac{dW}{d\tau} = -(2 \cdot L \cdot r_u + l - L) \cdot \frac{L \cdot l}{\Gamma \cdot V} \cdot \frac{u' \cdot X}{r_u}, \tag{57}$$

$$\frac{dw}{d\tau} = -(2 \cdot l \cdot r_u + L - l) \cdot \frac{L \cdot l}{\Gamma \cdot V} \cdot \frac{u' \cdot X}{r_u} < 0. \tag{58}$$

Moreover, the derivatives w.r.t. τ of total social welfare is

$$\frac{dU}{d\tau} = -2 \cdot L \cdot l \cdot \mathcal{E}_u \cdot \frac{u}{V} < 0. \tag{59}$$

Note that formulas (57)–(59) does not contain the second derivative of the production costs. So they are the same for the cases of linear and nonlinear production costs.

Further, in Free Trade,

$$U = \Gamma^2 \cdot \frac{u}{V} = \frac{\Gamma}{L} \cdot W = \frac{\Gamma}{l} \cdot w. \tag{60}$$

In terms of elasticities, formulas (57), (58) and especially (59) can be written, using (60), in more short form.

Corollary 3. *In Free trade, the elasticities w.r.t. τ of social welfare in each country are*

$$E_W = -\frac{l}{\Gamma^2} \cdot (2 \cdot L \cdot r_u + l - L) \cdot \frac{\mathcal{E}_u}{r_u}, \tag{61}$$

$$E_w = -\frac{L}{\Gamma^2} \cdot (2 \cdot l \cdot r_u + L - l) \cdot \frac{\mathcal{E}_u}{r_u} < 0, \tag{62}$$

the elasticity w.r.t. τ of total social welfare is

$$E_U = -\frac{2 \cdot L \cdot l}{\Gamma^2} \cdot \mathcal{E}_u = -\frac{2 \cdot L \cdot l}{(L+l)^2} \cdot \mathcal{E}_u \in (-\mathcal{E}_u, 0) \subseteq (-1, 0). \tag{63}$$

Due to Proposition 5, the total welfare and the welfare in the "small" country F decrease w.r.t. τ near Free Trade, see (63) and (62). As to the welfare in the "big" country H, the monotone the decrease/increase property depends on the sign of the expression $2 \cdot L \cdot r_u + l - L$, see (61). The example below shows that it is not possible to guarantee this sign.

Example. Let be $L = 3, l = 1$,

$$u(\xi) = 4 \cdot (1+\xi)^{0.5} - \xi - 4, \quad V(\eta) = c \cdot \eta + 1.$$

Then, in Free Trade, $Q = 4 \cdot X$, FOC ($\mathcal{E}_u = \mathcal{E}_V$, see (49)) is

$$\frac{2 \cdot (1+X)^{-0.5} - 1}{4 \cdot (1+X)^{0.5} - X - 4} \cdot X = \frac{c \cdot 4 \cdot X}{c \cdot 4 \cdot X + 1}.$$

Moreover,

$$r_u = \frac{(1+X)^{-1.5} \cdot X}{2 \cdot (1+X)^{-0.5} - 1}$$

and, due (61),

$$E_W \geq 0 \Longleftrightarrow r_u \leq \frac{1}{3}.$$

Let $c = \dfrac{21}{4}$, then

$$X = \frac{15}{49}, \quad \sqrt{1+X} = \frac{8}{7}, \quad r_u = \frac{35}{128} < \frac{1}{3}.$$

Let $c = \dfrac{5}{2}$, then

$$X = \frac{11}{25}, \quad \sqrt{1+X} = \frac{6}{5}, \quad r_u = \frac{55}{144} > \frac{1}{3}.$$

3.2 Autarky

In this section, we restrict our study by the case of linear production costs (31). Then we can use (32)–(37) and (45)–(48). We are interested in various cases of total autarky:[9]

1. τ_1 and τ_2 exist such that $1 < \tau_1 < \tau_2$ and

$$z(\tau) > 0, \quad \tau < \tau_1, \quad z(\tau_1) = 0, \quad Z(\tau) > 0, \quad \tau < \tau_2, \quad Z(\tau_2) = 0; \tag{64}$$

2. τ_1 and τ_2 exist such that $1 < \tau_1 < \tau_2$ and

$$Z(\tau) > 0, \quad \tau < \tau_1, \quad Z(\tau_1) = 0, \quad z(\tau) > 0 \quad \tau < \tau_2, \quad z(\tau_2) = 0; \tag{65}$$

3. τ_a exists such that

$$Z(\tau) > 0, \quad z(\tau) > 0, \quad \tau < \tau_a, \quad Z(\tau_a) = z(\tau_a) = 0. \tag{66}$$

[9] In equilibrium situation (see, e.g., [15]), due to trade balance, the moving from the trade to the total autarky is under the unique transport cost. Instead, in social optimality situation, several kinds of partial autarky can be before the total autarky.

The Proposition below describes the monotonicity of consumption, firm's sizes and welfare in each country, depending on the cases (64)–(66).

Proposition 6. *Under (7), in various cases of total autarky, the various results hold.*

1. *Let τ_1 and τ_2 exist such that $1 < \tau_1 < \tau_2$ and (64) holds. Then in $\tau = \tau_2$:*

$$\frac{dX}{d\tau} = \frac{dx}{d\tau} = \frac{dq}{d\tau} = 0, \quad \frac{dZ}{d\tau} = \frac{1}{\tau_2} \cdot \frac{u'(0)}{u''(0)} < 0, \quad \frac{dQ}{d\tau} = l \cdot \tau \cdot \frac{dZ}{d\tau} < 0,$$

$$\frac{dW}{d\tau} = -L \cdot l \cdot \frac{u'(0)}{V(Q)} \cdot \frac{dZ}{d\tau} = -\frac{dw}{d\tau} > 0.$$

2. *Let τ_1 and τ_2 exist such that $1 < \tau_1 < \tau_2$ and (65) holds. Then in $\tau = \tau_2$:*

$$\frac{dX}{d\tau} = \frac{dx}{d\tau} = \frac{dQ}{d\tau} = 0, \quad \frac{dz}{d\tau} = \frac{1}{\tau_2} \cdot \frac{u'(0)}{u''(0)} < 0, \quad \frac{dq}{d\tau} = L \cdot \tau_2 \cdot \frac{dz}{d\tau} < 0,$$

$$\frac{dW}{d\tau} = L \cdot l \cdot \frac{u'(0)}{V(q)} \cdot \frac{dz}{d\tau} = -\frac{dw}{d\tau} < 0.$$

3. *Let τ_a exists such that (66) holds. Then in $\tau = \tau_a$:*

$$\frac{dX}{d\tau} = \frac{dx}{d\tau} = 0, \quad \frac{dZ}{d\tau} = \frac{dz}{d\tau} = \frac{1}{\tau_2} \cdot \frac{u'(0)}{u''(0)} < 0, \quad \frac{dQ}{d\tau} = l \cdot \tau_2 \cdot \frac{dZ}{d\tau} = \frac{l}{L} \cdot \frac{dq}{d\tau} < 0,$$

$$\frac{dW}{d\tau} = \frac{dw}{d\tau} = 0.$$

4 Conclusion

In this paper, we study, in the monopolistic competition framework, the homogeneous model of international trade between the two countries. The utility function for each consumer is additively separable. We consider the situation of social optimality and consider two "limited" cases: free trade and autarky. Although the total social welfare decreases w.r.t. transport costs, we find two counter-intuitive results: when transport costs increase,

- near free trade, the social welfare *can increase* in the bigger country;
- near total autarky, the social welfare *can increase* as in the bigger country, as in smaller countries.

Note that it may be interesting to consider more complicated cases:

- non-additive utility function, cf. [9,16,35];
- non-linear production costs, cf. [13,14];
- heterogeneous case, cf. [19,20,29];
- marketing models, cf. [10–12].

Acknowledgments. The work was supported in part by the Russian Foundation for Basic Research, projects 18-010-00728 and 19-010-00910, by the program of fundamental scientific researches of the SB RAS, project 0314-2019-0018, and by the Russian Ministry of Science and Education under the 5-100 Excellence Programme.

References

1. Aizenberg, N., Bykadorov, I., Kokovin, S.: Optimal reciprocal import tariffs under variable elasticity of substitution, National Research University Higher School of Economics, Basic Research Program Working Papers, Series: Economics, WP BRP 204/EC/2018. https://doi.org/10.2139/ssrn.3291165
2. Antoshchenkova, I.V., Bykadorov, I.A.: Monopolistic competition model: the impact of technological innovation on equilibrium and social optimality. Autom. Remote Control **78**(3), 537–556 (2017)
3. Arkolakis, C., Costinot, A., Rodríguez-Clare, A.: New trade models, same old gains? Am. Econ. Rev. **102**(1), 94–130 (2012)
4. Arkolakis, C., Costinot, A., Rodríguez-Clare, A.: The elusive pro-Competitive effects of trade. Rev. Econ. Stud. **86**(1), 46–80 (2019)
5. Behrens, K., Murata, Y.: General equilibrium models of monopolistic competition: a new approach. J. Econ. Theory **136**(1), 776–787 (2007)
6. Brander, J., Krugman, P.: A 'Reciprocal Dumping' model of international trade. J. Int. Econ. **15**(3–4), 313–321 (1983)
7. Bykadorov, I.: Monopolistic competition model with different technological innovation and consumer utility levels. In: CEUR Workshop Proceeding, vol. 1987, pp. 108–114 (2017)
8. Bykadorov, I.: Monopolistic competition with investments in productivity. Optim. Lett. **13**(8), 1803–1817 (2019)
9. Bykadorov, I., Ellero, A., Funari, S., Kokovin, S., Pudova, M.: Chain store against manufacturers: regulation can mitigate market distortion. In: Kochetov, Y., Khachay, M., Beresnev, V., Nurminski, E., Pardalos, P. (eds.) DOOR 2016. LNCS, vol. 9869, pp. 480–493. Springer, Cham (2016). https://doi.org/10.1007/978-3-319-44914-2_38
10. Bykadorov, I., Ellero, A., Funari, S., Moretti, E.: Dinkelbach approach to solving a class of fractional optimal control problems. J. Optim. Theory Appl. **142**(1), 55–66 (2009)
11. Bykadorov, I., Ellero, A., Moretti, E.: Minimization of communication expenditure for seasonal products. RAIRO Oper. Res. **36**(2), 109–127 (2002)
12. Bykadorov, I., Ellero, A., Moretti, E., Vianello, S.: The role of retailer's performance in optimal wholesale price discount policies. Eur. J. Oper. Res. **194**(2), 538–550 (2009)
13. Bykadorov, I., Gorn, A., Kokovin, S., Zhelobodko, E.: Why are losses from trade unlikely? Econ. Lett. **129**, 35–38 (2015)
14. Bykadorov, I., Kokovin, S.: Can a larger market foster R&D under monopolistic competition with variable mark-ups? Res. Econ. **71**(4), 663–674 (2017)
15. Bykadorov, I., Ellero, A., Funari, S., Kokovin, S., Molchanov, P.: Painful Birth of Trade under Classical Monopolistic Competition, National Research University Higher School of Economics, Basic Research Program Working Papers, Series: Economics, WP BRP 132/EC/2016. https://doi.org/10.2139/ssrn.2759872
16. Bykadorov, I.A., Kokovin, S.G., Zhelobodko, E.V.: Product diversity in a vertical distribution channel under monopolistic competition. Autom. Remote Control **75**(8), 1503–1524 (2014)
17. Chamberlin, E.H.: The Theory of Monopolistic Competition: A Re-orientation of the Theory of Value. Harvard University Press, Cambridge (1933)
18. Chamberlin, E.H.: The Theory of Monopolistic Competition. Harvard University Press, Cambridge (1962)

19. Demidova, S.: Trade policies, firm heterogeneity, and variable markups. J. Int. Econ. **108**(8), 260–273 (2017)
20. Demidova, S., Rodríguez-Clare, A.: Trade policy under firm-level heterogeneity in a small economy. J. Int. Econ. **78**(1), 100–112 (2009)
21. Dhingra, S.: Trading away wide brands for cheap brands. Am. Econ. Rev. **103**(6), 2554–2584 (2013)
22. Dhingra, S., Morrow, J.: Monopolistic competition and optimum product diversity under firm heterogeneity. J. Polit. Econ. **127**(1), 196–232 (2019)
23. Dixit, A.K., Stiglitz, J.E.: Monopolistic competition and optimum product diversity. Am. Econ. Rev. **67**(3), 297–308 (1977)
24. Feenstra, R.C.: A homothetic utility function for monopolistic competition models, without constant price elasticity. Econ. Lett. **78**(1), 79–86 (2003)
25. Feldman, A.M., Serrano, R.: Welfare Economics and Social Choice Theory, 2nd edn. Springer, Brown University, Boston, Providence (2006). https://doi.org/10.1007/0-387-29368-X
26. Krugman, P.: Increasing returns, monopolistic competition, and international trade. J. Int. Econ. **9**(4), 469–479 (1979)
27. Krugman, P.: Scale economies, product differentiation, and the pattern of trade. Am. Econ. Rev. **70**(5), 950–959 (1980)
28. Melitz, M.J., Ottaviano, G.I.P.: Market size, trade, and productivity. Rev. Econ. Stud. **75**(1), 295–316 (2008)
29. Melitz, M.J.: The impact of trade on intra-industry reallocations and aggregate industry productivity. Econometrica **71**(6), 1695–1725 (2003)
30. Melitz, M.J., Redding, S.J.: Missing gains from trade? Am. Econ. Rev. **104**(5), 317–321 (2014)
31. Melitz, M.J., Redding, S.J.: New trade models, new welfare implications, new welfare implications. Am. Econ. Rev. **105**(3), 1105–1146 (2015)
32. Moulin, H.: Fair Division and Collective Welfare. MIT Press, Cambridge (2004)
33. Mrázová, M., Neary, J.P.: Together at last: trade costs, demand structure, and welfare. Am. Econ. Rev. **104**(5), 298–303 (2014)
34. Mrázová, M., Neary, J.P.: Not so demanding: demand structure and firm behavior. Am. Econ. Rev. **107**(12), 3835–3874 (2017)
35. Ottaviano, G.I.P., Tabuchi, T., Thisse, J.-F.: Agglomeration and trade revised. Int. Econ. Rev. **43**, 409–436 (2002)
36. Zhelobodko, E., Kokovin, S., Parenti, M., Thisse, J.-F.: Monopolistic competition in general equilibrium: beyond the constant elasticity of substitution. Econometrica **80**(6), 2765–2784 (2012)

Hamilton-Jacobi-Bellman Equations for Non-cooperative Differential Games with Continuous Updating

Ovanes Petrosian[1,2] and Anna Tur[1(✉)]

[1] St. Petersburg State University, 7/9, Universitetskaya nab., Saint-Petersburg 199034, Russia
petrosian.ovanes@yandex.ru, opetrosyan@hse.ru
[2] National Research University Higher School of Economics, Saint-Petersburg 194100, Russia
a.tur@spbu.ru

Abstract. This paper is devoted to a new class of differential games with continuous updating. It is assumed that at each time instant, players have or use information about the game defined on a closed time interval. However, as the time evolves, information about the game updates, namely, there is a continuous shift of time interval, which determines the information available to players. Information about the game is the information about motion equations and payoff functions of players. For this class of games, the direct application of classical approaches to the determination of optimality principles such as Nash equilibrium is not possible. The subject of the current paper is the construction of solution concept similar to Nash equilibrium for this class of differential games and corresponding optimality conditions, in particular, modernized Hamilton-Jacobi-Bellman equations.

Keywords: Differential games with continuous updating · Nash equilibrium with continuous updating · Nash equilibrium · Hamilton-Jacobi-Bellman equation

1 Introduction

This paper formulates a new approach to analyze differential games with uncertainties and unknowns in the players' future payoff structures. The approach [11,13] is used for constructing game theoretical models and defining solutions for conflict-controlled processes, where information about the process updates continuously in time. Existing differential game models often rely on the assumption of time-invariant game structures for the derivation of equilibrium solutions. However, many events in a considerably far future are intrinsically unknown.

The reported study was funded by RFBR according to the research project No. 18-00-00727 (18-00-00725).

I. Bykadorov et al. (Eds.): MOTOR 2019, CCIS 1090, pp. 178–191, 2019.
https://doi.org/10.1007/978-3-030-33394-2_14

Therefore, the behavior of players should as well be modeled using the assumption that they use only the truncated information about the game structure. It is supposed that players lack certain information about the motion equations and payoff functions on the whole-time interval on which the game is played. At each time instant information about the game structure updates, players receive information about motion equations and payoff functions. This new approach for the analysis of differential games via information updating provides a more realistic and practical alternative to the study of differential games. Existing differential game models defined on the closed time interval [1], infinite time interval [8], random time interval [16] do not take into account the fact that in many real-life processes, players at the initial instant do not know all the information about the game. Thus, existing approaches cannot be directly used to construct a range of real-life game-theoretic models.

This work aims to present optimality conditions in the form of Hamilton-Jacobi-Bellman equations for the solution concept similar to the feedback Nash equilibrium for a class of games with continuous updating. In the game models with continuous updating, it is assumed that players

1. in the current time instant $t \in [t_0, +\infty)$ have information about the motion equations and payoff functions on the truncated time interval $[t, t+\overline{T}]$ with length defined by the information horizon $\overline{T}$,
2. continuously or at any time instant $t \in [t_0, +\infty)$ receive updated information about the motion equations and payoff functions and as a result continuously adapt to the updated information (Fig. 1).

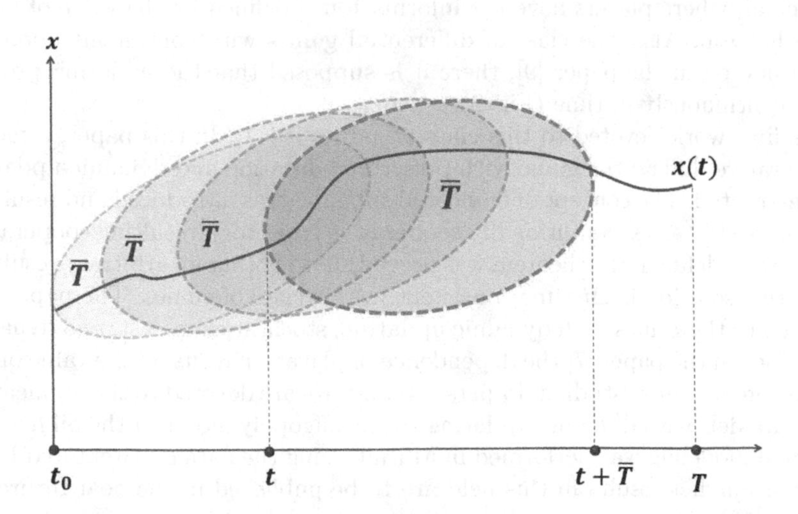

Fig. 1. Each blue oval shows the information available to players at instant t, namely $[t, t+\overline{T}]$, where $\overline{T}$ is the time horizon. (Color figure online)

Obviously, it is difficult to obtain Nash equilibrium due to the lack of fundamental approaches to control problems with continuously moving information horizon. Classical methods such as dynamic programming and Hamilton-Jacobi-Bellman equation [2] or Pontryagin Maximum Principal [17] do not allow to directly construct Nash equilibrium in problems with moving information horizon. Taking into account the assumptions described above the two main problems arise:

1. How to define a solution concept similar to Nash equilibrium for a class of games with continuous updating?
2. How to derive the corresponding optimality conditions?

Both questions are addressed in this work. The game model with continuous updating and the solution technique besides the current time parameter uses an additional one. It is used to take into account truncated information available to players. Feedback Nash equilibrium in the game model with continuous updating is defined using the so-called generalized feedback Nash equilibrium as a strategy profile depending not only on the current time t and state x but also on the additional time parameter. Special transformation is introduced to obtain Nash equilibrium with continuous updating as a strategy profile depending only on the current time t and state x. In order to define generalized Nash equilibrium, we introduce a new type of Hamilton-Jacobi-Bellman equations for the class of games with continuous updating.

Till now a class of games with dynamic updating was studied in the papers [7,11–15,21], where authors laid a foundation for further study of the class of games with dynamic updating. There it is assumed that the information about motion equations and payoff functions is updated in discrete time instants, and the interval, where players have the information is defined by the value of information horizon. Also, the class of differential games with continuous updating was considered in the paper [9], there it is supposed that the updating process evolves continuously in time (Fig, 2).

The first work devoted to this class of games is [11]. In this paper, a model of cooperative differential game with prescribed duration and dynamic updating was constructed. The concept of truncated subgame was introduced and resulting cooperative strategies, conditionally cooperative trajectory, resulting cooperative solution were defined, the theorem was proved showing that an arbitrary resulting cooperative solution is Δt-time consistent in this class of games. The paper [13] is devoted to the games with dynamic updating, stochastic forecast, and dynamic adaptation. In the paper [7] the dependence of players' payoffs on the value of the information horizon is studied. Papers [14] and [15] are devoted to the application of game models with dynamic updating to the oligopoly model of the oil market. Numerical modeling was performed in Matlab using the data on Brent and Light oil prices. Further results in this field are to be published in the near future. In the paper [12] the approach presented above was applied to the game models with infinite horizon. The paper [21] is devoted to the construction of a special class of Hamilton-Jacobi-Bellman equations for the noncooperative dynamic game model defining the various types of information structure. Obtained results can be used for constructing models, where players use different information structures. The

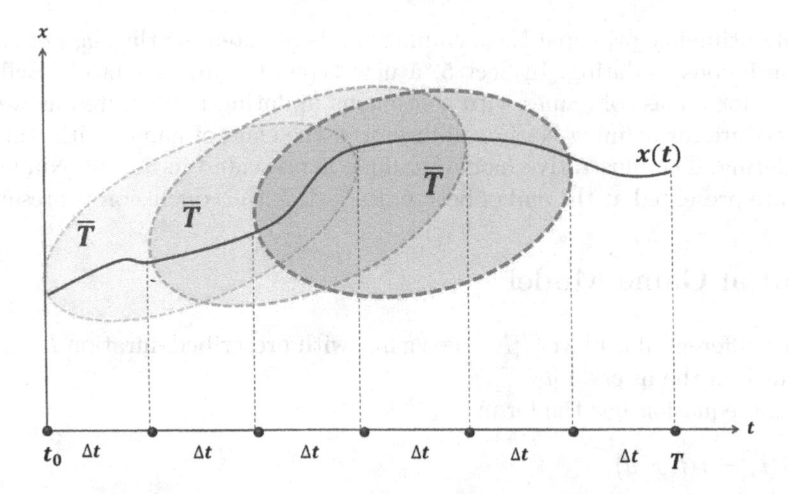

Fig. 2. Each blue oval shows the information available to players over the interval $[t_0 + j\Delta t, t_0 + (j+1)\Delta t]$, namely $[t_0 + j\Delta t, t_0 + j\Delta t + \overline{T}]$, $j = 0, \ldots, l$, $l = \frac{T - t_0}{\Delta t}$. (Color figure online)

results of the paper [9] are devoted to studying the class of linear-quadratic differential games with continuous updating. There the form of Nash equilibrium with continuous updating is presented and special convergence results are obtained for the case of dynamic and continuous updating.

The class of differential games with continuous updating has some similarities with the Model Predictive Control (MPC) theory worked out within the framework of numerical optimal control. We analyze [6,10,18,19] to get recent results in this area. MPC is a method of control when the current control action is achieved by solving at each sampling instant a finite horizon open-loop optimal control problem using the current state of an object as the initial state. This type of control is able to cope with strict limitations on controls and states, which is an advantage over other methods. There is, therefore, a wide application in petrochemical and related industries where key operating points are located close to the set of admissible states and controls. The main problem that is solved in MPC is the provision of movement along the target trajectory under the conditions of random perturbations and an unknown dynamical system. At each time step, the optimal control problem is solved for defining controls which will lead the system to the target trajectory. The class of games with continuous updating, on the other hand, solves the problem of modeling player behavior when information about the process updates dynamically. This means that the class of differential games with continuous updating does not use the target trajectory, but answers the question of composing a trajectory which will be used by players in case of using truncated information about the process and continuous updating.

The paper is structured as follows. In Sect. 2, the description of the initial differential game model is presented. In Sect. 3, the game model with continuous updating is defined as well as the concept of a strategy for it. In Sect. 4, classical optimality principal Nash equilibrium is adapted for the class of games with continuous updating. In Sect. 5, a new type of Hamilton-Jacobi-Bellman equations for a class of games with continuous updating is presented as well as the procedure for defining Nash equilibrium in the class of games with continuous updating. The illustrative model example is presented in Sect. 6. Numerical results are presented at the end of Sect. 6. In Sect. 7, the conclusion is presented.

2 Initial Game Model

Consider differential n-player ($|N| = n$) game with prescribed duration $\Gamma(x_0, T - t_0)$ defined on the interval $[t_0, T]$.

Motion equation has the form:

$$\begin{aligned} &\dot{x}(t) = f(t, x, u), \\ &x(t_0) = x_0, \\ &x \in \mathbb{R}^l, \; u = (u_1, \dots, u_n), \; u_i = u_i(t, x) \in U_i \subset \text{comp}\mathbb{R}^k, \; t \in [t_0, T]. \end{aligned} \tag{1}$$

Payoff function of player i is defined in the following way:

$$K_i(x_0, T - t_0; u) = \int_{t_0}^{T} g^i[t, x(t), u(t, x)]dt, \; i \in N, \tag{2}$$

where $g^i[t, x(t), u(t, x(t))]$, $f(t, x, u)$ are the integrable functions, $x(t)$ is the solution of Cauchy problem (1) with fixed $u(t, x) = (u_1(t, x), \dots, u_n(t, x))$. The strategy profile $u(t, x) = (u_1(t, x), \dots, u_n(t, x))$ is called admissible if the problem (1) has a unique and continuable solution. We use the conditions of existence, uniqueness and continuability of Filippov [5]:

1. right-hand side of motion equations $f(t, x, u)$ (1) is continuous on the set $[t_0, T] \times X \times U_1 \times \dots \times U_n$
2. right-hand side of motion equations $f(t, x, u)$ satisfies Lipschitz conditions for x with the constant $k_1 > 0$ uniformly regarding to u:
$$||f(t, x', u) - f(t, x'', u)|| \leq k_1 ||x' - x''||, \; \forall \, t \in [t_0, T], \; x', x'' \in X, u \in U$$
3. exists such a constant k_2 that function $f(t, x, u)$ satisfies condition:
$$||f(t, x, u)|| \leq k_2(1 + ||x||), \; \forall \, t \in [t_0, T], \; x \in X, u \in U$$
4. for any $t \in [t_0, T]$ and $x \in X$ set
$$G(x) = \{f(t, x, u) | u \in U\}$$
is a convex compact from R^l.

Using the initial differential game with prescribed duration of T, we construct the corresponding differential game with continuous updating.

3 Differential Game Model with Continuous Updating

Consider n-player differential game $\Gamma(x_0, t_0, \overline{T})$, defined on the interval $[t_0, t_0 + \overline{T}]$, where $0 < \overline{T} < +\infty$.

Motion equation has the form:

$$\begin{aligned} &\dot{x}_{t_0}(s) = f(s, x_{t_0}, u^{t_0}), \\ &x_{t_0}(t_0) = x_0, \\ &x_{t_0} \in \mathbb{R}^l,\ u^{t_0} = (u_1^{t_0}, \dots, u_n^{t_0}),\ u_i^{t_0} = u_i^{t_0}(s, x) \in U_i \subset \text{comp}\mathbb{R}^k. \end{aligned} \tag{3}$$

Payoff function of player i is defined in the following way:

$$K_i^{t_0}(x_0, t_0, \overline{T}; u^{t_0}) = \int\limits_{t_0}^{t_0+\overline{T}} g^i[s, x_{t_0}(s), u^{t_0}(s, x)]ds,\ i \in N, \tag{4}$$

where $x_{t_0}(s)$, $u^{t_0}(s, x)$ are trajectory and strategies in the game $\Gamma(x_0, t_0, \overline{T})$, $\dot{x}_{t_0}(s)$ is the derivative of s.

Subgame of Differential Game with Continuous Updating. Consider n-player differential game $\Gamma(x, t, \overline{T})$, $t \in [t_0, +\infty)$ defined on the interval $[t, t + \overline{T}]$, where $0 < \overline{T} < +\infty$.

Motion equation for the subgame $\Gamma(x, t, \overline{T})$ has the form:

$$\begin{aligned} &\dot{x}_t(s) = f(s, x_t, u^t), \\ &x_t(t) = x, \\ &x_t \in \mathbb{R}^l,\ u^t = (u_1^t, \dots, u_n^t),\ u_i^t = u_i^t(s, x) \in U_i \subset \text{comp}\mathbb{R}^k,\ \ s \in [t, t + \overline{T}]. \end{aligned} \tag{5}$$

Payoff function of player i for the subgame $\Gamma(x, t, \overline{T})$ has the form:

$$K_i^t(x, t; u^t) = \int\limits_{t}^{t+\overline{T}} g^i[s, x_t(s), u^t(s, x)]ds,\ i \in N, \tag{6}$$

where $x_t(s)$, $u^t(s, x)$ are trajectories and strategies in the game $\Gamma(x, t, \overline{T})$, $\dot{x}_t(s)$ is the derivative of s.

Differential game with continuous updating is developed according to the following rule:

Current time $t \in [t_0, +\infty)$ evolves continuously and as a result players continuously obtain new information about motion equations and payoff functions in the game $\Gamma(x, t, \overline{T})$.

Strategy profile $u(t, x)$ in the differential game with continuous updating has the form:

$$u(t, x) = u^t(t, x),\ t \in [t_0, +\infty), \tag{7}$$

where $u^t(s, x)$, $s \in [t, t + \overline{T}]$ are strategies in the subgame $\Gamma(x, t, \overline{T})$.

Trajectory $x(t)$ in the differential game with continuous updating is determined in accordance with

$$\begin{aligned} &\dot{x}(t) = f(t, x, u), \\ &x(t_0) = x_0, \\ &x \in \mathbb{R}^l, \end{aligned} \tag{8}$$

where $u = u(t, x)$ are strategies in the game with continuous updating (7) and $\dot{x}(t)$ is the derivative of t. We suppose that the strategy with continuous updating obtained using (7) is admissible or that the problem (8) has a unique and continuable solution. Corresponding conditions of existence, uniqueness and continuability of Filippov [5] are presented for the system (1)–(4.).

The essential difference between the game model with continuous updating and classic differential game with prescribed duration $\Gamma(x_0, T-t_0)$ is that players in the initial game are guided by the payoffs that they will eventually obtain on the interval $[t_0, T]$, but in the case of a game with continuous updating, at the time instant t they orient themselves on the expected payoffs (6), which are calculated based on the information defined on the interval $[t, t+\overline{T}]$ or the information that they have at the instant t.

4 Nash Equilibrium in Game with Continuous Updating

In the framework of continuously updated information, it is important to model the behavior of players. To do this, we use the concept of Nash equilibrium in feedback strategies. However, for the class of differential games with continuous updating, we would like to have it the following form:

- for any fixed $t \in [t_0, +\infty)$, $u^{NE}(t, x) = (u_1^{NE}(t, x), \dots, u_n^{NE}(t, x))$ coincides with the Nash equilibrium in the game (5), (6) defined on the interval $[t, t+\overline{T}]$ in the instant t.

However, direct application of classical approaches for definition of the Nash equilibrium in feedback strategies is not possible, consider two intervals $[t, t+\overline{T}]$ and $[t+\epsilon, t+\overline{T}+\epsilon]$, $\epsilon << \overline{T}$. Then according to the problem statement:

- $u^{NE}(t, x)$ in the instant t coincides with the feedback Nash equilibrium in the game defined on the interval $[t, t+\overline{T}]$,
- $u^{NE}(t+\epsilon, x)$ in the instant $t+\epsilon$ must coincide with the feedback Nash equilibrium in the game defined on the interval $[t+\epsilon, t+\epsilon+\overline{T}]$.

To construct such strategies, we consider a concept of generalized Nash equilibrium in feedback strategies as the principle of optimality

$$\widetilde{u}^{NE}(t, s, x) = (\widetilde{u}_1^{NE}(t, s, x), \dots, \widetilde{u}_n^{NE}(t, s, x)),\ t \in [t_0, T],\ s \in [t, t+\overline{T}], \tag{9}$$

which we are going to use further for construction of strategies $u^{NE}(t, x)$.

Definition 1. *Strategy profile* $\widetilde{u}^{NE}(t, s, x) = (\widetilde{u}_1^{NE}(t, s, x), \dots, \widetilde{u}_n^{NE}(t, s, x))$ *is a generalized Nash equilibrium in the game with continuous updating, if for any fixed* $t \in [t_0, +\infty)$ *strategy profile* $\widetilde{u}^{NE}(t, s, x)$ *is the feedback Nash equilibrium in the game* $\Gamma(x, t, \overline{T})$.

It is important to notice that the generalized feedback Nash equilibrium $\widetilde{u}^{NE}(t,s,x)$ for a fixed t is a function of s and x, where s is defined on the interval $[t, t+\overline{T}]$. Using generalized feedback Nash equilibrium it is possible to define a solution concept for a game model with continuous updating.

Definition 2. *Strategy profile $u^{NE}(t,x)$ is called the Nash equilibrium with continuous updating if it is defined in the following way:*

$$u^{NE}(t,x) = \widetilde{u}^{NE}(t,s,x)|_{s=t} = (\widetilde{u}_1^{NE}(t,s,x)|_{s=t}, \dots, \widetilde{u}_n^{NE}(t,s,x)|_{s=t}), \quad t \in [t_0, +\infty), \tag{10}$$

where $\widetilde{u}^{NE}(t,s,x)$ is the generalized feedback Nash equilibrium defined in Definition 1.

Trajectory $x^{NE}(t)$ corresponding to the Nash equilibrium with continuous updating $u^{NE}(t,x)$ can be obtained from the system

$$\begin{aligned} \dot{x}(t) &= f(t, x, u^{NE}), \\ x(t_0) &= x_0, \\ x &\in \mathbb{R}^l, \end{aligned} \tag{11}$$

Unlike the generalized feedback Nash equilibrium, $u^{NE}(t,x)$ does not contain feedback Nash equilibrium strategies for any $s \in [t, t+\overline{T}]$. Strategy profile $u^{NE}(t,x)$ only contains strategies of players that they perform according to the procedure described in Sect. 3, i.e. continuous updating procedure, where $s = t$. $u^{NE}(t,x)$ will be used as a solution concept in the game with continuous updating. The notion of Nash equilibrium with continuous updating $u^{NE}(t,x)$ is close to the subgame perfect Nash equilibrium. In the sense of the expected payoff (6) for any fixed current time t individual deviation from Nash equilibrium with continuous updating is not beneficial for the players or for any subgame of the game with continuous updating individual deviation is not beneficial due to the information structure.

5 Hamilton-Jacobi-Bellman Equations with Continuous Updating

To define strategy profile $u^{NE}(t,x)$, it is necessary to determine the generalized Nash equilibrium in feedback strategies $\widetilde{u}^{NE}(t,s,x)$ in the game with continuous updating $\Gamma(x_0, t_0, \overline{T})$. To do this, we will use a modernized version of dynamic programming. In the framework of this approach, the Bellman function $V^i(t,s,x)$ is defined as the payoff of player i in feedback Nash equilibrium $\widetilde{u}^{NE}(t,s,x)$ in the subgame starting at the instant s in the state x in the game defined on the interval $[t, t+\overline{T}]$:

$$V^i(t,s,x) = \int_s^{t+\overline{T}} g^i[\tau, x_t(\tau), \widetilde{u}^{NE}(t,\tau,x)]d\tau, \ i \in N \tag{12}$$

subject to

$$\dot{x}(\tau) = f(\tau, x, u),$$
$$x(s) = x. \tag{13}$$

The following theorem takes place:

Theorem 1. *$\widetilde{u}^{NE}(t,s,x)$ is the generalized Nash equilibrium in feedback strategies in the differential game with continuous updating $\Gamma(x_0, t_0, \overline{T})$, if there exist functions $V^i(t,s,x) : [t_0, +\infty) \times [t, t+\overline{T}] \times R \to R$, $i \in N$ continuously differentiable by s and x, satisfying the following system of partial differential equations:*

$$\begin{aligned} -V_s^i(t,s,x) &= \max_{\phi_i} \{ g^i(s,x,\widetilde{u}_{-i}^{NE}) + V_x^i(t,s,x) f(s,x,\widetilde{u}_{-i}^{NE}) \} \\ &= g^i(s,x,\widetilde{u}^{NE}) + V_x^i(t,s,x) f(s,x,\widetilde{u}^{NE}), \\ & \qquad V^i(t, t+\overline{T}, x) = 0, \quad i \in N, \end{aligned} \tag{14}$$

where $\widetilde{u}_{-i}^{NE}(\phi_i) = (\widetilde{u}_1^{NE}, \ldots, \phi_i, \ldots, \widetilde{u}_n^{NE})$.

Proof. According to the definition of generalized Nash equilibrium $\widetilde{u}^{NE}(t,s,x)$, $\widetilde{u}^{NE}(t,s,x)$ should be the feedback Nash equilibrium for any fixed t.

By fixing t in the formulation of the Theorem 1 and in particular in (14) we obtain classical sufficient conditions for feedback Nash equilibrium in the differential game with prescribed duration $[t, t+\overline{T}]$ presented in [1]. Therefore for any fixed t, the conditions for the definition of generalized Nash equilibrium are satisfied. The Theorem is proved.

We consider only the class of generalized Nash equilibrium such that for the Nash equilibrium with continuous updating the solution of the system (11) satisfies the conditions of existence, uniqueness, and continuability of Filippov [5]. In the case, if it is possible to obtain generalized Nash equilibrium $\widetilde{u}^{NE}(t,s,x)$ using equations (14), then by using the procedure (10) we obtain desired strategy profile $u^{NE}(t,x)$.

6 Differential Game of Investment in Public Goods

As an illustrative example consider a two-player differential game of investment in public goods. Two players are investing in a public stock of knowledge. It is assumed that knowledge is pure public good and every individual has access to it. The model of such type was firstly formulated in [4] (see also [3,20]). In these models, it was assumed that each player has a quadratic utility and that the cost of investment increases quadratically with the investment effort. Here we use other functions to define players' payoffs.

6.1 Initial Game Model

Let $x(t)$ be the stock of knowledge at time t and $u_i(t,x)$ is the investment of player i in public knowledge at time t. Assume that the stock of knowledge evolves according to the accumulation equation

$$\dot{x}(t) = -\beta x(t) + u_1(t,x) + u_2(t,x), \quad x(0) = x_0, \tag{15}$$

where β is the depreciation rate, $x_0 > 0$ is a given initial stock of knowledge. Let possible states at each time t satisfies the condition $x(t) > 0$. Assume that each individual i faces costs of investment which are state-dependent and given by function $\frac{u_i^2}{x}$. Suppose that the utility of each player is a linear function of knowledge stock consumption $q_i x(t)$, where $q_i > 0$. That is, the payoff function of each player is

$$K_i(x_0, T; u) = \int_0^T \left(q_i x(t) - \frac{u_i^2(t,x)}{x(t)} \right) dt. \tag{16}$$

In order to define the feedback Nash equilibrium $u^*(t,x)$ we use the sufficient conditions in the form of HJB equations:

$$\begin{aligned} &-V_t^i(t,x) = \max_{u_i}(q_i x - \tfrac{u_i^2}{x} + V_x^i(t,x)(-\beta x + u_i + u_j^*)), \quad i \neq j, \\ &V_i(T,x) = 0, \quad i = 1,2. \end{aligned} \tag{17}$$

Suppose that the Bellman function has the following form:

$$V^i(t,x) = a_i(t)x, \; i = 1,2.$$

The maximization problem in (17) yields a strategy for player i:

$$u_i^*(t,x) = \frac{x V_x^i(t,x)}{2} = \frac{a_i(t)x}{2}. \tag{18}$$

Substituting (18) into (17) we obtain the following system of differential equations for $a_i(t)$:

$$\begin{aligned} &\dot{a}_i(t) = \beta a_i(t) + \tfrac{a_i^2(t)}{4} - a_i(t)a_j(t) - q_i, \quad i \neq j, \quad i = 1,2, \\ &a_i(T) = 0. \end{aligned} \tag{19}$$

As an example consider the symmetric case when $q_1 = q_2 = q$, then the solution of (19) is

$$a_i(t) = \frac{2q(e^{v(t-T)} - 1)}{(\beta - v)e^{v(t-T)} - \beta - v}, \quad i = 1,2,$$

where $v = \sqrt{\beta^2 - 3q}$. Here we assume, that $\beta^2 > 3q$. The interval of validity to the solution of (19) is $(-\infty; T + \frac{1}{v} ln \frac{\beta+v}{\beta-v})$. Note that $[0;T] \subset (-\infty; T + \frac{1}{v} ln \frac{\beta+v}{\beta-v})$. Then the feedback Nash equilibrium strategies are

$$u_i^*(t,x) = \frac{q(e^{v(t-T)} - 1)}{(\beta - v)e^{v(t-T)} - \beta - v} x(t), \quad i = 1,2.$$

Resulting equilibrium trajectory is

$$x^*(t) = x_o e^{-\frac{1}{3}(\beta+2v)t} \left(\frac{(\beta - v)e^{vt} - (\beta + v)e^{vT}}{\beta - v - (\beta + v)e^{vT}} \right)^{\frac{4}{3}}.$$

The state constraint $x^*(t) > 0$ is satisfied for all $t \in [0; T]$ when $x_0 > 0$.

6.2 Game Model with Continuous Updating

Now consider this model as a game with continuous updating. It is assumed that information about motion equations and payoff functions is updated continuously in time. At every instant $t \in [0, +\infty)$ players have information only the interval $[t, t + \overline{T}]$. It means that because of the possibility of changing the number of investors or utilities of players at each time instant they can count for the stability of process only over period $\overline{T}$.

According to the Theorem 1 Hamilton-Jacobi-Bellman equations for generalized feedback Nash equilibrium are

$$\begin{aligned} &-V_s^i(t, s, x) = \max_{u_i}(q_i x - \tfrac{u_i^2}{x} + V_x^i(t, s, x)(-\beta x + u_i + \widetilde{u}_j^{NE})), \quad i \neq j, \\ &V_i(t, t + \overline{T}, x) = 0, \quad i = 1, 2. \end{aligned} \tag{20}$$

Suppose that the Bellman function is defined in the form:

$$V_i(t, s, x) = a_i(t, s)x.$$

Maximization problem in (20) yields a strategy for player i:

$$\widetilde{u}_i^{NE}(t, s, x) = \frac{x V_x^i(t, s, x)}{2} = \frac{x a_i(t, s)}{2}. \tag{21}$$

Substituting (21) into (20) we obtain the following system of differential equations for $a_i(t, s)$:

$$\begin{aligned} &\dot{a}_i(t, s) = \beta a_i(t, s) + \tfrac{a_i^2(t,s)}{4} - a_i(t, s)a_j(t, s) - q_i, \quad i \neq j, \quad i = 1, 2, \\ &a_i(t, t + \overline{T}) = 0. \end{aligned} \tag{22}$$

Taking into account the symmetry of players we obtain the solution:

$$a_i(t, s) = \frac{2q(e^{v(s-t-\overline{T})} - 1)}{(\beta - v)e^{v(s-t-\overline{T})} - \beta - v}, \quad i = 1, 2,$$

where $v = \sqrt{\beta^2 - 3q}$.

Finally we get the generalized feedback Nash equilibrium strategies:

$$\widetilde{u}_i^{NE}(t, s, x) = \frac{q(e^{v(s-t-\overline{T})} - 1)}{(\beta - v)e^{v(s-t-\overline{T})} - \beta - v} x(t).$$

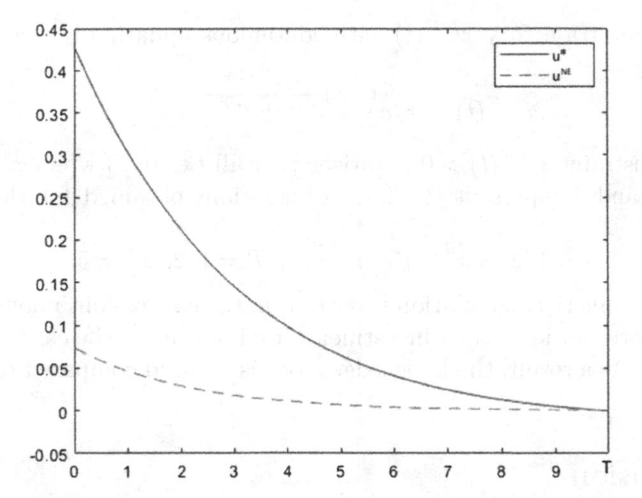

Fig. 3. Nash equilibrium with continuous updating $u^{NE}(t,x)$ (red dashed line), Nash equilibrium strategies in the initial game $u^*(t)$ (blue solid line). (Color figure online)

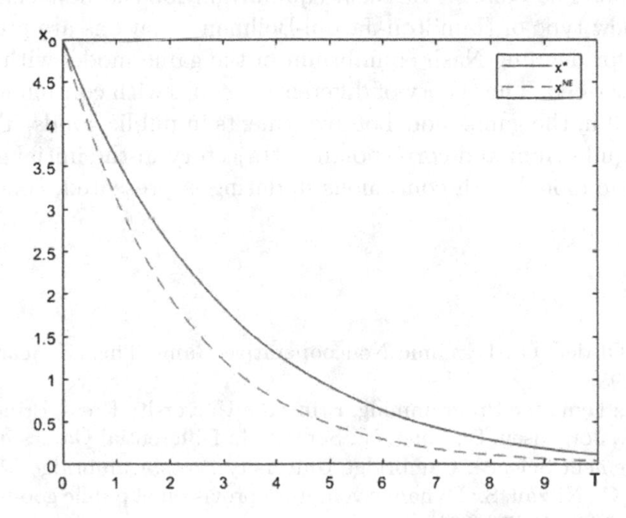

Fig. 4. Resulting equilibrium trajectory $x^{NE}(t)$ with continuous updating (red dashed line), resulting equilibrium trajectory $x^*(t)$ in the initial game (blue solid line). (Color figure online)

According to the procedure (10) we construct the feedback Nash equilibrium with continuous updating:

$$u_i^{NE}(t,x) = \widetilde{u}_i^{NE}(t,s,x)\mid_{s=t} = \frac{q(e^{-v\overline{T}}-1)}{(\beta-v)e^{-v\overline{T}}-\beta-v}x(t).$$

The equilibrium trajectory $x^{NE}(t)$ with continuous updating is

$$x^{NE}(t) = x_o e^{\left(-\beta + \frac{2q(e^{-v\overline{T}}-1)}{(\beta-v)e^{-v\overline{T}}-\beta-v}\right)t}.$$

The state constraint $x^{NE}(t) > 0$ is satisfied for all $t \in [0;T]$ when $x_0 > 0$.

Figures 3 and 4 represent the form of solutions obtained for the following parameters:

$$\beta = 1/2,\ q = 1/15,\ T = 10,\ \overline{T} = 1/2,\ x_0 = 5.$$

Results of numerical simulation show that in the case of continuous updating, players are more cautious, their investments are less than in the case of the initial game (Fig. 3). As a result, the knowledge stock is reduced compared to the initial game (Fig. 4).

7 Conclusion

A differential game model with continuous updating is presented and described in the first time. The concept of Nash equilibrium for the new class of games is defined. A new type of Hamilton-Jacobi-Bellman equations are presented and the technique for defining Nash equilibrium in the game model with continuous updating is described. The theory of differential games with continuous updating is demonstrated in the game model of investments in public goods. The comparison of Nash equilibrium and corresponding trajectory in the initial game model and in the game model with continuous updating is presented, conclusions are drawn.

References

1. Basar, T., Olsder, G.: Dynamic Noncooperative Game Theory. Academic Press, London (1995)
2. Bellman, R.: Dynamic Programming. Princeton University Press, Princeton (1957)
3. Dockner, E., Jorgensen, S., Long, N., Sorger, G.: Differential Games in Economics and Management Science. Cambridge University Press, Cambridge (2000)
4. Fershtman, C., Nitzan, S.: Dynamic voluntary provision of public goods. Eur. Econ. Rev. **35**(5), 1057–1067 (1991)
5. Filippov, A.: Introduction to the Theory of Differential Equations. Editorial URSS, Moscow (2004). (in Russian)
6. Goodwin, G., Seron, M., Dona, J.: Constrained Control and Estimation: An Optimisation Approach. Springer, New York (2005). https://doi.org/10.1007/b138145
7. Gromova, E., Petrosian, O.: Control of information horizon for cooperative differential game of pollution control. In: 2016 International Conference Stability and Oscillations of Nonlinear Control Systems (Pyatnitskiy's Conference) (2016)
8. Kleimenov, A.: Non-antagonistic Positional Differential Games. Science, Ekaterinburg (1993)
9. Kuchkarov, I., Petrosian, O.: On class of linear quadratic non-cooperative differential games with continuous updating. In: Khachay, M., Kochetov, Y., Pardalos, P. (eds.) MOTOR 2019. LNCS, vol. 11548, pp. 635–650. Springer, Cham (2019). https://doi.org/10.1007/978-3-030-22629-9_45

10. Kwon, W., Han, S.: Receding Horizon Control: Model Predictive Control for State Models. Springer, New York (2005)
11. Petrosian, O.: Looking forward approach in cooperative differential games. Int. Game Theory Rev. **18**, 1–14 (2016)
12. Petrosian, O.: Looking forward approach in cooperative differential games with infinite-horizon. Vestnik S.-Petersburg Univ. Ser. 10. Prikl. Mat. Inform. Prots. Upr. **4**, 18–30 (2016)
13. Petrosian, O., Barabanov, A.: Looking forward approach in cooperative differential games with uncertain-stochastic dynamics. J. Optim. Theory Appl. **172**, 328–347 (2017)
14. Petrosian, O., Nastych, M., Volf, D.: Non-cooperative differential game model of oil market with looking forward approach. In: Petrosyan, L.A., Mazalov, V.V., Zenkevich, N. (eds.) Frontiers of Dynamic Games, Game Theory and Management, St. Petersburg, 2017, Birkhäuser, Basel (2018)
15. Petrosian, O., Nastych, M., Volf, D.: Differential game of oil market with moving informational horizon and non-transferable utility. 2017 Constructive Nonsmooth Analysis and Related Topics (dedicated to the memory of V.F. Demyanov) (2017)
16. Petrosyan, L., Shevkoplyas, E.: Cooperative differential games with random duration. Vestnik Sankt-Peterburgskogo Universiteta. Ser 1. Matematika Mekhanika Astronomiya **4**, 18–23 (2000)
17. Pontryagin, L.: On theory of differential games. Successes Math. Sci. **26** (4(130)), 219–274 (1966)
18. Rawlings, J., Mayne, D.: Model Predictive Control: Theory and Design. Nob Hill Publishing, Madison (2009)
19. Wang, L.: Model Predictive Control System Design and Implementation Using MATLAB. Springer, New York (2005)
20. Wirl, F.: Dynamic voluntary provision of public goods: extension for nonlinear strategies. Eur. J. Polit. Econ. **12**(3), 555–560 (1996)
21. Yeung, D., Petrosian, O.: Cooperative stochastic differential games with information adaptation. In: International Conference on Communication and Electronic Information Engineering (2017)

10. Kwon, W., Han, S.: Receding Horizon Control: Model Predictive Control for State Models. Springer, New York (2005)
11. Petrosian, O.: Looking forward approach in cooperative differential games. Int. Game Theory Rev. 18, 1–14 (2016)
12. Petrosian, O.: Looking forward approach in cooperative differential games with infinite-horizon. Vestnik S.-Petersburg Univ. Ser. 10. Prikl. Mat. Inform. Protsessy Upr. 4, 18–30 (2016)
13. Petrosian, O., Barabanov, A.: Looking forward approach in cooperative differential games with uncertain-stochastic dynamics. J. Optim. Theory Appl. 172, 328–347 (2017)
14. Petrosian, O., Nastych, M., Volf, D.: Non-cooperative differential game model of oil market with looking forward approach. In: Petrosyan, L.A., Mazalov, V.V., Zenkevich, N. (eds.) Frontiers of Dynamic Games, Game Theory and Management, St. Petersburg, 2017. Birkhäuser, Basel (2018)
15. Petrosian, O., Nastych, M., Volf, D.: Differential game of oil market with moving informational horizon and non-transferable utility. In: Constructive Nonsmooth Analysis and Related Topics (dedicated to the memory of V.F. Demyanov) (2017)
16. Petrosian, O., Shevkoplyas, E.: [illegible] games with random time [illegible] Vestnik S.-Petersburg Univ. [illegible] (2000)
17. [illegible]
18. [illegible] Nob Hill Publishing, Madison (2009)
19. Wang, L.: Model Predictive Control System Design and Implementation Using MATLAB. Springer, New York (2009)
20. [illegible] for [illegible]
21. [illegible] Petrosian, O.: [illegible] differential games with information [illegible] International Conference on Communication and Electronic Information Engineering (2017)

Data Mining and Computational Geometry

On the Thinnest Covering of Fixed Size Containers with Non-euclidean Metric by Incongruent Circles

Alexander Kazakov[1], Anna Lempert[1](✉), and Quang Mung Le[2]

[1] Matrosov Institute for System Dynamics and Control Theory of SB RAS, 664033 Irkutsk, Russia
kazakov@icc.ru, lempert.a.a@gmail.com

[2] Irkutsk National Research Technical University (Baikal School of BRICS), 664074 Irkutsk, Russia
quangmungle2010@gmail.com
http://www.idstu.irk.ru, http://www.istu.edu

Abstract. The paper is devoted to the circle covering problem with unequal circles. The number of circles is given. Also, we know a function, which determines a relation between the radii of two neighboring circles. The circle covering problem is usually studied in the case when the distance between points is Euclidean. We assume that the distance is determined by means of some special metric, which, generally speaking, is not Euclidean. The special numerical algorithm is suggested and implemented. It based on optical-geometric approach, which is developed by the authors in recent years and previously used only for circles of the equal radius. The results of a computational experiment are presented and discussed.

Keywords: Circle covering problem · Non-Euclidean metric · Incongruent circles · Optical-geometric approach

1 Introduction

The circles packing (CPP) and covering (CCP) problems are well-known classical location problems. The aim of CPP is to pack a certain number of circles, each one with a maximal radius (not necessary the same for each circle) inside a container. The task of CCP is to answer the question: how large can be the acreage of a container that is completely covered with given circles. The shape of the container may be "simple" like a circle, a square, a rectangular, or, for example, consists of combination of line and arc segments. In most cases, we are talking about the single covering problem, which is considered in a large number of papers (see, for example, [6,13,25–27]).

It is considered that the problem of covering is more complicated than the packing one. Optimal packings of equal circles in a unit square were found up to 36 circles [19,20,24], and optimal coverings were proved only up to 12 circles [12,14].

I. Bykadorov et al. (Eds.): MOTOR 2019, CCIS 1090, pp. 195–206, 2019.
https://doi.org/10.1007/978-3-030-33394-2_15

Optimal packings in a unit circle are known up to 19 circles [9–11,21], and optimal coverings were obtained up to 11 circles [29].

There are a lot of computational results for the circle covering problem. In part, the best packings of equal circles in regular polygons were found by computational methods up to 10,000 circles, and the best coverings were found only up to 100 circles.

Algorithms for covering of simply connected sets by congruent circles employing quasi-differentiability of the objective function are presented in [15], heuristic and metaheuristic methods can be found in [1,2,32], algorithms of integer and continuous optimization are proposed by [5,22,23,25], geometric methods are suggested in [31]. A modification of feasible directions' method appears in [26], where optimal coverings are given for different $n \leq 100$.

There are considerably fewer works devoted to covering the plane with different circles. The article [30] states that this problem was first investigated in 1958 by Toth and Molnar. They presented a hypothesis about the lower bound of the covering density. Then, this hypothesis was proven by Toth in 1995 [28]. Florian and Heppes [8] established a sufficient condition for such a covering to be solid in the sense of [30]. Banhelyi et al. [3] suggested a special branch and bound algorithm for CCP with prescribed centres. It allows one to check if a given polygon is covered by a set of circles. Dorninger [7] presented an analytical description the general case (covering by unequal circles) in such a way that the conjecture can easily be numerically verified and upper and lower limits for the asserted bound can be gained.

Note that the most of known results are obtained for the case when a covered set is a subset of the Euclidean plane or a multi-dimensional Euclidean space. In the case of a non-Euclidean metric, covering and packing problems are relatively poorly studied.

In this paper, we expand a technique proposed in [16,17] for solving the problem of covering a simply connected container by unequal circles when the distance between two points is defined as minimal time of moving from one of them to another. This formulation appears in logistics when the time of moving is more important than the physical distance. The suggested algorithm is based on the fundamental physical principles of Fermat and Huygens, which makes it possible to use the non-Euclidean metric [16].

2 Formulation

Assume we are given a metric space X, a bounded domain $M \subset X$ with a continuous boundary ∂M, and n of covering circles $C_i(O_i, r_i)$ with centers $O_i = (x_i, y_i)$, $i = 1, ..., n$ and radii r_i. Let $0 < f(x,y) \leq \beta$ be a continuous function, which makes sense of the instantaneous speed of movement at every point of M. The distance in space X is determined as follows:

$$\rho(a,b) = \min_{\Gamma \in G(a,b)} \int\limits_{\Gamma} \frac{d\Gamma}{f(x,y)}, \tag{1}$$

where $G(a, b)$ is the set of all continuous curves, which belong to X and connect the points a and b. It is easy to see, that in this case the shortest route between two points is a curve, that requires the least time to be spent.

We have the following minimization problem:

$$r_1 \to \min, \tag{2}$$

$$r_i = i^k r_1, k \in R, \tag{3}$$

$$M \subseteq \bigcup_{i=1}^{n} C_i(O_i, r_i). \tag{4}$$

The objective function (2) minimizes the radius associated with the circles. Constraint (3) fixes the radius ratio, and (4) guarantees that each point of M belongs to at least one circle.

Note, if $k = 0$, we have the classic circle covering problem.

3 Solution Method

In this section, we propose a heuristic method for solving problem (2)–(4), based on the analogy between the propagation of the light wave and finding the minimum of integral functional (1). This analogy is a consequence of physical principles of Fermat and Huygens. This approach is described more detail in [16,18].

The following algorithm includes the basic steps: constructing the Dirichlet tessellation for the initial set of centers; moving O_i to the point O_i^* that is the center of the covering circle, which has the minimal radius for each part of the tessellation; revising radius ratio and returning to the first step with the new centers. Now we describe the idea in details:

Algorithm

1. Randomly generate initial coordinates of the circles centers $O_i \in M, i = \overline{1, n}$.
2. From $O_i, i = \overline{1, n}$, we initiate the light waves using the algorithm [16]. It allows us to divide set M on n segments M_i and to find their boundaries ∂M_i, $i = \overline{1, n}$. Note, that because of unequal radii we have to deal with different waves. The wave velocity v_i directly depends on the source number $v_i = i^k$.
3. Boundary ∂M_i of segment M_i is approximated by the closed polygonal line with nodes at the points $A_l, l = \overline{1, q}$.
4. From $A_l, l = \overline{1, q}$, we initiate the light waves using the algorithm [16] as well.
5. Every point $(x, y) \in D_i$, first reached by one of the light waves is marked (here and further on, we assume using an analytical grid for x and y). We memorize time $T(x, y)$, which is required to reach (x, y).
6. Find $\bar{O}_i = \arg\max_{(x,y) \in M_i} T(x, y)$. Then, the minimum radius of a circle that covers M_i, is given by

$$r_{i\,\min} = \max_{l=\overline{1,q}} \rho(\bar{O}_i, A_l).$$

Steps 3–6 are carried out independently for each segment $M_i, i = \overline{1,n}$. As a result we obtain a covering

$$P = \bigcup_{i=1}^{n} C_i(\bar{O}_i, r_{i\,\min}).$$

To ensure equality (3) we should include in the algorithm an additional procedure:

7. Fix radius of the first circle $r_{1\,\min}$ and calculate radii of other circles according to (3). Then we obtain set

$$S_1 = \left\{r_1^1 = r_{1\,\min}, r_2^1 = 2^k r_{1\,\min}, ..., r_2^1 = n^k r_{1\,\min}\right\}.$$

Similarly, we construct sets for the remaining circles

$$S_j = \{r_i^j\}, r_i^j = \frac{i^k}{j^k} r_{i\,\min}, \quad i, j = \overline{1,n}.$$

8. Find $j^* = \arg\min\limits_{j=1,\ldots,n} r_1^j$ and check inequalities $r_i^{j^*} \geq r_{i\,\min}$, $i = \overline{1,n}$. If the inequalities hold, then the current covering

$$\bar{P} = \bigcup_{i=1}^{n} C_i(\bar{O}_i, r_i^{j^*})$$

is memorized as a solution.
Otherwise, find $q^* = \arg\min\limits_{j=1,\ldots,n, j \neq j^*} r_1^j$ and so on. It is easy to see, that at least one of S_j satisfies (3). Then go to step 2 with $O_i = \bar{O}_i$, $i = \overline{1,n}$.
Steps 2–8 are being carried out as long as radii decreases.
9. The counter of initial coordinates generations $Iter$ is incremented. If $Iter$ becomes equal to some preassigned value, then the algorithm is terminated and $P^* = \bar{P}$ be a solution. Otherwise, go to step 1.

A drawback of the algorithm is that it does not guarantee a solution that globally minimizes the circles radii. This feature is inherited from the constructing of Voronoi diagram. We use multiple generating of initial positions (Step 1) to increase the probability of finding a global solution.

4 Computational Experiment

Testing of the algorithm proposed in the previous section was carried out using the PC of the following configuration: Intel (R) Core i5-3570K (3.4 GHz, 8 GB RAM) and Windows 10 operating system. The algorithm is implemented in C# using the Visual Studio 2013.

Example 1. This example illustrates how the proposed in the previous section algorithm works in the case of the Euclidean metric $f(x, y) \equiv 1$. The covered set is the unit square, $r_i = \sqrt{i}$. The number of random generations of initial positions $Iter = 25$. The best solutions are shown in Table 1. Here and further n is a number of circles, r_1 is the best radius of the first circle, r_{av} is the best average radius of the covering, R is the best of known radius of equal circles covering given from [26], $\Delta r = \frac{R - r_{av}}{R}$, t is executing time in seconds.

From Table 1 you can see that, despite the strict condition on the ratio of the radii, the average radius of covering with unequal circles is significantly less than with equal ones. From a practical point of view, this means that if we assume that the cost of opening a logistics facility is directly proportional to its service radius, then to minimize costs, it is preferable to use covering with unequal circles. In addition, the operating time of the proposed algorithm is not very long.

Table 1. Covering of a unit square by incongruent circles with Euclidean metric

n	r_1	r_{av}	R	$\Delta r(\%)$	t
2	0.70640003	0.60295014	0.55901699	−7.86	11
3	0.55769705	0.42467832	0.50389111	15.72	17
4	0.51618795	0.35932579	0.35355339	−1.63	25
5	0.46641399	0.30145928	0.32616058	7.57	34
6	0.44904231	0.27241293	0.29872706	8.81	47
7	0.43561451	0.25003547	0.27429189	8.84	55
8	0.42131935	0.23022136	0.26030011	11.56	71
9	0.41160175	0.21516574	0.23063693	6.71	86
10	0.40459856	0.20314885	0.21823351	6.91	102
11	0.39910399	0.19311224	0.21251602	9.13	109
12	0.39503544	0.18471806	0.20227589	8.68	122
13	0.38898972	0.17619841	0.19431237	9.32	146
14	0.38594560	0.16970016	0.18551055	8.52	184
15	0.37985769	0.16242701	0.17966176	9.59	202
16	0.37210886	0.15498321	0.16942705	8.53	235
17	0.36783828	0.14944036	0.16568093	9.80	251
18	0.36434208	0.14456755	0.16063966	10.01	283
19	0.36388468	0.14118052	0.15784198	10.56	311
20	0.36203423	0.13748712	0.15224681	9.69	353
21	0.35913406	0.13362306	0.14895379	10.29	346
22	0.35894706	0.13096140	0.14369318	8.86	381
23	0.35546027	0.12727314	0.14124482	9.89	403

(*continued*)

Table 1. (*continued*)

n	r_1	r_{av}	R	$\Delta r(\%)$	t
24	0.35454054	0.12466993	0.13830288	9.86	469
25	0.35313405	0.12203341	0.13354871	8.62	503
26	0.34935082	0.11871785	0.13176488	9.90	536
27	0.34901851	0.11669988	0.12863353	9.28	590
28	0.34738931	0.11435139	0.12731755	10.18	606
29	0.34618477	0.11224213	0.12555351	10.60	623
30	0.34505797	0.11024752	0.12203687	9.66	650
40	0.32937213	0.09278123	0.10546620	12.03	1106
50	0.32736524	0.08349368	0.09308878	10.31	1828
60	0.31629417	0.07430825	0.08434634	11.90	2503
70	0.31502516	0.06900192	0.07842673	12.02	3090
100	0.30895469	0.05743345	0.06481289	11.39	5631

Figure 1 shows the dependence of the first circle radius on the number of circles.

Example 2. Now we turn to non-Euclidean metrics. The covered set M is following

$$M = \{(x, y) : (x - 2.5)^2 + (y - 2.5)^2 \leq 4\}.$$

The radii ratio $r_i = \sqrt{i}$, the instantaneous speed of movement $f(x, y) = x + y + 1$ increases linearly with both coordinates. The number of random generations of initial positions $Iter = 25$. The best solutions are shown in Table 2.

Figure 5 shows the solutions associated with Table 2 for $n = 10, 15$.

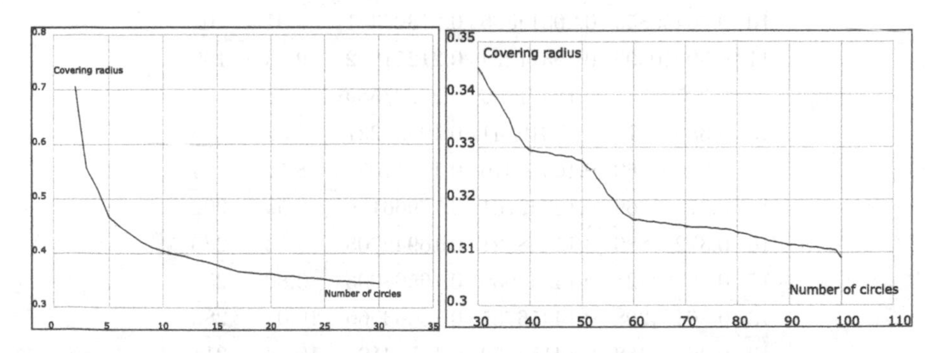

Fig. 1. The dependence of the first circle radius on the number of circles for $n \leq 30$ (left) and $30 \leq n \leq 100$ (right).

$$S_{10} = \{(3.680, 3.941), (3.979, 3.190), (1.959, 2.821), (2.318, 1.200), (1.769, 4.227), (0.889, 2.818), (3.372, 2.042), (1.108, 1.802), (2.623, 3.868), (1.262, 3.527)\},$$

$$S_{15} = \{(1.198, 3.984), (4.125, 2.721), (3.349, 4.050), (2.180, 3.441), (3.849, 3.491),$$
$$(1.158, 1.182), (3.030, 3.061), (0.722, 2.851), (3.500, 1.801), (0.990, 2.192),$$
$$(1.802, 2.924), (2.591, 4.102), (1.083, 3.459), (2.393, 1.621), (1.783, 4.072)\}.$$

Borovskikh [4] proved that in this case the wave fronts also have the form of a circle, as in the Euclidean metric, but the source of the wave (the center of the circle) is displaced. Recall that the radius here means the time of moving from center to the boundary of the circle (Fig. 2).

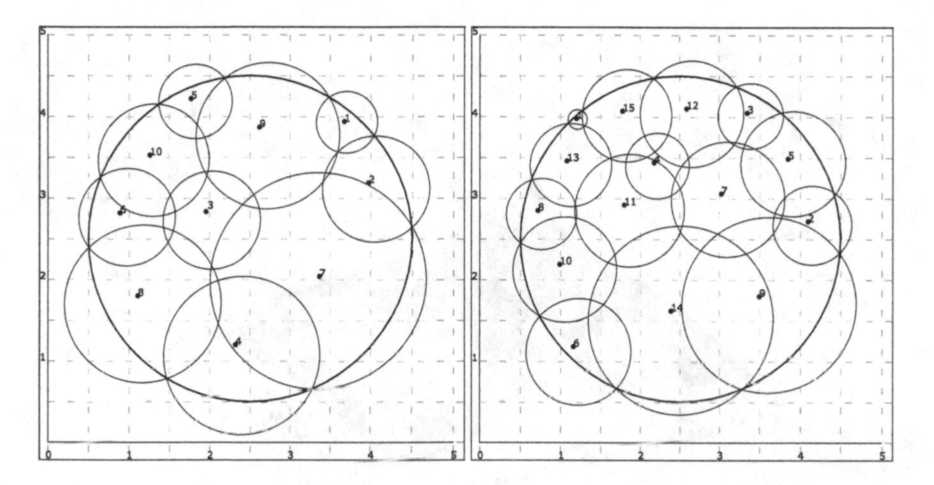

Fig. 2. The best covering of the circle by 10 (left) and 15 (right) incongruent circles with a linear metric

Table 2. The best coverings for Example 2

n	r_1	t
5	0.16338	51
10	0.07871	143
15	0.05336	301
20	0.04042	435
25	0.03266	568
30	0.02725	703

Example 3. The covered set M and the radii ratio r_i are the same as in the Example 2. The instantaneous speed of movement

$$f(x,y) = \frac{(x-2)^2 + (y-2.5)^2 + 4}{(x-2)^2 + (y-2.5)^2 + 1} + 0.5.$$

Figure 3 shows function $f(x,y)$ level lines. From the lowest to the highest point, the speed of the wave increases (Table 3).

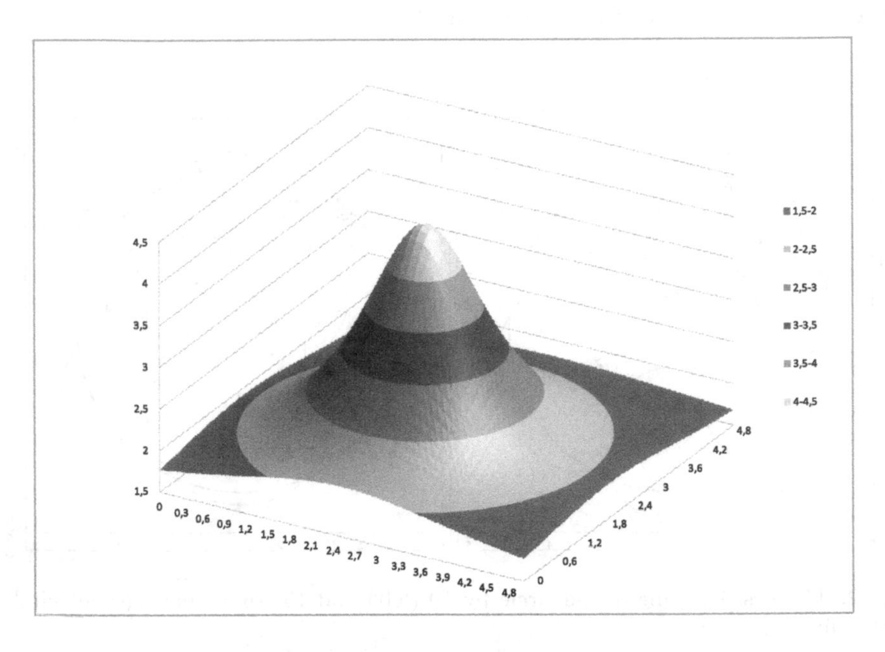

Fig. 3. Level lines of function $f(x,y)$

Table 3. The best coverings for Example 3

n	r_1	t
5	28.94242	65
10	14.03740	161
15	9.44260	290
20	7.06655	417
25	5.63607	605
30	4.74689	731

The best solutions are shown in Table 4.

Figure 4 shows the solutions associated with Table 2 for $n = 10, 15$. Note that here the boundary of the covering circle is a continuous closed line, and its shape is not known in advance.

$$S_{10} = \{(3.439, 4.212), (1.531, 4.158), (1.190, 1.189), (4.222, 2.320), (0.819, 2.161),$$
$$(3.692, 1.253), (3.841, 3.360), (1.162, 3.239), (2.443, 1.649), (2.508, 3.833)\},$$
$$S_{15} = \{(2.907, 3.727), (2.173, 0.639), (3.265, 4.122), (2.961, 2.584), (4.355, 2.760),$$
$$(0.831, 2.959), (2.682, 4.218), (1.920, 4.138), (1.420, 1.158), (1.300, 3.603),$$
$$(2.311, 3.064), (3.855, 1.976), (3.743, 3.507), (1.153, 2.169), (3.109, 1.213)\}.$$

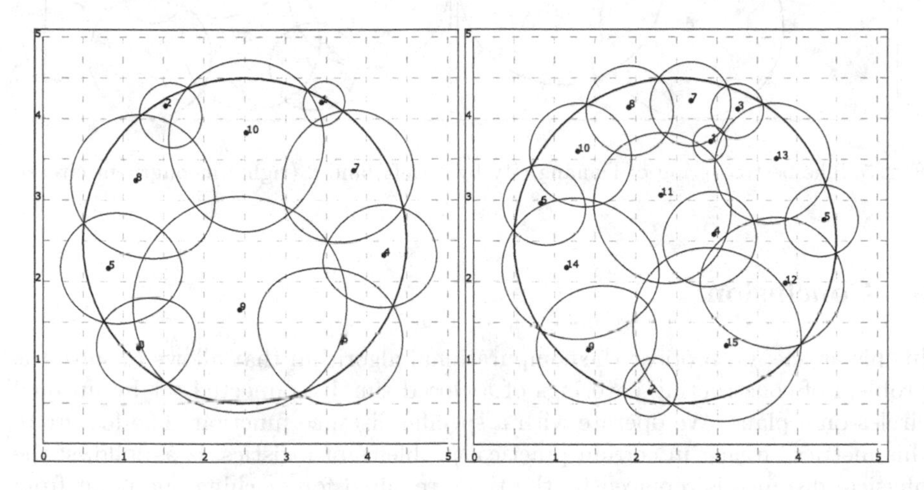

Fig. 4. The best covering of the circle by 10 (left) and 15 (right) incongruent circles

Table 4. The best covering of Danang city

Number of BS	The minimum radius of the service area of the first BS (km)	The total area of all service areas of the BS system (km^2)	Executed time (s)
2	10.616	1 062.155	19
3	7.049	936.535	32
4	5.625	994.186	55
5	4.444	930.543	81
6	3.777	941.137	106
7	3.263	936.537	139
8	3.034	1 041.076	185

Example 4. The problem of locating the system of base transceiver stations (BS) with minimal capacity covering the city of Da Nang (Vietnam) is considered. The Table 4 below presents the calculation results. In this case, it is proposed that the BS signal propagates in a straight line at the same speed throughout the city. It is easy to see, that the covering by 5 circles gives the minimal total area (Fig. 5).

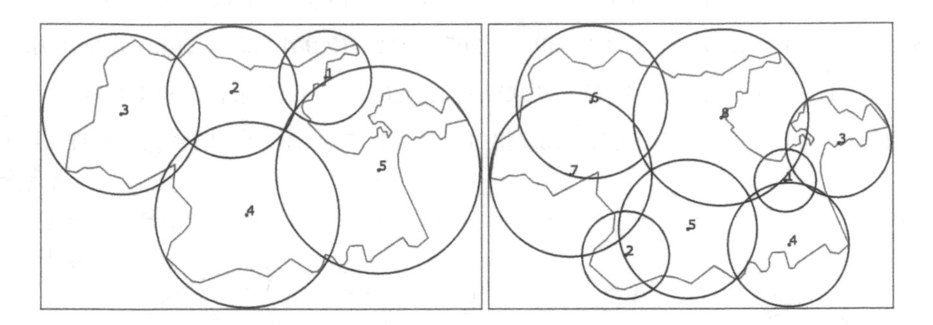

Fig. 5. The best covering of Danang city by 5 (left) and 8 (right) incongruent circles

5 Conclusion

In this article, we propose the computational algorithm that allows to solve the problem of constructing coverings of a closed simply-connected set by unequal circles on a plane. We operate with a specific distance function. The feature of this metrics, arising in certain practical problems of logistics, is as follows: the physical distance is replaced by the time, required for reaching one point from another.

The results of the computational experiment show that the algorithm is sensitive to the initial positions, but it is applicable to solve the considered problem. In this regard, further research will be aimed at creating algorithms with special procedure of initial positions generation and adaptive grids.

Acknowledgements. This work was partially supported by the Russian Foundation for Basic Research, research project No. 18-08-00604.

References

1. Al-Sultan, K., Hussain, M., Nizami, J.: A genetic algorithm for the set covering problem. J. Oper. Res. Soc. **47**(5), 702–709 (1996)
2. Azimi, Z., Toth, P., Galli, L.: An electromagnetism metaheuristic for the unicost set covering problem. Eur. J. Oper. Res. **205**(2), 290–300 (2010)
3. Bánhelyi, B., Palatinus, E., Lévai, B.: Optimal circle covering problems and their applications. CEJOR **23**(4), 815–832 (2015)

4. Borovskikh, A.: The two-dimensional eikonal equation. Siberian Math. J. **47**(2), 813–834 (2006)
5. Brusov, V., Piyavskii, S.: A computational algorithm for optimally covering a plane region. USSR Comput. Math. Math. Phys. **11**(2), 17–27 (1971)
6. Conway, J., Sloane, N.: Sphere Packing. Lattices and Groups. Springer Science and Business Media, New York (1999)
7. Dorninger, D.: Thinnest covering of the euclidean plane with incongruent circles. Anal. Geom. Metr. Spaces **5**(1), 40–46 (2017)
8. Florian, A., Heppes, A.: Solid coverings of the euclidean plane with incongruent circles. Discrete Comput. Geom. **23**(2), 225–245 (2000)
9. Fodor, F.: The densest packing of 19 congruent circles in a circle. Geom. Dedicata. **74**(2), 139–145 (1999)
10. Fodor, F.: The densest packing of 12 congruent circles in a circle. Beitr. Algebra Geom. **41**(2), 401–409 (2000)
11. Fodor, F.: The densest packing of 13 congruent circles in a circle. Contrib. Algebra Geom. **44**(2), 431–440 (2003)
12. Friedman, E.: Circles covering squares. http://www2.stetson.edu/~efriedma/circovsqu/. Accessed 7 Feb 2019
13. Heppes, A., Melissen, J.: Covering a rectangle with equal circles. Periodica Math. Hung. **34**, 65–81 (1997)
14. Heppes, A., Melissen, J.: Covering a rectangle with equal circles. Periodica Math. Hung. **34**(1–2), 65–81 (1997)
15. Jandl, H., Wieder, A.: A continuous set covering problem as a quasi differentiale optimization problem. Optim. J. Math. Program. Oper. Res. **19**(6), 781–802 (1988)
16. Kazakov, A., Lempert, A.: An approach to optimization in transport logistics. Autom. Remote Control **72**(7), 1398–1404 (2011)
17. Kazakov, A., Lempert, A.: On mathematical models for optimization problem of logistics infrastructure. Int. J. Artif. Intell. **13**(1), 200–210 (2015)
18. Kazakov, A., Lempert, A., Le, Q.: An algorithm for packing circles of two types in a fixed size container with non-euclidean metric. In: Supplementary Proceedings of the Sixth International Conference on Analysis of Images, Social Networks and Texts, vol. 1975, pp. 281–292. CEUR-WS (2017)
19. Kirchner, K., Wengerodt, G.: Die dichteste packung von 36 kreisen in einem quadrat. Beitr. Algebra Geom. **25**, 147–159 (1987)
20. Markot, M.: Interval methods for verifying structural optimality of circle packing configurations in the unit square. J. Comput. Appl. Math. **199**, 353–357 (2007)
21. Melissen, J.: Densest packings of eleven congruent circles in a circle. Geom. Dedicata. **50**, 15–25 (1994)
22. Melissen, J., Schuur, P.: Covering a rectangle with six and seven circles. Discrete Appl. Math. **99**, 149–156 (2000)
23. Nurmela, K.: Conjecturally optimal coverings of an equilateral triangle with up to 36 equal circles. Exp. Math. **9**(2), 241–250 (2000)
24. Nurmela, K., Ostergard, P.: Packing up to 50 circles in a square. Discrete Comput. Geom. **18**, 111–120 (1997)
25. Nurmela, K., Ostergard, P.: Covering a square with up to 30 equal circles. Technical report Res. rept A62., Lab. Technol. Helsinki Univ. (2000)
26. Stoyan, Y., Patsuk, V.: Covering a compact polygonal set by identical circles. Comput. Optim. Appl. **46**, 75–92 (2010)
27. Tarnai, T., Gaspar, Z.: Covering a square by equal circles. Elem. Math. **50**, 167–170 (1995)

28. Toth, G.: Covering the plane with two kinds of circles. Discrete Comput. Geom. **13**, 445–457 (1995)
29. Toth, G.: Thinnest covering of a circle by eight, nine, or ten congruent circles. Comb. Comput. Geom. **52**, 361–376 (2005)
30. Toth, L.F.: Solid circle-packings and circle-coverings. Studia Sci. Math. Hungar. **3**, 401–409 (1968)
31. Ushakov, V., Lebedev, P.: Algorithms of optimal set covering on the planar R^2. Vestn. Udmurtsk. Univ. Mat. Mekh. Komp. Nauki **26**(2), 258–270 (2016)
32. Watson-Gandy, C.D.T.: An electromagnetism metaheuristic for the unicost set covering problem. Eur. J. Oper. Res. **205**(2), 290–300 (2010)

The Problem K-Means and Given J-Centers: Polynomial Solvability in One Dimension

Alexander Kel'manov[1,2] and Vladimir Khandeev[1,2](✉)

[1] Sobolev Institute of Mathematics, 4 Koptyug Ave., 630090 Novosibirsk, Russia
{kelm,khandeev}@math.nsc.ru
[2] Novosibirsk State University, 2 Pirogova St., 630090 Novosibirsk, Russia

Abstract. We consider a problem of partitioning a finite set of points in Euclidean space into clusters so as to minimize the sum over all clusters of the intracluster sums of the squared distances between clusters elements and their centers. The centers of one part of the clusters are given as an input, while the centers of the other part of the clusters are defined as centroids (geometrical centers). It is known that in the general case this problem is strongly NP-hard. We prove constructively that the one-dimensional case of this problem is solvable in polynomial time. This result is based, first, on the proved properties of the problem optimal solution and, second, on the justified dynamic programming scheme.

Keywords: Minimum sum-of-squares clustering · Euclidean space · Strongly NP-hard problem · One-dimensional case · Polynomial-time algorithm

1 Introduction

The subject of this study is one strongly NP-hard problem of partitioning a finite set of points in Euclidean space into clusters. Our goal is to analyze the computational complexity of the problem in the one-dimensional case. The research is motivated by the openness of the specified mathematical question, as well as by the importance of the problem for some applications, in particular, for Data analysis, Data mining, Pattern recognition, Machine learning, and Big data processing.

The paper has the following structure. In Sect. 2, we give the problem formulation and establish a connection with a well-known problem that is the closest to we consider one. The next section presents auxiliary statements that reveal the structure of the optimal solution to the problem. These statements allow us to prove the main result. In Sect. 4, we prove our main result, i.e., the polynomial solvability of the problem in the one-dimensional case.

I. Bykadorov et al. (Eds.): MOTOR 2019, CCIS 1090, pp. 207–216, 2019.
https://doi.org/10.1007/978-3-030-33394-2_16

2 Problem Formulation, Its Sources and Related Problems

In the well-known clustering *K-Means* problem, an N-element set $\mathcal{Y}$ of points in d-dimension Euclidean space and a positive integer K are given. It is required to find a partition of the input set $\mathcal{Y}$ into non-empty clusters $\mathcal{C}_1, \ldots, \mathcal{C}_K$ minimizing the sum

$$\sum_{j=1}^{K} \sum_{y \in \mathcal{C}_k} \|y - \overline{y}(\mathcal{C}_k)\|^2,$$

where $\overline{y}(\mathcal{C}_k) = \frac{1}{|\mathcal{C}_k|} \sum_{y \in \mathcal{C}_k} y$ is the centroid of the k-th cluster.

Another common name of *K-Means* problem is *MSSC* (*Minimum Sum-of-Squares Clustering*). In statistics, this problem is known from the last century and is associated with Fisher (see, for example, [1,2]). In practice (in a wide variety of applications), this problem arises when there is the following hypothesis on a structure of some given numerical data. Namely, one has assumption that the set $\mathcal{Y}$ of sample (input) data contains K homogeneous clusters (subsets) $\mathcal{C}_1, \ldots, \mathcal{C}_K$, and in all clusters, the points are scattered around the corresponding unknown mean values $\overline{y}(\mathcal{C}_1), \ldots, \overline{y}(\mathcal{C}_K)$. However, the correspondence between points and clusters is unknown. Obviously, in this situation, for the correct application of classical statistical methods (hypothesis testing or parameter estimating) to the processing of sample data, at first it is necessary to divide the data into homogeneous groups (clusters). This situation is typical, in particular, for the above-mentioned applications.

The authors of [3] relatively recently proved *K-Means* strong NP-hardness. The paper [4] proposed the result on the problem solvability in exponential $\mathcal{O}(N^{dK+1})$ time. The authors of [5] proved that the problem is NP-hard even on a plain (when $d = 2$). However, even earlier in the last century, the authors of [6] proved the polynomial solvability of this problem on a line. The cited paper [6] presents an algorithm that implements a dynamic programming scheme. This algorithm running time is $\mathcal{O}(KN^2)$. This well-known algorithm relies on an exact polynomial algorithm for solving the well-known *Nearest neighbor search* problem [7]. In recent years, for the one-dimensional case of the *K-Means* problem, a number of exact algorithms with improved running time have been constructed. The paper [8] presents an overview of these algorithms and their properties.

The object of our research is the following problem that is close in its formulation to *K-Means* and is poorly studied.

Problem 1 (*K-Means and Given J-Centers*). *Given* an N-element set $\mathcal{Y}$ of points in d-dimension Euclidean space, a positive integer K, and a tuple $\{c_1, \ldots, c_J\}$ of points. *Find* a partition of $\mathcal{Y}$ into non-empty clusters $\mathcal{C}_1, \ldots, \mathcal{C}_K$, $\mathcal{D}_1, \ldots, \mathcal{D}_J$ such that

$$F = \sum_{k=1}^{K} \sum_{y \in \mathcal{C}_k} \|y - \overline{y}(\mathcal{C}_k)\|^2 + \sum_{j=1}^{J} \sum_{y \in \mathcal{D}_j} \|y - c_j\|^2 \to \min,$$

where $\overline{y}(\mathcal{C}_k)$ is the centroid of the k-st cluster.

On the one hand, Problem 1 may be considered as some modification of K-Means. On the other hand, the introduced notation allows us to call Problem 1 as *K-Means and Given J-Centers*.

Unlike *K-Means*, Problem 1 models an applied clustering problem in which for a part of clusters (i.e., for $\mathcal{D}_1, \ldots, \mathcal{D}_J$) the quadratic scatter data centers (i.e., $c_1, \ldots, c_J$) are known in advance, i.e., they are given as input instance. This applied problem is also typical for data analysis and its interpretation, pattern recognition, and machine learning. In particular, the two-cluster Problem 1, i.e., 1-*Mean and Given* 1-*Center*, is related to the solution of the applied signal processing problem. Namely, this two-clusters problem is related with the problem of joint detecting a quasi-periodically repeated pulse of unknown shape in a pulse train and evaluating this shape under Gaussian noise with given zero value (see [9–11]). In this two-cluster Problem 1, the zero mean corresponds to the cluster with the center specified at the origin. Apparently, the paper [9] mentioned this two-cluster Problem 1 for the first time. It should be noted that simpler optimization problems induced by the applied problems of noise-proof detection and discrimination of impulses of specified shapes are typical, in particular, for radar, electronic reconnaissance, hydroacoustics, geophysics, technical and medical diagnostics, and space monitoring (see, for example, [12–14]).

The strong NP-hardness of problem 1-*Mean and Given* 1-*Center* follows from [15,16]. From this result follows the strong NP-hardness of Problem 1 in the general case when K and J are the part of the input. Obviously, Problem 1 is not easier than *K-Means*. However, the solvability question of Problem 1 on a line until now remained open.

Our main result is the proof of Problem 1 polynomial solvability in the one-dimensional case.

3 Auxiliary Statements

Everywhere further we assume that $d = 1$. Let Problem 1D be the one-dimensional case of Problem 1.

Our proof is based on the few given below auxiliary statements, which reveal the structure of Problem 1 optimal solution.

Denote by $\mathcal{C}_1^*, \ldots, \mathcal{C}_K^*$, $\mathcal{D}_1^*, \ldots, \mathcal{D}_J^*$ the optimal clusters in Problem 1.

Lemma 1. *If in Problem 1D $c_m < c_\ell$, where $1 \le m \le J$, $1 \le \ell \le J$, then for each $x \in \mathcal{D}_m^*$ and $z \in \mathcal{D}_\ell^*$ the inequality $x \le z$ holds.*

Proof. Let there exist $x \in \mathcal{D}_m^*$ and $z \in \mathcal{D}_\ell^*$ such that $x > z$. Then, for the points x, z and c_m, c_ℓ we have

$$\begin{aligned}
&(x - c_m)^2 + (z - c_\ell)^2 - ((z - c_m)^2 + (x - c_\ell)^2) \\
&= ((x - z) + (z - c_m))^2 + (z - c_\ell) - ((z - c_m)^2 + ((x - z) + (z - c_\ell))^2) \\
&= 2(x - z)(x - c_m) - 2(x - z)(x - c_\ell) \\
&= 2(x - z)(c_\ell - c_m) > 0. \qquad (1)
\end{aligned}$$

Let $\mathcal{D}'_m = \mathcal{D}^*_m \setminus \{x\} \bigcup \{z\}$, $\mathcal{D}'_\ell = \mathcal{D}^*_\ell \setminus \{z\} \bigcup \{x\}$. Then by using (1) we get

$$\begin{aligned}
&\sum_{y\in\mathcal{D}^*_m} (y-c_m)^2 + \sum_{y\in\mathcal{D}^*_\ell} (y-c_\ell)^2 \\
&\quad = \sum_{y\in\mathcal{D}^*_m\setminus\{x\}} (y-c_m)^2 + \sum_{y\in\mathcal{D}^*_\ell\setminus\{z\}} (y-c_\ell)^2 + (x-c_m)^2 + (z-c_\ell)^2 \\
&\quad > \sum_{y\in\mathcal{D}^*_m\setminus\{x\}} (y-c_m)^2 + \sum_{y\in\mathcal{D}^*_\ell\setminus\{z\}} (y-c_\ell)^2 + (z-c_m)^2 + (x-c_\ell)^2 \\
&\quad = \sum_{y\in\mathcal{D}'_m} (y-c_m)^2 + \sum_{y\in\mathcal{D}'_\ell} (y-c_\ell)^2 ,
\end{aligned} \tag{2}$$

which contradicts our assumption about optimality of $\mathcal{C}^*_1, \ldots, \mathcal{C}^*_K$, $\mathcal{D}^*_1, \ldots, \mathcal{D}^*_J$. □

Lemma 2. *For the optimal solution of Problem 1D the following is true:*

(1) if $\overline{y}(\mathcal{C}^*_m) < c_\ell$, where $1 \le m \le K$, $1 \le \ell \le J$, then for each $x \in \mathcal{C}^*_m$ and $z \in \mathcal{D}^*_\ell$ the inequality $x \le z$ holds;

(2) if $\overline{y}(\mathcal{C}^*_m) > c_\ell$, where $1 \le m \le K$, $1 \le \ell \le J$, then for each $x \in \mathcal{C}^*_m$ and $z \in \mathcal{D}^*_\ell$ the inequality $x \ge z$ holds.

Proof. Let us prove the lemma for the case when $\overline{y}(\mathcal{C}^*_m) < c_\ell$. As in proof of Lemma 1, suppose that there exist such $x \in \mathcal{C}^*_m$ and $z \in \mathcal{D}^*_\ell$ for which $x > z$.

Then, in a similar way to (1), for the points x, z and $\overline{y}(\mathcal{C}^*_m)$, c_ℓ, we have

$$(x - \overline{y}(\mathcal{C}^*_m))^2 + (z - c_\ell)^2 > (z - \overline{y}(\mathcal{C}^*_m))^2 + (x - c_\ell)^2. \tag{3}$$

Let $\mathcal{C}'_m = \mathcal{C}^*_m \setminus \{x\} \bigcup \{z\}$, $\mathcal{D}'_\ell = \mathcal{D}^*_\ell \setminus \{z\} \bigcup \{x\}$. Then, in a similar way to (2), we get

$$\begin{aligned}
\sum_{y\in\mathcal{C}^*_m} (y-\overline{y}(\mathcal{C}^*_m))^2 + \sum_{y\in\mathcal{D}^*_\ell} (y-c_\ell)^2 &> \sum_{y\in\mathcal{C}'_m} (y-\overline{y}(\mathcal{C}^*_m))^2 + \sum_{y\in\mathcal{D}'_\ell} (y-c_\ell)^2 \\
&\ge \sum_{y\in\mathcal{C}'_m} (y-\overline{y}(\mathcal{C}'_m))^2 + \sum_{y\in\mathcal{D}'_\ell} (y-c_\ell)^2 ,
\end{aligned}$$

which contradicts our assumption about optimality of $\mathcal{C}^*_1, \ldots, \mathcal{C}^*_K$, $\mathcal{D}^*_1, \ldots, \mathcal{D}^*_J$.

The case when $\overline{y}(\mathcal{C}^*_m) > c_\ell$ is proved in the similar way. □

Lemma 3. *If in Problem 1D $\overline{y}(\mathcal{C}^*_m) < \overline{y}(\mathcal{C}^*_\ell)$, where $1 \le m \le K$, $1 \le \ell \le K$, then for each $x \in \mathcal{C}^*_m$ and $z \in \mathcal{C}^*_\ell$ the inequality $x \le z$ holds.*

Proof. As in proof of Lemmas 1 and 2, suppose that there exist such $x \in \mathcal{C}^*_m$ and $z \in \mathcal{C}^*_\ell$ for which $x > z$.

Then, in a similar way to (1) and (3), for the points x, z and $\overline{y}(\mathcal{C}^*_m)$, $\overline{y}(\mathcal{C}^*_\ell)$, we have

$$(x - \overline{y}(\mathcal{C}^*_m))^2 + (z - \overline{y}(\mathcal{C}^*_\ell))^2 > (z - \overline{y}(\mathcal{C}^*_m))^2 + (x - \overline{y}(\mathcal{C}^*_\ell))^2.$$

Let $\mathcal{C}'_m = \mathcal{C}^*_m \setminus \{x\} \bigcup \{z\}$, $\mathcal{D}'_\ell = \mathcal{D}^*_\ell \setminus \{z\} \bigcup \{x\}$. Then

$$\sum_{y \in \mathcal{C}^*_m} (y - \overline{y}(\mathcal{C}^*_m))^2 + \sum_{y \in \mathcal{D}^*_\ell} (y - \overline{y}(\mathcal{C}^*_\ell)^2 > \sum_{y \in \mathcal{C}'_m} (y - \overline{y}(\mathcal{C}^*_m))^2 + \sum_{y \in \mathcal{D}'_\ell} (y - \overline{y}(\mathcal{C}^*_\ell))^2$$
$$\geq \sum_{y \in \mathcal{C}'_m} (y - \overline{y}(\mathcal{C}'_m))^2 + \sum_{y \in \mathcal{D}'_\ell} (y - \overline{y}(\mathcal{C}'_\ell))^2 ,$$

which contradicts our assumption about optimality of $\mathcal{C}^*_1, \ldots, \mathcal{C}^*_K$, $\mathcal{D}^*_1, \ldots, \mathcal{D}^*_J$. □

Lemma 4. *If $d = 1$ in Problem 1, then for each $k \in \{1, \ldots, K\}$ and $j \in \{1, \ldots, J\}$ it is true that $\overline{y}(\mathcal{C}^*_k) \neq c_j$.*

Proof. Let there exist such $m \in \{1, \ldots, K\}$ and $\ell \in \{1, \ldots, J\}$ for which $\overline{y}(\mathcal{C}^*_m) = c_\ell$.

Consider arbitrary points $x \in \mathcal{C}_m$ and $z \in \mathcal{D}_\ell$. Let $\mathcal{C}'_m = \mathcal{C}^*_m \setminus \{x\} \bigcup \{z\}$, $\mathcal{D}'_\ell = \mathcal{D}^*_\ell \setminus \{z\} \bigcup \{x\}$. Then

$$\sum_{y \in \mathcal{C}^*_m} (y - \overline{y}(\mathcal{C}^*_m))^2 + \sum_{y \in \mathcal{D}^*_\ell} (y - c_\ell)^2 = \sum_{y \in \mathcal{C}'_m} (y - \overline{y}(\mathcal{C}^*_m))^2 + \sum_{y \in \mathcal{D}'_\ell} (y - c_\ell)^2. \quad (4)$$

Further, because $\mathcal{Y}$ is the set, $x \neq z$ and $\overline{y}(\mathcal{C}^*_m) \neq \overline{y}(\mathcal{C}'_m)$. Therefore for the centroids $\overline{y}(\mathcal{C}^*_m)$ and $\overline{y}(\mathcal{C}'_m)$ we have

$$\sum_{y \in \mathcal{C}'_m} (y - \overline{y}(\mathcal{C}^*_m))^2 = \sum_{y \in \mathcal{C}'_m} (y - \overline{y}(\mathcal{C}'_m) + \overline{y}(\mathcal{C}'_m) - \overline{y}(\mathcal{C}^*_m))^2$$
$$= \sum_{y \in \mathcal{C}'_m} (y - \overline{y}(\mathcal{C}'_m))^2 + 2 \sum_{y \in \mathcal{C}'_m} \left((y - \overline{y}(\mathcal{C}'_m)) \cdot (\overline{y}(\mathcal{C}'_m) - \overline{y}(\mathcal{C}^*_m))\right)$$
$$+ |\mathcal{C}'_m| \cdot (\overline{y}(\mathcal{C}'_m) - \overline{y}(\mathcal{C}^*_m))^2 . \quad (5)$$

Since $\sum_{y \in \mathcal{C}'_m} (y - \overline{y}(\mathcal{C}'_m)) = 0$, we have

$$\sum_{y \in \mathcal{C}'_m} \left((y - \overline{y}(\mathcal{C}'_m)) \cdot (\overline{y}(\mathcal{C}'_m) - \overline{y}(\mathcal{C}^*_m)\right) = 0,$$

and (5) implies

$$\sum_{y \in \mathcal{C}'_m} (y - \overline{y}(\mathcal{C}^*_m))^2 = \sum_{y \in \mathcal{C}'_m} (y - \overline{y}(\mathcal{C}'_m))^2 + |\mathcal{C}'_m| \cdot (\overline{y}(\mathcal{C}'_m) - \overline{y}(\mathcal{C}^*_m))^2$$
$$> \sum_{y \in \mathcal{C}'_m} (y - \overline{y}(\mathcal{C}'_m))^2. \quad (6)$$

Finally, by applying (6) to right part of (4), we get

$$\sum_{y \in \mathcal{C}^*_m} (y - \overline{y}(\mathcal{C}^*_m))^2 + \sum_{y \in \mathcal{D}^*_\ell} (y - c_\ell)^2 > \sum_{y \in \mathcal{C}'_m} (y - \overline{y}(\mathcal{C}'_m))^2 + \sum_{y \in \mathcal{D}'_\ell} (y - c_\ell)^2,$$

which contradicts our assumption about optimality of $\mathcal{C}^*_1, \ldots, \mathcal{C}^*_K$, $\mathcal{D}^*_1, \ldots, \mathcal{D}^*_J$. □

Lemma 5. *For each $k, j \in \{1, \ldots, K\}$ such that $k \neq j$, the following holds in Problem 1D: $\overline{y}(\mathcal{C}_k^*) \neq \overline{y}(\mathcal{C}_j^*)$.*

Proof. As in proof of Lemma 4, let us assume that there exist such $m, \ell \in \{1, \ldots, K\}$ for which $\overline{y}(\mathcal{C}_m^*) = \overline{y}(\mathcal{C}_\ell^*)$.

Consider arbitrary points $x \in \mathcal{C}_m$ and $z \in \mathcal{C}_\ell$. Let $\mathcal{C}'_m = \mathcal{C}_m^* \setminus \{x\} \bigcup \{z\}, \mathcal{C}'_\ell = \mathcal{C}_\ell^* \setminus \{z\} \bigcup \{x\}$. Then

$$\sum_{y \in \mathcal{C}_m^*} (y - \overline{y}(\mathcal{C}_m^*))^2 + \sum_{y \in \mathcal{C}_\ell^*} (y - c_\ell)^2 = \sum_{y \in \mathcal{C}'_m} (y - \overline{y}(\mathcal{C}_m^*))^2 + \sum_{y \in \mathcal{C}'_\ell} (y - c_\ell)^2. \quad (7)$$

In a similar way to (6), we have

$$\sum_{y \in \mathcal{C}'_m} (y - \overline{y}(\mathcal{C}_m^*))^2 > \sum_{y \in \mathcal{C}'_m} (y - \overline{y}(\mathcal{C}'_m))^2 \quad (8)$$

and

$$\sum_{y \in \mathcal{C}'_\ell} (y - \overline{y}(\mathcal{C}_\ell^*))^2 > \sum_{y \in \mathcal{C}'_\ell} (y - \overline{y}(\mathcal{C}'_\ell))^2. \quad (9)$$

Applying (8) and (9) to right part of (7), we get

$$\sum_{y \in \mathcal{C}_m^*} (y - \overline{y}(\mathcal{C}_m^*))^2 + \sum_{y \in \mathcal{C}_\ell^*} (y - \overline{y}(\mathcal{C}_\ell^*))^2 > \sum_{y \in \mathcal{C}'_m} (y - \overline{y}(\mathcal{C}'_m))^2 + \sum_{y \in \mathcal{C}'_\ell} (y - \overline{y}(\mathcal{C}'_\ell))^2,$$

which contradicts our assumption about optimality of $\mathcal{C}_1^*, \ldots, \mathcal{C}_K^*$, $\mathcal{D}_1^*, \ldots, \mathcal{D}_J^*$. □

Lemmas 1–5 establish the relative position of the optimal clusters $\mathcal{D}_1^*, \ldots, \mathcal{D}_J^*$ and $\mathcal{C}_1^*, \ldots, \mathcal{C}_K^*$ on a line.

The following theorem is true.

Theorem 1. *Let in Problem 1D points $y_1, \ldots, y_N$ of $\mathcal{Y}$ and points $c_1, \ldots, c_J$ be ordered so that*

$$y_1 < \ldots < y_N,$$

$$c_1 < \ldots < c_J.$$

Then optimal partition of $\mathcal{Y}$ into clusters $\mathcal{C}_1^*, \ldots, \mathcal{C}_K^*, \mathcal{D}_1^*, \ldots, \mathcal{D}_J^*$ corresponds to a partition of the positive integer sequence $1, \ldots, N$ into disjoint segments.

The validity of the theorem follows from Lemmas 1–5.

4 Polynomial Solvability of the Problem in 1D Case

The following theorem is the main result of the paper.

Theorem 2. *There exists a polynomial algorithm that finds the optimal solution of Problem 1D in $\mathcal{O}(KJN^2)$ time.*

Our proof of Theorem 1 is constructive. Namely, we justify an algorithm that implements a dynamic programming scheme and allows us to find an exact solution of Problem 1D in polynomial time.

The idea of the proof is as follows. Without loss of generality, we assume that the points $y_1, \ldots, y_N$ of $\mathcal{Y}$, as well as the points $c_1, \ldots, c_J$ are ordered as in Theorem 1.

Let $\mathcal{Y}_{s,t} = \{y_s, \ldots, y_t\}$, where $1 \leq s \leq t \leq N$, be a subset of $s - t + 1$ points of $\mathcal{Y}$ with numbers from s to t.

Let

$$f_{s,t}^j = \sum_{i=s}^{t} (y_i - c_j)^2, \quad j = 1, \ldots, J,$$

$$f_{s,t} = \sum_{i=s}^{t} (y_i - \overline{y}(\mathcal{Y}_{s,t}))^2,$$

where $\overline{y}(\mathcal{Y}_{s,t})$ is the centroid of the subset $\mathcal{Y}_{s,t}$.

In accordance with Theorem 1, let $\{i^*, i^* + 1, \ldots, N\}$ be the indexes of the elements of the optimal cluster containing the point y_N.

Note that for the optimal value F^* of the Problem 1D objective function it is true that

$$F^* = \begin{cases} \sum\limits_{k=1}^{K-1} \sum\limits_{y \in \mathcal{C}_k^*} (y - \overline{y}(\mathcal{C}_k))^2 + \sum\limits_{j=1}^{J} \sum\limits_{y \in \mathcal{D}_j^*} (y - c_j)^2 + f_{i^*,N}, \\ \qquad \text{if } \mathcal{C}_K^* = \{y_{i^*}, y_{i^*+1}, \ldots, y_N\}; \\ \sum\limits_{k=1}^{K} \sum\limits_{y \in \mathcal{C}_k^*} (y - \overline{y}(\mathcal{C}_k^*))^2 + \sum\limits_{j=1}^{J-1} \sum\limits_{y \in \mathcal{D}_j^*} (y - c_j)^2 + f_{i^*,N}^J, \\ \qquad \text{if } \mathcal{D}_J^* = \{y_{i^*}, y_{i^*+1}, \ldots, y_N\}. \end{cases} \tag{10}$$

Denote Problem 1D by $\langle K, J, N \rangle$. This problem is embedded in the family of subproblems $\{\langle k, j, n \rangle,\ k = 0, \ldots, K;\ j = 0, \ldots, J;\ n = 1, \ldots, N\}$, where $\langle k, j, n \rangle$ is the subproblem of searching the partition of $\{y_1, \ldots, y_n\}$ into nonempty clusters $\mathcal{C}_1, \ldots, \mathcal{C}_k$, $\mathcal{D}_1, \ldots, \mathcal{D}_j$ such that

$$\sum_{m=1}^{k} \sum_{y \in \mathcal{C}_m} (y - \overline{y}(\mathcal{C}_m))^2 + \sum_{\ell=1}^{j} \sum_{y \in \mathcal{D}_\ell} (y - c_\ell)^2 \to \min.$$

Let $F_{k,j}(n)$ be the optimal value of subproblem $\langle k, j, n \rangle$ objective function. Then in a similar way to (10),

$$F_{k,j}(n) = \begin{cases} \sum\limits_{m=1}^{k-1} \sum\limits_{y \in \mathcal{C}_m^*} (y - \overline{y}(\mathcal{C}_m))^2 + \sum\limits_{\ell=1}^{j} \sum\limits_{y \in \mathcal{D}_\ell^*} (y - c_\ell)^2 + f_{i^*,n}, \\ \qquad \text{if } \mathcal{C}_k^* = \{y_{i^*}, y_{i^*+1}, \ldots, y_n\}; \\ \sum\limits_{m=1}^{k} \sum\limits_{y \in \mathcal{C}_m^*} (y - \overline{y}(\mathcal{C}_m^*))^2 + \sum\limits_{\ell=1}^{j-1} \sum\limits_{y \in \mathcal{D}_\ell^*} (y - c_\ell)^2 + f_{i^*,n}^j, \\ \qquad \text{if } \mathcal{D}_j^* = \{y_{i^*}, y_{i^*+1}, \ldots, y_n\}, \end{cases} \tag{11}$$

where $\mathcal{C}_1^*, \ldots, \mathcal{C}_k^*$, $\mathcal{D}_1^*, \ldots, \mathcal{D}_j^*$ are the optimal clusters in subproblem $\langle k, j, n\rangle$, and $\{i^*, i^*+1, \ldots, n\}$ are the indexes of the elements of the optimal cluster containing the point y_n.

Let

$$F_{k,j}(n) = \begin{cases} 0, & \text{if } n = k = j = 0; \\ +\infty, & \text{if } n = 0;\ k = 0, \ldots, K;\ j = 0, \ldots, J;\ k + j \neq 0; \\ +\infty, & \text{if } k = -1;\ j = -1, \ldots, J;\ n = 0, \ldots, N; \\ +\infty, & \text{if } j = -1;\ k = -1, \ldots, K;\ n = 0, \ldots, N. \end{cases} \quad (12)$$

Then in accordance with (11) and Bellman's principle of optimality, we have

$$F_{k,j}(n) = \min_{i=1}^{n}\Big\{F_{k-1,j}(i-1) + f_{i,n}\Big\}, \quad k = 0, \ldots, K;\ j = 0, \ldots, J;\ n = 1, \ldots, N, \quad (13)$$

if $\mathcal{C}_k^* = \{y_{i^*}, y_{i^*+1}, \ldots, y_n\}$. If however $\mathcal{D}_j^* = \{y_{i^*}, y_{i^*+1}, \ldots, y_n\}$, then

$$F_{k,j}(n) = \min_{i=1}^{n}\Big\{F_{k,j-1}(i-1) + f_{i,n}^{j}\Big\}, \quad k = 0, \ldots, K;\ j = 0, \ldots, J;\ n = 1, \ldots, N. \quad (14)$$

Combining (13) and (14), we get that for $F_{k,j}(n)$ the following is true

$$F_{k,j}(n) = \min\Big\{\min_{i=1}^{n}\Big\{F_{k-1,j}(i-1) + f_{i,n}\Big\}, \min_{i=1}^{n}\Big\{F_{k,j-1}(i-1) + f_{i,n}^{j}\Big\}\Big\}, \quad k = 0, \ldots, K;\ j = 0, \ldots, J;\ n = 1, \ldots, N. \quad (15)$$

Thus, the optimal value of the Problem 1D objective function is found by the following formula

$$F^* = F_{K,J}(N),$$

and the values

$$F_{k,j}(n),\ k = -1, 0, 1, \ldots, K;\ j = -1, 0, 1, \ldots, J;\ n = 0, \ldots, N,$$

are calculated by the formula (12) and the recurrent formula (15). In general, the formulas (12), (15) implement the forward running of the algorithm.

Further, Bellman's principle of optimality implies that the optimal clusters $\mathcal{C}_1^*, \ldots, \mathcal{C}_K^*, \mathcal{D}_1^*, \ldots, \mathcal{D}_J^*$ may be found using the following recurrent rule, that implements the backward running of the algorithm.

The step-by-step rule looks as follows:

`Step 0.` $k := K$, $j := J$, $n := N$.

`Step 1.` If

$$\min_{i=1}^{n}\Big(F_{k-1,j}(i-1) + f_{i,n}\Big) \le \min_{i=1}^{n}\Big(F_{k,j-1}(i-1) + f_{i,n}^{j}\Big),$$

then

$$\mathcal{C}_k^* = \{y_{i^*}, y_{i^*+1}, \ldots, y_n\},$$

where

$$i^* = \arg\min_{i=1}^{n}\Big(F_{k-1,j}(i-1) + f_{i,n}\Big);$$

$k := k-1$; $n := i^* - 1$.
If, however,

$$\min_{i=1}^{n}\Big(F_{k-1,j}(i-1) + f_{i,n}\Big) > \min_{i=1}^{n}\Big(F_{k,j-1}(i-1) + f_{i,n}^j\Big),$$

then

$$\mathcal{D}_j^* = \{y_{i^*}, y_{i^*+1}, \ldots, y_n\},$$

where

$$i^* = \arg\min_{i=1}^{n}\Big(F_{k,j-1}(i-1) + f_{i,n}^j\Big);$$

$j := j-1$; $n := i^* - 1$.

Step 2. If $k > 0$ or $j > 0$, then go to Step 1; otherwise—the end of calculations.

The validity of this rule we have proved by induction.

From the above proving we have the following algorithm for Problem 1.

Algorithm $\mathcal{A}$.

Input: an N-element set $\mathcal{Y}$ of 1D points, a positive integer K, and a tuple $\{c_1, \ldots, c_J\}$ of points.

Step 1. Sort the points $y_1, \ldots, y_N$ and the points $c_1, \ldots, c_J$.

Step 2. Calculate the values $f_{s,t}^j$ and $f_{s,t}$.

Step 3. Find the optimal values of $F_{k,j}(n)$ using formulas (12) and (15).

Step 4. Find the optimal clusters $\mathcal{C}_1^*, \ldots, \mathcal{C}_K^*, \mathcal{D}_1^*, \ldots, \mathcal{D}_J^*$ using the backward rule.

Output: clusters $\mathcal{C}_1^*, \ldots, \mathcal{C}_K^*, \mathcal{D}_1^*, \ldots, \mathcal{D}_J^*$.

Finally, we have proved that the running time of the algorithm is $\mathcal{O}(KJN^2)$, that is, the algorithm is polynomial. Indeed, Step 1 requires $\mathcal{O}(N \log N)$ operations. Step 2 can be done in $\mathcal{O}(JN^2)$ time by using prefix sums. The running time of Step 3 is defined by the complexity of implementation of formula (15). This formula is calculated $\mathcal{O}(KJN)$ times and every calculation of $F_{k,j}(n)$ requires $\mathcal{O}(N)$ operations. Finally, Step 4 requires $\mathcal{O}((K+J)N)$ operations.

5 Conclusion

In the present paper, we have proved the polynomial solvability of the one-dimensional case of one strongly NP-hard problem of partitioning a finite set of points in Euclidean space. The construction of faster exact algorithms for this case seems to be the direction of future studies. The construction of approximate efficient algorithms with guaranteed accuracy bounds for the general case of the problem is also of great interest.

Acknowledgments. The study was supported by the Russian Foundation for Basic Research, projects 19-01-00308 and 18-31-00398, by the Russian Academy of Science (the Program of basic research), project 0314-2019-0015, and by the Russian Ministry of Science and Education under the 5–100 Excellence Programme.

References

1. Fisher, R.A.: Statistical Methods and Scientific Inference. Hafner, New York (1956)
2. MacQueen, J.B.: Some methods for classification and analysis of multivariate observations. In: Proceedings of the 5th Berkeley Symposium on Mathematical Statistics and Probability, vol. 1, pp. 281–297. Univ. of California Press, Berkeley (1967)
3. Aloise, D., Deshpande, A., Hansen, P., Popat, P.: NP-hardness of euclidean sum-of-squares clustering. Mach. Learn. **75**(2), 245–248 (2009)
4. Inaba, M., Katoh, N., Imai, H.: Applications of weighted Voronoi diagrams and randomization to variance-based clustering. In: Proceedings of the Annual Symposium on Computational Geometry, pp. 332–339 (1994)
5. Mahajana, M., Nimbhorkar, P., Varadarajan, K.: The Planar k-Means Problem is NP-Hard. Theoret. Comput. Sci. **442**, 13–21 (2012)
6. Rao, M.: Cluster analysis and mathematical programming. J. Amer. Stat. Assoc. **66**, 622–626 (1971)
7. Bellman, R.: Dynamic Programming. Princeton University Press, Princeton (1957)
8. Grønlund, A., Larsen, K.G., Mathiasen, A., Nielsen, J.S., Schneider, S., Song, M.: Fast exact k-Means, k-medians and bregman divergence clustering in 1D. CoRR arXiv:1701.07204 (2017)
9. Kel'manov, A.V., Khamidullin, S.A., Kel'manova, M.A.: Joint finding and evaluation of a repeating fragment in noised number sequence with given number of quasiperiodic repetitions. In: Book of abstract of the Russian Conference "Discrete Analysis and Operations Research" (DAOR-4), p. 185. Sobolev Institute of Mathematics SB RAN, Novosibirsk (2004). (in Russian)
10. Gimadi, E.Kh., Kel'manov, A.V., Kel'manova, M.A., Khamidullin, S.A.: A posteriori detection of a quasi periodic fragment in numerical sequences with given number of recurrences Siberian J. Ind. Math. **9**(1(25)), 55–74 (2006). (in Russian)
11. Gimadi, E.K., Kel'manov, A.V., Kel'manova, M.A., Khamidullin, S.A.: A posteriori detecting a quasiperiodic fragment in a numerical sequence. Pattern Recogn. Image Anal. **18**(1), 30–42 (2008)
12. Kel'manov, A.V., Khamidullin, S.A.: Posterior detection of a given number of identical subsequences in a guasi-periodic sequence. Comput. Math. Math. Phys. **41**(5), 762–774 (2001)
13. Kel'manov, A.V., Jeon, B.: A posteriori joint detection and discrimination of pulses in a quasiperiodic pulse train. IEEE Trans. Signal Process. **52**(3), 645–656 (2004)
14. Carter, J.A., Agol, E., et al.: Kepler-36: a pair of planets with neighboring orbits and dissimilar densities. Science **337**(6094), 556–559 (2012)
15. Kel'manov, A.V., Pyatkin, A.V.: On the complexity of a search for a subset of "similar" vectors. Doklady Math. **78**(1), 574–575 (2008)
16. Kel'manov, A.V., Pyatkin, A.V.: On a version of the problem of choosing a vector subset. J. Appl. Ind. Math. **3**(4), 447–455 (2009)

A Generalized Point-to-Point Approach for Orthogonal Transformations

Artyom Makovetskii[1(✉)], Sergei Voronin[1], Vitaly Kober[1,2], and Aleksei Voronin[1]

[1] Chelyabinsk State University, Chelyabinsk 454001, Russian Federation
artemmac@mail.ru
[2] CICESE, 22860 Ensenada, BC, Mexico

Abstract. The known Iterative Closest Point (ICP) algorithm utilizes point-to-point or point-to-plane approaches. The point-to-plane ICP algorithm uses points coordinates and normal vectors for aligning of 3D point clouds, whereas point-to-point approach uses point coordinates only. This paper proposes a new algorithm for orthogonal registration of point clouds based on a generalized point-to-point ICP algorithm for orthogonal transformations. The algorithm uses the known Horn's algorithm and combines point coordinates and normal vectors.

Keywords: Iterative Closest Points (ICP) · Rigid ICP · Point-to-point · Point-to-plane · Orthogonal transformations · Surface reconstruction

1 Introduction

The algorithm ICP (Iterative Closest Points) is the most important method of "alignment" of three-dimensional models based on the use of exclusively geometric characteristics. The alignment is a geometric transformation that connects two data sets (clouds) of points in R^3 in the best way with respect to the norm L_2. The algorithm is widely used to record data obtained with 3D scanners. The ICP algorithm, originally described by Besl and Mackay [1], and Chen and Medioni [2] consists of the following iteratively applied steps:

1. Selection of a subset of points in both clouds.
2. Determin the correspondence between the points of the selected subsets.
3. Compare weight coefficients obtained pairs.
4. Drop some pairs based on various criteria.
5. Select error metrics for pairs of points.
6. Minimize error metrics (variational subproblem of the ICP algorithm).

The work was supported by the Ministry of Education and Science of Russian Federation (grant N 2.1743.2017) and by the RFBR (grant N 18-07-00963).

I. Bykadorov et al. (Eds.): MOTOR 2019, CCIS 1090, pp. 217–231, 2019.
https://doi.org/10.1007/978-3-030-33394-2_17

The key element of the ICP algorithm [3] is the search for an orthogonal or affine transformation, which is the best in sense of a quadratic metric combining two clouds of points with a given correspondence between the points (variational subproblem of the ICP algorithm).

There are two main approaches to choice error metrics for pairs of points. The first approach (point-to-point) [1] uses the distance between the elements of a pair in R^3. The second approach (point-to-plane) [2] takes into account the distance between a point of the first cloud and the tangent plane to the corresponding point of the second cloud. For orthogonal transformations, the solution to this problem in closed form was obtained by Horn in [4] and [5]. In the first paper, the solution is based on the use of quaternions; in the second one, the solution is obtained with orthogonal matrices. Computational complexity of the solution is linear in time with respect to the number of pairs of points. The original ICP algorithm is widely used for registering rigid objects, but it does not work works well non-rigid objects. Variants of the ICP algorithm for affine transformations and non-rigid registration were proposed in [6–17].

It is known that the point-to-plane metric shows a better performance than that of the point-to-point metric in terms of accuracy and convergence rate [18]. In this paper we propose an approach to the ICP variational subproblem that utilizes the information about coordinates of both points and the normal vectors of the point clouds. The proposed approach is intermediate between common point-to-point and the point-to-plane methods. The results of numerical simulations show that the proposed method is more accurate than a common point-to-point ICP.

2 Formulation of the Variation Problem

Let $P = \{p^1, \ldots, p^s\}$ be a template point cloud, and $Q = \{q^1, \ldots, q^s\}$ be a target point cloud in R^3. Suppose that the relationship between points in P and Q is given in such a manner that for each point p^i exists a corresponding point q^i. Denote by $S(P)$ and $S(Q)$ the surfaces constructed from the clouds P and Q respectively; by $T_P(p^i)$ and $T_Q(q^i)$ we denote the tangent planes of $S(P)$ and $S(Q)$ at points p^i and q^i, respectively.

The ICP algorithm is commonly considered as a geometrical transformation for the rigid objects, mapping P to Q

$$Rp_i + T, \tag{1}$$

$$R = \begin{pmatrix} r_{11} & r_{12} & r_{13} \\ r_{21} & r_{22} & r_{23} \\ r_{31} & r_{32} & r_{33} \end{pmatrix}, \quad p^i = (p_1^i \quad p_2^i \ p_3^i)^t, \quad q^i = (q_1^i \quad q_2^i \quad q_3^i)^t, \tag{2}$$

where R is a rotation (orthogonal) matrix, T is a translation vector, $i = 1, \ldots, s$. Denote by $J_h(R, T)$ the following functional:

$$J_h(R, T) = \sum_{i=1}^{s} \| Rp^i + T - q^i \|^2 . \tag{3}$$

The point-to-point variational problem is formulated as follows:

$$arg \min_{R,T} J_h(R, T). \tag{4}$$

Denote by n_p^i and n_q^i the normal vectors to planes $T_P(p^i)$ and $T_Q(q^i)$, respectively, $i = 1, \ldots, s$. Consider the functional $J_g(R, T)$ as

$$J_g(R, t) = \sum_{1=1}^{s} \| Rp^i + T - q^i \|^2 + \lambda \sum_{1=1}^{s} \| Rn_p^i - n_q^i \|^2, \tag{5}$$

where λ is a parameter. Consider the generalized point-to-point variational problem as follows:

$$arg \min_{R,T} J_g(R, T). \tag{6}$$

2.1 Translation Vector Exclusion

Let us apply to all points of the cloud P the following transformation:

$$\begin{cases} (p')_1^i = p_1^i - \frac{1}{s}\sum_{1=1}^{s} p_1^i \\ (p')_2^i = p_2^i - \frac{1}{s}\sum_{1=1}^{s} p_2^i, \\ (p')_3^i = p_3^i - \frac{1}{s}\sum_{1=1}^{s} p_3^i \end{cases} \tag{7}$$

and the corresponding transformation to the points of the cloud Q,

$$\begin{cases} (q')_1^i = q_1^i - \frac{1}{s}\sum_{1=1}^{s} q_1^i \\ (q')_2^i = q_2^i - \frac{1}{s}\sum_{1=1}^{s} q_2^i, \\ (q')_3^i = q_3^i - \frac{1}{s}\sum_{1=1}^{s} q_3^i \end{cases} \tag{8}$$

where $i = 1, \ldots, s$. We get new clouds P' and Q' that are obtained from P and Q by translations. For clouds P' and Q' the functional (5) takes the form

$$J_g(R) = \sum_{1=1}^{s} \| Rp^i - q^i \|^2 + \lambda \sum_{1=1}^{s} \| Rn_p^i - n_q^i \|^2, \tag{9}$$

where p^i and q^i are the points of the clouds P' and Q', $i = 1, \ldots, s$. The variational problem (6) takes the following form:

$$arg \min_{R} J_g(R). \tag{10}$$

3 Reduction of the Variational Problem

We reduce of the variational problem (10) to such a problem that can be solved by the Horn's method [5]. Rewrite functional (9) as

$$\begin{aligned} J_g(R) &= \sum_{\imath=1}^{s} <Rp^i - q^i, Rp^i - q^i> + \lambda \sum_{\imath=1}^{s} <Rn_p^i - n_q^i, Rn_p^i - n_q^i> \\ &= \sum_{\imath=1}^{s} <Rp^i, Rp^i> - 2<Rp^i, q^i> + <q^i, q^i> \\ &\quad + \lambda \sum_{\imath=1}^{s} <Rn_p^i, Rn_p^i> - 2<Rn_p^i, n_q^i> + <n_q^i, n_q^i>, \end{aligned} \tag{11}$$

where $<\cdot,\cdot>$ denotes the inner product. Note that expressions $<q^i, q^i>$ and $<n_q^i, n_q^i>$ do not depend on R. Therefore, these expressions do not affect the variational problem (10). The functional $J_g(R)$ takes the form

$$\begin{aligned} J_g(R) &= \sum_{\imath=1}^{s} <Rp^i, Rp^i> - 2<Rp^i, q^i> \\ &\quad + \lambda \sum_{\imath=1}^{s} <Rn_p^i, Rn_p^i> - 2<Rn_p^i, n_q^i> \\ &= \sum_{\imath=1}^{s} <R^t Rp^i, p^i> - 2<Rp^i, q^i> \\ &\quad + \lambda \sum_{\imath=1}^{s} <R^t Rn_p^i, n_p^i> - 2<Rn_p^i, n_q^i>. \end{aligned} \tag{12}$$

Since R is an orthogonal matrix, we get that $R^t R = E$. The inner products $<p^i, p^i>$ and $<n_p^i, n_p^i>$ do not depend on R. It follows that the functional $J_g(R)$ takes the form

$$J_g(R) = -2\left(\sum_{\imath=1}^{s} <Rp^i, q^i> + \lambda \sum_{\imath=1}^{s} <Rn_p^i, n_q^i>\right). \tag{13}$$

The variational problem (10) can be rewritten as

$$arg \max_{R} \sum_{\imath=1}^{s} <Rp^i, q^i> + \lambda \sum_{\imath=1}^{s} <Rn_p^i, n_q^i>. \tag{14}$$

Note that the inner product $<Rp^i, q^i>$ can be expressed by the matrix trace

$$<Rp^i, q^i> = tr(R \cdot (p^i (q^i)^t)). \tag{15}$$

Denote the matrix $p^i(q^i)^t$ by M^i. It means that

$$\sum_{\imath=1}^{s} <Rp^i, q^i> = \sum_{\imath=1}^{s} tr(R \cdot (p^i(q^i)^t)) = \sum_{\imath=1}^{s} tr(R \cdot M^i)$$
$$= tr(\sum_{\imath=1}^{s}(R \cdot M^i)) = tr(R \cdot \sum_{\imath=1}^{s} M^i). \tag{16}$$

Let denote the matrix D_p as

$$D_p = \sum_{\imath=1}^{s} M^i. \tag{17}$$

Then we can write

$$\sum_{\imath=1}^{s} <Rp^i, q^i> = tr(R \cdot D_p). \tag{18}$$

Note that the following condition holds:

$$\sum_{\imath=1}^{s} <Rp^i, q^i> = tr(R \cdot D_p) = <R, (D_p)^t>. \tag{19}$$

Denote the matrix $n_p^i(n_q^i)^t$ by M_n^i, and let D_n be the following matrix:

$$D_n = \sum_{\imath=1}^{s} M_n^i. \tag{20}$$

In a similar way, we get

$$\sum_{\imath=1}^{s} <Rn_p^i, n_q^i> = tr(R \cdot D_n) = <R, (D_n)^t>. \tag{21}$$

The variational problem (14) can be rewritten as

$$arg \max_R \sum_{\imath=1}^{s} <Rp^i, q^i> + \lambda \sum_{\imath=1}^{s} <Rn_p^i, n_q^i>$$
$$= arg \max_R <R, (D_p)^t> + \lambda <R, (D_n)^t>. \tag{22}$$

Let D be the following matrix:

$$D = (D_p)^t + \lambda (D_n)^t. \tag{23}$$

Then the variational problem (22) takes the form

$$arg \max_R <R, D>. \tag{24}$$

Remark 1. The variational problem (24) can be solved by the Horn's method [5].

4 Closed Form Solution to the Variational Problem

The orthogonal matrix R in (24) can be computed in the following way [5]:

$$R = DC \begin{pmatrix} \frac{1}{\sqrt{\lambda_1}} & 0 & 0 \\ 0 & \frac{1}{\sqrt{\lambda_2}} & 0 \\ 0 & 0 & \frac{1}{\sqrt{\lambda_3}} \end{pmatrix} C^{\top}, \tag{25}$$

where C is an orthogonal matrix consisting of columns, that are eigenvectors of the matrix $D^{\top}$. Coefficients $\lambda_i, i = 1, 2, 3$, are eigenvalues of the matrix $D^{\top}D$.

4.1 Return from the Clouds P' and Q' to the Clouds P and Q

Let matrix R_* be a solution to the variational problem (24). Then R_* is also a solution to the variational problem (10)

$$arg \min_{R} \sum_{\imath=1}^{s} \| R(p')^i - (q')^i \|^2 + \lambda \sum_{\imath=1}^{s} \| Rn_p^i - n_q^i \|^2 . \tag{26}$$

Denote by v_p and v_q the following vectors:

$$v_p = (\sum_{\imath=1}^{s} p_1^i \ \sum_{\imath=1}^{s} p_2^i \ \sum_{\imath=1}^{s} p_3^i)^t, v_q = (\sum_{\imath=1}^{s} q_1^i \ \sum_{\imath=1}^{s} q_2^i \ \sum_{\imath=1}^{s} q_3^i)^t. \tag{27}$$

Rewrite (7) and (8) as

$$(p')^i = p^i - \frac{1}{s} v_p, \tag{28}$$

$$(q')^i = q^i - \frac{1}{s} v_q. \tag{29}$$

The functional in (26) can be rewritten taking into account (28) and (29) as

$$\sum_{\imath=1}^{s} \| R(p^i - \frac{1}{s} v_p) - (q^i - \frac{1}{s} v_q) \|^2 + \lambda \sum_{\imath=1}^{s} \| Rn_p^i - n_q^i \|^2$$
$$= \sum_{\imath=1}^{s} \| Rp^i - q^i + \frac{1}{s}(v_q - Rv_p) \|^2 + \lambda \sum_{\imath=1}^{s} \| Rn_p^i - n_q^i \|^2 . \tag{30}$$

In such a way, the matrix R_* minimizes both the left and the right sides of the Eq. (30). It means that the optimal translation vector T_* is defined as

$$T_* = \frac{1}{s}(v_q - R_* v_p). \tag{31}$$

The matrix R_* and the vector T_* are solutions to the variational problem (6).

5 Computer Simulation

We consider two variants of the ICP algorithm. The first one is the point-to-point ICP based on the Horn's algorithm. The second one is the generalized point-to-point ICP based on the proposed approach. Other parameters of the ICP algorithm are the same. The computational experiments are organized as follows. We apply to the points' cloud P a rigid geometrical transformation defined by an orthogonal matrix R_{true} and a translation vector T_{true}. The points' cloud Q is obtained from the point cloud P

$$Q = R_{true}P + T_{true}. \tag{32}$$

Remark 2. The information on the matrix R_{true} and the translation vector T_{true} is contained in the matrix M_{true} in homogenious coordinates.

We apply to the clouds P and Q the ICP algorithm based on the Horn method (H-PPt). The result of the H-PPt algorithm is a matrix M_h. Also we use for the clouds P and Q the ICP algorithm based on the proposed generalized point-to-point ICP algorithm (G-PPt). The result of the G-PPt algorithm is a matrix M_g. The figures show the clouds P and Q, clouds M_hP and Q, clouds M_gP and Q (here the coordinates of the points are homogeneous).

The value of the regularization parameter λ in all experiments is equal to 0.3.

The tested ICP algorithms were implemented in C++ in a conventional PC with an Intel Core i7-6700 3.4 GHz CPU and 8 GB of RAM with the same optimization and skill levels.

5.1 Experiments with Cube

Let us consider here the 3D model of the cube surface. The cloud P consists of 386 points.

1. Figure 1(a) shows the cloud P (yellow) and Q (red). The cloud Q is obtained from P by transformation M_{true}. Figure 1(b) shows the cloud M_hP (blue) and Q (red). Figure 1(c) shows the cloud M_gP (green) and Q (red).

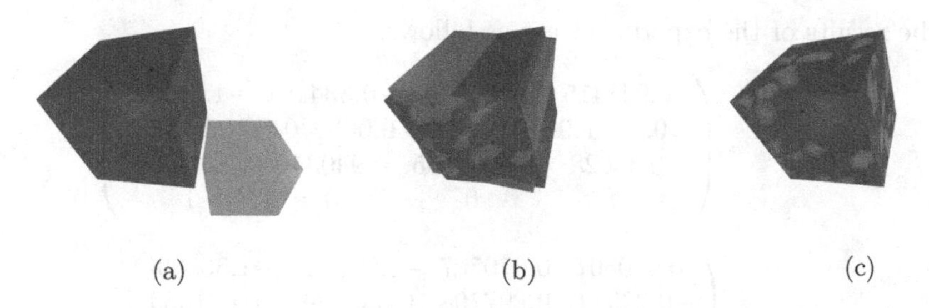

Fig. 1. (a) Test cloud P (yellow), resultant cloud Q (red); (b) alignment result of the algorithm H-PPt; (c) alignment result of the algorithm G-PPt. (Color figure online)

The results of the experiment are as follows:

$$M_{true} = \begin{pmatrix} 0.945116 & -0.229849 & 0.232219 & 2.72449 \\ 0.204079 & 0.970308 & 0.129821 & -2.766 \\ -0.255163 & -0.0753047 & 0.963961 & 1.01888 \\ 0 & 0 & 0 & 1 \end{pmatrix},$$

$$M_h = \begin{pmatrix} 0.972361 & -0.024245 & 0.232219 & 2.724490 \\ -0.006281 & 0.991518 & 0.129821 & -2.766001 \\ -0.233397 & -0.127691 & 0.963961 & 1.018876 \\ 0 & 0 & 0 & 1 \end{pmatrix},$$

$$M_g = \begin{pmatrix} 0.945116 & -0.229849 & 0.232219 & 2.724490 \\ 0.204078 & 0.970309 & 0.129821 & -2.766001 \\ -0.255163 & -0.075305 & 0.963961 & 1.018876 \\ 0 & 0 & 0 & 1 \end{pmatrix}.$$

The processing time of the H-PPt algorithm is 20 ms, the processing time of the proposed G-PPt algorithm is 20 ms.

2. Figure 2(a) shows the cloud P (yellow) and Q (red). The cloud Q is obtained from P by transformation M_{true}. Figure 2(b) shows the cloud M_hP (blue) and Q (red). Figure 2(c) shows the cloud M_gP (green) and Q (red).

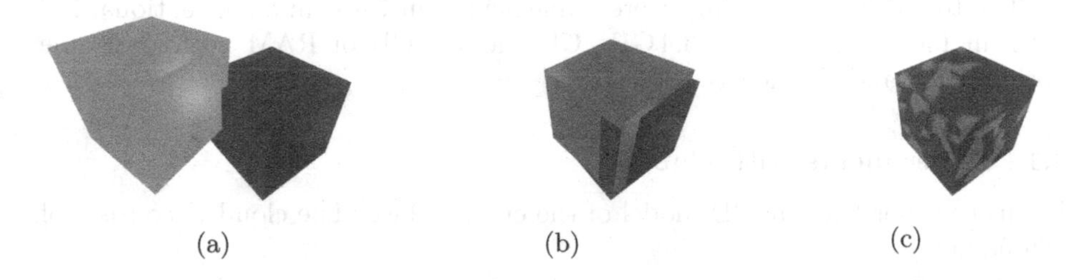

Fig. 2. (a) Test cloud P (yellow), resultant cloud Q (red); (b) alignment result of the algorithm H-PPt; (c) alignment result of the algorithm G-PPt. (Color figure online)

The results of the experiment are as follows:

$$M_{true} = \begin{pmatrix} 0.941415 & 0.0455463 & -0.334161 & -1.64269 \\ -0.0255698 & 0.997626 & 0.0639404 & -0.855317 \\ 0.33628 & -0.05165 & 0.940344 & 1.67539 \\ 0 & 0 & 0 & 1 \end{pmatrix},$$

$$M_h = \begin{pmatrix} 0.970507 & 0.970507 & -0.237248 & -1.55909 \\ -0.029447 & 0.997798 & 0.059429 & -0.691539 \\ 0.239268 & -0.05069 & 0.96963 & 1.62716 \\ 0 & 0 & 0 & 1 \end{pmatrix},$$

$$M_g = \begin{pmatrix} 0.941415 & 0.045546 & -0.33416 & -1.64269 \\ -0.02557 & 0.997626 & 0.06394 & -0.855317 \\ 0.336279 & -0.05165 & 0.940345 & 1.67539 \\ 0 & 0 & 0 & 1 \end{pmatrix}.$$

The processing time of the H-PPt algorithm is 14 ms, the processing time of the proposed G-PPt algorithm is 16 ms.

5.2 Experiments with Stanford Bunny

The cloud P consists of 34817 points.

1. Figure 3(a) shows the cloud P (yellow) and Q (red). The cloud Q is obtained from P by transformation M_{true}. Figure 3(b) shows the cloud $M_h P$ (blue) and Q (red). Figure 3(c) shows the cloud $M_g P$ (green) and Q (red).

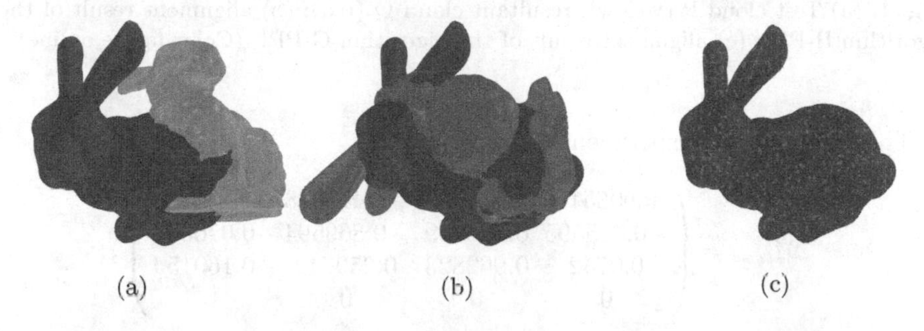

Fig. 3. (a) Test cloud P (yellow), resultant cloud Q (red); (b) alignment result of the algorithm H-PPt; (c) alignment result of the algorithm G-PPt. (Color figure online)

The results of the experiment are as follows:

$$M_{true} = \begin{pmatrix} 0.556337 & -0.778256 & -0.291217 & -0.281807 \\ 0.14472 & 0.43585 & -0.888308 & 0.442844 \\ 0.818257 & 0.452054 & 0.355109 & -0.490603 \\ 0 & 0 & 0 & 1 \end{pmatrix},$$

$$M_h = \begin{pmatrix} 0.824945 & -0.481528 & 0.295968 & -0.472856 \\ -0.351158 & -0.026325 & 0.935946 & 0.27693 \\ -0.442893 & -0.876035 & -0.190809 & 0.214402 \\ 0 & 0 & 0 & 1 \end{pmatrix},$$

$$M_g = \begin{pmatrix} 0.556337 & -0.778256 & -0.291217 & -0.281807 \\ 0.14472 & 0.43585 & -0.888308 & 0.442844 \\ 0.818257 & 0.452054 & 0.355109 & -0.490603 \\ 0 & 0 & 0 & 1 \end{pmatrix}.$$

The processing time of the H-PPt algorithm is 18463 ms, the processing time of the proposed G-PPt algorithm is 9105 ms.

2. Figure 4(a) shows the cloud P (yellow) and Q (red). The cloud Q is obtained from P by transformation M_{true}. Figure 4(b) shows the cloud $M_h P$ (blue) and Q (red). Figure 4(c) shows the cloud $M_g P$ (green) and Q (red).

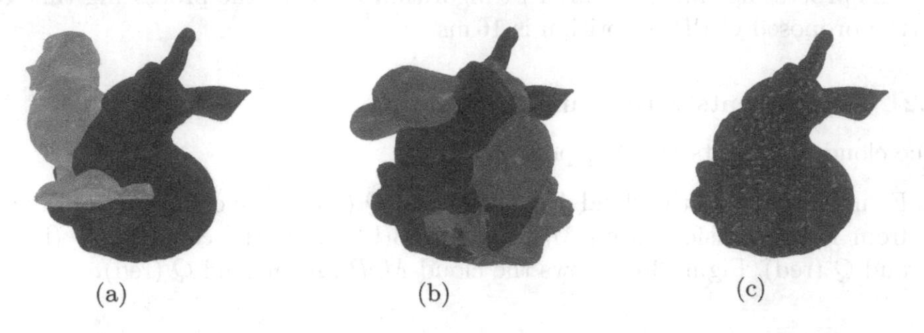

Fig. 4. (a) Test cloud P (yellow), resultant cloud Q (red); (b) alignment result of the algorithm H-PPt; (c) alignment result of the algorithm G-PPt. (Color figure online)

The results of the experiment are as follows:

$$M_{true} = \begin{pmatrix} 0.906518 & 0.0434237 & -0.419928 & 0.464897 \\ -0.415595 & 0.266619 & -0.869594 & -0.070553 \\ 0.0742 & 0.962823 & 0.259742 & -0.160154 \\ 0 & 0 & 0 & 1 \end{pmatrix},$$

$$M_h = \begin{pmatrix} 0.045591 & -0.831211 & 0.554085 & 0.697217 \\ -0.204511 & 0.535147 & 0.81963 & -0.546255 \\ -0.977802 & -0.150684 & -0.145594 & 0.372149 \\ 0 & 0 & 0 & 1 \end{pmatrix},$$

$$M_g = \begin{pmatrix} 0.906518 & 0.043424 & -0.419928 & 0.464897 \\ -0.415595 & 0.266619 & -0.869594 & -0.070553 \\ 0.0742 & 0.962823 & 0.259742 & -0.160154 \\ 0 & 0 & 0 & 1 \end{pmatrix}.$$

The processing time of the H-PPt algorithm is 16715 ms, the processing time of the proposed G-PPt algorithm is 6696 ms.

5.3 Experiments with Dragon

The cloud P consists of 22998 points.

1. Figure 5(a) shows the cloud P (yellow) and Q (red). The cloud Q is obtained from P by transformation M_{true}. Figure 5(b) shows the cloud $M_h P$ (blue) and Q (red). Figure 5(c) shows the cloud $M_g P$ (green) and Q (red).

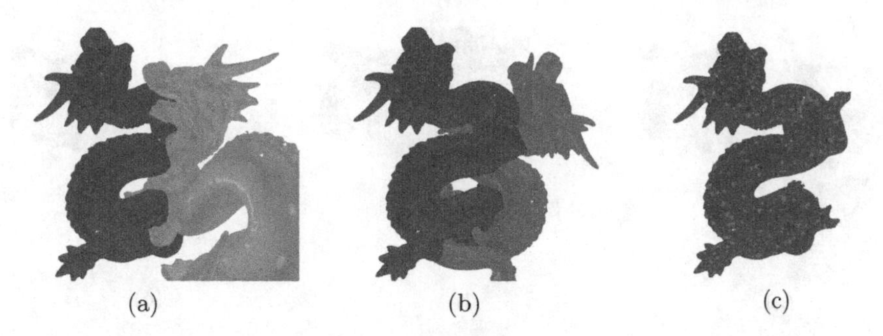

Fig. 5. (a) Test cloud P (yellow), resultant cloud Q (red); (b) alignment result of the algorithm H-PPt; (c) alignment result of the algorithm G-PPt. (Color figure online)

The results of the experiment are as follows:

$$M_{true} = \begin{pmatrix} 0.556762 & 0.0501222 & -0.829158 & -1.94604 \\ -0.739748 & 0.48399 & -0.467468 & 0.0424149 \\ 0.377873 & 0.873637 & 0.306545 & 1.07446 \\ 0 & 0 & 0 & 1 \end{pmatrix},$$

$$M_h = \begin{pmatrix} 0.71438 & 0.698225 & 0.046308 & -2.10059 \\ -0.582702 & 0.556931 & 0.59185 & -0.654277 \\ 0.387454 & -0.449789 & 0.804717 & 0.001504 \\ 0 & 0 & 0 & 1 \end{pmatrix},$$

$$M_g = \begin{pmatrix} 0.556762 & 0.050122 & -0.829158 & -1.94603 \\ -0.739748 & 0.48399 & -0.467468 & 0.042415 \\ 0.377873 & 0.873637 & 0.306545 & 1.07446 \\ 0 & 0 & 0 & 1 \end{pmatrix}.$$

The processing time of the H-PPt algorithm is 7022 ms, the processing time of the proposed G-PPt algorithm is 2647 ms.

2. Figure 6(a) shows the cloud P (yellow) and Q (red). The cloud Q is obtained from P by transformation M_{true}. Figure 6(b) shows the cloud M_hP (blue) and Q (red). Figure 6(c) shows the cloud M_gP (green) and Q (red).
The results of the experiment are as follows:

$$M_{true} = \begin{pmatrix} 0.938224 & -0.119789 & 0.324633 & 1.5113 \\ 0.340453 & 0.487297 & -0.804135 & 2.32942 \\ -0.0618666 & 0.864981 & 0.497976 & -2.92058 \\ 0 & 0 & 0 & 1 \end{pmatrix},$$

$$M_h = \begin{pmatrix} 0.746991 & 0.551314 & 0.371562 & 1.08177 \\ -0.213773 & 0.728377 & -0.650975 & 2.32943 \\ -0.629528 & 0.406842 & 0.661947 & -2.67112 \\ 0 & 0 & 0 & 1 \end{pmatrix},$$

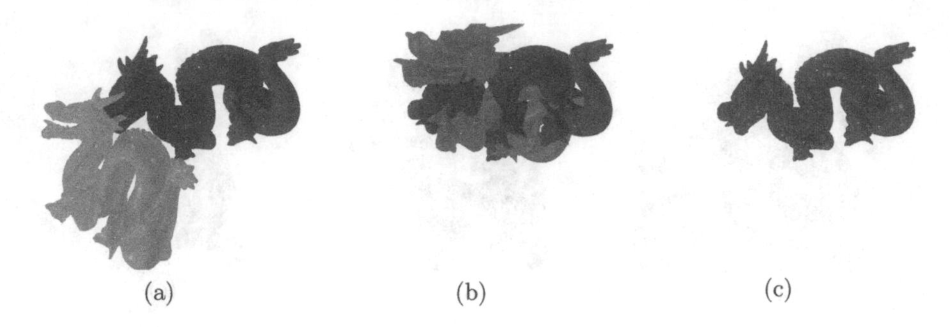

Fig. 6. (a) Test cloud P (yellow), resultant cloud Q (red); (b) alignment result of the algorithm H-PPt; (c) alignment result of the algorithm G-PPt. (Color figure online)

$$M_g = \begin{pmatrix} 0.938224 & -0.119789 & 0.324633 & 1.5113 \\ 0.340453 & 0.487297 & -0.804135 & 2.32942 \\ -0.061867 & 0.864981 & 0.497976 & -2.92058 \\ 0 & 0 & 0 & 1 \end{pmatrix}.$$

The processing time of the H-PPt algorithm is 3106 ms, the processing time of the proposed G-PPt algorithm is 6393 ms.

5.4 Experiments with Armadillo

The cloud P consists of 21259 points.

1. Figure 7(a) shows the cloud P (yellow) and Q (red). The cloud Q is obtained from P by transformation M_{true}. Figure 7(b) shows the cloud M_hP (blue) and Q (red). Figure 7(c) shows the cloud M_gP (green) and Q (red).

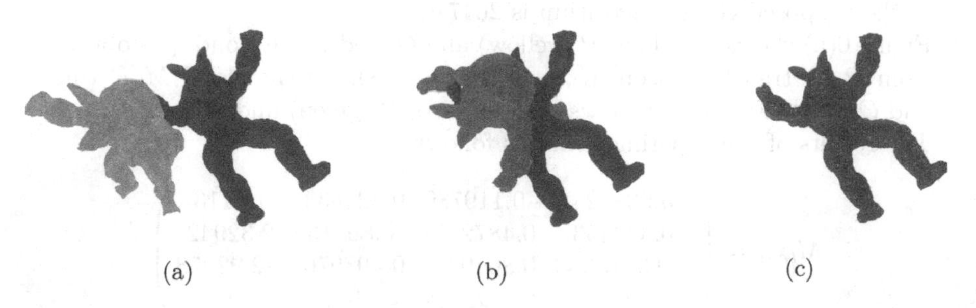

Fig. 7. (a) Test cloud P (yellow), resultant cloud Q (red); (b) alignment result of the algorithm H-PPt; (c) alignment result of the algorithm G-PPt. (Color figure online)

The results of the experiment are as follows:

$$M_{true} = \begin{pmatrix} 0.714482 & 0.69945 & -0.016872 & -1.96075 \\ -0.273847 & 0.257379 & -0.926695 & 0.930376 \\ -0.643834 & 0.666727 & 0.375435 & 0.465303 \\ 0 & 0 & 0 & 1 \end{pmatrix},$$

$$M_h = \begin{pmatrix} 0.859435 & 0.207441 & -0.467269 & -1.04969 \\ 0.356691 & 0.411477 & 0.838724 & 0.991471 \\ 0.366257 & -0.887499 & 0.279646 & 1.54364 \\ 0 & 0 & 0 & 1 \end{pmatrix},$$

$$M_g = \begin{pmatrix} 0.714482 & 0.69945 & -0.016872 & -1.96075 \\ -0.273847 & 0.257379 & -0.926695 & 0.930376 \\ -0.643834 & 0.666727 & 0.375435 & 0.465303 \\ 0 & 0 & 0 & 1 \end{pmatrix}.$$

The processing time of the H-PPt algorithm is 2675 ms, the processing time of the proposed G-PPt algorithm is 1834 ms.

2. Figure 8(a) shows the cloud P (yellow) and Q (red). The cloud Q is obtained from P by transformation M_{true}. Figure 8(b) shows the cloud M_hP (blue) and Q (red). Figure 8(c) shows the cloud M_gP (green) and Q (red).

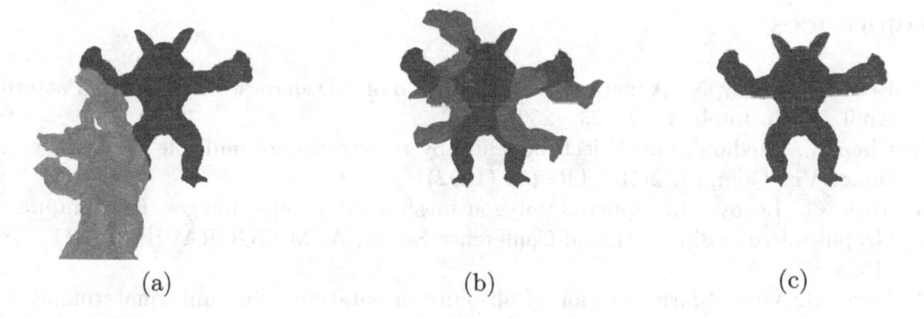

Fig. 8. (a) Test cloud P (yellow), resultant cloud Q (red); (b) alignment result of the algorithm H-PPt; (c) alignment result of the algorithm G-PPt. (Color figure online)

The results of the experiment are as follows:

$$M_{true} = \begin{pmatrix} 0.616773 & -0.680801 & 0.395097 & -0.352557 \\ -0.0448522 & 0.470728 & 0.881138 & 2.00338 \\ -0.785863 & -0.561183 & 0.259797 & 1.01968 \\ 0 & 0 & 0 & 1 \end{pmatrix},$$

$$M_h = \begin{pmatrix} 0.907628 & 0.416241 & -0.054365 & -1.20154 \\ -0.171592 & 0.486082 & 0.856901 & 2.10575 \\ 0.383103 & -0.768419 & 0.512606 & 1.04612 \\ 0 & 0 & 0 & 1 \end{pmatrix},$$

$$M_g = \begin{pmatrix} 0.616773 & -0.680801 & 0.395097 & -0.352557 \\ -0.044852 & 0.470728 & 0.881138 & 2.00338 \\ -0.785863 & -0.561183 & 0.259796 & 1.01968 \\ 0 & 0 & 0 & 1 \end{pmatrix}.$$

The processing time of the H-PPt algorithm is 1507 ms, the processing time of the proposed G-PPt algorithm is 1480 ms.

Remark 3. The proposed algorithm G-PPt shows best fidelity than standard point-to-point ICP algorithm (H-PPt). Also, H-PPt cannot work better than G-PPt, because G-PPt is reduced to H-PPt for $\lambda = 0$.

6 Conclusion

In this paper we proposed a new algorithm for orthogonal registration of point clouds based on a generalized point-to-point ICP algorithm. The algorithm uses the known Horn's algorithm and combines point coordinates and normal vectors. The computer simulation indicates that the proposed algorithm yields the best fidelity comparing with that of a common point-to-point ICP algorithm. It was shown that the common point-to-point ICP algorithm is a particular case of the proposed algorithm for $\lambda = 0$.

References

1. Besl, P., McKay, N.: A method for registration of 3-D shapes. IEEE Trans. Pattern Anal. Mach. Intell. **14**(2), 239–256 (1992)
2. Chen, Y., Medioni, G.: Object modeling by registration of multiple range images. Image Vis. Comput. **2**(10), 145–155 (1992)
3. Turk, G., Levoy, M.: Zippered polygon meshes from range images. In: Computer Graphics Proceedings. Annual Conference Series, ACM SIGGRAPH, pp. 311–318 (1994)
4. Horn, B.: Closed-form solution of absolute orientation using unit quaternions. J. Opt. Soc. Am. Ser. A **4**(4), 629–642 (1987)
5. Horn, B., Hilden, H., Negahdaripour, S.: Closed-form solution of absolute orientation using orthonormal matrices. J. Opt. Soc. Am. Ser. A **5**(7), 1127–1135 (1988)
6. Du, S., Zheng, N., Ying, S., You, Q., Wu, Y.: An extension of the ICP algorithm considering scale factor. In: Proceedings of the 14th IEEE International Conference on Image Processing, pp. 193–196 (2007)
7. Du, S., Zheng, N., Meng, G., Yuan, Z.: Affine registration of point sets using ICP and ICA. IEEE Signal Process. Lett. **15**, 689–692 (2008)
8. Du, S., Zheng, N., Ying, S., Liu, J.: Affine iterative closest point algorithm for point set registration. Pattern Recogn. Lett. **31**, 791–799 (2010)
9. Makovetskii, A., Voronin, S., Kober, V., Tihonkih, D.: An efficient point-to-plane registration algorithm for affine transformations. In: Proceedings SPIE, Applications of Digital Image Processing XL, vol. 10396, p. 103962J (2017)
10. Makovetskii, A., Voronin, S., Kober, V., Tihonkih, D.: Affine registration of point clouds based on point-to-plane approach. Procedia Eng. **201**, 322–330 (2017)

11. Makovetskii, A., Voronin, S., Kober, V., Voronin, A.: A non-iterative method for approximation of the exact solution to the point-to-plane variational problem for orthogonal transformations. Math. Methods Appl. Sci. **41**, 9218–9230 (2018)
12. Makovetskii, A., Voronin, S., Kober, V., Voronin, A., Tihonkih, D.: Point clouds registration based on the point-to-plane approach for orthogonal transformations. In: CEUR Workshop Proceedings, vol. 2210, pp. 236–242 (2018)
13. Makovetskii, A., Voronin, S., Kober, V., Voronin, A.: A point-to-plane registration algorithm for orthogonal transformations. In: Proceedings of the SPIE, Applications of Digital Image Processing XLI, vol. 10752, p. 107522R (2018)
14. Voronin, S., Makovetskii, A., Voronin, A., Diaz-Escobar, J.: A regularization algorithm for registration of deformable surfaces. In: Proceedings of the SPIE, Applications of Digital Image Processing XLI, vol. 10752, p. 107522S (2018)
15. Ruchay, A., Dorofeev, K., Kober, A.: Accurate reconstruction of the 3D indoor environment map with a RGB-D camera based on multiple ICP. In: CEUR Workshop Proceedings, vol. 2210, pp. 300–308 (2018)
16. Ruchay, A., Dorofeev, K., Kober, A., Kolpakov, V., Kalschikov, V.: Accuracy analysis of 3D object shape recovery using depth filtering algorithms. In: Proceedings of SPIE - The International Society for Optical Engineering, vol. 10752, p. 1075221 (2018)
17. Ruchay, A., Dorofeev, K., Kober, A.: Accuracy analysis of 3D object reconstruction using RGB-D sensor. In: CEUR Workshop Proceedings, vol. 2210, pp. 82–88 (2018)
18. Rusinkiewicz, S., Levoy, M.: Efficient variants of the ICP algorithm. In: Proceedings of the International Conference on 3-D Digital Imaging and Modeling, pp. 145–152 (2001)

11. [illegible] A fast iterative method for [illegible] problem for [illegible] (2014)
12. [illegible] Proceedings [illegible] 2016 [illegible]
13. [illegible]
14. [illegible]
15. [illegible]
16. [illegible]
17. [illegible] Proceedings [illegible]

Integer Programming

An Integer Programming Approach to the Irregular Polyomino Tiling Problem

Vadim M. Kartak[1] and Aigul Fabarisova[2](✉)

[1] Ufa State Aviation Technical University, Ufa, Russia
kvmail@mail.ru

[2] M. Akmullah Bashkir State Pedagogical University, Ufa, Russia
aygul_fab@mail.ru

Abstract. In this paper, new integer programming models of the problem of irregular polyomino tiling are introduced. We consider tiling of finite, square, NxN-sized structure with L-shaped trominoes without any restriction on their number. Each polyomino can be rotated 90°, so there are four orientations of the L-tromino. Developed models are effective for small-size instances. For medium- and large-size instances we suggest dividing the initial structure into several equally sized parts and combine the solutions of optimized tilings. We tried to apply new models to the existing information-theoretic entropy-based approach. We conducted computational experiments using IBM ILOG CPLEX package. The problem of irregular polyomino tiling can be applied to the design of phased array antennas where polyomino-shaped subarrays are used to reduce the cost of the array antenna and to reduce the undesired sidelobes radiation. Computational results along with antenna simulation results are presented in the paper.

Keywords: Optimization problems · Integer programming · Polyomino tiling · Phased array antenna

1 Introduction

In this paper, we consider the two-dimensional irregular polyomino tiling problem. Term polyomino was introduced by Golomb in 1954 as the shape made by connecting certain numbers of equal-sized squares, each joined together with at least one other square along an edge [1] (Fig. 1). Polyomino tiling has various fields of application from computer graphics [2,3] to mechanical engineering. One of them refers to the phased array antennas design. Phased array antennas consist of multiple stationary antenna elements, which are fed coherently and use variable phase or time-delay control at each element to scan a beam to given angles in space [4]. This phase controls and time-delay devices are the most important parts of phased array antennas, which make it possible to control the beam direction and to keep array pattern stationary. But for economical reasons, it is better to reduce the number of these devices in an antenna array.

I. Bykadorov et al. (Eds.): MOTOR 2019, CCIS 1090, pp. 235–243, 2019.
https://doi.org/10.1007/978-3-030-33394-2_18

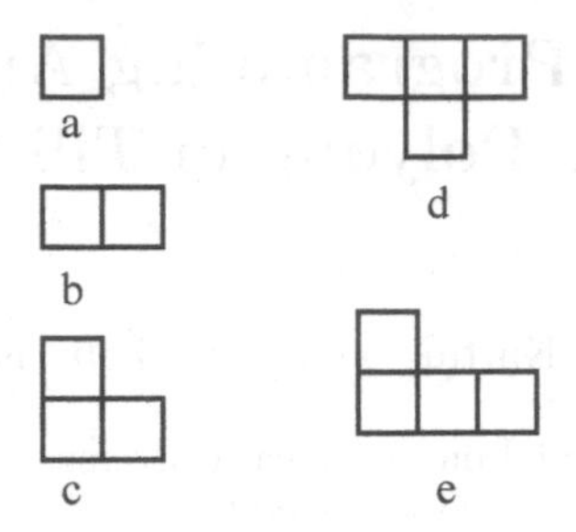

Fig. 1. Examples of polyominoes: monomino (a), domino (b), L-tromino (c), T-tetromino (d), L-tetromino (e).

To resolve this problem, the subarray technology can be used. This means that array elements are grouped into subarrays and controls are introduced at the subarray level. But using rectangular subarrays causes periodicity and radiate discrete sidelobes called the quantization sidelobes. Sidelobe is a beam that represents undesirable radiation in a direction different from the main direction of the antenna. Mailloux et al. argue that using irregular polyomino subarrays can result in a major decrease in sidelobes while presenting only a few tenths of a dB gain reduction compared to rectangular subarrays [5].

The problem of polyomino packing and tiling with polyominoes has been investigated by many researchers [6–10]. In our work, we consider the irregular polyomino tiling problem which has several approaches to its solution. Among recent publications in the field of irregular polyomino tiling we highlight the ones with the heuristic approach and the mathematical programming approach.

One of the many approaches to the polyomino tiling problem is using heuristic methods like the genetic algorithm. It was shown in Gwee and Lim studies [11]. Their algorithm was tested in the phased array antennas design by Chirikov et al. [12]. Who also presented their own approach called the Snowball algorithm which showed better results.

Another kind of approaches refers to the mathematical programming. For example, we used an integer programming approach to the problem of optimal layout [13]. There are examples of using ILP models to d-dimensional Orthogonal Packing Problem [14]. Karademir et al. used an integer programming in the phased array antenna design application. He formulated the irregular polyomino tiling problem as a nonlinear exact set covering model, where the irregularity of a tiling is incorporated into the objective function using the information-theoretic entropy concept [15].

This article is devoted to the new integer programming model for irregular polyomino tiling and to comparing this new model with the model built on the information-theoretic entropy-based approach.

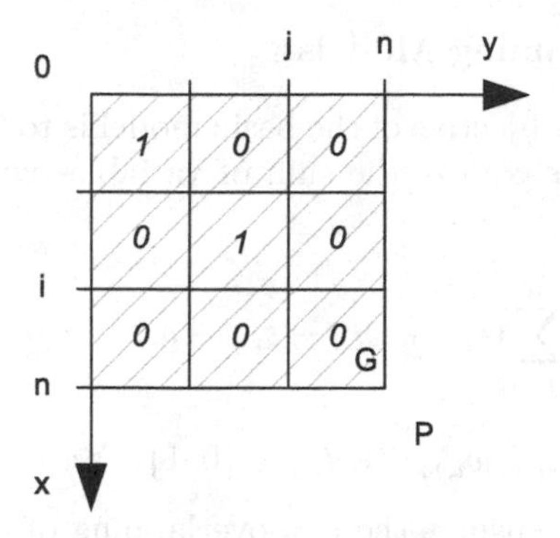

Fig. 2. NxN-sized structure

2 Model Formulation

We consider the tiling of the finite, square NxN-sized structure with L-shaped trominoes without any restriction on their number. L-tromino is a polyomino consisting of three equal squares joined at the edges to form a shape resembling the capital letter L. Each tromino can be rotated 90°. So there are four orientations of L-tromino. Let an NxN element structure be represented as the board G with the squares $G(i, j)$, $i \in \{1, \ldots, n\}, j \in \{1, \ldots, n\}$ of monomino size, where (i, j) is the coordinate of each square (Fig. 2). The problem is to place L-shaped trominoes on the given NxN-size board G.

To each orientation of an L-tromino, we put into correspondence a binary variable as illustrated in Fig. 3. Assume that each L-tromino has the center C in the corner square. The value of each tromino variable z, s, w, d depends on containing the center C in the corresponding coordinate of $G(i, j)$, thus:

$$z_{i,j}, w_{i,j}, s_{i,j}, d_{i,j} = \begin{cases} 1, C \in G(i,j) \\ 0, C \notin G(i,j) \end{cases} \tag{1}$$

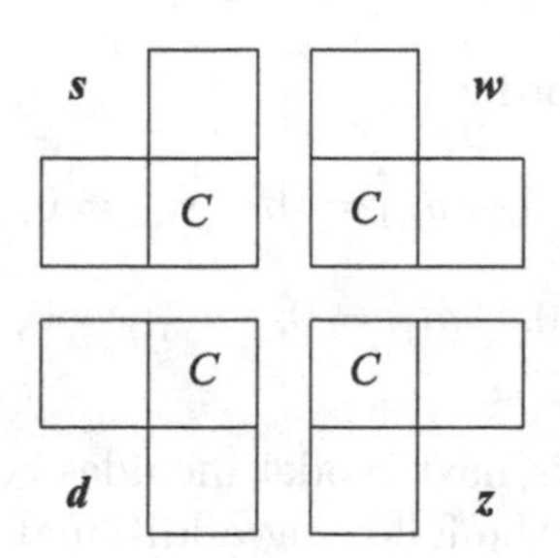

Fig. 3. Four orientations of L-tromino

2.1 Integer Programming Models

Basic Model T1. The objective of the basic model is to fill the whole structure with trominoes, i.e. to maximize the sum of variables while keeping the shape of given trominoes.

$$\sum_{i=1}^{n}\sum_{j=1}^{n}(z_{i,j} + w_{i,j} + s_{i,j} + d_{i,j}) \rightarrow max \tag{2}$$

$$z_{i,j}, w_{i,j}, s_{i,j}, d_{i,j} \in \{0, 1\} \quad \forall i, j$$

The following constraint ensures the non-overlapping of given trominoes for all cells of the structure except those on borders:

$$\begin{aligned} &w_{i,j} + w_{i+1,j} + w_{i,j-1} + z_{i,j} + z_{i-1,j} + z_{i,j-1} \\ &\quad + s_{i,j} + s_{i+1,j} + s_{i,j+1} + d_{i,j} + d_{i-1,j} + d_{i,j+1} \leq 1, \end{aligned} \tag{3}$$

$$i, j \in \{2, \ldots, n-1\}$$

Since some trominoes cannot be placed at the border of the structure, the following inequality reduces overlapping along the border:

Right border constraint:

$$s_{i,n} + s_{i+1,n} + d_{i,n} + d_{i-1,n} + w_{i,n-1} + z_{i,n-1} \leq 1, \quad i \in \{2, \ldots, n-1\} \tag{4}$$

Bottom border constraint:

$$d_{n-1,j} + w_{n,j} + w_{n,j-1} + z_{n-1,j} + s_{n,j} + s_{n,j+1} \leq 1, \quad j \in \{2, \ldots, n-1\} \tag{5}$$

Left border constraint:

$$w_{i,1} + w_{i+1,1} + z_{i,1} + z_{i-1,1} + s_{i,2} + d_{i,2} \leq 1, \quad i \in \{2, \ldots, n-1\} \tag{6}$$

Upper border constraint:

$$w_{2,j} + z_{1,j} + z_{1,j-1} + s_{2,j} + d_{1,j} + d_{1,j+1} \leq 1, \quad j \in \{2, \ldots, n-1\} \tag{7}$$

Additional constraints for borders:

$$z_{i,n} = 0, \quad w_{i,n} = 0, \quad d_{i,1} = 0, \quad s_{i,1} = 0, \quad i \in \{1, \ldots, n\}$$

$$z_{n,j} = 0, \quad d_{n,j} = 0, \quad s_{1,j} = 0, \quad w_{1,j} = 0, \quad j \in \{1, \ldots, n\}$$

Improved Model T2. This next model includes constraints described in the previous model T1 but with the following additional constraint:

$$w_{i,j} + z_{i,j} + z_{i-1,j} + s_{i,j} + s_{i,j+1} + d_{i-1,j+1} + d_{i-1,j} + d_{i,j+1} \leq 1, \tag{8}$$

$$i, j \in \{2, \ldots, n-1\}$$

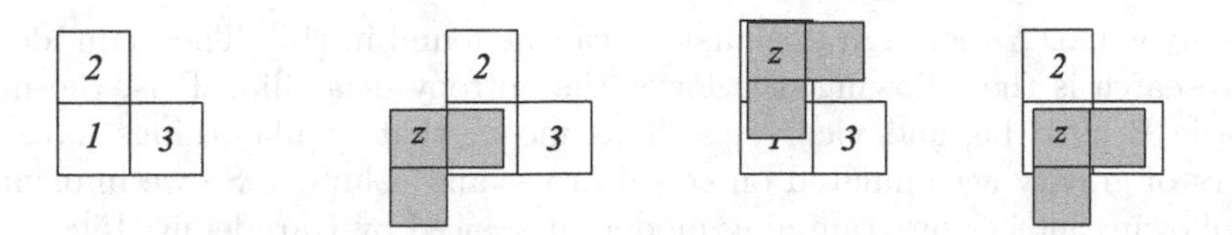

Fig. 4. Figure z coverig the first element

This constraint is not dominated by other constraints and improves the contiguous relaxation of our model. This constraint can be obtained as a mixed integer rounding cut by the following steps.

Let us find the sum of all possible ways to cover each element of the shape illustrated in Fig. 4, where the example of a figure z covering the first element is also shown.

$$\begin{aligned} &w_{i,j} + w_{i+1,j} + w_{i,j-1} + s_{i,j} + s_{i,j+1} + s_{i+1,j} \\ &\quad + z_{i,j} + z_{i,j-1} + z_{i-1,j} + d_{i,j} + d_{i,j+1} + d_{i-1,j} \leq 1 \end{aligned} \tag{9}$$

$$\begin{aligned} &w_{i-1,j} + w_{i-1,j-1} + w_{i,j} + s_{i,j} + s_{i-1,j} + s_{i-1,j+1} \\ &\quad + z_{i-1,j} + z_{i-2,j} + z_{i-1,j-1} + d_{i-1,j} + d_{i-2,j} + d_{i-1,j+1} \leq 1 \end{aligned} \tag{10}$$

$$\begin{aligned} &w_{i,j+1} + w_{i,j} + w_{i+1,j+1} + s_{i,j+1} + s_{i,j+2} + s_{i+1,j+1} \\ &\quad + z_{i,j} + z_{i,j+1} + z_{i-1,j+1} + d_{i,j+1} + d_{i,j+2} + d_{i-1,j+1} \leq 1 \end{aligned} \tag{11}$$

Now the sum of the three inequalities will be as follows:

$$3w_{i,j} + 2z_{i,j} + 2z_{i-1,j} + 2s_{i,j} + 2s_{i,j+1} + 2d_{i-1,j+1} + 2d_{i-1,j} + 2d_{i,j+1} \leq 3 \tag{12}$$

We already eliminated all entries with a coefficient less than 2. Then dividing by 2 with rounding half down gives the final constraint:

$$w_{i,j} + z_{i,j} + z_{i-1,j} + s_{i,j} + s_{i,j+1} + d_{i-1,j+1} + d_{i-1,j} + d_{i,j+1} \leq 1, \tag{13}$$

$$i, j \in \{2, \ldots, n-1\}$$

Each tromino contained in this inequality covers at least two elements of the shape shown in Fig. 4. Since two trominoes would cover at least one element twice, only one of such tromino can be used.

2.2 Entropy Based Models E1 and E2

In the previous two models, the notion of irregularity was not implemented in the objective of the model. The next model is an attempt to implement the entropy concept, where the polyomino's center C is considered as the gravity center for this polyomino. A thorough explanation can be found in [15]. The main idea of this research is the following: the lower the entropy of a tiling B is, the more periodic B must be, and vice versa. This means that regular tilings have the centers of gravity accumulated on certain rows and columns. So we implement the following integer programming model introduced by Karademir [15]:

$$-\sum_{i=1}^{n}\sum_{t=1}^{T}(\frac{t}{2T}\log_2(\frac{t}{2T})r_{i,t}) - \sum_{j=1}^{n}\sum_{t=1}^{T}(\frac{t}{2T}\log_2(\frac{t}{2T})c_{j,t}) \rightarrow max \quad (14)$$

Subject to

$$\sum_{t=1}^{T} t \cdot r_{i,t} = \sum_{j=1}^{n}(z_{i,j} + w_{i,j} + s_{i,j} + d_{i,j}) \quad i \in \{1,\dots,n\} \quad (15)$$

$$\sum_{t=1}^{T} t \cdot c_{j,t} = \sum_{i=1}^{n}(z_{i,j} + w_{i,j} + s_{i,j} + d_{i,j}) \quad j \in \{1,\dots,n\} \quad (16)$$

$$\sum_{t=1}^{T} r_{i,t} = 1 \quad i \in \{1,\dots,n\} \quad (17)$$

$$\sum_{t=1}^{T} c_{j,t} = 1 \quad j \in \{1,\dots,n\} \quad (18)$$

$$z_{i,j}, w_{i,j}, s_{i,j}, d_{i,j}, r_{i,t}, c_{j,t} \in \{0,1\} \quad \forall i,j,t$$

We combined this model with constraints (3–7) from model T1 and constraint (8) from model T2 to obtain two models E1 and E2 respectively.

3 Computational Results

The computational experiments are devoted to an evaluation of the quality of developed integer programming models. We have implemented described models in Python code using the API of IBM ILOG CPLEX 12.6 [16]. The machine with Intel(R) Core(TM) i5-2450M CPU @2.50GHz with 6 Gb RAM is used for all computations.

Table 1 shows computational results for models T1 and T2 without any segmentation used and no random shapes in the structure. It is clear that for some instances the second model T2 shows increasingly better results regarding the running time and the number of nodes created. The most significant fact is that CPLEX is capable of solving instances of board dimension beyond 32×32. The computational results for entropy based models are in Table 2, where X means that the time limit of 10000 s was exceeded. These models are more challenging for structure sizes greater than 16×16.

Table 1. Computational results for models T1 and T2 (no segmentation, no random).

Size (Rows * Cols)	Nodes				Total running time (s)	
	T1		T2		T1	T2
	Created	Processed	Created	Processed		
12 * 12	0	0	0	0	0.19	0.28
16 * 16	0	0	0	0	0.20	0.31
20 * 20	225	127	0	0	0.95	1.03
24 * 24	2284	1296	5418	4929	1.63	5.44
28 * 28	46566	24828	3686	2035	20.11	6.22
30 * 30	48036	24782	1738	1208	25.84	9.45
32 * 32	47336	24327	0	0	28.8	3.27
40 * 40	48202	24893	47496	25081	46.88	126.0
48 * 48	896182	481331	231651	124323	800.5	1439.36

Table 2. Computational results for models E1 and E2 (no segmentation, no random).

Size (Rows * Cols)	Nodes				Total running time (s)	
	E1		E2		E1	E2
	Created	Processed	Created	Processed		
12 * 12	2	0	2	0	0.64	0.52
16 * 16	50244	49894	15603	13137	468.16	X
20 * 20	–	–	–	–	X	X

The next set of computational experiments is aimed at evaluating the models regarding a practical application and at estimating the irregularity of obtained structures. The simulation of phased array antennas can show the irregularity of obtained structures and compare designed models. The structure with the size 30×30 was taken into consideration. In order to add more irregularity to the instances we decided to add randomly placed L-trominoes on the structure before optimization. The set of random coordinates was generated in advance and applied to all the models. Only the instances that could be solved within 5 min by each model were selected. The model E1 could not be solved with the structure of 30×30, therefore it was divided into four parts of equal size 15×15. Models T1 and T2 were not divided into segments. The second entropy model E2 was not tested due to the incapability of being solved even for the size of 16×16.

As seen from Table 3, no individual model shows a distinct advantage. The model based on the entropy approach E1 has more structures with 3 holes, while T1 and T2 show better optimization results.

Table 3. Computational results for models T1, T2, E1 (size 30 * 30).

N of instance	Number of random shapes	SLL (dB)						Number of holes		
		T1		T2		E1		T1	T2	E1
		r = 1.300	r = 1.818	r = 1.300	r = 1.818	r = 1.300	r = 1.818			
1	5	−25.76	−18.63	−27.99	−20.95	−25.79	−18.61	0	0	0
2	5	−27.11	−20.32	−26.38	−19.20	−25.99	−19.18	0	0	3
3	5	−26.09	−19.06	−27.32	−20.27	−27.47	−20.38	0	0	0
4	5	−27.46	−20.31	−25.54	−18.30	−26.00	−18.85	0	0	0
5	5	−26.96	−20.46	−27.00	−20.16	−25.19	−18.16	0	0	0
6	10	−25.07	−18.02	−25.22	−18.05	−26.22	−19.31	0	0	3
7	10	−27.08	−19.77	−27.27	−20.22	−25.22	−18.00	0	0	0
8	10	−27.41	−20.36	−25.59	−18.41	−26.40	−19.25	3	3	3
9	10	−26.22	−19.09	−27.64	−20.49	−27.28	−20.14	0	0	0
10	10	−26.38	−19.22	−27.15	−20.00	−26.16	−19.04	0	0	0

4 Conclusions

In this paper we study the problem of irregular polyomino tiling, considering tiling with L-trominoes. To solve this problem we propose an approach using integer programming and we present several models. We implemented ILP models as a Python program using CPLEX. One of the practical applications of this method could be in the phased array antennas design. Computational results show that the approach can be trusted. We intend to extend this approach to other complex shapes like octomino or pentomino.

Acknowledgments. This research was supported by DAAD grant and by the Russian Foundation for Basic Research, project No. 19-07-00895.

We thank our colleagues from the Institute of Numerical Mathematics (Dresden University of Technology) who provided insight and expertise that greatly assisted the research.

References

1. Golomb, S.: Polyominoes: Puzzles, Patterns, Problems and Packings, 2nd edn. Princeton University Press, Princeton (1996)
2. Ostromoukhov, V.: Sampling with polyominoes. ACM Trans. Graph. (SIGGRAPPH) **26**(3), 78:1–78:6 (2007)
3. Vanderhaeghe, D., Ostromoukhov, V.: Polyomino-based digital halftoning. In: IADIS International Conference on Computer Graphics and Visualization 2008, pp. 11–18. ADIS Press, Netherlands (2008)
4. Mailloux, R.: Phased Array Antenna Handbook, 2nd edn. Artech House, Norwood (2005)

5. Mailloux, R., Santarelli, S., Roberts, T., Luu, D.: Irregular polyomino-shaped subarrays for space-based active arrays. Int. J. Antennas Propag. **2009**, 9 (2009)
6. Conway, J.H., Lagarias, J.C.: Tiling with polyominoes and combinatorial group theory. J. Comb. Theory Ser. A **53**(2), 183–208 (1990)
7. Wolffram, J.: Packing polyominoes. Oper. Res. **91**, 168–171 (1992)
8. Bodlaender, H.L., Van Der Zanden, T.C.: On the exact complexity of polyomino packing. In: Ito, H., Leonardi, S., Pagli, L., Prencipe G. (eds.) 9th International Conference on Fun with Algorithms (FUN 2018), pp. 9:1–9:10. Dagstuhl Publishing (2018)
9. Aigrain, P., Beauquier, D.: Polyomino tilings, cellular automata and codicity. Theoret. Comput. Sci. **147**(1–2), 165–180 (1995)
10. Keating, K., Vince, A.: Isohedral polyomino tiling of the plane. Discrete Comput. Geom. **21**(4), 615–630 (1999)
11. Gwee, B.H., Lim, M.H.: Polyominoes tiling by a genetic algorithm. Comput. Optim. Appl. J. **6**(3), 273–291 (1996)
12. Chirikov, R., Rocca, P., Bagmanov, V., et al.: Algorithm for phased antenna array design for satellite communications. Vestnik UGATU **17**(4), 159–166 (2013)
13. Kartak, V.M., Marchenko, A.A., Petunin, A.A., Sesekin, A.N., Fabarisova, A.I.: Optimal and heuristic algorithms of planning of low-rise residential buildings. AIP Conf. Proc. **1895**(1), 110002 (2017). https://doi.org/10.1063/1.5007408
14. Kartak, V.M., Ripatti, A.V.: The minimum raster set problem and its application to the D-dimensional orthogonal packing problem. Eur. J. Oper. Res. **271**(1), 33–39 (2018). https://doi.org/10.1016/j.ejor.2018.04.046
15. Karademir, S., Prokopyev, O., Mailloux, R.: Irregular polyomino tiling via integer programming with application in phased array antenna design. J. Global Optim. **65**(2), 137–173 (2016)
16. IBM: IBM ILOG CPLEX. http://www.ibm.com/products/ilog-cplex-optimization-studio. Accessed 12 Feb 2019

Iterative Methods for Constructing Approximations to Optimal Coverings of Nonconvex Polygons

Pavel Lebedev[1,2](✉) and Vladimir Ushakov[1,2](✉)

[1] Krasovskii Institute of Mathematics and Mechanics, Ural Branch of the Russian Academy of Sciences, Yekaterinburg, Russia
pleb@yandex.ru, ushak@imm.uran.ru

[2] Ural Federal University named after the first President of Russia B.N. Yeltsin, Yekaterinburg, Russia

Abstract. The paper proposes algorithms for the iterative construction of optimal coverings of nonconvex flat figures using sets of circles. These algorithms are based on the procedures of dividing the figure into zones of influence of points that serve as the centers of the initial coverings and finding the Chebyshev centers of these zones. To generate the initial array of points, we use stochastic procedures based on the synthesis of optimal hexagonal grids and random vectors.

Keywords: Optimal covering · Chebyshev center · Voronoi diagram · Dirichlet domain · Nonconvex polygon

1 Introduction

In various areas of mathematics, it is necessary to approximate sets by tuples of elements of the same type. One of the simplest and most convenient ways to do this is to construct a set of congruent balls that reflect the geometry of the set. In the case of flat figures, one may talk of two types of approximation, namely, coverings (external approximations covering the set) and packings (internal approximations embedded in the set). For practical applications, coverings generally play a major role, since they ensure that, for each point of the figure, there is an element covering it. This is of great importance, for example, in the design of networks of communication centers, technical service centers or warehouses (for more details, see [1]). Earlier, theoreticians studied mostly coverings of convex figures of a regular shape (for example, circles [2] or regular polygons [3]). However, one often has to deal with nonconvex figures, especially in the problems of regional transport networking [4] and distribution networking in architecture [5, Chap. 3]. The necessity arises to develop algorithms that

The work was supported by the Decree no. 211 of the Government of the Russian Federation, contract no. 02.A03.21.0006.

I. Bykadorov et al. (Eds.): MOTOR 2019, CCIS 1090, pp. 244–254, 2019.
https://doi.org/10.1007/978-3-030-33394-2_19

allow us to construct coverings for figures of a rather general form, for example, arbitrary nonconvex simply connected polygons. This requires the use of computational geometry [6] and nonsmooth optimization [7] methods.

2 Optimal Covering Problem

Let M be a closed bounded simply connected set on the Euclidean plane. The union of $n \in \mathbf{N}$ disks $O(\mathbf{s}_i, r), i = \overline{1,n}$, of the same radius $r > 0$ is called a covering of the set M, if the condition

$$M \subseteq O(\mathbf{s}_1, r) \cup O(\mathbf{s}_1, r) \cup \ldots \cup O(\mathbf{s}_i, r)$$

holds. We assume that the covering of the set M for which the value of r is minimal, is optimal. Let us consider the problem of finding the optimal covering for the set M for given n.

Let us introduce the auxiliary notations.

Definition 1. *Let A and B be closed and bounded sets in $\mathbf{R}^2$. The Hausdorf semideviation of A from B [8] is the value*

$$h(A, B) = \max\{\rho(\mathbf{a}, B) \colon \ \mathbf{a} \in A\},$$

where $\rho(\mathbf{a}, B) = \min\{\|\mathbf{a} - \mathbf{b}\| \colon \ \mathbf{b} \in B\}$ is the Euclidean distance from the point $\mathbf{a}$ to the set B.

Definition 2. *The Chebyshev center of a closed bounded set $M \in \mathbf{R}^2$ is the point $\mathbf{c}(M)$ that satisfies the equality*

$$h(M, \{\mathbf{c}(M)\}) = \min\{h(M, \{\mathbf{x}\}) \colon \ \mathbf{x} \in \mathbf{R}^2\}. \tag{1}$$

For more information about the properties of the Chebyshev center, see the papers of A.L. Garkavi, for example, [9,10]. For any $M \in \mathrm{comp}(\mathbf{R}^2)$, the Chebyshev center $\mathbf{c}(M)$ exists, is unique, and belongs to the convex hull $\mathrm{co}\, M$ of the set M. The quantity (1) is called the Chebyshev radius $r(M)$ of the set M.

Definition 3. *A nonempty set consisting of at most n points in $\mathbf{R}^2$ is called an n-network [9,10] on the plane $\mathbf{R}^2$.*

Definition 4. *Denote by Σ_n the set of all n-networks of the space $\mathbf{R}^2$. An n-network S^* is called the best Chebyshev n-network of the set $M \in \mathbf{R}^2$ if*

$$h(M, S^*) = \min\{h(M, S) \colon \ S \in \Sigma_n\}.$$

The solution of the optimal covering problem for a figure M with an n-tuple is the construction of its best n-network. In the simplest case, for $n = 1$, the optimal coverage consists of a circle centered at $\mathbf{c}(M)$ with radius $r(M)$.

3 Algorithms for Solving the Problem

Consider a compact set M being nonconvex in the general case. We assume that some n-network S is originally given, possibly generated with the help of a random number generator or taken as a set of points from a lattice superimposed on the figure M. Based on this network, we can construct a new n-network $\widehat{S}$ for which $h(M, \widehat{S}) \leq h(M, S)$.

We introduce some definitions before describing the algorithms.

Definition 5. *The Voronoi cell* [11, Chap. 3] *of a point* $\mathbf{s}_i \in S$ *in an* n*-network* S *is the set*

$$W_i(S) = \left\{\mathbf{w} \in \mathbf{R}^2 \colon \ \|\mathbf{w} - \mathbf{s}_i\| = \min\{\|\mathbf{w} - \mathbf{s}_j\| \colon \ \mathbf{s}_j \in S\}\right\}.$$

Note that, by construction, the Voronoi cell is either a convex polygon or an unbounded convex part of the plane bounded by segments and rays of straight lines, or a half-plane.

Definition 6. *Let a compact set* $M \in \mathbf{R}^2$ *and a* n*-network* S *be given. The Dirichlet domain* [12] *of a point* $\mathbf{s}_i \in S$ *in the set* M *is the subset* $D_i(M, S) = M \cap W_i(S)$.

A new network $\widehat{S} = \{\widehat{\mathbf{s}}_i\}_{i=1}^n$ is constructed according to the scheme proposed in [13]:

$$\widehat{\mathbf{s}}_i = \begin{cases} \mathbf{c}\big(D_i(M, S)\big), \ D_i(M, S) \neq \emptyset, \\ \mathbf{s}_i, \ D_i(M, S) = \emptyset. \end{cases} \tag{2}$$

In [1] it was shown that scheme (2) yields a new n-network $\widehat{S}$, the Hausdorff semideviation $h(M, \widehat{S})$ of the set M from which is not large than $h(M, S)$. The authors constructed the coverings using formula (2) for some basic geometric shapes, such as a square and a circle, in [14, § 8]. The obtained results are close enough to the known optimal covering given in [15]. The paper [16] presents methods for calculating the Chebyshev center of a polygon, and [17] presents methods for constructing the Dirichlet domains.

We note that the construction of the Dirichlet domains for nonconvex sets is significantly hampered by the fact that they can be not only nonconvex but also unconnected. Therefore, for convenience, it is necessary to replace the set $D_i(M, S)$ by the set $H_i = \{\mathbf{h}_i^{(j)}\}_{j=1}^J$ with a finite number J of points such that the condition $\mathbf{c}\big(D_i(M, S)\big) = \mathbf{c}(H_i)$ holds. As the latter set, one can consider the union of points $\mathbf{h}_i^{(j)} \in D_i(M, S)$ satisfying one of the three conditions:

- $\mathbf{h}_i^{(j)}$ is a vertex of the polygon M;
- $\mathbf{h}_i^{(j)}$ belongs to the boundary ∂M of the polygon M and to at least one more Dirichlet domain $D_k(M, S)$, $k \neq i$;
- $\mathbf{h}_i^{(j)}$ belongs to at least two more Dirichlet domains $D_k(M, S)$ and $D_l(M, S)$, $k \neq i$, $l \neq i$, $k \neq l$.

By construction, the set H_i is the union of all vertices of the polygons that constitute the Dirichlet domain $D_i(M,S)$; therefore, their Chebyshev centers and radii coincide. Moreover, in the set H_i there is a set of at most three elements (i.e., a simplex on the plane) such that all of them lie on the circle $\partial O\left(\mathbf{c}(D_i(M,S)), r(D_i(M,S))\right)$ of the Chebyshev radius, with its center coinciding with $\mathbf{c}(D_i(M,S))$. Note that the points of H_i include, by construction, possible maxima of the function $f(\mathbf{x}) = h(\{\mathbf{x}\}, S)$ on the compact set M. Therefore, to find the radius of the covering circles for a given array of their centers, it suffices to calculate the maximum of $f(\mathbf{x})$ on $H_i, i = \overline{1,n}$; in more detail, see [12,18].

An important element of the algorithms under consideration is the construction of the Chebyshev center of the compact set H_i. In the general case, the Chebyshev center for a set can be constructed only by numerical methods. Therefore, it is important to know how close the value computed by the software package to the exact value. This is especially true for iterative algorithms where we need some termination conditions.

Theorem 1. *Let compact sets M and $M_0 \subseteq M$ be given in $\mathbf{R}^2$. If the Chebyshev center $\mathbf{c}(M_0)$ of the compact set M is found, then the estimate*

$$\|\mathbf{c}(M) - \mathbf{c}(M_0)\| \leq \sqrt{2h\left(M, \{\mathbf{c}(M_0)\}\right)\left(h\left(M, \{\mathbf{c}(M_0)\}\right) - r(M_0)\right)} \quad (3)$$

is true.

Proof. Suppose that estimate (3) does not hold. Denote the distance between the Chebyshev centers by $d = \|\mathbf{c}(M) - \mathbf{c}(M_0)\|$. Suppose that $d > 0$. Then we can draw through $\mathbf{c}(M)$ a perpendicular line to the segment $[\mathbf{c}(M), \mathbf{c}(M_0)]$. It divides the circle $\partial O(\mathbf{c}(M), r(M))$ in two halves. Moreover, according to the properties of the Chebyshev center of a closed set, there must be points of the set M on both halves of the circle $\partial O(\mathbf{c}(M), r(M))$ (otherwise, it would be possible to find a circle of radius less than $r(M)$, in which M can be embedded). Let $\mathbf{m}^*$ denote an arbitrary point that belongs to M and to a half of the circle $\partial O(\mathbf{c}(M), r(M))$ not containing $\mathbf{c}(M_0)$. Consider a triangle formed by the points $\mathbf{m}^*, \mathbf{c}(M)$, and $\mathbf{c}(M_0)$. By construction, the angle at the vertex $\mathbf{c}(M)$ is not less than $\pi/2$; hence, the cosine theorem implies the estimate

$$\|\mathbf{m}^* - \mathbf{c}(M_0)\|^2 \geq \|\mathbf{m}^* - \mathbf{c}(M)\|^2 + \|\mathbf{c}(M) - \mathbf{c}(M_0)\|^2,$$

which can be rewritten as

$$d^2 \leq \|\mathbf{m}^* - \mathbf{c}(M_0)\|^2 - \|\mathbf{m}^* - \mathbf{c}(M)\|^2.$$

It follows from the definition of the Chebyshev center that $\|\mathbf{m}^* - \mathbf{c}(M)\| = r(M)$. The definition of the Hausdorff deviation implies $\|\mathbf{m}^* - \mathbf{c}(M_0)\| \leq h\left(M, \{\mathbf{c}(M_0)\}\right)$. We get the estimate

$$d^2 \leq h\left(M, \{\mathbf{c}(M_0)\}\right)^2 - r(M)^2.$$

The embedding $M_0 \subseteq M$ implies the estimates $r(M_0) \leq r(M)$ and $r(M_0) \leq h(M, \{\mathbf{c}(M_0)\})$. Thus, we can write as follows

$$d^2 \leq h(M, \{\mathbf{c}(M_0)\})^2 - r(M_0)^2 = \big(h(M, \{\mathbf{c}(M_0)\}) + r(M_0)\big)\times$$
$$\times\big(h(M, \{\mathbf{c}(M_0)\}) - r(M_0)\big) \leq 2h(M, \{\mathbf{c}(M_0)\}) \cdot \big(h(M, \{\mathbf{c}(M_0)\}) - r(M_0)\big).$$

After extracting the square roots from the first and last part of the equality, we get estimate (3). □

To construct Dirichlet domains, it is required to construct the median perpendiculars to the segments connecting the points of S and look for their intersections with each other and with the sides of the polygon M. This procedure requires a lot of computational costs, especially if the set M is nonconvex. This is connected with the need to check whether a certain point belongs to the polygon, for which additional constructions should be done. Therefore, it is important to reduce the exhaustive search of points that can be vertices of the polygons contained in the Dirichlet domain.

Theorem 2. *Let a compact set M in $\mathbf{R}^2$ and a n-network S be given. Then two Dirichlet domains $D_i(M, S)$, $i \leq n$, and $D_j(M, S)$, $j \leq n$, may have common points only if*

$$\|\mathbf{s}_i - \mathbf{s}_j\| \leq 2\,h(M, S). \tag{4}$$

Proof. Consider an arbitrary point $\mathbf{m}^* \in D_i(M, S) \cap D_j(M, S)$. By the triangle inequality, we estimate

$$\|\mathbf{s}_i - \mathbf{s}_j\| \leq \|\mathbf{s}_i - \mathbf{m}^*\| + \|\mathbf{s}_j - \mathbf{m}^*\|.$$

Since $\mathbf{m}^*$ belongs to $D_i(M, S)$ and $D_j(M, S)$, we have $\|\mathbf{s}_i - \mathbf{m}^*\| = \|\mathbf{s}_j - \mathbf{m}^*\|$. Then we can write

$$\|\mathbf{s}_i - \mathbf{m}^*\| \geq \|\mathbf{s}_i - \mathbf{s}_j\|/2.$$

If the estimate (4) is not satisfied, then we have

$$\|\mathbf{s}_i - \mathbf{m}^*\| > h(M, S).$$

Since, by the definition of the Dirichlet domain, the point $\mathbf{s}_i$ is one of the closest to any point of $D_i(M, S)$, then

$$\rho(\mathbf{m}^*, S) > h(M, S).$$

We came to a contradiction. □

Theorem 3. *Let a compact set M in $\mathbf{R}^2$ and a n-network S be given. Then three Dirichlet domains $D_i(M, S)$, $i \leq n$, $D_j(M, S)$, $j \leq n$, and $D_k(M, S)$, $k \leq n$, may have a common point only if the condition*

$$\min\{\|\mathbf{s}_i - \mathbf{s}_j\|, \|\mathbf{s}_i - \mathbf{s}_k\|, \|\mathbf{s}_j - \mathbf{s}_k\|\} \leq \sqrt{3}\,h(M, S) \tag{5}$$

is true.

Proof. Consider a point $\mathbf{m}^* \in D_i(M,S) \cap D_j(M,S) \cap D_k(M,S)$. It follows from the definition of the Dirichlet domain that this point is equidistant from the points $\mathbf{s}_i$, $\mathbf{s}_j$, and $\mathbf{s}_k$. This means that the latter three points lie on a circle of radius r^* centered at $\mathbf{m}^*$. Consider the angles formed by the vectors $\mathbf{s}_i - \mathbf{m}^*$, $\mathbf{s}_j - \mathbf{m}^*$, and $\mathbf{s}_k - \mathbf{m}^*$. The sum of these three angles does not exceed 2π. Hence, the smallest of them does not exceed $2\pi/3$. Without loss of generality, we assume that

$$\alpha = \angle\big((\mathbf{s}_i - \mathbf{m}^*), (\mathbf{s}_j - \mathbf{m}^*)\big) \leq 2\pi/3. \tag{6}$$

By construction, we have

$$\|\mathbf{s}_i - \mathbf{s}_j\| = 2r^* \sin(\alpha/2).$$

Since the sine function increases on the interval $[0, \pi/2]$, then inequality (6) implies the estimate

$$\|\mathbf{s}_i - \mathbf{s}_j\| \leq 2r^* \sin\left((2\pi/3)/2\right) = 2r^* \sin(\pi/3) = \sqrt{3}r^*.$$

The definitions of the Dirichlet domain and the Hausdorff deviation imply the estimate $r^* \leq h(M,S)$; hence, the latter inequality implies (5). □

The results of Theorems 2 and 3 can be applied starting from the second cycle of correction of the network S according to (2). If the Hausdorff deviation h_0 of the set M from the previous network $S_0 = \{\mathbf{s}_i^{(0)}\}_{i=1}^n$ is known, then the following estimate for $h(M,S)$ can be used:

$$h(M,S) \leq h(M,S_0) - (\sqrt{2}-1)\frac{\min\limits_{i=\overline{1,n}} \left\|\mathbf{s}_i - \mathbf{s}_i^{(0)}\right\|}{h(M,S_0)},$$

the proof of which is given in [17].

4 Examples of Building Coverings

The software package for constructing approximations of optimal coverings by repeatedly improving the randomly generated n-network S using formula (2) has been developed in MATLAB. The original Chebyshev n-network was constructed on the basis of imposing a hexagonal lattice [19] on the figure M with stochastic changes in the coordinates of its nodes. Then exactly n elements were randomly selected among the arrays of points.

It is difficult to assess how close the result is to optimal. Therefore, the software package was running many times for each example.

Example 1. Consider a nonconvex octagon M defined by the array of its vertices

$$(-1,-1),\ (0,-0.5),\ (1,-1),\ (0.5,0),$$
$$(1,1),\ (0,0.5),\ (-1,1),\ (-0.5,0).$$

It is required to construct its optimal coverings by circles with the number of elements $n = 10$ and $n = 13$.

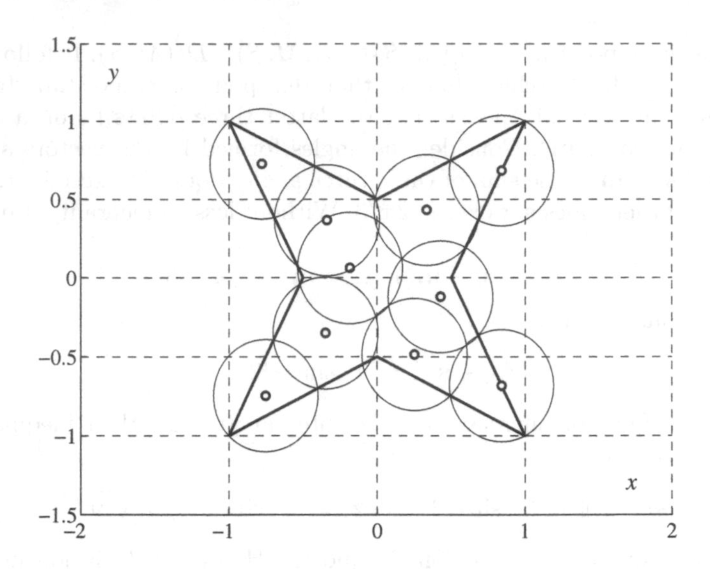

Fig. 1. Covering of the octagon M in Example 1 with the union of 10 disks.

The results are obtained by multiple launching the software package. For $n = 10$, the array of the disk centers is as follows:

$$S = \{(-0.7519, -0.7471),\ (-0.3473, -0.3501),\ (-0.1826, 0.0618),$$

$$(-0.7748, 0.7265),\ (0.2525, -0.4863),\ (0.4279, -0.1204),\ (-0.3372, 0.3662),$$

$$(0.3348, 0.4312),\ (0.8416, -0.6831),\ (0.8416, 0.6831)\}.$$

The best radius is found to be equal to $r \approx 0.3543$. The covering disks of 10 elements (thin lines), the tuple S of their centers (in the form of "bubbles"), and the set M (bold lines) are shown in Fig. 1.

For $n = 13$, the array of the disk centers is as follows:

$$S = \{(-0.8581, -0.7161),\ (-0.4428, -0.0605),\ (-0.5781, 0.4094),$$

$$(-0.7162, 0.8581),\ (-0.4039, -0.489),\ (-0.0612, -0.0823),\ (-0.2154, 0.4847),$$

$$(0.8318, 0.7309),\ (0.1726, -0.5599),\ (0.4694, -0.3213),$$

$$(0.2950, 0.2288),\ (0.4011, 0.4617),\ (0.7526, -0.8012)\}.$$

The best found radius is $r \approx 0.3174$. The covering disks of 13 elements, the tuple S of their centers, and the set M are presented in Fig. 2.

Example 2. Consider a nonconvex polygon M defined by the array of its vertices

$$(-1, -1),\ (-0.5, -0.5),\ (-0.5, -1),\ (0, -1),\ (0.5, -0.5),$$

$$(0.5, -1),\ (1, -1),\ (1, -0.5),\ (0.5, 0),\ (1, 0),$$

$$(1, 1),\ (0.5, 1),\ (0.5, 0.5),\ (0, 0.5),\ (0, 1),$$

$$(-0.5, 1),\ (-0.5, 0.5),\ (-1, 0.5),\ (-0.5, 0),\ (-1, 0).$$

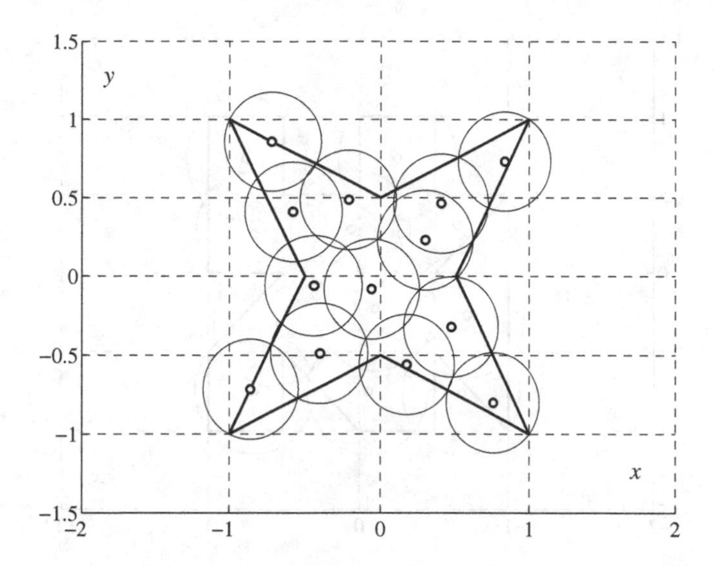

Fig. 2. Covering of the octagon M in Example 1 with the union of 13 disks.

It is required to construct its optimal coverings by circles with the number of elements $n = 11$ and $n = 12$.

The results are obtained by multiple launching the software package. For $n = 11$, the array of the disk centers is as follows:

$$S = \{(-0.6843 - 0.7603)\ (-0.7172 - 0.2667),\ (-0.65280.3088),\ (-0.25, 0.7346),$$
$$(-0.0569 - 0.7551),\ (-0.1041 - 0.2238),\ (0.05780.2759),\ (0.75, 0.7276),$$
$$(0.6924 - 0.75),\ (0.5668, -0.3155),\ (0.6924, 0.25)\}.$$

The best found radius is to be equal to $r \approx 0.3964$. The covering disks of 11 elements, the tuple S of their centers, and the set M are shown in Fig. 3.

For $n = 12$, the array of the disk centers is as follows:

$$S = \{(-0.0713, -0.6042),\ (-0.7171, -0.75),\ (-0.7171, -0.25),\ (-0.6311, 0.4199),$$
$$(-0.177, -0.927),\ (0.17740.2336),\ (-0.1682, -0.0258),\ (-0.25, 0.7171),$$
$$(0.75, -0.7171),\ (0.7301, 0.248),\ (0.4776, -0.3799),\ (0.7206, 0.75)\}.$$

The best radius is found $r \approx 0.3775$. The covering disks of 12 elements, the tuple S of their centers, and the set M are presented in Fig. 4.

The software complex has been launched for solving the problems around $10 \div 15$ times with the exit condition for the iteration cycle so as, after applying formula (2), the coordinates of the points must change by a quantity no more than $\Delta r = 10^{-4}$. The number of cycles varied within $200 \div 500$ and the runtime was from 10 to 20 min.

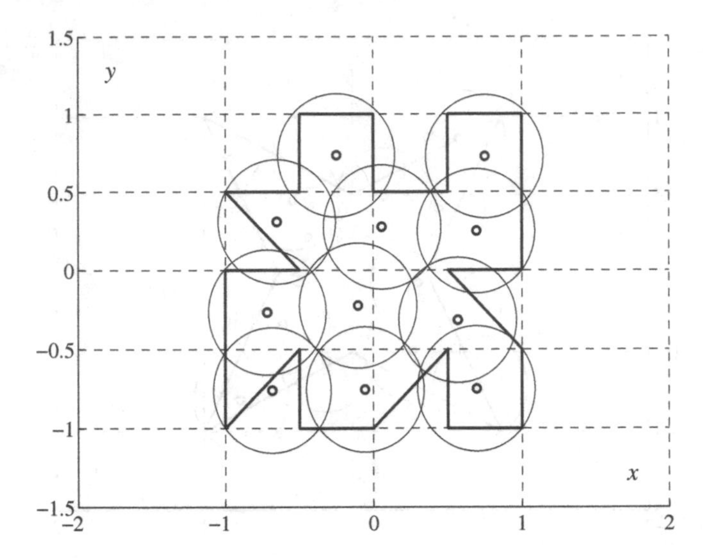

Fig. 3. Covering of the polygon M in Example 2 with the union of 11 disks.

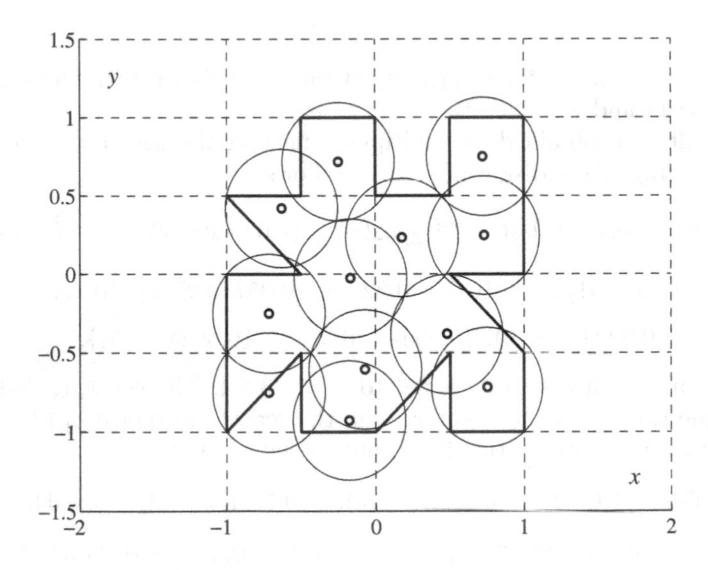

Fig. 4. Covering of the polygon M in Example 2 with the union of 12 disks.

5 Conclusions

The software complex for constructing optimal coverings of nonconvex flat sets M has been developed and tested by the authors for several examples. The centers of the covering circles are found as the Chebyshev centers of the zones of influence of points from the current array S of points. The accuracy of computing the Chebyshev center is estimated using Theorem 1. In turn, the zones of influence are constructed as the Dirichlet domains of the current array of points of the given figure M. To save computational costs of the search, we apply the results of Theorems 2 and 3. This significantly reduces the number of the considered intersections of the median perpendiculars to the segments with endpoints from S with each other and with the sides of the polygon M that can be included in the Dirichlet domains. When modeling the solution of problems, we considered various flat nonconvex figures (both having symmetry axes and irregular shaped) as a compact M which has to be covered. The multiple generation of the initial position of the points has been performed, including a combination of the previously found configuration close to optimal and a random vector array. The results are visualized and allow us to evaluate the pattern formed by the centers of the covering elements. This pattern differs significantly from the regular hexagonal lattice, which is optimal for covering the whole plane [19, Chap. III].

References

1. Kazakov, A.L., Lebedev, P.D.: Algorithms for constructing optimal n-networks in metric spaces. Autom. Remote Control **78**(7), 1290–1301 (2017)
2. Melissen, H.: Densest packings of eleven congruent circles in a circle. Geom. Dedicata. **50**(1), 15–25 (1994)
3. Heppes, A., Melissen, H.: Covering a rectangle with equal circles. Period. Math. Hung. **34**(1–2), 65–81 (1997)
4. Kazakov, A.L., Lempert, A.A., Bukharov, D.S.: On segmenting logistical zones for servicing continuously developed consumers. Autom. Remote Control **74**(6), 968–977 (2013)
5. Desyatov, V.G.: Proektirovanie Sistem Ob'ectov Obshchestvennogo Kompleksa Promyshlennykh Predpriyatiy (Systems Design for Public Service Objects of Industrial Plants). MARKhI, Moscow (1989)
6. Preparata, F.P., Shamos, M.I.: Computational Geometry: An Introduction. Springer, New York (1988)
7. Dem'yanov, V.F., Vasil'ev, L.V.: Nondifferentiable Optimization. Nauka, Moscow (1981); Springer, New York (1985)
8. Hausdorff, F.: Set Theory. Chelsea Publishing Co., New York (1962); Komkniga, Moscow (2006)
9. Garkavi, A.L.: On the existence of an optimal network and best diameter of a set in a Banach space. Usp. Mat. Nauk **15**(2), 210–211 (1960). (in Russian)
10. Garkavi, A.L.: The best possible net and the best possible cross-section of a set in a normed space. Am. Math. Soc. Transl. II. Ser. **39**, 111–132 (1964); transl. from Izv. Akad. Nauk SSSR, Ser. Mat. **26**(1), 87–106 (1962)
11. Mestetskiy, L.M.: Continuous Morphology of Binary Images. Figures, skeletons, Circular. Fizmatlit, Moscow (2009). (in Russian)
12. Brusov, V.S., Piyavskii, S.A.: A computational algorithm for optimally covering a plane region. USSR Comput. Math. Math. Phys. **11**(2), 17–27 (1971)
13. Lebedev, P.D., Uspenskii, A.A., Ushakov, V.N.: Algorithms of the best approximations of the flat sets by the union of circles. Vestn. Udmurt. Univ. Mat. Mekh. Komp'yut. Nauki (4), 88–99 (2013). (in Russian)

14. Ushakov, V.N., Lakhtin, A.S., Lebedev, P.D.: Optimization of the Hausdorff distance between sets in Euclidean space. Proc. Steklov Inst. Math. **291**(1), S222–S238 (2015)
15. https://www2.stetson.edu/~efriedma/packing.html
16. Ushakov, V.N., Lebedev, P.D.: Algorithms for the construction of an optimal cover for sets in three-simensional Euclidean space. Proc. Steklov Inst. Math. **293**(1), S225–S237 (2016)
17. Ushakov, V.N., Lebedev, P.D.: Algorithms of optimal set covering on the planar $\mathbf{R}^2$. Vestn. Udmurt. Univ. Mat. Mekh. Komp'yut. Nauki **26**(2), 258–270 (2016). (in Russian)
18. Piyavskii, S.A.: On optimization of networks. Izv. Akad. Nauk SSSR Tekh. Kibern. (1), 68–80 (1968). (in Russian)
19. Fejes Tóth, L.: Lagerungen in der Ebene, auf der Kugel und im Raum. Springer, Berlin (1953); Fizmatlit, Moscow (1958)

Polyhedral Attack on the Graph Approximation Problem

R. Yu. Simanchev[1,2](✉), I. V. Urazova[2], and Yu. A. Kochetov[3]

[1] Omsk Scientific Center of SB RAS, Marksa Ave., 15, Omsk 644024, Russia
[2] Omsk State University, Mira Ave., 55a, Omsk 644077, Russia
osiman@rambler.ru, urazovainn@mail.ru
[3] Sobolev Institute of Mathematics,
Akademika Koptyuga pr. 4, Novosibirsk 630090, Russia
jkochet@math.nsc.ru

Abstract. In the clique partition problem (CPP), we need to find a spanning family of pairwise vertex-disjoint cliques of minimum total weight in a complete edge-weighted graph. In this paper, we consider the special case of the CPP, the so-called graph approximation problem (GAP), where the weights of edges are 1 or −1. It is one of the most computationally difficult cases of the CPP. We present our polyhedral approach to this problem based on the facet inequalities and the branch and cut framework. Computational experiments on the randomly generated instances indicate simple and hard classes of the GAP and maximal dimension for exact and an approximate solution with a given accuracy.

Keywords: Branch and cut · Facet inequality · Local search · Rounding

1 Introduction

Let us consider a simple graph D. We say that a simple graph is *a matroidal graph* or *M-graph* if each of its connected components is a complete graph [23]. In the graph approximation problem (GAP), we need to find an M-graph H with the minimal symmetric difference on the edge sets for graphs H and D. In other words, we wish to transform graph D into a vertex-disjoint union of cliques by a minimum number of edge modifications. This optimization problem seems to have been first studied by Harary in 1955 [11]. Now this problem is known as the graph correlation clustering [14], cluster editing [20], and transitive graph projection [19]. The problem is NP-hard [15] and some nontrivial polynomially solvable cases are found [6,25]. Approximation algorithms with performance guarantee are reviewed in [14].

The GAP is a special case of the clique partition problem (CPP) [7]. It is an important problem in cluster analysis when we need to divide a set of objects into disjoint clusters and optimize a measure of intergroup relations. Recently, there are many results devoted to the approximation methods for the problem

I. Bykadorov et al. (Eds.): MOTOR 2019, CCIS 1090, pp. 255–265, 2019.
https://doi.org/10.1007/978-3-030-33394-2_20

including heuristics and metaheuristics [3,13]. Interest to this problem deals with its real–life applications in community detection [5], group technology [10], and aggregation of binary relations (ABRP) [16,24]. The last application will be studied here as well.

In this paper, we are interested in the exact solution methods for the CPP and GAP based on the polyhedral approach. Up to now, the branch and cut method is the most powerful framework for this goal. Properties of the polyhedron and structure of the facet inequalities play a crucial role in this method. The first results in this direction were obtained in [7,8] where the integer linear programming formulation is presented for the CPP. Note that the GAP has the same set of constraints and slightly different objective function. The cutting plane methods are based on the facet inequalities. The first classes of facets so-called triangle inequalities, 2-partition inequalities, 2-chorded cycle inequalities, and 2-chorded wheel inequalities are described in [8]. For the aggregation of binary relations problem, this approach can find an optimal solution for the special type of voting graphs with 158 vertices. Later, new facet inequalities were discovered in [9,18,21] and computational complexity of the corresponding separation problems are studied [18,22]. Comparison of lower bounds obtained by some facets is illustrated for the CPP in [18]. In the weighted cluster editing problem, we consider the case similar to GAP but with arbitrary weights on the edges. In [2], a special reduction procedure is suggested to reduce the dimension of the problem. For the graphs from real-world applications and random graphs with large cliques, this reduction procedure combined with the branch and cut method can find the optimal solutions for the large scale instances with up to 1000 vertices. Nevertheless, in the general case, the branch and cut approach is still not so effective and allows us to solve the small size instances only [13].

In this paper, we apply a new polyhedral attack to the GAP in the general setting. We use the branch and cut method with triangle inequalities and 1-parachutes inequalities [18,21] for improving the lower bounds. These inequalities are the most useful in the polyhedral approach for the CPP [2,7,18]. We guess that they are a *golden mean* in known classes of facets between the degree of approximation to the polytope and the cost of finding the cuts. We use randomized local search heuristic to collect 1-parachutes inequalities and select a small subset of them for including into the current set of the problem constraints. We believe that this trick is crucial for the separation problem and used for the GAP for the first time.

The paper is organized as follows. Section 2 contains the basic concepts and some known results on the polyhedral structure of the CPP and GAP. Section 3 provides a brief description of our branch and cut method including the lower and upper bounds of the optimum, and randomized local search heuristic for discovering facet inequalities. The results of computational experiments are presented in Sect. 4. The first part of the experiments is devoted to studying of the cutting plane procedure. The second part is devoted to the branch and cut method for the GAP. We indicate the difficult and simple cases of this problem. The last Sect. 5 concludes the paper.

2 Notations and Basic Concepts

A set P is called a polytope if it is the convex hull of a finite set of points in the Euclidean space. The linear inequality $a^T x \leq a_0$ where $a \neq 0$, is called the valid inequality for the polytope P if it is satisfied for all points from the polytope. The valid inequality is called the support inequality if there exists at least one point x from the polytope which satisfies the equality $a^T x = a_0$. Each support inequality generates a face $\{x \in P | a^T x = a_0\}$ for the polytope P. It is called a facet if it is the maximal face by inclusion. It is easy to see that each facet has the dimension which is less by 1 than the dimension of the polytope. The support inequality which produces a facet is called the faceted inequality.

Polyhedron in the Euclidean space is the set of all solutions for the system of linear equations and inequalities if it is bounded. By the Weyl-Minkowski theorem, there is a polyhedron for each polytope and vice versa.

For arbitrary subgraph D of a complete simple graph K_n with n vertices, we denote by VD and ED the sets of its vertices and edges respectively. For edge $e \in ED$, we will use the notation uv if vertices u and v from VD incident to the edge e. The operations of union and intersection of graphs will be understood in the edge sense.

For the graph K_n, we consider the Euclidean space R^E with dimension $\frac{n^2-n}{2}$. For each edge of the graph, we define an axis in this space. The incident vector for a spanning graph $D \subseteq K_n$ is a vector $x^D \in R^E$ where $x^D_e = 1$ if $e \in ED$ and $x^D_e = 0$ otherwise. Thus, we have the one–to–one correspondence between spanning subgraphs in K_n and vertices of the 0-1 cube in R^E. Now we can introduce the polytope for the CPP as the following

$$P_n = conv\{x^H \in R^E \mid H \text{ is } M\text{-graph} \}$$

and the CPP is the minimization problem over the vertices of the polytope P_n:

$$\min\{c^T x \mid x \quad \text{is a vertex of} \quad P_n\}, \tag{1}$$

where the vector c defines the weights for the edges in the graph K_n.

If $x \in R^E$ and $E' \subseteq E$, then $x(E')$ is the linear function $\sum_{e \in E'} x_e$. Now the GAP can be presented as the minimization problem over the vertices of the polytope P_n:

$$\min\{|ED| + x(E\bar{D}) - x(ED) \mid x \in P_n \cap Z^E\}, \tag{2}$$

where $E\bar{D}$ is the complement of the set ED and Z^E is the 0-1 lattice in R^E [21].

Each $(0,1)$-vector $x \in R^E$ is the incidence vector of the M-graph if and only if it satisfies the system

$$\begin{aligned} -x_{uv} + x_{uw} + x_{vw} &\leq 1, \\ x_{uv} - x_{uw} + x_{vw} &\leq 1, \\ x_{uv} + x_{uw} - x_{vw} &\leq 1, \end{aligned} \tag{3}$$

$$x_{uv} \geq 0, \forall\ uv \in E, \tag{4}$$

where $u, v, w \in V$ are all triples of pairwise different vertices [8,21]. The polyhedron defined by the system (3)–(4) is denoted by M_n. It may contain non-integer vertices [21]. Thus, the inclusion $P_n \subset M_n$ is strict one and each integer vertex of M_n is an incidence vector of the M-graph. The constraints (3) will be called the triangle inequalities [7].

In [8], some classes of faceted inequalities for P_n are presents. In particular, all constraints (3), (4) are faceted. Computational experiments for the ABRP indicate that we can get an optimal solution by the cutting plane method using constraints (3), (4) and the 2-partition inequalities only [8].

In [21], a new class of support inequalities for P_n was introduced. Let $U = \{u_1, u_2, \ldots, u_k\}$ and $W = \{v_1, v_2, \ldots, v_p\}$ are non-empty subsets of the set V and $U \cap W = \emptyset$, $k \geq 1$, $p \geq 2$. We denote by T_i, $i = 1, 2, \ldots, k$, the star in K_n with center at the vertex u_i and rays u_iv_j, $j = 1, 2, \ldots, p$. By K_p, we denote a clique on the set W. We define $T = \cup_{i=1}^{k} T_i$. The graph $T \cup K_p$ we called k-parachute. We associate with this graph the following inequality

$$x(ET) - x(EK_p) \leq \frac{k^2 + k}{2}.$$

The inequality induced by the k-parachute $T \cup K_p$ is a support inequality if and only if $p \geq k$ and it is a faceted inequality if and only if $k = 1$ [21]. Thus, the 1-parachutes will be important in our exact method.

3 The Branch and Cut Method

We implement the classical branch and cut a framework to solve the problem. Below we describe some specific features of this implementation. The 0-1 cube in R^E is taken as the initial relaxation [7] to reduce the number of triangle constraints (3). In each node of branching tree we get the optimal solution for linear programming relaxation and choose either a fractional coordinate, say x_u, or an integer one that does not participate in the formation of the current node. Then we create two new nodes in the branching tree for $x_u = 0$ and $x_u = 1$, respectively. For each node, we generate at most $n^2 - n + 1$ cutting planes to get lower bound.

3.1 Lower Bound

Our cutting plane procedure uses triangle inequalities (3) and inequalities generated by 1-parachutes only. At the initial step, we ignore all triangle inequalities and use the 0-1 cube as the initial relaxation. If the optimal solutions are an integer, we check all triangle inequalities. If these inequalities are not violated then the current optimum is the exact solution of the problem. Otherwise, we look for inequalities that strictly separate it from the P_n. Using these inequalities, we form another relaxation of the convex hull of M-graphs from the current face. All the constraints of the previous relaxation which are not basic (except for the constraints of the unit cube) are discarded. We limit the running time for each node of the branching tree by a small threshold. As a result, we get either an exact solution or a lower bound on the weights of the M-graphs contained in the current node. The IBM ILOG CPLEX package is used as the linear programming solver.

Now we show how to discover the 1-parachutes inequalities to cut off the current optimum $\bar{x}$ of the linear programming relaxation. It is a very important problem in the polyhedral approach which is called the separation problem. Let the point $\bar{x}$ satisfies all the triangle inequalities. Is there a 1-parachute that strictly separates this point from the polytope P_n? In other words, we want to find a 1-parachute $T \cup K_p$ with the following property

$$\sum_{e \in ET} \bar{x}_e - \sum_{e \in EK_p} \bar{x}_e > 1.$$

To this end, we introduce new 0-1 variables:

$x_u \in \{0,1\}$ defines the center of the star T;
$y_u \in \{0,1\}$ defines the vertices in the clique K_p;
$z_{uv} \in \{0,1\}$ defines the edges in the clique K_p;
$t_{uv} \in \{0,1\}$ defines the rays in the star T.

Now the separation problem can be presented as the integer linear program:

$$\max \frac{1}{2}\Big(\sum_{u,v \in V} \bar{x}_{uv} t_{uv} - \sum_{u,v \in V} \bar{x}_{uv} z_{uv}\Big) \tag{5}$$

$$s.t. \;\; z_{uv} \geq y_u + y_v - 1, \;\; u,v \in V; \tag{6}$$

$$z_{uv} \leq y_u, \;\; z_{uv} \leq y_v, \;\; u,v \in V; \tag{7}$$

$$t_{uv} \geq x_u + y_v - 1, \;\; u,v \in V; \tag{8}$$

$$t_{uv} \leq x_u, \;\; t_{uv} \leq y_v, \;\; u,v \in V; \tag{9}$$

$$x_u + y_u \leq 1, \;\; u \in V; \tag{10}$$

$$\sum_{u \in V} x_u = 1; \tag{11}$$

$$\sum_{u \in V} y_u \geq 1; \tag{12}$$

$$x_u, y_u, z_{uv}, t_{uv} \in \{0,1\}, \;\; u,v \in V. \tag{13}$$

Constraints (6), (7) guarantee the equality $z_{uv} = y_u y_v$ and constraints (8), (9) guarantee the equality $t_{uv} = x_u y_v$. Constraints (10) prohibit the center of the star T to be in the clique K_p. Constraint (11) allows one center only for the star T. Constraint (12) indicates that the clique K_p is not empty.

Since the separation problem for 1-parachutes is NP-complete [18,22], we use the local search heuristic for it. We generate a starting solution for the problem (5)–(13) by a greedy heuristic and define the Flip+Swap neighborhood [1,12]. This set of neighboring solutions consists of all feasible solutions for the problem which can be obtained by two rules: (1) we can add or remove one vertex from the clique, (2) we can replace one vertex in the current 1-parachute. It is a well–known neighborhood for the facility location problems [12,17]. To reduce the running time of the search, we use the randomized neighborhood [4] with parameter κ. Each element of the Flip+Swap neighborhood is included in this randomized neighborhood with probability κ independently from other elements.

Our computational experiments show that this local search heuristic discovers a lot of cuts. If we add all of them into the current relaxation of the problem, the number of constraints grows rapidly. Thus, we apply two strategies. In the first case, we add at most q inequalities $a^T x \leq 1$ with the largest values $h(\bar{x}, a) = a^T \bar{x} - 1$. In the second case, we add at most q inequalities with the largest values $d(\bar{x}, a) = \frac{a^T \bar{x} - 1}{\|a\|}$. In our computational experiments, the comparison of these strategies is carried out with a time limit for the lower bound.

3.2 Upper Bound

To find near–optimal incumbent solution at the initial step of the branch and cut method, we apply local search heuristic again [13]. In each node of the branching tree, we try to improve it by the rounding procedures. We used two methods of rounding the point $\bar{x}$. Let $G_c \subset K_n$ be the spanning subgraph induced by the set of edges $EG_c = \{e \in E \mid c_e < 0\}$ and $T_{\bar{x}}$ be a subgraph in K_n induced by the set of edges $ET_{\bar{x}} = \{e \in E \mid \bar{x}_e = 1\}$. In the first rounding method, we complement each connected component of the graph $G_c \cap T_{\bar{x}}$ to a clique. In the second rounding method, we do the same, but the graph $S_{\bar{x}}$ with the set of edges $ES_{\bar{x}} = \{e \in E \mid \bar{x}_e > 0\}$ is used instead of the graph $T_{\bar{x}}$. Note that these procedures are applied for non-integer solutions and for integer solutions that are not an M-graph. Moreover, if $\bar{x}$ is integer then $T_{\bar{x}} = S_{\bar{x}}$.

3.3 Approximate Solution with a Given Accuracy

In our computational experiments, we apply this branch and cut method for finding approximate solutions with a priori accuracy also. The M-graph x^* is called the α-approximate solution of the problem, $\alpha \geq 0$, if $c^T x^* \leq (1 + \alpha) c^T x^{opt}$, where x^{opt} is the optimal solution of the problem. At each iteration of the method, we compare the current optimum $\bar{x}$ and the incumbent solution x^*. If $(1 + \alpha) c^T \bar{x} > c^T x^*$, then x^* is an α-approximate solution.

4 Computational Experiments

The computational experiments are conducted on the random test instances for the GAP and real–world instances for the ABRP taken from [7]. For the GAP, we generate the coefficients of the objective function as follows. For a given value of density β, each edge of the graph K_n has weight -1 with probability β and weight $+1$ with probability $1-\beta$. As we will see later, the value of density is crucial for our exact method.

In the first computational experiment, we study the cutting plane procedure to improve the lower bounds by 1-parachutes inequalities. We wish to identify the best values of parameter κ for solving the separation problem by local search. For graph K_{100}, five values of κ are tested: 0.25, 0.4, 0.5, 0.6 and 0.75. The best values for the objective function (5) are obtained for $\kappa = 0.6$ and $\kappa = 0.75$. Moreover, these values are the best for other random test instances with n from 20 to 80. We tested 5 random graphs for each dimension and added the 1-parachute inequalities one by one to constraints (3)–(4). Again, we see the best growth of the lower bounds in average for these values. Thus, we use $\kappa = 0.6$ in the rest part of our experiments.

In the second computational experiment, we try to select the best combination of the strategies from Sect. 3.1 to select the most strong cutting planes and identify the number of inequalities for including into the linear program. We tested three values of the parameter $q = \frac{n}{10}, \frac{n}{5}, \frac{n}{2}$ and two strategies: $h(\bar{x}, a) = a^T\bar{x} - 1$ and $d(\bar{x}, a) = \frac{a^T\bar{x}-1}{\|a\|}$, where a is vectors of coefficients of the 1-parachute inequality. Comparison of these combinations was carried out under the time limit equals 10 min. The best combination shows the highest value of the objective function. In average, the combination $(q = \frac{n}{2}; d(\bar{x}, a))$ is the best and we use it in all our experiments.

The third computational experiment is devoted to the branch and cut method. We have tested it on the ABRP benchmarks from [7] and new benchmarks. The structure of the objective function is as follows. Let H_i, $i = 1, 2, \dots, p$, is a family of M-graphs in K_n. The weight of an edge $e \in E$ is defined as $c_e = p - 2|\{k \in \{1, \dots, p\} \mid e \in EH_k\}|$. The graphs H_i are called characteristics and the ABRP is NP-hard [24]. All data in these instances have real content and concern politics, business, and social processes. The maximum dimension of the instances from [7] is $n = 158$. Our algorithm is similar to the algorithm from [7]. The main difference lies in solving the separation problem and the choice of cuts. It is not clear why the authors did not attempt to solve the problems of a higher dimension. For all instances, the exact solution was obtained in a sufficiently short time and without the branching stage. In [7], some the ABRP instances with random data were also solved, but their dimension did not exceed $n = 34$. We solved all these random instances as well. Moreover, all the characteristics of the method almost completely coincide with [7]. We also considered several new instances of the ABRP with random data. All of them were solved without branching. Our computational results for new instances of the ABRP are given in Table 1.

Table 1. The ABRP with random data

n	Number of characteristics of the ABRP	f_{opt}	Time (sec.)	Number of iterations
160	3	−7175	1033	5
160	4	−7732	1399	7
200	3	−7174	1971	5
200	4	−8572	1525	7
300	3	−7175	1756	5
300	4	−3209	7210	19

Further investigation of the branch and cut method was carried out on the GAP instances. We study the influence of the graph density on the running time of the method and ability to find exact and α-approximate solutions. The computational results are presented in Table 2. For each instance, the running time of the method is limited by three hours. The first column of the table indicates the type of the instance and its dimension. The second column shows the density of the graph. The third column presents the approximation parameter α. The fourth column (FC) contains the results for solving the GAP without branching, that is, by pure the cutting plane method. The fifth column (BFC) presents the running time of the branch and cut method. Finally, the last sixth column shows the objective function value of the α-approximate solution obtained. As we can see, the cutting plane method can solve the problem instances with small dimensions only, $n \leq 30$. The dash means that the problem is not solved in 3 h.

The most important result of our empirical study deals with the density of random graphs. We observe that this parameter defines the computational difficulty of the GAP instances for exact branch and cut method. In Table 3, we show our results for small, mean, and high density. Column f_{opt} in the table presents the objective function value for the optimal solution of the problem. Column *Time* indicates the running time of the branch and cut method in seconds. Column FC shows the number of calls for the local search heuristic to solve the separation problem for 1-parachutes, 0 means that we use the triangle inequalities only in each node of the branching tree. Column FBC presents the total number of nodes in the branching tree. Again, the running time of the method is limited by three hours for each instance. As we can see, all instances with small and high density are easy for our exact method. The case $0.2 \leq \beta \leq 0.6$ is the most difficult.

Table 2. Computational results for the GAP

Problem	Density	α	FC (sec.)	BFC (sec.)	α-approximate solution
$f1-20$	0,47	0,00	20,95	7,18	58
$f2-20$	0,56	0,00	17,55	9,55	65
$f3-20$	0,46	0,00	13,21	7,19	53
$f4-20$	0,52	0,00	9,21	9,3	59
$f5-20$	0,48	0,00	18,39	12,2	58
$f1-30$	0,53	0,00	2493,49	1785,3	155
$f2-30$	0,46	0,00	1318,22	652,29	138
$f3-30$	0,52	0,00	1157,13	693,32	151
$f4-30$	0,49	0,00	1818,56	1398,12	148
$f5-30$	0,51	0,00	1595,42	1271,09	152
$f1-40$	0,49	0.05	–	1540,13	330
$f2-40$	0,51	0.05	–	1600,08	327
$f3-40$	0,51	0.05	–	1671,02	330
$f4-40$	0,49	0.05	–	1586,45	315
$f5-40$	0,51	0.05	–	1652,34	321
$f1-50$	0,26	0.10	–	3534,12	267
$f2-50$	0,25	0.10	–	2976,31	263
$f3-50$	0,27	0.10	–	4985,07	279
$f4-50$	0,25	0.10	–	3109,43	269
$f5-50$	0,27	0.10	–	3988,16	276
$f1-60$	0,25	0.15	–	9720,38	398
$f2-60$	0,26	0.15	–	10681,14	396
$f3-60$	0,26	0.15	–	9307,46	407
$f4-60$	0,26	0.15	–	10142,09	407
$f5-60$	0,24	0.15	–	9223,27	366
$f1-70$	0,27	0.15	–	13980,09	565
$f2-70$	0,25	0.15	–	13320,54	548
$f3-70$	0,26	0.15	–	14400,05	562
$f4-70$	0,24	0.15	–	11,520,31	517
$f5-70$	0,24	0.15	–	12180,24	519

Table 3. The influence of density

n	Density	f_{opt}	Time (sec.)	FC	BFC	n	Density	f_{opt}	Time (sec.)	FC	BFC
50	0,95	71	4	0	3	60	0,2	–			
50	0,85	132	8	0	5	60	0,15	256	882	3	396
50	0,7	375	35	2	16	60	0,1	160	798	7	355
50	0,6	–				60	0,05	83	2840	31	220
50	0,3	–				70	0,95	161	20	0	4
50	0,2	3	569	7	344	70	0,85	338	73	0	13
50	0,15	174	868	21	350	70	0,7	788	347	0	26
50	0,1	111	1896	54	419	70	0,6	–			
50	0,05	64	379	10	351	70	0,2	–			
60	0,95	92	12	0	4	70	0,1	237	4401	6	700
60	0,85	202	26	0	9	70	0,05	99	2859	23	508
60	0,7	537	59	0	17	80	0,95	202	7	0	7
60	0,6	–				80	0,85	504	232	0	19

5 Conclusion

In this paper, we present computational results for the polyhedral attack to the ABRP and GAP instances. We design the exact branch and cut method where the most strong cutting plane inequalities are generated and collected by local search heuristic. We show that our method can find an optimal solution for previously studied benchmarks of the ABRP and new high dimensional graphs. For the GAP instances, we observe that the efficiency of the approach is depended on the graph density. The randomly generated instances with small and high density are easy for the method. Instances with mean density are the most difficult. Moreover, the method cannot solve the GAP without branching for $n > 30$ and can do that for the ABRP. Note, that both problems are NP-hard, have the same polytope and differ only in the structure of the objective functions. Thus, we need additional research in this direction to understand the influence of this aspect to the polyhedral methods.

Acknowledgment. This research was supported by the Russian Foundation for Basic Research, project 18-07-00599.

References

1. Alekseeva, E., Kochetov, Yu., Plyasunov, A.: Complexity of local search for the p-median problem. Eur. J. Oper. Res. **191**, 736–752 (2008)
2. Bocker, S., Briesemeister, S., Klau, G.W.: Exact algorithms for cluster editing: evaluation and experiments. Algorithmica **60**, 316–334 (2011)
3. Brimberg, J., Janicijevic, S., Mladenovic, N., Urosevic, D.: Solving the clique partitioning problem as a maximally diverse grouping problem. Optim. Lett. **1**, 1123–1135 (2017)
4. Diakova, Z., Kochetov, Yu.: A double VNS heuristic for the facility location and pricing problem. Electron. Notes Discrete Math. **39**, 29–34 (2012)
5. Fortunato, S.: Community detection in graphs. Phys. Rep. **486**(3), 75–174 (2010)
6. Fridman, G.Š.: A graph approximation problem. Upravlyaemye Sistemy **8**, 73–75 (1971). (in Russian)
7. Grotschel, M., Wakabayashi, Y.: A cutting plane algorithm for a clustering problem. Math. Program. **B**(45), 59–96 (1989)
8. Grotschel, M., Wakabayashi, Y.: Facets of the clique partitioning polytope. Math. Program. **47**, 367–387 (1990)
9. Grotschel, M., Wakabayashi, Y.: Composition of facets of the clique partitioning polytope. In: Bodendiek, R., Henn, R. (eds.) Topics in Combinatorics and Graph Theory, pp. 271–284. Physica-Verlag, Heidelberg (1990). https://doi.org/10.1007/978-3-642-46908-4_31
10. Ham, I., Hitomi, K., Yoshida, T.: Group Technology: Applications to Production Management. Kluwer, Dordrecht (1988)
11. Harary, F.: On the notion of balance of a signed graph. Mich. Math. J. **2**, 143–146 (1955)
12. Iellamo, S., Alekseeva, E., Chen, L., Coupechoux, M., Kochetov, Yu.: Competitive location in cognitive radio networks. 4OR **13**(1), 81–110 (2015)

13. Kochenberger, G., Glover, F., Alidaee, B.: Clustering of microarray data via clique partitioning. J. Comb. Optim. **10**, 77–92 (2005)
14. Il'ev, V., Il'eva, S., Kononov, A.: Short survey on graph correlation clustering with minimization criteria. In: Kochetov, Y., Khachay, M., Beresnev, V., Nurminski, E., Pardalos, P. (eds.) DOOR 2016. LNCS, vol. 9869, pp. 25–36. Springer, Cham (2016). https://doi.org/10.1007/978-3-319-44914-2_3
15. Křivánek, M., Morávek, J.: NP-hard problems in hierarchical-tree clustering. Acta Informatica **23**, 311–323 (1986)
16. Marcotorchino, F., Michaud, P.: Heuristic approach to the similarity aggregation problem. Methods Oper. Res. **43**, 395–404 (1981)
17. Mladenović, N., Brimberg, J., Hansen, P., Moreno-Pérez, J.A.: The p-median problem: a survey of metaheuristic approaches. Eur. J. Oper. Res. **179**(3), 927–939 (2007)
18. Oosten, M., Rutten, J.H.G.C., Spieksma, F.C.R.: The clique partitioning problem: facets and patching facets. Networks **38**(4), 209–226 (2001)
19. Rahmann, S., Wittkop, T., Baumbach, J., Martin, M., Truß, A., Böcker, S.: Exact and heuristic algorithms for weighted cluster editing. In: 6th Annual International Conference on Computational Systems Bioinformatics, vol. 6, pp. 391–401. Imperial College Press, London (2007)
20. Shamir, R., Sharan, R., Tsur, D.: Cluster graph modification problems. Discrete Appl. Math. **144**(1–2), 173–182 (2004)
21. Simanchev, R.Yu., Urazova, I.V.: On the polytope faces of the graph approximation problem. J. Appl. Ind. Math. **9**(2), 283–291 (2015)
22. Simanchev, R.Yu., Urazova, I.V.: Separation problem for k-parashutes. In: Proceedings of the DOOR 2016, Vladivostok, Russia, 19–23 September, CEUR-WS, vol. 1623, pp. 109–114 (2016)
23. Tyshkevich, R.I.: Matroidal decompositions of a graph. Diskretnaya Matematika **1**(3), 129–139 (1989). (in Russian)
24. Wakabayashi, Y.: Aggregation of binary relations: algorithmic and polyhedral investigations. Ph.D. thesis, Universitat Augsburg, West Germany (1986)
25. Zahn, C.T.: Approximating symmetric relations by equivalence relations. J. Soc. Ind. Appl. Math. **12**(4), 840–847 (1964)

Analysis of Integer Programming Model of Academic Load Distribution

Lidia Zaozerskaya(✉)

Sobolev Institute of Mathematics, Novosibirsk, Russia
zaozer@ofim.oscsbras.ru

Abstract. The bicriteria problem of academic load distribution (ALD) and its integer linear programming (ILP) model are considered. Earlier it was shown that the search for a feasible solution to this problem is NP-hard and the cardinality of the complete set of alternatives is polynomial. For the ALD problem, the problem of finding a Pareto-optimal solution can be formulated as a weighted bin packing problem with color constraints and lower bounds on the load of the bins. In this problem, the number of bins is given and items have volume and color. For each bin, there is an upper bound on the number of different colors and this bound depends on the bin volume. For each item, coefficients of the efficiency of placing in any bin are set. In this paper, we study the ILP model for finding a Pareto-optimal solution. Parametric families of ALD instances are constructed and the L-coverings of these instances are studied. These instances have a small duality gap, in particular, it can be equal to one. We investigate the complexity of solving these families by the Land and Doig algorithm for some known branching rules. It is shown, that the iterations number grows exponentially with the number of bins.

Keywords: Teachers assignment · Weighted bin packing · General assignment problem · Integer linear programming · Land and Doig Algorithm · L-partition

1 Introduction

Distribution of academic load among teachers at an university department is one of the actual organization problems of the learning process. Problem formulations can be diverse and largely depend on the requirements of a particular university. In [1–3], some optimization models for this problem are proposed. In [1], the average load of a teacher at the department is minimized. The presented mixed integer programming model is a special variant of the fixed charge transportation problem. The NP-hardness of this problem is also shown. For the equivalent combinatorial statement of the problem, the branch and bound algorithm has been proposed there as well. However, the use of this model can lead to a solution

Supported by the program of fundamental scientific researches of the SB RAS No. I.5.1., project No. 0314-2019-0019.

I. Bykadorov et al. (Eds.): MOTOR 2019, CCIS 1090, pp. 266–279, 2019.
https://doi.org/10.1007/978-3-030-33394-2_21

whereby one course can be distributed arbitrarily among several teachers, which is contrary to the practice of Russian universities. In [2,3], each training course consists of parts that correspond to some type of academic load, such as lectures, seminars, exams, etc. Each part is indivisible and can be assigned to only one teacher.

In [3], the bicriteria ALD problem was considered in which it is required to reduce the diversity of courses assigned to each teacher and to maximize the total effectiveness of the distribution. Shown that finding a feasible solution to this problem is NP-hard and the cardinality of the full set of alternatives is polynomial. Note that the bicriteria model of the ALD problem can be interpreted as a supply management problem with the discrete sizes of the batches [4].

Finding a Pareto-optimal solution to the ALD problem can be described using the terminology of the bin packing problem with color constraints. In considered problem, all items have volume and color and any bin has an upper and a lower bounds on total volume of items loaded into it. For any bin, there is an upper bound on the number of different colors. This bound is directly proportional to the volume of a bin. In addition, for each item, there is a set of the effectiveness coefficients (or weights) of placing it in any bin. Any placement of items is characterized by a total weight which should be maximized. Bin packing problems with color constraints are studied, for example, in [5,6]. For these problems, the number of bins should be minimized in contrast to the considered problem in which the number of bins is given.

In this paper, we investigate the ILP model of finding a Pareto-optimal solution to the ALD problem. We propose two parametric families of instances that have the duality gap equal to one. It is shown that these families are difficult for the Land and Doig algorithm with some branching rules. Namely, it is shown, that the iterations number grows exponentially with the number of bins. On the basis of these families, a generalized family of instances for the ALD problem was constructed. These instances have a small duality gap and are difficult for the Land and Doig algorithm with the considered branching rules.

For the Land and Doig algorithm, families of difficult instances are known for the one-dimensional knapsack problem, the set covering problem, the set packing problem, the supply management problem, etc [7–10]. Analysis of difficult instances properties may be useful to increase the efficiency of integer programming algorithms. In addition, such families can be used for the theoretical study of other algorithms, for example, evolutionary ones [11].

To analyze the behavior of some other ILP algorithms when solving the proposed families, we study relaxation polytopes of proposed instances. We use the regular partitions approach, suggested by A.A. Kolokolov for analysis and solving the integer programming (IP) problems [12,13]. It is based on using some special partitions of space $\mathbb{R}^n$, in particular, L-partition. The fractional covering of a problem is defined as a special subset of the relaxation polytope of the problem. All points of the fractional covering must be excluded from the relaxation polytope in a process of solving the problem when some well-known algorithms are applied. The L-covering cardinality is a characteristic of fractional covering of the IP problems and plays an important role in analysis of efficiency of many algorithms based on continuous optimization techniques. In particular, it is an upper bound on the iterations number for a class of cutting plane algorithms introduced in [13] which includes the first Gomory algorithm [14,15].

The paper is organized as follows: in Sect. 2, we give the formulation of the ALD problem. In Sect. 3, we construct parametric families of instances for this problem and their generalization and study solving these families by the Land and Doig algorithm. In Sect. 4, we give the necessary information about the L-partition method and analyze the L-structure of the instances of constructed families.

2 Problem Formulation

The ALD problem and its model from [3] are described below. Let $I = \{1, ..., m\}$ be a set of teachers. The values a_i, c_i set the upper and the lower bounds of the possible load for teacher i where $i \in I$. Let $J = \{1, ..., n\}$ be a set of training courses. We denote by t_j a number of individual units of j-th course. Let b_j^k be volume (i.e., the hours number) of k-th unit of j-th course where $k \in K_j = \{1, ..., t_j\}$.

We denote by l_{ij}^k efficiency coefficient of assignment of unit k of course j to teacher i for all i, j, k. Let s_j be a maximum number of units of one training course that can be assigned to one teacher ($s_j \leq t_j$). Parameters s_j are introduced to avoid the formation of monotonous load for the teachers.

It is necessary to assign each unit of any training course to a certain teacher so that a total load of each teacher satisfies the given bounds. The first goal of optimization is to minimize the number of the assigned courses for a teacher with the highest upper bound on his/her allowable load. For other teachers, an individual upper bound on the courses number is set. This value is proportional to the upper bound on the allowable load of the teacher. The second optimization goal is to maximize the total efficiency of assignments of units to the teachers.

We introduce Boolean variables x_{ij} and z_{ij}^k where $i \in I$, $j \in J$, $k \in K_j$. Here $x_{ij} = 1$ if teacher i gives course j and $x_{ij} = 0$, otherwise; $z_{ij}^k = 1$ if teacher i gives k-th unit of course j and $z_{ij}^k = 0$, otherwise. Denote the vector of variables z_{ij}^k by Z and the vector of variables x_{ij} by X.

Let $a_{max} = \max_{i \in I} a_i$ and i_{max} be the index where this maximum is reached. Let $p_i = \frac{a_i}{a_{max}}$ for $i \in I$. We introduce a non-negative integer variable y. It is the upper bound on the number of courses assigned to teacher i_{max}.

The ALD problem can be formulated as a bicriteria ILP problem as follows

$$\text{minimize } y \tag{1}$$

$$\text{maximize } L(Z) = \sum_{i=1}^{m} \sum_{j=1}^{n} \sum_{k=1}^{t_j} l_{ij}^k z_{ij}^k \tag{2}$$

subject to

$$c_i \leq \sum_{j=1}^{n} \sum_{k=1}^{t_j} b_j^k z_{ij}^k \leq a_i, \quad i \in I, \tag{3}$$

$$\sum_{i=1}^{m} z_{ij}^k = 1, \quad j \in J, \; k \in K_j, \tag{4}$$

$$x_{ij} \leq \sum_{k=1}^{t_j} z_{ij}^k \leq s_j x_{ij}, \quad i \in I, \; j \in J, \tag{5}$$

$$\sum_{j=1}^{n} x_{ij} \leq p_i y + q, \quad i \in I, \tag{6}$$

$$y \geq 0, \; y \in \mathbb{Z}, \; x_{ij}, z_{ij}^k \in \{0, 1\}, \; i \in I, \; j \in J, \; k \in K_j. \tag{7}$$

Conditions (3) describe the allowable ranges of the total load of each teacher. Equalities (4) show that each unit of any course should be assigned to just one teacher. Inequalities (5) describe the relationship of the variables x_{ij} and z_{ij}^k. The variable $x_{ij} = 1$ if and only if there exists index k such that $z_{ij}^k = 1$, i.e., course j is assigned to teacher i if and only if at least one unit of this course is assigned to him/her. Restrictions (6) set the greatest number of courses for each teacher, depending on the value of his/her maximum permissible load. Here $q \in [0, 1)$ is a constant that controls the rounding of the values on the right side. Optimization criterion (1) means the minimization of the number of courses assigned to the teacher with the highest upper bound on allowable load. Criterion (2) maximizes the value of function $L(Z)$, i.e., the total efficiency of the load distribution.

It was shown that finding a feasible solution of (1)–(7) is NP-hard [3].

The values of function (1) belong to set $\{1, 2, ..., n\}$, i.e., the cardinality of the full set of alternatives does not exceed n.

Let integer parameter y_{max} takes values from 1 to n. The problem of finding a Pareto-optimal solution is (2)–(7) with the additional constraint

$$y \leq y_{max}. \tag{8}$$

Notice that when problems (2)–(8) are being solved in order of decreasing of y_{max}, then some values y_{max} can be skipped [4].

3 Analysis of Parametric Families of ALD Problem

We construct some families of instances of the ALD problem and study the branch and bound algorithm (Land and Doig scheme) for solving these instances.

3.1 Family $F1(n)$

Let α, β be positive integers and $\alpha \geq 3$. Consider the ALD instances with n teachers and n courses with the following input data. All courses consist of one unit of volume α, i.e., $t_j = 1$ for all j. Since index k takes a single value, we will omit it in the notation of variables and parameters, i.e., $b_j = b_j^1$, $l_{ij} = l_{ij}^1$, $z_{ij} = z_{ij}^1$ for all i, j. For teachers, the following lower and upper bounds on the possible load are set: $c_i = \alpha - 1$, $a_i = \alpha$ for $i < n$ and $c_n = \alpha$, $a_n = \alpha + 1$. Also, the efficiency coefficients are given: $l_{ij} = \beta$, $l_{nj} = \alpha + \beta$ for $i < n$ and all j. By $F1(n)$ we denote the family of these instances.

From $s_j \leq t_j$, it follows that $s_j = 1$ and conditions (5) transform into $x_{ij} = z_{ij}$ for all i, j. Next, we replace the variables vector (Z, X) with $Z = (z_{ij})_{n\times n}$. Let $q \in [\frac{1}{\alpha},\ 1)$ for $y_{max} = 1$ and $q \in [0,\ 1]$, otherwise.

The problem of finding a Pareto-optimal solution may be written as

$$\text{maximize } L(\bar{Z}) = \beta \sum_{i=1}^{n-1} \sum_{j=1}^{n} z_{ij} + (\alpha + \beta) \sum_{j=1}^{n} z_{nj} \tag{9}$$

subject to

$$\alpha - 1 \leq \sum_{j=1}^{n} \alpha z_{ij} \leq \alpha, \quad i < n, \tag{10}$$

$$\alpha \leq \sum_{j=1}^{n} \alpha z_{nj} \leq \alpha + 1, \tag{11}$$

$$\sum_{i=1}^{n} z_{ij} = 1, \quad j \in J, \tag{12}$$

$$z_{ij} \in \{0, 1\}, \ i \in I, \ j \in J. \tag{13}$$

In fact, constraints (6) are converted to

$$\sum_{j=1}^{n} z_{ij} \leq \frac{\alpha}{\alpha + 1} y_{max} + q, \quad i < n,$$

$$\sum_{j=1}^{n} z_{nj} \leq y_{max} + q.$$

For $y_{max} = 1$, these constraints are a consequence of the right-hand inequalities from (10), (11). Therefore these restrictions are not included in the ILP model for $F1(n)$. When $y_{max} > 1$, this property is kept for all $q \in [0,\ 1)$.

Note that problem (9)–(13) can be considered as an instance of the generalized assignment problem with additional lower bounds on the total volume of items placed into each knapsack.

Denote by Z^* and $\widetilde{Z}$ the optimal solutions of problem (9)–(13) and its linear relaxation which is obtained if restrictions (13) are replaced by $z_{ij} \geq 0$ for all

i, j. From (10)–(12) it follows that the matrix Z^* contains exactly one element equal to one in each row and each column. Another elements of Z^* are zeros. The number of such solutions is $n!$ and $L(Z^*) = n\beta + \alpha$.

The maximum of the objective function of the linear relaxation will be achieved when teacher n is fully loaded. Really this teacher has the maximum efficiency coefficients and the highest upper bound on the allowable load. The other teachers have efficiency coefficients equal to each other and with a lower value. Thus condition $\sum_{j=1}^{n} \tilde{z}_{nj} = 1 + \frac{1}{\alpha}$ is satisfied for any $\widetilde{Z}$. From (12), it follows that the sum of all $\tilde{z}_{ij}$ is equal to n. So there is the left-hand inequality from (10) which is satisfied as equality for some index i. For the remaining indexes, the right-hand inequalities are satisfied as equalities. Consequently, $\widetilde{Z}$ satisfies the conditions of a balanced transportation problem. If the continuous optimal solution $\widetilde{Z}$ is obtained by the simplex method, then it is a vertex of the polytope of the linear relaxation and this solution has the following property: it is impossible to construct a cycle for non-zero elements of the plan or for subsets of these elements.

The matrix $\widetilde{Z}$ can have a different structure. For any of them, condition $\sum_{j=1}^{n} \tilde{z}_{nj} = 1 + \frac{1}{\alpha}$ is satisfied and there are $j \in J$ such that $\tilde{z}_{nj} = \frac{1}{\alpha}$ and $i \neq n$ such that $\tilde{z}_{ij} = \frac{\alpha-1}{\alpha}$ (see Fig. 1). Clearly, $L(\widetilde{Z}) = n\beta + \alpha + 1$. The duality gap for the instances from $F1(n)$ is equal to 1.

$$\begin{pmatrix} 0 & 0 & \dots & 0 & 0 & \frac{\alpha-1}{\alpha} \\ 1 & 0 & \dots & 0 & 0 & 0 \\ 0 & 1 & \dots & 0 & 0 & 0 \\ \dots & \dots & \dots & \dots & \dots & \dots \\ 0 & 0 & \dots & 1 & 0 & 0 \\ 0 & 0 & \dots & 0 & 1 & \frac{1}{\alpha} \end{pmatrix} \qquad \begin{pmatrix} 0 & 0 & \dots & 0 & \frac{1}{\alpha} & \frac{\alpha-1}{\alpha} & 0 \\ 0 & 0 & \dots & \frac{1}{\alpha} & \frac{\alpha-1}{\alpha} & 0 & 0 \\ 0 & 0 & \dots & \frac{\alpha-1}{\alpha} & 0 & 0 & 0 \\ \dots & \dots & \dots & \dots & \dots & \dots & \dots \\ \frac{1}{\alpha} & \frac{\alpha-1}{\alpha} & \dots & 0 & 0 & 0 & 0 \\ \frac{\alpha-1}{\alpha} & 0 & \dots & 0 & 0 & 0 & 0 \\ 0 & 0 & \dots & 0 & 0 & \frac{1}{\alpha} & 1 \end{pmatrix}.$$

Fig. 1. Examples of $\widetilde{Z}$ with a minimum and maximum number of non-integer elements.

3.2 Analysis of Land and Doig Algorithm

We study the branch and bound algorithm (Land and Doig scheme) for solving instances from $F1(n)$. Denote this algorithm by LD. It is based on the sequential partition of the feasible set into subsets and on a calculation for each subset bounds on the objective function. These bounds are used to exclude subsets that do not contain optimal solutions for the integer problem. The bound is equal to the optimal value of the objective linear relaxation function and calculated by the simplex method. Algorithm LD constructs a binary search tree. Variable z_{kl} corresponding to a fractional element of the optimal solution to the linear relaxation is chosen as a branching variable. At the current iteration, the subset of feasible solutions is divided into two subsets by adding one of the constraints $z_{kl} = 0$ (left branch) or $z_{kl} = 1$ (right branch). Such variables are called fixed, and other variables are called free. Different branching rules are applied when solving ILP problems (see, for example, [16–18]). The number of iterations of algorithm LD determined as the number of solved linear programming (LP) problems.

Let $A1$ be algorithm LD with a branching rule for which $k \neq n$ holds for any branching variable z_{kl} when solving instances from $F1(n)$. Suppose the variables of the problem are ordered as follows: $z_{11}, ..., z_{1n}, z_{21}, ..., z_{2n}, ..., z_{n1}, ..., z_{nn}$ then, for example, the rules are possible: the first fractional element, the fractional element that is closest to one, the fractional element with the lowest value l_{ij}.

Theorem 1. *Algorithm* $A1$ *requires at least* $2^n - 1$ *iterations to solve any instance from* $F1(n)$.

Proof. Let us analyze a part of the search tree that occurs during the running of the algorithm $A1$. We will be interested in branches (i.e. paths from the tree root to the current node) that have the length not less than $n-1$. Each such branch corresponds to not less than $n-1$ fixed variables.

Let $z_{kl} = 1$. From (10), it follows that variables $\tilde{z}_{kj}$ equal to 0 for $j \neq l$ in any feasible solution of the corresponding LP subproblem. From (12), it results that $\tilde{z}_{il} = 0$ for $i \neq k$. So, variables of row k and column l of matrix Z cannot be selected below for branching. The LP subproblem has an integer optimal solution Z^* and $L(Z^*) = n\beta + \alpha$ when assigning $n-1$ variables equal to ones. The tree branch breaks off.

Now we prove that all other nodes of the search tree at the level $n-1$ have the upper bounds on the objective function equal to $L(\widetilde{Z}) = n\beta + \alpha + 1$.

Let $n-1$ variables from some row of the matrix Z be fixed at 0. Then we obtain an optimal continuous solution $\widetilde{Z}$, in which the single non-zero element in this row is equal to $\frac{a-1}{a}$ or 1. Assigning $n-1$ zeros in some column j leads to an optimal continuous solution $\widetilde{Z}$ in which $\tilde{z}_{nj} = 1$.

Suppose that $n-1$ fixed zeros are located in arbitrary rows, excluding the row n. Note that for the linear relaxation of problem (9)–(13), the value $n(n-1)(n-1)!$ is the number of optimal solutions having the structure shown in Fig. 1 left. Here $n(n-1)$ is the number of location variants of fractional elements and $(n-1)!$ is the number of location variants of ones. If some variable z_{kl} $(k < n)$ is equal to 0, then $(n-1)!$ matrices $\widetilde{Z}$ of specified type are excluded from the relaxation set, namely, those for which $\tilde{z}_{kl} = \frac{\alpha-1}{\alpha}$. If $n-1$ zeros are fixed, then the number of excluded solutions is $(n-1)(n-1)!$. Therefore, there are at least n optimal solutions with the specified structure for such LP subproblem and $L(\widetilde{Z}) = n\beta + \alpha + 1$.

Consider the general case when t ones and $n-t-1$ zeros are assigned where $1 \leq t \leq n-2$. If t variables are fixed at 1 then the linear relaxation of problem (9)–(13) is converted to a similar LP subproblem of dimension $(n-t) \times (n-t)$. As follows from the case discussed above, the assignment of $n-t-1$ zeros does not break the branch.

Thus, when solving instances from $F1(n)$, algorithm $A1$ builds the search tree of depth at least $n-1$, and this corresponds to solve at least 2^n-1 LP problems. The theorem is proved.

Remark 1. Consider algorithm LD with any branching rule by which a branching variable is selected only from the last row of the matrix Z when solving instances from $F1(n)$. As an example, we give the following rule: the fractional element with the highest coefficient l_{ij}. In this case, we can show that algorithm LD requires n^2+n-1 iterations to solve problems from $F1(n)$.

3.3 Families $F2(n)$ and $G(2n)$

For the ALD problem, consider the family $F2(n)$ of instances which differ from the instances from $F1(n)$ only by the parameters. We have $c_i=\alpha$, $a_i=\alpha+1$ for $i<n$ and $c_n=\alpha-1$, $a_n=\alpha$, $l_{ij}=\alpha+\beta$ and $l_{nj}=\beta$ for $i<n$ and all j. Here $\alpha\geq 3$, $\beta>0$ are integers.

The ILP problem to find a Pareto-optimal solution takes the following form

$$\text{maximize } L(Z)=(\alpha+\beta)\sum_{i=1}^{n-1}\sum_{j=1}^{n} z_{ij}+\beta\sum_{j=1}^{n} z_{nj} \tag{14}$$

subject to

$$\alpha\leq\sum_{j-1}^{n}\alpha z_{ij}\leq\alpha+1,\ \ i<n, \tag{15}$$

$$\alpha-1\leq\sum_{j=1}^{n}\alpha z_{nj}\leq\alpha, \tag{16}$$

$$\sum_{i=1}^{n} z_{ij}=1,\ \ j\in J, \tag{17}$$

$$z_{ij}\in\{0,1\},\ i\in I,\ j\in J. \tag{18}$$

Instances from $F1(n)$ and $F2(n)$ have the same set of optimal solutions. Linear relaxations of instances from $F2(n)$ also have several optimal solutions, but these solutions have another structure. For example, one of the structures with the minimum number of fractional elements is shown in Fig. 2.

It is clear that $L(Z^*)=(n-1)\alpha+n\beta$, $L(\widetilde{Z})=(n-1)\alpha+n\beta+1$. The duality gap is still equal to one.

Denote by $A2$ the algorithm LD with a branching rule for which condition $k\neq n$ holds for any branching variable z_{kl} when solving instances from $F2(n)$. As examples of such rule, we can mention the following: the fractional element with the highest value of l_{ij}, the fractional element that is closest to zero or the first fractional element.

Theorem 2. *Algorithm $A2$ requires at least 2^{n-1} iterations to solve any instance from family $F2(n)$.*

$$\begin{pmatrix} 1 & 0 & \dots & 0 & \frac{1}{\alpha} \\ 0 & 1 & \dots & 0 & 0 \\ \dots & \dots & \dots & \dots & \dots \\ 0 & 0 & \dots & 1 & 0 \\ 0 & 0 & \dots & 0 & \frac{\alpha-1}{\alpha} \end{pmatrix}.$$

Fig. 2. Example of structure of LP solution $\widetilde{Z}$ for instances of $F2(n)$.

Proof. Initially, we analyze the optimal solution $\widetilde{Z}$ of the linear relaxation of problem (14)–(18) with $n-1$ fixed variables. Consider several cases.

Let $n-1$ variables from column j_0 of matrix Z be fixed to zeros. From (17) it follows that $z_{nj_0} = 1$, and from (16) it follows that $z_{nj} = 0$ for $j \neq j_0$. The other variables have the efficiency coefficients equal to $\alpha+\beta$. From here we obtain that $L(\widetilde{Z}) = (n-1)(\alpha+\beta)+\beta$ for any feasible solution to the LP problem. If the optimal solution of the LP problem is integer, then this branch breaks off.

Let $n-1$ zeros are fixed in several columns of matrix Z. Then there is an optimal solution $\widetilde{Z}$ with the minimum number of fractional components, and $L(\widetilde{Z}) = L(Z^*)+1$. This can show as in the proof of Theorem 1.

Let a variable is fixed to 1, for example, $z_{kl} = 1$ for $k \neq n$. From (17) it follows that $z_{il} = 0$, $i \neq k$, and these variables will not be selected further for branching. Let $n-1$ ones are fixed. Then the linear relaxation has an optimal solution, which, up to a permutation of the columns, has the structure shown in Fig. 2. Solving the problem continues on this branch.

Consider the general case when t ones and $n-t-1$ zeros are assigned where $1 \leq t \leq n-2$. For simplicity, we put $z_{kk} = 1$ for $k \leq t$. In addition, we assume that the other elements are equal to zeros in the first t rows of $\widetilde{Z}$, i.e., the corresponding variables are not selected for branching at the next iterations. In this case, the assignment of t ones means the transition to a subproblem of dimension $(n-t)\times n$. Let $z_{ij} = 0$ for $i \leq t$ and all j in this problem then we have the subproblem of type (14)–(18) of dimension $(n-t)\times(n-t)$. It can be shown as in the proof of Theorem 1 that the assignment of no more than $n-t-1$ zeros does not break this branch.

Therefore, all branches of the search tree have a length at least $n-1$. This tree has at least 2^n-1 nodes, i.e., the number of the solved LP subproblems is at least 2^n-1.

Now we only need to consider the case when the branching variable z_{kl} is selected from the row in which some variable is already fixed to 1. If $z_{kl} = 0$, then the LP subproblem is solvable and its dimension is keep and we have $L(\widetilde{Z}) = L(Z^*)+1$. If $z_{kl} = 1$ then the LP subproblem has no solution because a constraint from (15) is false with $i = k$. For the search tree of depth $n-1$, this means the insertion of the fragments corresponding to the just mentioned branchings. So, the number of iterations of algorithm $A2$ can only to increase.

The theorem is proved.

Remark 2. Consider algorithm LD with a branching rule, in which a branching variable is selected only from the last row of the linear relaxation solution to an instance from $F2(n)$, for example, it is a fractional element that is closest to one or a fractional element with a minimum value of l_{ij}. Then, to solve instances from $F2(n)$, algorithm LD requires at least $2n+1$ iterations.

Using $F1(n)$ and $F2(n)$, we construct a family $G(2n)$. These instances have $2n$ teachers and $2n$ courses with a single unit of volume α. Other input data is given below

$$(l_{ij})_{2n\times 2n} = \begin{pmatrix} L_1 & H \\ H & L_2 \end{pmatrix} \quad c = (c_1, c_2) \quad a = (a_1, a_2).$$

Here L_1, L_2 are the efficiency matrices of dimension $n \times n$; c_1, a_1 and c_2, a_2 are vectors of the lower and the upper bounds on the allowable load of teachers for families $F1(n)$ and $F2(n)$. H is a $(n \times n)$-matrix and all of its elements are equal to -1. For $G(2n)$, the duality gap is equal to 2. Easy to show that this problem is difficult for algorithm LD with branching rules from algorithms $A1$ and $A2$.

4 Analysis of L-Covering

At first, we give the necessary information about the method of regular partitions [13]. Let $\mathbb{Z}^n$ be the set of all integer points of space $\mathbb{R}^n$. The L-partition of space $\mathbb{R}^n$ is defined as follows. Points $z \in \mathbb{Z}^n$ constitute the separate L-classes, i.e., the elements of partition. Points $x,\ y \notin \mathbb{Z}^n$ $(x \succ y)$ belong to same fractional L-class if no $z \in \mathbb{Z}^n$ exist such that $x \succ z \succ y$ is holds. Here $\succ$ are the symbol of the lexicographical order. Denote by X/L the quotient set induced by L-partition for a set $X \subset \mathbb{R}^n$. Set X/L is called L-structure of set X and its elements are called L-classes. It is known that any fractional L-class V from X/L can be represented as:

$$V = X \cap \{x \mid x_1 = d_1, ..., x_{r-1} = d_{r-1}, d_r < x_r < d_r + 1\}, \tag{19}$$

where d_j is integer for $j = 1, ..., r$ and $r \leq n$.

We consider the following problem of finding the lexicographically maximal integer element of a set Ω

$$\text{find} \quad \text{lexmax}\ (\Omega \cap \mathbb{Z}^n) \tag{20}$$

where Ω be some closed subset from $\mathbb{R}^n$.

Assume that the relaxation of this problem

$$\text{find} \quad \text{lexmax}\ \Omega$$

is solvable.

The fractional covering of problem (20) is defined as a set

$$\Omega_* = \{x \in \Omega \mid x \succ z \text{ for all } z \in (\Omega \cap \mathbb{Z}^n)\}.$$

The quotient set Ω_*/L is called the L-covering of problem (20), and $|\Omega_*/L|$ is called the L-covering cardinality.

Let us investigate the L-coverings of instances from family $F1(n)$. Denote by M the polytope of the linear relaxation of an instance from $F1(n)$. This polytope is described by conditions (10)–(12) and inequalities $z_{ij} \geq 0$ for all i, j. Vector Z denotes all variables z_{ij} ordered in an arbitrary fixed way. As before, Z^* and $\widetilde{Z}$ denote optimal solutions to the problem and its linear relaxation. Note that $L(\widetilde{Z}) = L(Z^*) + 1$. Let us introduce a new variable $z_0 = L(Z)$. It is clear that z_0 has integer value when vector Z is integer. Now we have the following lexicographical formulation of $F1(n)$

$$\text{find} \quad \text{lexmax}(\hat{M} \cap \mathbb{Z}^{n \times n+1})$$

where $\hat{M} = \{(z_0, Z) \in \mathbb{R}^{n \times n+1} \mid z_0 = L(Z),\ Z \in M\}$.

Let $\hat{M}_*$ be the fractional covering of this problem. In [14], the cardinality of the L-covering of the lexicographic optimization problem has been described through the L-structure of the relaxation polytope of the problem in the formulation with the objective function. We put $M_{L(\widetilde{Z})} = \{Z \in M \mid L(Z) = L(\widetilde{Z})\}$. For problem (9)–(13), the duality gap is equal to one and the lexicographical maximal element of set $\{Z \in M \mid L(Z) = L(Z^*)\}$ is integer. So above mentioned relation has the form

$$|\hat{M}_*/L| = |M_{L(\widetilde{Z})}/L| + 1. \tag{21}$$

The first sum term is the number of L-classes of set $\{Z \in \hat{M} \mid L(Z) = L(\widetilde{Z}\})$. The second term corresponds to the L-class with $L(Z^*) < z_0 < L(\widetilde{Z})$.

Theorem 3. *When the order of variables is z_{11}, ..., z_{1n}, z_{21}, ..., z_{2n}, ..., z_{n1}, ..., z_{nn}, the following estimate holds for $n \geq 4$*

$$|\hat{M}_*/L| \geq 1.7 \cdot n!.$$

Proof. As follows from (21), it suffices to estimate the number of optimal solutions of the linear relaxation of the problem (9)–(13) belonging to different L-classes. For simplicity, we will consider only solutions with the minimum number of fractional components. For a given order of variables, it is convenient to consider the solution Z as a matrix.

Property (19) of the L-partition implies that any fractional class is determined by the index r of its first fractional component and the values of the first $r-1$ integer components. Let the first fractional element of $\widetilde{Z}$ belongs to row i where $i < n$. According to Fig. 1, other elements of this row are equal to zeros, i.e., the number of variants of the fractional element location is equal to n. All variables of the preceding rows take integer values. Also from (10), it follows that each such row contains a single element equal to one and this element cannot belong to the same column as the first fractional element. Consequently, the number of such L-classes is equal to $P(i) = n(n-1)...(n-i+1)$ for $i > 1$ and $P(1) = n$. Thus, we have

$$|\hat{M}_*/L| > |M_{L(\tilde{Z})}/L| = \sum_{i=1}^{n-1} P(i) = n! \sum_{i=1}^{n-1} \frac{1}{(n-i)!} \geq 1.7 \cdot n!.$$

The theorem is proved.

From this theorem, it follows that the upper bound on the number of iterations of a dual fractional cutting plane processes [13] including the first Gomory algorithm is exponential for family $F1(n)$. The result holds for a lower bound on the number of iterations of such processes based on totally regular cuts [15]. Note that the ILP problems, possessing exponential L-coverings, are difficult also for the L-class enumeration method [19].

The following theorem is easy to prove.

Theorem 4. *Consider the order of variables in which the variables z_{nj} are located in the first n places. Then the L-covering cardinality does not exceed* $\frac{1}{2}(n-1)(n+2)$.

In this case the number of iterations for above mentioned cutting plane process is a polynomial.

Similar statements hold for the family $F2(n)$.

5 Conclusion

We have considered the problem of academic load distribution among teachers as the bicriteria ILP problem and the single-criterion problem of searching for a Pareto optimal solution. We constructed parametric families of problems with duality gap equal to one and two. We showed that the proposed problems are difficult with certain branching rules for the Land and Doig algorithm, although with other branching rules the instances lose their complexity. The generalization of families remains difficult for the same branching rules. The study of the fractional covering of the problems showed that the cardinality of the L-covering varies depending on the order of variables from exponential to polynomial. This means that the constructed families of instances have the corresponding complexity for the L-class enumeration algorithm and some dual cutting plane algorithms. In the future, it is of interest to study more complex branching rules for the Land and Doig algorithm when solving the considered problem.

References

1. Hultberg, T.H., Cardoso, D.M.: The teacher assignment problem: a special case of the fixed charge transportation problem. Eur. J. Oper. Res. **101**, 463–473 (1997). https://doi.org/10.1016/S0377-2217(96)00082-3
2. Sultanova, S.N., Tarkhov, S.V.: Modeli i algoritmy podderzhki prinyatiya resheniy pri raspredelenii uchebnoy nagruzki prepodavateley (Models and algorithms of decision support in the distribution of academic load of teachers). Vestnik UGATU **7**(3), 107–114 (2006). (in Russian)
3. Zaozerskaya, L., Plankova, V., Devyaterikova, M.: Modeling and solving academic load distribution problem. In: CEUR Workshop Proceedings, Proceedings of the School-Seminar on Optimization Problems and their Applications (OPTA-SCL 2018), pp. 438–445. CEUR (2018)
4. Zaozerskaya, L.A., Plankova, V.A.: Researching and solving a bicriteria supply management problem with the given volumes of batches. J. Phys: Conf. Ser. **1210**, 012164 (2019). https://doi.org/10.1088/1742-6596/1210/1/012164
5. Peeters, M., Degraeve, Z.: The co-printing problem: a packing problem with a color constraint. Oper. Res. **52**(4), 623–638 (2004)
6. Kondakov, A., Kochetov, Y.: A core heuristic and the branch-and-price method for a bin packing problem with a color constraint. In: Eremeev, A., Khachay, M., Kochetov, Y., Pardalos, P. (eds.) OPTA 2018. CCIS, vol. 871, pp. 309–320. Springer, Cham (2018). https://doi.org/10.1007/978-3-319-93800-4_25
7. Jeroslow, R.: Trivial integer programs unsolvable by branch-and-bound. Math. Program. **6**(1), 105–109 (1974)
8. Kolpakov, R., Posypkin, M.: Asimptoticheskaya otsenka slozhnosti metoda vetvey i granits s vetvleniyem po drobnoy peremennoy dlya zadachi o rantse (Asymptotic estimate on the complexity of the branch-and-bound method with branching by a fractional variable for the knapsack problem). Diskret. Anal. Issled. Oper. **15**(1), 58–81 (2008). (in Russia)
9. Saiko, L.: Issledovaniye moshchnosti L-nakrytiy nekotorykh zadach o pokrytii (Investigation of cardinality of L-coverings for some set covering problems). In: Discrete Optimization and Analysis of Complex Systems, pp. 76–97. VTs SO AN SSSR, Novosibirsk (1989). (in Russia)
10. Zaozerskaya, L.: Analysis of fractional covering of some supply management problems. J. Math. Model. Algorithms **5**(2), 201–213 (2006). https://doi.org/10.1007/s10852-005-9016-z
11. Borisovsky, P.A., Eremeev, A.V.: A study on performance of the (1+1)-evolutionary algorithm. In: De Jong, K., Poli, R., Rowe, J. (eds.) Foundations of Genetic Algorithms, vol. 7, pp. 271–287. Morgan Kaufmann, Burlington (2003). An Imprint of Elsevier Science
12. Kolokolov, A.A.: Regular cuts by solving integer optimization problems. Upravlyaemye Sistemy, Institute Math. SB AS USSR **21**, 18–25 (1981). (in Russian)
13. Kolokolov, A.A.: Regular partitions and cuts in integer programming. In: Korshunov, A.D. (ed.) Discrete Analysis and Operations Research. Mathematics and Its Applications, vol. 355, pp. 59–79. Springer, Dordrecht (1996). https://doi.org/10.1007/978-94-009-1606-7_6
14. Kolokolov, A.A., Zaozerskaya, L.A.: Finding and analysis of estimation of the number of iterations in integer programming algorithms using the regular partitioning method. Russ Math. **58**(1), 35–46 (2014). https://doi.org/10.3103/S1066369X14010046

15. Kolokolov, A.A., Zaozerskaya, L.A.: Analysis of some cutting plane algorithms of integer programming. In: Stukach, O. (ed.) Dynamics of Systems, Mechanisms and Machines (Dynamics), Omsk, Russia, 15–17 November 2016, pp. 1–5. IEEE (2016). http://ieeexplore.ieee.org/document/7819028/
16. Kellerer, H., Pferschy, U., Pisinger, D.: Knapsack Problems. Springer, Heidelberg (2003). https://doi.org/10.1007/978-3-540-24777-7
17. Balas, E., Carrera, M.C.: A dynamic subgradient-based branch-and-bound procedure for set covering. Oper. Res. **44**(6), 875–890 (1996)
18. Caprara, A., Toth, P., Fischetti, M.: Algorithms for the set covering problem. Ann. Oper. Res. **98**, 353–371 (2000)
19. Kolokolov, A.A.: Some L-class enumeration algorithms for integer programming problems. In: Proceedings of the of 3rd IFIP WG-7.6 Working Conference on Optimization - Based Computer - Aided Modelling and Design, pp. 256–260. IITA, Prague, Czech Republic (1995)

Mathematical Programming

Regularization and Matrix Correction of Improper Linear Programming Problems

Vladimir Erokhin(✉), Sergey Sotnikov, Andrey Kadochnikov, and Alexey Vaganov

Mozhaisky Military Space Academy, St. Petersburg, Russia
{erohin_v_i,sergey_v.sotnikov,andrey_p.kadochnikov, alexey_a.vaganov}@mail.ru

Abstract. The results of the study of improper linear programming problems are presented, in which the duality theory is essentially used and the approaches of I.I. Eremin (correction of incompatible constraints) and A.N. Tikhonov (creation of compatible systems of constraints equivalent in accuracy to given incompatible constraints). The problem of a stable solution to an approximate (and, possibly, improper) pair of mutually dual linear programming problems with a coefficient matrix of size $m \times n$ is reduced to a Mathematical Programming problem of dimension $m + n + 2$. The necessary and sufficient conditions for the existence of a solution and constructive formulas for its calculation are obtained. Computational experiments were carried out on a model Linear Programming problem with an approximate matrix and vectors of the right-hand side and the objective function, demonstrating the convergence of the obtained solutions to the normal solutions to direct and dual problems with a decrease in the level of data error.

Keywords: Improper linear programming problems · Duality theory · Matrix correction · Tikhonov regularization

1 Introduction

Consider a pair of mutually dual Linear Programming problems (LPs)

$$L(A, b, c) : c^\top x \to \max, \text{s.t. } Ax = b, x \geqslant 0,$$
$$L^*(A, b, c) : u^\top b \to \min, \text{s.t. } u^\top A \geqslant c^\top,$$

where $A \in \mathbb{R}^{m \times n}$; $b, u \in \mathbb{R}^m$; $c, x \in \mathbb{R}^n$.

I. Bykadorov et al. (Eds.): MOTOR 2019, CCIS 1090, pp. 283–293, 2019.
https://doi.org/10.1007/978-3-030-33394-2_22

The feasible sets, optimal values, and the sets of optimal solutions to these problems are defined as

$$\mathcal{X}(A,b) \triangleq \{x \,|\, Ax = b, x \geqslant 0\},\ \mathcal{U}(A,c) \triangleq \{u \,|\, u^\top A \geqslant c^\top\},$$
$$\ell \triangleq \sup_{x \in \mathcal{X}(A,b)} c^\top x,\ \ell^* \triangleq \inf_{u \in \mathcal{U}(A,b)} u^\top b,$$
$$\mathcal{X}_{\text{opt}}(A,b,c) \triangleq \{x \in \mathcal{X}(A,b) \,|\, c^\top x = \ell\},$$
$$\mathcal{U}_{\text{opt}}(A,b,c) \triangleq \{u \in \mathcal{U}(A,c) \,|\, u^\top b = \ell^*\}.$$

Since both *proper* and *improper* problems will be mentioned below, it is important to recall the relevant definitions and the main facts of duality theory. It is convenient to present this information in the form of a theorem (see, for example, [1,2]).

Theorem 1. *The solvability or unsolvability of problems $L(A,b,c)$ and $L^*(A,b,c)$ are completely characterized by the following four cases.*

1. *$\mathcal{X}(A,b) \neq \varnothing$, $\mathcal{U}(A,c) \neq \varnothing$. Then $L(A,b,c)$ and $L^*(A,b,c)$ are both solvable. They are called* ***proper*** *problems and it is true that $-\infty < \ell = \ell^* < +\infty$.*
2. *$\mathcal{X}(A,b) = \varnothing$, $\mathcal{U}(A,c) \neq \varnothing$. Then ℓ is not defined, $\ell^* = -\infty$ and the problems are both unsolvable. $L(A,b,c)$ is called an* ***improper problem of the first type****, while $L^*(A,b,c)$ is called an* ***improper problem of the second type****.*
3. *$\mathcal{X}(A,b) \neq \varnothing$, $\mathcal{U}(A,c) = \varnothing$. Then $\ell = +\infty$, ℓ^* is not defined and the problems are both unsolvable. $L(A,b,c)$ is called an* ***improper problem of the second type****, while $L^*(A,b,c)$ is called an* ***improper problem of the first type****.*
4. *$\mathcal{X}(A,b) = \varnothing$, $\mathcal{U}(A,c) = \varnothing$. Then $L(A,b,c)$ and $L^*(A,b,c)$ are both unsolvable and are called an* ***improper problems of the third type****.*

Remark 1. In the cases 2–4, the sets $\mathcal{X}_{opt}(A,b,c)$ and $\mathcal{U}_{opt}(A,b,c)$ are not defined.

Assume that the parameters A, b, c are subject to perturbations that either make the solutions of $L(A,b,c)$ and $L^*(A,b,c)$ unstable and far from hypothetical exact solutions or these problems become improper. In this case, it is reasonable to apply regularization and parameter correction procedures, which can be formalized, for example, in the form of the search for a Tikhonov regularized solution to an approximate pair of mutually dual LPs.

Let us clarify the problem's statement. Suppose that there exist matrix $A_0 \in \mathbb{R}^{m\times n}$ and vectors $b_0 \in \mathbb{R}^m$, $c_0 \in \mathbb{R}^n$ such that problems $L(A_0,b_0,c_0)$ and $L^*(A_0,b_0,c_0)$ are proper. Define the matrix A_0 and vectors b_0, c_0 as *exact*, the matrix A and vectors b, c as *approximate*, and the corresponding LP problems as problems with the precise and approximate data. Let us assume that the following conditions are satisfied: $\|A - A_0\| \leqslant \mu$, $\|b - b_0\| \leqslant \delta_b$, $\|c - c_0\| \leqslant \delta_c$, where $\mu, \delta_b, \delta_c > 0$ are some constants known a priori, and the symbol $\|\cdot\|$ denotes Euclidean matrix norm (the symbol will also be used to denote an Euclidean vector norm through the paper). Following the logic of works [3,4], we consider the **problem** $R(A, b, c, \mu, \delta_b, \delta_c)$:

$$\min_{x_1,u_1,A_1,b_1,c_1} \|x_1\|^2 + \|u_1\|^2,$$
$$\text{s.t. } A_1 \in \mathbb{R}^{m\times n}, b_1 \in \mathbb{R}^m, c_1 \in \mathbb{R}^n, x_1 \in \mathbb{R}^n, u_1 \in \mathbb{R}^m,$$
$$\|A - A_1\| \leqslant \mu, \|b - b_1\| \leqslant \delta_b, \|c - c_1\| \leqslant \delta_c,$$
$$x_1 \in \mathcal{X}_{opt}(A_1, b_1, c_1), u_1 \in \mathcal{U}_{opt}(A_1, b_1, c_1).$$

The specified problem allows the following interpretation: find vectors x_1, u_1, b_1, c_1 and matrix A_1 such that the pair of problems $L(A, b, c)$ and $L^*(A, b, c)$ (maybe improper) with the approximate matrix A and vectors b, c is mapped to the corresponding solvable problems $L(A_1, b_1, c_1)$ and $L^*(A_1, b_1, c_1)$. Matrix A_1 should be close to matrix A and vectors b_1 and c_1 should be close to vectors b and c accordingly. Euclidean norm is the proximity measure of matrices and the Euclidean norm for vectors is the proximity measure of vectors. Numerical characteristic of proximity for the mentioned objects are μ, δ_b, δ_c. Vectors x_1 and u_1 are solutions to these problems, and the sum of the squares of their Euclidean norms is minimal.

Using a technique similar to that used by A.N. Tikhonov, for justification of the method of stable solution to approximate systems of linear algebraic equations, it can be shown that

$$\mu, \delta_b, \delta_c \to 0 \Rightarrow A_1 \to A_0, b_1 \to b_0, c_1 \to c_0, x_1 \to x_0, u_1 \to u_0,$$

where x_0 and u_0 are the solutions to the problems $L(A_0, b_0, c_0)$ and $L^*(A_0, b_0, c_0)$ respectively, and the sum of the squares of the Euclidean norms of the vectors x_0 and u_0 is minimal. From now on vectors that exhibit those characteristics will be called a *normal solution* to a pair of mutually dual LP problems. That is, vectors x_1 and u_1 are a stable approximation of the normal solution to a pair of mutually dual LP problems with exact data.

If problems $L(A, b, c)$ and $L^*(A, b, c)$ are improper, then problem $R(A, b, c, \mu, \delta_b, \delta_c)$ can be also interpreted as a correction problem, introduced in studies of I.I. Eremin and his followers:

Problem $C(A, b, c, \mu, \delta_b, \delta_c)$:

$$\min_{x,u,H,h_b,h_c} \|x\|^2 + \|u\|^2,$$
$$\text{s.t. } H \in \mathbb{R}^{m\times n}, h_b \in \mathbb{R}^m, h_c \in \mathbb{R}^n, x \in \mathbb{R}^n, u \in \mathbb{R}^m,$$
$$\|H\| \leqslant \mu, \|h_b\| \leqslant \delta_b, \|h_c\| \leqslant \delta_c,$$
$$x \in \mathcal{X}_{opt}(A + H, b + h_b, c + h_c), u \in \mathcal{U}_{opt}(A + H, b + h_b, c + h_c).$$

It is obvious that problems R and C are solvable (not solvable) simultaneously, and the solutions are interrelated: $A_1 = A + H$, $b_1 = b + h_b$, $c_1 = c + h_c$, $x_1 = x$, $u_1 = u$. Hence, the problems R and C are equivalent.

This paper is devoted to the study of the problem R, which includes the justification of the necessary and sufficient conditions for the existence of a solution, reduction to the auxiliary problem of Mathematical Programming, constructive formulas, and computational experiments.

The work continues the research performed by the authors earlier and devoted to both Tikhonov's method of solving approximate systems of linear algebraic equations [5,6] and the problem of matrix correction and regularization of improper LPs [7–14].

It should be noted that a detailed review of papers devoted to the construction of stable solutions to improper LPs could be the subject of a separate article. We will only mention work [15] and contemporary works close to the subject of the article [16–20]. §§12–13 of the monography [1] and the work [21] should be noted as fundamental and novel papers dedicated to matrix correction of improper problems in LP. The multiparametric approximation was considered as a correction approach for the improper problems in those papers. The term "matrix correction" was first introduced in the paper [22] and also problems of matrix correction for improper problems in LP were formulated as problems of Mathematical Programming.

2 Mathematical Tools

Consider a "technical" problem, which is the *inverse* LP. As an example of problems of this kind, we can mention one recent work [23], devoted to the study of the problem of the minimal (in the Euclidean norm) change (correction) of the vector of the objective function that guarantees the optimality of a given vector taken from the feasible set of vectors.

The problem considered below is inverse because the solutions to the primal and dual LPs are given, whereas the coefficient matrix is unknown.

Problem $C_1(x, u, b, c)$: *Given vectors* $x, c \in \mathbb{R}^n$, $u, b \in \mathbb{R}^m$ *with* $x, u \neq 0$ *and* $x \geqslant 0$, *find a matrix* $A \in \mathbb{R}^{m\times n}$ *of a minimal Euclidean norm such that* $x \in \mathcal{X}_{opt}(A, b, c)$, $u \in \mathcal{U}_{opt}(A, b, c)$.

The solution to problem $C_1(x, u, b, c)$ is given by the following statement, based on duality theory (see Theorem 1) and the results of papers [7,13].

Theorem 2. *[13][1] Given* $x, u \neq 0$, *a solution to problem* $C_1(x, u, b, c)$ *exists if and only if*

$$c^\top x = u^\top b = \alpha.$$

The solution $\hat{A}$ *to problem* $C_1(x, u, b, c)$ *with the minimal Euclidean norm is unique and given by the formula*

$$\hat{A} = \frac{bx^\top}{x^\top x} + \frac{ug^\top}{u^\top u} - \alpha\frac{ux^\top}{x^\top x \cdot u^\top u},$$

[1] The author of the theorem is Alexander Krasnikov.

where

$$g = (g_j) \in \mathbb{R}^n,\ g_j = \begin{cases} 0 \ \mathit{if}\ c_j \leqslant 0 \ \mathit{and}\ x_j = 0, \\ c_j \ \mathit{otherwise}. \end{cases}$$

Moreover,

$$\|\hat{A}\| = \frac{\|b\|^2}{\|x\|^2} + \frac{\|g\|^2}{\|u\|^2} - \frac{\alpha^2}{\|x^2\| \cdot \|u\|^2}.$$

Let's consider one more "technical" problem.

Problem $C_2(A, b, c, x, u, w_b, w_c)$: *Given vectors* $x, c \in \mathbb{R}^n$, $u, b \in \mathbb{R}^m$ *with* $x \geqslant 0$, *matrix* $A \in \mathbb{R}^{m\times n}$, *real scalars* $w_b, w_c > 0$, *find a matrix* $H \in \mathbb{R}^{m\times n}$, *vectors* $h_b \in \mathbb{R}^m$, $h_c \in \mathbb{R}^n$ *being a solution to optimization problem*

$$\Theta(H, h_b, h_c) \triangleq \|H\|^2 + \|h_b\|^2 + \|h_c\|^2 \to \min\left(= \hat{\Theta}\right),$$

such that $x \in \mathcal{X}_{opt}(A+H, b+w_b h_b, c+w_c h_c)$, $u \in \mathcal{U}_{opt}(A+H, b+w_b h_b, c+w_c h_c)$.

The solution to problem $C_2(A, b, c, x, u, w_b, w_c)$ is given by the following statement.

Theorem 3. *The solution to problem* $C_2(A, b, c, x, u, w_b, w_c)$ *for any data* A, b, c, x, u *and* $w_b, w_c > 0$ *exists, is unique and defined by formulas*

$$\hat{H} = \frac{(b - Ax)\, x^\top}{x^\top x + w_b^2} + \frac{u g^\top}{u^\top u + w_c^2} - \alpha \frac{u x^\top}{(x^\top x + w_b^2)\,(u^\top u + w_c^2)}, \tag{1}$$

$$\hat{h}_b = -\left(\frac{w_b\,(b - Ax)}{x^\top x + w_b^2} + \frac{\omega u}{u^\top u + w_c^2} - \alpha \frac{w_b u}{(x^\top x + w_b^2)\,(u^\top u + w_c^2)}\right), \tag{2}$$

$$\hat{h}_c = -\left(\frac{w_c g}{u^\top u + w_c^2} + \frac{\upsilon x}{x^\top x + w_b^2} - \alpha \frac{w_c x}{(x^\top x + w_b^2)\,(u^\top u + w_c^2)}\right), \tag{3}$$

$$\hat{\Theta} = \Theta(\hat{H}, \hat{h}_b, \hat{h}_c) = \frac{\|b - Ax\|^2 + \upsilon^2}{\|x\|^2 + w_b^2} + \frac{\|g\|^2 + \omega^2}{\|u\|^2 + w_c^2} - \frac{\alpha^2}{(\|x\|^2 + w_b^2)(\|u\|^2 + w_c^2)}, \tag{4}$$

where

$$\zeta = w_c^2 \cdot x^\top x + w_b^2 \cdot u^\top u + w_b^2 w_c^2, \tag{5}$$

$$\alpha = \frac{w_c^2\,(c^\top x - u^\top Ax)\,(x^\top x + w_b^2) + w_b^2\,(u^\top b - u^\top Ax)\,(u^\top u + w_c^2)}{\zeta}, \tag{6}$$

$$\upsilon = \frac{w_c}{\zeta} \cdot \left(w_b^2\,(c^\top x - u^\top Ax) + x^\top x\,(c^\top x - u^\top b)\right), \tag{7}$$

$$\omega = \frac{w_b}{\zeta} \cdot \left(w_c^2\,(u^\top b - u^\top Ax) + u^\top u\,(u^\top b - c^\top x)\right), \tag{8}$$

$$g = (g_j) \in \mathbb{R}^n,\ g_j = \begin{cases} 0 \ \mathit{if}\ \left(c - A^\top u\right)_j \leqslant 0 \ \mathit{and}\ x_j = 0, \\ \left(c - A^\top u\right)_j \ \mathit{otherwise}. \end{cases} \tag{9}$$

Proof. Consider the **problem** $\tilde{C}_1(\tilde{x}, \tilde{u}, \tilde{b}, \tilde{c})$, which is a modification of the problem $C(x, u, b, c)$: *Given: vectors* $x, c \in \mathbb{R}^n$, $u, b \in \mathbb{R}^m$ *with* $x \geqslant 0$, *real scalars* $w_b, w_c > 0$ *and the vectors*

$$\begin{aligned} \tilde{x} = \begin{bmatrix} x \\ w_b \end{bmatrix} \in \mathbb{R}^{n+1}, \tilde{u} = \begin{bmatrix} u \\ w_c \end{bmatrix} \in \mathbb{R}^{m+1}, \\ \tilde{b} = \begin{bmatrix} b - Ax \\ v \end{bmatrix} \in \mathbb{R}^{m+1}, \tilde{c} = \begin{bmatrix} c - A^\top u \\ \omega \end{bmatrix} \in \mathbb{R}^{n+1}, \end{aligned} \tag{10}$$

where $v, \omega \in \mathbb{R}$ *are some parameters.*

Find: a matrix

$$G = \begin{bmatrix} \tilde{H} & -\tilde{h}_b \\ -\tilde{h}_c^\top & 0 \end{bmatrix} \in \mathbb{R}^{(m+1)\times(n+1)}, \textit{ where } \tilde{H} \in \mathbb{R}^{m\times n}, \tilde{h}_b \in \mathbb{R}^m, \tilde{h}_c \in \mathbb{R}^n, \tag{11}$$

having minimal Euclidean norm and such that

$$\tilde{x} \in \mathcal{X}_{opt}\left(G, \tilde{b}, \tilde{c}\right), \tilde{u} \in \mathcal{U}_{opt}\left(G, \tilde{b}, \tilde{c}\right). \tag{12}$$

Suppose that problems $C_2(A, b, c, x, u, w_b, w_c)$ and $\tilde{C}_1(\tilde{x}, \tilde{u}, \tilde{b}, \tilde{c})$ are solvable. In this case, the following equivalences take place:

$$x \in \mathcal{X}_{opt}\left(A + \tilde{H}, b + w_b\tilde{h}_b, c + w_c + \tilde{h}_c\right) \Leftrightarrow \begin{bmatrix} x \\ w_b \end{bmatrix} \in \mathcal{X}_{opt}\left(G, \tilde{b}, \tilde{c}\right), \tag{13}$$

$$u \in \mathcal{U}_{opt}\left(A + \tilde{H}, b + w_b\tilde{h}_b, c + w_c + \tilde{h}_c\right) \Leftrightarrow \begin{bmatrix} x \\ w_b \end{bmatrix} \in \mathcal{U}_{opt}\left(G, \tilde{b}, \tilde{c}\right). \tag{14}$$

From formulas (13)–(14), it follows that the optimal solutions to problems $C_2(A, b, c, x, u, w_b, w_c)$ and $\tilde{C}_1(\tilde{x}, \tilde{u}, \tilde{b}, \tilde{c})$ are connected by the relations

$$\hat{H} = \tilde{H},\ \hat{h}_b = \tilde{h}_b,\ \hat{h}_c = \tilde{h}_c, \tag{15}$$

which, in turn, allow to write

$$\|G\|^2 = \|\tilde{H}\|^2 + \|\tilde{h}_b\|^2 + \|\tilde{h}_c\|^2 = \|\hat{H}\|^2 + \|\tilde{h}_b\|^2 + \|\tilde{h}_c\|^2 = \hat{\Theta}. \tag{16}$$

Thus, to prove the theorem, it suffices to verify that the solution to the problem $\tilde{C}_1(\tilde{x}, \tilde{u}, \tilde{b}, \tilde{c})$ exists for any values of parameters A, b, c, x, u and $w_b, w_c > 0$, further obtain constructive formulas for parameters $\tilde{H}$, $\tilde{h}_b$, $\tilde{h}_c$ and to apply formulas (15), (16).

Let's note that due to (10) inequations $\tilde{x} \neq 0$, $\tilde{u} \neq 0$ hold for any x, u, among which $x = 0$, $u = 0$. By Theorem 2, a solution to a problem $\tilde{C}_1(\tilde{x}, \tilde{u}, \tilde{b}, \tilde{c})$ exists if and only if the following relation is carried out:

$$\tilde{c}^\top \tilde{x} = \tilde{u}^\top \tilde{b} = \alpha. \tag{17}$$

The condition (17), by virtue of (10), is equivalent to the following system of equations:

$$c^\top x - u^\top Ax + w_b\omega = \alpha, \tag{18}$$

$$u^\top b - u^\top Ax + w_c v = \alpha. \tag{19}$$

This system contains two uncertain parameters v and ω. The fulfillment of condition (17) can be achieved for any values of A, b, c, x, u and $w_b, w_c > 0$ with the appropriate set of these parameters. Thus, by Theorem 2, the matrix W providing the fulfillment of conditions (12), exists at any values of the indicated above parameters. Also, by Theorem 2, for all values of the referred above parameters the corresponding matrix $\hat{W}$ with minimal Euclidean norm exists and is unique:

$$\hat{W} = \begin{bmatrix} S & p \\ q^\top & \theta \end{bmatrix} = \frac{\tilde{b}\tilde{x}^\top}{\tilde{x}^\top \tilde{x}} + \frac{\tilde{u}\tilde{g}^\top}{\tilde{u}^\top \tilde{u}} - \alpha \frac{\tilde{u}\tilde{x}^\top}{\tilde{x}^\top \tilde{x} \cdot \tilde{u}^\top \tilde{u}}, \tag{20}$$

$$\|\hat{W}\|^2 = \frac{\|\tilde{b}\|^2}{\|\tilde{x}\|^2} + \frac{\|\tilde{g}\|^2}{\|\tilde{u}\|^2} - \frac{\alpha^2}{\|\tilde{x}\|^2 \cdot \|\tilde{u}\|^2}, \tag{21}$$

where $S \in \mathbb{R}^{m \times n}$, $p \in \mathbb{R}^m$, $q \in \mathbb{R}^n$, $\theta \in \mathbb{R}$, vectors $\tilde{x}$, $\tilde{u}$, $\tilde{b}$ are determined by formulas (10) for the data A, b, c, x, u, w_b, w_c, and vector $\tilde{g} \in \mathbb{R}^{n+1}$ is determined as

$$\tilde{g} = \begin{bmatrix} g \\ \omega \end{bmatrix}, \tag{22}$$

where vector g is determined by the formula (9) for the given A, x, u and c. Using block representation (10) for vectors $\tilde{x}$, $\tilde{u}$, $\tilde{b}$, $\tilde{g}$ from (20)–(22), one can get

$$\begin{aligned} \hat{W}(v, \omega, \alpha) = {} & \frac{\begin{bmatrix} b - Ax \\ v \end{bmatrix} \cdot \begin{bmatrix} x^\top & w_b \end{bmatrix}}{x^\top x + w_b^2} + \frac{\begin{bmatrix} u \\ w_c \end{bmatrix} \cdot \begin{bmatrix} g^\top & \omega \end{bmatrix}}{u^\top u + w_c^2} - \\ & -\alpha \frac{\begin{bmatrix} u \\ w_c \end{bmatrix} \cdot \begin{bmatrix} x^\top & w_b \end{bmatrix}}{(x^\top x + w_b^2) \cdot (u^\top u + w_c^2)}, \end{aligned} \tag{23}$$

$$\begin{aligned} \|\hat{W}(v, \omega, \alpha)\|^2 = {} & \frac{\|b - Ax\|^2 + v^2}{\|x\|^2 + w_b^2} + \frac{\|g\|^2 + \omega^2}{\|u\|^2 + w_c^2} - \\ & - \frac{\alpha^2}{\left(\|x\|^2 + w_b^2\right) \cdot \left(\|u\|^2 + w_c^2\right)}. \end{aligned} \tag{24}$$

The block representation (23) allows to write down a condition $\theta = 0$ which is necessary for the transformation of matrix $\hat{W}$ into matrix G which is a solution to problem $\tilde{C}_1(\tilde{x}, \tilde{u}, \tilde{b}, \tilde{c})$:

$$\theta = \frac{w_b v}{x^\top x + w_b^2} + \frac{w_c \omega}{u^\top u + w_c^2} - \frac{w_b w_c \alpha}{(x^\top x + w_b^2) \cdot (u^\top u + w_c^2)} = 0. \tag{25}$$

Conditions (18), (19) and (25) form a system of linear algebraic equations concerning variables v, ω, and α, which can be rewritten in the following vector-matrix form:

$$\begin{bmatrix} w_c & 0 & -1 \\ 0 & w_b & -1 \\ \frac{w_b}{x^\top x + w_b^2} & \frac{w_c}{u^\top u + w_c^2} & -\frac{w_b}{x^\top x + w_b^2} \cdot \frac{w_c}{u^\top u + w_c^2} \end{bmatrix} \cdot q = \begin{bmatrix} u^\top Ax - u^\top b \\ u^\top Ax - c^\top x \\ 0 \end{bmatrix}, \quad (26)$$

where $q = [\upsilon \quad \omega \quad \alpha]^\top$.

Denote by Q the matrix of system (26). The solution $q^* = \left[\upsilon^* \ \omega^* \ \alpha^*\right]^\top$ to the system (26) exists and is unique for any x, u, w_b, w_c, such that $w_b, w_c \neq 0$. This can be seen by analyzing the range of values of the determinant of matrix Q:

$$0 < \det(Q) = \frac{w_c^2 \cdot x^\top x + w_b^2 \cdot u^\top u + w_b^2 w_c^2}{w_c^2 \cdot x^\top x + w_b^2 \cdot u^\top u + w_b^2 w_c^2 + x^\top x \cdot u^\top u} \leqslant 1.$$

Therefore, a solution to problem $\tilde{C}_1(\tilde{x}, \tilde{u}, \tilde{b}, \tilde{c})$ exists, is unique and given by the matrix $\hat{W}(\upsilon^*, \omega^*, \alpha^*)$, that is, $\hat{W}(\upsilon^*, \omega^*, \alpha^*) = G$. Solving the system (26), and comparing the expressions for υ^*, ω^*, α^* with formulas (5)–(8), we conclude that conditions $\upsilon = \upsilon^*, \omega = \omega^*, \alpha = \alpha^*$ are satisfied. Taking into account this fact, one can deduce formulas (1)–(3) from (9) and (23), and (4) from (16) and (24).

3 Main Result

Problem $C_3(A, b, c, \mu, \delta_b, \delta_c)$: *Given vectors $c \in \mathbb{R}^n$, $b \in \mathbb{R}^m$, matrix $A \in \mathbb{R}^{m \times n}$, real scalars $\mu, \delta_b, \delta_c > 0$. Find a vectors $x \in \mathbb{R}^n$, $u \in \mathbb{R}^m$, real scalars w_b,w_c being the solution to optimization problem*

$$\|x\|^2 + \|u\|^2 \to \min,$$

such that

$$\begin{gathered} x \in \mathcal{X}_{opt}(A + H, b + w_b h_b, c + w_c h_c), \\ u \in \mathcal{U}_{opt}(A + H, b + w_b h_b, c + w_c h_c), \\ \|H\| \leqslant \mu, \|w_b h_b\| \leqslant \delta_b, \|w_c h_c\| \leqslant \delta_c, \end{gathered}$$

where H, h_b and h_c are the solutions to the problem $C_2(A, b, c, x, u, w_b, w_c)$.

Let x^*, u^*, w_b^*, w_c^* be a solution to the problem $C_3(A, b, c, \mu, \delta_b, \delta_c)$, and H^*, h_b^*, h_c^* be a solution to the problem $C_2(A, b, c, x^*, u^*, w_b^*, w_c^*)$. Based on Theorem 3, we can prove the following assertion.

Theorem 4. *Problem $R(A, b, c, \mu, \delta_b, \delta_c)$ has a solution in the form: $A_1 = A + H^*$, $b_1 = b + w_b^* h_b^*$, $c_1 = c + w_c^* h_c^*$, $x_1 = x^*$, $u_1 = u^*$. For a sufficiently small μ, δ_b, δ_c, this solution is unique.*

Numerical Example (taken from [24], the vector c_0 is changed). Let the exact data for the problems L, L^* have the form

$$A_0 = \begin{bmatrix} 2 & -2 & 1 & 1 & 1/2 \\ 1 & 1 & 1 & 0 & 1/2 \\ 3 & -1 & 2 & 1 & 1 \end{bmatrix}, b_0 = \begin{bmatrix} 2 \\ 1 \\ 3 \end{bmatrix}, c_0 = \left[-1 \ -1 \ 1 \ 1 \ 1\right]^\top,$$

$$x_0 = \left[0 \ 1/14 \ 0 \ 17/14 \ 13/7\right]^\top, u_0 = \left[1/3 \ 1/3 \ 2/3\right]^\top.$$

The approximate data have the form $A = A_0 + \frac{\mu}{\|\Delta A\|} \cdot \Delta A$, $b = b_0 + \frac{\delta_b}{\|\Delta b\|} \cdot \Delta b$, $c = c_0 + \frac{\delta_c}{\|\Delta c\|} \cdot \Delta c$, where

$$\Delta A = \begin{bmatrix} 0.017 & 0.051 & -0.044 & -0.304 & 0.001 \\ 0.449 & 0.347 & 0.239 & 0.339 & 0.073 \\ -0.074 & -0.028 & 0.483 & 0.157 & -0.471 \end{bmatrix}, \Delta b = \begin{bmatrix} -0.031 \\ -0.343 \\ 0.158 \end{bmatrix},$$

$$\Delta c = \begin{bmatrix} 0.342 & -0.391 & -0.186 & 0.214 & -0.363 \end{bmatrix}^\top,$$

$$\varepsilon = (\mu = \delta_b = \delta_c) = 0.5, 0.1, 0.05, 0.01, ..., 0.000005.$$

The computations show that the problem $L(A, b, c)$ is an improper problem of LP of the 1st type, and the problem $L^*(A, b, c)$ is an improper problem of LP of the 2nd type. The results of the numerical solution to the problem $C_3(A, b, c, \mu, \delta_b, \delta_c)$ obtained in the Matlab® environment using the `fmincon` solver are shown in Table 1 and Fig. 1. Note that the corresponding solutions to problem $R(A, b, c, \mu, \delta_b, \delta_c)$ (in particular, matrix A_1, vectors b_1, c_1) are easily obtained by applying Theorem 4.

Table 1. The results of solving two $C_3(A, b, c, \mu, \delta_b, \delta_c)$ problems.

ε	0.5	0.1
$x_1(\varepsilon)$	0.00000000	0.00000000
	0.04244959	0.06084441
	0.00000000	0.00000000
	1.29983478	1.21814139
	1.66263558	1.82468616
$u_1(\varepsilon)$	0.38460944	0.34637553
	0.31771598	0.32949911
	0.60208196	0.64607356
$w_b^*(\varepsilon)$	1.37492315	1.69882886
$w_c^*(\varepsilon)$	1.09418157	1.61027491
$\|x_0 - x_1(\varepsilon)\|$	0.21445630	0.03435590
$\|u_0 - u_1(\varepsilon)\|$	0.08393048	0.02467541

As can be seen from the presented table and graph, the method for solving an approximate (and, possibly, improper) pair of dual LPs, considered in this paper, allows us to obtain a stable approximation to the normal solution to an exact pair of dual LPs.

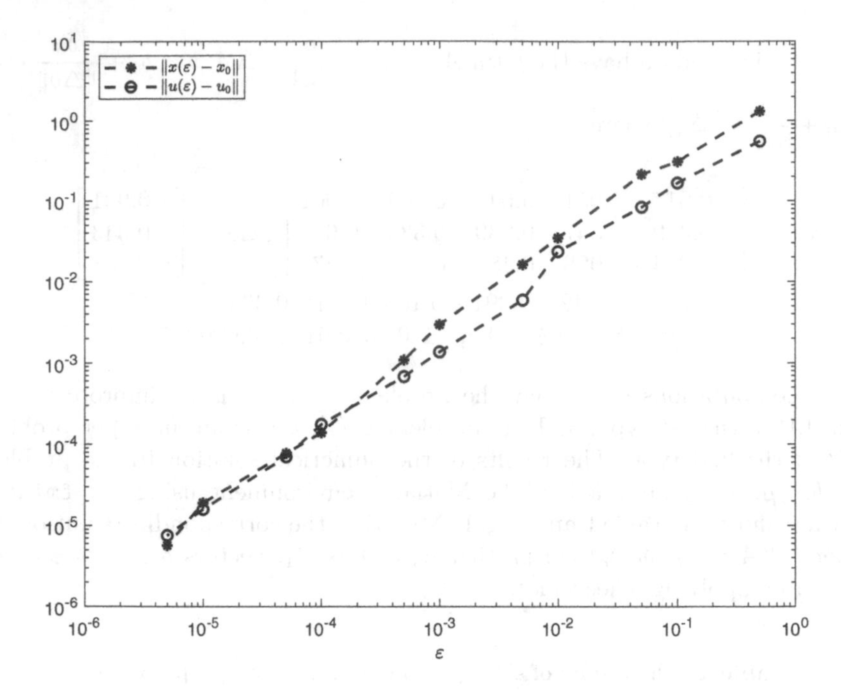

Fig. 1. The results of solving a series of $C_3(A, b, c, \mu, \delta_b, \delta_c)$ problems.

References

1. Eremin, I.I., Mazurov, V.D., Astaf'ev, N.N.: Improper Problems of Linear and Convex Programming. Nauka, Moscow (1983). (in Russian)
2. Vasil'ev, F.P., Ivanitskii, A.Yu.: Linear Programming. Faktorial Press, Moscow (2008). (in Russian)
3. Tikhonov, A.N.: Approximate systems of linear algebraic equations. USSR Comput. Math. Math. Phys. **20**(6), 10–22 (1980). (in Russian)
4. Tikhonov, A.N., Arsenin, V.Ja.: Methods for Solving ill-posed Problems, 3rd edn. Nauka, Moscow (1986). (in Russian)
5. Volkov, V.V., Erokhin, V.I.: Tikhonov solutions of approximately given systems of linear algebraic equations under finite perturbations of their matrices. Comput. Math. Math. Phys. **50**(4), 589–605 (2010). https://doi.org/10.1134/S0965542510040032
6. Erokhin, V.I., Volkov, V.V.: On A. N. Tikhonov's regularized least squares method. Vestn. St.-Peterburg Univ. Prikl. Mat. Inform. Protsessy Upr. **13**(1), 4–16 (2017). (in Russian). https://doi.org/10.21638/11701/spbu10.2017.101
7. Erokhin, V.I.: Matrix correction of a dual pair of improper linear programming problems. Comput. Math. Math. Phys. **47**(4), 564–578 (2007). https://doi.org/10.1134/S0965542507040033
8. Erokhin, V.I., Krasnikov, A.S., Khvostov, M.N.: Matrix corrections minimal with respect to the euclidean norm for linear programming problems. Autom. Remote Control **73**(2), 219–231 (2012). https://doi.org/10.1134/S0005117912020026

9. Erokhin, V.I., Krasnikov, A.S., Khvostov, M.N.: On sufficient conditions for the solvability of linear programming problems under matrix correction of their constraints. Tr. Inst. Mat. Mekh. **19**(2), 144–156 (2013). (in Russian)
10. Erokhin, V.I.: On some sufficient conditions for the solvability and unsolvability of matrix correction problems of improper linear programming problems. Tr. Inst. Mat. Mekh. **21**(3), 110–116 (2015). (in Russian)
11. Erokhin, V.I., Krasnikov, A.S., Volkov, V.V., Khvostov, M.N.: Matrix correction minimal with respect to the euclidean norm of a pair of dual linear programming problems. In: CEUR Workshop Proceedings 9th, pp. 196–209 (2016)
12. Erokhin, V.I.: A stable solution of linear programming problems with the approximate matrix of coefficients. In: 2017 Constructive Nonsmooth Analysis and Related Topics (dedicated to the memory of V.F. Demyanov) (CNSA), St. Petersburg, Russia, pp. 88–90. IEEE (2017). https://doi.org/10.1109/CNSA.2017.7973953
13. Volkov, V.V., Erokhin, V.I., Krasnikov, A.S., et al.: Minimum-euclidean-norm matrix correction for a pair of dual linear programming problems. Comput. Math. Math. Phys. **57**(11), 1757–1770 (2017). https://doi.org/10.1134/S0965542517110148
14. Erokhin, V.I., Razumov, A.V., Krasnikov, A.S.: Tikhonov's solution of approximate and improper LP problems. In: IX Moscow International Conference on Operations Research (ORM 2018), Proceedings, vol. I, pp. 131–136. MAKS Press, Moscow (2018). https://doi.org/10.29003/m211.ORM2018_v1
15. Eremin, I.I., Makarova, D.A., Schultz, L.V.: Questions of stability and regularization of improper linear programming problems. In: Proceedings of the Ural State University, vol. 30, pp. 43–62 (2004)
16. Gorelik, V., Zolotova, T.: Approximation of the improper linear programming problem with restriction on the norm of the correction matrix of the left-hand side of the constraints. In: 2018 IX International Conference on Optimization and Applications (OPTIMA 2018), pp. 14–24. DEStech Transactions on Computer Science and Engineering (2018). https://doi.org/10.12783/dtcse/optim2018/27918
17. Shary, S.P.: Interval Regularization for Imprecise Linear Algebraic Equations. arXiv preprint arXiv:1810.01481 (2018)
18. Artem'eva, L.A., Vasil'ev, F.P., Potapov, M.M.: Extragradient method for correction of inconsistent linear programming problems. Comput. Math. Math. Phys. **58**(12), 1919–1925 (2018). https://doi.org/10.1134/S0965542518120163
19. Popov, L.D.: Methods of interior points, adapted to improper linear programming problems. Tr. Inst. Mat. Mekh. **24**(4), 208–216 (2018). (in Russian). https://doi.org/10.21538/0134-4889-2018-24-4-208-210
20. Skarin, V.D.: Method of penalty functions and regularization in the analysis of improper problems of convex programming. Tr. Inst. Mat. Mekh. **24**(3), 187–199 (2018). (in Russian). https://doi.org/10.21538/0134-4889-2018-24-3-187-199
21. Vatolin, A.A.: Approximation of improper linear programming problems using euclidean norm criterion. USSR Comput. Math. Math. Phys. Fiz. **24**(12), 1907–1908 (1984). (in Russian)
22. Gorelik, V.A.: Matrix correction of a linear programming problem with inconsistent constraints. Comput. Math. Math. Phys. **41**(12), 1615–1622 (2001)
23. Amirkhanova, G.A., Golikov, A.I., Evtushenko, Yu. G.: On an inverse linear programming problem. Proc. Steklov Inst. Math. **295**(1), 21–27 (2016)
24. Agayan, G.M., Ryutin, A.A., Tikhonov, A.N.: The problem of linear programming with approximate data. USSR Comput. Math. Math. Phys. **24**(5), 14–19 (1984). (in Russian)

Discrete Time Lyapunov-Type Convergence Conditions for Recurrent Sequences in Optimization and Subgradient Method for Weakly Convex Functions

Evgeni Nurminski[1] and Natalia Shamray[2](✉)

[1] School of Natural Sciences, Far Eastern Federal University, Vladivostok, Russia
nurminski.ea@dvfu.ru

[2] Institute of Automation and Control Processes, Vladivostok, Russia
shamray@dvo.ru

Abstract. We present here the set of conditions for recursive optimization-like processes which guarantee their convergence to a given solution set. These conditions simplify convergence studies for such processes by essentially reducing them to the analysis of the processes behavior at arbitrary small vicinity of points outside the solution set. They also implicitly implement a rather complicated part of the logic of convergence proofs when there is no strict monotonicity of Lyapunov function along the process trajectory.

Keywords: Recurrent sequences · Optimization methods · Convergence conditions

1 Introduction

We consider recursive algorithms which can be described by relations

$$x^{k+1} = \Phi_k(x^k), k = 0, 1, \dots, \tag{1}$$

where $\Phi_k(\cdot)$ can be called recursive operators $\Phi_k : E \to E$. This model is typical for algorithms for solution of optimization problems $\min\{f(x), x \in X\}$ for x from finite-dimensional euclidean space E. This model of optimization process stimulated studies of general conditions to ensure the convergence of recursive optimization algorithms.

One of the general techniques to analyze the convergence of numerical algorithms of multidimensional optimization at the early stages of development of such algorithms was well-developed Lyapunov theory for asymptotic stability of ordinary differential equations (ODE), see f.i. [4] and works cited there. This idea is based first on similarities between recursive optimization processes and finite difference versions of ODE, and second on similarities between the notion

I. Bykadorov et al. (Eds.): MOTOR 2019, CCIS 1090, pp. 294–303, 2019.
https://doi.org/10.1007/978-3-030-33394-2_23

of asymptotic stability of ODE and convergence of optimization algorithms. This technique still remains popular, see f.i. [14] and references there. The latest developments in this area include the use of Lyapunov technique for the estimation of the rate of convergence.

Nevertheless, there is an essential difference between ODE and optimization processes, as the latter are of essentially recursive nature and their rate of convergence is defined in terms of the number of function/gradient calculations, not in terms of anything like running time for ODE. Moreover, optimization algorithms, in fact, have very little in common with methods of numerical integration where the main goal is to keep accurate discrete approximation of the whole trajectory of ODE.

Therefore the different convergence theories for convergence of optimization algorithms were developed which use convergence indicators, analogies to Lyapunov functions but with different convergence conditions of essentially finite nature. Within this framework, the algorithm for solving a particular problem is considered as a rule for construction of a sequence of approximate solutions $\{x^k\}$, which has to converge to a certain set $X_\star$ which by definition is a set of desirable solutions. Typically this set is defined by optimality conditions (for optimization problems), and the like. We refer to the case when the sequence $\{x^k\}$ has at least one cluster point in $X_\star$ as *weak* convergence.

The most general form of these conditions belongs probably to Zangwill [16] who formulated necessary and sufficient (!) conditions for weak convergence of recursive models of optimization algorithms. These conditions were relaxed by Huard [6], Hu [5] and probably by others. Other forms of convergence conditions were proposed by [8,13]. Eremin [3] suggested the algorithmic model based on Fejer processes, where the Euclidean distance to solution set is used as the Lyapunov function for convex feasibility problems and others. For the state-of-the-art development of this model see [1].

Even if these approaches were quite successful in many areas the problem of finding the relevant and practically computable convergence indicator which behaves monotonically on algorithm trajectory remains the hardest part of the analysis of advanced algorithms. It is especially true for non-differentiable and stochastic optimization where even descent directions are hard to find. So the new approach was suggested in the early publication [11] which was further extended in the number of follow-up works, see [9] for the references. As these papers are published mainly in Russian and are almost impossible to read in translation the present paper is intended to popularize this technique for proving convergence of non-monotone algorithms. It also provides some extension to this technique to deal with algorithms for approximate solutions of optimization problems and alike.

2 Notations and Preliminaries

Let E denotes a finite-dimensional vector space with the inner product xy for $x, y \in E$, and the standard Euclidean norm $\|x\| = \sqrt{xx}$. The one-dimensional E

is denoted nevertheless as $\mathbb{R}$ and $\mathbb{R}_\infty = \mathbb{R} \cup \{\infty\}$. The unit ball in E is denoted as $B = \{x : \|x\| \leq 1\}$. The linear vector space of bounded closed convex subsets of E with Minkowski addition and standard multiplication by real numbers is denoted as $\mathcal{C}(E)$. For any $X \subset E$ we denote its interior as $\text{int}(X)$. The closure of a set X is denoted as $\text{cl}\{X\}$ and its boundary as ∂X. The distance function $\rho(x, X)$ between point x and set $X \subset E$ is defined as $\rho(x, X) = \inf_{y \in X} \|x - y\|$. The norm of a set X is defined as $\|X\| = \sup_{x \in X} \|x\|$.

3 Convergence Theory

Zangwill [16] very early suggested the following conditions under which an optimizing sequence $\{x^k\}$ converges to a bounded solution set $X_\star$. Namely, convergence is guaranteed if the sequence $\{x^k\}$ has the following properties:

Z1 The sequence $\{x^k\}$ is bounded.
Z2 There exists continuous function $V(x) : E \to \mathbb{R}$ such that
  Z2.1 $V(x^\star) = 0$ for any $x^\star \in X_\star$ and $V(x) > 0$ otherwise;
  Z2.2 if $\{x^k\}$ has a cluster point $x' \notin X_\star$ then this sequence has another cluster point x'' such that $V(x'') < V(x')$.

We can call function $V(x)$ a Lyapunov function by analogue with Lyapunov stability theory for ordinary differential equations.

The following theorem holds.

Theorem 1. *For the sequence $\{x^k\}$ to have a cluster point $x^\star \in X_\star$ it is necessary and sufficient to have conditions* **Z1, Z2** *satisfied.*

Proof. The necessity is obvious for $V(x) = \text{dist}(x, X_\star)$ where $\text{dist}(\cdot, \cdot)$ is the distance function (see Sect. 2 for definition). The sufficiency is also practically obvious: consider a sub-sequence $\{x^{k_t}, t = 0, 1, \dots\}$ such that $\{x^{k_t}, t = 0, 1, \dots\}$ converges to certain $\bar{x}$ such that

$$\lim_{t \to \infty} V(x^{k_t}) = \liminf_{n \to \infty} V(x^m) = V(\bar{x}) = \bar{V} > 0$$

due to continuity of V on any bounded closed set containing $\{x^k, k = 0, 1, \dots\}$. Then $\bar{x} \in X_\star$ and $\bar{V} = 0$ otherwise using **Z2.1** we arrive to contradiction.

These conditions are very easy to apply to descent algorithms for optimization problems with unique solutions since in these cases the objective functions of optimization problems normally substitutes the Lyapunov function and conditions **Z1, Z2** are easy to check. However, in non-differentiable optimization, it is not so easy to guarantee monotone property of optimization sequences and in non-convex cases it is hardly possible to construct a Lyapunov function with zero value on a potential solution sets.

Therefore the more specific conditions are to be used which were suggested first in [11] with their more advanced variant from [10]. We formulate them here as follows.

A1 The sequence $\{x^k\}$ is bounded;
A2 If $\{x^{n_t}\} \to x' \notin X_\star$ then there exists such $\epsilon > 0$ that for any t

$$m_t = \inf\{s : \|x^{n_t} - x^s\| > \epsilon,\ s > n_t\} < \infty; \tag{2}$$

A3 There exists continuous function $V(x) : E \to \mathbb{R}$ such that:
A3.1 The set $V_\star = \{V(x^\star), x^\star \in X_\star\}$ has everywhere dense complement;
A3.2 For any sequences $\{x^{n_t}\}, \{x^{m_t}\}$, defined in condition **A2**

$$\limsup_{t\to\infty} V(x^{m_t}) = V'' < V' = \lim_{t\to\infty} V(x^{n_t}) = V(x'); \tag{3}$$

A4 If $\{x^{n_t}\} \to x^\star \in X_\star$ then $\|x^{n_t+1} - x^{n_t}\| \to 0$ when $t \to \infty$.

Theorem 2. *If conditions* **A1–A4** *are satisfied then every cluster point of the sequence* $\{x^k\}$ *belongs to* $X_\star$.

Proof. Assume, contrary to the statement of the theorem, that the sequence $\{x^k\}$ has a cluster point $x' \notin X_\star$. It means that there is a sub-sequence $\{x^{n_t}\}$ which converges to x' when $t \to \infty$. Then according to **A2** there is an $\epsilon > 0$ such that $\|x^{n_t} - x^{m_t}\| > \epsilon$ for some $m_t > n_t$ and $\|x^{n_t} - x^s\| \le \epsilon$ for $n_t < s < m_t$.

By condition **A3.2** there is a continuous function $V(x)$ such that

$$\limsup_{t\to\infty} V(x^{m_t}) = V'' < V' = V(x') = \lim_{t\to\infty} V(x^{n_t}).$$

It can be assumed without any loss of generality that the sequence $\{V(x^{m_t})\}$ converges to V'' when $t \to \infty$. Denote $\varepsilon = \frac{1}{3}(V' - V'') > 0$. Then

$$V(x^{m_t}) < V'' + \varepsilon = V_a < V' \tag{4}$$

and

$$V(x^{n_t}) > V' - \varepsilon = V_b > V_a \tag{5}$$

for all t large enough.

As $V_a + \gamma < V_b$ for any $\gamma \in (0, \varepsilon)$ it is then clear that the sequence $\{V(x^k)\}$ down-crosses the interval $(V_a + \gamma, V_b)$ infinitely many times and hence it up-crosses the same interval infinitely many times as well.

It means that there are infinite sub-sequences $\{p_t\}, \{q_t\}, t = 1, 2, \ldots$ such that

$$V(x^{p_t}) \le V_a + \gamma < V_b \le V(x^{q_t}), \quad p_t < q_t, \quad t = 1, 2, \ldots \tag{6}$$

and

$$V_a + \gamma < V(x^s) < V_b, \quad p_t < s < q_t, \quad t = 1, 2, \ldots. \tag{7}$$

It can be assumed without any loss of generality that $\{x^{p_t}\}$ converges to a certain $\tilde{x}$ with $\{V(x^{p_t})\} \to V(\tilde{x}) \le V_a + \gamma$. If $\tilde{x} \in X_\star$ then according to **A4** $|V(x^{p_t+1}) - V(x^{p_t})| \to 0$ and eventually $V(x^{p_t+1})$ becomes smaller than V_b and hence $p_t < p_t + 1 < q_t$. It implies that $V(x^{p_t}) < V_a + \gamma < V(x^{p_t+1})$ and therefore

$V(x^{p_t}) \to V_a + \gamma$. Due to **A3.1** the correction $\gamma \in (0, \varepsilon)$ can be chosen such that $\lim_{t\to\infty} V(x^{p_t}) = V_a + \gamma \notin V_\star$ and hence $\tilde{x} \notin X_\star$. But in this case

$$r_t = \min_{s>p_t}\{\|x^{p_t} - x^s\| > \delta > 0\} \in (p_t, q_t) \tag{8}$$

for δ small enough and

$$V(x^{r_t}) > V(x^{p_t}) \text{ for all } t \text{ large enough} \tag{9}$$

or

$$\limsup_{t\to\infty} V(x^{r_t}) \geq \lim_{t\to\infty} V(x^{p_t}) = V(\tilde{x}); \tag{10}$$

which contradicts **A3.2** and hence proves the theorem.

We complete the paper with the model proof of convergence for quasi-gradient method of minimization of weakly convex functions.

4 Subgradient Method for Weakly Convex Functions

One of the frequently used classes of non-convex non-differentiable functions is the class of weakly convex functions. They are becoming popular in the area of non-convex optimization, mostly because the set of such functions contains common differentiable ones without convexity requirements. This class is also closed with respect to maximum operation, at least finite and in many cases continuous. The review of applications and algorithmic approaches for solving optimization problems with weakly convex functions can be found for instance in [2].

The definition of weakly convex function [12] goes like following.

Definition 1. *Function $f : E \to \mathbb{R}$ is called weakly convex if for any x there exists a set $\partial f(x)$ such that for all y*

$$f(y) - f(x) \geq g(y - x) + r(x, y) \tag{11}$$

for arbitrary $g \in \partial f(x)$ where the reminder term is such that $r(x, y)/\|x - y\| \to 0$ when $y \to x$ locally uniformly in x.

By *locally uniform* we mean that for any $\epsilon > 0$ there is $\delta > 0$ such that $r(z, y)/\|z - y\| \leq \epsilon$ when $\|z - x\| \leq \delta$, $\|y - x\| \leq \delta$. By analogy with convex functions the set $\partial f(x)$ can be called a sub-differential of f at point x. It is immediately can be seen that $\partial f(x)$ is a convex set and that due to locally uniform smallness of $r(x, y)$ with respect to $\|x - y\|$ the sub differential $\partial f(x)$ considered as a multi-function of x is upper-semicontinious. As in convex case the directional derivative $f'(x, d)$ of weakly convex $f(\cdot)$ at point x in the direction d can be computed from the sub-differential $\partial f(x)$ as

$$f'(x, d) = \lim_{\delta\to+0} \delta^{-1}(f(x + \delta d) - f(x)) = \sup_{g\in\partial f(x)} gd = (\partial f(x))_d \tag{12}$$

It remains to notice that the similar, but slightly less general, notion of generalized differentiability was suggested in [7] and later studied in [15, see Definition 8.3, page 301] under the name of the regular sub-gradient.

From the point of view of optimization theory the necessary condition for the minimization problem $\min_x f(x) = f(x^\star)$ is the expected $0 \in \partial f(x^\star)$. Therefore it is quite natural to consider a sub-gradient method

$$x^{k+1} = x^k - \lambda_k g^k, k = 0, 1, \ldots \tag{13}$$

with certain initial point x^0 and step multipliers $\lambda_k > 0$ for minimization problems with weakly convex functions. The subgradient g^k used in this recursive relationship is picked up in an arbitrary way from subdifferential $\partial f(x^k)$. Analysis of convergence of the algorithm (13) is complicated however by the fact that the direction g^k is not generally a descent direction and moreover contrary to the convex case $\partial f(x)$ is not a monotone set-valued mapping. It excludes monotonicity of two common convergence indicators for the optimizing trajectories $\{x^k\}$: values of objective function $f(x^k)$ and distances $\operatorname{dist}(x^k, X_\star)$ to the solution set $X_\star$. Hence traditional means to prove convergence of (13) are hardly suitable and we use this algorithm to demonstrate application of the convergence conditions **A1**–**A4**. To simplify the demonstration we focus on key steps **A2, A3**, which are the most difficult conditions to prove.

We need also one auxiliary result on the "averaging" effect of iterations between n_t and m_t from **A2**. In fact, the ability to take this effect into account is one of the attractive features of condition **A3**. The effect can be described by the following lemma.

Lemma 1. *Let D is a closed convex bounded subset of E with* $\operatorname{dist}(0, D) = \delta > 0$ *and $\{\sigma_k\} \subset (0, 1]$ is a sequence of scalar weights such that $\sigma_k \to 0$ when $k \to \infty$ and $\sum_k \sigma_k = \infty$. Then there is such N that for any sequence $\{x^k\} \subset D$ there is $m \le N$ that*

$$y^m x^{m+1} \ge \delta^2/2, \tag{14}$$

where $\{y^k\}$ is a Cesare-like averaging of $\{x^k\}$:

$$y^1 = x^1; \quad y^{k+1} = (1 - \sigma_k) y^k + \sigma_k x^{k+1}, k = 1, 2, \ldots \tag{15}$$

The number N depends on D and on the sequence of weights $\{\sigma_k\}$.

Proof. Assume contrary to the lemma statement that there is a sequence $\{x^k\}$ such that $y^{k+1} x^k \le \delta^2/2$ for all $k = 1, 2, \ldots$ where the sequence $\{y^k\}$ is defined by (15).

Notice that in any case $\{y^k\} \subset D$ and therefore $\delta \le \|y^k\| \le \|D\|$. Then

$$\begin{aligned}
\|y^{k+1}\|^2 &\le \|(1 - \sigma_k) y^k + \sigma_k x^{k+1}\|^2 \\
&= \|y^k\|^2 + 2\sigma_k y^k (x^{k+1} - y^k) + \sigma_k^2 \|x^{k+1} - y^k\|^2 \\
&\le \|y^k\|^2 + 2\sigma_k y^k (x^{k+1} - y^k) + \sigma_k^2 \|x^{k+1} - y^k\|^2 \\
&\le \|y^k\|^2 + 2\sigma_k (\delta^2/2 - \|y^k\|^2) + 4\sigma_k^2 \|D\|^2 \\
&\le \|y^k\|^2 + 2\sigma_k (\delta^2/2 - \delta^2) + 4\sigma_k^2 \|D\|^2 \\
&\le \|y^k\|^2 - \sigma_k \delta^2 + 4\sigma_k^2 \|D\|^2 \le \|y^k\|^2 - \sigma_k \delta^2/2,
\end{aligned} \tag{16}$$

for sufficiently large k such that $-\sigma_k\delta^2 + 4\sigma_k^2\|D\|^2 \leq -\sigma_k\delta^2/2$. The universal lower bound K for such k which does not depend on a sequence $\{x^k\}$ may be defined from the inequality

$$\sup_{k\geq K} \sigma_k^2 \leq \delta^2/8\|D\|^2. \tag{17}$$

Summing up the last inequality in (16) with respect to $k > K$ obtain

$$\begin{aligned}\delta^2 \leq \|y^{K+L}\|^2 &\leq \|y^K\|^2 - \delta^2/2\sum\nolimits_{t=0}^{L}\sigma_{K+t}\\ &\leq \|D\|^2 - \delta^2/2\sum\nolimits_{t=0}^{L}\sigma_{K+t} \to -\infty\end{aligned}$$

when $L \to \infty$. As this is impossible the upper bound for L follows from the inequality

$$\sum_{s=0}^{\hat{L}} \sigma_{K+s} \leq 2(\|D\|^2 - \delta^2)/\delta^2. \tag{18}$$

Summing estimates (17) and (18) together we may ensure that for any $N \geq K+L$ with K and L defined by their respective lower bounds there must be $m \leq N$ such that $y^m x^{m+1} \geq \delta^2/2$. Notice that both estimates (17) and (18) do not depend upon sequence $\{x^k\}$ which finally proves the lemma.

With this lemma, we are ready to show that conditions **A2, A3** holds providing that the other conditions are already satisfied. So we assume that there is a sub-sequence $\{x^{n_t}\} \to x' \notin X_\star$, where $X_\star$ is a set of stationary points $x^\star : 0 \in \partial f(x^\star)$. Due to convexity of $\partial f(x')$ and upper semi-continuity of $\partial f(\cdot)$ it implies that

$$0 \notin P = \mathrm{co}(\partial f(x), x \in U_1 = \{x : \|x' - x\| \leq 4\epsilon\}) \tag{19}$$

for $\epsilon > 0$ but small enough. If we assume that $\{x^k\}_K \subset U_1$ for K large enough, then all $\{g^k\}_K \subset P$ and hence by separation theorem there is a vector p such that $g^k p \geq \delta > 0$. Then

$$x^{k+1}p = x^k p - \lambda_k g^k p \leq x^k p - \lambda_k\delta \to -\infty \text{ when } k \to \infty. \tag{20}$$

which is contradiction. Therefore the sequence $\{x^k\}$ must leave U_1 and hence must leave $U_2 = \{x : \|x^{n_t} - x\| \leq \epsilon\} \subset U_1$ where t is large enough that $x^{n_t} \in U_3 = \{x : \|x' - x\| \leq \epsilon\}$. It means that **A2** is satisfied and the sequence $\{m_t\}$ such that

$$m_t = \inf\{s : \|x^{n_t} - x^s\| > \epsilon,\ s > n_t\} \tag{21}$$

is well defined. Now we use the objective function $f(x)$ as Lyapunov function and demonstrate that **A3** is satisfied.

To simplify notations we assume $n_t = 0,\ \rho_\tau = \lambda_{n_t+\tau},\ \tau = 0, 1, \ldots$. Then t iteration of (13) can be represented as (Fig. 1)

$$x^t - x^0 = (\sum_{\tau=0}^{t} \rho_\tau) z^t$$

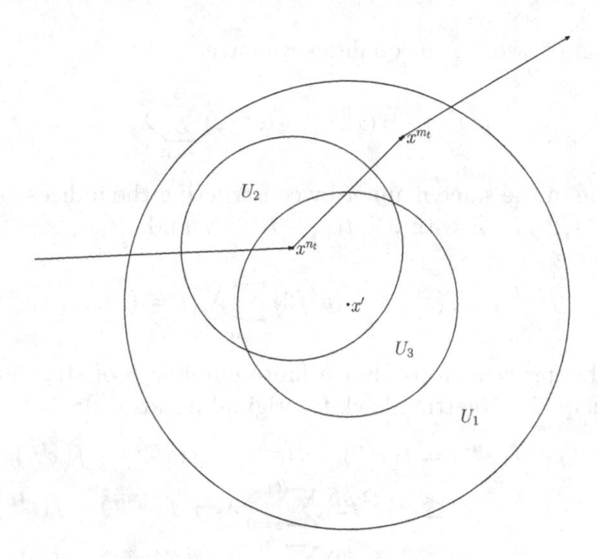

Fig. 1. U_1, U_2, U_3 neighborhoods of x' and x^{n_t}. $x^{n_t} \to x', \|x^{m_t} - x^{n_t}\| > \epsilon$

where z^t can be obtained by the averaging process

$$z^{\tau+1} = (1 - \sigma_\tau)z^\tau + \sigma_\tau g^\tau, \ \tau = 1, 2, \dots, t$$

where $\sigma_\tau = \rho_\tau / (\sum_{t=0}^{\tau} \rho_t)$. Of course $\sigma_\tau \to 0$ when $\tau \to \infty$ and it is easy to check that $\sum_\tau \sigma_\tau = \infty$. Also as $\lambda_{t+1}/\lambda_t \to 1$ then $\sigma_\tau \to 1/\tau$ and according to Lemma 1 there is $t_0 \leq N$ such that

$$0 < \delta^2 \leq z^{t_0} g^{t_0+1} = g^{t_0+1}(x^{t_0} - x^0)/(\sum_s \rho_s)$$

or

$$f(x^{t_0}) - f(x^0) \leq g^{t_0+1}(x^{t_0} - x^0) + r(x^0, x^{t_0}) \leq -\delta^2(\sum_{s=0}^{t_0} \lambda_s) + r(x^0, x^{t_0}).$$

The reminder term can be estimated from above as

$$|r(x^{t_0}, x^0)| \leq (\delta^2/\|P\|^2)\|x^{t_0} - x^0\| \leq (\delta^2/2\|P\|)(\sum_{s=0}^{t_0} \lambda_s)\|P\| \leq (\delta^2/2)\sum_{s=0}^{t_0} \lambda_s$$

which gives

$$f(x^{t_0}) - f(x^0) \leq -(\delta^2/2)\sum_{s=0}^{t_0} \lambda_s.$$

By substituting x^0 with x^{t_0} we can determine in the same manner t_1 such that

$$f(x^{t_1}) - f(x^{t_0}) \leq -(\delta^2/2)\sum_{s=t_0}^{t_1} \lambda_s.$$

After summation of two last inequalities we have

$$f(x^{t_1}) - f(x^0) \le -(\delta^2/2) \sum_{s=0}^{t_1} \lambda_s$$

and can continue in the same manner by constructing the indices $t_2, t_3, \dots$ as far as $t_l \le m_k,\ l = 1, 2, \dots, L$ with $0 < t_{l+1} - t_l \le N$ and

$$f(x^{t_l}) - f(x^0) \le -(\delta^2/2) \sum_{s=0}^{t_l} \lambda_s, l = 1, 2, \dots \tag{22}$$

Clear enough this process stops after a finite number L of steps and we obtain the following inequality (getting back to original notations):

$$\begin{aligned} f(x^{m_t}) - f(x^{n_t}) &= f(x^{t_L}) - f(x^{n_t}) + f(x^{m_t}) - f(x^{t_L}) \\ &\le -(\delta^2/2) \sum\nolimits_{s=0}^{t_L} \lambda_s + f(x^{m_t}) - f(x^{t_L}) \\ &= -(\delta^2/2) \sum\nolimits_{s=0}^{m_t} \lambda_s + f(x^{m_t}) - f(x^{t_L}) \\ &\quad +(\delta^2/2) \sum\nolimits_{s=0}^{m_t} \lambda_s - (\delta^2/2) \sum\nolimits_{s=0}^{t_L} \lambda_s. \end{aligned}$$

As $0 < m_t - t_L \le N$ both $|f(x^{m_t}) - f(x^{t_L})| \to 0$ and

$$(\delta^2/2) | \sum_{s=0}^{m_t} \lambda_s - \sum_{s=0}^{t_L} \lambda_s | = (\delta^2/2) | \sum_{s=t_L+1}^{m_t} \lambda_s | \to 0.$$

when $t \to \infty$.

At the same time

$$\epsilon < \|x^{m_t} - x^{n_t}\| \le \sum_{s=n_t}^{m_t} \lambda_s \|g^s\| \le \|D\| \sum_{s=n_t}^{m_t} \lambda_s$$

so $\sum_{s=n_t}^{m_t} \lambda_s \ge \epsilon / \|D\|$ and finally we obtain the key inequality

$$f(x^{m_t}) - f(x^{n_t}) \le -\frac{\delta^2}{2\|D\|}\epsilon + \xi_t$$

where $\xi_t \to 0$ when $t \to \infty$. By passing to the limit we obtain **A3**:

$$\limsup_{t \to \infty} f(x^{m_t}) \le f(x') - \frac{\delta^2}{2\|D\|}\epsilon < f(x')$$

and therefore confirms convergence of (13) at least to stationary points.

Notice that namely the analysis of the cumulative effect of iterations in the vicinity of $x' \in X_\star$ allowed to demonstrate asymptotic local monotonicity of $\{f(x^k)\}$ aka Lyapunov function of this algorithm.

Acknowledgments. For E. Nurminski this work was supported by the Ministry of Science and Education of Russian Federation, project 1.7658.2017/6.7. N. Shamray acknowledges the support for this work from the RFBR grant 18-29-03071.

References

1. Berdnikova, E.A., Eremin, I.I., Popov, L.D.: Distributed Fejer processes for systems of linear inequalities and problems of linear programming. Autom. Remote Control **65**(2), 168–183 (2004). https://doi.org/10.1023/B:AURC.0000014714.97496.79
2. Drusvyatskiy, D.: The proximal point method revisited. ArXiv e-prints, December 2017
3. Eremin, I.: The application of the method of Fejer approximations to the solution of problems of convex programming with non-smooth constraints. USSR Comput. Math. Math. Phys. **9**(5), 225–235 (1969). https://doi.org/10.1016/0041-5553(69)90163-3. http://www.sciencedirect.com/science/article/pii/0041555369901633
4. Evtushenko, Y., Zhadan, V.: Application of the method of Lyapunov functions to the study of the convergence of numerical methods. USSR Comput. Math. Math. Phys. **15**(1), 96–108 (1975). https://doi.org/10.1016/0041-5553(75)90138-X. http://www.sciencedirect.com/science/article/pii/004155537590138X
5. Hu, X.: An overall study of convergence conditions for algorithms in nonlinear programming. Acta Mathematicae Applicatae Sinica **9**(2), 97–103 (1993). https://doi.org/10.1007/BF02007433
6. Huard, P.: Extensions of Zangwill's theorem. In: Huard, P. (ed.) Point-to-Set Maps and Mathematical Programming. Mathematical Programming Studies, vol. 10, pp. 98–103. Springer, Heidelberg (1979). https://doi.org/10.1007/BFb0120846
7. Kruger, A.Y., Mordukhovich, B.S.: Extremal points and the euler equation in nonsmooth problems of optimization. Doklady Academia Nauk BSSR **24**, 684–687 (1980)
8. Meyer, R.R.: The validity of a family of optimization methods. SIAM J. Control **8**(1), 41–54 (1970). https://doi.org/10.1137/0308003
9. Norkin, V.I.: Necessary and sufficient conditions for convergence of iteration algorithms of nonlinear programming. Cybernetics **23**(4), 497–504 (1987). https://doi.org/10.1007/BF01078907
10. Nurminski, E.A.: Numerical Methods of Convex Optimization. Nauka, Moscow (1991)
11. Nurminskii, E.A.: Convergence conditions for nonlinear programming algorithms. Cybernetics **8**(6), 959–962 (1972). https://doi.org/10.1007/BF01068520
12. Nurminskii, E.A.: The quasigradient method for the solving the nonlinear programming problems. Cybernetics **9**(1), 145–150 (1973). https://doi.org/10.1007/BF01068677
13. Polak, E.: Computational methods in optimization: a unified approach. Math. Sci. Eng. (1971). https://books.google.ru/books?id=YNaetHz0J4oC
14. Polyak, B., Shcherbakov, P.: Lyapunov functions: an optimization theory perspective. IFAC-PapersOnLine **50**(1), 7456–7461 (2017). https://doi.org/10.1016/j.ifacol.2017.08.1513, http://www.sciencedirect.com/science/article/pii/S2405896317320955, 20th IFAC World Congress
15. Tyrrell Rockafellar, R., Wets, R.J.-B.: Variational Analysis, Grundlehren der mathematischen Wissenschaften, vol. 317. Springer, Heidelberg (1998). https://doi.org/10.1007/978-3-642-02431-3
16. Zangwill, W.I.: Convergence conditions for nonlinear programming algorithms. Manag. Sci. **16**(1), 1–13 (1969). https://doi.org/10.1287/mnsc.16.1.1

Methods for Matrix Games with Mixed Strategies and Quantile Payoff Function

Leonid D. Popov[1,2(✉)]

[1] Krasovskii Institute of Mathematics and Mechanics UB RAS, Ekaterinburg, Russia
popld@imm.uran.ru
[2] Ural Federal University, Ekaterinburg, Russia

Abstract. Walsh–Vries approach to matrix games with mixed strategies is considered. According to this approach, the payment function is defined not as the mathematical expectation of a random gain in a long series of parties, but as its quantile (VaR-estimate) for a given level of risk. The properties of such games are studied, and the methods for their solution are suggested.

Keywords: Matrix games · Mixed strategies · VaR-criterion · Linear and parametric programming

1 Introduction

The matrix game is the simplest mathematical model of the antagonistic conflict of two persons each of which has a finite set of possible strategies [1–3]. It is completely determined by its payoff matrix

$$A = \begin{pmatrix} a_{11} & a_{12} & \dots & a_{1n} \\ a_{21} & a_{22} & \dots & a_{2n} \\ & & \dots & \\ a_{m1} & a_{m2} & \dots & a_{mn} \end{pmatrix}.$$

Each row of this matrix corresponds to one of the admissible strategies of the first player (from his strategy set $\mathcal{I} = \{1, \dots, m\}$), and each of its columns corresponds to one of the admissible strategies of the second player (from his strategy set $\mathcal{J} = \{1, \dots, n\}$). The elements of the matrix a_{ij} determine the winning of the first player and, at the same time, the losses of the second one, if the players use the strategies i and j, respectively. Row and column indexes are called pure strategies of the players.

Following the principle of guaranteed result, the first player seeks a row index i_0, such that

$$\max_i \min_j \{a_{i,j}\} = \min_j \{a_{i_0,j}\} \ \big(=: v_*\big).$$

I. Bykadorov et al. (Eds.): MOTOR 2019, CCIS 1090, pp. 304–318, 2019.
https://doi.org/10.1007/978-3-030-33394-2_24

Using this strategy, he gets win at least v_* (this is the lower value of the game). Analogously, the second player is interested in finding a column index j_0, such that

$$\min_j \max_i \{a_{i,j}\} = \max_i \{a_{i,j_0}\} \ (=: v^* \geq v_*).$$

Using this strategy, he gets loss no more than v^* (this is the upper value of the game). The selected row and column are called minimax pure strategies of the players.

Usually $v^* > v_*$. If the lower and upper values of the game coincide, then the game has an equilibrium (in pure strategies), and the player who deviates from his minimax strategy alone will only worsen his result and improve the result of the opponent. Unfortunately, the matrix game in pure strategies has rarely equilibrium, so if the players play it only once, then the result can't be predicted.

When, however, the players have to repeat the game many times (potentially an infinite number of times), they can use so-called mixed strategies. It means that to disorient the opponent, the players can alternate their pure strategies in every new party according to some random rule and aim not at winning (or losses) in a one party, but at some criterion like average return in a long series of such parties.

Let us denote by

$$X = \{x \in \mathbf{R}^m_+ : \ x_1 + x_2 + \ldots + x_n = 1\}$$

the set of all mixed strategies of the first player (all of them are the probability distributions on the set $\mathcal{I}$), and by

$$Y = \{y \in \mathbf{R}^n_+ : \ y_1 + y_2 + \ldots + y_n = 1\}$$

the set of all mixed strategies of the second player (all of them are the probability distributions on the set $\mathcal{J}$). If the players choose their strategies $x \in X$ and $y \in Y$ respectively, then the income of the first player $\xi(x, y)$ (and the losses of the second player) is random value with discrete distributions function

$$F(z; x, y) = \mathbf{P}\big(\xi(x, y) < z\big) = \sum_{ij:a_{ij}<z} \mathbf{P}\big(\xi(x, y) = a_{ij}\big) = \sum_{ij:a_{ij}<z} x_i y_j.$$

The classic approach to mixed strategies exploits the average payoff, which is equal to

$$\mathbf{E}(x, y) := x^T A y = \sum_{i,j} a_{ij} x_i y_j.$$

In this instance, the first player looks for his optimal mixed strategy $\bar{x} \in X$, such that

$$\max_{x\in X} \min_{y\in Y} x^T A y = \min_{y\in Y} \bar{x}^T A y \ \big(=: \bar{v}_*\big).$$

Similarly, the second player is interested in finding his optimal mixed strategy $\bar{y} \in Y$, for which

$$\min_{y\in Y} \max_{x\in X} x^T A y = \max_{x\in X} x^T A \bar{y} \ \big(=: \bar{v}^*\big).$$

By the von Neumann theorem, every matrix game in mixed strategies and average income always has an equilibrium [1], so the average result of the game is known in advance.

Unfortunately, the classic approach does not take into account the risk of too much deviation of the random payoff from its average (i.e., its variance). This deficiency was overcome in [4], where games with a median criterion were considered, and in [5,6], where a quantile criterion corresponding to a certain specified level of confidence was also investigated. Among the closest topics in making decision area, we note the works [7–11].

In this paper we continue a line [5,6] and deal with quantile criterion. Let the strategies $x \in X$ and $y \in Y$ be chosen, and the acceptable risk level $\alpha \in (0;1)$ be given close to 0. The quantile (or VaR-estimate) of random payoff $\xi(x,y)$ of the game corresponding to this level is defined as the greatest value $\bar{\xi}_\alpha(x,y)$ such that

$$\mathbf{P}(\xi(x,y) < \bar{\xi}_\alpha(x,y)) < \alpha.$$

This means that the random income $\xi(x,y)$ of the first player will take unfavorable to him values, namely, less than $\bar{\xi}_\alpha(x,y)$, with probability less than α, that is quite rare, and conversely, the value of $\xi(x,y)$ will be greater or equal to the VaR-estimate with probability greater than $1-\alpha$, that is, almost always.

Thus, to maximize quantile of his payoff the first player must seek his strategy $\bar{x}_\alpha \in X$ such that

$$\min_{y\in Y} \mathrm{VaR}_\alpha(\bar{x}_\alpha, y) = \max_{x\in X}\min_{y\in Y} \mathrm{VaR}_\alpha(x,y) \ \big(=: \breve{v}_*\big). \tag{1}$$

Similarly, the second player seeks the opponent strategy $\bar{y}_\alpha \in X$ such that

$$\max_{x\in X} \mathrm{VaR}_\alpha(x, \bar{y}_\alpha) = \min_{y\in Y}\max_{x\in X} \mathrm{VaR}_\alpha(x,y) \ \big(=: \breve{v}^* \geq \breve{v}_*\big); \tag{2}$$

here $\mathrm{VaR}_\alpha(x,y) = \bar{\xi}_\alpha(x,y)$.

Below, to find $\breve{v}^*$ and $\breve{v}_*$, we develop some methods other than the methods proposed in [6]. The distinction is discussed in the next section.

2 Two Different Classes of the Methods

In what follows, it will be convenient to range various elements of matrix A in ascending order. We denote the elements of the resulting array as

$$\xi_1 < \xi_2 < \ldots < \xi_{N-1} < \xi_N, \quad N \leq mn. \tag{3}$$

Here, each ξ_k matchs one or more equal elements of the payout matrix. So we have a very interesting detailed grid. It is easy to see that our var-score always coincides with one of the elements of this grid. Namely,

$$\mathrm{VaR}_\alpha(x,y) = \max\{\xi_k :\ \mathbf{P}\big(\xi(x,y) < \xi_k\big) < \alpha\}.$$

As a consequence,

$$\mathbf{P}\big(\xi(x,y) < z\big) \geq \alpha \qquad \forall z > \mathrm{VaR}_\alpha(x,y).$$

Due to a general ideas from [5], instead of direct solving the problems (1), (2), it is natural to consider the following array of the inverse instances: for every ξ_k from the series (3) to find a minimum of corresponding acceptable risk

$$\alpha_k = \min_{x\in X}\max_{y\in Y} \mathbf{P}\big(\xi(x,y) < \xi_k\big). \tag{4}$$

These elements form an auxiliary array

$$0 = \alpha_1 \leq \alpha_2 \leq \cdots \leq \alpha_N \leq 1.$$

To solve (1), (2) now it is sufficiently to find index K such that $\alpha_K < \alpha \leq \alpha_{K+1}$.

Remark that to find α_k from (4) one can use linear programming methods.

Since series (3) may be too long, in [6] a more compact grid was constructed which approximates the initial one. Let S denote an arbitrary subset of the rows of the matrix A, $|S|$ denote the number of elements of S. The elements of the compact grid are

$$\beta_k = \max_{S:|S|=k}\left[\min_j \max_{i\in S}\{a_{ij}\}\right], \quad k = 1,\ldots,m.$$

For any β from interval $\beta_{k-1} < \beta \leq \beta_k$ a series of simple estimates were proposed like that

$$1/k \leq \max_{x\in X}\min_{y\in Y} \mathbf{P}(\xi(x,y) \geq \beta) \leq 1 - (k-1)/m$$

and others.

Though usually $m < N$, and pure strategies are the more simple object for investigation than mixed strategies, nevertheless it is not so easy to work with a triplet of max-min operations.

Counterpoise, we return to a series of exact games (4) and try to construct a more efficient algorithm to solve them. For this purpose, we generate a series of linear programs which differ each other only in a few coefficients of constraint matrix. Then we adopt standard simplex-technology to analyze these programs in a consecutive mode, as it takes place in parametric programming. Also, we give some additional estimates for optimal values in (4). These estimates help us to drop some points of the grid (3).

3 Description of the Auxiliary Matrix Games

Let us associate with (3) the following auxiliary series

$$Q_1 \;\leq\; Q_2 \;\leq\; \ldots \;\leq\; Q_N \;\leq\; Q_{N+1}, \tag{5}$$

composed of Boolean matrices of special type

$$Q_k = \begin{pmatrix} 0\,1\,0\ldots0 \\ 1\,0\,1\ldots0 \\ \ldots \\ 0\,1\,0\ldots1 \end{pmatrix}_{m\times n}.$$

These matrices consist only of 0 and 1 and are built according to the payoff matrix so that units are in those positions, where $a_{ij} < \xi_k$. The rest of the positions are 0. Besides, let us add into consideration the zero matrix Q_1 and the matrix Q_{N+1} which is completely filled with 1.

Using the new notation, we can rewrite $F(\xi_k; x, y) = x^T Q_k y$. Therefore (recall that α is very small),

$$\Big(\mathrm{VaR}_\alpha(x,y) = \xi_k\Big) \iff \Big(x^T Q_k y < \alpha \le x^T Q_{k+1} y\Big).$$

Thus, since all ξ_k are strictly increasing, we can first find

$$K_\alpha(x,y) = \max\{k: \text{ from any } k, \text{ such that } x^T Q_k y < \alpha\},$$

and only then take $\mathrm{VaR}_\alpha(x,y) = \xi_{K_\alpha(x,y)}$.

Similarly, we can replace tasks (1), (2) with reformulated search tasks

$$K_* := \min_{y\in Y} K_\alpha(\bar{x}_\alpha, y) = \max_{x\in X}\min_{y\in Y} K_\alpha(x,y), \tag{6}$$

$$K^* := \max_{x\in X} K_\alpha(x, \bar{y}_\alpha) = \min_{y\in Y}\max_{x\in X} K_\alpha(x,y). \tag{7}$$

Here $\check{v}_* = \xi_{K_*}$ and $\check{v}^* = \xi_{K^*}$. Of course, $K_* \le K^*$.

Along with matrices from the set (5) let us consider a series of auxiliary matrix games with mixed strategies

$$\Gamma(Q_1),\ \ \Gamma(Q_2),\ \ \ldots,\ \ \Gamma(Q_N). \tag{8}$$

In contrast to the classical setting, we assume that in these games, players have switched roles, so that the first player minimizes his winnings, and the second one maximizes his loss. Each such game, however, have an equilibrium, and its lower and upper values coincide.

We denote the values of the auxiliary games by $-\infty < v_k < +\infty$ and the corresponding optimal strategies of the players by $\bar{x}(k) \in X$ and $\bar{y}(k) \in Y$, respectively. We repeat once more that

$$v_k = \min_{x\in X}\max_{y\in Y} x^T Q_k y = \max_{y\in Y}\min_{x\in X} x^T Q_k y = \bar{x}(k)^T Q_k \bar{y}(k), \quad 1 \le k \le K+1.$$

Remark 1. Note that the first several matrices from series (5) will have at least one zero-row and therefore the corresponding matrix games will have zero value and a trivial Wald-like solution.

In the general case, when all the rows of the matrix Q_k contain at least one unit, a solution of the introduced auxiliary game can be found by reducing this game to an equivalent dual pair of linear programming problems of the form

$$0 < \frac{1}{v_k} = \max\{\|x\|_1 : \ Q_k^T x \le e, \ \ x \ge 0\} \ = \ \min\{\|y\|_1 : \ Q_k y \ge e, \ y \ge 0\}. \quad (9)$$

Here $(1/v_k)\bar{x}(k)$ is a solution of the first (primal) problem and $(1/v_k)\bar{y}(k)$ is a solution of the second (dual) one; $e = (1, 1, \ldots, 1)^T$ is a vector of suitable dimension, composed of units; $\|z\|_1 = |z_1| + |z_2| + \cdots + |z_s|$ for $z \in \mathbf{R}^s$.

Lemma 1. *The sequence of numbers v_k does not decrease.*

Proof. As already mentioned, for positive v_{k+1} we have

$$0 < \frac{1}{v_{k+1}} = \max\{\|x\|_1 : Q_{k+1}^T x \le e, \ x \ge 0\}.$$

The optimal vector $\bar{\bar{x}}$ of this problem is nonnegative and $Q_{k+1} \ge Q_k$. Therefore the same optimal vector turns out to be feasible for the preceding problem

$$\frac{1}{v_k} = \max\{\|x\|_1 : \ Q_k^T x \le e, \ x \ge 0\}.$$

So the optimal value of the latter cannot be less than $1/v_{k+1}$. In other words, we have $v_{k+1} \ge v_k$. ◇

The Lemma 1 states that in the given sequence one can single out the number $1 \le \bar{K} < N$ such that

$$v_{\bar{K}} < \alpha \le v_{\bar{K}+1}. \quad (10)$$

Recall that, for the indices $k = 1, \ldots, N$, the inequalities hold

$$\bar{x}(k)^T Q_k y \le v_k = \bar{x}(k)^T Q_k \bar{y}(k) \le x^T Q_k \bar{y}(k), \quad \forall x \in X, \ y \in Y. \quad (11)$$

The question arises: how are these quantities related to the solutions of problems (1), (2)?

Lemma 2. *The inequality $\bar{K} \ge K^*$ is valid.*

Proof. Let us return to the inequalities (10), (11). From the definition of index $\bar{K}$ it follows that

$$\alpha \le v_{\bar{K}+1} \le x^T Q_{\bar{K}+1} \bar{y}(\bar{K}+1) \quad \forall x \in X.$$

Consequently,

$$K_\alpha(x, \bar{y}(\bar{K}+1)) \le \bar{K} \quad \forall x \in X.$$

Hence

$$\max_{x \in X} K_\alpha(x, \bar{y}(\bar{K}+1)) \le \bar{K}$$

and, respectively,

$$K^* = \min_{y \in Y} \max_{x \in X} K_\alpha(x, y) \le \max_{x \in X} K_\alpha(x, \bar{y}(\bar{K}+1)) \le \bar{K},$$

that is what we seek. ◇

Lemma 3. *The inequality $\bar{K} \leq K_*$ holds.*

Proof. Once again, let us turn to the inequalities (10), (11). By the definition of the index $\bar{K}$, we have one more relation

$$\alpha > v_{\bar{K}} \geq \bar{x}(\bar{K})^T Q_{\bar{K}} y \quad \forall y \in Y.$$

Consequently,

$$K_\alpha(\bar{x}(\bar{K}), y) \geq \bar{K} \quad \forall y \in Y.$$

Hence

$$\min_{y \in Y} K_\alpha(\bar{x}(\bar{K}), y) \geq \bar{K}$$

and, respectively,

$$K_* = \max_{x \in X} \min_{y \in Y} K_\alpha(x, y) \geq \min_{y \in Y} K_\alpha(\bar{x}(\bar{K}+1), y) \geq \bar{K},$$

that is what we ask. ◇

Theorem 1. *Any mixed matrix game (1), (2) with quantile-criterion always has an equilibrium. The value of this game coincides with $\xi_{\bar{K}}$, where the index $\bar{K}$ is determined by the relations (11) . More over, the strategy $\bar{x}(\bar{K})$ is optimal for the first player and the strategy $\bar{y}(\bar{K}+1)$ is optimal for the second one.*

Proof. This assertion follows from the Lemmas 1, 2 and the fact that $K_* \leq K^*$.

4 How to Find the Index $\bar{K}$

Let us discuss now, how we can solve the matrix games (8) with the help of the primal simplex method taking into account a discrete variation of the coefficients of its constraint matrix [12].

Note that the first game $\Gamma(Q_0)$ has a trivial solution. Also, the first few games may have the auxiliary matrices Q_k with null rows too. Such games may be solved elementary by Wald's principle.

Only when each row of the matrix Q_k has at least one unit, the task becomes nontrivial.

In this case, we can find a solution of the game $\Gamma(Q_k)$ by solving the linear programming problem (9). Reduce the latter to standard form

$$\max\{z = x_1 + x_2 + \cdots + x_m : \ Q_k^T x + u = e, \ \ x \geq 0, \ u \geq 0\}, \qquad (12)$$

where u is a vector of slack variables.

To solve the linear program formulated above, let us apply the so-called revised primal simplex method. Recall, that this variant of the primal simplex method works only with inverse basis matrix and not with a simplex-tableau as a whole. As a starting basis, we can take one including all slack variables.

Suppose that we have already solved the nontrivial game $\Gamma(Q_k)$ with index k. Proceeding to the next problem $\Gamma(Q_{k+1})$, we see that the coefficients of (12) have changed (one or several additional units have added to this constraint matrix). How can we take these changes into account in such a way that the optimal inverse basis matrix obtained for the previous linear program will be useful for further calculations of the simplex method?

If the modified columns of Q_k do not belong to the optimal basis of the previous step, then the calculations may be continued without any troubles. The troubles appear only if the modified column is a basic one. In this case, it is convenient to convert this column into an artificial one. To do this, we replace the corresponding coefficient of the objective function by $-M$, where M is very large constant. Simultaneously we add the new content of this column to the constraint matrix. Thus the number of structural variables in (12) will increase by one. After that, the simplex method can continue execution from the previous basis, since the basis matrix and the matrix inverse to it remain unchanged, while the basic solution itself remains feasible. All now needs to be done is to calculate the new reduced costs (i.e., new simplex estimates of the nonbasic variables).

Calculations may be continued according to the usual scheme of the primal simplex method. After finding the next optimal basic solution, it will turn out that all artificial variables have left it (both the former and the newly created ones).

Thus, in the transition from one auxiliary matrix game to another, in the current linear programming problem, new structural variables may appear, while some of the former structural variables may become artificial. At the end of each optimization cycle, the artificial variables become equal to zero, and we can remove them from the current linear program.

The outlined technique can be used not only for the transition from matrices Q_s with a small index to matrices Q_r with a large index but also for a transition in the opposite direction.

Let us illustrate the proposed technique by a small numerical example.

Example 1
Consider the game $\Gamma(Q)$, where $Q = E$, E is the identity (3×3)-matrix. The corresponding linear program (12) is given by the following simplex tableau:

Basis	x_1	x_2	x_3	u_1	u_2	u_3	$r.h.s.$
z	-1	-1	-1	0	0	0	0
	1	0	0	1	0	0	1
	0	1	0	0	1	0	1
	0	0	1	0	0	1	1

Let the variables x_1, x_2, x_3 be basic. The corresponding basis matrix coincides with the identity one. To make sure that this basis is optimal, let us exclude the basic variables from the objective function:

Thus, the value of the game $\Gamma(Q)$ is equal to 3, and the optimal strategy of the first player is $\bar{x} = (1/3; 1/3; 1/3)$.

Basis	x_1	x_2	x_3	u_1	u_2	u_3	$r.h.s.$
z	0	0	0	1	1	1	3
x_1	1	0	0	1	0	0	1
x_2	0	1	0	0	1	0	1
x_3	0	0	1	0	0	1	1

Let the auxiliary matrix of the next game differ from the previous one by one unit into position (3.1). The corresponding simplex tableau changes as follows (here, the extra unit is highlighted):

Basis	x_1	x_2	x_3	u_1	u_2	u_3	$r.h.s.$
z	−1	−1	−1	0	0	0	0
	1	0	**1**	1	0	0	1
	0	1	0	0	1	0	1
	0	0	1	0	0	1	1

In order not to start calculations with the slack basis again, we will slightly expand the number of the variables of the program, namely, return the previous state of the modified column to the table. We denote this column with a new name x_3^*. Emphasize that x_3^* will play the role of an artificial variable, that is, it will get a large penalty constant $M > 0$ in the corresponding position of the objective function:

Basis	x_1	x_2	x_3^*	x_3	u_1	u_2	u_3	$r.h.s.$
z	−1	−1	M	−1	0	0	0	0
	1	0	0	**1**	1	0	0	1
	0	1	0	0	0	1	0	1
	0	0	1	1	0	0	1	1

Now we can continue the calculation using the previous optimal basis, taking into account that both the base matrix and the matrix inverse to it are unchanged, as well as the transformed right-hand sides:

Basis	x_1	x_2	x_3^*	x_3	u_1	u_2	u_3	$r.h.s.$
z	-1	-1	M	-1	0	0	0	0
x_1	1	0	0	1	1	0	0	1
x_2	0	1	0	0	0	1	0	1
x_3^*	0	0	1	1	0	0	1	1

To determine new reduced costs let us eliminate the basic unknowns from the objective function:

Basis	x_1	x_2	x_3^*	x_3	u_1	u_2	u_3	$r.h.s.$
z	0	0	0	$-M$	1	1	$-M$	$2-M$
x_1	1	0	0	1	1	0	0	1
x_2	0	1	0	0	0	1	0	1
x_3^*	0	0	1	1	0	0	1	1

Next, we apply the usual rules of the simplex method to chose an incremental variable and a blocking constraint. These will be the column x_3 and the row x_3^* respectively (they are shown below in bold):

Basis	x_1	x_2	x_3^*	x_3	u_1	u_2	u_3	$r.h.s.$
z	0	0	0	$-M$	1	1	$-M$	$2-M$
x_1	1	0	0	**1**	1	0	0	1
x_2	0	1	0	**0**	0	1	0	1
x_3	**0**	**0**	**1**	**1**	**0**	**0**	**1**	**1**

The variable x_3 enters a basis and the artificial variable x_3^* leaves it. Now we can remove the variable x_3^* from the tableau. After the corresponding pivot transformation we get:

Basis	x_1	x_2	x_3	u_1	u_2	u_3	$r.h.s.$
z	0	0	0	1	1	0	2
x_1	1	0	0	1	0	-1	0
x_2	0	1	0	0	1	0	1
x_3	0	0	1	0	0	1	1

Thus, we obtain an optimal basic solution of the modified game in one step of the procedure for improving the previous basic solution. The value of the updated game is equal to 1/2, this value corresponds to the optimal strategy $\bar{x} = (0; 1/2; 1/2)$ of the first player.

Remark 2. When we pass from one auxiliary matrix game to another one with a greater number of units in the payoff matrix, the value of an auxiliary game grow quite quickly, so very few steps will be required to achieve a given level of risk. It also means that a nontrivial solution of the original VAR-problem with small $\alpha \approx 0$ can occur only if a payoff matrix is of sufficiently large size $m = O(\alpha^{-1})$.

Remark 3. The process of constructing auxiliary games (8) requires a preliminary ordering of the elements of the payoff matrix A in ascending order. However, this regulating can be done gradually, step by step. Initially, to form Q_2 it is sufficient to find only the smallest elements of matrix A and only then expand the list of ordered elements gradually, as the index of subsequent auxiliary linear program grows.

Next, we present two lemmas which give us lower and upper estimates for the value of the auxiliary games. These lemmas can help us in the implementation of our algorithm above.

Let us introduce the notation

$$w_K^R(i) = \sum_{j=1}^{n} q_{ij}^K, \quad 1 \le i \le m,$$

for the numbers of units in each of the rows of the matrices $Q_K = (q_{ij}^K)$. We also introduce the notation

$$w_K^C(j) = \sum_{i=1}^{m} q_{ij}^K, \quad 1 \le j \le n,$$

for the numbers of units in each of the columns of the matrices $Q_K = (q_{ij}^K)$.

Lemma 4. *The value of the auxiliary game $\Gamma(Q_K)$ satisfies the two-side inequality*

$$\frac{1}{m} \cdot \min_{1 \le i \le m} \{w_K^R(i)\} \le v_K \le \frac{1}{n} \cdot \max_{1 \le j \le n} \{w_K^C(j)\}.$$

Proof. As it is well-known, to find the value of the game $\mathbf{G}(Q_K)$, we can solve not only the linear program (9) but another one as well, namely

$$v_K = \min\{\lambda : \ \lambda e \ge Q_K^T x, \ e^T x = 1, \ x \ge 0\}.$$

From the system of inequalities and equations of this program, it follows that, if λ and x are feasible, then

$$m\lambda = \lambda e^T e \ge e^T Q_K^T x = \sum_{i=1}^{m} \sum_{j=1}^{n} q_{ij}^K x_i = \sum_{i=1}^{m} w_K^R(i) x_i$$
$$\ge \min_{1 \le i \le m} \{w_K^R(i)\} \sum_{i=1}^{m} x_i = \min_{1 \le i \le m} \{w_K^R(i)\}.$$

Thus, in any case,

$$\lambda \geq \frac{1}{m} \cdot \min_{1 \leq i \leq m} \{w_K^R(i)\},$$

that is what we want.

The second estimate may be obtained by repeating of similar arguments for the dual problem

$$v_K = \max\{\lambda : \ \lambda e \leq Q_K y, \ e^T y = 1, \ y \geq 0\}.$$

The proof is complete. ◇

Next, let us consider the transition from the game $\Gamma(Q_K)$ to the game $\Gamma(Q_{K+1})$. Recall that the matrix Q_{K+1} contains one or more additional units as compared to the matrix Q_K. Denote by $W_{K+1} \subset \{1, \ldots, m\} \times \{1, \ldots, n\}$ the set of all pairs of indices (i, j) such that $q_{ij}^K \neq q_{ij}^{K+1}$. These are the positions in the matrix Q_{K+1} such that consist of the additional units in question. Also set

$$I_{K+1}(j) = \{i : \ (i, j) \in W_{K+1}\}, \quad J_{K+1}(i) = \{j : \ (i, j) \in W_{K+1}\}.$$

Lemma 5. *If the game $\Gamma(Q_K)$ is already resolved and the optimal strategies $\bar{x}$ and $\bar{y}$ of the players are known, then the two-side inequality holds*

$$v_K + \min_{1 \leq i \leq m} \sum_{j \in J_{K+1}(i)} \bar{y}_j \leq v_{K+1} \leq v_K + \max_{1 \leq j \leq n} \sum_{i \in I_{K+1}(j)} \bar{x}_i.$$

Proof. Let us again turn to the linear program

$$v_K = \min\{\lambda : \ \lambda e \geq Q_K^T x, \ e^T x = 1, \ x \geq 0\}.$$

Obviously, the properties of the optimal strategies imply

$$v_K = \max_{1 \leq j \leq n} \sum_{i=1}^{m} q_{ij}^K \bar{x}_i.$$

Using this strategy in the next game, we have

$$\begin{aligned} v_{K+1} \leq \lambda &= \max_{1 \leq j \leq n} \sum_{i=1}^{m} q_{ij}^{K+1} \bar{x}_i = \max_{1 \leq j \leq n} \left(\sum_{i=1}^{m} q_{ij}^K \bar{x}_i + \sum_{i=1}^{m} (q_{ij}^{K+1} - q_{ij}^K) \bar{x}_i \right) \\ &\leq \max_{1 \leq j \leq n} \sum_{i=1}^{m} q_{ij}^K \bar{x}_i + \max_{1 \leq j \leq n} \sum_{i=1}^{m} (q_{ij}^{K+1} - q_{ij}^K) \bar{x}_i \\ &= v_K + \max_{1 \leq j \leq n} \sum_{i \in I_{K+1}(j)} \bar{x}_i. \end{aligned}$$

To obtain the second estimate, we can repeat similar arguments for the dual problem

$$v_K = \max\{\lambda : \ \lambda e \leq Q_K y, \ e^T y = 1, \ y \geq 0\}.$$

The proof is complete. ◇

The above lemmas can help us to choose a good initial problem for starting the calculations and simplifying the subsequent parametric analysis.

Example 2
Suppose, that some commercial firm needs to select a decision from the given finite set of alternatives, to maximize its profit. A discrete random factor with unknown law of the probability distribution also influences on the future result. The following matrix consists of the initial data about the possible profit

$$A = \begin{pmatrix} -3 & 2 & 0 & 1 & 0 & 0 & 0 & 0 & 1 & -1 & 1 \\ 1 & 0 & 0 & 0 & -4 & 0 & 1 & -1 & 0 & 0 & 0 \\ 0 & 0 & -2 & 0 & 1 & 1 & 0 & 0 & 1 & 0 & 1 \\ 0 & 0 & 1 & 0 & 0 & -1 & 0 & 1 & -3 & 0 & 1 \\ 1 & 1 & 0 & 1 & -1 & 0 & 0 & 0 & 0 & 1 & -3 \\ 0 & 0 & 1 & 0 & 0 & 5 & -2 & 1 & 0 & 0 & 2 \\ 0 & -3 & 0 & -1 & 1 & 0 & 2 & 0 & 1 & 0 & 0 \\ 0 & 0 & 1 & 1 & -1 & 0 & 0 & 1 & 0 & -3 & 0 \\ 0 & 0 & -1 & 0 & 1 & -3 & 1 & 0 & 1 & 1 & 0 \\ 1 & 0 & 0 & 1 & 0 & 0 & 1 & -2 & 1 & 0 & 0 \\ 0 & 3 & 0 & -2 & 1 & 0 & 0 & 0 & 2 & 0 & 1 \end{pmatrix}.$$

Its rows correspond to the possible variants of the firm's decision, and the columns correspond to various realizations of the random factor.

If the number of the game parties to be played is sufficiently large (potentially infinite), then we can focus not on winning in a one party, but on some average return. However, negative winnings mean the ruin of the firm and are unacceptable (at least the probability of such an outcome should be small).

The application of the classical approach associated with maximization of average outcome leads to minimax strategies $\bar{x}$ = (0.039; 0.002; 0.014; 0.137; 0.147; 0.166; 0.087; 0.052; 0.189; 0.107; 0.059) for the first player and $\bar{y}$ = (0.074; 0.097; 0.014; 0.136; 0.006; 0.068; 0.27; 0.179; 0.019; 0.06; 0.074) for the second player. These strategies provide a positive average outcome $\mathbf{E}(\bar{x}, \bar{y}) = 0.140$ for the first player. Note that the corresponding ruin probability equal to $\mathbf{P}(\xi(\bar{x}, \bar{y}) < 0) = 0.186$.

Next, let us find the strategy of the company, which minimize the probability of ruin. We construct an auxiliary 0–1 matrix with the units located in the positions of the outcomes associated with ruin. The matrix looks as

$$Q = \begin{pmatrix} 1 & 0 & 0 & 0 & 0 & 0 & 0 & 0 & 0 & 1 & 0 \\ 0 & 0 & 0 & 0 & 1 & 0 & 0 & 1 & 0 & 0 & 0 \\ 0 & 0 & 1 & 0 & 0 & 0 & 0 & 0 & 0 & 0 & 0 \\ 0 & 0 & 0 & 0 & 0 & 1 & 0 & 0 & 1 & 0 & 0 \\ 0 & 0 & 0 & 0 & 1 & 0 & 0 & 0 & 0 & 0 & 1 \\ 0 & 0 & 0 & 0 & 0 & 0 & 1 & 0 & 0 & 0 & 0 \\ 0 & 1 & 0 & 1 & 0 & 0 & 0 & 0 & 0 & 0 & 0 \\ 0 & 0 & 0 & 0 & 1 & 0 & 0 & 0 & 0 & 1 & 0 \\ 0 & 0 & 1 & 0 & 0 & 1 & 0 & 0 & 0 & 0 & 0 \\ 0 & 0 & 0 & 0 & 0 & 0 & 0 & 1 & 0 & 0 & 0 \\ 0 & 0 & 0 & 1 & 0 & 0 & 0 & 0 & 0 & 0 & 0 \end{pmatrix}.$$

The solution of the problem (12) with matrix Q gives us the optimal strategy of the first player $\bar{x} = (0.143; 0; 0.143; 0.143; 0.143; 0.143; 0.143; 0; 0; 0.143; 0)$ and the inequality $\mathbf{P}(\xi(\bar{x}, y) < 0) \leq 0.1428$ for all $y \in Y$. It is less than what was obtained with the classical approach to the problem.

The next auxiliary matrix is

$$Q' = \begin{pmatrix} 1 & 0 & 1 & 0 & 1 & 1 & 1 & 1 & 0 & 1 & 0 \\ 0 & 1 & 1 & 1 & 1 & 1 & 0 & 1 & 1 & 1 & 1 \\ 1 & 1 & 1 & 1 & 0 & 0 & 1 & 1 & 0 & 1 & 0 \\ 1 & 1 & 0 & 1 & 1 & 1 & 1 & 0 & 1 & 1 & 0 \\ 0 & 0 & 1 & 0 & 1 & 1 & 1 & 1 & 1 & 0 & 1 \\ 1 & 1 & 0 & 1 & 1 & 0 & 1 & 0 & 1 & 1 & 0 \\ 1 & 1 & 1 & 1 & 0 & 1 & 0 & 1 & 0 & 1 & 1 \\ 1 & 1 & 0 & 0 & 1 & 1 & 1 & 0 & 1 & 1 & 1 \\ 1 & 1 & 1 & 1 & 0 & 1 & 0 & 1 & 0 & 0 & 1 \\ 0 & 1 & 1 & 0 & 1 & 1 & 0 & 1 & 0 & 1 & 1 \\ 1 & 0 & 1 & 1 & 0 & 1 & 1 & 1 & 0 & 1 & 0 \end{pmatrix}$$

Solving the dual problem to (12) with matrix Q', we find the optimal strategy of the second player $\bar{y} = (0; 0; 0; 0; 0.167; 0.333; 0; 0; 0.167; 0.167; 0.167)$ and get the inequality $\mathbf{P}(\xi(x, \bar{y}) \leq 0) \geq 0.666$ for all $x \in X$. Thus, $\bar{x}$ is optimal strategy for the firm that guarantees its non-bankruptcy for any $y \in Y$ with any acceptable risk level from interval $0.143 < \alpha < 0.6667$.

5 Conclusion

Matrix games in mixed strategies are considered, in which the payoff function is defined not as the mathematical expectation of a random return in a long series of game parties, but as its guaranteed VaR-estimate at a given risk level. The properties of such games are studied, and novel methods for their solution additional to known ones are proposed.

References

1. Von Neumann, J., Morgenstern, O.: The Theory of Games and Economic Behavior. Princeton University Press, Princeton (1944)
2. Karlin, S.: Mathematical Methods in Games, Programming and Economics. In two vols. Dover Publications Inc., New York (1959). Reprinted 1992
3. Luce, R.D., Raiffa, H.: Games and Decisions. Introduction and Critical Survey. Dover Publication Inc., New York (1957). Reprinted 1989
4. Walsh, J.E.: Discrete two-person game theory with median payoff criterion. Opsearch (J. Oper. Res. Soc. India) **6**(2), 83–97 (1969)
5. Walsh, J.E., Kelleher, G.J.: Generally applicable two-person percentile game theory. Opsearch (J. Oper. Res. Soc. India) **8**(2), 143–152 (1971)
6. De Vries, H.: Quantile criteria for the selection of strategies in game theory. Int. J. Game Theory **3**(2), 105–114 (1974)
7. Kibzun, A.I., Kan, Y.S.: Stochastic Programming Problems with Probability and Quantile Functions. Wiley, Chichester (1996)
8. Uryasev, S., Rockafellar, R.T.: Conditional value-at-risk: optimization approach. In: Uryasev, S., Pardalos, P.M. (Eds.) Stochastic Optimization: Algorithms and Applications, pp. 411–435. Kluwer, London (2001)
9. Kan, Y.S., Krasnopol'skaya, A.N.: Selection of a fixed-income portfolio. Autom. Remote Control **67**(4), 598–605 (2006). https://doi.org/10.1134/S0005117906040072
10. Pankov, A.R., Platonov, E.N., Semenikhin, K.V.: Minimax optimization of investment portfolio by quantile criterion. Autom. Remote Control **64**(7), 1122–1137 (2003)
11. Konyukhovskiy, P.V., Malova, A.S.: Game-theoretic models of collaboration among economic agents. Contrib. Game Theory Manage. **6**, 211–221 (2013)
12. Murtagh, B.: Advanced Linear Programming: Computation and Practice. MacGraw-Hill Intern., New York (1981)

Simulation of Flow Regimes of Non-isothermal Liquid Films

Liudmila Prokudina(✉) and Dmitrii Bukharev(✉)

South Ural State University, Pr. Lenina 76, Chelyabinsk, Russia
prokudinala@susu.ru, needleinspace@gmail.com

Abstract. For moderate Reynolds numbers, a nonlinear partial differential equation of the free surface state of a non-isothermal liquid film is presented. The algorithm was developed and the program was written in Matlab R2017b using the Symbolic Math Toolbox module. The wave characteristics of the liquid film under heat and mass transfer are calculated. The flow regimes of a vertical liquid film with a maximum perturbation growth rate are distinguished, and the effect of temperature gradients and surface viscosity on them is investigated.

Keywords: Liquid film · Non-linear mathematical model · Instability · Increment

1 Introduction

Studies of thin viscous liquid layers (liquid film) flows are carried out both theoretically [1–3,9,11,12,18,21] and experimentally [2,4–8,10]. The relevance and practical significance of these studies are associated with the implementation of liquid film flows in numerous devices, for example, in evaporators, absorbers, distillation columns, crystallizers, refrigeration equipment, as well as liquid film is the basis of many technological processes in chemical, petrochemical, food and other industries [14,15]. Low thermal resistance and a large contact surface at low specific fluid flow rates make liquid film a very effective tool in the process of interfacial heat and mass transfer. In addition, in many cases there is an additional intensification of transport processes due to wave formation. Various physical and chemical factors, such as temperature effects (temperature gradients) and the presence of insoluble surfactants (surface viscosity) on the free surface of the film affect the wave characteristics of the liquid film and the process of wave formation [5,13,14,19,20,22]. Experimental studies of the flow regimes of liquid films [14,16] show that the regimes with the maximum value of the increment are the most stable with respect to small perturbations. The maximum value of the increment and the corresponding wave number determine the optimal flow regime of the liquid film. Mathematical modeling of optimal flow regimes of liquid films allows us to determine the influence of various physical and chemical factors on the corresponding wave characteristics.

The novelty of the study is related to the effect of temperature gradients and surface viscosity on the wave characteristics of the liquid film.

I. Bykadorov et al. (Eds.): MOTOR 2019, CCIS 1090, pp. 319–328, 2019.
https://doi.org/10.1007/978-3-030-33394-2_25

2 Mathematical Model

We consider the flow of a thin viscous liquid layer with a free surface under the action of gravity on a solid impermeable surface with a temperature T in the coordinate system OXY (Fig. 1), where the OX axis is directed in the direction of the layer flow and the OY is directed perpendicular to the liquid layer. Presence of insoluble surfactants on the free surface of the liquid film was also taken into account. Liquid film is described by the system of Navier-Stokes equations and the continuity equation with boundary conditions [20, 22].

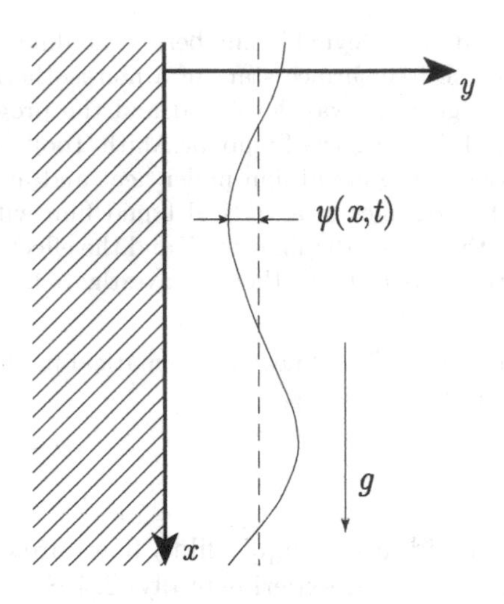

Fig. 1. Flow of liquid film.

The nonlinear mathematical model of the state of the free surface of the liquid film for moderate Reynolds numbers has the form [20]:

$$\begin{aligned}&\left(a_7\frac{\partial}{\partial x}+a_{13}\right)\frac{\partial\psi}{\partial t}+a_1\frac{\partial^4\psi}{\partial x^4}+a_4\frac{\partial^3\psi}{\partial x^3}+a_6\frac{\partial^2\psi}{\partial x^2}+a_{11}\frac{\partial\psi}{\partial x}\\&+a_{14}\psi\frac{\partial\psi}{\partial x}+a_{16}\psi\frac{\partial^2\psi}{\partial x^2}+a_{17}\psi\frac{\partial^2\psi}{\partial x\partial t}+a_{21}\frac{\partial\psi}{\partial x}\frac{\partial\psi}{\partial t}+a_{22}\left(\frac{\partial\psi}{\partial x}\right)^2\\&+a_{26}\frac{\partial\psi}{\partial x}\frac{\partial^2\psi}{\partial x^2}+a_{28}\psi\frac{\partial^3\psi}{\partial x^3}+a_{30}\frac{\partial\psi}{\partial x}\frac{\partial^3\psi}{\partial x^3}+a_{34}\psi\frac{\partial^4\psi}{\partial x^4}+a_{37}\psi^2\frac{\partial\psi}{\partial x}\\&+a_{39}\psi^2\frac{\partial^2\psi}{\partial x^2}+a_{40}\psi^2\frac{\partial^2\psi}{\partial x\partial t}+a_{44}\psi\frac{\partial\psi}{\partial x}\frac{\partial\psi}{\partial t}+a_{45}\psi\left(\frac{\partial\psi}{\partial x}\right)^2\\&+a_{49}\frac{\partial\psi}{\partial x}\frac{\partial^2\psi}{\partial x^2}+a_{51}\psi\frac{\partial\psi}{\partial x}\frac{\partial^3\psi}{\partial x^3}+a_{55}\psi^2\frac{\partial^4\psi}{\partial x^4}+a_{58}\psi^2\frac{\partial^3\psi}{\partial x^3}=0\end{aligned}\tag{1}$$

The coefficients of Eq. (1) include physical and chemical parameters: Re—Reynolds number, F_x—Froude number, σ—surface tension, M—temperature gradients, N—surface viscosity.

$$\begin{aligned}
&a_1 = -\frac{Re\sigma}{3}, a_4 = -\frac{Re^2 F_x N}{2},\\
&a_6 = -\frac{ReF_y}{3} - \frac{ReM}{2} + \frac{3}{40}Re^3 F_x^2, a_7 = \frac{5}{24}Re^2 F_x,\\
&a_{11} = -ReF_x, a_{13} = -1, a_{14} = -2ReF_x,\\
&a_{16} = -ReF_y - ReM + \frac{9}{20}Re^3 F_x^2, a_{17} = \frac{5}{6}Re^2 F_x,\\
&a_{21} = a_{17}, a_{22} = a_{16}, a_{26} = -\frac{1}{2}Re^2 F_x N, a_{28} = 3a_{26},\\
&a_{30} = a_{34} = -Re\sigma, a_{37} = -ReF_x,\\
&a_{39} = -ReF_y - \frac{ReM}{2} + \frac{9}{8}Re^3 F_x^2,\\
&a_{40} = \frac{5}{4}Re^2 F_x, a_{44} = 2a_{40}, a_{45} = -2ReF_y, a_{49} = -Re^2 F_x N,\\
&a_{51} = -2Re\sigma, a_{55} = \frac{1}{2}a_{54}, a_{58} = -\frac{3}{2}Re^2 F_x N.
\end{aligned} \tag{2}$$

We consider the linear part of Eq. (1). Using the solution type $\psi(x,t) = A\, exp\, i\,(k_x x - \omega t)$, we obtain the dispersion equation

$$\omega\,(a_7 k_x + i) + a_1 k_x^4 - a_4 i k_x^3 - a_6 k_x^2 + a_{11} i k_x = 0. \tag{3}$$

Let us split real and imaginary parts of Eq. (3) and get the solutions for increment ω_i and phase velocity c_r

$$\omega_r = \frac{Y - XZ}{1 + Z^2}, \tag{4}$$

$$\omega_i = X + \omega_r Z, \tag{5}$$

$$c_r = \frac{\omega_r}{k_x}, \tag{6}$$

where $X = a_1 k_x^4 - a_6 k_x^2$, $Y = a_4 k_x^3 - a_{11} k_x$, $Z = a_7 k_x$.

Numerical study of liquid film flows for the range of Reynolds numbers [1,15] in the framework of Eq. (3) allows to solve a number of important problems:

1. Finding areas of instability of the flow of non-isothermal liquid films under the influence of various physical and chemical factors, such as temperature gradients and surface viscosity.

2. Selection of optimal flow regimes of liquid films. Optimal modes are flow regimes of liquid film with a maximum value of the increment. Optimal flow regimes are needed for correct operation of heat and mass transfer devices.
3. Calculation of phase velocity and wavelength for wave numbers corresponding to optimal flow regimes.

We present a model of the algorithm for calculating the wave characteristics.

1: **for** $Re = 5$ **to** 15 **do**
2: Compute equation coefficients $a_1, a_2, ..., a_{58}$
3: Construct symbolic expressions for ω_i, c_r
4: Get first derivatives for ω_i, c_r
5: Compute roots of derivatives
6: **print** $k_{x_{max}}, \omega_{i_{max}}, c_{r_{min}}$
7: **end for**

3 Computational Experiments

Unstable modes of liquid films are characterized by positive values of increment. On the increment curve (Fig. 2), the following points could be noted: the maximum value of the increment and the zero value of the increment with the corresponding wave numbers. The set of points k_{max} for the studied range of Reynolds numbers form the curve of the maximum growth of disturbances (Fig. 3) and the set of points k_0—neutral stability curve.

For the free flow of a liquid film of water, Table 1 presents maximum values of the increment, wave number and wavelength.

Phase velocity of the liquid film in the region of instability varies within the limits $2 \leq c_r \leq 3$, which corresponds to the experimental data of the authors [4–7]. Figure 4 shows values of minimal phase velocity that correspond to the regime with the maximum value of increment.

Table 1. Optimal regimes of the flow

Re	$\omega_{i_{max}}$	k_{max}	λ
5	0.015357	0.072459	86.713455
6	0.023877	0.083407	75.331313
7	0.033763	0.093321	67.328672
8	0.044279	0.102114	61.531077
9	0.054640	0.109758	57.245706
10	0.064189	0.116292	54.029236
11	0.072502	0.121807	51.583077
12	0.079388	0.126422	49.700115
13	0.084838	0.130264	48.234194
14	0.088956	0.133455	47.080929
15	0.091903	0.136103	46.164965

Table 2. Optimal flow regimes of film falling over the heated surface

Re	$\omega_{i_{max}}$	k_{max}	λ
5	0.030210	0.085813	73.219488
6	0.046217	0.098381	63.865784
7	0.063997	0.109499	57.381366
8	0.081920	0.119092	52.759059
9	0.098542	0.127194	49.398544
10	0.112905	0.133926	46.915380
11	0.124586	0.139461	45.053350
12	0.133577	0.143985	43.637913
13	0.140113	0.147671	42.548412
14	0.144543	0.150675	41.700353
15	0.147236	0.153122	41.033778

Table 3. Optimal flow regimes of film with insoluble surfactants

Re	$\omega_{i_{max}}$	k_{max}	λ
5	0.029069	0.084120	74.693091
6	0.043593	0.095363	65.887001
7	0.059045	0.104714	60.003557
8	0.073901	0.112144	56.027958
9	0.086995	0.117757	53.357396
10	0.097706	0.121748	51.608198
11	0.105892	0.124356	50.525784
12	0.111732	0.125823	49.936514
13	0.115556	0.126373	49.719269
14	0.117740	0.126199	49.787755
15	0.118636	0.125465	50.079185

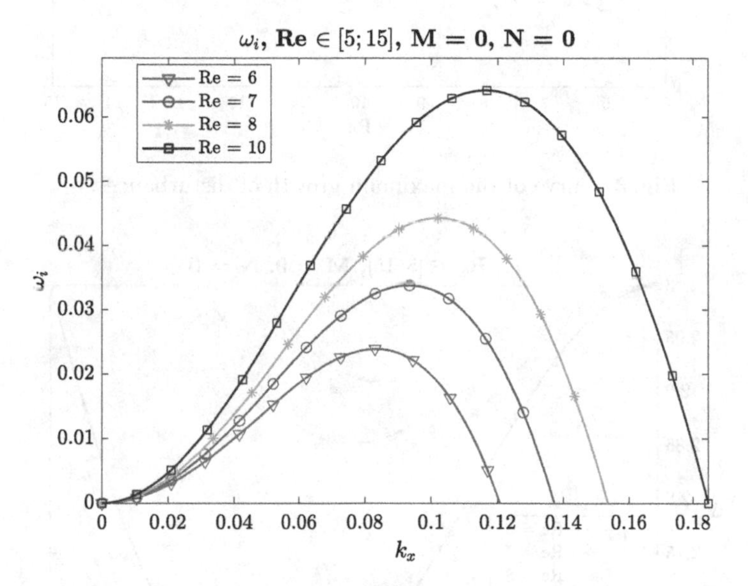

Fig. 2. Increment of liquid film.

We investigate the effect of temperature gradients arising during the flow of the liquid film over the heated surface. This flow is characterized by increase in values of the increment (Fig. 5) and a decrease in the values of phase velocity (Fig. 6). Table 2 shows wave characteristics of optimal film flow regimes. In region of instability of the liquid film, depending on the magnitude of the temperature gradients, rupture of film is possible. In Fig. 7 critical values of the temperature gradients are presented, where possible gaps in film structure could lead to emergency modes [17].

The presence of insoluble surfactants (oils, fats) on the free surface of film leads to stabilization of the flow. The increment value is significantly reduced (Fig. 8), while phase velocity increased (Fig. 9). Table 3 shows wave characteristics of optimal water film flow with insoluble surfactants present.

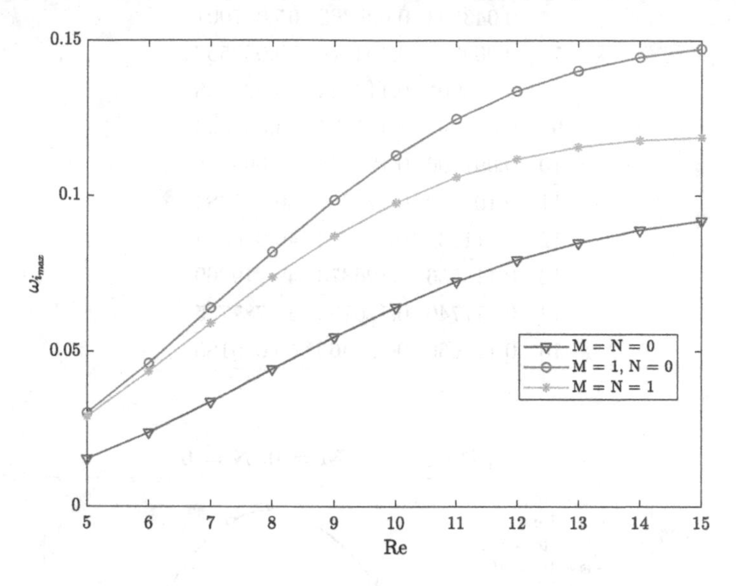

Fig. 3. Curve of the maximum growth of disturbances.

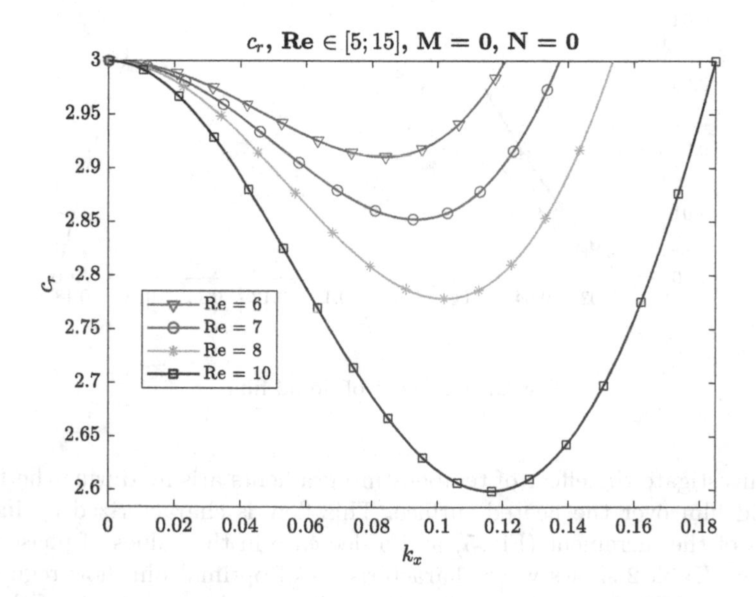

Fig. 4. Phase velocity of liquid film.

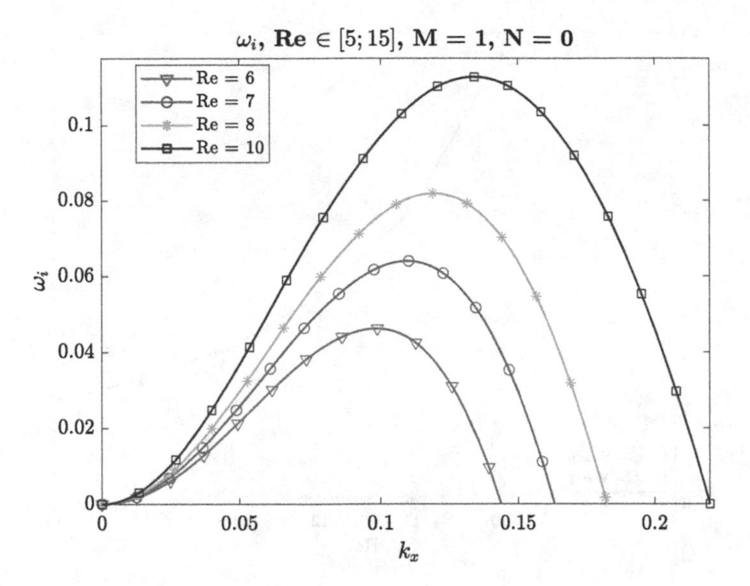

Fig. 5. Increment of liquid film flowing over the heated surface.

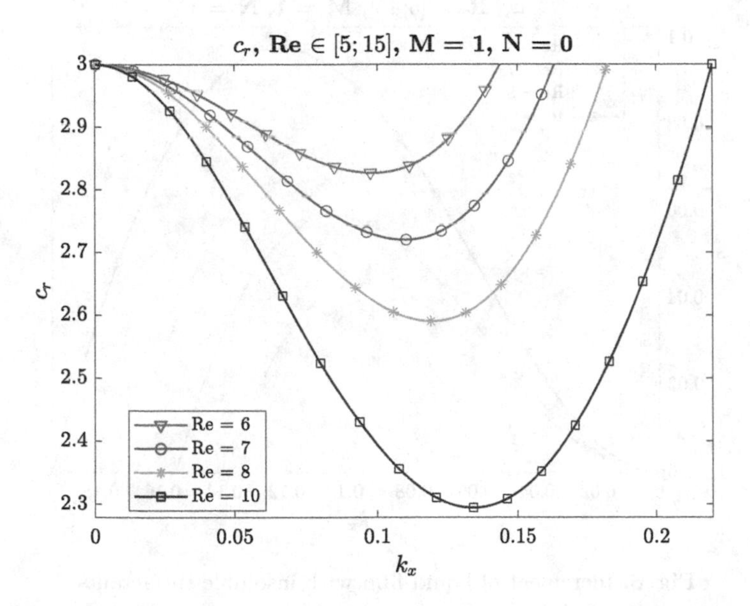

Fig. 6. Phase velocity of liquid film flowing over the heated surface.

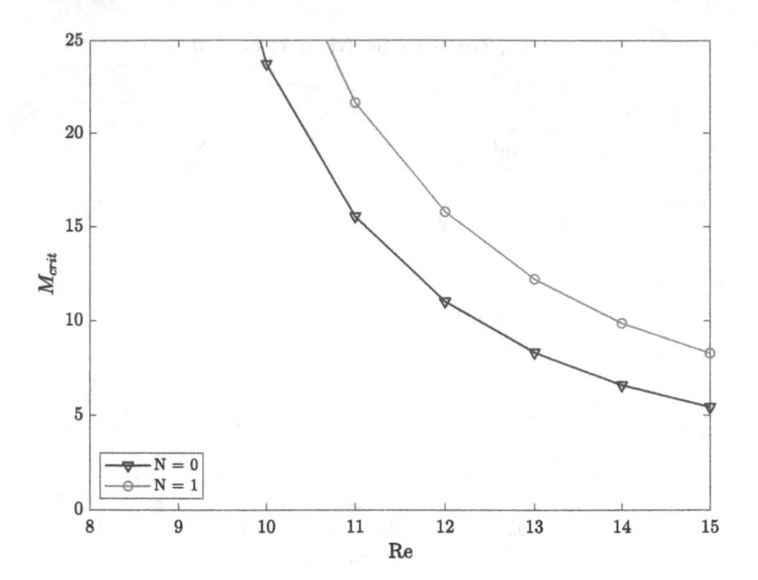

Fig. 7. Critical values of temperature gradients.

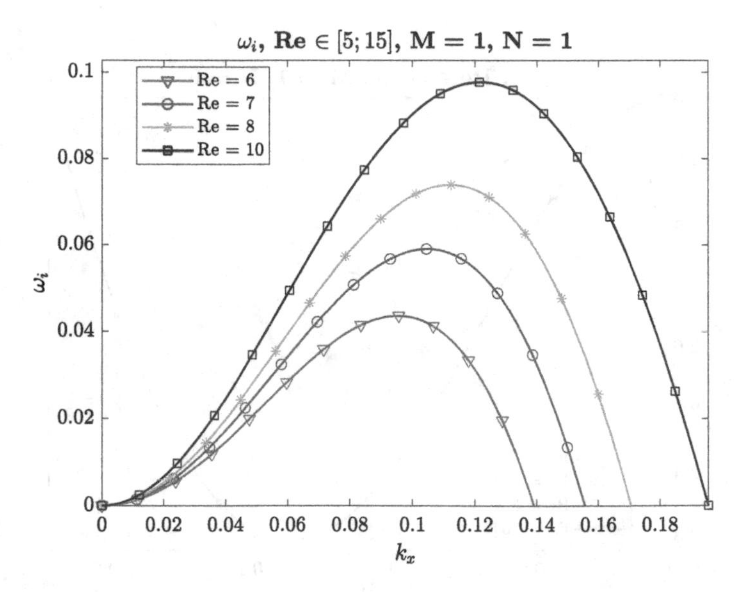

Fig. 8. Increment of liquid film with insoluble surfactants

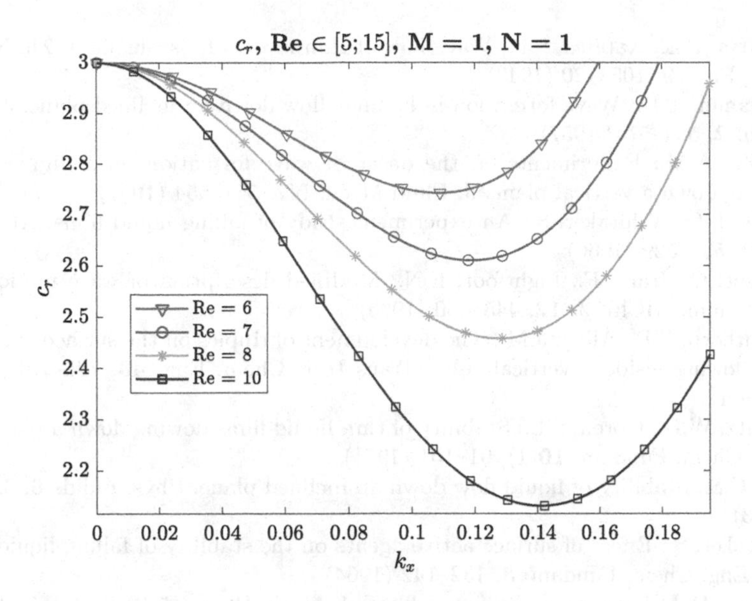

Fig. 9. Phase velocity of liquid film with insoluble surfactants

4 Conclusion

In the framework of free surface state differential Eq. (1), numerical simulation of unstable flow regimes of a vertical liquid water film at moderate Reynolds numbers is carried out.

Optimal flow regimes of the liquid film characterized by the maximum increment and the minimum phase velocity are revealed.

The destabilizing effect of temperature gradients on the wave characteristics of the liquid film was studied. Critical values of the temperature gradients that lead to destruction of film are calculated.

Presence of insoluble surfactants on the free surface of film leads to the appearance of surface viscosity forces that stabilize the film flow. Wave characteristics of the liquid film with the combined effect of temperature gradients and surface viscosity are calculated.

Results of computational experiments are aimed at improving the technologies in liquid films and operation of film devices.

The main contribution of this article is pointing out optimal liquid flow regimes. That includes adding insoluble surfactants into the liquid film in order to achieve more stable flow when liquid films are used in different heated devices.

References

1. Kapitsa, P.L.: Wave flow of thin layers of viscous fluid. Zh. Eksper. Teor. Fiz. **18**, 3–28 (1948)

2. Kapitsa, P.L., Kapitsa, S.P.: Wave flow of thin layers of viscous fluid. Zh. Eksper. Teor. Fiz. **19**, 105–120 (1949)
3. Benjamin, T.B.: Wave formation in laminar flow down an inclined plane. J. Fluid Mech. **2**, 554–574 (1957)
4. Binnie, A.M.: Experiments on the onset of wave formation on a film of water flowing down a vertical plane. J. Fluid Mech. **N2**, 551–554 (1957)
5. Jones, L.O., Whitaker, S.: An experiment study of falling liquid films. AICHE J. **12**(3), 525–529 (1966)
6. Massot, C., Irani, F., Lightfoot, E.N.: Modified description of wave motion in a falling film. AIChE J. **12**, 445–450 (1966)
7. Stainthorp, F.P., Allen, J.M.: The development of ripples on the surface of a liquid film flowing inside a vertical tube. Trans Inst. Chem. Eng. **43**, 785–791 (1965). London
8. Krantz, W.B., Goren, S.L.: Stability of thin liquid films flowing down a plane. Ind. Eng. Chem. Fundam. **10**(1), 91–101 (1971)
9. Yih, C.S.: Stability of liquid flow down an inclined plane. Phys. Fluids **6**, 321–334 (1963)
10. Whitaker, S.: Effect of surface active agents on the stability of falling liquid films. Ind. Eng. Chem. Fundam. **3**, 132–142 (1964)
11. Benney, D.J.: Long waves on liquid films. J. Math. Phys. **45**(2), 150–155 (1966)
12. Krylov, V.S., Vorotilin, V.P., Levich, V.G.: The theory of wave motion of thin liquid films. Theor. Found. Chem. Technol. **3**(4), 499–507 (1969). (in Russian)
13. Filippov, A.G., Saltanov, G.A., Kukushkin, A.N.: Fluid Flow and Heat and Mass Transfer in the Presence of Surfactants. Energoizdat, Moscow (1988). (in Russian)
14. Kholpanov, L.P., Shkadov, V.Ya.: Hydrodynamic and Heat and Mass Transfer with Free Surface. Nauka, Moscow (1990). (in Russian)
15. Gogonin, I., Shemagin, I.: Heat Exchange at Film Condensation and Film Boiling in the Components of NPS. Energoatomizdat, Moscow (1993). (in Russian)
16. Alekseenko, S.V., Nakoryakov, V.E.: Pokusaev Wave Flow of Liquid Films. Begell House, New York (1994)
17. Prokudina, L.A., Vyatkin, G.P.: Instability of a nonisothermal liquid film. Dokl. Phys. **43**(10), 652–654 (1998)
18. Vlachogiannis, M., Bontozoglou, V.: Observation of solitary wave dynamics film flows. J. Fluid Mech. **435**, 191–215 (2001)
19. Burmistrova, O.A.: Stability of vertical liquid film with consideration of the Marangoni effect and heat exchange with the environment. Appl. Mech. Tech. Phys. **55**(3), 17–25 (2014)
20. Prokudina, L.A.: Influence of surface tension inhomogeneity on the wave flow of a liquid film. J. Eng. Phys. Thermophys. **87**, 165–173 (2014)
21. Prokudina, L.A.: Nonlinear evolution of perturbations in a thin fluid layer during wave formation. J. Exp. Theor. Phys. **118**(3), 480–488 (2014)
22. Prokudina, L.A.: Nonlinear development of the marangoni instability in liquid films. J. Eng. Phys. Thermophys. **89**(4), 921–928 (2016)

Counterexamples in the Theory of α-Sets

Vladimir Ushakov[1,2], Aleksandr Ershov[1,2(✉)], and Maksim Pershakov[2]

[1] N.N. Krasovskii Institute of Mathematics and Mechanics, 16 S.Kovalevskaya Str., 620108 Ekaterinburg, Russia
ale10919@yandex.ru

[2] Ural Federal University, 19 Mira Str., 620002 Ekaterinburg, Russia

Abstract. The paper considers α-sets that are the generalization of convex sets. This concept was introduced by V.N. Ushakov in the 2000s to classify the reachable sets of controlled systems according to the degree of their nonconvexity. Since then a lot of properties of such sets have been discovered and proven. However, not all the "natural" properties are fulfilled. We have proved two "unnatural" properties for such sets in the paper. Firstly, we provide an example of a non-self-intersecting curve, a connected segment of which is "less convex" than the entire curve in terms of α-sets. Secondly, we show that there is an α-curve which is not representable as a graph of the function for all $\alpha > 0$.

Keywords: α-Set · Nonconvexity · α-Curve

1 Introduction

The α-sets are the generalization of the convex sets along with E. Michael's paraconvex sets [1], and with the weakly convex sets according to Vial, Efimov and Stechkin [2]. Note that each generalized-convex set of these classes is associated with a numerical parameter, a measure of nonconvexity, and these measures satisfy certain relationships [2,3].

The following notations are used [4]:
$\operatorname{co} M$ is the convex hull of set M;
$\langle x_*, x^* \rangle$ is the scalar product of x_* and x^* from $\mathbb{R}^n$;
$||x_*|| = \langle x_*, x_* \rangle^{1/2}$ is the standard norm (generated by the scalar product) in the Euclidean space;
$\angle(x_*, x^*) = \arccos \dfrac{\langle x_*, x^* \rangle}{||x_*|| \cdot ||x^*||} \in [0, \pi]$ is the angle between vectors x_* and x^*;
$\operatorname{con} M = \{y = \lambda x : \lambda \geqslant 0, x \in M\}$ is the cone in $\mathbb{R}^n$ spanned by set M and with the vertex at zero.

Projection p^* of the point x^* onto set M is the closest to x^* point from M.

The reported study was funded by RFBR according to the research project no. 18-01-00018 mol_a. The work was supported by Act 211 Government of the Russian Federation, contract no. 02.A03.21.0006.

I. Bykadorov et al. (Eds.): MOTOR 2019, CCIS 1090, pp. 329–340, 2019.
https://doi.org/10.1007/978-3-030-33394-2_26

Definition 1. *Let A be a closed set in the n-dimensional Euclidean space $\mathbb{R}^n$ and $z^* \in \mathbb{R}^n \backslash A$. By $\Omega_A(z^*)$ we denote the set of all projections of point z^* onto A, and by $H_A(z^*) = \mathrm{con}(\mathrm{co}\,\Omega_A(z^*) - z^*)$ we denote the cone spanned by $\mathrm{co}\,\Omega_A(z^*) - z^* = \{z - z^* : z \in \mathrm{co}\,\Omega_A(z^*)\}$.*

Let us define the function $\alpha_A(z^*) = \max\limits_{h_*, h^* \in H_A(z^*)} \angle(h_*, h^*) \in [0, \pi]$ and suppose that $\alpha_A = \sup\limits_{z^* \in \mathbb{R}^n \backslash A} \alpha_A(z^*) \in [0, \pi]$.

Then the set A is called α-set, where $\alpha = \alpha_A$.

2 Results Statement

The α-sets have the following properties (Lemmas 1 and 2).

Lemma 1. *[5] Let $a \leqslant c < d \leqslant b$, $f \in C[a, b]$, $\Gamma = \{(x, y) : y = f(x), a \leqslant x \leqslant b\}$, $\gamma = \{(x, y) : y = f(x), c \leqslant x \leqslant d\}$.*

Then $\alpha_\Gamma = \sup\limits_{z^* \in \mathbb{R}^2 \backslash \Gamma} \alpha_\Gamma(z^*) \geqslant \alpha_\gamma = \sup\limits_{z^* \in \mathbb{R}^2 \backslash \gamma} \alpha_\gamma(z^*)$.

Lemma 1 shows that any single connected segment γ of the graph Γ of the continuous function has the measure of nonconvexity α_γ not exceeding α_Γ. In the next section, the counterexample 1 is given, which implies that this lemma cannot be applied to an arbitrary non-self-intersecting curve.

Lemma 2. *[6] Let scalar function $f(\cdot)$, defined on the closed set $M \subset \mathbb{R}^n$, be Lipschitz with constant $L \geqslant 0$. Then the sets* $\mathrm{hypo}\, f(\cdot)$, $\mathrm{epi}\, f(\cdot)$, *and* $\mathrm{gr}\, f(\cdot)$ *are β_α-sets, where $\beta_\alpha \leq \alpha$, $\alpha = 2 \arctan L$.*

The following new result (counterexample 2) indicates that the converse is wrong in some sense. It will be constructively proved in Sect. 4, and the following theorem can be formulated.

Theorem 1. *There exists a such non-self-intersecting curve $\gamma \subset \mathbb{R}^2$ with an arbitrarily small $\alpha_\gamma > 0$ that it is not a graph of some function in any rectangular coordinate system.*

3 Counterexample 1

Let's consider a non-self-intersecting curve consisting of two arcs of circles and its simply connected segment consisting of one semicircle (Fig. 1).

Let's describe these curves in detail and define the curve $\Gamma = \gamma \cup \omega$, where

$$\gamma = \{(x, y) : x^2 + y^2 = 1, y \leq 0\},$$

$$\omega = \left\{(x, y) : \left(x - \frac{1}{2}\right)^2 + \left(y - \frac{1}{4}\right)^2 \leq \frac{5}{16}, y \leq 0, x \geq \frac{1}{2}\right\}.$$

In other words, γ is the lower half of the unit radius circle centered at $O = (0, 0)$, where the end points of this curve are points $A = (-1, 0)$ and $B = (1, 0)$.

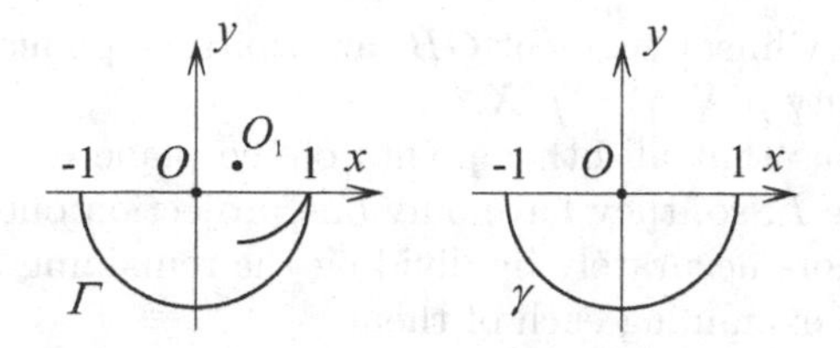

Fig. 1. Curve Γ and its single connected segment γ.

The curve ω is an arc of the circle with radius $\frac{\sqrt{5}}{4}$ centered at $O_1 = \left(\frac{1}{2}, \frac{1}{4}\right)$ with the endpoints at $B = (1, 0)$ and $C = \left(\frac{1}{2}, \frac{1-\sqrt{5}}{4}\right)$.

The purpose of the counterexample 1 is to prove that $\alpha_\Gamma < \alpha_\gamma$.

It is obvious that $\alpha_\gamma = \alpha(\gamma)(O) = \pi$. It is more difficult to calculate the value of α_Γ. Firstly, a bisector [7] β of the curve Γ (Fig. 2) will be constructed.

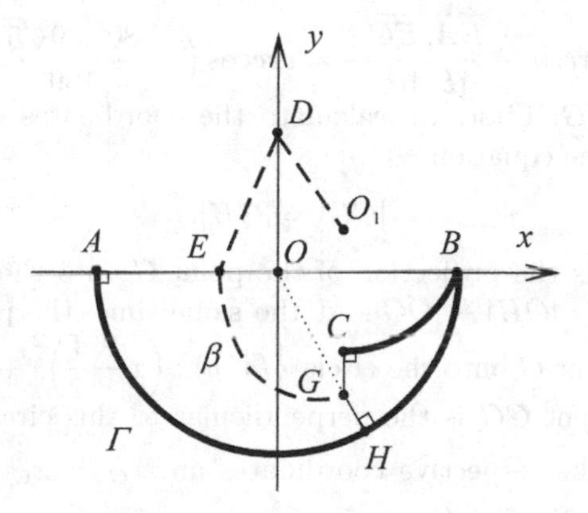

Fig. 2. Bisector β.

The bisector β is represented by the raw $DK = \left\{(x, y) : x = 0, y \geq \frac{2+\sqrt{5}}{4}\right\}$ at a large distance from the curve Γ. The coordinates of the point $D = \left(0, \frac{2+\sqrt{5}}{4}\right)$ can be easily calculated from the equality of the lengths $|AD| = |CD| = |BD| = \frac{1}{4}\sqrt{25 + 4\sqrt{5}}$. Next, the bisector is divided into two segments DO_1 and DE. The point $E = \left(-\frac{3+\sqrt{5}}{24}, 0\right)$ is the last bisector's point, for which $\rho(E, \gamma) = \rho(E, A)$. Next, an arbitrary point F of the bisector β satisfies the equality of the distances $\rho(F, \gamma) = |FC|$ on the curvilinear segment

EG. On the last curvilinear segment GB, an arbitrary point X of the bisector β satisfies the equality $\rho(X,\gamma) = \rho(X,\omega)$.

It is quite obvious that all other points of the plane do not belong to the bisector of the curve Γ, so, they have only one projection onto Γ. It is possible to prove this fact more accurately by dividing the remaining parts of the plane into the sectors and examining each of them.

Now we estimate the function $\alpha_\Gamma(X)$, where X is a point of the bisector β, on each segment of the bisector.

(1) The raw DK. It's obvious that $\max\limits_{X\in DK} \alpha_\Gamma(X) = \alpha_\Gamma(D) = \angle(\overrightarrow{DA}, \overrightarrow{DB}) =$
$$\arccos \frac{\langle \overrightarrow{DA}, \overrightarrow{DB} \rangle}{|DA|^2} = \arccos\left(\frac{128\sqrt{5}-255}{545}\right) \approx 1.513 < \pi.$$
(2) The segment $DO_1\backslash\{D\}$. It's obvious that $\max\limits_{X\in DO_1\backslash\{D\}} \alpha_\Gamma(X) = \alpha_\Gamma(O_1) =$
$$\angle(\overrightarrow{O_1C}, \overrightarrow{O_1B}) = \arccos \frac{\langle \overrightarrow{O_1C}, \overrightarrow{O_1B} \rangle}{|O_1C|^2} = \arccos\left(\frac{\sqrt{5}}{5}\right) \approx 1.107.$$
(3) The segment $DE\backslash\{D\}$. On this segment, $\max\limits_{X\in DE\backslash\{D\}} \alpha_\Gamma(X) = \alpha_\Gamma(E) =$
$$\angle(\overrightarrow{EA}, \overrightarrow{EC}) = \arccos \frac{\langle \overrightarrow{EA}, \overrightarrow{EC} \rangle}{|EA|^2} = \arccos\left(-\frac{80+9\sqrt{5}}{109}\right) \approx 2.735.$$
(4) The segment EG. First, we calculate the coordinates of the point $G = (x_G, y_G)$ from the equation
$$|GC| = |GH|, \tag{1}$$

where the point H is the projection of the point G onto the semicircle γ, and consequently $|GH| = |OH| - |OG|$. At the same time, the point C is also the projection of the point G onto the circle $\left\{(x,y) : \left(x-\frac{1}{2}\right)^2 + \left(y-\frac{1}{4}\right)^2 \le \frac{5}{16}\right\}$. Therefore, the segment CG is the perpendicular to this circle at the point C, which means that the respective coordinates are $x_G = x_C = \frac{1}{2}$. In order to calculate y_G, rewrite the Eq. (1) in the coordinate form:
$$\frac{1-\sqrt{5}}{4} - y_G = 1 - \sqrt{\frac{1}{4} + y_G^2}.$$

Hence, $G = \left(\frac{1}{2}, -\frac{\sqrt{5}}{4}\right)$.

Now, rewrite the equation of the curve EG in the coordinate form. Let arbitrary point $F = (x,y)$ belong to the segment EG of the bisector, which implies that the equation $\rho(F,\gamma) = |FC|$ holds. This equation has the following coordinate form:
$$1 - \sqrt{x^2+y^2} = \sqrt{\left(x-\frac{1}{2}\right)^2 + \left(y - \frac{1-\sqrt{5}}{4}\right)^2}$$

or, after transformation,
$$3x^2 + (\sqrt{5}-1)xy + \frac{5+\sqrt{5}}{2}y^2 - \frac{3+\sqrt{5}}{4}x + \frac{1+\sqrt{5}}{4}y - \frac{7+3\sqrt{5}}{32} = 0. \tag{2}$$

We provide this equation in the canonical form. The coefficients of the corresponding quadratic form

$$a_{11}x^2 + 2a_{12}xy + a_{22}y^2$$

are $a_{11} = 3$, $a_{12} = \sqrt{5} - 1$, and $a_{22} = \dfrac{5+\sqrt{5}}{2}$. It is known that, in order to exclude the mixed product, the required rotation angle of the coordinate plane can be calculated from the equation

$$\tan(2\varphi) = \frac{2a_{12}}{a_{11} - a_{22}}$$

or, equally,

$$\tan^2\varphi + \frac{a_{11} - a_{22}}{a_{12}} \tan\varphi - 1 = 0.$$

A suitable solution of this equation is $\varphi = \arctan\left(\dfrac{1-\sqrt{5}}{2}\right)$. Accordingly, the Eq. (2) will not contain the mixed product in the new coordinate system Ox_1y_1, associated with the old coordinate system Oxy by the change of coordinates

$$\begin{cases} x = \cos\varphi \cdot x_1 - \sin\varphi \cdot y_1, \\ y = \sin\varphi \cdot x_1 + \cos\varphi \cdot y_1, \end{cases}$$

or

$$\begin{cases} x = \sqrt{\dfrac{5-\sqrt{5}}{10}} \cdot x_1 - \sqrt{\dfrac{5+\sqrt{5}}{10}} \cdot y_1, \\ y = \sqrt{\dfrac{5+\sqrt{5}}{10}} \cdot x_1 + \sqrt{\dfrac{5-\sqrt{5}}{10}} \cdot y_1. \end{cases}$$

In the new coordinates, the Eq. (2) takes the form:

$$4x_1^2 + \frac{3+\sqrt{5}}{2}y_1^2 + \frac{\sqrt{5+2\sqrt{5}}}{2}y_1 = \frac{7+3\sqrt{5}}{32}.$$

By extracting the full squares, we obtain the following canonical equation of an ellipse:

$$4x_1^2 + \left(\frac{1+\sqrt{5}}{2}\right)^2 \cdot \left(y_1 + \frac{\sqrt{10-2\sqrt{5}}}{8}\right)^2 = \left(\frac{1+\sqrt{5}}{4}\right)^2.$$

It can be written in the parametric form:

$$\begin{cases} x_1 = \dfrac{1+\sqrt{5}}{8}\cos\tau, \\ y_1 = -\dfrac{\sqrt{10-2\sqrt{5}}}{8} + \dfrac{1}{2}\sin\tau, \end{cases} \qquad \tau \in [0, 2\pi].$$

Accordingly, the curve EG can be parameterized in the original coordinates as follows:

$$\begin{cases} x = x_F(\tau) = \sqrt{\dfrac{5-\sqrt{5}}{10}} \cdot \dfrac{1+\sqrt{5}}{8}\cos\tau - \sqrt{\dfrac{5+\sqrt{5}}{10}}\Big(-\dfrac{\sqrt{10-2\sqrt{5}}}{8} + \dfrac{1}{2}\sin\tau\Big), \\ y = y_F(\tau) = \sqrt{\dfrac{5+\sqrt{5}}{10}} \cdot \dfrac{1+\sqrt{5}}{8}\cos\tau + \sqrt{\dfrac{5-\sqrt{5}}{10}}\Big(-\dfrac{\sqrt{10-2\sqrt{5}}}{8} + \dfrac{1}{2}\sin\tau\Big), \end{cases}$$

$$\tau \in \Big[-\arctan\Big(\frac{9-\sqrt{5}}{2}\Big) + \pi, \arctan\Big(\frac{1+\sqrt{5}}{2}\Big) + \pi\Big].$$

Thus, $\alpha_\Gamma(F) = \angle(\overrightarrow{FC}, \overrightarrow{OF}) = \dfrac{\langle\, \overrightarrow{FC}, \overrightarrow{OF}\,\rangle}{|FC|\cdot|OF|} = f(\tau)$, where

$$f(\tau) = \frac{\Big(\frac{1}{2} - x_F(\tau)\Big)\cdot(-x_F(\tau)) + \Big(\frac{1-\sqrt{5}}{4} - y_F(\tau)\Big)\cdot(-y_F(\tau))}{\sqrt{\Big(\frac{1}{2} - x_F(\tau)\Big)^2 + \Big(\frac{1-\sqrt{5}}{4} - y_F(\tau)\Big)^2}\cdot\sqrt{x_F^2(\tau) + y_F^2(\tau)}}.$$

By analyzing $f'(\tau)$ in the interval $\Big[-\arctan\Big(\frac{9-\sqrt{5}}{2}\Big)+\pi, \arctan\Big(\frac{1+\sqrt{5}}{2}\Big)+\pi\Big]$, we find that the maximum of the function $f(\tau)$ is attained at the point $\tau_0 = -\arctan\Big(\frac{9-\sqrt{5}}{2}\Big) + \pi$.

So, we have established that

$$\max_{F\in EG} \alpha_\Gamma(F) = \alpha_\Gamma(E) = \arccos\Big(-\frac{80+9\sqrt{5}}{109}\Big) \approx 2.735.$$

(5) Let us consider the last curvilinear segment GB. We present its equation in the coordinate form. Assume that $X \in GB$. Then $\rho(X,\gamma) = \rho(X,\omega)$. Since $\rho(X,\gamma) = |XO_1| - \dfrac{\sqrt{5}}{4}$, and $\rho(X,\omega) = 1 - |XO|$, the equation of the curve GB will have the following form:

$$\sqrt{\Big(x-\frac{1}{2}\Big)^2 + \Big(y-\frac{1}{4}\Big)^2} - \frac{\sqrt{5}}{4} = 1 - \sqrt{x^2+y^2}$$

or

$$\frac{17+8\sqrt{5}}{4}x^2 - xy + (5+2\sqrt{5})y^2 - (2+\sqrt{5})x - \frac{2+\sqrt{5}}{2}y - \frac{9+4\sqrt{5}}{4} = 0.$$

After the substitution

$$\begin{cases} x = \dfrac{2}{\sqrt{5}}x_1 - \dfrac{1}{\sqrt{5}}y_1, \\ y = \dfrac{1}{\sqrt{5}}x_1 + \dfrac{2}{\sqrt{5}}y_1, \end{cases}$$

and the segregation of the complete squares, we get the canonical equation of the ellipse:

$$(4+2\sqrt{5})\Big(x_1-\frac{\sqrt{5}}{8}\Big)^2+\Big(\frac{21}{4}+2\sqrt{5}\Big)y_1^2=\frac{82+37\sqrt{5}}{32}.$$

This equation has the following parametric form:

$$\begin{cases} x_1=\dfrac{\sqrt{5}}{8}+\dfrac{4+\sqrt{5}}{8}\cos\tau,\\ y_1=\dfrac{\sqrt{4+2\sqrt{5}}}{4}\sin\tau. \end{cases}$$

The segment GB of the bisector has the following parameterizations in the original coordinates:

$$\begin{cases} x=x(\tau)=\dfrac{2}{\sqrt{5}}\cdot\Big(\dfrac{\sqrt{5}}{8}+\dfrac{4+\sqrt{5}}{8}\cos\tau\Big)-\dfrac{1}{\sqrt{5}}\cdot\dfrac{\sqrt{4+2\sqrt{5}}}{4}\sin\tau,\\ y=y(\tau)=\dfrac{1}{\sqrt{5}}\cdot\Big(\dfrac{\sqrt{5}}{8}+\dfrac{4+\sqrt{5}}{8}\cos\tau\Big)+\dfrac{2}{\sqrt{5}}\cdot\dfrac{\sqrt{4+2\sqrt{5}}}{4}\sin\tau, \end{cases}$$

$$\tau\in\Big[\arctan(2\sqrt{11+5\sqrt{5}})-\pi,-\arctan\Big(\frac{2}{11}\sqrt{10\sqrt{5}-4}\Big)\Big].$$

Let us introduce the function

$$g(\tau)=\alpha_\Gamma(X)=\angle(\overrightarrow{OX},\overrightarrow{XO_1})=\arccos\frac{\langle\,\overrightarrow{OX},\overrightarrow{XO_1}\,\rangle}{|OX|\cdot|XO_1|}$$

$$=\arccos\frac{x\Big(\dfrac{1}{2}-x\Big)+y\Big(\dfrac{1}{4}-y\Big)}{\sqrt{x^2+y^2}\cdot\sqrt{\Big(\dfrac{1}{2}-x\Big)^2+\Big(\dfrac{1}{4}-y\Big)^2}}.$$

By analyzing its derivative $g'(t)$ (note that $g'\Big(-\dfrac{\pi}{2}\Big)=0$, $g''\Big(-\dfrac{\pi}{2}\Big)<0$), we can conclude that

$$\sup_{X\in GB\setminus\{B\}}\alpha_\Gamma(X)=g\Big(-\arctan\Big(\frac{2}{11}\sqrt{10\sqrt{5}-4}\Big)\Big)=\angle(\overrightarrow{OB},\overrightarrow{BO_1})$$

$$=\arccos\Big(-\frac{2}{\sqrt{5}}\Big)\approx 2.678.$$

So, we have constructed an example of the non-self-intersecting curve Γ, for which $\alpha_\Gamma=\arccos\Big(-\dfrac{80+9\sqrt{5}}{109}\Big)\approx 2.735$, but its simply connected segment γ has a value $\alpha_\gamma=\pi$.

4 Counterexample 2 (Proof of Theorem 1)

With the purpose of convenience, the following definition will be used.

Definition 2. *Any curve γ, for which $\alpha_\gamma = \alpha$, will be called α-curve.*

As an α-curve, which can not be represented as a graph, we consider a spiral σ parametrically defined as follows:

$$\sigma = \{(x, y) : x = e^{k\varphi} \cos\varphi, y = e^{k\varphi} \sin\varphi, -\infty < \varphi < \infty)\} \cup \{(0, 0)\},$$

where the point $O = (0, 0)$ is added to σ with the purpose of the closure, since we are able to calculate its measure α of the nonconvexity only for the closed sets.

The curve σ cannot be represented as a graph of the function for some k.

Let us prove that having chosen some parameter k, we can ensure that α_σ takes any value from the interval $(0, \pi)$.

Let us take an arbitrary point $P_0 = (x_0, y_0) \in \sigma$ corresponding to the parameter value $\varphi = \varphi_0$, and formulate the tangent equation at the point P_0:

$$\begin{cases} x = x_0 + x'(\varphi_0)t, \\ y = y_0 + y'(\varphi_0)t, \end{cases} \quad t \in (-\infty, \infty).$$

Since

$$x'(\varphi_0) = ke^{k\varphi_0} \cos\varphi_0 - e^{k\varphi_0} \sin\varphi_0 = kx_0 - y_0,$$

$$y'(\varphi_0) = ke^{k\varphi_0} \sin\varphi_0 - e^{k\varphi_0} \cos\varphi_0 = ky_0 + x_0,$$

then the tangent equation can be represented in the form:

$$\begin{cases} x = x_0 + (kx_0 - y_0)t, \\ y = y_0 + (ky_0 + x_0)t, \end{cases} \quad t \in (-\infty, \infty).$$

By using this form, we can write out the parametrization of the normal to curve σ at the same point P_0:

$$\begin{cases} x = x_0 + (x_0 + ky_0)t, \\ y = y_0 + (y_0 - kx_0)t, \end{cases} \quad t \in (-\infty, \infty).$$

Having expressed the parameter t in both equations, we obtain the canonical normal equation:

$$\frac{x - x_0}{x_0 + ky_0} = \frac{y - y_0}{y_0 - kx_0}. \tag{3}$$

Let $P_1 = (x_1, y_1)$ be the some other point on the curve σ. Then we can draw a normal passing through the point P_1 to the curve σ described by the same kind equation, namely, the equation

$$\frac{x - x_1}{x_1 + ky_1} = \frac{y - y_1}{y_0 - kx_1}. \tag{4}$$

Suppose that these two normals intersect at some point $P = (x, y)$, and this point belongs to the bisector. If points P_0 and P_1 are the projections of the point P onto σ, then the following equality should be true:

$$\sqrt{(x - x_0)^2 + (y - y_0)^2} = \sqrt{(x - x_1)^2 + (y - y_1)^2},$$

which after the transformation will take the following form:

$$2x(x_1 - x_0) + 2y(y_1 - y_0) = x_1^2 + y_1^2 - x_0^2 - y_0^2. \tag{5}$$

The Eqs. (3)–(5) are the necessary conditions for the point P to belong to the bisector of the curve σ. The solution of the system (3)–(5) contains the bisector of curve σ.

From (3) and (4) it follows that the coordinates of point $P = (x, y)$ are expressed through the coordinates of the points P_0 and P_1 as follows:

$$x = \frac{k}{1+k^2} \frac{(x_1^2 + y_1^2)(x_0 + ky_0) - (x_0^2 + y_0^2)(x_1 + ky_1)}{x_1 y_0 - x_0 y_1},$$

$$y = \frac{k}{1+k^2} \frac{(x_1^2 + y_1^2)(y_0 - kx_0) - (x_0 62 + y_0^2)(y_1 - kx_1)}{x_1 y_0 - x_0 y_1}.$$

Including them into Eq. (5), we get the following equation

$$\frac{2k}{(1+k^2)(x_1 y_0 - x_0 y_1)} \Big(-2(x_0^2 + y_0^2)(x_1^2 + y_1^2) + (x_0 x_1 + y_0 y_1)(x_0^2 + y_0^2 + x_1^2 + y_1^2) \Big)$$

$$= \frac{1-k^2}{1+k^2}(x_1^2 + y_1^2 - x_0^2 - y_0^2).$$

Since $x_0 = e^{k\varphi_0} \cos \varphi_0$, $y_0 = e^{k\varphi_0} \sin \varphi_0$, $x_1 = e^{k\varphi_1} \cos \varphi_1$, $y_1 = e^{k\varphi_1} \sin \varphi_1$, then our equation can be represented as follows:

$$2k\big(2e^{(\varphi_1 - \varphi_0)} - \cos(\varphi_1 - \varphi_0)(1 + e^{2k(\varphi_1 - \varphi_0)})\big)(1 - k^2)(e^{2k(\varphi_1 - \varphi_0)} - 1)\sin(\varphi_1 - \varphi_0).$$

Denote $\beta = \varphi_1 - \varphi_0$. Then the last equation is converted to the form:

$$\frac{2k}{1-k^2} \cdot \frac{2e^{k\beta} - \cos\beta(1 + e^{2k\beta})}{e^{2k\beta} - 1} = \sin\beta. \tag{6}$$

If P_0 and P_1 are the projections of the point P onto σ, then the sections PP_0 and PP_1 can not intersect the curve σ by the definition of the projection. We assume that $\varphi_1 > \varphi_0$. Based on this additional information, we conclude that the points from the bisector of the curve σ correspond to the smallest positive root of the Eq. (6).

The dependence of the roots of the Eq. (6) on parameter k is shown in Fig. 3.

Here, we will omit the cumbersome asymptotic analysis of the Eq. (6). However, note that since the roots of the Eq. (6) continuously depend on its coefficients, then, by decomposing them into the Taylor series we can prove that one of its roots $\beta \to \pi$ as $k \to \infty$ and $\beta \to 2\pi$ as $k \to 0$. We can also notice that

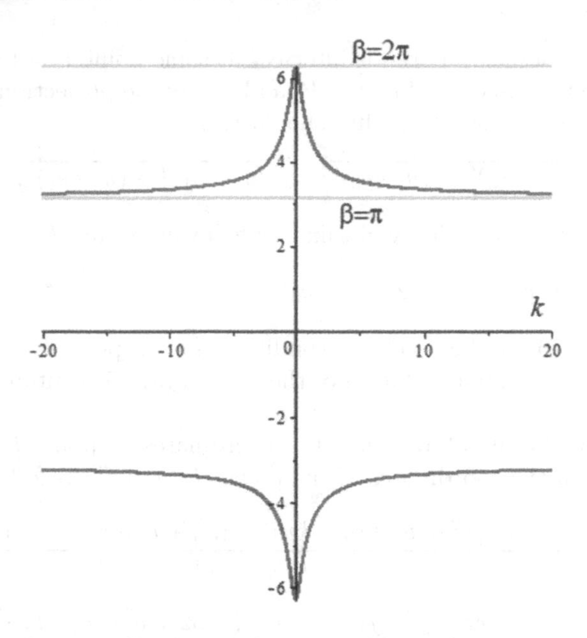

Fig. 3. The roots of the Eq. (6).

$\beta_0 = 0$ is a root of Eq. (6) for any value of parameter k. This root corresponds to the two coincident projections.

Consider the quadrilateral OP_0PP_1 (Fig. 4).

By direct calculation, it is easy to show that $\angle(\overrightarrow{OP_0}, \overrightarrow{PP_0}) = \angle(\overrightarrow{OP_1}, \overrightarrow{PP_1}) = \arccos \dfrac{\langle \overrightarrow{OP_0}, \overrightarrow{PP_0} \rangle}{|OP_0| \cdot |PP_0|} = \arccos \sqrt{1+k^2}$. Indeed, $\overrightarrow{OP_0} = (x_0, y_0)$ and the normal Eq. (3) implies that $\overrightarrow{PP_0} = t \cdot (x_0 + ky_0, y_0 - kx_0)$, wherein $\langle \overrightarrow{OP_0}, \overrightarrow{PP_0} \rangle = t(x_0^2 + y_0^2)$, and $|OP_0| \cdot |PP_0| = t(x_0^2 + y_0^2)\sqrt{1+k^2}$.

Next, note that $\angle OP_0P = \pi - \angle(\overrightarrow{OP_0}, \overrightarrow{PP_0})$ can be considered as an adjoining corner, and $\angle OP_1P = \angle(\overrightarrow{OP_1}, \overrightarrow{PP_1})$ can be considered as a cross corner. Therefore, $\angle OP_0P + \angle OP_1P = \pi$. Since the sum of the angles in the quadrangle OP_0PP_1 equals $\angle P_1OP_0 + \angle OP_0P + \angle P_0PP_1 + \angle OP_1P = 2\pi$, then

$$\alpha_\sigma(P) = \angle P_0PP_1 = \pi - \angle P_1OP_0 = \pi - (2\pi - (\varphi_1 - \varphi_0)) = \beta - \pi.$$

Taking into account the asymptotic of angle β at $k \to 0$ and $k \to \infty$, as well as the continuous dependence of β on k, we see that $\alpha_\sigma(P) \to 0$ at $k \to \infty$ and $\alpha_\sigma(P) \to \pi$ at $k \to 0$. The value of $\alpha_\sigma(P)$ depends only on the value of the parameter k. It follows that $\alpha_\sigma = \alpha_\sigma(P)$ for any point P of the bisector σ.

So, α_σ equals to any value from the interval $(0, \pi)$ with the proper choice of the parameter k. Theorem 1 has been proved.

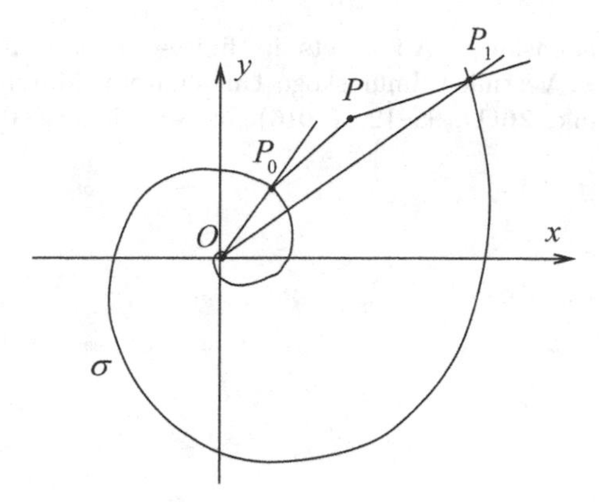

Fig. 4. Curve σ.

5 Conclusion

Using the counterexample 1, we have shown that a simply connected segment γ of α-curve Γ may be less convex than the entire curve. However, the value of α_γ is quite close to π in the constructed example. In this regard, the hypothesis arises that if α_Γ is sufficiently small then it is impossible to identify the simply connected less convex segment. If so, then it is interesting to find the largest critical value of α_Γ for which this property is still relevant.

In the second counterexample, we have constructed an α-curve, which can not be represented as a graph of the function, regardless of the coordinate rotation. However, the constructed curve has one end at the point $(0,0)$. It can be supposed that the α-curve, without the ends and with a sufficiently small α, can still be represented as a graph of the function in some coordinate system.

References

1. Michael, E.: Paraconvex sets. Mathematica Scandinavica **7**(2), 312–315 (1959)
2. Ivanov, G.E.: Weakly Convex Sets and Functions: Theory and Applications. Fizmatlit, Moscow (2006)
3. Ershov, A.A., Pershakov, M.V.: On the relation of alpha-sets with other generalizations of convex sets. In: Treskova, P.P., Kirsanova, A.I. (eds.) VI Information School for Young Scientists, Yekaterinburg, pp. 143–150 (2018). https://doi.org/10.32460/ishmu-2018-6-0017
4. Polovinkin, E.S., Balashov, M.V.: Elements of Convex and Strongly Convex Analysis. Fizmatlit, Moscow (2007)
5. Ushakov, V.N., Uspenskii, A.A., Ershov, A.A.: Alpha sets in finite-dimensional euclidean spaces and their applications in control theory. Vestn. Saint Petersburg Univ. Appl. Math. Comput. Sci. Control Process. **14**(3), 261–272 (2018). https://doi.org/10.21638/11702/spbu10.2018.307
6. Ushakov, V.N., Uspenskii, A.A.: Theorems on the separability of α-sets in Euclidean space. Proc. Steklov Inst. Math. **299**(suppl. 1), 277–291 (2017). https://doi.org/10.1134/S0081543817090255

7. Ushakov, V.N., Uspenskii, A.A.: α-sets in finite-dimensional Euclidean spaces and their properties. Vestnik Udmurtskogo Universiteta. Matematika. Mekhanika. Komp'yuternue Nauki **26**(1), 95–120 (2016). https://doi.org/10.20537/vm160109. (in Russian)

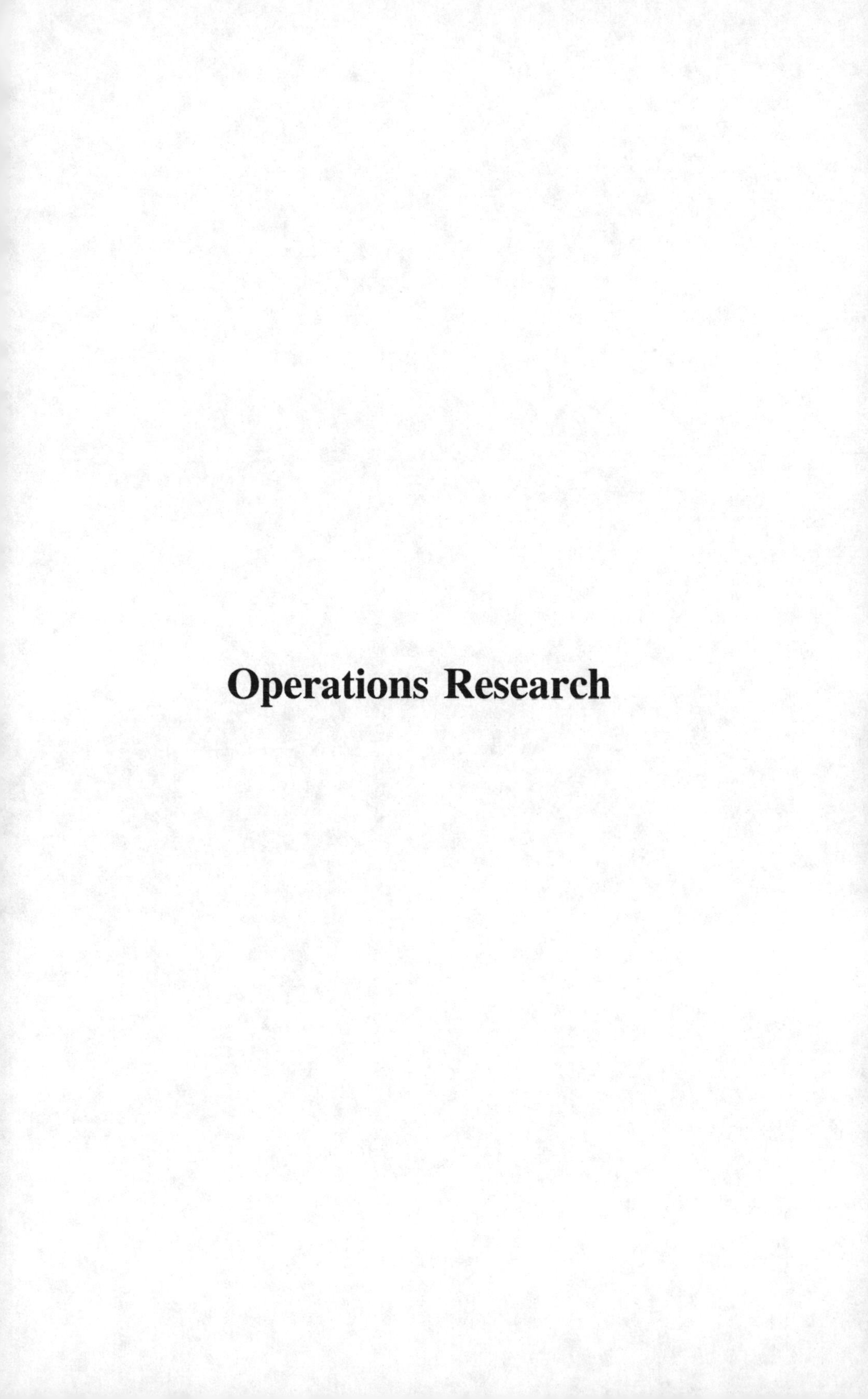

Operations Research

Using Machine Learning Algorithm for Diagnosis of Stomach Disorders

Yedilkhan Amirgaliyev[1,2], Shahriar Shamiluulu[1,2], Timur Merembayev[1,3(✉)], and Didar Yedilkhan[1]

[1] Institute Information and Computational Technologies CS MES RK, Almaty, Kazakhstan
amir_ed@mail.ru

[2] Suleyman Demirel University, Kaskelen, Kazakhstan
shahriar.shamiluulu@sdu.edu.kz

[3] International Information Technology University, Almaty, Kazakhstan
timur.merembayev@gmail.com

Abstract. Medicine is one of the rich sources of data, generating and storing massive data, begin from description of clinical symptoms and end by different types of biochemical data and images from devices. Manual search and detecting biomedical patterns is complicated task from massive data. Data mining can improve the process of detecting patterns. Stomach disorders are the most common disorders that affect over 60% of the human population. In this work, the classification performance of four non-linear supervised learning algorithms i.e. Logit, K-Nearest Neighbour, XGBoost and LightGBM for five types of stomach disorders are compared and discussed. The objectives of this research are to find trends of using or improvements of machine learning algorithms for detecting symptoms of stomach disorders, to research problems of using machine learning algorithms for detecting stomach disorders. Bayesian optimization is considered to find optimal hyperparameters in the algorithms, which is faster than the grid search method. Results of the research show algorithms that base on gradient boosting technique (XGBoost and LightGBM) gets better accuracy more 95% on the test dataset. For diagnostic and confirmation of diseases need to improve accuracy, in the article, we propose to use optimization methods for accuracy improvement with using machine learning algorithms.

Keywords: Stomach disorder · Machine learning algorithm · Decision support system · Bayesian optimization

1 Introduction

Computer applications and tools are being used in almost every field to assist the work on a daily basis and the medicine is not an exception to that [1]. Currently, there are various Machine Learning (ML) methods that are being applied for disease diagnosis. It is strongly believed that it will be more widely used in

I. Bykadorov et al. (Eds.): MOTOR 2019, CCIS 1090, pp. 343–355, 2019.
https://doi.org/10.1007/978-3-030-33394-2_27

biomedical systems in this century [2]. This is because of the high complexity included in the clinical data. The aim of the work is, to perform a comparative study between four ML methods on stomach disorders and identify the top ones according to performance metrics which can be incorporated into the clinical decision support system. Gastrointestinal diseases are considered to be one of the most common disorders which affect more than 46% of the human population, where over 60% of the population is affected by stomach disorders [3]. Generally, it is easier to find medical data related to digestive system disorders and ethical reports are more likely to be accepted and permission is more likely to be granted to conduct research. Due to the aforementioned reasons, five stomach diseases were selected: stomach flu, gastroesophageal reflux (heartburn), gastritis, peptic ulcer, and stomach cancer [4]. Currently, there is continuous on-going research in the field of medical diagnosis and treatment. Tremendous work has been done by applying the supervised learning methods i.e., Neural Networks, Regression models, and Support Vector Machines for diagnosis diseases like diabetes, heart attack, cancer, and kidney diseases [2,5,6]. Further, the methods of background information and implications in the medical field are described.

2 Background Information and Implications

This section provides brief information on four ML algorithms i.e., Logistic Regression (Logit), K-Nearest Neighbour, XGBoost and LightGBM. The two last algorithms are similar and built under a gradient boosting method. LightGBM differs from XGBoost by specific features, especially the process of creation trees. Parameters for tuning models are almost similar for both algorithms. Each section also describes the implications of algorithms and provides plausible outcomes.

2.1 Logistic Regression Model

In statistics, the Logit has wide range implications in medical areas which are generally used to create models for the classification of the attributes that might determine the happening of the resulting outcome. The distinctive feature of the Logit is that the resulting outcome is dichotomous. Generally, patients data is being used to develop a proper logistic regression model by identifying the important attributes in the data, which are important in predicting the given outcome. As a result, the created model can be used to classify newly provided patient data through placing in the Logit model to calculate the probability $P(Y_i)$ of a given outcome [2].

There are several research studies that use regression models for classification and prediction in the biomedical field. In this work [7] authors applied the logistic regression method to predict the probability of a fail outcome in the Tuberculosis treatment course that might be used to determine the level of patients' supervision and support. They proved that the developed model-based of Logit achieved 95% prediction accuracy based on optimal sensitivity and specificity. In another study [8] researchers talk about dichotomous diagnostic tests and advantages of using logistic regression models in terms of sensitivity, specificity, and likelihood ratios (LRs). The exerted model allows calculating the LRs of diagnostic test results which is conditional on these covariates with an accuracy of 90%. The intended logistic regression approach proves an efficient method to determine the performance of tests at the level of the individual patient risk profile and to examine the effect of patient's characteristics on diagnostic test features. In another study, the authors applied regression models in the biomechanics field.

2.2 Metric Algorithm K-Nearest Neighbour

K nearest neighbour (KNN) is a very simple, the most popular, highly efficient and effective pattern recognition algorithm. KNN is a direct classifier, where a choice is classified based on the class of their nearest neighbour. The marked data is provided to the algorithm for the training process when the training algorithm gets a new object, it is assigned to a class that is most common among k neighbours of the element whose classes are already known.

KNN is also widely used in medicine for various types of tasks. In the article [9], authors propose a new algorithm that is based on KNN with a genetic algorithm for detecting heart disease.

In the article [10] considers a combination of decision tree algorithms and the classifier of k-nearest neighbors as an assessment of the selected features for the diagnosis of Parkinson's disease at an early stage. Training data included: speech with several types of sound recordings and Parkinson Handwriting sample datasets used to evaluate the proposed model. In this study, authors used one of the approaches to solving the problem of automatic segmentation of discrete speech signal for the diagnosis of disease, in [11] authors consider one of the new approaches to solving this problem. For this purpose, a new type of information functions called TAC-coefficients (throat and acoustic correlation coefficients) is used, which provides sufficient accuracy, the efficiency of segmentation for diagnosis of Parkinson's disease.

2.3 Algorithms Based on Gradient Boosting: XGBoost and LightGBM

In this study, we propose to use algorithms based on the idea of gradient boosting: XGBoost and LightGBM.

XGBoost main features are easy parallelization and impressive prediction accuracy compared to other methods. XGBoost is an efficient and scalable version of the gradient boosting method [12], which has proven itself in several recent machine learning competitions. The authors reported that XGBoost is an ensemble of classification and regression trees that can capture non-linear dependents. The idea of the algorithm is to add a classifier in interaction and next iteration the classifier has been trained on how to improve accuracy base on a trained ensemble of trees. In the article [13], several machine learning algorithms are used to determine the rule for predicting the diagnosis of influenza by combining clinical signs and symptoms in patients and further assessing the accuracy of the prediction model developed using the machine learning algorithm. The XGBoost algorithm showed AUC results: 0.87 on the train dataset, 0.79 in the test dataset.

LightGBM is a relatively new library [14] and is not widely used in the scientific area [15], but it is widely used in machine learning competitions. The main problem faced by gradient boosting algorithms is that for each function they need to scan all data instances in order to evaluate the gain in information about all possible separation points, which takes a very long time when the dimension of the object is high and the size of the data is large. In the article, the authors proposed a new gradient boosting algorithm, which contains two new techniques: a gradient sampling on one side and an exclusive set of functions for solving a large number of data instances and a large number of functions, respectively. Experimental and theoretical results according to have shown that using the LightGBM algorithm can significantly outpace XGBoost in computing speed.

3 The Issue of Tuning Hyper Parameters in Models

One relative disadvantage of these algorithms (XGBoost and LightGBM) is a large number of hyperparameters that are provided to the end-user, which in turn can affect the problems of the practical use of machine learning algorithms in clinics for diagnosing diseases. Therefore, there is a great attraction for automatic approaches that can optimize the performance of any given learning algorithm for the problem in question. Training time can take considerable time with large amounts of data. To maximize the predictive power of gradient boosting models, you must manually configure the hyperparameters or use automated methods, such as those based on Bayesian optimization.

Bayesian optimization is an effective method for global optimization of objective functions $f : X \to R$, where $X \subset R^d$. Where we decide:

$$x^* = \underset{x \in \mathcal{X}}{\operatorname{argmax}} f(x) \tag{1}$$

Where X is a compact and convex set. Often, you can access only the perturbed estimates of the function $f()$, which further complicates the optimization. Below we provide a brief description of Bayesian optimization. A more detailed formulation of the Bayes optimization problem is described in [16].

There are two basic steps that must be made when performing Bayesian optimization. The first is to select prior functions that will express assumptions about the function being optimized. To do this, we will pre-select the Gaussian process, due to its flexibility and controllability. Secondly, it is necessary to choose the data collection function that is used to construct the utility function from the posterior model, which allows us to determine the next point to be estimated.

The properties of the Gaussian distribution allow us to calculate predictive averages and variances in closed form [17]. It is determined by the mean function

$\mu(x)$ and the covariance function $k(x, x')$. A sample of the Gaussian process is a function, given in the form:

$$f(x) \sim GP(\mu(x), k(x, x')) \tag{2}$$

where the value of the function at an arbitrary point x is a random variable distributed according to Gauss. Without any loss in generality, it can be assumed that the previous middle function is a zero function, which makes the Gaussian process a completely definable covariance function. A popular choice for the covariance function is the square exponential function, represented as:

$$k(x, x') = (-\frac{1}{2\theta}||x - x'||^2) \tag{3}$$

where θ is the length scale parameter. We assume that the length scale is isotropic in our method. Other popular covariance functions include the Mattern kernel, a rational quadratic kernel. In Bayesian optimization, there is the concept of a utility function as a receiving function. The data collection function helps us achieve the optimum of the base function by examining areas where the uncertainty about the function is high and exploring areas where the expected function values can be higher.

Data collection functions can be defined either using criteria based on improvement, or using criteria based on confidence. The upper confidence limit of the predictive GP distribution as a function of data collection. However, paper [16] suggests using a combination of these data collection functions.

The upper limit of the reliability of the Gaussian process:

$$\alpha_t(x) = \mu_{t-1}(x) + \beta_t^{\frac{1}{2}} \sigma_{t-1}(x) \tag{4}$$

where $\beta_t = 2 + 2 \cdot d \cdot log(t^2 \cdot d \cdot b \cdot r\sqrt{log(\frac{4da}{\delta})}), \sum_{t \geq 1} \pi_t^{-1} = 1, \pi_t \geq 0$

a, b are constants, and d is the dimension of the problem and are given as $a > 0, b > 0, d > 0, r > 0, \delta \in (0, 1), t \geq 1$. The constants a, b are related to the Lipschitz constant of the objective function $f(x)$.

Now we have discussed the use of a prior over smooth functions using the kernel Mattern, we will focus our attention on computing Bayesian optimization. The role of the data collection function is to find the optimal value. Typically, data collection functions are defined in such a way that high data acquisition corresponds to potentially high values of the target function. Maximization function is used to select the next point at which to evaluate the function. We consider maximizing the probability of improvement $f(x^+)$, where $x^+ = \text{argmax}_{x \in \mathcal{X}} f(x_i)$. The improvement function is defined as:

$$I(x) = max\{0, f(x) - f(x^+)\} \tag{5}$$

The new point will be found by maximizing the expected improvement:

$$x = \underset{x}{\text{argmax}}\, E(max\{0, f(x)_{i+1} - f(x^+)\}|D_i + 1) \tag{6}$$

The expected improvement is similar with an analytical solution:

$$E(I) = \begin{cases} (\mu(x) - f(x^+))\Phi(Z) + \sigma(x)\phi(Z), & if\, \sigma(x) > 0 \\ 0, & if\, \sigma(x) = 0. \end{cases} \quad (7)$$

where $Z = \dfrac{\mu(x) - f(x^+)}{\sigma(x)}$. $\Phi(.)$ and $\phi(.)$ are the cumulative distribution and probability distribution functions respectively.

Bayesian optimization is a powerful tool for machine learning, where often the problem is not in getting data, but in getting tags. In many ways, this is similar to the usual active learning, but instead of obtaining training data for classification or regression, it allows us to develop structures to effectively solve new types of learning problems. Nevertheless, Bayesian optimization is also a fairly recent addition to machine learning algorithms, and not yet sufficiently studied in user applications.

4 Materials and Methods

4.1 Problem Description

The objective of this study is to find a better algorithm for machine learning, which will improve the diagnosis of the disease of gastric disorders and provide a result in an acceptable time. In the previous section, we gave a description of the selected algorithms for research on data. We have identified two main points from these studies; the first one is, used XGBoost and LightGBM gradient boosting algorithms, these algorithms performed well on data science (Kaggle) competition platforms. LightGBM did not use for similar medical data. LightGBM did not use for similar medical data. The second one is, proving that moderate ML algorithms can outperform by performing several genuine preprocessing on clinical data. The third is, 90% of the work, which is being done in disease diagnosis, especially in the areas related to health are not open source and being kept secret. The objective is to implement the mentioned algorithms from scratch and make them suitable for integration in any clinical decision support system.

4.2 Description of Data

In the retrospective analysis study, the medical data related to five digestive systems disorders is considered. The medical data were collected from 1999 to 2014 in the process of routine endoscopic practice for over 1000 subjects, in two hospitals i.e., Samatya and Frunze located in different countries i.e., Turkey and Kyrgyzstan. Patients data selected for those who were confirmed as having mentioned stomach disorders. The disorders prevalence in the dataset is 65% that is close to WHO statistics.

In the present study, the patient's data has 26 independent attributes of different types described in Table 1. The attributes are organized into three groups;

the first group shows the patient's history, the second group shows symptoms, and the last group shows the lab test results. The correlation and significance levels for the attributes are provided Table 1. Prior to applying any ML algorithm, the dataset has undergone several normalizations and standardization changes. Only those patients data who gave their permission for use had been used in the research in this retrospective study and those patients kept strictly anonymous and confidential.

Table 1. Medical data attributes.

Attributes	Value ranges	Comp. groups
History		
Dental probs.	[0.01/1.00]	Comp 1
Sleep disorders	[0.01/1.00]	Comp 1
Constipation	[0.01/1.00]	Comp 1
Age	[0 to 100]	Comp 1
NSAIDs	[0.01/0.5/1.00]	Comp 3
Appetite	[0.01/1.00]	Comp 3
Stress	[0.01/1.00]	Comp 4
Gender	[0/1]	Comp 4
Breakfast	[0.01/1.00]	Comp 4
Smoking	[0.01/1.00]	Comp 5
Alcohol	[0.01/1.00]	Comp 5
Symptoms		
Swelling	[0.01/1.00]	Comp 1
Burning	[0.01/1.00]	Comp 1
Souring	[0.01/1.00]	Comp 1
Abdom. pain	[0.01/1.00]	Comp 2
Nausea	[0.01/1.00]	Comp 2
Weakness	[0.01/1.00]	Comp 2
Vomiting	[0.01/1.00]	Comp 2
Diarrhea	[0.01/1.00]	Comp 2
Weightloss	[0.01/1.00]	Comp 3
Lab tests		
Leukocytes (mcL)	[4.0/10.0]	Comp 2
Hemoglobin (g/dl)	[9.0/17.0]	Comp 3
Stool blood test	[0.00/1.00]	Comp 3
CLO test	[0.00/1.00]	Comp 4

4.3 Disease Groups

The defined disease groups metrics are shown in Table 3. The disease groups are related to five stomach disorders, that are commonly occur in the Middle East and Central Asian populations with a prevalence of 65%. There are two final disease conditions for the ML models in order to perform classification, which is 0 representing patients without disorder and 1 representing patients with the defined condition suffering from any stomach disorder. Also, several types of disorders were detected in several patients at the same time, these types of diseases were matched into separate classes for subsequent diagnosis of dual diseases in patients. Figure 1 shows a histogram of the distribution of diseases and such dual diseases as gastritis - reflux and cancer - an ulcer has 16 and 33 patients, respectively (Table 2).

Table 2. Basic statistics for selected features of dataset.

	Age	HPTEST	WBC	HMG	PLT	SBT	RBC	Stomach diseases
Count	1041	1041	1041	1041	1041	1041	1041	1041
Mean	33.99	0.19	6.05	12.62	160.89	0.07	4.57	1.85
Std	17.40	0.39	1.56	0.90	13.58	0.25	0.41	1.99
Min	6	0	4.2	9.8	148	0	3.5	0
25%	22	0	4.8	12.2	152	0	4.3	0
50%	32	0	5.4	12.5	155	0	4.6	1
75%	48	0	7.3	13.2	162	0	4.8	4
Max	78	1	9.8	16.2	220	1	5.6	6

Table 3. Stomach disorders condition metrics.

Stomach disorders	Model classification metrics
Cancer	[0.00/1.00]
Ulcer	[0.00/1.00]
Gastritis	[0.00/1.00]
Reflux	[0.00/1.00]
Flu	[0.00/1.00]
Gastritis and Reflux	[0.00/1.00]
Cancer and Ulcer	[0.00/1.00]

To determine how the various features are related to each other, we constructed two plots: correlation and pair plot, Figs. 2 and 3 respectively. The pair plot allows us to see a distribution of individual variables and the relationship between two variables. Paired charts are an excellent method for determining trends for further data analysis. To determine the dependencies between features, 5 features (Age, WBC, HMG, PLT, RBC) have been selected which values are not binary. In Fig. 2 it is possible to separate 5–6 features that have a correlation above 60%, this gives us the opportunity to reduce the dimension of features in the dataset.

4.4 Performance Measure Metrics

For the comparison of the ML algorithms, several performance metrics are used. The selected metrics are briefly described in this section. The experimental results

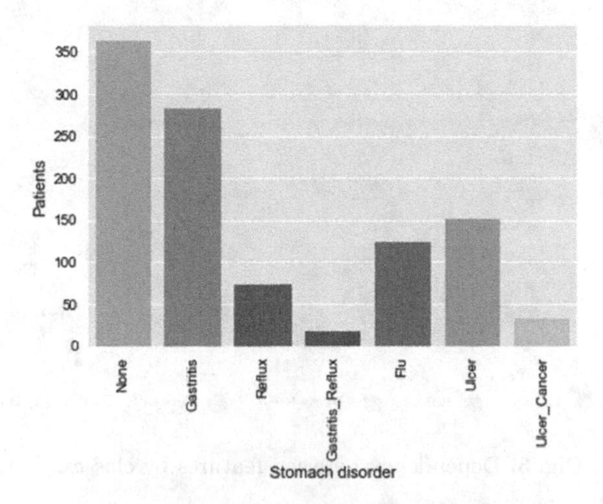

Fig. 1. The histogram of stomach diseases distribution.

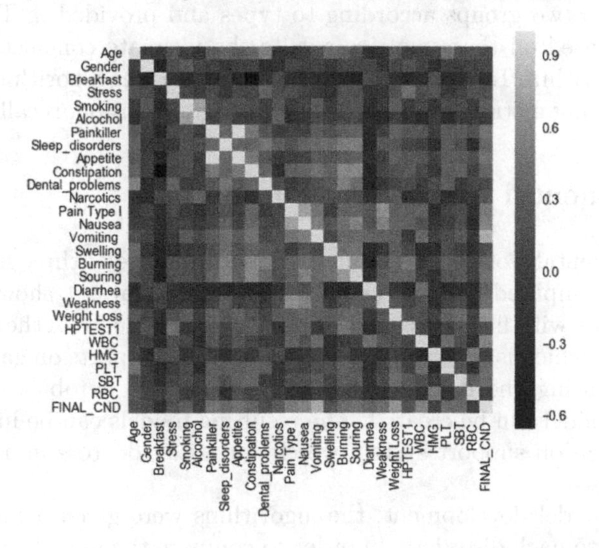

Fig. 2. Correlation matrix between features.

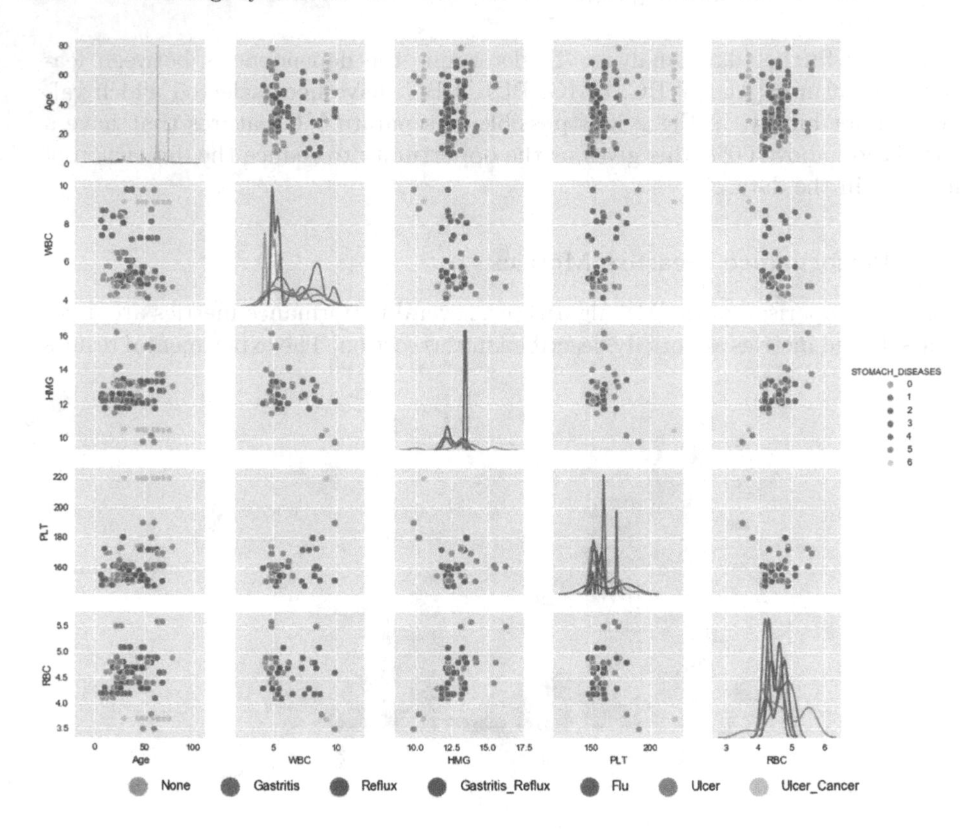

Fig. 3. Dependency between features by classes.

are divided into two groups according to types and provided in Tables 4 and 5 respectively. Based on these metrics results the ultimate conclusions are made about each algorithm. To estimate the accuracy of learning algorithms for selected data, the following metrics were selected: accuracy, precision, recall, and f1.

5 Experimental Result and Discussions

In this experimental work, the classification performance of three non-linear ML algorithms is compared and discussed. The medical dataset shown in Table 1, contains patients with five stomach disorders. The prevalence of the disease in the dataset is 65%, which is with the accordance of WHO reports on gastrointestinal disorders. By using the dataset with such prevalence a robust and accurate classification model can be created. The resulting models can be integrated into any clinical decision support system which can assist doctors in a more precise disease diagnosis.

After the model development, the algorithms were given a task to classify patients with stomach disorders. In order to compare the models' performances, we were concentrated on ten metrics shown in Tables 4 and 5. The underlined metric values indicate which technique is good with respect to which measure.

During the model's simulations, the 5-fold cross-validation used, where the dataset divided into 5 randomly equal parts, out of which 3 parts used for training and the remaining part used to test the model. A learning rate was fixed to 0.3 for assessing the performance of the models

As demonstrated in Table 4, the proposed LightGBM model has the best performance in terms of testing accuracy. The Logit showed lower performance results as compared to other models. There is no single champion model in classification problems and the best model in terms of accuracy is different from one case to another. Thus we recommend that one should try all alternative models to determine which one will perform best for an underlined data set.

Table 4. Classification performance of models for test dataset.

	Logit	KNN	XGBoost	LightGBM
Accuracy	0.947	0.958	0.971	0.980
Precision	0.930	0.944	0.816	0.959
Recall	0.920	0.958	0.971	0.980
F1	0.925	0.958	0.971	0.980

As shown in the Table 5, all models are non-linear but the Logit is more moderate in terms of complexity. The three methods i.e., KNN, XGBoost and LightGBM outperforms Logit according to marginal error making a more accurate learning process.

Table 5. Classification performance of models for test dataset.

	Logit	KNN	XGBoost	LightGBM
Runtime (sec)	0.05	0.15	0.81	0.14
Marginal error	0.013	0.003	0.003	0.003
Complexity	Moderate	Complex	Complex	Complex
Type	Non-linear	Non-linear	Non-linear	Non-linear

As can be seen from Table 5, the Logit algorithm is still the best result in terms of execution time. LightGBM algorithm suits by two indicators. This fact requires confirmation of the authors of the article [14], where the author claims that the proposed algorithm shows that the LightGBM shows the best indicators of accuracy and speed of execution.

6 Conclusion

In recent years computer-based disease diagnosis by using machine learning methods have played an important role in improving the quality of medical services. In addition, using computer-based disease diagnosis makes the diagnosis more reliable and therefore increases patient satisfaction. In this study, we developed predictive models using four machine learning algorithms to diagnose stomach disorders. In this study, we studied the capabilities of machine learning algorithms for the diagnosis of diseases of the stomach. The study showed that there is no single method that clearly surpasses all methods in all problem situations. Therefore, one recommendation from this study is to try all alternative models to determine which one will perform best for particular clinical data set. However, the performances varied slightly between models, XGBoost and LightGBM achieved better prediction results (test dataset 97% and 98% respectively). One novelty of this paper was LightGBM, which has never been implemented in medical or diagnosis problems. According to results, LoghGBM model tree performed quite competitively with other algorithms. The results of our study suggest that stomach disease can be classified with an accuracy of approximately 94% with all four machine learning methods. Which is really good in real-life problems and we can comfortably suggest that computer-based disease diagnosis can use these four implemented algorithms in their diagnosis problems.

In addition to testing the selected machine learning algorithms on medical data, we highlight a problem of tuning hyperparameters for algorithms. We considered the promising direction of Bayesian optimization for the tuning parameters, but this method also has disadvantages, such as Gaussian processes are not always the best or the simplest solution but even when it is, you need to be very careful when developing a kernel. It can go through a lot of iteration without improvement. These problems are exacerbated by the increasing of dataset dimension—more dimensionality means that more samples are required to cover a space, therefore more hyperparameters also need to be tuned.

These studies will continue in this direction - optimization and tuning hyperparameters for machine learning algorithms with a practical bias on big dataset dimensions and data specifics.

Acknowledgement. The work was supported by the program-targeted funding projects of the Ministry of Education and Science of Republic of Kazakhstan AP05132648 and BR05236699.

References

1. Ramesh, A.: Artificial intelligence in medicine. Ann. R. Coll. Surg. Engl. **86**, 334 (2004)
2. Chaudhry, B.: Systematic review: impact of health information technology on quality, efficiency, and costs of medical care. Ann. Intern. Med. **144**, 742–752 (2006)
3. WHO, World health statistics 2010, World Health Organization (2010)
4. Yamada, T.: Textbook of Gastroenterology. Wiley - Blackwell, Hoboken (2009)

5. Alkim, E.: A fast and adaptive automated disease diagnosis method with an innovative neural network model. Neural Networks. **33**, 88–96 (2012)
6. Çomak, E.: A decision support system based on support vector machines for diagnosis of the heart valve diseases. Comput. Biol. Med. **37**, 21–27 (2007)
7. Kalhori, S., Nasehi, M., Zeng, X.J.: A logistic regression model to predict high risk patients to fail in tuberculosis treatment course completion. Int. J. Appl. Mathe. **40**, 102–107 (2010)
8. Janssens, A., Deng, Y., Borsboom, G., Eijkemans, M., Habbema, J., Steyerberg, E.: A new logistic regression approach for the evaluation of diagnostic test results. Med. Decis. Making **25**, 168–177 (2005)
9. Akhiljabbar, M., Deekshatulu, B.L., Chandrac, P.: Classification of heart disease using K-Nearest Neighbor and Genetic algorithm. Procedia Technol. **10**, 85–94 (2013)
10. Gupta, D., et al.: Optimized cuttlefish algorithm for diagnosis of Parkinson's disease. Cogn. Syst. Res. **52**, 36–48 (2018)
11. Mussabayev, R., Kalimoldayev, M., Amirgaliyev, Y., Mussabayev, T.: Automatic speech segmentation using throat-acoustic correlation coefficients. Open Eng. **6**(1), 335–346 (2016)
12. Chen, T., Guestrin, C.: XGBoost: a scalable tree boosting system. In: Proceedings of the 22nd ACM SIGKDD International Conference on Knowledge Discovery and Data Mining - KDD 2016, pp. 785–794 (2016)
13. Hung, S., Hsieh, C., Chen, K.: Predicting influenza infection by clinical feature-based machine learning algorithms. Ann. Emerg. Med. **72**(4 Supplement), S58 (2018)
14. Ke, G.L., et al.: LightGBM: a highly efficient gradient boosting decision tree. In: Advances in Neural Information Processing Systems (2017)
15. Merembayev, T., Yunussov, R., Amirgaliyev, Y.: Machine learning algorithms for stratigraphy classification on uranium deposits. Procedia Comput. Sci. **150**, 46–52 (2019)
16. Brochu, E., Cora, V.M., de Freitas, N.: A tutorial on Bayesian optimization of expensive cost functions, with application to active user modeling and hierarchical reinforcement learning. pre-print (2010). arXiv:1012.2599
17. Williams, C., Rasmussen, C.: Gaussian Processes for Machine Learning, p. 4. MIT Press, Cambridge (2006)

The Convergecast Scheduling Problem on a Regular Triangular Grid

Adil Erzin[1,2(✉)] and Roman Plotnikov[1]

[1] Sobolev Institute of Mathematics, SB RAS, Novosibirsk 630090, Russia
{adilerzin,prv}@math.nsc.ru
[2] Novosibirsk State University, Novosibirsk 630090, Russia

Abstract. The problem of conflict-free data aggregation in an arbitrary graph is NP-hard. On a square unit grid, in each node of which a sensor is located, the problem is polynomially solvable. For the case when the graph is a regular triangular grid, the upper bound on the length of the schedule of conflict-free data aggregation was previously known. In this paper, the refined estimates are given for the length of the schedule of conflict-free data aggregation on a triangular grid, as well as polynomially solvable cases are found and algorithms for constructing optimal and approximate schedules are proposed.

Keywords: Conflict-free data aggregation scheduling · Triangular grid

1 Introduction

In wireless sensor networks (WSNs), the data collected by the sensors is transmitted to the center, which is called the *base station* (BS) [2]. In this case, the data transmission is carried out over the edges of the communication graph, which connects the sensors [6]. If the information obtained by an arbitrary vertex can be generalized and a single aggregated data packet is then sent, then this process of data transmission to the BS is called *aggregation*. In the TDMA (Time Division Multiple Access) standard, time is discrete. It is divided into such equal *time rounds* (or *slots*) that the duration of one round is sufficient to transmit a data packet along every edge of the graph. The aggregation time, equal to the number of time rounds, during which the aggregated data from all sensors will fall into the BS, is the most important criterion in many networks.

The communication graph is usually synthesized based on the criterion of minimum transmission energy consumption [6]. Therefore, it is highly sparse, and not all nodes (sensors) can transmit information directly to the BS. Packages from most sensors go through other sensors.

A. Erzin thanks the Russian Foundation for Basic Research, grant 19–47–540007 and the program of fundamental scientific researches of the SB RAS, project 0314–2019–0014 (contribution: Sects. 1–3), and R. Plonikov thanks the Russian Science Foundation, grant 18–71–00084 (contribution: Sects. 4, 5), for financial support.

I. Bykadorov et al. (Eds.): MOTOR 2019, CCIS 1090, pp. 356–368, 2019.
https://doi.org/10.1007/978-3-030-33394-2_28

The aggregation time in a WSN depends on various constraints caused by *conflicts.* Thus, in most WSNs, the sensor cannot receive and transmit a data packet at the same time, and it cannot simultaneously receive or transmit more than one packet. Violation of these conditions is called a *conflict of the first type.* The transmission energy consumption depends on the transmission distance to a power of 2–6, so this is a very power-consuming operation. Therefore, for the reason of energy efficiency, each sensor sends a packet only once during the whole aggregation session. This means that packets are transmitted along the edges of some desired *aggregation tree* (AT) rooted in the BS, and an arbitrary vertex must first receive packets from all its children (in the AT) and only after that can send an aggregated packet to its parent vertex. Moreover, in most WSNs transmitters share one common radio frequency. Therefore, if more than one transmitter is operating in the sensor's receiving area, then due to the *interference* of radio waves, the receiver cannot get the data packet intended for it [2]. The situation when more than one transmitter is operating in the reception area is the *second type of conflict.*

In the problem of conflict-free data aggregation, it is necessary to find an AT, as well as a schedule (a time round, when each sensor transmits) of conflict-free data transmission of the minimum length [1,2,13]. This problem is known as *Convergecast Scheduling Problem* (CSP) and it is NP-hard even for the case when AT is given [7].

The CSP is intensively investigated. To construct an approximate solution, a number of heuristic algorithms have been proposed [1,2,9,11–13,17]. For some of them, guaranteed accuracy estimates were found in terms of the degree and radius of the communication graph [14,18]. To assess the quality of other heuristics, numerical experiments were carried out [2,11,16]. Special cases of the problem are also considered. For example, when conflicts arise only between children of a common parent in AT. This situation occurs when sensors use different radio frequencies to transmit data [9,11,16]. Such a problem is also NP-hard in general, but in the case when AT is known, it is solved in polynomial time.

In [8], a special case of communication graph is considered in the form of a unit square grid with a sensor in each node, and the transmission distance of each sensor is 1. A polynomial algorithm for constructing an *optimal* solution to this problem is proposed. In [4,7] a similar grid graph is considered with an arbitrary transmission distance $d \geq 2$. On the one hand, an increase of d may reduce the length of the schedule. On the other hand, the number of conflicts of the second type is increasing (due to interference). Several methods are proposed for constructing a schedule of conflict-free data transmission with a guaranteed estimate of accuracy depending on d. In particular, for the case when $d = 2$ a proposed algorithm on the $(n+1) \times (m+1)$ grid with the BS at the origin $(0, 0)$ builds a schedule whose length does not exceed $(n + m)/2 + 3$. Later in [3] the *optimality* of the constructed schedule was proved.

In this paper, we consider a special case of the CSP on a regular *triangular* grid, in which all the vertices are to the left and above the BS. This is due to the fact that for the problem of energy-efficient sensor coverage of the area, regular covers are often used, in which the sensors are located at the nodes of a regular grid, in particular, a triangular grid [19]. A polynomial algorithm is proposed for constructing an *optimal* schedule in the case when there is exactly one most remote vertex (MRV) from the BS is in the grid. If there is more than one MRV, then the proposed algorithm builds a feasible schedule. For this case, another polynomial algorithm is also proposed, which becomes more accurate than the first algorithm with an increase in the number of the MRVs.

The paper is organized as follows. In Sect. 2, the CSP problem is formulated for an arbitrary graph. Section 3 is devoted to the consideration of the problem on a complete triangular grid. The linear-time algorithm HCA is proposed for constructing a feasible schedule and it is proved that in the case under consideration it builds an *optimal* schedule. The incomplete grid is considered in Sect. 4, where a linear-time complexity algorithm SCA is proposed for constructing an approximate solution and the accuracy of this algorithm is estimated. Section 5 concludes the paper.

2 Formulation of the CSP

A communication graph is specified in which the vertices are images of the sensors, and two vertices are linked by an edge if the transmission between them can be carried out in both directions. Among the vertices of the graph, we select the BS to which it is necessary to transmit data from all the vertices of the graph. Let's suppose that:

- time is discrete, and a data packet can pass each edge during a one-time slot;
- the sensor cannot simultaneously receive and transmit, as well as receive or transmit more than one packet. Otherwise, a conflict of the first type arises;
- each sensor during the aggregation session transmits a data packet only once (except the sink which always can only receive messages), i.e. once a sensor sends a message, it can no longer be a destination of any transmission;
- the subset of vertices can transmit at the same time unless there is a conflict of the second type associated with the interference of radio waves.

A schedule satisfying these properties is called *feasible*. Thus, a feasible schedule is a conflict-free schedule with additional constraints associated with energy savings. In the CSP, it is required to find a feasible schedule (i.e. to find a transmission time slot for each vertex), the length of which is minimal.

As noted above, the CSP in general, as well as in many special cases, is NP-hard [2, 7]. However, in some special cases, the problem is polynomially solvable. This is the case, for example, when the communication graph is a square unit grid, and the transmission distance is 1 [8] or 2 [3]. In this paper, we are interested in the case when the communication graph is a regular *triangular* grid.

3 Complete Triangular Grid

Let's consider a grid graph (Fig. 1a), in each node of which (x, y), $x = 0, 1, \ldots, n$, $y = 0, 1, \ldots, m$, except the vertex $(0, 0)$, is a sensor. At the node (0,0) is the BS.

Two different nodes (x_i, y_i) and (x_j, y_j), for which conditions $x_i \geq 0$, $y_i \geq 0$, $x_j \geq 0$, $y_j \geq 0$ and $|x_i - x_j| + |y_i - y_j| = 1$, or $x_i - x_j + y_i - y_j = 0$ and $|y_i - y_j| = 1$ are fulfilled, are connected by an edge (i, j) (see Fig. 1*a*). All vertices with coordinates (x, y), $x = 0, 1, \ldots, n$, $y = 0, 1, \ldots, m$ are included in this grid, therefore we call it *complete*. The transmission distance of each vertex is 1. This means that each vertex hears only adjacent vertices. Every sensor should send the collected data to the BS, which is placed at the origin (yellow vertex in Fig. 1). In this case, the data packet is transmitted to the parent node, aggregation of the received data occurs there, and then one packet is sent further. The time slot number when the last packet arriving at the BS is the aggregation time or the *length of the schedule*. Denote the minimum length of the schedule by $L(n, m)$.

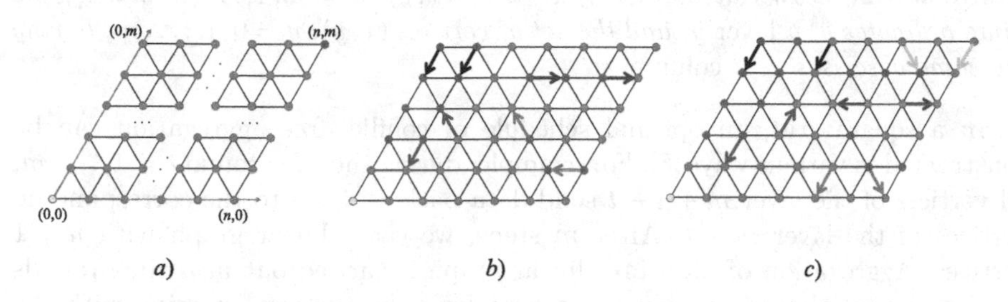

Fig. 1. (*a*) Grid graph; (*b*) Conflict transmissions; (*c*) Conflict-free transmissions. (Color figure online)

The restrictions indicated in the previous section, of course, must also be satisfied here. Thus, Fig. 1*b* shows examples of invalid transmissions (transmissions shown by arrows of the same color cannot be performed simultaneously, because the red vertices hear more than one transmitter). And Fig. 1*c* shows examples of conflict-free transmissions (arrows of the same color), they can be carried out during the same time round.

3.1 Preliminary Considerations

Definition 1. *The* distance *to the vertex is the minimum number of edges in the path connecting it to the base station.*

Then the vertex (x, y) is at a distance $x + y$ from the BS. The length of the schedule cannot be less than $n + m$, since the most remote vertex (red vertex in Fig. 1*a*) is at a distance of $n + m$, and the packet from this vertex cannot arrive at the BS before the time slot $n + m$. Due to the fact that the BS (as, indeed, any other vertex) can receive no more than one packet during one time slot, the fairly obvious

Property 1. If at least two vertices of an arbitrary graph are located at a distance R from the BS, then the aggregation time cannot be less than $R + 1$.

Lemma 1. *The minimal length of the schedule of conflict-free aggregation $L(n, m)$ on a complete triangular grid cannot be less than $n + m + 1$.*

Proof. The most remote vertex (MRV) (n, m) is at a distance of $n + m$. If it does not transmit during time slot 1, then the length of the schedule will be at least $n + m + 1$. If it transmits during the first time round, then both adjacent to it vertices at distance $n + m - 1$ cannot transmit because of the emerging conflict. This means that they can transmit no earlier than during the 2nd time round. The distance from both vertices $(n - 1, m)$ and $(n, m - 1)$ is equal to $n+m-1$. Therefore, (according to Property 1), the aggregation time from these vertices is at least $n + m$. Therefore, the total aggregation time cannot be less than $n + m + 1$. The lemma is proved.

Definition 2. *Let's call the set of vertices (x, y), $x = 0, 1, \ldots, n$, having the same ordinates y, a* layer y, *and the set of vertices (x, y), $y = 0, 1, \ldots, m$, having the same abscissas x, a* column x.

In a square grid, an optimal schedule of conflict-free aggregation can be constructed in various ways [8]. For example, during the time round $t = 1, \ldots, m$, all vertices of the layer $m + 1 - t$ send data packets down to the corresponding vertices of the layer $m - t$. After m steps, we get a linear graph with $n + 1$ vertices. Aggregation of data in a linear graph is carried out in n time rounds by the sequential transmission of packets from the vertices, starting with the most remote one. The length of the constructed schedule is equal to $n + m$ and coincides with the lower bound. Therefore, this is the optimal schedule.

In a triangular grid, the use of such an algorithm is unacceptable, since the simultaneous transmission of neighboring vertices of one layer or one column leads to conflicts. To transmit packets from all vertices of one layer, at least two time rounds are required, and as a result, a similar algorithm will construct a schedule of at least $2m + n$ length. In the next subsection, we present the polynomial *Hexagonal Corridor Algorithm* (HCA), which builds an *optimal* schedule on a complete grid, whose length coincides with the lower bound $n + m + 1$.

3.2 The HCA

Definition 3. *We call a* hexagonal corridor *(HC) the subgrid of a complete triangular grid consisting of hexagons (each of which consists of six triangles) touching each other, in which the two common sides of two adjacent hexagons coincide and are at an angle of* $120°$ *to the horizontal (the angle is counted from the horizontal axis counterclockwise). In Fig. 2, the HCs are highlighted in one color.*

Definition 4. *We call the HC* hollow *if vertices at the centers of the hexagons constituting HC are removed.*

Property 2. Two vertices of a hollow HC with positive ordinates located at the same distance from the BS or at a distance differing by more than 1 can transmit data packets along the edges of the hollow HC simultaneously without conflict between them. If not all ordinates of two nodes are positive, then when passing along two paths of the same length, a conflict will arise at the junction point of the paths.

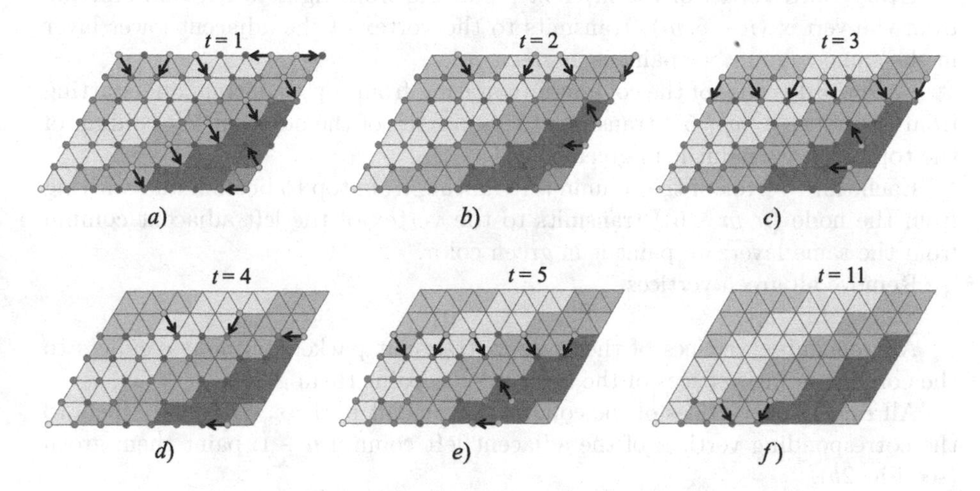

Fig. 2. Several time iterations of the HCA. (Color figure online)

We use this property to build a schedule of conflict-free data aggregation of minimum length. To do this, we construct a central hollow HC, which includes the MRV (n, m) (in Fig. 2, this is the green corridor). During each time slot, nodes that do not belong to the central corridor will transmit packets towards the central HC. In Fig. 2 arrows show transmissions of several time rounds. For convenience, the transmitting vertices are colored green. If a vertex has transmitted a packet, then it is excluded and is not considered in the following time rounds, since each vertex must transmit a packet only once. Since the distance to the MRV limits the length of the schedule from below, we will build a schedule in which the transmissions along the longest path are carried out without extra delay. If we want to build a schedule of length $n + m + 1$, then during the transmission along the longest path, a delay of no more than one time round is possible. That is, only once the most remote current vertex can be silent. Moreover, it is necessary that the distance to the two MRVs of the central corridor differ by at least 2. Otherwise, the conflict will inevitably arise at the junction point of these paths. We give a formal description of the HCA (without loss of generality, we assume that $n \geq m$).

Algorithm HCA (see Fig. 2)
Step 1. $t := 1$;
Green colored vertex $(n-1, m)$ sends a packet to the vertex (n, m).
Green colored vertex $(n-2, m)$ transmits to the vertex $(n-3, m)$.

Each internal vertex $(n-k, m-k)$, $k = 1, \ldots, m-1$ of the central (green) corridor transmits to the vertex $(n-k+1, m-k-1)$; paint it green.

Every third vertex of the layer m (counting from right to left and starting from the vertex $(n-5, m)$) is colored green; it transmits to the vertex of the adjacent lower layer and the adjacent right column.

Every third vertex of the layer m (counting from right to left and starting from the vertex $(n-6, m)$) transmits to the vertex of the adjacent lower layer in the same column; we paint it in green color.

Each third vertex of the column n (counting from top to bottom and starting from the node $(n, m-5)$) transmits to the vertex of the adjacent left column of the top layer; we paint it in green color.

Each third vertex of the column n (counting from top to bottom and starting from the node $(n, m-6)$) transmits to the vertex of the left adjacent column from the same layer; we paint it in green color.

Remove all green vertices.

Step 2. $t := 2$;

All remaining vertices of the layer m transmit packets without conflicts to the corresponding vertices of the layer $m-1$; paint them green.

All remaining vertices of the column n transmit packets without conflicts to the corresponding vertices of the adjacent left column $n-1$; paint them green (see Fig. 2*b*).

Remove all green vertices.

Step 3. $t := t + 1$;

Starting on the right from the vertices of the central corridor, we paint all the vertices of the layer $m - [t/2]$, the transmission of which to the corresponding vertices of the layer $m - [t/2] - 1$ does not lead to conflicts, in green.

We paint all the vertices of the $n - [t/2]$ column not belonging to the central corridor, the transmission of which to the corresponding vertices of the $n - [t/2] - 1$ column does not lead to conflicts, in green (see Fig. 2*c*).

Remove all green vertices.

$t := t + 1$;

We paint all the remaining vertices of the layer $m - [t/2]$ in green; they transmit to the corresponding vertices of the layer $m - [t/2] - 1$.

Paint the node of the column $n - [t/2] + 2$ in green; it passes the packet horizontally to the left.

We paint all the remaining vertices of the column $n - [t/2]$ in green; they transmit to the corresponding vertices of the column $n - [t/2] - 1$.

If $t < 2m$, then **goto** Step 3.

Step 4.

Carry out the aggregation in a linear graph with $n - m$ vertices by passing the packets to the left, starting with the right-most remaining vertex $(n - m, 0)$.

Stop.

Theorem 1. *The HCA builds an optimal schedule of conflict-free data aggregation in a complete triangular grid of dimension* $(n+1) \times (m+1)$ *with BS at the origin* $(0, 0)$*, whose length coincides with the lower bound* $n + m + 1$.

Proof. In the HCA, the MRV (n, m) begins to transmit with a delay of 1 time round. During the first time slot, a hollow central HC is created. In subsequent time rounds, two vertices of a hollow central HC, the distance between which is at least 2, simultaneously transmit packets. Therefore, according to Property 2, such transmissions do not lead to conflicts and, consequently, to delays.

It remains to show that the most remote current vertex (vertices that have already transmitted packets are deleted), starting from the second time slot, transmits packets without delay. To do this, make sure that all the vertices outside the central (green) corridor do not interfere with the transmission of two vertices along the edges of the hollow central corridor. Indeed, during the second round, the vertex (n, m) transmits the packet along with the remaining vertices of layer m and column n. As a result, during the first 2 rounds, all vertices of layer m will transmit packets, and in column n there will remain vertices $(n, m-1)$ and $(n, m-2)$ (Fig. 2*b*). Next, for every two time rounds, packets are transmitted from all the vertices of the current upper layer (including the vertices of central HC). Vertices located to the right transmit packets as early as possible. All vertices below the central corridor belonging to the same rightmost column transmit packets during two consecutive time rounds. Moreover, the vertices located above transmit as early as possible (Fig. 2*c*). As a result, at each subsequent time slot, both the most distant nodes of the central corridor located at least 2 edges apart from each other transmit packets.

After the time slot $t = 2m$ we get the situation shown in Fig. 2*f*, after which it remains to aggregate in a linear graph consisting of $n-m$ vertices (not counting the BS). For this, obviously, $n - m$ time rounds are enough. As a result, for the linear time, we built a schedule of non-conflict data aggregation, the length of which coincides with the lower bound $n + m + 1$. The theorem is proved.

In conclusion of this section, we note that the HCA also works in the case of $n = m$, constructing a schedule of length $2n + 1$, as well as in cases where the BS is located in some other places. However, if the BS is located, for example, in the lower right corner of the triangular grid, then the situation changes fundamentally. The fact is that in this case, for example, when $n = m$, the number of MRV is $2n + 1$, and the distance to them is equal to n. Obviously, in this case, layer-by-layer aggregation can be performed in $3n$ time rounds. The trivial lower bound for the length of the conflict-free aggregation schedule is $n + 1$. This case will be the subject of another paper.

4 Incomplete Triangular Grid

Let, as before, the BS is on the lower left, but there is more than one MRV in the grid (in Fig. 4 there are five MRVs at the distance $n + m - 4$). In this case, from the complete grid, all vertices are removed, the distance to which from the BS is greater than $n + m - 3$. We call such a grid *incomplete*. If we save the transmission schedule that was built by the HCA for the complete grid, for the remaining vertices, it is obvious that we will get a feasible schedule of length $n + m + 1$. However, if the incomplete grid has k MRVs, then the trivial lower bound for the length of the schedule is $n + m - k + 2$, and, therefore, the constructed schedule may not be optimal.

Let's see whether it is possible to find a more accurate lower bound for the length of the schedule, depending on the number of MRVs k. Let the distance to the MRVs be R, so each of MRV can transmit a packet in R time rounds. However, the BS can receive no more than one packet at a time. If packets from all MRVs arrive at different vertices adjacent to the BS at time $R-1$, then at time R can transmit the packet to the BS only one adjacent vertex. The remaining vertices will transmit packets in turn, and the last adjacent vertex will transmit the packet to the BS at time $R+k$. However, some conflicts can be solved earlier (further away from the BS). For this, it is necessary that several paths from MRVs join earlier. Then, due to the conflicts, the transmission time along the shared path (after the merge) will increase, but the number of the longest paths will decrease. Suppose a transmission from the MRVs is organized in such a way that packets from them come in BS at different time slots $R, R+1, \ldots, R+a$. Let us estimate the value of a. A packet from the some MRV may arrive at the BS at time R if it is transmitted along a path that does not intersects with other paths from the MRVs (left path in Fig. 3). What is the maximum number of MRVs that packets from them can be delivered to the BS at time round $R+1$? This, of course, depends on R. The vertex adjacent to the BS transmits at time $R+1$. Consequently, packets can come into it from two vertices, at time R (say, from vertex i) and at time $R-1$ (from vertex j) (the second left branch of the paths in Fig. 3). Then, only one packet can arrive at the vertex j at time $R-2$. And at the vertex i, one packet can arrive at time $R-1$, and another - at time $R-2$. Figure 3 shows the maximum possible number of MRVs in fragment of AT, when $R = 4$, and $a = 2$. The number inside the circle corresponds to the time round when the vertex sends a message.

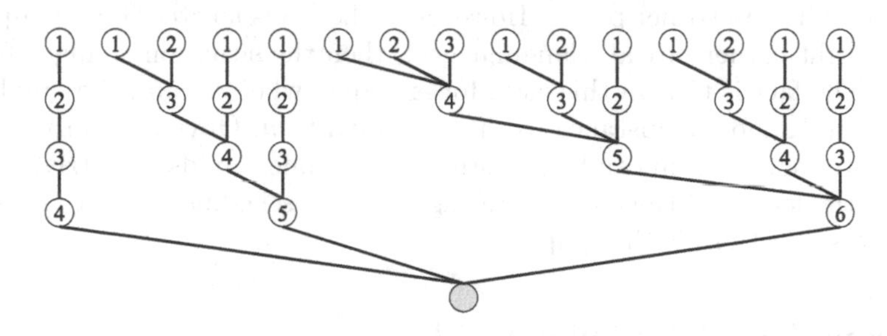

Fig. 3. Illustration of an AT fragment for transmitting packets from the maximum number of MRVs during $R + a$ time rounds ($R = 4$, $a = 2$).

If in the AT there are several fragments similar to those shown in Fig. 3, then a will increase. But it can be argued that if the number of MRVs is at least 2, then $a \geq 1$. If the number of MRVs is at least 6, and $R \leq 4$, then $a \geq 2$. If the number of MRVs is not less than 16, but $R \leq 4$, then $a \geq 3$. That is, a grows slowly, and the lower bound of the length of the schedule is slightly differs from $R+1$ (especially for large R).

Let $n \geq m$. To accelerate the aggregation, let us not use the HCs, but select the straight corridors (SCs), inclined to the horizontal at an angle of 60°, consisting of four triangles in the layers (in Fig. 4 different SCs are painted in different colors). If we delete internal vertices in SC, leaving only the boundary vertices, then we call this SC *hollow*. In the hollow SC, two nodes located on different sides of the corridor can transmit simultaneously. The paths that go along opposite sides of the corridor intersect at the layer $y = 0$. In this case, a conflict may arise if the right vertex sends to the left, and the left sends down-left, and the distance between them is no more than 2 (see Figs. 4*e* and 4*f*). If a conflict arises, then suppose that the left vertex sends, and the right one waits. Then the current MRV will always be the rightmost vertex.

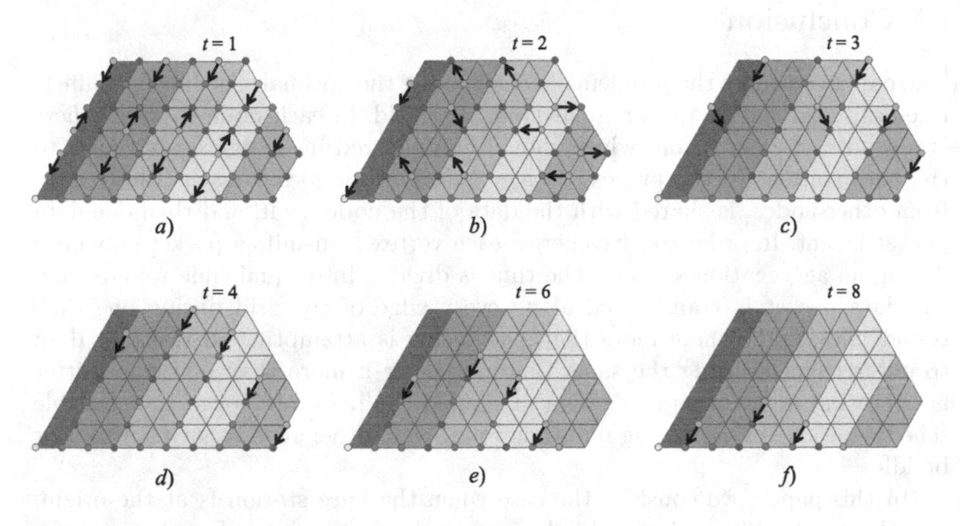

Fig. 4. An example of the operation of the SCA. The length of the schedule is $n+m-1$. (Color figure online)

Without describing the operation of the *Straight Corridor Algorithm* (SCA) in detail, we illustrate it in Fig. 4 where the number of MRVs is 5. In order for vertices lying on different sides of hollow SC to be able to transmit messages simultaneously, it is necessary that the internal vertices of SC transmit packets earlier than the MRVs located on SC boundaries. To organize this, two first time slots are enough. Starting from time round 3, only the MRVs prevent each other from transmitting without conflicts. Not all, but only half of them, because the internal MRVs in SCs have already sent the packets and deleted. For example, in Figs. 4*e* and 4*f* due to conflicts the nodes circled in red cannot transmit.

Lemma 2. *If in an incomplete grid the number of the MRVs equals k, then SCA builds a schedule of length, not more than* $n + m - k + 5$.

Proof. Starting from the third time slot, the remaining (non-internal) MRVs can transmit without delay. If a conflict arises between the two rightmost MRVs, then in the SCA the right vertex is waiting. It needs to wait no more than two time rounds. During this time, all possible conflicts between the vertices located to the left can be solved, giving the first opportunity to transmit those MRVs, which are to the left. Then in total, 4 *additional* time rounds are enough to transmit packets from all MRVs. The distance to the MRVs is $n+m-k+1$. The lemma is proved.

So, we proved that the length of the schedule constructed by SCA exceeds the length of the optimal schedule by no more than 4 time rounds. If we use HCA, then the length of the schedule is equal to $n+m+1$. Therefore, for $k \geq 4$, the SCA is preferable to HCA.

5 Conclusion

The paper considers the problem of constructing the optimal schedule of conflict-free data aggregation on a regular triangular grid, in each node of which there is a sensor, the data from which should be delivered in an aggregated form to the base station. In the process of aggregation, data goes to each transit node from other nodes, is shared with the data of the node itself, and then one data packet is sent. In order to save energy, each vertex transmits a packet only once during an aggregation session. The time is divided into equal time rounds, and the data packet is transmitted along every edge of the grid during one time round. Situations where more than one vertex is attempting to transmit data to one recipient during the same time round, or if more than one transmitter is operating in the recipient's area are called conflicts. Moreover, in a feasible schedule, each vertex during one time round can either receive, or transmit, or be idle.

In this paper, we consider the case when the base station is at the origin, and the regular triangular grid is located in the first quarter. For the case of a complete grid (when it contains all points with coordinates (x, y), $x = 0, \ldots, n$, $y = 0, \ldots, m$), an algorithm HCA of linear complexity is proposed for constructing an optimal schedule whose length coincides with the lower bound. This is the first polynomial algorithm known to us that builds an optimal solution to such a problem. If the grid is incomplete (vertices are removed from the complete grid, the distance to which exceeds a certain integer), then we cannot guarantee the optimality of the schedule under construction. For this case, the HCA is also applicable, but with a sufficiently large number of MRVs, another algorithm (SCA) developed by us builds a shorter schedule.

In further studies, we plan to consider various options for the location of the BS. In some cases, nothing new happens, and in some situation changes dramatically. For example, if the BS is in the lower right corner of the grid and, for example, $n = m$, then the distance to the MRVs becomes equal to n (hence, the trivial lower bound for the length of the schedule is $n+1$), and the number of MRVs is $2n+1$. Building a feasible schedule of length less than $2n$ is not so easy.

References

1. Cheng, C.-T., Tse, C.K., Lau, F.C.M.: A delay-aware data collection network structure for wireless sensor networks. IEEE Sens. J. **11**(3), 699–710 (2011)
2. De Souza, E., Nikolaidis, I.: An exploration of aggregation convergecast scheduling. Ad Hoc Netw. **11**, 2391–2407 (2013)
3. Erzin, A.: Solution of the convergecast scheduling problem on a square unit grid when the transmission range is 2. In: Battiti, R., Kvasov, D.E., Sergeyev, Y.D. (eds.) LION 2017. LNCS, vol. 10556, pp. 50–63. Springer, Cham (2017). https://doi.org/10.1007/978-3-319-69404-7_4
4. Erzin, A., Plotnikov, R.: Conflict-free data aggregation on a square grid when transmission distance is not less than 3. In: Fernández Anta, A., Jurdzinski, T., Mosteiro, M.A., Zhang, Y. (eds.) ALGOSENSORS 2017. LNCS, vol. 10718, pp. 141–154. Springer, Cham (2017). https://doi.org/10.1007/978-3-319-72751-6_11
5. Erzin, A., Plotnikov, R.: Using VNS for the optimal synthesis of the communication tree in wireless sensor networks. Electron. Notes Discrete Math. **47**, 21–28 (2015)
6. Erzin, A., Plotnikov, R., Mladenovic, N.: Variable neighborhood search variants for min-power symmetric connectivity problem. Comput. Oper. Res. **78**, 557–563 (2017)
7. Erzin, A., Pyatkin, A.: Convergecast scheduling problem in case of given aggregation tree. the complexity status and some special cases. In: 10th International Symposium on Communication Systems, Networks and Digital Signal Processing (CSNDSP), no. 16. IEEE-Xplore, Prague (2016)
8. Gagnon, J., Narayanan, L.: Minimum latency aggregation scheduling in wireless sensor networks. In: 11th International Symposium on Algorithms and Experiments for Sensor Systems, Wireless Networks and Distributed Robotics, pp. 152–168. Wroclaw, Poland (2014)
9. Ghods, F., Yousefi, H., Mohammad, A., Hemmatyar, A., Movaghar, A.: MC-MLAS: multi-channel minimum latency aggregation scheduling in wireless sensor networks. Comput. Netw. 57, 3812–3825 (2013)
10. Hansen, P., Kuplinsky, J., De Werra, D.: Mixed graph colorings. Math. Methods Oper. Res. **45**, 145–160 (1997)
11. Incel, O.D., Ghosh, A., Krishnamachari, B., Chintalapudi, K.: Fast data collection in tree-based wireless sensor networks. IEEE Trans. Mobi. Comput. **11**(1), 86–99 (2012)
12. Li, H., Hua, Q.-S., Wu, C., Lau, F.C.M.: Minimum-latency aggregation scheduling in wireless sensor networks under physical interference model. HKU CS Tech Report TR-2010-07 (2010). https://doi.org/10.1145/1868521.1868581
13. Malhotra, B., Nikolaidis, I., Nascimento, M.A.: Aggregation convergecast scheduling in wireless sensor networks. Wireless Netw. **17**, 319–335 (2011)
14. Nguyen, T.D., Zalyubovskiy, V., Choo, H.: Efficient time latency of data aggregation based on neighboring dominators in WSNs. In: IEEE Globecom, pp. 6133827 (2011)
15. Plotnikov, R., Erzin, A., Mladenovic, N.: Variable neighborhood search-based heuristics for min-power symmetric connectivity problem in wireless networks. In: Kochetov, Y., Khachay, M., Beresnev, V., Nurminski, E., Pardalos, P. (eds.) DOOR 2016. LNCS, vol. 9869, pp. 220–232. Springer, Cham (2016). https://doi.org/10.1007/978-3-319-44914-2_18
16. Plotnikov, R., Erzin, A., Zalyubovskiy, V.: Convergecast with unbounded number of channels. In: MATEC Web of Conferences, vol. 125, pp. 03001 (2017). https://doi.org/10.1051/matecconf/201712503001

17. Wang, P., He, Y., Huang, L.: Near optimal scheduling of data aggregation in wireless sensor networks. Ad Hoc Netw. **11**, 1287–1296 (2013)
18. Xu, X., Li, X.-Y., Mao, X., Tang, S., Wang, S.: A delay-efficient algorithm for data aggregation in multihop wireless sensor networks. IEEE Trans. Parallel Distrib. Syst. **22**, 163–175 (2011)
19. Zalyubovskiy, V., Erzin, A., Astrakov, S., Choo, H.: Energy-efficient area coverage by sensors with adjustable ranges. Sensors **9**(4), 2446–2460 (2009)

On an Applied Problem of Vector Optimization

Igor Kandoba[1,2(✉)] and Alexander Uspenskii[1,2]

[1] Krasovsky Institute of Mathematics and Mechanics,
Ural Branch of the Russian Academy of Sciences, Ekaterinburg, Russia
{kandoba,uspen}@imm.uran.ru
[2] Institute of Natural Sciences and Mathematics, Ural Federal University named after the first President of Russia B. N. Yeltsin, Ekaterinburg, Russia

Abstract. The paper is devoted to the construction and investigation of mathematical models of economic processes in a local product market. The problem of optimization of prices at outlets of an autonomous network of wholesale under additional restrictions is in focus. The mathematical model of this problem belongs to the class of linear problems of vector optimization. The main properties of the multicriteria problem are studied. An optimal plan is defined. The necessary and sufficient conditions for optimality are established. The theorem of the existence and uniqueness of the optimal plan is formulated. A finite iterative procedure for the problem solution is developed on the base of the obtained theoretical results. The suggested numerical algorithm is based on specific variations of model parameters. The results are illustrated by examples of numerical solutions of some intuitive economic problems with using model data.

Keywords: Multicriteria problem · Optimal plan · Existence and uniqueness · Iterative procedure

1 Introduction

The paper is devoted to the construction and investigation of mathematical models of economic processes in an autonomous commodity market. The problem of price optimization at interconnected outlets of a certain uniform good under additional restrictions is in focus. The mathematical model of this problem belongs to the class of linear problems of vector optimization. A large number of works (see, for example, [4,6,8,11,14] are devoted to the investigation of such problems. The questions discussed here deal with earlier investigations conducted by the authors [12]. In this paper, it is managed to remove some restrictions on an admissible solution of the problem, which were taken into account in [12].

Properties of the multicriteria optimization problem are examined in the research. A finite iterative procedure for solving the problem is developed on the base of established properties. This procedure takes into account the specific

I. Bykadorov et al. (Eds.): MOTOR 2019, CCIS 1090, pp. 369–380, 2019.
https://doi.org/10.1007/978-3-030-33394-2_29

properties of the problem. The numerical algorithm is based on special variations of parameters of the mathematical model. The findings are illustrated by examples of numerical solving of some model problems with economic content.

2 Price Optimization Problem in an Autonomous Wholesale Market

A local autonomous market of wholesale trades of certain uniform goods (for example, of some energy resource) is considered. It is supposed that this market consists of a finite number of interconnected outlets. The only connection between two various outlets can be defined. This connection is interpreted as a canal of the transportation of goods from one outlet to another using one of possible transport methods chosen in advance (for example, power lines, road haulage, railway transport, etc.). It is assumed that the transportation of goods, possibly transit transportation, from each outlet to any other outlet is feasible in the considered market. Costs of the transportation of goods from one outlet of the market to any other outlet are known. Each outlet of the market is characterized by the goods price.

The problem of setting optimal prices at the network points consists in the following. Suppose that costs of the goods transportation from one outlet to any other are fixed. Under the assumption that the prices at some outlets are fixed, it is necessary to find maximal possible price values at other outlets in such a way that the following condition is fulfilled. For any pair of interconnected outlets, the price at each of them should not exceed the price at another one plus the cost of transportation between them.

In our opinion, such a condition creates objective reasons for those consumers who are geographically or economically "fastened" to a concrete outlet of the market to acquire goods exactly at this outlet. Such reasons form a grounded base for the long-term planning of goods deliveries to the outlets of the market.

3 Mathematical Model of the Market

Under the assumptions above, the market can be interpreted as a connected undirected graph (see Fig. 1). Each node of this graph is associated with some outlet of the market and all edges of the graph are interpreted as the connections between outlets corresponded to nodes.

Let $V = \{\boldsymbol{v}_i \in \mathbb{R}^2 \mid i = 1, 2, \ldots, n\}$ denote a finite set of points from two-dimensional Euclidean space $\mathbb{R}^2$, and let E denote the set of segments that connect some pairs of different points from the set V. Thus, $E = \{\boldsymbol{e}_{ij} = [\boldsymbol{v}_i, \boldsymbol{v}_j] \mid 1 \leq i \leq n, 1 \leq j \leq n : i \neq j\}$. Note that, in the general case, there exist two different points $\boldsymbol{v}_k$ and $\boldsymbol{v}_s$ in V such that they are not connected directly, i.e., the set E does not contain elements $\boldsymbol{e}_{ks}$ and $\boldsymbol{e}_{sk}$ (see Fig. 1).

Let us consider a flat arcwise connected nonoriented graph $G = (V, E)$ [2]. This graph consists of n interconnected nodes connected by edges $\boldsymbol{e}_{ij} \in E$

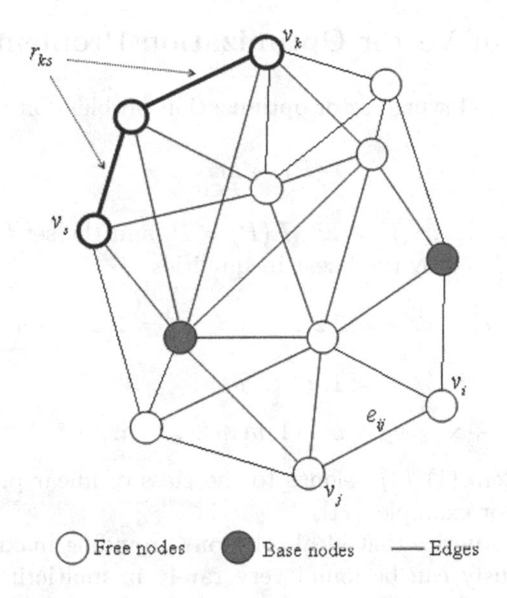

Fig. 1. An example of a flat graph with twelve interconnected nodes.

(see Fig. 1). It is known [2] that any pair of different nodes can be connected by a simple chain in a arcwise connected graph. A path that connects two arbitrary nodes of the graph G is called a chain [2]. A chain that consists of different nodes is called a simple chain. Any chain is identified with the set of its nodes $\boldsymbol{r}_{i_0 i_l} = \{\boldsymbol{v}_{i_0}, \boldsymbol{v}_{i_1}, \boldsymbol{v}_{i_2}, \ldots, \boldsymbol{v}_{i_l}\}$ that are consequently connected by edges of the graph G. Thus, for a simple chain, we have $\boldsymbol{v}_{i_j} \neq \boldsymbol{v}_{i_k}$ for every $j, k = 0, 1, 2, \ldots, l : j \neq k$. The first node $\boldsymbol{v}_{i_0}$ in the chain $\boldsymbol{r}_{i_0 i_l}$ is called the initial node of the chain $\boldsymbol{r}_{i_0 i_l}$, and the last node $\boldsymbol{v}_{i_l}$, the end node of the chain $\boldsymbol{r}_{i_0 i_l}$. The length of the chain is the number of edges of the graph that connect the nodes defining the chain. In this notation, the length of the chain $\boldsymbol{r}_{i_0 i_l}$ is equal to l.

Let us assign numerical characteristics $p_i \geq 0$ and $c_{ij} > 0$ to all nodes $\boldsymbol{v}_i$ and edges $\boldsymbol{e}_{ij}$ of the graph G. Here, p_i is the goods price, c_{ij} are the costs of transportation connection. Let us make the following assumptions regarding the transportation costs c_{ij}. First, there is a constant $c_0 > 0$ such that $c_{ij} \geq c_0$ for all $i, j = 1, 2, \ldots, n$. Second, let $c_{ks} = \overline{c} > 0$ for those pairs of values of indices k and s for which there is no any corresponding element in E (there is no any edge $\boldsymbol{e}_{ks}$ in the graph G). In addition, $\overline{c} >> c_{\max}$, $c_{\max} = \max\{c_{ij} \mid \boldsymbol{e}_{ij} \in E\}$. Third, $c_{ij} = c_{ji}$. Let us fix the values of all parameters c_{ij} and values of a part of parameters $p_k = s_k < +\infty$ $(k = m+1, \ldots, n, m < n)$. All other parameters p_i $(i = 1, \ldots, m)$ are nonnegative variables. The nodes corresponding to variable goods prices are called free nodes, and the nodes corresponding to fixed prices, basic nodes. Here, it is supposed that the first m $(1 \leq m < n)$ nodes are considered as free ones (otherwise, the nodes of the graph can be renumbered).

4 Statement of Vector Optimization Problem

Let us consider the following vector optimization problem on the graph G:

$$F(P) \to \max_{P \in U} \tag{1}$$

Here, $P = (p_1, p_2, \ldots, p_n)^\top \in \mathbb{R}^n$, $F(P) = P$, and the set $U \subseteq \mathbb{R}^n$ of admissible solutions is defined by the linear inequalities

$$p_i \le p_j + c_{ij} \qquad \forall i = 1, 2, \ldots, m, \quad \forall j = 1, 2, \ldots, n : j \ne i, \tag{2}$$

$$p_i \ge 0 \qquad \forall i = 1, 2, \ldots, n, \tag{3}$$

$$p_j = s_j < +\infty \quad \forall j = m+1, m+2, \ldots, n. \tag{4}$$

Optimization problem (1)–(4) belongs to the class of linear problems of vector optimization (see, for example, [21]).

It is necessary to notice that ideal solutions providing maximal values to all criteria simultaneously can be found very rarely in multicriteria optimization problems. But if it is found then, naturally, this solution should be considered as an optimal one. Definitions of vector preferences with the help of effective solutions or Pareto optimal solutions lead to common and well-designed methods for solving vector optimization problems. In what follows, vector inequalities $P \ge P^\star$ are understood coordinate-wise.

Definition 1. *A vector $P^\star = (p_1^\star, p_2^\star, \ldots, p_n^\star)^\top \in U$ is called a Pareto optimal solution of problem (1)–(4), if there is no any vector $P = (p_1, p_2, \ldots, p_n)^\top \in U$ such that $P \ge P^\star, P \ne P^\star$.*

From formulas (2)–(4) it follows that the objective space U in problem (1)–(4) is convex and compact. Hence, the set U_P of Pareto optimal solutions is not empty. It is known [18] that the Pareto set U_P has the property of external stability (for any $P \in U$, there exists an estimate $P^\star \in U_P$ such that $P^\star \ge P$), which is formulated as the following definition.

Definition 2. *A vector $P^\star = (p_1^\star, p_2^\star, \ldots, p_n^\star)^\top \in U$ is called an ideal solution of problem (1)–(4) (an optimal plan for problem (1)–(4)) if*

$$P^\star \ge P \quad (p_i^\star \ge p_i \quad \forall i = 1, 2, \ldots, n) \tag{5}$$

for every $P = (p_1, p_2, \ldots, p_n)^\top \in U$.

The goal is to find an ideal solution (see Definition 2) that maximizes the prices at all outlets simultaneously under specified conditions (2)–(4).

5 Main Properties of Optimization Problem

Note that problem (1)–(4) degenerates under condition $m = n$, since in this case optimal values of the parameters p_i are reached at infinity ($p_i = +\infty, \quad i = 1, 2, \ldots, n$). Therefore, the existence of base nodes in the graph G is of principal importance. In this case ($m < n$), the following statements are true.

Lemma 1. *The set $U \subseteq \mathbb{R}^n$ defined by system of linear inequalities (2)–(4) is bounded, i.e.* $\exists\, \overline{p} > 0 : p_i \leq \overline{p} \quad \forall i = 1, 2, \ldots, n$.

Proof. Let us consider an arbitrary node $\boldsymbol{v}_{i_0}$.

There exists a simple chain $\boldsymbol{r}_{i_0 i_l} = \{\boldsymbol{v}_{i_0}, \boldsymbol{v}_{i_1}, \boldsymbol{v}_{i_2}, \ldots, \boldsymbol{v}_{i_l}\}$ connected this node with a base node $\boldsymbol{v}_{i_l}$. Obviously, the length of this chain satisfies inequality $l \leq n - 1$. Hence, taking into account (2), the following inequalities

$$p_{i_0} \leq p_{i_1} + c_{i_0 i_1} \leq p_{i_2} + c_{i_1 i_2} + c_{i_0 i_1} \leq p_{i_3} + c_{i_2 i_3} + c_{i_1 i_2} + c_{i_0 i_1} \leq \ldots \leq p_{i_l} + \sum_{j=0}^{l-1} c_{i_j i_{j+1}}$$

are fulfilled. Taking into account (4), we obtain

$$p_{i_0} \leq s_l + (n-1)c_{\max}. \tag{6}$$

Let $\overline{p} = (n-1)c_{\max} + \max\{s_i \mid i = m+1, \ldots, n\}$. Then, inequality (6) leads to $p_i \leq \overline{p} \quad \forall i = 1, 2, \ldots, n$.

The lemma is proved.

The structure of conditions (2) provides a necessary condition of optimality of the plan $P^\star$.

Lemma 2. *(Necessary condition of optimality)*

Let $P^\star = (p_1^\star, p_2^\star, \ldots, p_n^\star)^\top \in U$ be an optimal plan for problem (1)–(4). Then,

$$\forall i = 1, 2, \ldots, m \ \ \exists j = 1, 2, \ldots, n, j \neq i : \ \ p_i^\star = p_j^\star + c_{ij}. \tag{7}$$

Proof. Assume the contrary.

Let $\exists\, 1 \leq i \leq m : \ \forall j = 1, 2, \ldots, n, j \neq i \Rightarrow p_i^\star < p_j^\star + c_{ij}$.

Let us define $\Delta p_i = \min\{p_j^\star + c_{ij} - p_i^\star \mid j = 1, 2, \ldots, m : \ j \neq i\}$ and consider the vector $P_\Delta = (p_1^\star, \ldots, p_{i-1}^\star, p_i^\star + \Delta p_i, p_{i+1}^\star, \ldots, p_n^\star)^\top$. It is easy to see that $P_\Delta \in U$. This contradicts to the optimality of $P^\star$ since $p_i^\star + \Delta p_i > p_i^\star$.

The lemma is proved.

Let us introduce the additional definitions.

Definition 3. *A chain $\boldsymbol{r}_{i_0 i_l} = \{\boldsymbol{v}_{i_0}, \boldsymbol{v}_{i_1}, \boldsymbol{v}_{i_2}, \ldots, \boldsymbol{v}_{i_l}\}$ connecting the nodes $\boldsymbol{v}_{i_0}$ and $\boldsymbol{v}_{i_l}$ of the graph G is called a limit chain for $P = (p_1, p_2, \ldots, p_n)^\top \in U$ if* $p_{i_k} = p_{i_{k+1}} + c_{i_k i_{k+1}} \quad \forall k = 0, 1, \ldots, l-1$.

Definition 4. *A limit chain $\boldsymbol{r}_{i_0 i_l} = \{\boldsymbol{v}_{i_0}, \boldsymbol{v}_{i_1}, \boldsymbol{v}_{i_2}, \ldots, \boldsymbol{v}_{i_l}\}$ is called a basic limit chain for $P = (p_1, p_2, \ldots, p_n)^\top \in U$ if the end node $\boldsymbol{v}_{i_l}$ of the chain $\boldsymbol{r}_{i_0 i_l}$ is a base node.*

One can specify the following useful properties of a limit chain.

First, it is easy to prove that any limit chain $\boldsymbol{r}_{i_0 i_l} = \{\boldsymbol{v}_{i_0}, \boldsymbol{v}_{i_1}, \boldsymbol{v}_{i_2}, \dots, \boldsymbol{v}_{i_l}\}$ for any plan P of problem (1)–(4) is a simple chain. Obviously, it is enough to show that the nodes $\boldsymbol{v}_{i_0}$ and $\boldsymbol{v}_{i_l}$ are different. Indeed, assuming the opposite, i.e. $\boldsymbol{v}_{i_0} = \boldsymbol{v}_{i_l}$, it is easy to get a contradiction with the positivity of the parameters $c_{i_j i_{j+1}}$ $(j = 0, 1, \dots, l-1)$ corresponded to edges connecting the nodes of this chain. This follows from the equalities

$$\begin{aligned} p_{i_0} &= p_{i_1} + c_{i_0 i_1} = p_{i_2} + c_{i_1 i_2} + c_{i_0 i_1} \\ &= p_{i_3} + c_{i_2 i_3} + c_{i_1 i_2} + c_{i_0 i_1} = \dots = p_{i_l} + \sum_{j=0}^{l-1} c_{i_j i_{j+1}}. \end{aligned}$$

If $\boldsymbol{v}_{i_0} = \boldsymbol{v}_{i_l}$ then the equality $p_{i_0} = p_{i_0} + \sum_{j=0}^{l-1} c_{i_j i_{j+1}}$ means that $\sum_{j=0}^{l-1} c_{i_j i_{j+1}} = 0$. This contradicts to the positivity of $c_{i_j i_{j+1}}$.

Second, it is evident that the limit chain $\boldsymbol{r}_{i_0 i_l} = \{\boldsymbol{v}_{i_0}, \boldsymbol{v}_{i_1}, \boldsymbol{v}_{i_2}, \dots, \boldsymbol{v}_{i_l}\}$ is an oriented chain in the sense that, unlike the original limit chain $\boldsymbol{r}_{i_0 i_l}$, the chain $\boldsymbol{r}_{i_l i_0} = \{\boldsymbol{v}_{i_l}, \boldsymbol{v}_{i_{l-1}}, \boldsymbol{v}_{i_{l-2}}, \dots, \boldsymbol{v}_{i_0}\}$ is not a limit chain.

Theorem 1. *(Existence and uniqueness of an ideal solution)*
Problem (1)–(4) has a unique ideal solution (the optimal plan $P^\star$).

Proof. We prove the theorem by two steps. First, we prove the existence of a stationary point in problem (1)–(4) that meets necessary condition of optimality (7). Then we prove the uniqueness of such point and show that this point is an ideal solution.

1. The existence of a stationary point.

Let us construct the sequence of points $\Sigma = \{P^{(t)} = (p_1^{(t)}, p_2^{(t)}, \dots, p_n^{(t)})^\top\} \subset \mathbb{R}^n$ as follows. The initial point is defined as $P^{(t)} = (0, 0, \dots, 0, s_{m+1}, \dots, s_n)^\top \in U$. The subsequent elements of the sequence are determined recursively:

$$\begin{aligned} &P^{(t+1)} = P^{(t)} + \Delta P^{(t)}, \\ &\Delta P^{(t)} = (\Delta p_1^{(t)}, \Delta p_2^{(t)}, \dots, \Delta p_n^{(t)})^\top \in \mathbb{R}^n : \\ &\Delta p_i^{(t)} = 0 \;\; \forall i = 1, 2, \dots, n: \;\; i \neq k, \;\; 1 \leq k \leq m, \\ &\Delta p_k^{(t)} = \Delta \overline{p}, \\ &t = 0, 1, 2, \dots, \end{aligned} \tag{8}$$

where

$$\begin{aligned} &k = Arg\ \max_{j=1,2,\dots,m}\{\min\{p_i^{(t)} + c_{ij} - p_j^{(t)} \mid c_{ij} < \overline{c},\ i = 1, 2, \dots, n\}\}, \\ &\Delta \overline{p} = \min\{p_i^{(t)} + c_{ik} - p_k^{(t)} \mid c_{ik} < \overline{c},\ i = 1, 2, \dots, n\} \geq 0. \end{aligned} \tag{9}$$

Thus, each successor $P^{(t+1)}$ of the sequence Σ is a result of such variation of the antecedent $P^{(t)}$ that leads to the increase of the value of only one coordinate of the point $P^{(t)}$. According to (9), the number k of this coordinate and the increment value $\Delta p_k^{(t)}$ are defined in the following way. The number k corresponds to a free node $\boldsymbol{v}_k$ of the graph G such that the current value of its characteristic $p_k^{(t)}$ can be increased more than at other free nodes. Hence, the value $\Delta p_k^{(t)}$ is defined as a maximal possible value that meets the condition $P^{(t+1)} = P^{(t)} + \Delta P^{(t)} \in U$. As a result, we obtain the sequence satisfying the conditions $P^{(t+1)} \in U, \quad P^{(t+1)} \geq P^{(t)} \ \forall t = 0, 1, 2, \ldots$.

Due to the completeness of the space $\mathbb{R}^n$ and the boundedness of the objective set $U \subseteq \mathbb{R}^n$, there exists $\lim_{t\to\infty} P^{(t)} = P^\star$. Herewith, $\lim_{t\to\infty} \Delta P^{(t)} = (0, 0, \ldots, 0)^\top$ by definition (8), (9) of the sequence Σ.

Let us prove that $P^\star$ is a unique optimal plan for problem (1)–(4).

2. The optimality and uniqueness of plan $P^\star$.

It is easy to show that the plan $P^\star$ meets necessary condition of optimality (7). Otherwise, according to rule (9), one can find a vector $\Delta P \neq 0$ such that $P^\star + \Delta P \geq P^\star$. This contradicts to the fact that $P^\star$ is the limit of the sequence Σ.

Now, we show that the plan $P^\star$ satisfies condition (5), i.e., it is an optimal plan for problem (1)–(4). Suppose the opposite. Let there is a point $\widetilde{P} = (\widetilde{p}_1, \widetilde{p}_2, \ldots, \widetilde{p}_n)^\top \in U$ such that $\exists\, 1 \leq i_0 \leq m : \widetilde{p}_{i_0} > p^\star_{i_0}$. Obviously, by virtue of condition (7), for each free node $\boldsymbol{v}_{j_0}$ $(1 \leq j_0 \leq m)$ of the graph G, there is a limit base chain $\boldsymbol{r}_{j_0 j_l} = \{\boldsymbol{v}_{j_0}, \boldsymbol{v}_{j_1}, \boldsymbol{v}_{j_2}, \ldots, \boldsymbol{v}_{j_l}\}$ for $P^\star$ connecting a node $\boldsymbol{v}_{j_0}$ with some base node $\boldsymbol{v}_{j_l}$. Consider the limit base chain $\boldsymbol{r}_{i_0 i_l} = \{\boldsymbol{v}_{i_0}, \boldsymbol{v}_{i_1}, \boldsymbol{v}_{i_2}, \ldots, \boldsymbol{v}_{i_l}\}$. Then, the conditions $p^\star_{i_j} = p^\star_{i_{j+1}} + c_{i_j i_{j+1}}, \quad \widetilde{p}_{i_j} \leq \widetilde{p}_{i_{j+1}} + c_{i_j i_{j+1}}$ $(j = 0, 1, \ldots, l-1)$ leads to the inequalities $\widetilde{p}_{i_j} > p^\star_{i_j} \quad \forall j = 0, 1, 2, \ldots, l$. This contradicts to the fact that $\widetilde{p}_{i_l} = s_{i_l} = p^\star_{i_l}$ for the base node $\boldsymbol{v}_{i_l}$.

Let us prove the uniqueness of an optimal plan for problem (1)–(4). Suppose the opposite, i.e. there are two different optimal plans $\widetilde{P}^\star, \overline{P}^\star$ for this problem. Then, $\widetilde{P}^\star \geq P$ and $\overline{P}^\star \geq P \ \forall P \in U$. Thus, $\widetilde{P}^\star \geq \overline{P}^\star$ and $\overline{P}^\star \geq \widetilde{P}^\star$. Hence, $\widetilde{P}^\star = \overline{P}^\star$.

The theorem is proved.

The proof of Theorem 1 implies the fact that necessary optimality condition (7) of the plan $P^\star$ is also a sufficient condition for its optimality in problem (1)–(4).

6 Solution Algorithm for Vector Optimization Problem

The well-known rather general methods [5,7,9,16,17] for solving vector optimization problems can be used for constructing the optimal plan $P^\star$ for problem (1)–(4).

In addition, it should be noted that constructing an optimal plan for multicriteria problem (1)–(4) can be reduced to solving a linear programming problem with one scalar criterion. In particular, it is known [15] that there exists a vector $H = (h_1, h_2, \ldots, h_n)^\top \in \mathbb{R}^n$: $h_i > 0 \quad \forall i = 1, 2, \ldots, n$, such that the problem

$$f(P) = \sum_{i=1}^{m} h_i p_i \to \max_{P \in U} \tag{10}$$

is equivalent to problem (1)–(4).

Koopmans [13] was the first who obtained the analogous result of reducing a specialized linear problem with a vector criterion to a standard linear programming problem by means of the convolution technique. Charnes and Cooper [3] obtained the analogous result with the help of duality theorems for linear programming problems. Later, the equivalence of a linear vector optimization problem and a linear programming problem whose objective function is presented as the sum of scalar coordinates with positive coefficients was established by Bod, Focke, Isermann, Podinovskii, and Steuer [18].

Thus, an optimal plan for problem (1)–(4) can be found as a solution of standard problem of linear programming (10). The traditional methods [19] for solving linear programming problems (for example, the simplex-method) can be used. However, the implementation of this approach is connected with the inevident construction of the vector H, which provides the equivalence of original multicriteria problem (1)–(4) and auxiliary one-criterion problem (10).

In this paper, another solving method, which essentially takes into account the specific properties of problem (1)–(4), is proposed. In essence, the proof of Theorem 1 contains a numerical algorithm for solving the vector optimization problem. A modification of this algorithm is described below. This modification is an iterative procedure that allows to obtain both exact and approximate solutions of problem (1)–(4) for a finite number of steps.

We make the following simple observations. Let us choose an arbitrary free node $\boldsymbol{v}_k$ $(1 \leq k \leq m)$ of the graph G (see Fig. 2).

Let $\{k_1, k_2, \ldots, k_s\}$ denote the set of indices of the nodes connected with this node (see Fig. 2). Let us select from constraints (2) the inequalities that include the parameter p_k:

$$p_k \leq p_{k_j} + c_{kk_j} \quad \forall j = 1, 2, \ldots, s, \tag{11}$$

$$p_{k_j} \leq p_k + c_{k_j k} \quad \forall j = 1, 2, \ldots, s. \tag{12}$$

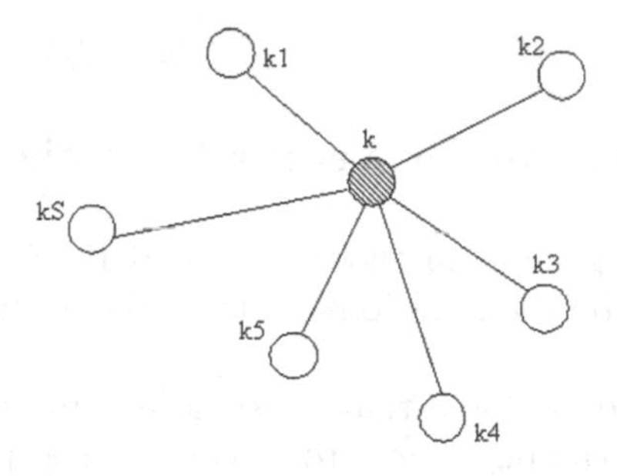

Fig. 2. The k-th node of the graph and its "neighborhood".

From (11) it follows that, for fixed values of the variables $p_{k_1}, p_{k_2}, \ldots, p_{k_s}$, the maximal admissible value $\overline{p}_k$ of the variable p_k can be calculated by the following rule

$$\overline{p}_k = \min\{p_{k_j} + c_{kk_j} \mid j = 1, 2, \ldots, s\}. \tag{13}$$

Thus, relations (11)–(13) reflect the following tendency. The increase of the value of one variable gives the possibility to increase the values of other variables. This tendency entirely corresponds to the purposes of solving problem (1)–(4).

These observations form the basis for constructing a numerical algorithm for calculating the vector $P^\star = (p_1^\star, p_2^\star, \ldots, p_n^\star)^\top$ that is the optimal plan for problem (1)–(4). The main idea consists in the iterative maximum increment of price values p_k $(k = 1, 2, \ldots, m)$ at free nodes of the graph from some admissible initial values until this increment is possible.

The following iterative numerical procedure for calculating the optimal plan $P^\star$ is suggested.

1. Let us fix the initial approximation $P^{(0)}$ for the plan $P^\star$ as

$$P^{(t)} = (0, 0, \ldots, 0, s_{m+1}, s_{m+2}, \ldots, s_n)^\top,$$

where $t = 0$. Obviously, $P^{(t)} \in U$.

2. The values of all components $p_k^{(t)}$ $(k = 1, 2, \ldots, m)$ of the current plan

$$P^{(t)} = (p_1^{(t)}, p_2^{(t)}, \ldots, p_m^{(t)}, s_{m+1}, s_{m+2}, \ldots, s_n)^\top$$

that correspond to the free nodes of the graph G are sequentially redefined by formula (13).

3. The next approximation $P^{(t+1)}$ for the plan $P^\star$ is defined as

$$\begin{aligned} P^{(t+1)} &= (p_1^{(t+1)}, p_2^{(t+1)}, \ldots, p_m^{(t+1)}, s_{m+1}, s_{m+2}, \ldots, s_n)^\top \\ &= (\overline{p}_1^{(t)}, \overline{p}_2^{(t)}, \ldots, \overline{p}_m^{(t)}, s_{m+1}, s_{m+2}, \ldots, s_n)^\top. \end{aligned}$$

4. The terminal condition for the calculations is checked:

$$\Delta p^{(t+1)} = \max\{p_i^{(t+1)} - p_i^{(t)} \mid i = 1, 2, \ldots, m\} \le \varepsilon. \tag{14}$$

Here, $\varepsilon \ge 0$ is an accuracy of the calculation of $P^\star$ defined in advance. If condition (14) is fulfilled, then we accept $P^\star \approx P^{(t+1)}$. Otherwise, the procedure of calculating the next approximation of the optimal plan $P^\star$ described above is repeated from the second step.

One can show, for example, using the method of mathematical induction by the number of free nodes in the graph G, that any required accuracy (condition (14)) can be reached in a finite number of iterations.

Note that iterative procedures are often used for solving problems in mathematical economics. Like in this research, they are applied to problems of pricing. Here, one can refer to the Walras "tatonnement" process [10,20]. It is a well-known iterative procedure in economic theory [1] for defining balance prices at a competitive market.

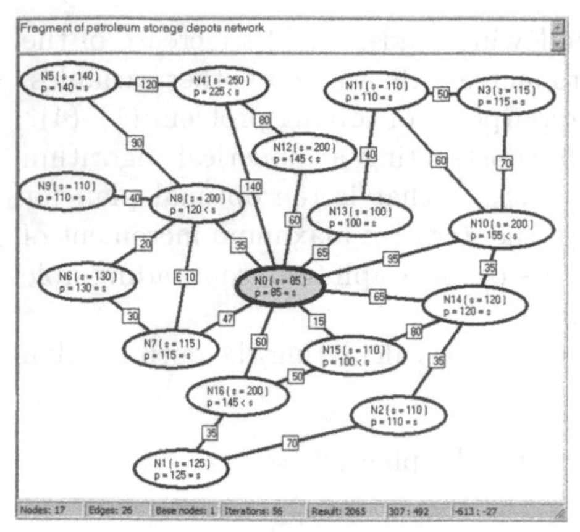

Fig. 3. Results of optimization of prices at petroleum storage depots under restrictions (15).

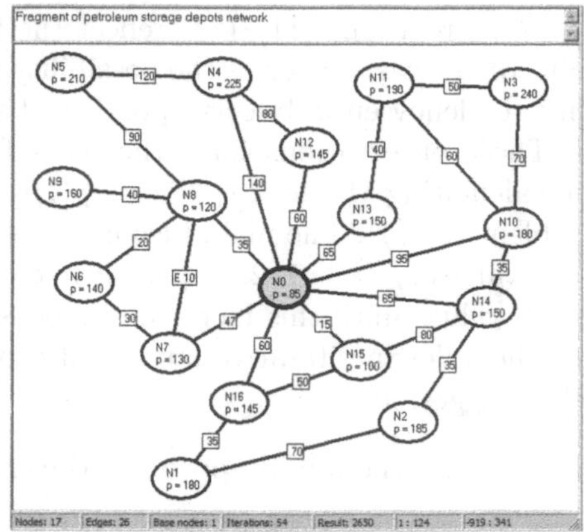

Fig. 4. Results of optimization of prices at petroleum storage depots without restrictions (15).

7 Results of Numerical Modeling

The intuitive examples of problem (1)–(4) are considered in [12]. Model data on local petroleum products markets in Sverdlovsk region are used in these examples. The results of the numerical solution of this problem given in [12] were obtained with taking into account additional restrictions, which simplify the problem, on admissible values of variables p_i $(i = 1, 2, \ldots, m)$ corresponded to free nodes of the graph G:

$$p_i \leq s_i < +\infty \quad i = 1, 2, \ldots, m. \tag{15}$$

In this case, in many free nodes of the considered graph the optimal (maximal) values of the variables p_i can reach the corresponding limit values s_i $(i = 1, 2, \ldots, m)$ specified by restrictions (15).

Figure 3 presents the optimization results with taking into account restrictions (15) for one of intuitive examples of problem (1)–(4) considered in [12]. In this example, the graph G consists of seventeen nodes connected by twenty six edges. One of these nodes is a base node. In Fig. 3, the base node $N0$ is highlighted by color and a thicker ellipse. As seen in Fig. 3, for ten of sixteen free nodes, the maximal values of prices coincide with the limit values given by restrictions (15) corresponded to these nodes. For the same intuitive example, the optimization results with the help of the suggested algorithm without taking into account restrictions (15) in problem (1)–(4) are presented in Fig. 4.

In both cases, the exact solutions of the problem were obtained. Here, it was supposed that $\varepsilon = 0$ in condition (14) for stopping the iterative procedure. Note that a fewer number of iterations was required to determine the optimal plan $P^\star$ for this problem without taking into account restrictions (15), see Figs. 3 and 4.

8 Conclusion

The results of computational experiment for model examples of problem (1)–(4) emphasize the effectiveness of the suggested algorithm for solving the vector optimization problem under consideration.

On the basis of the mathematical model of the intuitive economic problem in question, the statements of other applied management problems (for example, the dispatching problem) are possible.

References

1. Arrow, K., Debreu, G.: Existence of an eguilibrium for a completive economy. Econometrica **22**, 256–290 (1954)
2. Asanov, M.: Discrete Optimization. UralSCIENCE, Ekaterinburg (1988)
3. Charnes, A., Cooper, W.: Management Models and Industrial Applications of Linear Programming, vol. 1. Wiley, New York (1961)
4. Doumpos, M., Zopounidis, C.: Multicriteria Analysis in Finance. SOR. Springer, Cham (2014). https://doi.org/10.1007/978-3-319-05864-1
5. Ehrgott, M., Puerto, J., Rodríguez-Chía, A.M.: Primal-dual simplex method for multiobjective linear programming. J. Optim. Theory Appl. **134**(3), 483–497 (2007). https://doi.org/10.1007/s10957-007-9232-y
6. Ehrgott, M.: Multicriteria Optimization. Springer, Heidelberg (2005). https://doi.org/10.1007/3-540-27659-9
7. Evans, J., Steuer, R.: A revised simplex method for multiple objective programs. Math. Program. **5**(1), 54–72 (1973)
8. Ginchev, I., Guerraggio, A., Rocca, M.: From scalar to vector optimization. Appl. Math. **51**(1), 5–36 (2006). https://doi.org/10.1007/s10492-006-0002-1
9. Hamel, A., Lohne, A., Rudloff, B.: Benson type algorithms for linear vector optimization and applications. J. Optim. Theory Appl. **59**(4), 811–836 (2007)
10. Intriligator, M.: Mathematical Optimization and Economic Theory. Prentice-Hall Inc., Englewood Cliffs (1971)
11. Johannes, J.: Vector Optimization, Theory, Applications, and Extensions. Springer-Verlag, Heidelberg (2011). https://doi.org/10.1007/978-3-642-17005-8
12. Kandoba, I., Kotelnik, K., Uspenskii, A.: On one problem of vector optimization of pricing in an autonomous wholesale market, pp. 1–21. Interim report IR-04-026, IIASA, May 2004. http://pure.iiasa.ac.at/id/eprint/7420/1/IR-04-026.pdf
13. Koopmans, T.: Analysis of production as an efficient combination of activities. In. Activity Analysis of Production and Allocation, pp. 33–97 (1951)
14. Lohne, A.: Vector Optimization with Infimum and Supremum. Vector Optimization. Springer Science+Business Media, Heidelberg (2011). https://doi.org/10.1007/978-3-642-18351-5
15. Nogin, V.: Multi-criteria decision making. UTAS, S. Peterburg (2007)
16. Rudloff, B., Ulus, F., Vanderbei, R.: A parametric simplex algorithm for linear vector optimization problems. Math. Program. **163**, 213–242 (2017)
17. Ruzika, S., Wiecek, M.M.: Approximation methods in multiobjective programming. J. Optim. Theory Appl. **126**(3), 473–501 (2005)
18. Steuer, R.: Multiple Criteria Optimization: Theory, Computation and Application. Wiley, New York (1986)

19. Vasilyev, F.: Metody optimizacii [Optimization Methods]. Faktorial Press, Moscow (2002)
20. Walras, L., Jaffe, W.: Elements of Pure Economics, or The Theory of Social Wealth. Richard D. Irwin Inc., Homewood (1954). American Economic Association and the Royal Economic Society
21. Wierzbicki, A.: Reference point methods in vector optimization and decision support, pp. 1–40. Interim report IR-98-017, IIASA, April 1998. http://pure.iiasa.ac.at/id/eprint/5631/1/IR-98-017.pdf

Net Present Value Maximization in Inventory Management System

Svetlana A. Malakh(✉) and Vladimir V. Servakh(✉)

Sobolev Institute of Mathematics, Novosibirsk, Russia
malahsveta@mail.ru, svv_usa@rambler.ru

Abstract. The paper researches the model of profit maximization for a commercial company, taking into account the intensity of the sale of goods, the cost of purchase, the cost of delivery, the cost of storage and the cost of sale of goods. The alternative investments of available capital are also taken into account. It is shown that the profit function, depending on the period of delivery of goods, has a single maximum point. A model of the problem of the profit maximization in multi-product systems with limited working capital has been built and an algorithm for solving it has been developed.

Keywords: Logistics · NP-hardness · Pseudo-polymonial algorithm

1 Introduction

Let us consider the problem of the activity optimization of a trading company that buys goods on a commodity exchange and retails them. The company has a modern warehouse. It is equipped with a warehouse management system. The latest technology and efficient organization of the warehouse allowed the company to reduce the costs substantially. Classical models based on cost minimization, such as [1–5], do not always reflect the situation adequately. The high mobility of the economy, great competition and the introduction of modern logistics systems have created the need to develop new models. These models are aimed at the efficient use of capital and receiving a better return on their investments [6–9]. It is especially important when optimizing the procurement of the goods for which the ratio of volume/price is a small value. They are, for example, radio components, medicines, etc [10, 11]. Therefore, the rate of capital circulation is greater importance for the company. Delivery and storage costs are becoming less important. Using such models gives the company a good chance to use automation to optimize procurement processes.

There is another problem: the limitation of the working capital of a company. The automated system determines the optimal volume of procurement of goods. Sometimes it happens that the total value of the requested goods may be greater than the available money to invest. In this situation, a company has to either reduce the order or take a loan. The loan changes the value of money. Therefore,

I. Bykadorov et al. (Eds.): MOTOR 2019, CCIS 1090, pp. 381–389, 2019.
https://doi.org/10.1007/978-3-030-33394-2_30

the optimal volume of procurement should be changed. Order reduction also leads to additional costs: additional delivery costs and lost profits due to unmet demand. A company needs to solve the problem of adjusting a multi-product order taking into account the local limit on working capital.

In paragraph 2, we consider the model of inventory management, which takes into account the alternative use of capital. Paragraph 3 describes the model of the problem of optimizing current procurement in multi-product systems with limited working capital and with the possibility of using loans. An algorithm for solving this problem has been developed.

2 Task of Maximizing Net Present Value

This paragraph discusses the classical model of inventory management and its development, taking into account the factors of the modern economy. The focus is on the efficient use of capital when there are alternative opportunities for its use. We will consider the model under the assumption that there is a guaranteed possibility of an alternative risk-free capital allocation at the interest rate of r_0. Then money at different points in time will have different value. To compare receipts received in different time, the following approach is used. If at the moment t_1 some capital K_1 is available, then it can always be placed on the market under the current market interest rate of r_0. With this placement by the time t_2 capital will increase to $K_1(1+r_0)^{t_2-t_1}$. And vice versa, if at the moment t_2 capital K_2 is needed, then at the moment t_1 it is enough to have it in quantity $K_2/(1+r_0)^{t_2-t_1}$. Therefore, to compare the capital of K_1 in time point t_1 and K_2 at time point t_2, it is necessary to compare the values of K_1 and $K_2/(1+r_0)^{t_2-t_1}$. This operation is called the coercion operation by the time point t_1 or discounting.

The process of purchasing and selling goods will be considered as in the classical model: we buy goods in the amount of v at the price of β and sell it with the intensity of λ at the price of c. Delivery costs are defined by the function $\alpha + \beta v$, where α are the fixed costs including the cost of purchase and product delivery. The cost of storing a unit of goods per unit of time is denoted by c_{xp}.

The implementation time of the purchased goods will be $T = \frac{v}{\lambda}$. The intensity of the proceeds from the sale is $c\lambda$. Taking discounting into account, the intensity of the cash flow is expressed by the function $\frac{c\lambda}{(1+r_0)^t}$. For the period of the sale of goods $[0, T]$ total revenues discounted to the initial point in time will be equal to

$$Q(T) = \int_0^T \frac{c\lambda}{(1+r_0)^t} dt = \frac{c\lambda}{ln(1+r_0)} \cdot \frac{(1+r_0)^T - 1}{(1+r_0)^T},$$

and the cost of storing the goods, taking into account discounting, is

$$Z(T) = \int_0^T \frac{(T\lambda - t\lambda)c_{xp}}{(1+r_0)^t} dt = \frac{c_{xp}\lambda}{ln^2(1+r_0)} \cdot \left(T \cdot ln(1+r_0) + \frac{1}{(1+r_0)^T} - 1\right).$$

Thus, for the specified period $[0,T]$ net present value (NPV) will be received in the amount of $NPV = Q(T) - (\alpha + \beta v) - Z(T)$. The task requires to maximize profit per unit, which is expressed by the following function:

$$U(T) = \frac{Q(T) - (\alpha + \beta T\lambda) - Z(T)}{T} \rightarrow max.$$

The Fig. 1 shows an example of the function $U(T)$ depending from the period of delivery of goods with the following input data $\lambda = 1$, $\alpha = 2$, $\beta = 0,25$, $c = 1$, $r_0 = 0,1$, $c_{xp} = 0,03$. The optimal value of the specific net profit is $0,109$ and is reached at $T = 7,02$.

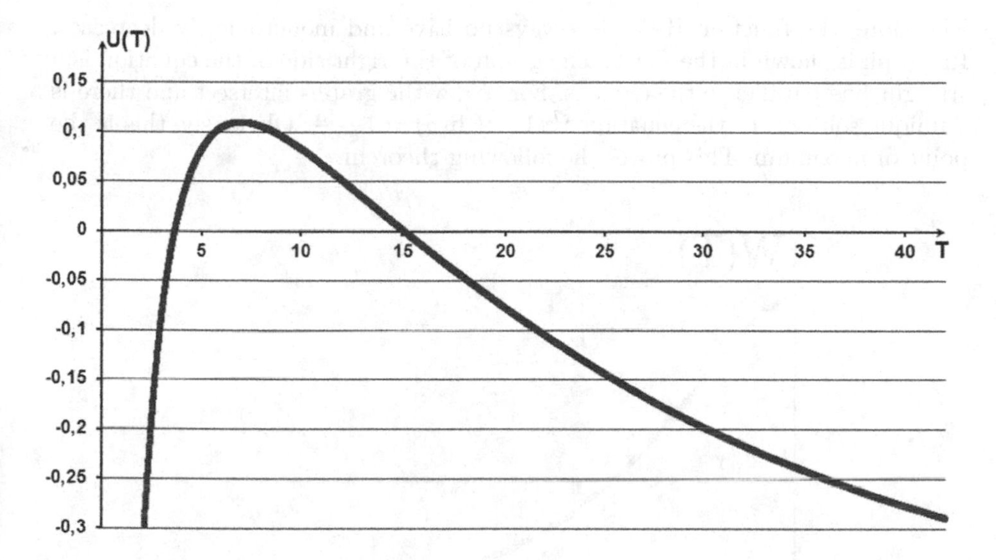

Fig. 1. The dependence of the specific profit from T

Let us investigate the extremal values of this function $U(T)$. Denote

$$\gamma = \frac{c\lambda}{ln(1+r_0)} + \frac{c_{xp}\lambda}{ln^2(1+r_0)} > 0, \qquad \delta = \frac{1}{1+r_0}, \quad 0 < \delta < 1.$$

Then

$$U(T) = \frac{1}{T}(\gamma(1-\delta^T) - \alpha - \beta T\lambda) - const,$$

where $const = \frac{c_{xp}\lambda}{ln^2(1+r_0)}$.

$$\lim_{T\to+0} U(T) = -\infty, \qquad \lim_{T\to+\infty} U(T) = -\beta\lambda.$$

Let us investigate the zeros of the derivative

$$U'(T) = \frac{-\gamma\delta^T \ln\delta T - \gamma(1-\delta^T) + \alpha}{T^2} = 0.$$

We get the following equation

$$\delta^T(1 - T\ln\delta) = 1 - \frac{\alpha}{\gamma}.$$

We will solve this graphically. Denote $W(T) = \delta^T(1 - T\ln\delta)$.

$$W(0) = \delta^0 = 1;$$

$$W'(T) = \delta^T\ln\delta - \delta^T T\ln^2\delta - \delta^T\ln\delta = -\delta^T T\ln^2\delta < 0;$$

$$\lim_{T\to\infty} W(T) = 0.$$

Therefore, the function $W(T)$ is always positive and monotonically decreases. Its graph is shown in the Fig. 2. The graph of the right side of the equation is a straight line parallel to the Ox axis. For $\alpha < \gamma$ the graphs intersect and there is a unique solution to the equation $\delta^T(1 - T\ln\delta) = 1 - \frac{\alpha}{\gamma}$. Obviously, this is the point of maximum. This proves the following theorem.

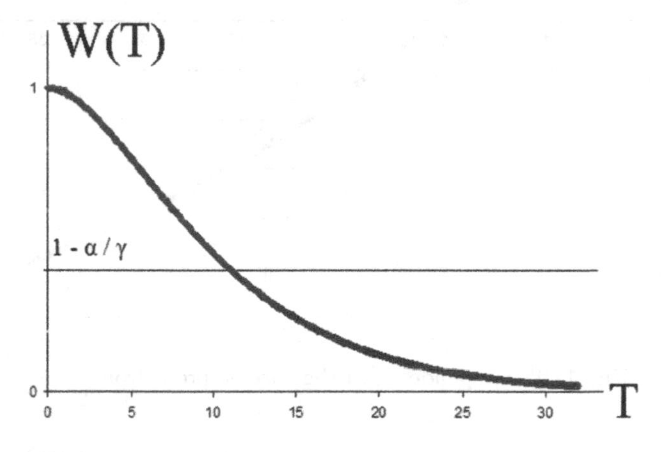

Fig. 2. Changing the function $W(T)$ depending from T

Theorem. *For* $\alpha < \frac{c\lambda}{ln(1+r_0)} + \frac{c_{xp}\lambda}{ln^2(1+r_0)}$ *the profit function has a single point of maximum. For* $\alpha \geq \frac{c\lambda}{ln(1+r_0)} + \frac{c_{xp}\lambda}{ln^2(1+r_0)}$ *the profit function monotonously increases on all the interval* $(0, \infty)$, *and the profit value is always negative and less than* $-\beta\lambda - \frac{c\lambda}{\ln(1+r_0)}$.

Thus, under certain conditions, there is a period of delivery of products in which the money invested is used as efficiently as possible, bringing the greatest profit. The point of maximum is found as the single solution to the equation $\delta^T(1 + T\ln\delta) = 1 - \frac{\alpha}{\gamma}$.

3 Problem with Working Capital Limits

The approach described above is the mathematical basis of the models implemented in practice in chain trading companies. In multi-product systems, it is determined which goods have to be imported and in what amounts, taking into account current demand, purchase and sale prices, delivery and storage costs. The criterion is to maximize the profit margins. This model allows taking into account various additional restrictions on the insurance stocks of goods, time lags, and so on. The dates and volumes of orders are calculated automatically by the robot program. The company's employees do not participate in the process of calculation.

Difficulties can arise when, during the order formation, the total price of the ordered goods may be higher than the available money for investment. First, the value of a purchase order fluctuates, as different types of goods are imported at different points in time. Secondly, the amount of working capital may be reduced. This may happen in the period of payment of taxes or because of the immobilization of current assets. In this situation, you have to either reduce the order or take a loan. Reducing the order leads to additional costs. A loan reduces profits due to interest payments. In the task presented below, we have built a model and developed an algorithm that makes it possible to minimize costs if the amount of working capital is limited.

Let us describe the task parameters:

N – the number of types of goods;
α_i – the cost of one consignment of goods i;
β_i – the purchase price per unit of output i;
$\alpha_i + \beta_i v_i$ – the cost of ordering and consignment of goods i with a volume of v_i;
λ_i – the intensity of sale of goods i;
c_i – sale price per unit i;
c_i^{xp} – the cost of storing a unit of good i per unit of time;
r_0 – the rate of alternative risk-free liquid capital allocation.

At the current time, for each item i with zero balance in stock, it is required to find a delivery period of T_i, when the profit margins

$$U_i(T_i) = \frac{1}{T_i}\left(\int_0^{T_i} \frac{c_i\lambda_i}{(1+r_0)^t}dt - \alpha_i - \beta_i T_i\lambda_i - \int_0^{T_i} \frac{(T_i\lambda_i - t\lambda_i)c_i^{xp}}{(1+r_0)^t}dt\right)$$

are maximum.

We make some natural assumptions that do not affect the generality of the model under consideration. Orders cannot be made at any time. The discreteness of planning in time is natural for economic tasks. If you select the appropriate unit of measurement, you can consider only integer moments. We will not take into account temporary lags during shipment and delivery of goods since this does not affect the essence of the task. Therefore, we will assume that the payment and delivery of goods are carried out when its balance is equal to zero.

Under these conditions, it suffices to consider only integer values of T_i. Indeed, if T_i is not an integer, once $\lfloor T_i \rfloor$ you will have to make another delivery, since the remaining goods $(T_i - \lfloor T_i \rfloor)\lambda_i$ are not enough until the next delivery. The integer optimum $U_i(T_i)$ is denoted by T_i^*. The corresponding amount of imported goods equal to $v_i^* = T_i^* \lambda_i$.

The next order is being formed at the current moment $t \in Z^+$. We are buying only the products that are running out at this point and there is a non-zero demand. For the convenience of presentation, we number them with numbers $1, 2, \dots, n$. From the database we take the current parameters for these positions, then for each product $i = 1, 2, \dots, n$ we find the integer optima T_i^* and the corresponding value of the delivered volumes $v_i^* = T_i^* \lambda_i$. To purchase them you need

$$\sum_{i=1}^{n} (\alpha_i + \beta_i v_i^*)$$

units of capital. Where the necessary financial resources are available, the order is paid and the goods are delivered.

3.1 Problem with Order Reduction

Problems arise when the current capital K is not enough to pay the whole order. In practice, three options are possible: order reduction, bank short-term loan, commodity loan. The last two options differ only in the way they are organized. As for the maximization of NPV, there is no difference between them, a part of the income has to be given. We first look at the problem of reducing the order, and then we generalize it in case we need to use loans.

We introduce the variable $x_i \le v_i^*$ – the order quantity of the product $i = 1, 2, \dots, n$. Order reduction to the level of x_i leads to a loss of profit in the amount of $H_i(x_i) = U_i\left(\frac{v_i^*}{\lambda_i}\right) - U_i\left(\frac{x_i}{\lambda_i}\right)$. To build an algorithm for finding the optimal solution of the problem, the following statement is important.

Statement 1. On the interval $x_i \in (0, v_i^*)$, the function $H_i(x_i)$ decreases monotonically.

The proof follows from the type of the function $U(T)$ and its investigation in Sect. 2.

Statement 2. Variables x_i can take only values from the set $\{\lambda_i, 2\lambda_i, \dots, v_i^*\}$.

The arguments are the same as at the end of the previous paragraph. If x_i is not a multiple of λ_i, then at some moment the remainder will be $x_i - \lfloor \frac{x_i}{\lambda_i} \rfloor \lambda_i$ and it will not be enough until the next delivery. By removing this residue, the objective function can be improved. Also, x_i cannot be equal to zero, since demand must be satisfied at least in the amount of λ_i to the nearest delivery. Thus, if $\sum_{i=1}^{n} (\alpha_i + \beta_i v_i^*) > K_0$, then you need to solve the problem of reducing the application to minimize costs. We get the following model:

$$\sum_{i=1}^{n} H_i(x_i) \to \min,$$

$$\sum_{i=1}^{n} (\alpha_i + \beta_i x_i) \leq K_0,$$

$$x_i \in \{\lambda_i, 2\lambda_i, 3\lambda_i, \ldots, v_i^*\}, i = 1, 2, \ldots, n.$$

Note that in this model $K_0 \geq \sum_{i=1}^{n} (\alpha_i + \beta_i \lambda_i)$, since minimum demand must be satisfied before the next delivery. The value $\sum_{i=1}^{n} (\alpha_i + \beta_i \lambda_i)$ is denoted by K_{min}.

A dynamic programming scheme is used to solve this problem. Denote by $\varphi(m, k)$ the optimal value of the objective function for the current capital k and the subset of products $\{1, 2, \ldots, m\}$, where $m = 1, 2, \ldots, n$, $k = K_{min}, \ldots, K$. Purchase of m in the amount of x_m is allowed if $x_m \in \{\lambda_m, 2\lambda_m, 3\lambda_m, \ldots, v_m^*\}$ and $\sum_{i=1}^{m-1} (\alpha_i + \beta_i \lambda_i) + \alpha_m + \beta_m x_m \leq k$. The set of admissible values of x_m is denoted by $D(m, k)$.

We look through all the values of $x_m \in D(m, k)$ to calculate $\varphi(m, k)$. If the m product is purchased in the amount of x_m, then we get a subtask:

$$\sum_{i=1}^{m-1} H_i(x_i) - H_m(x_m) \to \min,$$

$$\sum_{i=1}^{m-1} (\alpha_i + \beta_i x_i) \leq k - (\alpha_m + \beta_m x_m),$$

$$x_i \in \{\lambda_i, 2\lambda_i, 3\lambda_i, \ldots, v_i^*\}, i = 1, 2, \ldots, m - 1.$$

The optimal solution to this problem is $\varphi(m-1, k-\alpha_m-\beta_m x_m)$. The optimal solution to this problem is

$$\varphi(m, k) = \min_{x_m \in D(m,k)} \{H_m(x_m) + \varphi(m - 1, k - \alpha_m - \beta_m x_m)\}.$$

To find $\varphi(n, K)$ this recurrence relation must be implemented in a double cycle $m = 2, 3, \ldots, n$, $k = K_{min}, \ldots, K$, with initial conditions $\varphi(1, k) = H_m(x_1)$, where $x_1 = \lfloor \frac{k-\alpha_1}{\beta_1 \lambda_1} \rfloor \lambda_1$. Restoration of the optimal solution is carried out by the reverse course of the standard scheme. Note that the implementation of the algorithm requires the integer values of $\alpha_m + \beta_m x_m$, which can be ensured by the selection of units of capital measurement. The complexity of the algorithm depends pseudo-polynomially on the length of the input data record and amounts to $O(K \cdot \sum_{m=1}^{n} T_m^*)$ operations, where K is the available capital in the selected units, T_m^* – the optimal period of delivery of goods m.

3.2 Forming Orders Using Credits

Let us generalize the written model in case of using credits. The loan rate of r is known. The variable D will denote the size of the loan. As before, $K_{\min} = \sum_{i=1}^{n} (\alpha_i + \beta_i \lambda_i)$ is the minimum required amount of working capital. We also introduce the value $K_{\max} = \sum_{i=1}^{n} (\alpha_i + \beta_i \lambda_i T_i^*)$ – the amount of funds sufficient to fully support the application.

In case the minimum required the amount of working capital $K_{\min}$ is more than cash capital K, the loan amount cannot be less than the value of $K_{\min} - K$. The maximum loan value D does not exceed $K_{\max} - K$. The costs of loan reduced to the current point in time will be $\frac{D(1+r)}{1+r_0} - D$. Size available capital will be equal to $K + D$. We obtain the following model:

$$\sum_{i=1}^{n} H_i(x_i) + \frac{D(1+r)}{1+r_0} - D \rightarrow \min,$$

$$\sum_{i=1}^{n} (\alpha_i + \beta_i x_i) \le K + D,$$

$$\max\{0, K_{\min} - K\} \le D \le K_{\max} - K,$$

$$x_i \in \{\lambda_i, 2\lambda_i, 3\lambda_i, \ldots, v_i^*\}.$$

If the variable D is fixed the value of $\frac{D(1+r)}{1+r_0} - D$ will be constant and we will get the problem described in the previous section with the initial capital $K + D$. We solve it by the algorithm described there and find all $\varphi(m, k)$ for $m = 1, 2, \ldots, n$ and $k = K_{min}, \ldots, K_{max}$. After that, it remains to go through all the integer values of $D \in [\max\{0, K_{min} - K\}, K_{max} - K]$ and find the minimum

$$\min_{D \in [\max\{0, K_{min} - K\}, K_{max} - K]} \left\{ \varphi(n, K + D) - D + \frac{D(1+r)}{1+r_0} \right\}.$$

The complexity of the algorithm is $O(K_{\max} \cdot \sum_{i=1}^{n} T_i^*)$ operations. Calculations on real data allow us to show the relevance of the model with credits. Often the optimum for D is reached within the interval $[\max\{0, K_{min} - K\}, K_{max} - K]$, that is, the application is still reduced, but its part is paid for credit.

4 Conclusion

The paper investigates the problem of maximizing profits, taking into account the alternative use of capital. The theorem on the uniqueness of the maximum point of the profit function is proved. A model of the task of generating a profit maximization application has been built, taking into account the limitation of working capital, its properties have been identified and justified, an algorithm for solving the problem based on a dynamic programming scheme has been proposed and implemented. An algorithm for solving the problem of forming an application with the possibility of using loans is proposed and implemented. The models and algorithms were tested when solving the problem of bidding for the purchase and sale of goods for large wholesale and retail trading company in the pharmaceutical market.

References

1. Wilson, R.H.: A scientific routine for stock control. Harvard Bus. Rev. **13**, 116–128 (1934)
2. Bukan, J., Koenigsberg, E.: Scientific inventory management. Science, Moscow (1967)
3. Pervozvansky, A.A.: Mathematical models in production management. The main editors of the physical and mathematical literature of the publishing house. Science, Moscow (1975)
4. Ryzhikov, Y.I.: Queue theory and inventory management. Peter, St. Petersburg (2001)
5. Hadley, J., Whitin, T.: Analysis of inventory management systems. Science, Moscow (1969)
6. Brodetsky, G.L.: Inventory Management. Eksmo, Moscow (2007)
7. Brodetsky, G.L.: Optimization models of inventory management systems taking into account the time value of money with restrictions on the size of capital. Logistics Supply Chain Manag. **2**, 70–88 (2007)
8. Brodetsky, G.L.: Multi-inventory management: a new approach to optimizing solutions. Logistics Today **1**, 34–45 (2014)
9. Burlakova, N.I., Servakh, V.V.: Optimization of multi-product purchases with limited working capital. In: Abstracts of the XV All-Russian Conference "Mathematical Programming and Applications", Ekaterinburg, p. 74 (2015)
10. Burlakova, N.I., Servakh, V.V.: Maximizing net present profit in the problem of inventory management. In: Proceedings Scientific Papers of the VIII International School-Symposium Analysis, Management, Modeling, Development, Simferopol, pp. 61–62 (2014)
11. Burlakova, N.I., Polantseva, I.A., Servakh, V.V.: Procurement optimization with possibility of alternative use of capital. In: Abstracts of the XVI Baikal International School-Seminar "Optimization Methods and Their Applications", Irkutsk, ISEM SB RAS, p. 32 (2014)

Constructive Heuristics for Min-Power Bounded-Hops Symmetric Connectivity Problem

Roman Plotnikov[1(✉)] and Adil Erzin[1,2]

[1] Sobolev Institute of Mathematics, Novosibirsk, Russia
prv@math.nsc.ru
[2] Novosibirsk State University, Novosibirsk, Russia

Abstract. We consider a problem of constructing an energy-efficient bounded diameter communication spanning tree when the vertices are located on a plane, and the energy required to transmit a message between a pair of vertices is proportional to the squared distance between them. For this NP-hard problem, we have developed several approximate heuristic algorithms. The results of a posteriori analysis of solutions constructed by the proposed algorithms are presented.

Keywords: Energy efficiency · Approximation algorithms · Symmetric connectivity · Bounded hops

1 Introduction

Due to the prevalence of wireless sensor networks (WSNs) in human life, the different optimization problems aimed to increase their efficiency remain actual. Since usually WSN consists of elements with the non-renewable power supply, one of the most important issues related to the design of WSN is prolongation of its lifetime by minimizing the energy consumption of its elements. A significant amount of sensor's energy is consumed on communication. Therefore, modern sensors often can adjust their transmission ranges changing the transmitter's power. Herewith, usually, the energy consumption of a network's element is assumed to be proportional to d^s, where $s \geq 2$ and d is the transmission range [1].

The problem of the optimal power assignment in WSN is well-studied. The most general problem is the Range Assignment Problem where the goal is to find a strongly connected subgraph in a given directed graph has been considered in [2,3]. Its subproblem, the Minimum Power Symmetric Connectivity Problem (MPSCP), was first studied in [4]. The authors proved that the Minimum Spanning Tree (MST) is a 2-approximate solution to this problem. Also, they

The research of R. Plotnikov is supported by the Russian Science Foundation (project 18-71-00084, Sections 3, 4, and 5). The research of A. Erzin is supported by the Russian Science Foundation (project 19-71-10012, Sections 1 and 2).

I. Bykadorov et al. (Eds.): MOTOR 2019, CCIS 1090, pp. 390–407, 2019.
https://doi.org/10.1007/978-3-030-33394-2_31

proposed a polynomial-time approximation scheme with a performance ratio $1 + \ln 2 + \varepsilon \approx 1.69$ and a 15/8-approximation polynomial algorithm. In [5] a greedy heuristic Incremental Power: Prim (IPP) was proposed. IPP is similar to the Prim's algorithm for MST constructing. A Kruscal-like heuristic Incremental Power: Kruscal was studied in [6]. Both of these incremental power heuristics have been proposed for the Minimum Power Asymmetric Broadcast Problem, but they are suitable for the MPSCP too. It is proved in [7] that these algorithms have an approximation ratio 2, and it was shown in the same paper that in practice they yield a significantly more accurate solution than MST. Also, in a series of papers different heuristic algorithms have been proposed for the MPSCP and the experimental studies have been done: local search procedures [7–9], methods based on iterative local search [10], hybrid genetic algorithm that uses a variable neighborhood descent as mutation [11], variable neighborhood search [12], and variable neighborhood decomposition search [13].

Another measure of WSN's efficiency is a transmission delay, the required time for transmitting a message from one sensor to another. As a rule, the delay is proportional to the number of hops (edges) between two nodes of a network. In the general case, when the network is represented as a directed arc-weighted graph and the goal is to find a strongly connected subgraph with minimum total power consumptions and bounded path length, the problem is called the Min-Power Bounded-Hops Strong Connectivity Problem. In [3] a special case of this problem, when sensors are spread equidistantly on the line, was considered. In [14] the approximation algorithms with guaranteed estimates have been proposed for the Euclidean case of this problem. The bi-criteria approximation algorithm for the general case (not necessarily Euclidian) has been proposed in [15]. The authors of [16] propose improved constant factor approximation for the planar Euclidian case of the problem.

In this paper, we consider the symmetric case of the Min-Power Bounded-Hops Strong Connectivity Problem, when the network is represented as an undirected edge-weighted graph. Such a problem is known as the Min-Power Bounded-Hops Symmetric Connectivity Problem (MPBHSCP) [15]. We also assume that sensors are positioned on the Euclidian plane. Energy consumption for the data transmission from one node to another is assumed to be proportional to the squared distance between them. This problem is still NP-hard [17], and, therefore, the approximation heuristic algorithms that construct the near-optimal solution in a short time, are required for it.

Although the MPBHSCP is known to be NP-hard, to the best of our knowledge, none research has been done to find the efficient in practice approximation algorithms. This paper is aimed to fill this gap. We propose six different constructive heuristics for the approximation solution of the MPBHSCP. We employ the ideas of the most natural and widely spread heuristics for the Bounded-Diameter Minimum Spanning Tree (BDMST). We conducted an extensive numerical experiment where these algorithms have been compared. In small size cases, we compared our algorithms with CPLEX, that was run with our mixed integer linear programming (MILP) formulation, which is also proposed in this paper.

The rest of the paper is organized as follows. In Sect. 2 the problem is formulated, in Sect. 3 descriptions of the proposed algorithms are given, Sect. 4 contains results and analysis of an experimental study, and Sect. 5 concludes the paper.

2 Problem Formulation

Mathematically, the MPBHSCP can be formulated as follows. Given a connected edge-weighted undirected graph $G = (V, E)$ and an integer $D \geq 2$, find such spanning tree T^* in G, which is the solution to the following problem:

$$W(T) = \sum_{i \in V} \max_{j \in V_i(T)} c_{ij} \to \min_T,$$

$$dist_T(u, v) \leq D \; \forall u, v \in V,$$

where $V_i(T)$ is the set of vertices adjacent to the vertex i in the tree T, $c_{ij} \geq 0$ is the weight of the edge $(i, j) \in E$ and $dist_T(u, v)$ is the number of edges in a path between the vertices $u \in V$ and $v \in V$ in T.

Obviously, this problem may not have any feasible solution. In this paper, we consider a planar Euclidian case, where an edge weight equals the squared distance between the corresponding points, and G is a complete graph. Therefore, a solution always exists. Also, we assume that the sensors are randomly uniformly distributed on a square with fixed side. Therefore, the density of a network grows with increase of the number of its elements.

3 Heuristic Algorithms

We propose several heuristic algorithms that construct an approximate solution to the MPBHSCP. Many of these algorithms use ideas that previously have been applied to the solution of the BDMST. As well as it is done in many efficient heuristic algorithms for the BDMST, we use a *center-based* approach, where the center (one vertex if D is even or two vertices if it is odd) is chosen, and after that, a tree is constructed taking care of the depth of each vertex (the number of edges in the path from the center). The main difference between the algorithms applied to the BDMST and our methods is a calculation of the objective function increment after the modifications of a partial solution. An objective function of the MST problem is additive, that is, adding (or removing) an edge increases (or decreases) the objective value exactly by the weight of an edge, which is not held for the objective function of the MPSCP: if one wants to calculate the change of an objective function value for the MPSCP after adding or removing an edge, then he has to take into account the weights of all adjacent edges of a tree.

Let us define the notations that will be used further. For convenience purposes, we will construct a directed tree rooted in a center. If the center contains two vertices then one of them will be referred to as a root, and the second one—as its child. Let's call the number of edges between a vertex and center in a tree the *depth* of a vertex and call the maximum depth in a tree the *height* of a tree. Let $V_T \subset V$ be the set of vertices of T, and let E_T be the set of edges of T. Let $Par_T(v) \in V_T$ be the parent vertex of $v \in V_T$. If $v \notin V_T$ then $Par_T(v) = \emptyset$. Let $depth_T(v)$ be the depth of the vertex v in T. If $v \notin V_T$ then $depth_T(v) = -1$. Let $Pow(v, u) = Pow(u, v)$ be the power consumption of the direct communication between the vertices u and v. As it was mentioned before, we assume that $Pow(u, v)$ is a squared Euclidian distance between the vertices u and v. Of course, these values may be calculated in advance since the positions of the nodes are known. Let $Pow_T(v)$ be the total power consumption for the communication of the vertex v in T. Let $N_T(v) \subset V_T\{v\}$ be the set of neighbors of $v \in V_T$ in T. Then, $Pow_T(v) = \max_{u \in N_T(v)} Pow(u, v)$, and the total power consumption of T (the objective function), is $W(T) = \sum_{v \in V_T} Pow_T(v)$.

3.1 Prim-Like Heuristics

Many of known greedy approaches for the BDMST use the Prim's strategy [18] for tree building. Starting from a tree with one vertex, these algorithms repeatedly add a new edge that connects a non-tree vertex with a vertex in a tree satisfying the requirement on the diameter. Herewith, criteria of choosing a new non-tree vertex may vary while the in-tree vertex is always chosen greedily. A way of choosing the center vertex (or two center vertices in a case of odd value of D), which is rather essential, may vary too. The general scheme of the Prim-Like Heuristic (PLH) is presented in Appendix A, Algorithm 1. Below we consider three different heuristics that are based on the Prim's strategy: Min-Power Center-Based Tree Construction, Min-Power Randomized Tree Construction, and Min-Power Center-Based Least Sum-of-Costs. The difference between these algorithms lies in the different implementations of the methods $ChooseFirstCenters$, $ChooseSecondCenter$, and $ChooseEachVertex$.

Min-Power Center-Based Tree Construction. The first algorithm based on the PLH is Min-Power Center-Based Tree Construction (MPCBTC) which is similar to the Center-Based Tree Construction (CBTC) [19] for the BDMST. In this algorithm, the procedure $ChooseFirstCenters$ chooses each vertex, that is, the algorithm starts n times with each vertex selected as a center. The method $ChooseSecondCenter(v_0)$ returns the vertex $v_1 = \mathsf{argmin}_{v \in V \setminus \{v_0\}} Pow_T(v, v_0)$. And, finally, the method $ChooseEachVertex(U, V_0, wBestNeighbor)$ finds such vertex $u \in U$ that $wBestNeighbor(u)$ (the minimum contribution to the objective function of addition an arc between u and its neighbor to the tree) is minimum. CBTC is known to perform worse with a decrease of the maximum number of hops and increase of the density of the points, since the nodes that lie far from the center (let's call them *far nodes*) often have the maximum allowable depth and, therefore, once added, they cannot be connected with any other node. So, far nodes cannot be connected with any node in their proximity without violating the restriction on the number of hops, and they are forced to be connected with a tree by long arcs. Obviously, in MPCBTC, as well as in CBTC, the closest to the center nodes are added sooner, and in a case of large density and small D MPCBTC will have the same disadvantage as CBTC: far nodes will be connected with a tree using long edges. Due to this fact solution obtained by MPCBTC should appear extremely inefficient for the cases when n is large and D is small. The time complexity of MPCBTC is $O(n^3)$ since it is repeated n times for each vertex chosen as a center, and each iteration has an $O(n^2)$ time complexity.

Min-Power Randomized Tree Construction. One simple approach aimed to overcome the mentioned disadvantage of the CBTC is the Randomized Tree Construction (RTC) proposed in [19]. As well as CBTC, RTC chooses a center vertex (or two center vertices if D is odd), then it iteratively selects a vertex outside a tree and connects it with some vertex in a tree. But, in contrast to MPCBTC, each time the vertex is taken at random. The process is repeated n times, and the best tree is returned. We adapted this algorithm to the MPBHSCP. Let's denote our heuristic as Min-Power Randomized Tree Construction (MPRTC). Since this algorithm is also based on the PLH, the only procedures that should be mentioned are the special implementations of the subroutines $ChooseFirstCenters$, $ChooseSecondCenter$, and $ChooseEachVertex$, which are extremely simple in this case: the method $ChooseFirstCenters$ n times chooses a vertex $v \in V$ at random, as well as it is done in RTC [19]. The both methods $ChooseSecondCenter$ and $ChooseEachVertex$ take a vertex $v \in U$ at random (where U is a set of non-tree vertices, see Appendix A, Algorithm 1). This circumstance theoretically should cause better results of MPRTC comparing with MPCBTC on high-dense graphs constructed on uniformly spread set of points, because on each step of MPRTC the constructed partial solution consists of a random subset of V. Because of the fact that MPRTC is repeated n times with different randomly chosen center, its time complexity is $O(n^3)$.

Min-Power Center-Based Least Sum-of-Costs. Another greedy algorithm for the BDMST was proposed in [21], it is called Center-based Least Sum-of-Costs. In similar manner to the CBTC and RTC, it constructs a tree iteratively adding a vertex and an edge to the current tree. The difference of this algorithm from the mentioned above heuristics is that it chooses a vertex outside a tree with the minimum sum of costs of edges that connect it with other non-tree vertices. We employed a similar strategy and called the proposed algorithm Min-Power Center-based Least Sum-of-Costs (MPCBLSoC). But instead of minimizing the sum of the edge weights, we minimized the sum of the vertices power costs in a star-like subgraph with a center in a given vertex what is more suitable for the MPBHSCP. As well as the methods described above, MPCBLSoC is based on PLH. In this case, the methods $ChooseFirstCenters$, $ChooseSecondCenter$, and $ChooseEachVertex$ have the same implementation: given an already constructed partial tree T, there is selected a such vertex $v \in V \setminus V_T$, that a star-graph on remaining vertices rooted in v has a minimum total power. The algorithm that chooses the *best* star-graph center is called *FindBestStarCenter*, and its pseudo-code is given in Appendix A, Algorithm 2. Thus, from the one hand, since *ChooseFirstCenters* returns a single vertex, the algorithm MPCBLSoC contains one iteration. But, from the other hand, *FindBestStarCenter* runs in time $O(n^2)$, and, therefore, the total computational complexity of MPCBLSoC is $O(n^3)$.

3.2 Min-Power Center-Based Recursive Clustering

Authors of [20] suggest another greedy heuristic called Center-based Recursive Clustering (CBRC) for the BDMST. This algorithm starts with a spanning star-tree rooted in the center, selected in such a way that the sum of edge weights is minimum. Then the leaves, whose depth is less than $\lfloor D/2 \rfloor$, are reorganized into a cluster with a center in some node. At each iteration, the leaves are reattached to a center if this improves the solution and the restriction on the number of hops is held. We called our implementation for the MPBHSCP Min-Power Center-based Recursive Clustering (MPCBRC). As a center choosing subroutine the previously described algorithm *FindBestStarCenter* is used. The pseudo-code of MPCBRC is presented in Appendix A, Algorithm 3. Each iteration of the algorithm takes $O(n^2)$ operations because of the complexity of *FindBestStarCenter*, and, since there are $O(n)$ iterations, the algorithm runs in time $O(n^3)$.

3.3 Min-Power Quadrant Center-Based Heuristic

One of the most efficient heuristics applied to the BDMST in planar Euclidian case with uniformly distributed vertices consists of recursive splitting the given region into equal parts (*quadrants*) and search of their centers [20]. We implemented a variant of the similar approach to the MPBHMSCP and called it Min-Power Quadrant Center-based Heuristic (MPQCH). The pseudo-code of this algorithm is given in Appendix A, Algorithm 4. As well as in some of the previous heuristics, it starts with choosing a center by the algorithm *FindBestStarCenter*. But this time in order to reach central symmetry we select one start center despite the parity of D. Then inside the main loop, the region is iteratively split into the squared cells of equal size. For each cell, its center is chosen by the algorithm *FindBestStarCenter* and then it is added to the tree with an edge that connects it with a center of a previous iteration's cell that contains it, or with v_0 at the first iteration. After each iteration, the number of cells four times greater than the number of cells in the previous iteration, that is, each cell consists of four cells of the next iteration. At each iteration, the height of a constructed tree is increased by 1, and, since *stepsCount* is bounded by $\lfloor D/2 \rfloor$, the diameter constraint is not violated.

In our implementation, for the speed purposes, a regular rectangular grid of size $qsize \times qsize$ is initially set on the given region, and a corresponding grid cell is assigned to each vertex. Then, due to this grid, during the main loop the subset of vertices that belong to each cell $c \in C$ are found in constant time. Actually, $qsize$ is a parameter of the algorithm, and the greater value of $qsize$ allows to obtain better solution but increases the running time. The computational complexity of the algorithm is $O(qsize^2 + \min\{\lfloor D/2 \rfloor, \log(qsize)\}n^2)$.

3.4 Min-Power Iterative Refinement

Another efficient approach for building a spanning tree with bounded diameter is the following. At first, a tree without restriction on diameter is constructed. Then, it is iteratively modified in such way that the depths of its vertices decrease. This procedure is performed until the restriction on tree diameter is satisfied. The iterative algorithm that reduces the diameter of an input spanning tree for the BDMST has been proposed in [22]. We propose the heuristic for the MPBHSCP called Min-Power Iterative Refinement (MPIR), which is based on the similar idea. The pseudo-code of this algorithm is presented in Appendix A, Algorithm 5. At first, a center v_0 is chosen by the *FindBestStarCenter* subroutine. Then, a near-optimal solution for an unbounded problem rooted in v_0 is constructed by IPP [5]. If D is odd, then the most remote neighbor of v_0 in T is selected as second center. The algorithm works with a set of vertices U whose depth exceeds $\lfloor D/2 \rfloor$. For each $u \in U$ the best removing of an edge from the path from u to v_0 and subsequent adding another edge that decreases a depth of u and minimally increases the value of objective function are found. The best of such edge exchanges among all vertices of U is performed. After each modification of a tree depth of some vertices in U may be decreased, therefore, the vertices whose depth is less than $\lfloor D/2 \rfloor$ are removed from U. The time complexity of the algorithm is $O(n^3)$.

4 Simulation

We have implemented all the described algorithms in C++ programming language and run them on the data sets that are given in Beasley's OR-Library for the Euclidian Steiner Problem (http://people.brunel.ac.uk/~mastjjb/jeb/orlib). These test cases present the random uniformly distributed points in the unit square. For the same size 15 different instances are provided. The experiment was launched on the Intel Core i5-4460 3.2 GHz processor with 8 Gb RAM. For the algorithm MPQCH we chose $qsize = n$ since choosing such value does not slow down the algorithm much, while the objective function on the obtained solution becomes significantly greater than, for example, in the case $qsize = \sqrt{n}$.

We have tested our algorithms on two groups of test instances: small and large. The first group contains the instances when n is 20, 30, 40, or 50, and D is 4, 5, or 6. In Appendix B we propose a MILP model of the BHMPSCP that uses the ideas of integer linear programming (ILP) model of the BDMST from [23]. We have launched CPLEX with MILP models of small size instances. The results are presented in Appendix C, Table 1. In this table, in the first two columns, correspondingly, diameter of a tree D and problem size n are presented, the third column contains the numbers of instances in the OR-Library. In the next 7 columns the objective function values on solutions obtained by different algorithms are shown, and the last column shows the running time of CPLEX in seconds. We bounded the calculation time for CPLEX by 1 h. In cases when CPLEX constructed an optimal solution within the allotted time, the corresponding objective function value is marked by an asterisk, in other cases CPLEX built some feasible solution, not necessarily optimal. The least objective function values are marked bold. It is seen that MPRTC often appeared to be the best in these small size cases. On the 1st instance with $n = 50$, and $d = 5$, MPRTC outperforms CPLEX. In all tested cases the objective value on the best solution constructed by the new fast heuristics never exceeds the optimal value by more than 40%, and, in most cases, this gap is significantly less: for example, it is often less than 15% in cases when $n = 20$.

Within the second group we tested 4 variants of size: $n = 100$, 250, 500, and 1000, 15 instances for each. We also took different values of D for each size. The results of the experiment are presented in Appendix C, Table 2. For each algorithm and each tested pair of n and D the average objective value (av), average time in seconds (time), and standard deviation (err) are shown. In average, when the diameter bound is low, the best solution is constructed by MPRTC. With large values of D MPIR constructs the best solution. Note that MPBTC and MPCBLCoS results are very poor when D is small, but with large values of D their average objective values are close to minimum. MPBTC and MPRTC appeared to be the most time consuming on large size cases, while MPQCH always runs significantly faster than other algorithms. Besides, MPQCH performance almost does not depends on D. Most probably, this is because the maximum diameter of the constructed solution is much less than D,—this gives us a possibility for the further improvements of this algorithm.

Of course, CPLEX is not applicable in reasonable time for the large size instances. In order to obtain a lower bound of the objective function we launched CPLEX on the LP-relaxation of the problem, but the obtained bounds appeared to be too small, so we decided not to include them into the table.

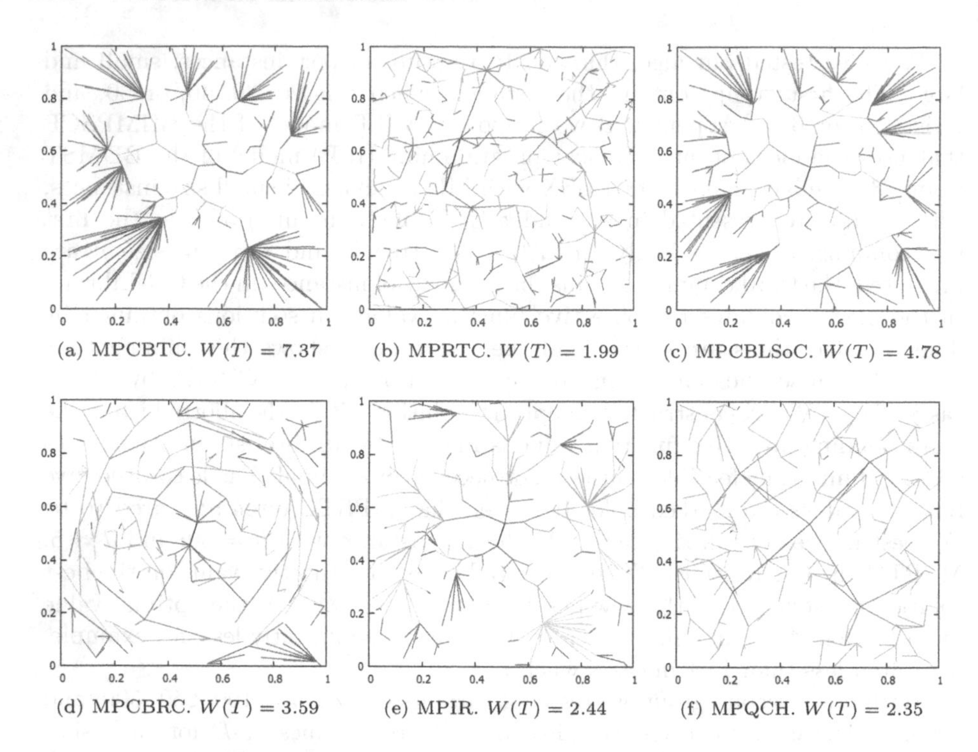

(a) MPCBTC. $W(T) = 7.37$ (b) MPRTC. $W(T) = 1.99$ (c) MPCBLSoC. $W(T) = 4.78$

(d) MPCBRC. $W(T) = 3.59$ (e) MPIR. $W(T) = 2.44$ (f) MPQCH. $W(T) = 2.35$

Fig. 1. Algorithms results on the same instance. $D = 15, n = 250$ (Color figure online)

As an illustration, we also present in Fig. 1 the solutions that were obtained by different algorithms on the same instance when $D = 15, n = 250$. For the convenience, the edges that remote from a center by an equal distance (i.e., hops count) are colored in the same color. Since the diameter bound is odd in this case, there are two centers (linked by a black edge) in solutions constructed by all algorithms except MPQCH, which always builds a tree with only one center. The difference of solutions obtained by the algorithms is seen in these figures. The diameter bound is still not enough for MPCBTC and MPCBLSoC to construct good solutions: in both cases the edges of a backbone (i.e., a subgraph induced by all the tree's non-leaf vertices) are often too short and there are many leaves far from a center that are coincident with long edges (colored in red). MPCBRC constructs a tree with big amount of long edges in backbone, since the backbone vertices are always chosen as center of the current set of leaves during the tree construction. MPIR result contains a lot of vertices with a large degree that are coincident with rather long edges, that slightly deteriorate solution. The remained two algorithms, MPRTC and MPQCH, that performed the best, have the following common features: (1) the number of vertices increases with increasing of their depths; (2) the average edge weight decreases with increase of the depth. MPRTC always chooses a vertex at random, and, in average, the distance to the closest in-tree vertex becomes less while the constructed tree size grows. MPQCH constructs a tree whose backbone vertices are located close to the quadrants geometric centers. Note that MPQCH built a tree with maximum depth equal 6 while the depth upper bound is 7. This allows to improve solution in this case: each of the longest edges that connect a center with its four children could be replaced by two shorter edges with intermediate vertex that is located close to edge's geometric center. We assume that such modification will significantly improve the solution, and we plan to implement it in future.

5 Conclusion

In this paper, the NP-hard Min-Power Bounded Hops Symmetric Connectivity Problem is considered. We propose six different constructive heuristics for its approximation solution. As the main ideas of our approaches, we used some of the known heuristics that were previously developed for the Bounded-Diameter Minimum Spanning Tree problem. We implemented all the proposed algorithms and conducted a numerical experiment on different randomly generated test instances. The simulation shows that in cases with large diameter the algorithm Min-Power Iterative Refinement yields the best results, while the usage of the Min-Power Randomized Tree Construction is more preferable when the diameter is small. If one needs to obtain a solution of rather good quality in shortest time, then the Min-Power Quadrant Center-based Heuristic could be the best choice. Besides, the experiment results show that the Min-Power Quadrant Center-based Heuristic can be significantly improved. As the experiment results on small size instances show, the best solution obtained by the proposed heuristics always differs from the optimal one by not more than 40% in terms of objective function. In the future, we plan to develop different variants of local search and other metaheuristics that appeared to be efficient for the Bounded-Diameter Minimum Spanning Tree problem, such as variable neighborhood search, genetic algorithm, and ant colony optimization, where the trees obtained by different algorithms proposed in this paper will serve as start solutions.

Appendix A: Algorithm Descriptions

Algorithm 1. Prim-Like Heuristic

$C[.] \leftarrow ChooseFirstCenters()$; $W^* \leftarrow \infty$;
for all $v_0 \in C[.]$ **do**
  $V_0 \leftarrow \{v_0\}$; $U \leftarrow V \setminus \{v_0\}$;
  $depth_T[.] \leftarrow$ an array of size n filled with -1;
  $bestNeighbor[.] \leftarrow$ an array that stores the best neighbor in V_0 for each vertex in U;
  $wBestNeighbor[.] \leftarrow$ an array that stores the total power increase if the vertex will be connected with its best neighbor;
  $depth(v_0) \leftarrow 0$; $T \leftarrow (v_0, \emptyset)$;
  if D is odd **then**
    $v_1 \leftarrow ChooseSecondCenter(v_0)$; $depth_T(v_1) \leftarrow 0$;
    Add a vertex v_1 and an edge (v_0, v_1) to T;
  end if
  $V_0 \leftarrow V_T$;
  for all $u \in U$ **do**
    $bestNeighbor(u) \leftarrow \text{argmin}_{v \in V_T}\{Pow(u, v)\}$; $wBestNeighbor(u) \leftarrow Pow(u, bestNeighbor(u))$;
  end for
  while U is not empty **do**
    $u \leftarrow ChooseEachVertex(U, V_0, wBestNeighbor)$;
    Add a vertex u and an edge $(u, bestNeighbor(u))$ to T;
    $depth_T(u) \leftarrow depth_T(bestNeighbor(u) + 1)$;
    $Pow_T(u) \leftarrow Pow(u, bestNeighbor(u))$;
    $Pow_T(bestNeighbor(u)) \leftarrow \max\{Pow_T(bestNeighbor(u)), Pow(u, bestNeighbor(u))\}$;
    $U \leftarrow U \setminus \{u\}$;
    for all $v \in U$ **do**
      $w \leftarrow Pow(bestNeighbor(u), v) + \max\{0, Pow(bestNeighbor(u), v) - Pow_T(v)\}$;
      if $w < wBestNeighbor(v)$ **then**
        $wBestNeighbor(v) \leftarrow w$; $bestNeighbor(v) \leftarrow bestNeighbor(u)$;
      end if
    end for
    if $depth_T(u) < \lfloor D/2 \rfloor$ **then**
      $V_0 \leftarrow V_0 \cup \{u\}$;
      for all $v \in U$ **do**
        $w \leftarrow Pow(u, v) + \max\{0, Pow(u, v) - Pow_T(v)\}$;
        if $w < wBestNeighbor(v)$ **then**
          $wBestNeighbor(v) \leftarrow w$; $bestNeighbor(v) \leftarrow u$;
        end if
      end for
    end if
  end while
  if $W(T) < W^*$ **then**
    $W^* \leftarrow W(T)$; $T^* \leftarrow T$;
  end if
end for
return T^*;

Algorithm 2. FindBestStarCenter

```
Input: U ⊂ V;
Output: center ∈ U;
center ← ∅;
minCost ← ∞;
for all u ∈ U do
  leavesCostSum ← 0;
  centerCost ← 0;
  for all v ∈ U \ u do
    leavesCostSum ← leavesCostSum + Pow(u, v);
    centerCost ← max(Pow(u, v), centerCost);
  end for
  if centerCost + leavesCostSum < minCost then
    center ← u;
    minCost ← centerCost + leavesCostSum;
  end if
end for
return center;
```

Algorithm 3. Min-Power Center-based Recursive Clustering

```
v0 ← FindBestStarCenter(V);
V0 ← {v0};
T ← a star-graph rooted in v0;
U ← V \ {v0};
depthT[.] ← an array of size n that stores a depth for each vertex in a tree, filled
with -1;
depth(v0) ← 0;
if D is odd then
  v1 ← FindBestStarCenter(V0);
  depthT(v1) ← 0;
  Add a vertex v1 and an edge (v0, v1) to T;
end if
while U is not empty do
  U0 ← {v ∈ U : depthT(v) < ⌊D/2⌋}
  center ← FindBestStarCenter(U0);
  if center == ∅ then
    break;
  end if
  U ← U \ {center};
  for all u ∈ U do
    Set powerIncrease ← {power increase after reassigning a parent of u from
ParT((u) to center};
    if powerIncrease < 0 then
      T ← (T \ {(u, ParT((u))}) ∪ {(u, center)};
      depthT(u) = depthT(center) + 1;
    end if
  end for
end while
```

Algorithm 4. Min-Power Quadrant Center-based Heuristic

$v_0 \leftarrow FindBestStarCenter(V)$;
$T \leftarrow (\{v_0\}, \emptyset)$;
$U \leftarrow V \setminus \{v_0\}$;
Construct rectangular grid of size $qsize \times qsize$ on a given square;
$stepsCount \leftarrow \min(\lfloor D/2 \rfloor, \log_2(qsize))$;
$cellCenter$—an array of size n that stores a cell center for each vertex;
Fill $cellCenter$ with v_0 (initially the whole square is a single cell and the root is a center);
for all $step \in \{1, ..., stepsCount\}$ **do**
  Split grid into $2^{step} \times 2^{step}$ cells C of equal size;
  for all $c \in C$ **do**
    $U_c \leftarrow$ vertices of U located in c;
    $center \leftarrow FindBestStarCenter(U_c)$;
    $T \leftarrow T \cup \{(bestCenter, cellCenter(center))\}$;
    $U \leftarrow U \setminus \{center\}$;
    for all $u \in U_c \setminus \{center\}$ **do**
      $cellCenter(u) \leftarrow center$;
    end for
  end for
end for

Algorithm 5. Min-Power Iterative Refinement

$v_0 \leftarrow FindBestStarCenter(V)$;
Construct spanning tree T rooted in v_0 by IPP;
$V_0 \leftarrow \{v_0\}$;
$U \leftarrow V \setminus \{v_0\}$;
$depth(v_0) \leftarrow 0$;
if D is odd **then**
  $v_1 \leftarrow$ most remote neighbor of v_0 in T;
  $depth_T(v_1) \leftarrow 0$;
  Add a vertex v_1 and an edge (v_0, v_1) to T;
end if
Calculate the values of $depth_T$;
$U \leftarrow \{v \in V \setminus \{s\} : depth_T(v) > h\}$;
while U is not empty **do**
  $bestChild \leftarrow \emptyset$;
  $bestParent \leftarrow \emptyset$;
  $minPowerIncrease \leftarrow \infty$;
  Mark all vertices in U as not considered;
  for all $u \in U$ **do**
    $C \leftarrow \{u\} \cup \{v \in V : depth_T(v) > 1 \;\&\; v$ is predecessor of u in $T\}$
    for all $c \in$ {not considered elements of C} **do**
      if c is considered **then**
        continue;
      end if
      Mark c as considered;
      $P \leftarrow \{v \in V : depth_T(v) < \min(\lfloor D/2 \rfloor - 1, depth_T(c) - 1)\}$;
      for all $p \in P$ **do**
        $powerIncrease \leftarrow$maximum power costs change of vertices c, $Par_T((c)$, and p after assigning p as a parent of c in T;
        if $powerIncrease < minPowerIncrease$ **then**
          $minPowerIncrease \leftarrow powerIncrease$;
          $bestChild \leftarrow c$;
          $bestParent \leftarrow p$;
        end if
      end for
    end for
  end for
  $T \leftarrow T \setminus (\{(bestChild, Par_T((bestChild))\}) \cup \{(bestChild, bestParent)\}$;
  Decrease $Level_T$ for all the vertices in the branch rooted in $bestChild$ by $Level_T(bestChild) - depth_T(bestParent) - 1$;
  $U \leftarrow U \setminus \{v \in U : depth_T(v) \leq \lfloor D/2 \rfloor\}$;
end while

Appendix B: MILP Formulation

We propose a MILP formulation of the BHMPSCP that is generally based on the ILP formulation of the BDMST from [23]. Let us construct such directed graph $G' = (V, A)$ from the given graph $G = (V, E)$, that each edge $(i, j) \in E$ corresponds to two oppositely directed arcs $(i, j) \in A$ and $(j, i) \in A$ with the same weight $c_{ij} = c_{ji}$. Let us introduce the following variables:

- $u_{il} = 1$ if the depth of vertex i in a tree equals l and $u_{il} = 0$ otherwise, $i = 1, ..., n; l = 0, ..., \lfloor D/2 \rfloor$;
- $x_{ij} = 1$ if the arc $(ij) \in A$ belongs to the tree and $x_{ij} = 0$ otherwise, $i, j = 1, ..., n$;
- $C_i \geq 0$—maximum weight of an arc adjacent to the vertex i in the tree, $i = 1, ..., n$;
- $r_{ij} = 1$ if $(i, j) \in E$, and the vertices i and j are the centers in the solution to the problem, $i, j = 1, ..., n$. These variables are used only in the case when D is odd.

We propose two different formulations depending on the parity of D. If D is even then the problem is formulated in a following way:

$$\sum_{i=1}^{n} C_i \to \min \tag{1}$$

$$x_{ij} c_{ij} \leq C_i, i, j = 1, ..., n \tag{2}$$

$$x_{ij} c_{ij} \leq C_j, i, j = 1, ..., n \tag{3}$$

$$\sum_{l=1}^{\lfloor D/2 \rfloor} u_{il} = 1, i = 1, ..., n \tag{4}$$

$$\sum_{i:(i,j) \in A} x_{ij} = 1 - u_{j0}, j = 1, ..., n \tag{5}$$

$$x_{ij} \leq 1 - u_{jl} + u_{i,l-1}, i, j = 1, ...n : (i, j) \in A, l = 1, ..., \lfloor D/2 \rfloor \tag{6}$$

$$\sum_{i=1}^{n} u_{i0} = 1 \tag{7}$$

In this formulation, constraints (2)–(3) bound the maximum weight of the arc in a tree for each vertex i by the corresponding variable C_i which is used in the minimized function (1). Equation (4) guarantee that for each vertex the only value of depth is assigned. Equations (5) imply that each vertex, except the center, has the only incoming arc and only the center has the depth assigned to 0. Inequalities (6) reflect the fact that for each in-tree arc (i, j), the depth of i is less by 1 than the depth of j. Equation (7) ensures that the tree has the only center.

Formulation for the case when D consists of the same minimized function (1) and constraints (2)–(6). But it does not contain the equality (7) since in this case solution should have two centers. In addition, the following constraints are included to the problem formulation in the case when D is odd:

$$r_{ij}c_{ij} \leq C_i, i, j = 1, ..., n \tag{8}$$

$$r_{ij}c_{ij} \leq C_j, i, j = 1, ..., n \tag{9}$$

$$\sum_{j:(i,j)\in E} r_{ij} = u_{i0}, j = 1, ..., n \tag{10}$$

$$\sum_{i=1}^{n} u_{i0} = 2 \tag{11}$$

In this formulation, the inequalities (8)–(9) bound below the appropriate variables C_i by the weight of the edge that connects two centers. Equations (10) imply that only the centers (i.e., the vertices with zero depth) are connected by the special edge that is defined by the variables r_{ij}. And, finally, the equality (11) guarantees that there are two centers in solution.

Appendix C: Tables with Experiment Results

Table 1. Comparison of the experiment's results obtained by different heuristics with CPLEX.

D	n	nr	Objective							Time
			MPCBTC	MPRTC	MPCBLSoC	MPCBRC	MPQCH	MPIR	CPLEX	CPLEX
4	20	1	1.83	**1.38**	1.66	1.86	1.50	2.06	1.28*	8.55
		2	1.86	1.54	1.77	3.15	**1.32**	2.04	1.24*	6.83
		3	1.59	**1.02**	1.63	1.40	1.07	1.69	0.94*	7.50
	30	1	2.06	**1.75**	2.43	3.11	1.92	2.33	1.51*	175.00
		2	2.11	**1.77**	2.11	3.94	2.28	2.63	1.58*	79.47
		3	2.15	2.16	2.51	2.83	**1.91**	3.93	1.70*	91.25
	40	1	3.14	**1.88**	3.16	3.21	1.99	3.57	1.69	3620.21
		2	3.01	**2.15**	3.66	2.94	2.28	4.16	1.67	3620.71
		3	4.09	2.46	4.17	4.89	**2.42**	4.90	1.97	3613.03
	50	1	4.85	**2.77**	3.59	4.63	3.10	7.12	2.85	3602.64
		2	3.89	**2.80**	3.63	6.32	3.27	6.21	2.28	3600.38
		3	4.51	**2.83**	4.68	4.55	2.99	5.09	2.42	3602.34
5	20	1	1.49	**1.22**	1.61	1.97	1.50	1.53	1.08*	39.94
		2	1.62	**1.25**	1.35	1.89	1.32	2.01	1.04*	17.25
		3	1.29	**0.97**	1.60	1.51	1.07	1.60	0.77*	16.61
	30	1	2.35	**1.54**	2.43	2.51	1.92	2.16	1.15	3606.57
		2	2.22	**1.54**	2.22	2.06	2.28	2.52	1.28	2344.61
		3	1.99	**1.71**	1.86	2.43	1.91	3.05	1.31	3613.91
	40	1	2.94	**1.58**	2.91	3.09	1.99	3.47	1.37	3614.71
		2	2.51	**1.79**	2.49	3.13	2.28	3.22	1.53	3610.20
		3	3.56	**2.22**	3.30	3.31	2.42	4.12	1.69	3600.80
	50	1	3.86	**2.44**	3.59	4.11	3.10	7.76	2.16	3620.42
		2	4.41	**2.43**	4.37	6.24	3.27	6.42	2.35	3612.53
		3	4.10	**2.25**	3.88	3.28	2.99	5.09	2.04	3620.67
6	20	1	1.15	**1.11**	1.25	1.86	1.22	1.17	0.97*	12.41
		2	1.17	1.16	1.24	1.89	**1.12**	1.26	0.97*	15.14
		3	0.82	0.81	0.91	1.06	**0.75**	0.82	0.69*	12.63
	30	1	1.65	**1.31**	1.40	2.38	1.59	1.54	1.00*	222.75
		2	1.73	1.38	2.26	1.31	1.38	**1.28**	1.02*	127.82
		3	1.48	**1.35**	1.64	2.22	1.60	1.66	1.08*	156.61
	40	1	2.34	1.38	2.93	1.95	**1.32**	1.40	1.07	3605.49
		2	1.70	1.66	1.55	2.94	**1.45**	1.68	1.06	3601.24
		3	2.69	**1.64**	2.79	3.09	1.65	1.93	1.31	3608.30
	50	1	2.88	2.00	2.40	3.15	**1.89**	7.75	1.53	3614.82
		2	2.77	**1.77**	2.63	3.69	2.03	5.73	1.49	3611.65
		3	2.66	**1.77**	2.72	5.78	1.79	2.69	1.41	3600.79

Table 2. Comparison of the experiment's results obtained by different heuristics.

n	D	MPCBTC			MPRTC			MPCBLSoC			MPCBRC			MPQBH			MPIR		
		av	err	time	av	err	time	av	err	time	av	err	time	av	err	time	av	err	time
100	5	8.17	0.47	0	**3.6**	0.13	0	8.8	0.66	0	8.41	0.79	0	5.04	0.17	0	12.1	0.48	0
	10	3.38	0.21	0	1.88	0.06	0	3.5	0.31	0	3.07	0.29	0	2.06	0.07	0	**1.84**	0.14	0
	15	1.87	0.16	0	1.75	0.07	0	1.62	0.18	0	2.39	0.17	0	2.06	0.07	0	**1.19**	0.05	0
	25	0.92	0.05	0	1.74	0.07	0	0.96	0.03	0	2.36	0.17	0	2.06	0.07	0	**0.89**	0.02	0
250	10	13.2	0.93	0.07	**2.32**	0.07	0.09	14.1	1.56	0.02	6.98	1.28	0.02	2.44	0.04	0	5.41	0.43	0.04
	15	8.17	0.65	0.08	**2**	0.03	0.1	7.94	0.97	0.02	3.85	0.29	0.02	2.47	0.04	0	2.6	0.46	0.05
	20	4.3	0.5	0.08	2.03	0.05	0.11	3.49	0.42	0.02	3.31	0.25	0.02	2.47	0.04	0	**1.48**	0.22	0.04
	40	0.96	0.05	0.1	2.03	0.05	0.11	**0.91**	0.02	0.02	3.32	0.22	0.02	2.47	0.04	0	1.04	0.26	0.02
500	15	26	1.89	0.65	**2.26**	0.03	1	26.6	2.5	0.14	6.24	0.78	0.14	2.62	0.03	**0.03**	5.1	0.49	0.33
	30	6.37	0.52	0.78	2.2	0.04	1.04	4.23	0.61	0.14	3.88	0.27	0.15	2.62	0.03	**0.03**	**1.41**	0.17	0.34
	45	1.87	0.19	0.8	2.2	0.04	0.94	1.1	0.09	0.13	3.89	0.25	0.15	2.62	0.03	**0.03**	**1.04**	0.11	0.23
	60	0.91	0.04	0.95	2.2	0.04	1.04	0.89	0.01	0.14	3.88	0.27	0.16	2.62	0.03	**0.04**	**0.857**	0.04	0.16
1000	20	50.4	1.98	6.27	**2.45**	0.04	13.5	49.4	3.01	1.13	6	0.57	1.23	2.81	0.02	**0.15**	5.26	0.43	2.94
	40	14.6	1.39	8.4	2.43	0.03	14.6	8.87	1.02	1.16	4.52	0.32	1.32	2.81	0.02	**0.16**	**1.52**	0.16	3.07
	60	4.02	0.33	9.88	2.42	0.03	15.2	1.25	0.09	1.16	4.52	0.32	1.34	2.81	0.02	**0.16**	**1.12**	0.11	2.43
	100	**0.81**	0.02	11.7	2.44	0.02	14.1	0.9	0	1.15	4.52	0.32	1.29	2.81	0.02	**0.16**	**0.85**	0.05	1.25

References

1. Rappaport, T.S.: Wireless Communications: Principles and Practices. Prentice Hall, Upper Saddle River (1996)
2. Clementi, A.E.F., Penna, P., Silvestri, R.: Hardness results for the power range assignment problem in packet radio networks. In: Hochbaum, D.S., Jansen, K., Rolim, J.D.P., Sinclair, A. (eds.) APPROX/RANDOM -1999. LNCS, vol. 1671, pp. 197–208. Springer, Heidelberg (1999). https://doi.org/10.1007/978-3-540-48413-4_21
3. Kirousis, L.M., Kranakis, E., Krizanc, D., Pelc, A.: Power consumption in packet radio networks. Theore. Comput. Sci. **243**(1–2), 289–305 (2000)
4. Călinescu, G., Măndoiu, I.I., Zelikovsky, A.: Symmetric connectivity with minimum power consumption in radio networks. In: Baeza-Yates, R., Montanari, U., Santoro, N. (eds.) Foundations of Information Technology in the Era of Network and Mobile Computing. ITIFIP, vol. 96, pp. 119–130. Springer, Boston, MA (2002). https://doi.org/10.1007/978-0-387-35608-2_11
5. Cheng, X., Narahari, B., Simha, R., Cheng, M.X., Liu, D.: Strong minimum energy topology in wireless sensor networks: NP-completeness and heuristics. IEEE Trans. Mob. Comput. **2**(3), 248–256 (2003)
6. Chu, T., Nikolaidis, I.: Energy efficient broadcast in mobile ad hoc networks. In: Proceedings AD-HOC Networks and Wireless (2002)
7. Park, J., Sahni, S.: Power assignment for symmetric communication in wireless networks. In: Proceedings of the 11th IEEE Symposium on Computers and Communications (ISCC), Washington, pp. 591–596. IEEE Computer Society, Los Alamitos (2006)
8. Althaus, E., Calinescu, G., Mandoiu, I.I., Prasad, S.K., Tchervenski, N., Zelikovsky, A.: Power efficient range assignment for symmetric connectivity in static ad hoc wireless networks. Wireless Netw. **12**(3), 287–299 (2006)

9. Erzin, A., Plotnikov, R., Shamardin, Y.: On some polynomially solvable cases and approximate algorithms in the optimal communication tree construction problem. J. Appl. Ind. Math. **7**, 142–152 (2013)
10. Wolf, S., Merz, P.: Iterated local search for minimum power symmetric connectivity in wireless networks. In: Cotta, C., Cowling, P. (eds.) EvoCOP 2009. LNCS, vol. 5482, pp. 192–203. Springer, Heidelberg (2009). https://doi.org/10.1007/978-3-642-01009-5_17
11. Erzin, A., Plotnikov, R.: Using VNS for the optimal synthesis of the communication tree in wireless sensor networks. Electron. Notes Discrete Math. **47**, 21–28 (2015)
12. Erzin, A., Mladenovic, N., Plotnikov, R.: Variable neighborhood search variants for Min-power symmetric connectivity problem. Comput. Oper. Res. **78**, 557–563 (2017)
13. Plotnikov, R., Erzin, A., Mladenovic, N.: VNDS for the min-power symmetric connectivity. Optim. Lett. (2018). https://doi.org/10.1007/s11590-018-1324-0
14. Clementi, A.E.F., Ferreira, A., Penna, P., Perennes, S., Silvestri, R.: The minimum range assignment problem on linear radio networks. In: Paterson, M.S. (ed.) ESA 2000. LNCS, vol. 1879, pp. 143–154. Springer, Heidelberg (2000). https://doi.org/10.1007/3-540-45253-2_14
15. Calinescu, G., Kapoor, S., Sarwat, M.: Bounded-hops power assignment in ad hoc wireless networks. Discrete Appl. Math. **154**(9), 1358–1371 (2006)
16. Carmi, P., Chaitman-Yerushalmi, L., Trabelsi, O.: On the bounded-hop range assignment problem. In: Dehne, F., Sack, J.-R., Stege, U. (eds.) WADS 2015. LNCS, vol. 9214, pp. 140–151. Springer, Cham (2015). https://doi.org/10.1007/978-3-319-21840-3_12
17. Clementi, A.E.F., Penna, P., Silvestri, R.: On the power assignment problem in radio networks. In: Electronic Colloquium on Computational Complexity (ECCC), (054) (2000)
18. Prim, R.C.: Shortest connection networks and some generalizations. Bell Syst. Tech. J. **36**, 1389–1401 (1957)
19. Julstrom, B.A.: Greedy heuristics for the bounded diameter minimum spanning tree problem. J. Exp. Algorithmics **14**(1), 1–14 (2009)
20. Patvardhan, C., Prem Prakash, V., Srivastav, A.: Fast heuristics for large instances of the Euclidean bounded diameter minimum spanning tree problem. Informatica **39**, 281–292 (2015)
21. Nghia, N.D., Binh, H.T.T.: Heuristic Algorithms for Solving Bounded Diameter Minimum Spanning Tree Problem and Its Application to Genetic Algorithm Development, Greedy Algorithms, Witold Bednorz. IntechOpen (2008). https://doi.org/10.5772/6345
22. Deo, N., Abdalla, A.: Computing a diameter-constrained minimum spanning tree in parallel. In: Bongiovanni, G., Petreschi, R., Gambosi, G. (eds.) CIAC 2000. LNCS, vol. 1767, pp. 17–31. Springer, Heidelberg (2000). https://doi.org/10.1007/3-540-46521-9_2
23. Gruber, M., Raidl, G.R.: A new 0–1 ILP approach for the bounded diameter minimum spanning tree problem. In: Proceedings of the 2nd International Network Optimization Conference (2005)

Identification of the Optimal Set of Informative Features for the Problem of Separating of Mixed Production Batch of Semiconductor Devices for the Space Industry

G. Sh. Shkaberina, V. I. Orlov, E. M. Tovbis, and L. A. Kazakovtsev(✉)

Reshetnev Siberian State University of Science and Technology, prosp. Krasnoyarskiy Rabochiy 31, Krasnoyarsk 660031, Russia
levk@bk.ru

Abstract. In this paper, we investigate the problem of separation of a mixed production batch of semiconductor devices of space application into homogeneous production batches. The results of the mandatory testing for each item contain a large number of parameters. Many optimization models and algorithms were developed for solving this clustering problem in the most efficient way. However, due to a rather high data dimensionality, such algorithms take significant computational resources. We analyzed methods of reducing the dimensionality of the data set with the use of factor analysis based on Pearson matrix in order to improve the accuracy of the separation. We investigated efficiency of the proposed method for separating a mixed lot of semiconductor devices which consists of two, three, four and seven homogeneous batches, with various methods of selection and rotation of factors. It was shown on real data that with any orthogonal rotation, with an increasing number of homogeneous batches in the sample, the clustering accuracy decreases. Moreover, it was impossible to identify a universal clustering model with a limited number of factors for dividing a mixed lot composed from an arbitrary number of homogeneous batches. Thus, the use of the multidimensional data was shown to be inevitable.

Keywords: Clustering · Factor analysis · Semiconductor devices

1 Introduction

In order to supply space equipment with highly reliable electronic components, specialized testing centers conduct a variety of tests for each installed semiconductor device. Electronic component base (ECB) designed for installation in spacecraft equipment, along with the input testing is subjected to additional rejection tests, including a selective destructive physical analysis (DPA). DPA allows us to confirm the good quality of the batches of ECB, or to identify the

I. Bykadorov et al. (Eds.): MOTOR 2019, CCIS 1090, pp. 408–421, 2019.
https://doi.org/10.1007/978-3-030-33394-2_32

batches, having defects due to manufacturing technology and are not detected during conventional rejection tests and additional non-destructive testing. In order to be able to transfer the results of DPA of several devices to the entire batch of semiconductor devices, the following requirement is put forward for the ECB intended for installation in space equipment: all devices from the same batch must be made from the same raw materials. Equipment manufacturers for general consumption (not designed solely for use in spacecraft) can not guarantee the implementation of this requirement. Therefore the problem of automatic grouping of semiconductor devices by production batches is very relevant.

It was shown [1], that the problem of allocation of homogeneous batches can be further reduced to a problem of cluster analysis. Authors [1] consider k-means, p-median and other optimization models for solving such a problem. Each group (cluster) must represent a homogeneous batch. To solve the problem of identifying homogeneous batches, in papers [2–4], the application of the clustering optimization algorithm k-means is proposed. In [5], authors consider the clustering method based on the EM algorithm which maximizes the log likelihood function. A model of separation of homogeneous production batches based on a mixture of Gaussian distributions was proposed in [6]. In [7], authors propose using ensembles of optimization models (k-means, k-medoids, k-medians), EM, as well as their optimized versions. In [1], authors consider the application of genetic optimization algorithms with greedy heuristic procedures, in combination with the EM algorithm for the separation of homogeneous batches of electronic devices. The advantage of the new algorithms over classical clustering algorithms for multidimensional data is shown.

In this paper, the initial data are represented by multidimensional sets (arrays) of parameters of electronic radio components (ERC), measured as the results of several hundred mandatory non-destructive tests [8]. In order to reduce the dimensionality of the input parameter sets for clustering devices into homogeneous batches, we propose the application of factor analysis methods. The aim of factor analysis is to find a simple structure that would accurately reflect and reproduce the real dependencies existing in nature [9]. Factor analysis is based on the definition of the factor model

$$X_i = \sum_{j=1}^{m} a_{ij} F_j + u_i \tag{1}$$

where X_i is a vector of values of measured parameter ($i = 1, \ldots, n$), F_j are primary factors ($j = 1, \ldots, m$), a_{ij} are coefficients named factor loadings, u_i are characteristic (specific) factors describing the part of the parameter that is not included in any primary factor. If $m < n$, the reduction of the original problem dimensionality takes place. By reducing the dimension of the data in the article we mean reducing the number of input variables due to the introduction of factors.

The quality improvement is achieved both by more coordinated functioning of radio elements with identical characteristics (from a single production batch), and by improving the quality and reliability of the results of destructive testing, for which it is possible to select elements from each production batch [1]. This paper is devoted to the problem of reducing the dimension of the original data for the corresponding problems of cluster analysis and attempts to find an optimal set of the informative features used in such cluster analysis optimization problems.

2 Data and Preprocessing

As an example of real data, in this paper we consider a sample consisting of seven different homogeneous batches. The sample is deliberately composed of homogeneous batches, some of which are extremely difficult to separate by known methods of cluster analysis.

One of the largest samplings, which the specialized test center was faced with, is presented in this paper. The total number of all devices in all batches is 3987: batch 1 contains 71 devices, 116 in batch 2, 1867 in batch 3, 1250 in batch 4, 146 in batch 5, 113 in batch 6, 424 in batch 7. Each batch contains information about 205 input measured parameters of the device. Input parameters for which the data vector contains only zero values or for which the number of non-zero values does not exceed 10% were excluded from consideration. For further processing, 67 input parameters remain to be considered.

At the first step, the analysis of the input parameters showed that the considered set of parameters can be divided into three groups:

1. parameters for which the histograms represent the normal Gaussian distribution (In21 - In28, In39 - In46, In92 - In107);
2. parameters for which the histograms represent a Gaussian distribution with frequency gaps (In84 - In91);
3. parameters for which the histogram does not correspond to Gaussian distributions (In 57 - In64, In75 - In82, In10-In20).

For each group, the histograms of observed frequencies and graphs of adjustment of distributions are given on the example of several input parameters (Figs. 1, 2 and 3).

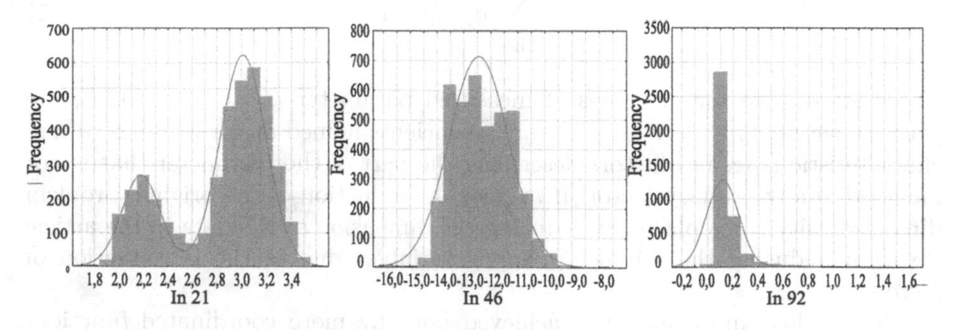

Fig. 1. Histogram of observed frequencies and graphs of the fit of the distributions. Normal Gaussian distribution

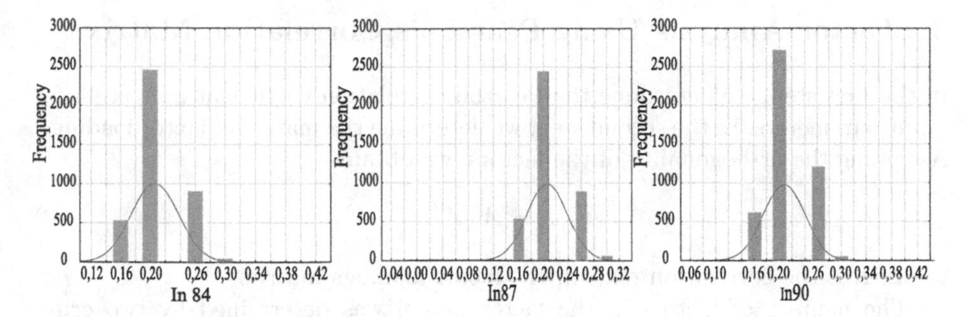

Fig. 2. Histogram of observed frequencies and graphs of the fit of the distributions. Gaussian distribution with frequency gaps

In the second step, the parameters were normalized according to the Eq. (2),

$$a_{i,k} = \frac{a^*_{i,k} - \overline{a^*_k}}{\delta^{max}_k - \delta^{min}_k} \tag{2}$$

where a^*_{ik} is the value of the measuared parameter before normalization, $\overline{a^*_k}$ are average values of the parameter, δ^{min}_k and δ^{max}_k are the lower and upper bounds of the parameter drift, respectively. The drift means the amount of change of parameters of ERC arising during the additional non-destructive testing, simulating extreme operating conditions. This method of normalization by the drift bounds was proposed in [1]. It is shown experimentally that this method of normalization gives a separation by production batches with a much smaller number of errors.

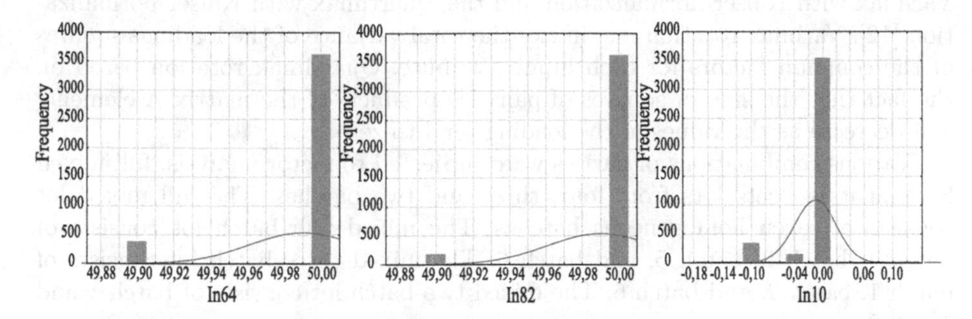

Fig. 3. Histogram of observed frequencies and graphs of the fit of the distributions. The histogram does not correspond to Gaussian distribution

3 Factor Analysis Using Pearson's Correlation Matrix

In the first step, we determine the Pearson correlation coefficient matrix [9] for input parameters. In the second step, we determine the matrix of factor loadings. Assuming the orthogonality of the factors, we obtain

$$R = A \cdot A^T \tag{3}$$

where R is the correlation matrix, A - factor loadings matrix.

The number of factors in the factor model was determined by two criteria. The first of them, the Kaiser criterion [10], selects factors with eigenvalues greater than one. However, the number of sufficient factors also depends on the total share of variance reproduced by these factors. The second of them, Cattel screening criterion [11], selects factors by scree plot based on eigenvalues of factors. The number of factors defined at the point on the chart where the decrease of eigenvalues from left to right slows down as much as possible. Since the Kaiser criterion selects factors with eigenvalues greater than one, and the Cattel screening criterion involves visual observation of the scree plot, there is no need to use any software to calculate these criteria.

Also, to simplify the factor structure, rotation is used to find one of the possible coordinate systems in the space of factors. The consequence of this is the maximization of high correlations and the minimization of low correlations. The problem of rotation is formulated as follows [9]: need to find the transformation matrix T corresponding to:

$$A^{\cdot} = A \cdot T \qquad R = A \cdot A^T = A^{\cdot} \cdot A^{\cdot T} \tag{4}$$

The following methods of orthogonal rotation are used in this paper: the Varimax with Kaiser normalization and the Quartimax with Kaiser normalization [12]. Varimax rotation maximizes the total variance of the loadings squares of the common factors for each input attribute. Quartimax rotation based on the fact that the sum of squares of pairwise products of the matrix A elements will decrease as the values of the loading tend to zero.

Various combinations of parties were subjected to factor analysis: full mixed lot and its subsets lots from four, three and two batches. The full mixed lot consists of seven homogeneous batches. The mixed four-batch lot consists of batch 1, batch 2, batch 5, and batch 6. The mixed three-batch lot consists of batch 1, batch 2, and batch 6. The mixed two-batch lot consists of batch 1 and batch 2.

In this paper, the number of factors was determined by the Kaiser criterion, and the total proportion of variance reproduced by these factors should be at least 70%.

4 Computational Experiments with Various Compositions of the Mixed Lot

To extract factors, we used the principal components method, the principal factor method with multiple R-square, principal axes method, maximum likelihood factors method, iterated communalities method (MINRES) and centroid method [9]. In further consideration, we used principal components method since it describes the maximum variance of input parameters.

For the whole mixed lot, the method based on Cattel criterion recommends to select 4 factors in the model, and this number does not change with any rotation (Fig. 4). According to Kaiser criterion, taking into account the total share of variance of at least 70%, there are five factors selected. Uberla [9] recommends in cases of dispute to select a larger number, therefore we allocate 5 factors for further consideration. Factor 1 corresponds to the highest loadings on the parameters In92-In107. This factor describes 22.779–23.954% of total variance. Factor 2 corresponds to the highest loadings on the parameters In58-In64, In76-In82. This factor describes an additional 19.335–21.265% of the total variance. Factor 3 corresponds to highest loadings on the parameters In39-In46, This factor describes an additional 12.300–14.776% of total variance. Factor 4 (parameters In10, In11, In13, In14, In18) describes 9.003–9.375% of total variance. Factor 5 (parameters In21 - In28) describes 6,781% (unrotated), 11.928% (Varimax) and 11.993% (Quartimax) of total variance. Regardless of the rotation method, the final solution has a cumulative percent of the total variance 75.794% (Table 1).

Table 1. Rotation of factor structure. Full mixed lot

Factor	Eigenvalues			Percent of the total variance (%)			Cumulative percent of the total variance (%)		
	Varimax	Quartimax	Unrotated	Varimax	Quartimax	Unrotated	Varimax	Quartimax	Unrotated
Factor 1	15.26	15.37	16.05	22.78	22.94	23.95	22.78	22.94	23.95
Factor 2	12.95	13.11	14.25	19.34	19.56	21.27	42.12	42.5	45.22
Factor 3	8.29	8.24	9.9	12.38	12.3	14.78	54.49	54.8	60
Factor 4	6.28	6.03	6.04	9.38	9	9.02	63.87	63.8	69.01
Factor 5	7.99	8.04	4.54	11.93	11.99	6.78	75.79	75.79	75.79

The total number of devices in a mixed lot composed of four batches is 446. For further processing 62 input parameters remain. The Cattel criterion, regardless of the rotation, recommends to select 4 factors in the model (Fig. 5), however, according to the Kaiser criterion, taking into account the total percentage of variance at least 70%, we allocate 6 factors. Substantial loadings on the Factor 1 appear for the parameters In21 - In28, In39 - In46. This factor describes 23.304–38.622% of total variance. Factor 2 shows substantial loadings for the parameters In58-In64, it describes in additional 13.220–17.761% of the total variance. Factor 3 has substantial loadings for In91-In107, Factor 4 for In79 - In82, Factor 6 for In57, In78. Regardless of the rotation method, the final solution has a cumulative percent of the total variance equal to 70.364% (Table 2).

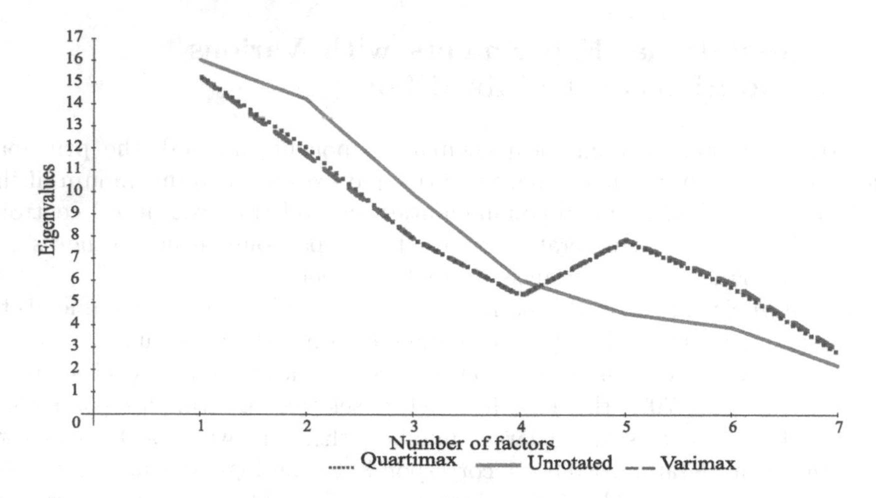

Fig. 4. Scree plot for whole mixed lot. Adv.Grapher

Table 2. Rotation of factor structure. Four-batch mixed lot

Factor	Eigenvalues			Percent of the total variance (%)			Cumulative percent of the total variance (%)		
	Varimax	Quartimax	Unrotated	Varimax	Quartimax	Unrotated	Varimax	Quartimax	Unrotated
Factor 1	15.07	16.92	23.95	24.3	27.3	38.62	24.3	27.3	38.62
Factor 2	11.01	10.36	8.2	17.76	16.7	13.22	42.07	44	51.84
Factor 3	12.27	11.41	6.96	19.8	18.41	11.23	61.86	62.41	63.07
Factor 4	2.42	2.07	1.9	3.9	3.34	3.06	65.76	65.75	66.13
Factor 5	1.41	1.48	1.38	2.28	2.38	2.22	68.05	68.13	68.35
Factor 6	15.07	16.92	23.95	24.3	27.3	38.62	24.3	27.3	38.62

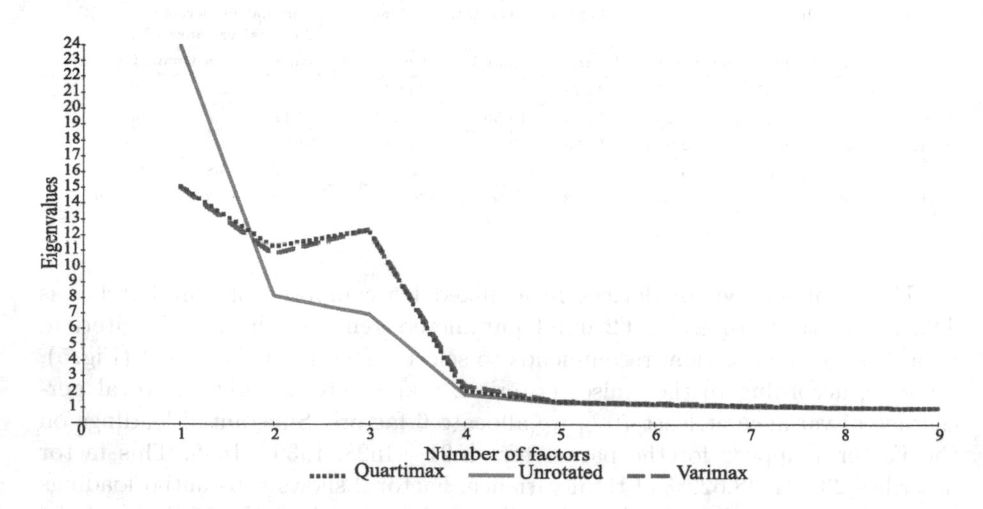

Fig. 5. Scree plot for four-batch mixed lot. Software - Adv.Grapher

The total number of devices in a mix of three batches is 300. The Cattel criterion, regardless of the rotation, recommends selecting 3 factors in the model (Fig. 6). According to the Kaiser criterion, taking into account the total percentage of variance at least 70%, we also allocate 3 factors. Substantial loadings on the Factor 1 appear for the parameters In21-In28, In39-In46. This factor describes 37.09–46.39% of total variance. Factor 2 has substantial loadings for In92-In107 and describes 22.03–26.61% of total variance. Factor 3 has substantial loadings for In84-In91 and describes in addition 9.192-13.905% of the total variance. Regardless of the rotation, the total solution has a cumulative percentage of the total variance 77.61% (Table 3).

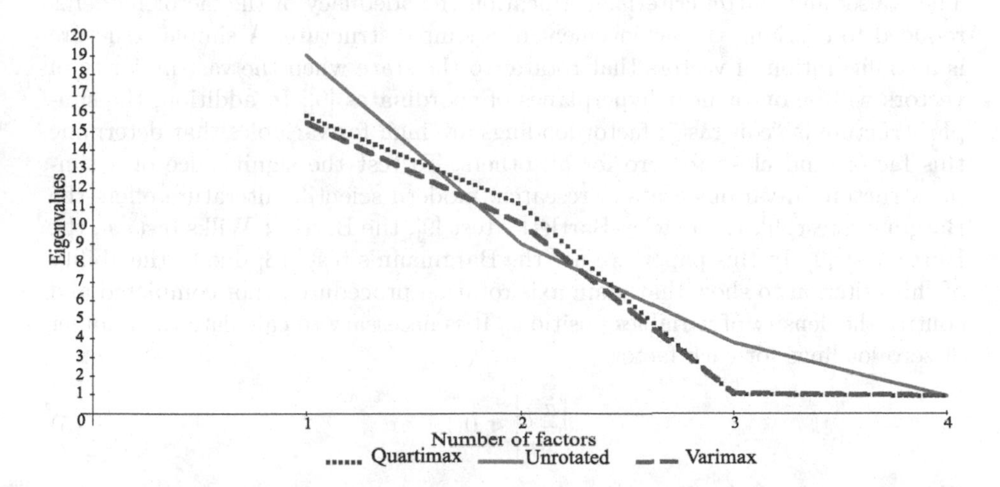

Fig. 6. Scree plot for three-batch mixed lot. Adv.Grapher

Table 3. Rotation of factor structure. Three-batch mixed lot

Factor	Eigenvalues			Percent of the total variance (%)			Cumulative percent of the total variance (%)		
	Varimax	Quartimax	Unrotated	Varimax	Quartimax	Unrotated	Varimax	Quartimax	Unrotated
Factor 1	15.21	15.89	19.02	37.09	38.74	46.38	37.09	38.74	46.38
Factor 2	10.91	10.89	9.03	26.61	26.57	22.03	63.7	65.31	68.41
Factor 3	5.7	5.04	3.77	13.91	12.3	9.19	77.61	77.61	77.61

The number of devices in the simplest mixed lot of two batches is 187. According to the Kaiser criterion, taking into account the total percentage of variance at least 70%, we allocate 2 factors. Factor 1 shows the highest loadings for the parameters In21 - In28, In39 - In46 and describes 45.41–66.20% of the total variance. Factor 2 shows the highest loadings for the In92-In95, In100-In102, In106, and describes in addition 7.28–28.07% of total variance. Regardless of the rotation, the solution has a cumulative percentage of the total variance 73.48% (Table 4).

Table 4. Rotation of factor structure. Two-batch mixed lot

Factor	Eigenvalues			Percent of the total variance (%)			Cumulative percent of the total variance (%)		
	Varimax	Quartimax	Unrotated	Varimax	Quartimax	Unrotated	Varimax	Quartimax	Unrotated
Factor 1	18.62	26.61	27.14	45.41	64.91	66.2	45.41	64.91	66.2
Factor 2	11.51	3.51	2.99	28.07	8.57	7.28	73.48	73.48	73.48

5 Adequacy of the Factor Model

Verification of the factors number sufficiency in the model was performed using The Kaiser and Cattel criteria. Verification the adequacy of the factor model is reduced to checking the achievement of a simple structure. A simple structure is a configuration of vectors that rotates to the state when the vast majority of vectors will be on or near hyperplanes of coordinates [9]. In addition, the simple structure is "contrast": factor loadings are high for variables that determine this factor, and close to zero for all others. To test the significance of a simple structure in various areas of research, modern scientific literature offers the Bargmann test [9], the Lawley-Bartlett's test [9], the Bartlett-Wilks test [9], the Burt's test [9]. In this paper, we use the Bargmann's test [13] due to the ability of this criterion to show that main axis rotation procedure is not completed and control the density of variables positions. It is necessary to calculate the number of zero loadings for each factor:

$$\left|\frac{a_{ij}}{h_i}\right| < 0.1 \tag{5}$$

where a_{ij} are factor loadings on each parameter, h_i - square root of communality (communality refers to the variance of a parameter due to common factors). If the number of zero loadings is not lower than the table value, the simple structure is considered to be achieved.

For the full mixed lot Bargmann test is satisfied for 3 of 5 factors in case of unrotated structure and for all factors in case of rotation with $\alpha <= 0,05$ (where α is a level of significance). For four-batch mixed lot test is satisfied for 3 from 6 factors in case of unrotated structure and for 4 from 6 factors in case of rotated structure with $\alpha <= 0.25$. For three-batch mixed lot test satisfied just for 1 factor in case of unrotated structure and for 2 factors in case of rotated structure with $\alpha <= 0,25$. And for two-batch mixed lot test is satisfied in one case with $\alpha <= 0.25$ (Table 5).

Analysis of the percentage of zero loadings shows, that with increasing the number of batches and at any rotation the number of cases for which test Bargmann is satisfied also increases.

Factor values obtained by orthogonal rotations described above are considered as input data for clustering algorithms. Clustering was performed with Deductor Studio Academic tool. EM algorithm applied with lower bound of likelihood = 0.2, level of accuracy = 10^{-5}, maximum of iterations = 100. Self-organizing Kohonen maps (SOM) [14] applied with linear initialization with

Table 5. Bargmann test

		Factor no.	Table value for $\alpha <= 0,05$	Table value for $\alpha <= 0.25$	The number of zero loading	Percent of zero loading
Full mixed lot (67 parameters)	Unrotated	1	17	14	9	13%
		2			6	8%
		3			18	27%
		4			31	46%
		5			26	39%
	Varimax	1			43	64%
		2			26	39%
		3			33	49%
		4			33	49%
		5			42	63%
	Quartimax	1			41	61%
		2			24	36%
		3			33	49%
		4			33	49%
		5			43	64%
Four-batch mixed lot (62 parameters)	Unrotated	1	20	17	3	5%
		2			8	13%
		3			14	23%
		4			45	73%
		5			36	58%
		6			51	82%
	Varimax	1			14	23%
		2			10	16%
		3			18	29%
		4			44	71%
		5			37	60%
		6			51	82%
	Quartimax	1			14	23%
		2			9	15%
		3			17	27%
		4			52	84%
		5			36	58%
		6			50	81%
Three-batch mixed lot (41 parameters)	Unrotated	1	9	7	0	0%
		2			0	0%
		3			8	20%
	Varimax	1			7	17%
		2			0	0%
		3			10	24%
	Quartimax	1			7	17%
		2			0	0%
		3			15	37%
Two-batch mixed lot (41 parameters)	Unrotated	1	6	4	0	0%
		2			3	7%
	Varimax	1			1	2%
		2			0	0%
	Quartimax	1			0	0%
		2			5	12%

eigenvalues, bubble neighborhood function, significance level = 0,1%. The clustering accuracy for considered mixed lots with different orthogonal rotations is presented in Table 6.

The analysis of Table 6 showed that for any orthogonal rotations and clustering algorithms, the clustering accuracy increases with a decrease the number of homogeneous batches in the sample from 39% up to 98%.

Clustering results on three-batch and two-batch mixed lots are shown in Figs. 7 and 8, respectively. Separating batches takes place exclusively on Factor 1 in both cases.

Table 6. Clustering results

		Unrotated		Varimax		Quartimax	
		EM	SOM	EM	SOM	EM	SOM
Two-batch mixed lot							
Batch 1 n = 71	Number of hits	71 (100%)	71 (100%)	71 (100%)	42 (59%)	71 (100%)	71 (100%)
	Number of errors	0 (0%)	0 (0%)	0 (0%)	29 (41%)	0 (0%)	0 (0%)
Batch 2 n = 116	Number of hits	113 (97%)	113 (97%)	104 (90%)	114 (98%)	116 (100%)	116 (100%)
	Number of errors	3 (3%)	3 (3%)	12 (10%)	2 (2%)	0 (0%)	0 (0%)
Three-batch mixed lot							
Batch 1 n = 71	Number of hits	71 (100%)	36 (51%)	71 (100%)	41 (58%)	71 (100%)	40 (56%)
	Number of errors	0 (0%)	35 (49%)	0 (0%)	30 (42%)	0 (0%)	31 (44%)
Batch 2 n = 116	Number of hits	113 (97%)	110 (95%)	110 (95%)	60 (52%)	111 (96%)	116 (100%)
	Number of errors	3 (3%)	6 (5%)	6 (5%)	56 (48%)	5 (4%)	0 (0%)
Batch 6 n = 113	Number of hits	106 (94%)	102 (90%)	93 (82%)	102 (90%)	98 (87%)	91 (81%)
	Number of errors	7 (6%)	11 (10%)	20 (18%)	11 (10%)	15 (13%)	22 (19%)
Four-batch mixed lot							
Batch 1 n = 71	Number of hits	70 (99%)	71 (100%)	70 (99%)	71 (100%)	70 (99%)	71 (100%)
	Number of errors	1 (1%)	0 (0%)	1 (1%)	0 (0%)	1 (1%)	0 (0%)
Batch 2 n = 116	Number of hits	108 (93%)	108 (93%)	108 (93%)	116 (100%)	108 (93%)	86 (74%)
	Number of errors	8 (7%)	8 (7%)	8 (7%)	0 (0%)	8 (7%)	30 (26%)
Batch 5 n = 146	Number of hits	146 (100%)	68 (41%)	146 (100%)	116 (79%)	146 (100%)	38 (26%)
	Number of errors	0 (0%)	78 (59%)	0 (0%)	20 (21%)	0 (0%)	108 (74%)
Batch 6 n = 113	Number of hits	107 (95%)	107 (95%)	107 (95%)	107 (95%)	108 (96%)	103 (91%)
	Number of errors	6 (5%)	6 (5%)	6 (5%)	6 (5%)	5 (4%)	10 (9%)
Full mixed lot							
Batch 1 n = 71	Number of hits	68 (96%)	71 (100%)	70 (99%)	71 (100%)	63 (89%)	71 (100%)
	Number of errors	3 (4%)	0 (0%)	1 (1%)	0 (0%)	8 (11%)	0 (0%)
Batch 2 n = 116	Number of hits	108 (93%)	106 (91%)	60 (52%)	114 (98%)	81 (70%)	113 (97%)
	Number of errors	8 (7%)	10 (9%)	56 (48%)	2 (2%)	35 (30%)	3 (3%)
Batch 3 n = 1867	Number of hits	487 (35%)	1337 (72%)	618 (33%)	1453 (78%)	699 (37%)	781 (42%)
	Number of errors	1380 (65%)	530 (28%)	1249 (67%)	414 (22%)	1168 (63%)	1086 (58%)
Batch 4 n = 1250	Number of hits	537 (43%)	721 (58%)	462 (37%)	583 (47%)	467 (37%)	571 (46%)
	Number of errors	713 (57%)	529 (42%)	788 (63%)	667 (53%)	783 (63%)	679 (54%)
Batch 5 n = 146	Number of hits	121 (83%)	79 (54%)	135 (92%)	102 (70%)	133 (91%)	73 (50%)
	Number of errors	25 (17%)	67 (46%)	11 (8%)	44 (30%)	13 (9%)	73 (50%)
Batch 6 n = 113	Number of hits	107 (95%)	113 (100%)	107 (95%)	113 (100%)	105 (93%)	113 (100%)
	Number of errors	6 (5%)	0 (0%)	6 (5%)	0 (0%)	8 (7%)	0 (0%)
Batch 7 n = 424	Number of hits	314 (74%)	369 (87%)	256 (60%)	421 (99%)	255 (60%)	284 (70%)
	Number of errors	110 (26%)	55 (13%)	168 (40%)	3 (1%)	169 (40%)	140 (30%)

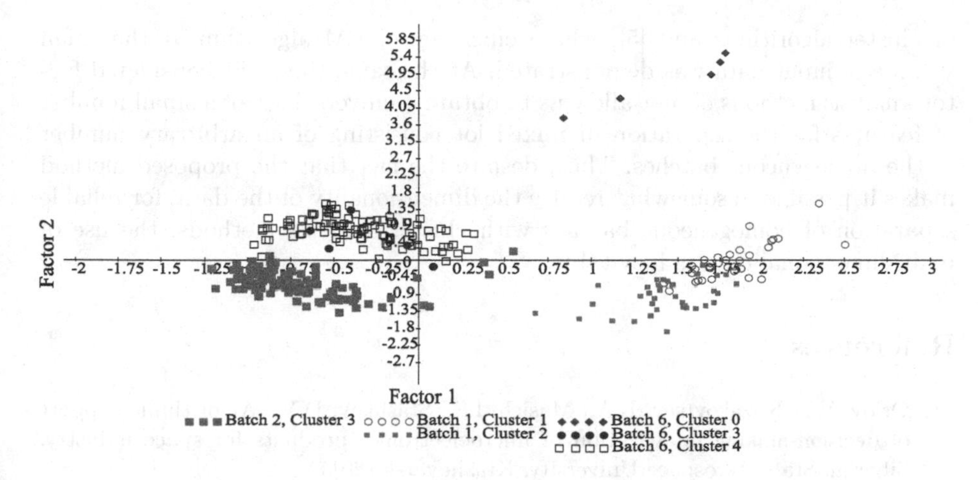

Fig. 7. Clustering results for three-batch mixed lot

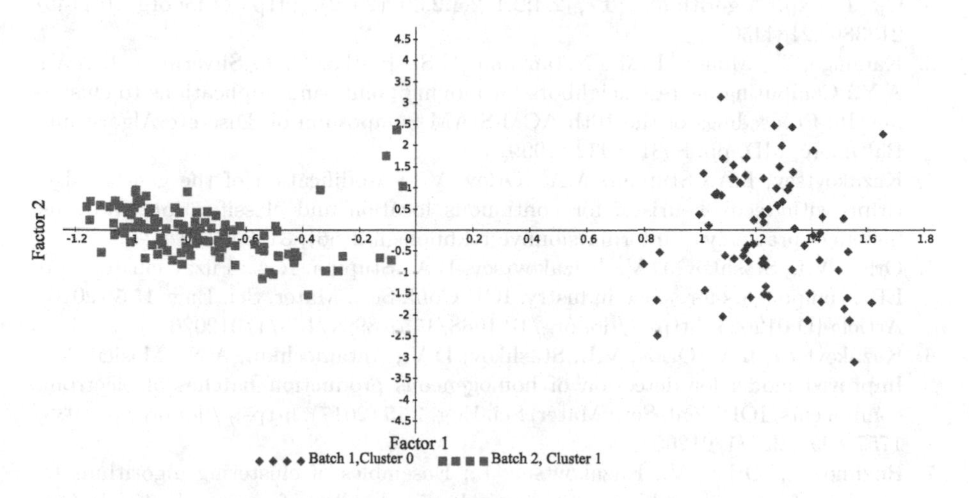

Fig. 8. Clustering results for two-batch mixed lot

6 Conclusions

The possibility of using factor analysis for the separation of a mixed lot, consisting of an arbitrary number of homogeneous batches of electronic radio components, has been proposed and described in the paper. Thus, the use of the factor model is appropriate to improve the accuracy of batch separation, regardless of the clustering algorithm used. It is shown, that the optimal number of the selected factors depends on the number of considered devices in the mixed lot, as well as on the input measured parameters of the device in a given sample. Regardless of the type of orthogonal rotation, the clustering accuracy decreases with the increase of the number of homogeneous batches in the mixed lot. A similar result was shown earlier in [6,7] when using the ensemble approach of cluster algorithms and [5], where efficiency of EM algorithm at the small volume of input data was demonstrated. At the same time, the considered factor analysis methods do not allow us to obtain a universal set of a small number of features for the separation of mixed lot consisting of an arbitrary number of the homogeneous batches. Thus, despite the fact that the proposed method makes it possible to somewhat reduce the dimensionality of the data, for reliable separation of homogeneous batches with cluster analysis methods, the use of multidimensional data is inevitable.

References

1. Orlov, V.I., Kazakovtsev, L.A., Masich, I.S., Stashkov, D.V.: Algorithmic support of decision-making on selection of microelectronics products for space industry. Siberian State Aerospace University, Krasnoyarsk (2017)
2. Ackermann, M.R., Märtens, M., Raupach, C., Swierkot, K., Lammersen, C., Sohler, C.: J. Exp. Algorithmics **17**, 2.4:2.1–2.4:2.30 (2012). https://doi.org/10.1145/2133803.2184450
3. Kanungo, T., Mount, D.M., Netanyahu, N.S., Piatko, C.D., Silverman, R., Wu, A.Y.: Computing nearest neighbors for moving points and applications to clustering. In: Proceedings of the 10th ACM-SIAM Symposium on Discrete Algorithms, Baltimore, MD, pp. S931–S932 (1999)
4. Kazakovtsev, L.A., Stupina, A.A., Orlov, V.I.: Modification of the genetic algorithm withgreedy heuristic for continuous location and classification problems. Sistemy upravleniya i informatsionnye tekhnologii **2**(56), 31–34 (2014)
5. Orlov, V.I., Stashkov, D.V., Kazakovtsev, L.A., Stupina, A.A.: Fuzzy clustering of EEE components for space industry. IOP Conf. Ser.: Mater. Sci. Eng. **155** (2016). Article ID 012026. https://doi.org/10.1088/1757-899x/155/1/012026
6. Kazakovtsev, L.A., Orlov, V.I., Stashkov, D.V., Antamoshkin, A.N., Masich, I.S.: Improved model for detection of homogeneous production batches of electronic components. IOP Conf. Ser.: Mater. Sci. Eng. **255** (2017). https://doi.org/10.1088/1757-899x/255/1/012004
7. Rozhnov, I., Orlov, V., Kazakovtsev, L.: Ensembles of clustering algorithms for problem of detection of homogeneous production batches of semiconductor devices. In: 2018 School-Seminar on Optimization Problems and their Applications, OPTA-SCL 2018, vol. 2098, pp. 338–348. CEUR-WS (2018)
8. Kazakovtsev, L.A., Antamoshkin, A.N.: Greedy heuristics method for location problems. Vestnik SibGAU **16**(2), 317–325 (2015)
9. Uberla, K.: Factorenanalyse. Springer, Heidelberg (1977). https://doi.org/10.1007/978-3-642-61985-4
10. Kaiser, H.F., Dickman, K.: Sample and population score matrices and sample correlation matrices from an arbitrary population correlation matrix. Psychometrika **27**(2), 179–182 (1962). https://doi.org/10.1007/bf02289635
11. Cattel, R.B.: The scree test the number of factors. Multivar. Behav. **1**, 245–276 (1966). https://doi.org/10.1207/s15327906mbr0102_10

12. Harman, H.: Modern Factor Analysis. The University of Chicago Press, Chicago (1967). https://doi.org/10.1002/bimj.19700120119
13. Bargmann, R.: Signifikanz Untersuchungen der einfachenStruktur in der Faktoren-Analyse. Mitteilungsblatt für Mathematische Statistik **1**, 1–24 (1955)
14. Kohonen, T.: Self-organized formation of topologically correct feature maps. Biol. Cybern. **43**(1), 59–69 (1980). https://doi.org/10.1007/bf00337288

A Cost Minimizing at Laser Cutting of Sheet Parts on CNC Machines

Anastasia Tavaeva[1,2], Alexander Petunin[2,3](✉), Stanislav Ukolov[2], and Vladimir Krotov[4]

[1] Production association "Urals Optical and Mechanical Plant", Yekaterinburg, Russia
a.f.tavaeva@urfu.ru
[2] Ural Federal University, Yekaterinburg, Russia
aapetunin@gmail.com, s.s.ukolov@urfu.ru
[3] Institute of Mathematics and Mechanics, UBr RAS, Yekaterinburg, Russia
[4] Joint-Stock Company "Technocomproject", Yekaterinburg, Russia
wikrot@mail.ru

Abstract. The problem of cost minimizing at laser cutting of sheet parts on CNC machines is considered in this paper. As an objective function the cost function of cutting process is used. The model of exact cost function calculation is presented depending on the number of frames (commands) in the NC program. The each command is written using G-code. In order to most correctly construct optimal cutting path the accurate value of objective function basic parameters should be calculated. To this end, the accurate calculation methodologies of basic parameters values are presented. The methodologies relate to calculation of cost parameters and cutting speed. Based on proposed methodology the subsystem of cutting cost calculation was developed by using .Net Framework technology. In order to solve optimization problem the special cutting techniques are used. There are some multi-contour and multi-segment cutting techniques. In this paper special cutting techniques for common geometrical types of contours widely used in blank production are presented. In order to verify the proposed methodologies on practice the computational experiments which show a statistically significant improvement of the objective function value compared with using standard cutting techniques are presented.

Keywords: CNC laser cutting machines · Thermal cutting · Tool path optimization · Cost of cutting process · Cutting techniques

1 Introduction

One of the complex optimization problem arising in technical applications is the cutting path optimization problem for CNC sheet cutting machines. This problem belongs to the class of NP-hard problems of continuous-discrete optimization equivalent to some types of traveling salesman problem with additional

I. Bykadorov et al. (Eds.): MOTOR 2019, CCIS 1090, pp. 422–437, 2019.
https://doi.org/10.1007/978-3-030-33394-2_33

restrictions that do not allow the use of known algorithms to solve them. As an objective function of the problem, the cost of parts cutting process for the resulting cutting path is considered.

Recently the CNC sheet cutting machines are widely used in order to manufacture sheet metal products. In particular, such machines include thermal cutting machines (laser, flame and plasma cutting). During development of NC programs there is need to take into account some important technological features and constraints arising in the process of part sheet cutting on CNC equipment.

Before cutting of part contour the piercings must be selected (Fig. 1). Piercings are operations where the laser cutting tool initiates the material. Piercings is selected according to the material type, its thickness and cutting parameters. In order to avoid material beading and part deformation the piercings must be selected by some distance from contour.

During thermal cutting the "burning out" and "sweeping" of material are occurred. Due to the fact the contour of parts and cutting tool path are not matched. The cutting tool is moved by equidistant curve of contour (Fig. 1).

The precedence constraint is taken into account which is due to the features of portal type machine [15,19]. If the contour is fully cut, one detaches from the rest of the material and can possibly shift its position, and thus it will be impossible to continue cutting in this area [2]. The constraint ties to inner-outer contour relation which means that an inner contour needs to be completely cut before the outer contour is cut. Figure 1 presents example of cutting precedence of contour by number 1–3.

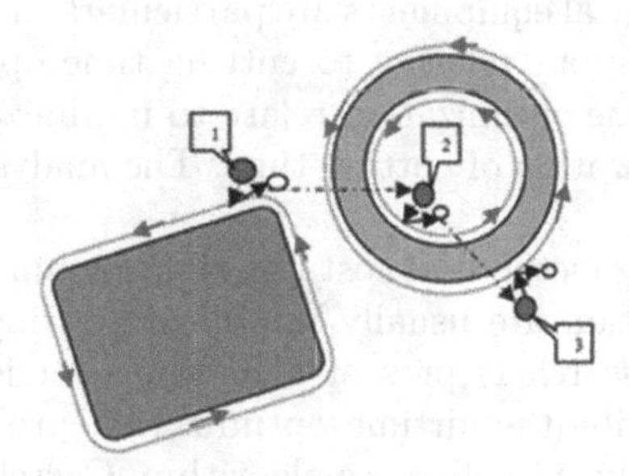

● **The piercing;** ○ **the point of tool switching off; —— the cutting motion; - - - the air motion**

Fig. 1. Cutting scheme example of two parts using standard cutting technique

The objective function (cutting cost F_{cost}) is calculated by [16]:

$$F_{cost} = L_{on} \cdot C_{on} + L_{off} \cdot C_{off} + N_{pt} \cdot C_{pt} \rightarrow \min \tag{1}$$

L_{off} is length of air tool path; L_{on} is length of working tool path; C_{off} is cost of air tool path unit; C_{on} is cost of working tool path unit; N_{pt} is numbers of piercing; C_{pt} is cost of one piercing.

In general when using different types ($j = \overline{1,k}$) of piercing the F_{cost} is calculated by:

$$F_{cost} = L_{on} \cdot C_{on} + L_{off} \cdot C_{off} + \sum_{j=1}^{k} L_{pt}^{j} \cdot C_{pt}^{j} \rightarrow \min \qquad (2)$$

The problem of cost function minimization is considered as *generalized travelling salesman problem (GTSP) with restrictions* [16,17]. The formalization of minimization problem of cutting time and cost for CNC sheet cutting machines is presented in [16].

In [3,9,20] the following classes of cutting tool routing problems for CNC sheet cutting machines are allocated:

- The Traveling Salesman Problem - TSP;
- The Generalized Traveling Salesman Problem - GTSP;
- The Continuous Cutting Problem - CCP;
- The End Point Cutting Problem - ECP;
- Intermittent Cutting Problem - ICP;
- Based on conception of contours cutting by segment [20] the new class of optimization tool routing problem is presented: Segment Continuous Cutting Problem (SCCP).

The detailed analysis of existing methods, which are solving the optimization problem of cutting tool route, was presented in [3]. A few algorithms of tool routing for other technological equipments are particularly described in [7,8]. In these articles there are questions relating to cutting time optimization. The present methods of cutting time optimization relate to minimization of idle moves time and slightly to minimization of cutting time. The analysis of current methods is provided below.

Analysis of existing methods of cost function minimizing showed that airtime and length of air motion are usually minimized during cutting path optimizing. The following researchers present algorithms for idle moves optimization. Yang et al. [28] describe the airtime optimization problem in leather cutting. They proposed the hybrid intelligence algorithm. Castelino et al. [4] describe an algorithm for airtime minimizing by optimally connecting the tool path. They consider heuristic methods that are used in order to obtain the optimal or near optimal solutions. Murzakaev et al. [14] consider problem of idle moves length minimization. The model is presented for standard cutting technique. In order to minimize cutter idle moves length the three metaheuristics (Simulated Annealing, Threshold Accepting, Great Deluge Algorithm) were chosen. They propose the generalized scheme of problem solving. The algorithm of idle moves minimizing is proposed by Chen et al. [1]. They divide into two sub-optimal problems (pattern cutting order and entry/exit cutting point) and solve ones using an ant colony optimization algorithm (min-max ant system). Lee and Kwon [11] consider tool path problem and proposed two-step genetic algorithm. The aim is to minimize the idle moving of cutting tool. They combine global search for piercing optimization and local search for part sequencing. The two stage of sequential optimization procedure for nesting and cutting sequence are presented by Sherif et al. [22]. The objectives are maximizing the material utilization and minimizing the cutter idle moves. They consider simulated annealing algorithm in order to find the near optimal cutting tool path.

Analysis of existing methods shows the deficiency of research in the field of piercing numbers N_{pt} reducing and length of cutting tool motion L_{on} in solving the optimization problem (2). In [12] authors consider the problem of N_{pt} reducing in thermal cutting of sheet material in terms of graph theory. It should be noted that the precedence constrain was not taken into account and the intersection of the existing cuts is allowed. The problem of cutting path optimization in terms of cutting of parts group with one of pierce point is considered in [6]. The last stage of solving the problem (cutter routing optimization at idle moves) is reduced to the TSP. The problem of cutter routing optimization at CNC machines is formulated and the mathematical model of total cutting time minimization is proposed by using standard and special cutting techniques by Faizrahmanov et al. [5]. Verhoturov et al. [27] present "chain" cutting technique in order to minimize the numbers of pierce points.

The one of cutting time and cost minimization methods is application of special cutting techniques. In order to optimize the cutting parameters and to observe the necessary cutting requirements the some special cutting techniques are used. There are "chain" cutting [23], common cut [18], partial cutting of contour with the subsequent completion of the contour cutting after cutting the contour of another part and some others. Petunin and Krotov [18] proposed the classification of various cutting techniques used to form the cutting tool path. The cutting techniques are classified into three main classes: standard cutting, multi-contour cutting and multi-segment cutting technique. Every contour is cut with pierce point by using *standard cutting technique*. The numbers of pierce points equal numbers of contours. The several contours are cut in one cutting segment with one pierce point by using *multi-contour cutting technique*. For example, the multi-contour cutting includes "chain" cutting, common cut. The several cutting segments are cut with several pierce points by using *multi-segment technique*.

Analysis of optimization problem solutions shows that there are no or are negligible considered the questions about cutting cost optimization in addressed articles. To this end the methodology of exact calculation of cutting cost objective function is conducted in this article. In optimization problem (2) there are difficulties in calculating the basic parameters C_{on}, C_{off}, C_{pt} depending on many factors in order to exact computation of F_{cost} and construct the exact optimal cutting path. For CNC laser cutting machines C_{on}, C_{off}, C_{pt} depend on the type of laser used in CNC machine, type and thickness of treatment material. The selected factors depend on analytically or tabular functions. The formulas of C_{on}, C_{off}, C_{pt} calculation and their values may significantly differ for various CNC sheet cutting machines. The analysis of existing methods for cutting cost calculation shown the insufficiency of research in the field. The time per pierce in laser cutting process is calculated in [10]. In [13] the laser cutting cost is compared with water jet, plasma and oxygen cutting costs when treatment sheet material of 1.0114 (thicknesses $\Delta = 3$–10 mm). In [21] the assessment of plasma and CO_2 laser cutting machines operating cost is performed. But it should be noted that the calculation of cost parameters remains outside the scope of present researches, hence the calculation of C_{on}, C_{off}, C_{pt} values and consequently exact calculation of cost function F_{cost} value are actual problems today, which are solved in this article. The methodology of cost parameters calculation in objective function F_{cost} is developed.

As seen from (2) F_{cost} also depends on L_{on}, L_{off} and N_{pt}, in turn L_{on} depends on value of working tool speed V_{on}. The value of V_{on} is usually constant parameter which is programming during the NC program development, but as the practical shows [19,26] the value of actual working speed of cutting tool is varied by various technological factors and parameters of NC programs. Consequently problem of accurate calculation of cost function (2) in optimization problem of tool path routing is arisen. In order to solve the encountered problem the need of correction parameters calculation for V_{on} values is emerged. It should be noted that the question of exact calculation of V_{on} values remains open, then there is a need of research in order to calculate the correction parameters of V_{on} values and consequently cutting cost in this article for CNC laser cutting machines.

It is observed that L_{on}, L_{off} and N_{pt} are interdependent. In some cases the reduction of N_{pt} leads to some increase of total cutting tool path length value L_{on} due to cutting motion between contours. Wherein the length of air path L_{off} is reduced.

The problem of cost function minimization (2) during treatment of figured parts from sheet material at CNC cutting machines is solved by the optimization of parameters L_{on}, L_{off} and N_{pt}. As the practice shows the length of cutting tool motion L_{on} and numbers of pierce points N_{pt} have the greatest impact on the cutting cost compared with the length of air tool motion L_{off}. Depending on the thickness and type of material the C_{pt} can reach up to 33% from C_{on} and at the same time can exceed the C_{off} by three orders [24]. Consequently the most interest are the methods of solving the problem aimed at the minimizing L_{on} and N_{pt}.

The article is organized as follows. The model of F_{cost} calculation and basic parameters C_{on}, C_{off}, C_{pt} is presented in Sect. 2.1. Exact calculation of V_{on} values and values of correction coefficients for V_{on} is given in Sect. 2.2. Based on the exact computation of objective functions the cutting path and cutting cost are evaluated as a true. The results of the computational experiment are presented in Sect. 3.

2 Exact Calculation of Cost Function F_{cost} in the Cutter Path Optimization Problem

2.1 Model of Basic Cost Parameters C_{on}, C_{off}, C_{pt} Calculation

The most important economic characteristic of the developed NC program quality is the cost F_{cost} of cutting parts at CNC machine. F_{cost} includes the costs of electricity and expendable materials, maintenance of a CNC machine and other operating costs incurred during cutting. The problem of exact calculation of cost function F_{cost} in optimizing of cutting tool route related with search of adequate value F_{cost}, which calculation depends on basic parameters C_{on}, C_{off}, C_{pt}. The allotted parameters in turn depend on values of L_{on}, L_{off} and N_{pt}. The functional dependence C_{on}, C_{off}, C_{pt} on type and thickness of material, laser type in CNC machine, cost of expendable materials, cost of laser and technological gases, depreciation of equipment can be set either table functions or analytically. Frequently the cutting cost is not often considered in blank production or is calculated based on normative is not dependent on values of C_{on}, C_{off}, C_{pt}. Obviously that necessity of cutting cost calculation arises on manufacturer which provides cutting material service for the third-party firm. As a rule during cutting cost calculation only L_{on} is taken into account which usually equals compound perimeter of cutting parts edge contours that leads to inaccurate cutting cost calculation. Subsequently the calculation methodology of cutting cost parameters is actual problem today.

The calculation methodology of cutting cost parameters in optimization problem of cutting path applied to CNC laser cutting machine (laser type is CO_2) is considered. In order to calculate C_{on} the following notations for cost parameters calculation on 1 m of cutting tool motion are entered: $C_{exp.mat}$ - the cost of expendable materials (for example, adjudge, protective glass, gas tubes); C_{tech} - the cost of technological gas (nitrogen or oxygen depending on processed material); C_{las} - the cost of cutting gas (when working on a gas flow laser); C_{elec}^{on} - the cost of electricity; C_{salary}^{on} - the cost related with salary of accompanying personnel; C_A^{on} - amortization of equipment. In general C_{on} is calculated by:

$$C_{on} = C_{elec}^{on} + C_{tech} + C_{las} + C_{exp.mat} + C_{salary}^{on} + C_A^{on} \tag{3}$$

In order to calculate $C_{elec}^{on}, C_{tech}, C_{las}, C_{exp.mat}, C_{salary}^{on}, C_A^{on}$ the additional notations are entered: t_{on} - the time spent on 1 m of cutting tool motion, h.; P_{on} - the electricity costs for 1 h of CNC laser machine work on cutting motion, kW/h; V_{tech} - technological gas consumption, m^3/h; V_{las} - laser gas consumption, m^3/h; C_{elec} - electricity cost per 1 kW; C_{lasM^3} - the cost of 1 m^3 laser gas; C_{techM^3} - the cost of 1 m^3 technological gas; $C_{exp.mat.Unit}$ - the cost of expendable materials unit; $t_{exp.mat.Term}$ - serviceable life of expendable materials; C_{salary} - cost of 1 h work of accompanying personnel; A – amortization of 1 h work of CNC laser cutting machine; N - useful life of equipment; C_{equip} - initial cost of the CNC laser cutting machine.

$C_{elec}^{on}, C_{tech}, C_{las}, C_{exp.mat}, C_{salary}^{on}, C_A^{on}$ are calculated by:

$$C_{elec}^{on} = P_{on} t_{on} C_{elec} \tag{4}$$

$$C_{tech} = V_{tech} C_{techM^3} t_{on} \tag{5}$$

$$C_{las} = V_{las} C_{lasM^3} t_{on} \tag{6}$$

$$C_{exp.mat} = \frac{C_{exp.mat.Unit} t_{on}}{t_{exp.mat.Term}} \tag{7}$$

$$C_{salary}^{on} = C_{salary} t_{on} \tag{8}$$

$$C_A^{on} = \frac{1}{N} \frac{C_{equip}}{1920} t_{on} \tag{9}$$

In order to calculate C_{off} the following notations are entered: P_{off} – the electricity costs for 1 h of CNC laser machine work on air motion, kW/h; t_{off} – the time spent on 1 m of air tool motion, h. Consequently C_{off} is calculated by:

$$C_{off} = P_{off} t_{off} C_{elec} + C_{salary} t_{off} + \frac{1}{N} \frac{C_{equip}}{1920} t_{off} \tag{10}$$

In order to calculate C_{pt} the following notations for cost parameters calculation on 1 pierce point are entered: $C_{exp.mat}^{pt}$ - the cost of expendable materials (for example, adjudge, protective glass, gas tubes); C_{tech}^{pt} - the cost of technological gas (nitrogen or oxygen depending on processed material); C_{las}^{pt} - the cost of cutting gas (when working on a gas flow laser); C_{elec}^{pt} - the cost of electricity; C_{salary}^{pt} - the cost related with salary of accompanying personnel; C_A^{pt} - amortization of equipment. In general C_{pt} is calculated by:

$$C_{pt} = C_{elec}^{pt} + C_{exp.mat}^{pt} + C_{las}^{pt} + C_{tech}^{pt} + C_{salary}^{pt} + C_A^{pt} \tag{11}$$

In order to calculate $C_{elec}^{pt}, C_{exp.mat}^{pt}, C_{las}^{pt}, C_{tech}^{pt}, C_{salary}^{pt}, C_A^{pt}$ the additional notations are entered: P_{pt} – the electricity costs for 1 pierce point, kW/h; t_{pt} – the time spent on 1 pierce point, h. Consequently C_{elec}^{pt}, $C_{exp.mat}^{pt}, C_{las}^{pt}, C_{tech}^{pt}, C_{salary}^{pt}, C_A^{pt}$ are calculated by:

$$C_{elec}^{pt} = P_{pt} t_{pt} C_{elec} \tag{12}$$

$$C_{exp.mat}^{pt} = V_{tech} C_{techM^3} t_{pt} \tag{13}$$

$$C_{las}^{pt} = V_{las} C_{lasM^3} t_{pt} \tag{14}$$

$$C^{pt}_{exp.mat} = \frac{C_{exp.mat.Unit} t_{pt}}{t_{exp.mat.Term}} \tag{15}$$

$$C^{pt}_{salary} = C_{salary} t_{pt} \tag{16}$$

$$C^{pt}_{A} = \frac{1}{N} \frac{C_{equip}}{1920} t_{pt} \tag{17}$$

During calculation of C_{pt} and C_{on} the following parameters C^{pt}_{las} and C_{las} must be taken into account when processing of material at flow-through gas laser machines. The parameter C^{pt}_{tech} must be considered during calculation of F_{cost} when technological gas is applied.

Consequently, F_{cost} can be written as follows:

$$\begin{aligned} F_{cost} = L_{on}\left(C^{on}_{elec} + C_{tech} + C_{las} + C_{exp.mat} + C^{on}_{salary} + C^{on}_{A}\right) + L_{off} C_{off} \\ + N_{pt}\left(C^{pt}_{elec} + C^{pt}_{exp.mat} + C^{pt}_{las} + C^{pt}_{tech} + C^{pt}_{salary} + C^{pt}_{A}\right) \end{aligned} \tag{18}$$

The main expendable materials for gas laser include: swivel mirrors, focusing lenses, protective glasses, nozzles, adjusting units, gas tubes. The main expendable materials for fiber laser are nozzles, protective glasses, focusing lenses. And for the case of using solid-state lasers, the expendable materials are optical pumping lamps, protective glasses, mirrors, a quantron, an active element. In [24] the values of cost parameters C_{on}, C_{off}, C_{pt} are presented by taken into account the above parameters (3)–(17) for CNC laser cutting machine by example ByStar 3015. For each type of material the parameters C_{on}, C_{off}, C_{pl} are calculated by taken into account that $V_{on} = const$. As the practical shown [19, 26] the value of actual working speed of cutting tool is varied by various technological factors and parameters of NC programs. Consequently problem of accurate calculation of cost function (2) in optimization problem of tool path construction is arisen. In order to solve the encountered problem the need of correction parameters calculation for V_{on} values is emerged.

Based on proposed methodology the subsystem "Cutting cost calculation" was developed by using Net. Framework technology. A subsystem may be integrated with existing CAM software. Figure 2 presents interface of developed subsystem. In order to calculate F_{cost} the values of basic cost parameters C_{on}, C_{off}, C_{pt} are added into database in XML format.

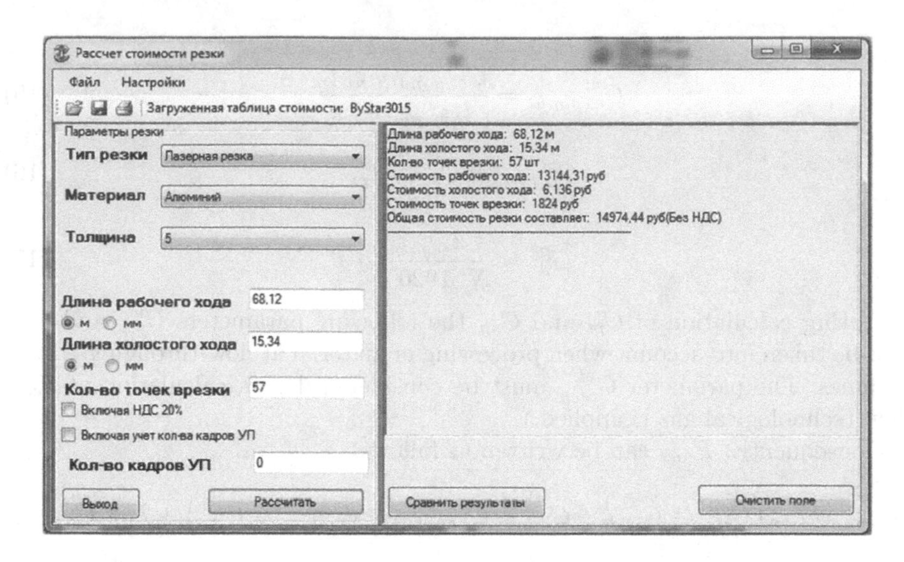

Fig. 2. The interface of "Cutting cost calculation" subsystem

2.2 Accurate Calculation of V_{on} in Objective Function from the Example of the CNC Laser Cutting Machine ByStar 3015

The inaccuracy of the actual cutting time and cost calculation is due to the fact that V_{on}, which is programmed as constant value in NC program, is actually varied by various technological factors. It was found that increasing of frames numbers in NC program for various sets of parts, which have the same total perimeter of the contours, the actual V_{on} is decreased [19,26]. The reasons why NC program can contain a large numbers of frames is mainly due to the contours of complex geometry (for example, splines) when converting from CAD systems to a CAM are divided into a large numbers of geometric primitives due to the difference of geometric file formats (for example, on segments of straight lines and circular arcs), i.e. approximated by simple geometric primitives. The difference in formats is due to the fact that almost all CNC systems are equipped with only linear and circular interpolators. As a rule the approximation of a complex geometry reduces to a linear approximation.

The functional dependence of V_{on} should be determined by science-based table functions or analytically. However in practice $V_{on} = const$ and in this case the accuracy of objective function calculation during cutting path optimization is not provided The algorithmization of objective function (2) calculation based on science–based determination of function parameters is requirement for the development of cutting tool path optimization algorithms. The cutting tool route is optimal only if the objective function is adequate calculated. For this reason the exact parameters values of objective function (2) must be calculated.

Some practical results on determining dependence of cutting speed on number of NC program commands are given below. Based on received results the objective function (2) can exactly calculated and exact optimal cutting path can constructed.

The research was conducted for following materials: 1.0114 (thickness $\Delta = 1$–10 mm) and AWAIMg3 (thickness $\Delta = 1$–5 mm). In order to conduct experiments the 150 NC programs for cutting of various types of parts with numbers of frames $n = \overline{10, 5000} (n \in \mathbb{N})$ for 1.0114 and 150 NC programs for cutting of various types of parts with numbers of frames $n = \overline{10, 2000} (n \in \mathbb{N})$ for AWAIMg3 were developed.

The statistical materials were processed by using "Mathcad". Based on received results the following upshots were made:

1. The actual average speed of cutting tool speed V_{act} is monotonically decreasing function depending on frames numbers of NC program (Fig. 3);
2. The predetermined cutting tool speed V_{on} coincides with the actual average speed when the numbers of frames reaches a certain threshold value N. When the frames numbers $n < N$, then the actual speed is greater than predetermined cutting tool speed, if the frames numbers of NC program is arisen ($n > N$) then the actual speed is less than predetermined cutting tool speed of NC program (in the experiments the reduction of average actual cutting tool speed value compared with predetermined cutting tool speed in NC program is 70%);
3. The threshold value N is varied for different thickness and grade of material.

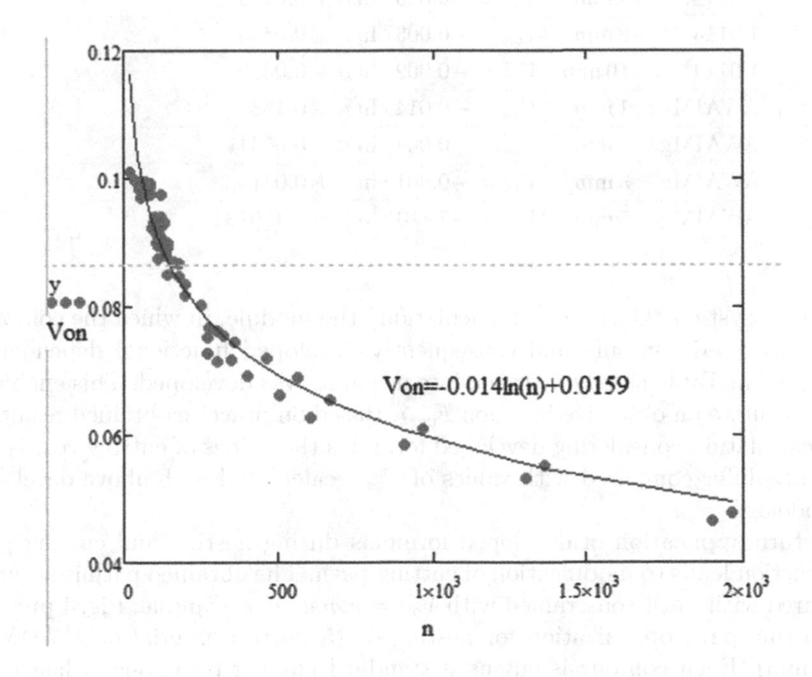

Fig. 3. Change of the real cutting tool speed for AWAIMg3, $\Delta = 1$ mm

In order to present the results of computational experiments the following notations are introduced: n – number of NC program commands; V_{act} - the actual speed of cutting tool; N – the number of commands when; $\sum \varepsilon_n^2$ - the deviation squares sum of the original data from the values of the approximation functions at these points.

When approximation the actual speed dependence presented on point chart on the number of commands with approximating curves in “Mathcad” $\sum \varepsilon_n^2 \to 0$ for all values of studied grade materials and thickness are achieved using logarithmic approximation function. Figure 3 presents following results for material of AWAIMg3 with $\Delta = 1$ mm. Similar results were obtained for AWAIMg3 with $\Delta = 1$–5 mm and 1.0114 with $\Delta = 1$–10 mm. The generalized formulas for calculating of cutting tool speed by example CNC laser cutting machine ByStar3015 are presented in Table 1.

Table 1. Generalized table of formulas for calculating of cutting tool speed by example CNC laser cutting machine ByStar3015

Material	Δ	Formulas for cutting tool speed calculation
1.0114	1 mm	$V_{on} = -0.025 \cdot \ln n + 0.25$
1.0114	2 mm	$V_{on} = -0.015 \cdot \ln n + 0.1711$
1.0114	3 mm	$V_{on} = -0.009 \cdot \ln n + 0.1062$
1.0114	3.5 mm	$V_{on} = -0.006 \cdot \ln n + 0.0759$
1.0114	4 mm	$V_{on} = -0.006 \cdot \ln n + 0.0709$
1.0114	8 mm	$V_{on} = -0.003 \cdot \ln n + 0.0443$
1.0114	10 mm	$V_{on} = -0.002 \cdot \ln n + 0.0359$
AWAIMg3	1 mm	$V_{on} = -0.014 \cdot \ln n + 0.1589$
AWAIMg3	2 mm	$V_{on} = -0.004 \cdot \ln n + 0.0641$
AWAIMg3	3 mm	$V_{on} = -0.001 \cdot \ln n + 0.0315$
AWAIMg3	5 mm	$V_{on} = -7 \cdot 10^{-4} \cdot \ln n + 0.0182$

For subsystem “Cutting cost calculation” the module, in which the complexity of processed contours and consequently developed functional dependences (presented in Table 1) may be taken into account, was developed. This enable to exact calculate an objective function F_{cost}. Based on practice obtained results of F_{cost} calculation considering developed formulas the values of cutting cost is significantly differ compared with values of F_{cost} calculated with above developed methodology.

In turn application of developed formulas during nesting and cutting path construction leads to modification of cutting path. The obtained path is accurate compared with path constrained with $V_{on} = const$. For example, Fig. 4 presents the cutting path optimization for nesting of 15 parts (material of AWAIMg3, $\Delta = 1$ mm). Each contour is cut used standard cutting technique (when number of piercing equals number of parts contours). In order to reduce acceptable solutions set the acceptable piercings set is limited with finite aggregate consists of 55 points (these points are green squares in Fig. 4, in turn points of tool switching off are X). The blue arrows are idle moves of cutter. The number of NC program commands for this nesting is $n = 120$. For the case of Fig. 4(a) $V_{on} = const = 0.1\,\mathrm{m/s}$, for the case of Fig. 4(b) $V_{on} = -0.014 \cdot \ln n + 0.1589$.

Based on proposed results (Fig. 4) the accurate calculation of objective function ensures not only the exact computation of function extremum value but also the correct results of optimal cutting path search taking into account parts complexity.

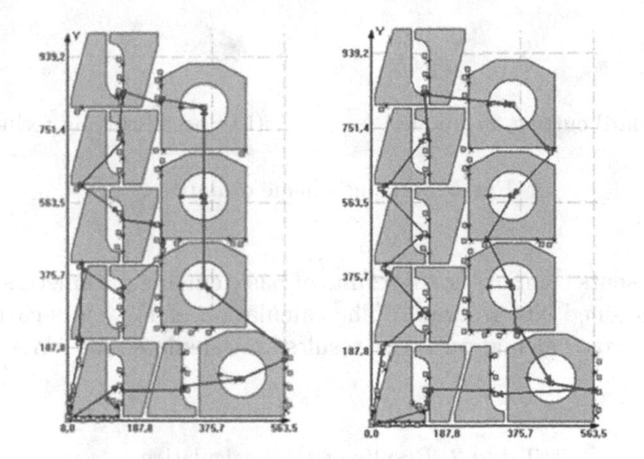

(a) Standard cutting technique (b) Special cutting technique

Fig. 4. Optimal cutting path (Color figure online)

3 Computational Experiments

The proposed methodology of F_{cost} calculation taking into account dependence of cutting speed V_{on} on parts complexity is useful during practice technological problems solving in terms of optimal cutting route planning on CNC thermal machines. There is example of cutting route planning below with reduction of F_{cost} for shaped parts taken into account application of special cutting techniques with thermal deformation reduction developed in [25]. The conditions of thermal deformation reduction during nesting and cutting route planning are considered in [15].

Based on algorithms presented on [25] the cutting tool route for nesting is automatically built at CAD/CAM "SIRIUS". In order to evaluate the effectiveness of developed special cutting techniques two nestings are obtained by using standard (Fig. 5(a)) and special (Fig. 5(b)) cutting techniques for various types of parts.

Figure 5(a) presents the cutting tool path built for various geometrical types of parts including circles using standard cutting techniques. Figure 5(b) presents the cutting tool path built for various geometrical types of parts including circles using special cutting techniques.

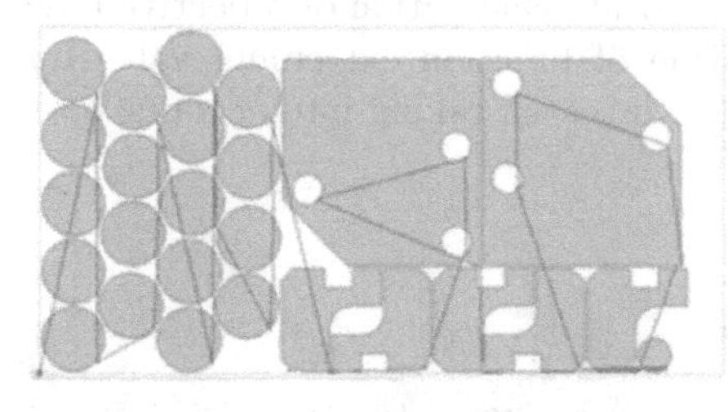

(a) Standard cutting technique

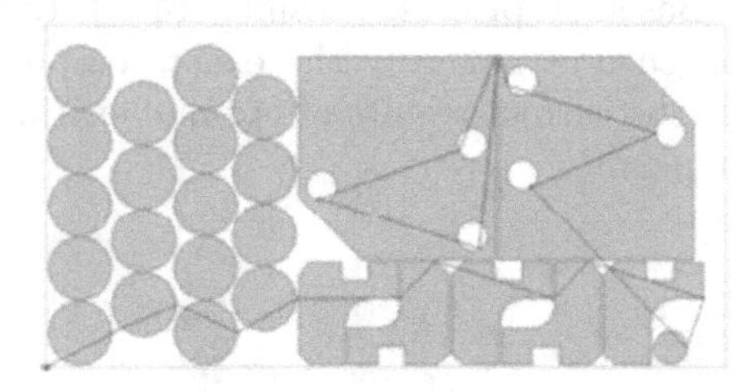

(b) Special cutting technique

Fig. 5. Cutting scheme example

Table 2 presents computational results of basic cutting parameters and values of F_{cost} for obtained NC programs. The calculation of F_{cost} was carried out by using "Cutting cost calculation". The results are calculated for AWAIMg3 $\Delta = 1$ and 5 mm.

Table 2. Results of F_{cost} calculation

Material	Δ	Technique	Fig.	N_{pt}	L_{off}, m	L_{on}, m	n	F_{cost}, rub	F^{n}_{cost}, rub	%
AWAIMg3	1 mm	Standard	Fig. 5(a)	32	16.43	36.82	130	809.5	866.5	6.6
		Special	Fig. 5(b)	18	8.6	36.97		757.1	814.3	7
	5 mm	Standard	Fig. 5(a)	32	16.43	36.82		13121.1	16062.2	18.3
		Special	Fig. 5(b)	18	8.6	36.97		12748.1	15701.2	18.8

The following notations in Table 2 are used: n – numbers of frames in NC program; F_{cost} - the cutting cost calculated taking into account that $V_{on} = const$; F^{n}_{cost} - the cutting cost calculated taking into account that $V_{on} = var$ and depends on the frames numbers of NC program; % - value of difference between F_{cost} and F^{n}_{cost}.

The results presented in Table 2 indicate that the basis cutting parameters are reduced by using special cutting techniques compared with standard cutting. In turn the difference between F_{cost} and F^{n}_{cost}. reaches to 18%.

4 Conclusion

In this paper the following results were obtained:

1. In order to exact calculate objective function and consequently to construct exact tool path the methodology of objective function F_{cost} and basic cost parameters calculation is presented for CNC laser cutting machines. Due to the problem of exact F_{cost} values calculation have arisen on many production factories and based on proposed methodology the subsystem "Cutting cost calculation" were developed used Net. Framework technology, which may be integrated with existing CAM software;
2. The functional dependencies on number of NC program commands for V_{on} are developed. These dependencies ensure exact calculation of objective function F_{cost} and exact tool optimization path construction. For subsystem "Cutting cost calculation" the module of exact cutting cost computation of cutting cost was developed used functional dependencies for V_{on};
3. In order to evaluate the developed results the computational experiments have been conducted taking into account previously proposed special cutting techniques compared with standard cutting. The cutting route is constructed with taking into account the thermal deformation reduction. The correct F_{cost} values calculation is carried out with developed above methodology given the $V_{on} = var(F^n_{cost})$ and $V_{on} = const(F_{cost})$. The results shown that the cutting cost is reduced by using special cutting techniques compared with standard cutting. In turn the difference between F_{cost}. and F^n_{cost} reaches to 18%.

Acknowledgments. The work was supported by the Russian Foundation for Basic Research (grant №19-01-00573).

References

1. Chen, M., Li, X., Tang, K.: Optimal air-move path generation based on MMAS algorithm. Int. J. Prod. Res. **52**(24), 7310–7323 (2014). https://doi.org/10.1080/00207543.2014.922713
2. Dewil, R., Vansteenwegen, P., Cattrysse, D.: Sheet metal laser cutting tool path generation: dealing with overlooked problem aspects. Key Eng. Mater. **639**, 517–524 (2015). https://doi.org/10.4028/www.scientific.net/KEM.639.517
3. Dewil, R., Vansteenwegen, P., Cattrysse, D.: A review of cutting path algorithms for laser cutters. Int. J. Adv. Manuf. Technol. **87**(5–8), 1865–1884 (2016). https://doi.org/10.1007/s00170-016-8609-1
4. D'Souza, K.C.R., Wright, P.K.: Tool-path optimization for minimizing airtime during machining. J. Manuf. Syst. **22**, 173–180 (2003)
5. Faizrahmanov, R., Murzakaev, R., Burylov, A., Pristupov, V.: The minimization of total cutting time taking into account technologies for CNC machines. Elektrotekhnika **11**, 7–12 (2016)
6. Frolovskii, V.: The automation of NC program design in thermal cutting at CNC machines. Inf. Technologii v proektirovanii i proizvodstve **4**, 63–66 (2005)

7. Ghiani, G., Improta, G.: The laser-plotter beam routing problem. J. Oper. Res. Soc. **52**(8), 945–951 (2001). https://doi.org/10.1057/palgrave.jors.2601161
8. Grötschel, M., Jünger, M., Reinelt, G.: Optimal control of plotting and drilling machines: a case study. Zeitschrift für Oper. Res. **35**(1), 61–84 (1991). https://doi.org/10.1007/BF01415960
9. Hoeft, J., Palekar, U.S.: Heuristics for the plate-cutting traveling salesman problem. IIE Trans. **29**(9), 719–731 (1997). https://doi.org/10.1023/A:1018582320737
10. Konnikova, E., Kyklina, A.: The specific of time per pierce determining when calculating of production cost using laser cutting technology. Economica I ypravlenie narodnim hozyaistvom **30**(2), 41–49 (2016)
11. Lee, M.K., Kwon, K.B.: Cutting path optimization in CNC cutting processes using a two-step genetic algorithm. Int. J. Prod. Res. **44**(24), 5307–5326 (2006). https://doi.org/10.1080/00207540600579615
12. Manber, U., Israni, S.: Pierce point minimization and optimal torch path determination in flame cutting. J. Manuf. Syst. **3**(1), 81–89 (1984). https://doi.org/10.1016/0278-6125(84)90024-4
13. Milcin, A., Shabanov, M., Lysych, M., Romanov, V.: The modern cutting methods of sheet materials. Actualnye napravleniya nauchnykh issledovanii XXI: teoriya i praktika **3**(2-1), 83–87 (2015)
14. Murzakaev, R., Shilov, V., Burylov, A.: Application of metaheuristic algorithms to minimize idle running length of cutting tool. J. PNRPU Bull. Electrotechnics Inf. Technol. Control Syst. **14**, 123–136 (2015)
15. Petunin, A.: About some strategies of tool routing when designing NC programs applied to thermal cutting machines. Bulletin UGATU, Upravlenie, VTiT **35**(2), 280–286 (2009)
16. Petunin, A.: The two problems of cutting tool routing for CNC figured cutting machines. Intell. Technol. Inf. Process. Manage. **1**, 215–220 (2014)
17. Petunin, A., Chentsov, A., Chentsov, P.: About one problem of cutting tool routing during sheet cutting. Model. Anal. Inform. Sist. **22**, 278–297 (2015)
18. Petunin, A., Krotov, V.: About classification of cutting techniques for CNC machines and problem of tool path optimization. In: Materialovedenie, pp. 466–475. Mashinostroenie, Energetika (2015)
19. Petunin, A., Tavaeva, A.: Optimization of tool route for CNC shape cutting machines provided that working stroke speed is not constant value. Fundamental Res. **6**, 56–62 (2015)
20. Petunin, A.A.: Modeling of tool path for the CNC sheet cutting machines. AIP Conf. Proc. **1690**(1), 060002 (2015). https://doi.org/10.1063/1.4936740
21. Shaparev, A., Savina, A.: Technological and economical comparison of plasma and CO2 laser cutting of sheet material. Nauka I sovremennost **9**(3), 181–189 (2016)
22. Sherif, S.U., Jawahar, N., Balamurali, M.: Sequential optimization approach for nesting and cutting sequence in laser cutting. J. Manuf. Syst. **33**(4), 624–638 (2014). https://doi.org/10.1016/j.jmsy.2014.05.011
23. Tavaeva, A., Petunin, A.: The some problems of cutting tool optimization applied to CNC thermal cutting machines. Bull. BSTU named after V.G. Shukhov **2**(9), 147–153 (2017). https://doi.org/10.12737/article_59a93b0b29fa13.40976330
24. Tavaeva, A., Petunin, A.: The accurate calculation of parts treatment cost from sheet metal on the CNC laser cutting machine in problem of tool routing optimization. J. Model. Optim. Inf. Technol. (2019). https://moit.vivt.ru/?p=7489&lang=ru. Accessed 15 Feb 2019

25. Tavaeva, A., Petunin, A., Krotov, V.: About effectiveness of special cutting techniques in developing of NC programs for CNC thermal cutting machines. In: Proceedings of the Workshop on Computer Science and Information Technologies, CSIT 2017, vol. 1, pp. 221–227 (2017)
26. Tavaeva, A., Kurennov, D.: Cost minimizing of cutting process for CNC thermal and water-jet machines. AIP Conf. Proc. **1690**(1), 020003 (2015). https://doi.org/10.1063/1.4936681
27. Verhoturov, M., Tarasenko, P.: Mathematical provision of tool path problem at flat shape nesting based on "chained" cutting. Bulletin UGATU, Upravlenie, VTiT **27**(2), 123–130 (2008)
28. Yang, W.B., Zhao, Y.W., Jie, J., Wang, W.L.: An effective algorithm for tool-path airtime optimization during leather cutting. Adv. Mater. Res. **102–104**, 373–377 (2010). https://doi.org/10.4028/www.scientific.net/AMR.102-104.373

A Local Branching MIP Heuristic for a Real-World Curriculum-Based Course Timetabling Problem

Pasquale Avella[1], Maurizio Boccia[2], Sandro Viglione[1], and Igor Vasilyev[3](✉)

[1] DING – Dipartimento di Ingegneria, Università del Sannio, Piazza Roma 21, 82100 Benevento, Italy
{avella,sandro.viglione}@unisannio.it

[2] DIETI – Dipartimento di Ingegneria Elettrica e delle Tecnologie dell'Informazione, Universitá degli Studi di Napoli "Federico II", Naples, Italy
maurizio.boccia@unina.it

[3] Matrosov Institute of System Dynamics and Control Theory, Siberian Branch of Russian Academy of Sciences, Lermontov str. 134, 664033 Irkutsk, Russia
vil@icc.ru

Abstract. Automated timetabling is a challenging area in the timetabling and scheduling theory and practice, intensively addressed in research papers in the last two decades. There are three main classes of problems, which are usually studied: school timetabling, course timetabling and examination timetabling. In this report, we address a case study of the Curriculum-Based Course Timetabling (CB-CTT) problem, arising at Engineering Department of Sannio University. In general, the problem consists of finding a feasible weekly assignment of course lectures to rooms and time periods while respecting a wide range of constraints, which have to be either strictly satisfied (hard constraints) or satisfied as much as possible (soft constraints). The case study here addressed here has many special requirements due to local organizational rules. We were able to model the complex requirements by an Integer Programming formulation. The solution approach consists of using an MIP solver, integrated with two local branching heuristics tailored for the problem. The effectiveness of the proposed approach is illustrated by the computational results on two real instances.

Keywords: Automated timetabling · Curriculum-based course timetabling · Integer programming · Heuristic

I. Bykadorov et al. (Eds.): MOTOR 2019, CCIS 1090, pp. 438–451, 2019.
https://doi.org/10.1007/978-3-030-33394-2_34

1 Introduction

Automated timetabling is a challenging area in the timetabling and scheduling theory and practice, intensively addressed in research papers in the last two decades. The problem statements, its classifications, the-state-of-the-art approaches, successful applications can be found in many surveys: Burke et al. [9], Schaerf [34], Burke and Petrovic [14], McCollum [29], Lewis [25], Kristiansen [23], Babaei et al. [4], Pillay [33], Bettinelli [7]. There are three main classes of problems, which are usually studied: school timetabling, course timetabling and examination timetabling.

In this paper we address a case study of the Curriculum-Based Course Timetabling (CB-CTT) problem, arising at the Engineering Department of Sannio University, a small university located in Southern Italy. In general, the problem consists of finding a feasible weekly assignment of course lectures to rooms and to time periods whilst respecting a wide range of constraints, which have to be either strictly satisfied (hard constraints) or satisfied as much as possible (soft constraints). A formal definition of CB-CTT problem, has been provided by Di Gaspero et al. [17] and McCollum et al. [30] in the Second International Timetabling Competition (ITC-2007), along with a set of benchmark instances. Several variants and extensions of this problem have been proposed by Bonutti et al. [8].

The CB-CTT problem is known to be NP-hard as it was shown in Burke et al. [11]. Real-world instances, proposed by Di Gaspero et al. [17] and Bonutti et al. [8], are very hard to solve to optimality, so great attention has been paid to developing heuristic approaches, as: Tabu search by Clark et al. [16], Lü et al. [27,28]; Simulated annealing of Geiger [20], Bellio et al. [6], Tarawneh et al. [36]; combination of different approaches in hybrid algorithms of Müller [31], Shaker [35], Bellio et al. [6], Kiefer et al. [22].

For deeper investigation of CB-TTT problem, Integer Linear Programming (ILP) formulations were proposed and investigated by many researchers. In general case of ITC-2007 problem, Burke et al. [10,13] proposed a compact ILP formulation, called *Monolithic.* Such formulation even if not leading to an optimal solution of nontrivial instances, was successfully used in heuristic approaches to get lower bounds [13]. It was also a base for the branch-and-cut algorithm of Burke et al. [12], where some small instances were solved to optimality. There are many attempts to develop exact algorithms for CB-TTT problem, but they are only able to improve lower bounds without proving the optimality (see, for example, the divide-and-conquer approach of Hao and Benlic [21], the column generation method of Cacchiani et al. [15], Benders decomposition in Bagger [5]). Exact approaches and lower bounds for different university timetabling problems have been also presented in [3,26,32].

Problem statements and benchmark instances proposed in [8,17] are quite general and can be applied to many real-world cases. Anyway, it is almost impossible to state a CB-CTT problem that fits all local rules and requirements of every university or department. In this paper we focus on a case study at the Engineering Department of Sannio University, which is very challenging as it includes many special requirements which are not usually met in the general CB-CTT problem. We remark that this problem also differs substantially from the one previously considered in Avella et al. [3] more than ten years ago, as pointed out in Sect. 2. Therefore we develop a new ILP formulation for this case and show preliminary computational results of using a commercial ILP solver and some simple heuristics on real instances. Finally we remark that the overall approach was implemented using the modelling language JuMP/Julia [18], which has turned out to be a very effective tool for fast prototyping real problems.

The remainder of the paper is organized as follows. In Sect. 2 we give a formal statement of the problem. In Sect. 3 we outline the ILP formulation of the case study. In Sect. 4 the ILP-based solution approach is outlined and finally in Sect. 5 we provide computational results for the case study.

2 Problem Statement

In this section, we describe the problem statement, highlighting the differences from the ITC-2007 instances and from those previously considered in [3]. First, let us introduce the main notations used further in the text.

- Let $T = \{1, \ldots, \bar{t}\}$ be the discrete time horizon, i.e. a set of time periods in which lectures can be given.
- Let $D = \{1, \ldots, \bar{d}\}$ be the teaching days of the week. For any $d \in D$, let τ_d and ι_d denote, respectively, the first time slots of the morning and afternoon session in day d.
- Let $C = \{1, \ldots, \bar{c}\}$ be a set of courses. For any $c \in C$, let n_c be the number of teaching hours (lectures) to be scheduled per week and let $n^c_{\min}$ and $n^c_{\max}$ be, respectively, the minimum and maximum daily number of teaching hours (no less than $n^c_{\min}$ and no more than $n^c_{\max}$ lectures have to be assigned to the day d, if the course c is scheduled in d).
- Let Π be a set of pairs of incompatible courses, i.e. if $(c_1, c_2) \in \Pi$ then the courses c_1 and c_2 cannot be taught in the same day.
- Let $R = \{1, \ldots, \bar{r}\}$ be a set of rooms, which is divided with subsets of close rooms $R_1, \ldots, R_K$, such that $\cup_{k=1}^{K} R_k = R$ and $R_k \cap R_l = \emptyset$ if $k \neq l$. Usually, the rooms from different subsets are located in different buildings.
- Let $G = \{1, \ldots, \bar{g}\}$ be a set of curricula. A curriculum is a group of courses having common students, which we call a class to be consistent with [3]. Let $C_g \subset C$ denote the courses of class $g \in G$ and $l_{\max}$ be the maximum daily number of teaching hours allowed for any class $g \in G$ and m_g is the maximum number of days to be involved in teaching for class $g \in G$.
- Let $S = \{1, \ldots, \bar{s}\}$ be a set of teachers and, for any $s \in S$, let $\mathcal{C}_s \subset C$ be the subset of courses taught by teacher s. Let k_s denote the maximum weekly number of teaching days allowed for the teacher s.

The problem consists of assigning course lectures to rooms and time periods, while satisfying hard constraints and minimizing the violation of soft constraints. The hard constraints are:

(i) For each course $c \in C$: n_c hours a week must be scheduled.
(ii) For each class $g \in G$: class g cannot attend more than one course at time $t \in T$.
(iii) For each teacher $s \in S$: teacher s cannot teach more than one course at time $t \in T$.
(iv) For each room $r \in R$: room r cannot host more than one course at time $t \in T$.
(v) If a course $c \in C$ is scheduled in day $d \in D$, it should take between $n^c_{\min}$ and $n^c_{\max}$ hours.
(vi) No class can attend more than l_{max} teaching hours a day.
(vii) For each class $g \in G$: the class g cannot attend more than m_g teaching days a week.
(viii) For each teaching day $d \in D$: the class g has a "lunch break" at a time slot between $\iota_d - 1$ or ι_d.
(ix) Each pair of courses $c_1, c_2 \in \Pi$ cannot be scheduled in the same day.
(x) For each course $c \in C$, the timetable should be "compact". If two hours of the same course c are scheduled in day d, they have to be assigned to adjacent time periods. In other words, let t_1 and t_3 ($t_1 < t_3$) be time periods belonging to the same day. If course c is assigned to the time periods t_1 and t_3, the same course should be scheduled at every time period between t_1 and t_3 as well, to guarantee compactness.
(xi) For each class $g \in G$, the timetable should be "compact": for each class, empty periods between any two courses are not allowed, except for the lunch break.
(xii) All the hours of a course $c \in C$ scheduled in a day $d \in D$ should be located in the same room $r \in R$.
(xiii) During a day, a class can move only between the rooms in the same building.
(xiv) A teacher $s \in S$ cannot give lectures for more than k_s days a week.
(xv) Due to the availability of equipment and capacity of the room, course $c \in C$ can be assigned to a subset of rooms $R_c \subseteq R$.
(xvi) A room $r \in R$ is available in a subset of time slots $T_r \subseteq T$.
(xvii) A teacher $s \in S$ is available in a subset of the time slots $\mathcal{T}_s \subseteq T$.

The soft constraints whose violation have to minimized are defined as follows:

1. The number of times a class has to move to another room in the same day is penalized with p_1.
2. The number of classes that attend courses in last two hours in the afternoon is penalized with p_2.

The overall problem is depicted in Fig. 1. Due to many special requirements, the problem does not fit with the problem stated in ITC-2007 [17], the main difference being that the soft constraints in [17] are indeed hard in our problem. More precisely these constraints are those named Minimum Working Days, Curriculum Compactness, Room Stability and Room Capacity respectively in [17].

Several generalizations are presented in [8], but they do not suit our case. For example, they do not consider requirements (vii)–(ix), (xi), (1), (2). The most similar problem is the one considered in [3]. The main difference is that it requires that all the courses of a class in one day must be scheduled either in the morning or in the afternoon session and the teacher's preferences are the soft constraints. Moreover, it does not contain requirements (viii), (xi), that the class timetable must be compact with the time window during the lunch break, which have a big impact on problem formulation presented in the next section.

3 Integer Linear Programming Formulation

To define an Integer Linear Programming (ILP) formulation for the timetabling problem stated above, we introduce the following integer variables:

- $x_{crt} = 1$ if course $c \in C$ is scheduled in room $r \in R$ at time $t \in T$, $x_{crt} = 0$ otherwise;
- $u_{cd} = 1$ if course $c \in C$ is assigned to the day $d \in D$, 0 otherwise;
- $y_{gd} = 1$ if class $g \in G$ has lectures on the day $d \in D$, 0 otherwise;
- $v_{gd} = 1$ if class $g \in G$ has lectures in the morning of day $d \in D$, 0 otherwise;
- $w_{gd} = 1$ if class $g \in G$ has lectures in the afternoon of day $d \in D$, 0 otherwise;
- $\psi_{sd} = 1$ if $d \in D$ is a teaching day for teacher $s \in S$, 0 otherwise.
- $\phi_{grd} = 1$ if class g has a lecture in room r on day $d \in D$, 0 otherwise.
- ω_{gd} equals to the number of moves of class $g \in G$ on day $d \in D$ between different rooms.

With these variables, a formulation that meets both basic and local requirements is:

$$\min\ p_1 \sum_{g \in G} \sum_{d \in D} \omega_{gd} + p_2 \sum_{c \in C} \sum_{r \in R} \sum_{(d-1) \in D} (x_{cr\tau_d - 2} + x_{cr\tau_d - 1}) \tag{1}$$

$$\sum_{r \in R} \sum_{t \in T} x_{crt} = n_c, \qquad c \in C \tag{2}$$

$$\sum_{c \in C_g} \sum_{r \in R} x_{crt} \leq 1, \qquad g \in G,\ t \in T \tag{3}$$

$$\sum_{c \in C_s} \sum_{r \in R} x_{crt} \leq 1, \qquad s \in S,\ t \in T \tag{4}$$

$$\sum_{c \in C} x_{crt} \leq 1, \quad r \in R, t \in T \tag{5}$$

$$\sum_{r \in R} \sum_{\tau_d \leq t < \tau_{d+1}} x_{crt} \geq n^c_{\min} u_{cd}, \quad c \in C, d \in D \tag{6}$$

$$\sum_{r \in R} \sum_{\tau_d \leq t < \tau_{d+1}} x_{crt} \leq n^c_{\max} u_{cd}, \quad c \in C, d \in D \tag{7}$$

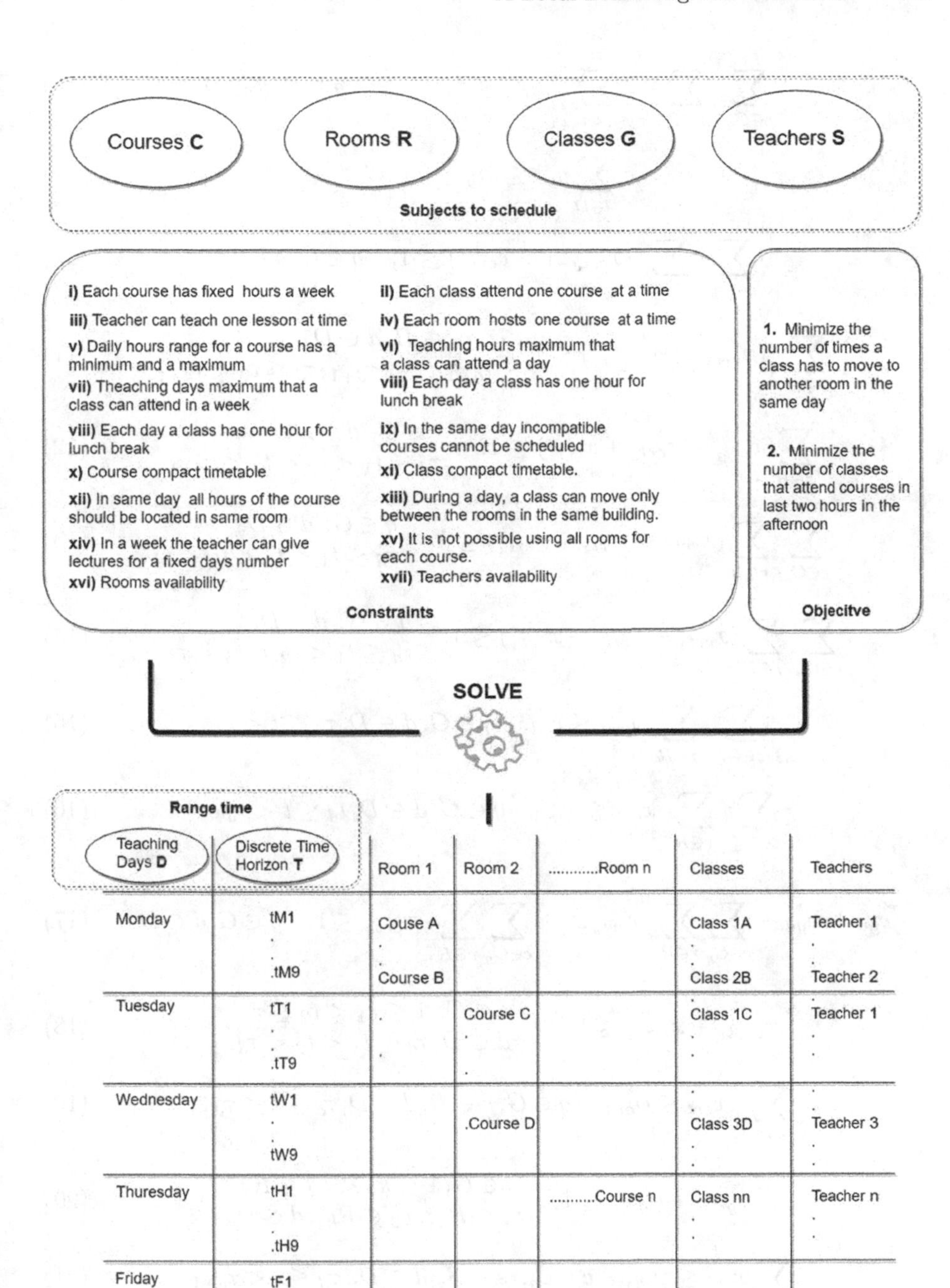

Fig. 1. Timetabling problem

$$\sum_{c \in C_g} \sum_{r \in R} \sum_{\tau_d \le t < \tau_{d+1}} x_{crt} \le l_{\max} y_{gd}, \quad g \in G, d \in D \tag{8}$$

$$\sum_{d \in D} y_{gd} \le m_g, \quad g \in G \tag{9}$$

$$\sum_{c \in C_g} \sum_{r \in R} (x_{cr\iota_d - 1} + x_{cr\iota_d}) \le 1, \quad g \in G, d \in D \tag{10}$$

$$\sum_{r \in R} (x_{c_1 r t_1} + x_{x_{c_2 r t_2}}) \le 1, \quad \begin{array}{l} c_1 c_2 \in \Pi, d \in D, \\ \tau_d \le t_1 < \tau_{d+1}, \tau_d \le t_2 < \tau_{d+1} \end{array} \tag{11}$$

$$\sum_{r \in R} (x_{crt_1} - x_{crt_2} + x_{crt_3}) \le 1, \quad \begin{array}{l} c \in C,\ d \in D, \\ \tau_d \le t_1 < t_2 < t_3 < \tau_{d+1} \end{array} \tag{12}$$

$$\sum_{c \in C_g} \sum_{r \in R} (x_{crt_1} - x_{crt_2} + x_{crt_3}) \le 1, \quad \begin{array}{l} g \in G,\ d \in D, \\ \tau_d \le t_1 < t_2 < t_3 < \iota_d \end{array} \tag{13}$$

$$\sum_{c \in C_g} \sum_{r \in R} (x_{crt_1} - x_{crt_2} + x_{crt_3}) \le 1, \quad \begin{array}{l} g \in G,\ d \in D, \\ \iota_d \le t_1 < t_2 < t_3 < \tau_{d+1} \end{array} \tag{14}$$

$$\sum_{c \in C_g} \sum_{r \in R} x_{crt} \le v_{gd}, \quad g \in G, d \in D, \tau_d \le t < \iota_d \tag{15}$$

$$\sum_{c \in C_g} \sum_{r \in R} x_{crt} \le w_{gd}, \quad g \in G, d \in D, \iota_d \le t < \tau_{d+1} \tag{16}$$

$$v_{gd} + w_{gd} - \sum_{c \in C_g} \sum_{r \in R} x_{cr\iota_d - 1} - \sum_{c \in C_g} \sum_{r \in R} x_{cr\iota_d} \le 1, \quad g \in G, d \in D \tag{17}$$

$$x_{cr_1 t_1} + x_{cr_2 t_2} \le 1, \quad \begin{array}{l} c \in C,\ 1 \le r_1 < r_2 \le \bar{r}, \\ d \in D, \tau_d \le t_1 < t_2 < \tau_{d+1} \end{array} \tag{18}$$

$$\sum_{c \in C_g} x_{crt} \le \phi_{grd}, \quad g \in G, r \in R, d \in D, \tau_d \le t < \tau_{d+1} \tag{19}$$

$$\phi_{gr_1 d} + \phi_{gr_2 d} \le 1, \quad \begin{array}{l} g \in G, 1 \le k_1 < k_2 \le K, \\ r_1 \in R_{k_1}, r_2 \in R_{k_2}, d \in D \end{array} \tag{20}$$

$$\sum_{r \in R} x_{crt} \le \psi_{sd}, \quad c \in \mathcal{C}_s,\ s \in S,\ d \in D,\ \tau_d \le t < \tau_{d+1} \tag{21}$$

$$\sum_{d \in D} \psi_{sd} \le k_s, \quad s \in S \tag{22}$$

$$\sum_{r \in R} \phi_{grd} - \omega_{gd} \le 1, \quad g \in G, d \in D \tag{23}$$

$$
\begin{aligned}
x_{crt} &\in \{0,\ 1\}, c \in C,\ r \in R,\ t \in T \\
y_{gd} &\in \{0,\ 1\},\ g \in G,\ d \in D \\
u_{gd} &\in \{0,\ 1\},\ g \in G,\ d \in D \\
w_{gd} &\in \{0,\ 1\},\ g \in G,\ d \in D \\
u_{cd} &\in \{0,\ 1\},\ c \in C,\ d \in D \\
\psi_{sd} &\in \{0,\ 1\},\ s \in S,\ d \in D \\
\omega_{gd} &\in \mathbb{Z}_+,\quad g \in G,\ d \in D
\end{aligned}
\tag{24}
$$

The objective function (1) minimizes the violation of the soft constraints (1) and (2) on p. x.

Constraints (2) require that the number of weekly hours for each course c is n_c (requirement (i)). Constraints (3) require that a class g cannot attend more than one course at time t (requirement (ii)). Requirement (iii) – a teacher cannot teach more than one course at time t – is defined by constraints (4). Requirement (iv) – a room r cannot host more than one course at time t – is defined by constraints (5).

Constraints (6) and (7) require that, if course c is scheduled in day d, i.e. if $u_{cd} = 1$, the number of daily hours of course c ranges between n^c_{min} and n^c_{max} (requirement (v)). Inequality (8) limits the maximum number of lectures for a class (requirement (vi)). The maximum number of teaching days (requirement (vii)) for a group is defined by constraints (9). The lunch breaks (requirement (viii)) are insured by constraints (10). Constraints (11) set that each incompatible pair of courses $c_1 c_2 \in \Pi$ cannot be scheduled in the same day (requirement (ix)).

Constraints (12) guarantee compactness for course c, i.e. they require that the time slots assigned to the day d be adjacent (requirement (x)). The class timetable compactness (requirement (xi)) is enforced by constraints (13)–(17), actually they ensure that only one window is possible during the lunch break. Constraints (18) require that all the teaching hours of course c scheduled in the day d be assigned to the same room r (requirement (xii)). The requirement (xiii), that a class can move only between the rooms in the same building is expressed by constraints (19)–(20). Constraints (21)–(22) limit the number of working days for each teacher (requirement (xiv)). Constraints (23) binds the number of class moves during a day. Finally, constraints (24) define the integrality of variables.

The requirements concerning capability/availability of rooms and teachers (requirements (xv)–(xvii)) are not expressed explicitly in the formulation (1)–(24). However such requirements can be easily given by fixing the corresponding variables, i.e.

$$x_{crt} = 0, c \in C, r \in R \setminus R_c,\ t \in T \tag{25}$$

$$x_{crt} = 0, c \in C, r \in R,\ t \in T \setminus T_r \tag{26}$$

$$x_{crt} = 0,\ s \in S, c \in C_s,\ r \in R,\ t \in T \setminus T_s \tag{27}$$

This formulation is based on those presented in [3], which was further more generalized and formalized as a "monolitic" formulation in [10,13]. In comparison with monolitic formulation, we have exactly the same variables x, u and constraints (3)–(5), but others are more or less different due to different sets of soft and hard constraints and local requirements.

4 Solution Approach

The problem turned out to be much harder to solve than the one addressed more than ten years ago in [3] as it contains much more complex requirements, which turn into a much larger size of the formulation.

State-of-the-art MIP solvers are not able to solve it to optimality, so we used a MIP solver as a heuristic tool, setting a short time limit (10 min). More specifically, we first set the MIP solver parameters to increase the time spent in MIP heuristics looking for good feasible solutions. On the other hand we introduced a customized Local Branching heuristic able to significantly improve the solutions provided by the MIP solver.

The idea is based on wide-spread approach of fixing some elements of timetable and solving a reduced problem, it is also known as the local branching heuristic. Some examples of this idea can be found in [13,16,24].

The local branching heuristics are MIP heuristics integrating local search and the outcomes of a MIP solver [19]. Given a feasible solution provided by the built-in heuristics of a MIP solver, they attempt to improve it by exploring a neighborhood of the current solution and iterating the procedure until the neighbor does not return any better feasible solution. In general-purpose local branching heuristics, neighbors for 0-1 IP problems are defined by introducing an additional constraint imposing that the new solution must not differ from the current one in more than k variables.

For our problem we introduce two local branching heuristics, defining two neighbors, strictly tailored for the problem:

Local Branching (LB) Strategy 1. Given a feasible solution $\bar{x}_{crt}$, which defines a feasible timetable, we define a neighborhood consisting of all the solutions in which a course keeps the assignment to the time slot, but the room can be changed. This is obtained by fixing corresponding variables and eliminating them from the formulation. Then we run the MIP solver over the reduced problem, within a prescribed time limit.

Local Branching (LB) Strategy 2. For a given a feasible solution $\bar{x}_{crt}$, we define a neighborhood consisting of all the solutions in which, a course keeps the assignment to the room and to the time slot of a day, but the day can be changed.

As shown in Fig. 2, we run Cplex solver for 10 min on the overall formulation to get an initial upper bound UB_{ini} and a lower bound LB. Starting from the current solution UB_{ini} we run the two MIP heuristics described above and we choose UB_{fin} as the better between the two results. In our computational experience we also tried to run the two heuristics in sequence, but we could not get any significantly improvement. The time limit of the first stage is set to

10 min to make the algorithm usable as a decision tool to carry out a what-if analysis on the availability/unavailability of some rooms. Moreover, as shown in Fig. 2, a longer time limit does not lead to significantly better solutions.

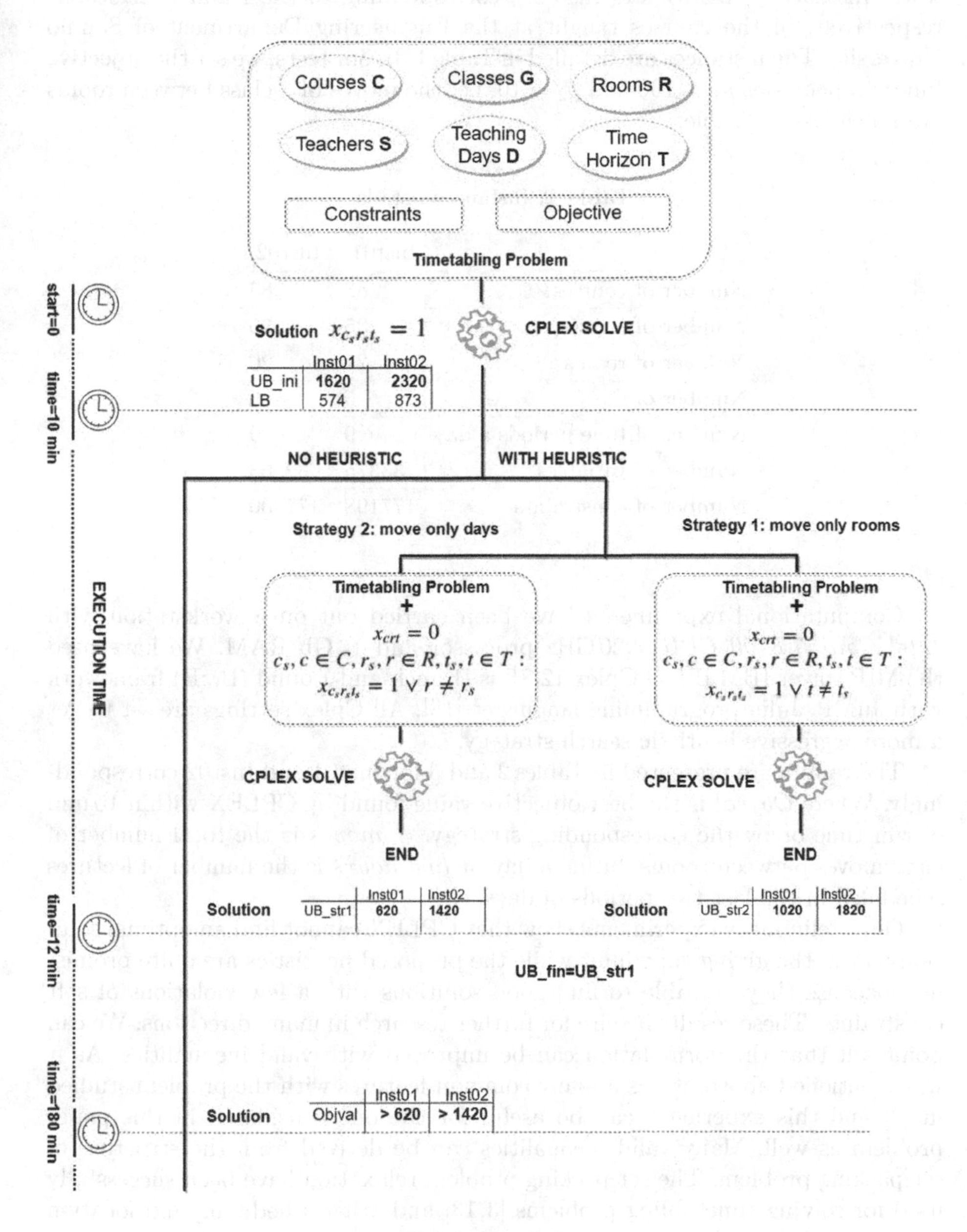

Fig. 2. Heuristic

5 Computational Results

In this section we give the preliminary computational experiments on two real world instances – Inst01 and Inst02 – corresponding to the I and II semester, respectively, of the courses taught at the Engineering Department of Sannio University. The instances are detailed in Table 1. In our tests, we set the objective function penalties $p_1 = 100$ and $p_2 = 10$, i.e. the moves of a class between rooms are much less desirable.

Table 1. Instances details

	Inst01	Inst02
Number of courses	82	83
Number of classes	35	35
Number of rooms	21	20
Number of days	5	5
Number of time periods a day	9	9
Number of variables	33375	32663
Number of constraints	177198	177900

Computational experiments have been carried out on a workstation with *Intel Core i7-8700 CPU*, 3.20 GHz processor and 16 Gb RAM. We have used the MIP solver IBM ILOG Cplex 12.8[1] as Branch-and-Bound (B&B) framework with JuMP/Julia programming language[2] [18]. All Cplex settings are set to get a more aggressive heuristic search strategy.

The results are presented in Tables 2 and 3 for Inst01 and Inst02 correspondingly. Where *Objval* is the best objective value found by CPLEX within 10 min of run time or by the corresponding strategy. # *moves* is the total number of class moves between rooms during a day, # *last hours* is the number of lectures scheduled in the last two periods of days.

Our preliminary experiments show that CPLEX cannot find an optimal solution within the giving time limit while the proposed heuristics are quite promising, because they are able to find good solutions with a few violations of soft constraints. These results inspire for further research in many directions. We can point out that this formulation can be improved with valid inequalities. As it was mentioned above, it has a many common features with the problem studied in [3] and this experience can be useful for the one considered in this paper problem as well. Many valid inequalities can be derived from the structure of set packing problem. The set packing problem relaxation have been successfully used for solving timetabling problems [3,13] and other scheduling and location

[1] https://www.ibm.com/products/ilog-cplex-optimization-studio.
[2] https://julialang.org/.

problems [1,2,37,38]. In fact, a study of the problem polytope structure could lead to derive new valid inequalities which are able to improve the lower bounds as well as the outcomes of the MIP heuristics.

Table 2. Inst01 results

	Objval	# moves	# last hours
CPLEX	1620	16	2
Strategy 1	1020	10	2
Strategy 2	620	6	2

Table 3. Inst02 results

	Objval	# moves	# last hours
CPLEX	2320	23	2
Strategy 1	1820	18	2
Strategy 2	1420	14	2

References

1. Avella, P., Boccia, M., Mannino, C., Vasilyev, I.: Time-indexed formulations for the runway scheduling problem. Transp. Sci. **51**(4), 1196–1209 (2017). https://doi.org/10.1287/trsc.2017.0750
2. Avella, P., Boccia, M., Vasilyev, I.: A branch-and-cut algorithm for the multilevel generalized assignment problem. IEEE Access **1**, 475–479 (2013). https://doi.org/10.1109/ACCESS.2013.2273268
3. Avella, P., Vasil'ev, I.: A computational study of a cutting plane algorithm for university course timetabling. J. Sched. **8**(6), 497–514 (2005). https://doi.org/10.1007/s10951-005-4780-1
4. Babaei, H., Karimpour, J., Hadidi, A.: A survey of approaches for university course timetabling problem. Comput. Ind. Eng. **86**, 43–59 (2015). https://doi.org/10.1016/j.cie.2014.11.010. Applications of computational intelligence and fuzzy logic to manufacturing and service systems
5. Bagger, N.C.F., Sørensen, M., Stidsen, T.R.: Benders' decomposition for curriculum-based course timetabling. Comput. Oper. Res. **91**, 178–189 (2018). https://doi.org/10.1016/j.cor.2017.10.009
6. Bellio, R., Ceschia, S., Di Gaspero, L., Schaerf, A., Urli, T.: A simulated annealing approach to the curriculum-based course timetabling problem. In: Proceedings of the 6th Multidisciplinary International Conference on Scheduling: Theory and Applications, MISTA 2013, Belgium, pp. 314–317, January 2013
7. Bettinelli, A., Cacchiani, V., Roberti, R., Toth, P.: An overview of curriculum-based course timetabling. TOP **23**(2), 313–349 (2015). https://doi.org/10.1007/s11750-015-0366-z

8. Bonutti, A., De Cesco, F., Di Gaspero, L., Schaerf, A.: Benchmarking curriculum-based course timetabling: formulations, data formats, instances, validation, visualization, and results. Ann. Oper. Res. **194**(1), 59–70 (2012). https://doi.org/10.1007/s10479-010-0707-0
9. Burke, E., Jackson, K., Kingston, J.H., Weare, R.: Automated university timetabling: the state of the art. Comput. J. **40**(9), 565–571 (1997). https://doi.org/10.1093/comjnl/40.9.565
10. Burke, E.K., Mareček, J., Parkes, A.J., Rudová, H.: Penalising patterns in timetables: novel integer programming formulations. In: Kalcsics, J., Nickel, S. (eds.) Operations Research Proceedings 2007, pp. 409–414. Springer, Heidelberg (2008). https://doi.org/10.1007/978-3-540-77903-2_63
11. Burke, E.K., Mareček, J., Parkes, A.J., Rudová, H.: A supernodal formulation of vertex colouring with applications in course timetabling. Ann. Oper. Res. **179**(1), 105–130 (2010). https://doi.org/10.1007/s10479-010-0716-z
12. Burke, E.K., Mareček, J., Parkes, A.J., Rudová, H.: A branch-and-cut procedure for the Udine Course Timetabling problem. Ann. Oper. Res. **194**(1), 71–87 (2012). https://doi.org/10.1007/s10479-010-0828-5
13. Burke, E.K., Marečdek, J., Parkes, A.J., Rudová, H.: Decomposition, reformulation, and diving in university course timetabling. Comput. Oper. Res. **37**(3), 582–597 (2010). https://doi.org/10.1016/j.cor.2009.02.023. Hybrid metaheuristics
14. Burke, E.K., Petrovic, S.: Recent research directions in automated timetabling. Eur. J. Oper. Res. **140**(2), 266–280 (2002). https://doi.org/10.1016/S0377-2217(02)00069-3
15. Cacchiani, V., Caprara, A., Roberti, R., Toth, P.: A new lower bound for curriculum-based course timetabling. Comput. Oper. Res. **40**(10), 2466–2477 (2013). https://doi.org/10.1016/j.cor.2013.02.010
16. Clark, M., Henz, M., Love, B.: QuikFix-a repair-based timetable solver. In: Proceedings of the 7th International Conference on the Practice and Theory of Automated Timetabling (2009). http://www.comp.nus.edu.sg/henz/publications/ps/PATAT2008.pdf
17. Di Gaspero, L., Schaerf, A., Mccollum, B.: The 2nd International Timetabling Competition (itc2007): Curriculum-based course timetabling (track 3. Technical report, School of Electronics, Electrical Engineering and Computer Science, Queens University, Belfast (UK), ITC-2007 (2007). http://www.cs.qub.ac.uk/itc2007/
18. Dunning, I., Huchette, J., Lubin, M.: JuMP: a modeling language for mathematical optimization. SIAM Rev. **59**(2), 295–320 (2017)
19. Fischetti, M., Lodi, A.: Local branching. Math. Program. **98**(1), 23–47 (2003). https://doi.org/10.1007/s10107-003-0395-5
20. Geiger, M.J.: Applying the threshold accepting metaheuristic to curriculum based course timetabling. Ann. Oper. Res. **194**(1), 189–202 (2012). https://doi.org/10.1007/s10479-010-0703-4
21. Hao, J.K., Benlic, U.: Lower bounds for the ITC-2007 curriculum-based course timetabling problem. Eur. J. Oper. Res. **212**(3), 464–472 (2011). https://doi.org/10.1016/j.ejor.2011.02.019
22. Kiefer, A., Hartl, R.F., Schnell, A.: Adaptive large neighborhood search for the curriculum-based course timetabling problem. Ann. Oper. Res. **252**(2), 255–282 (2017). https://doi.org/10.1007/s10479-016-2151-2
23. Kristiansen, S., Stidsen, T.: A Comprehensive Study of Educational Timetabling - A Survey. Department of Management Engineering, Technical University of Denmark, Lyngby (2013)

24. Lach, G., Lübbecke, M.E.: Curriculum based course timetabling: new solutions to Udine benchmark instances. Ann. Oper. Res. **194**(1), 255–272 (2012). https://doi.org/10.1007/s10479-010-0700-7
25. Lewis, R.: A survey of metaheuristic-based techniques for university timetabling problems. OR Spectr. **30**(1), 167–190 (2008). https://doi.org/10.1007/s00291-007-0097-0
26. Lindahl, M., Mason, A.J., Stidsen, T., Srensen, M.: A strategic view of university timetabling. Eur. J. Oper. Res. **266**(1), 35–45 (2018). https://doi.org/10.1016/j.ejor.2017.09.022
27. Lü, Z., Hao, J.K.: Adaptive tabu search for course timetabling. Eur. J. Oper. Res. **200**(1), 235–244 (2010). https://doi.org/10.1016/j.ejor.2008.12.007
28. Lü, Z., Hao, J.K., Glover, F.: Neighborhood analysis: a case study on curriculum-based course timetabling. J. Heuristics **17**(2), 97–118 (2011). https://doi.org/10.1007/s10732-010-9128-0
29. McCollum, B.: A perspective on bridging the gap between theory and practice in university timetabling. In: Burke, E.K., Rudová, H. (eds.) PATAT 2006. LNCS, vol. 3867, pp. 3–23. Springer, Heidelberg (2007). https://doi.org/10.1007/978-3-540-77345-0_1
30. McCollum, B.: Setting the research agenda in automated timetabling: the second international timetabling competition. INFORMS J. Comput. **22**(1), 120–130 (2010). https://doi.org/10.1287/ijoc.1090.0320
31. Müller, T.: ITC 2007 solver description: a hybrid approach. Ann. Oper. Res. **172**(1), 429 (2009). https://doi.org/10.1007/s10479-009-0644-y
32. Phillips, A.E., Waterer, H., Ehrgott, M., Ryan, D.M.: Integer programming methods for large-scale practical classroom assignment problems. Comput. Oper. Res. **53**, 42–53 (2015). https://doi.org/10.1016/j.cor.2014.07.012
33. Pillay, N.: A review of hyper-heuristics for educational timetabling. Ann. Oper. Res. **239**(1), 3–38 (2016). https://doi.org/10.1007/s10479-014-1688-1
34. Schaerf, A.: A survey of automated timetabling. Artif. Intell. Rev. **13**(2), 87–127 (1999). https://doi.org/10.1023/A:1006576209967
35. Shaker, K., Abdullah, S., Alqudsi, A., Jalab, H.: Hybridizing meta-heuristics approaches for solving university course timetabling problems. In: Lingras, P., Wolski, M., Cornelis, C., Mitra, S., Wasilewski, P. (eds.) RSKT 2013. LNCS (LNAI), vol. 8171, pp. 374–384. Springer, Heidelberg (2013). https://doi.org/10.1007/978-3-642-41299-8_36
36. Tarawneh, H., Ayob, M., Ahmad, Z.: A hybrid simulated annealing with solutions memory for curriculum-based course timetabling problem. J. Appl. Sci. **13**, 262–269 (2013). https://doi.org/10.3923/jas.2013.262.269
37. Vasilyev, I., Klimentova, X., Boccia, M.: Polyhedral study of simple plant location problem with order. Oper. Res. Lett. **41**(2), 153–158 (2013). https://doi.org/10.1016/j.orl.2012.12.006
38. Vasilyev, I., Avella, P., Boccia, M.: A branch and cut heuristic for a runway scheduling problem. Autom. Remote Control **77**(11), 1985–1993 (2016). https://doi.org/10.1134/S0005117916110084

Optimal Control and Applications

Iterative Method with Exact Fulfillment of Constraints in Optimal Control Problems

Alexander Sergeevich Buldaev(✉) and Ivan Dmitrievich Burlakov

Buryat State University, Ulan-Ude, Russia
buldaev@mail.ru

Abstract. A new approach is proposed for constructing a relaxation sequence of admissible controls in the class of optimal control problems with constraints. The approach is based on the construction of a system of non-local conditions for improving the admissible control in the form of a fixed point problem of the control operator. To build the conditions for improving the admissible control, we apply the transition to an auxiliary optimization problem based on the well-known principle of extension. Sufficient conditions for the optimality of admissible control and the existence of a minimizing sequence of admissible controls in the considered class of problems with constraints are substantiated. A comparative analysis of the computational efficiency of the proposed iterative method of fixed points with the exact implementation of constraints in model and test optimal control problems is carried out.

Keywords: Controlled system with constraints · Conditions for improving control · Fixed point problem

1 Introduction

A known method for solving optimal control problems with constraints is to reduce to auxiliary problems without constraints on the basis of the extension principle [1] in the form of a penalty functional, a regular or modified Lagrange functional. The process of solving auxiliary problems, as a rule, is based either on the implementation of necessary optimality conditions such as the maximum principle [2–6], or on constructing a relaxation sequence of controls using local methods to improve control such as gradient methods [4–6]. Moreover, at each iteration of the control improvement, the exact execution of the constraints of the initial problem is not guaranteed.

A new approach is proposed for constructing a relaxation sequence of controls based on the principle of extension with the help of constructed systems of conditions for a non-local improvement of control with the exact fulfillment

This work was carried out with the financial support of the Ministry of Education and Science, project 1.5049.2017BC; RFBR project 18-41-030005-r-a.

I. Bykadorov et al. (Eds.): MOTOR 2019, CCIS 1090, pp. 455–469, 2019.
https://doi.org/10.1007/978-3-030-33394-2_35

of the constraints of the original problem. The construction of such conditions is carried out with the help of non-standard increment formulas for functionals of an auxiliary problem that does not contain residual terms of extensions. Such formulas allow us to interpret the conditions of non-local improvement of admissible control as the problem of a fixed point of the control operator. This allows us to apply the developed theory and fixed point methods for the effective search for admissible improvement controls.

The fixed point approach under consideration is the development and extension of a non-local approach to improving control, originally developed for linear and linear-quadratic in the state of optimal control problems [6]. The methods of fixed points were constructed and justified in classes of nonlinear optimal control problems [7–13]. In this paper, the fixed point approach is developed for problems with constraints.

2 Statement of the Problem with Restrictions

We consider a class of optimal control problems with terminal, phase, and mixed constraints, including non-fixed time, reducible to the following general form:

$$\dot{x}(t) = f(x(t), u(t), \omega, t), x(t_0) = x^0, \tag{1}$$

$$u(t) \in U, \omega \in W, t \in T = [t_0, t_1],$$

$$\Phi_0(\sigma) = \varphi_0(x(t_1), \omega) + \int_T F_0(x(t), u(t), \omega, t)dt \to \inf_{\sigma \in \Omega}, \tag{2}$$

$$\Phi_1(\sigma) = \varphi_1(x(t_1), \omega) = 0, \tag{3}$$

in which $x(t) = (x_1(t), ..., x_n(t))$ is the state vector, $u(t) = (u_1(t), ..., u_m(t))$ is the vector of the control functions, $\omega = (\omega_1, ..., \omega_l)$ is the vector of the control parameters. The sets $U \subseteq R^m$ and $W \subseteq R^l$ are closed and convex. The interval T is fixed. As the available control functions, we consider a set V of piecewise continuous functions on T with values in the set U: $V = \{u \in PC(T) : u(t) \in U, t \in T\}$. $\sigma = (u, \omega)$ is an available control with values in the set $\Omega = V \times W$. The functions $\varphi_0(x, \omega)$ and $\varphi_1(x, \omega)$ are continuously differentiable on $R^n \times W$, the functions $F_0(x, u, \omega, t)$, $f(x, u, \omega, t)$ and their partial derivatives with respect to x, u, ω are continuous in the set of arguments on the set $R^n \times U \times W \times T$. The function $f(x, u, \omega, t)$ satisfies the Lipschitz condition by x in $R^n \times U \times W \times T$ with the constant $L > 0$: $\|f(x, u, \omega, t) - f(y, u, \omega, t)\| \leq L \|x - y\|$.

The conditions guarantee the existence and uniqueness of the solution $x(t, \sigma)$, $t \in T$ of the system (1) for any available control $\sigma \in \Omega$. The available control $\sigma \in \Omega$ is called admissible if the functional constraint (3) is satisfied. We denote the set of admissible controls by $D = \{\sigma \in \Omega : \Phi_1(\sigma) = \varphi_1(x(t_1), \omega) = 0\}$.

The problem of improving admissible control in the class of problems (1)–(3) is considered in the following general formulation: for a given admissible control $\sigma^I \in D$ is required to find an admissible control $\sigma \in D$ with the condition $\Delta_\sigma \Phi_0(\sigma^I) = \Phi_0(\sigma) - \Phi_0(\sigma^I) \leq 0$.

3 The Fixed Point Problem Based on the Extension Functional

We consider the auxiliary problem without constraints based on the extension functional

$$\dot{x}(t) = f(x(t), u(t), \omega, t), x(t_0) = x^0, \tag{4}$$

$$u(t) \in U, \omega \in W, t \in T = [t_0, t_1],$$

$$J(\sigma) = \varphi(x(t_1), \omega) + \int_T F(x(t), u(t), \omega, t)dt \to \inf_{\sigma \in \Omega}, \tag{5}$$

in which the extension functional is determined by the condition

$$J(\sigma) \leq \Phi_0(\sigma), \sigma \in D \subset \Omega.$$

As an example of the extension functional, the penalty functional for violation of the restriction (3) can be considered. In particular, the functional with a square penalty

$$M^{\mu}(\sigma) = \Phi_0(\sigma) + \mu \Phi_1^2(\sigma) \to \inf_{\sigma \in \Omega}, \mu > 0. \tag{6}$$

Another topical example of the extension functionality is the regular Lagrange functional

$$L^{\lambda}(\sigma) = \Phi_0(\sigma) + \lambda \Phi_1(\sigma) \to \inf_{\sigma \in \Omega}, \lambda \in R. \tag{7}$$

Pontryagin function with conjugate variable $\psi \in R^n$ and the standard conjugate system in the problem (4), (5) have the form

$$H(\psi, x, u, \omega, t) = \langle \psi, f(x, u, \omega, t) \rangle - F(x, u, \omega, t),$$

$$\dot{\psi}(t) = -H_x(\psi(t), x(t), u(t), \omega, t), t \in T, \quad \psi(t_1) = -\varphi_x(x(t_1), \omega). \tag{8}$$

For available control $\sigma \in \Omega$ denote $\psi(t, \sigma)$, $t \in T$ is a solution of the standard conjugate system (8) with $x(t) = x(t, \sigma)$ and arguments u, ω, corresponding to the control components σ.

Consider the problem of improving the available control in the tasks (4), (5): for a given available control $\sigma^I \in \Omega$ need to find available control $\sigma \in \Omega$ with the condition $\Delta_{\sigma} J(\sigma^I) = J(\sigma) - J(\sigma^I) \leq 0$. In accordance with [10], the projection conditions for improving the available of control $\sigma^I \in \Omega$ based on the projection operator has the following form.

Next, we use the following notation for a particular increment an arbitrary vector function $g(y_1, ..., y_l)$ with respect to y_{s_1}, y_{s_2}

$$\begin{aligned} &\Delta_{y_{s_1} + \Delta y_{s_1}, y_{s_2} + \Delta y_{s_2}} g(y_1, ..., y_l) \\ &= g(y_1, ..., y_{s_1} + \Delta y_{s_1}, ..., y_{s_2} + \Delta y_{s_2}, ..., y_l) - g(y_1, ..., y_l). \end{aligned}$$

In addition, we denote $\Delta x(t) = x(t, u) - x(t, u^I)$, $\Delta u(t) = u(t) - u^I(t)$, $\Delta \omega = \omega - \omega^I$.

P_Y - projection operator onset $Y \subset R^k$ in the Euclidean norm

$$P_Y(z) = \arg\min_{y \in Y}(\|y - z\|), z \in R^k.$$

We introduce a modified differential-algebraic conjugate system including an additional phase variable $y(t) = (y_1(t), ..., y_n(t))$,

$$\dot{p}(t) = -H_x(p(t), x(t), u(t), \omega, t) - r(t), \tag{9}$$

$$\langle H_x(p(t), x(t), u(t), \omega, t) + r(t), y(t) - x(t)\rangle = \Delta_{y(t)} H(p(t), x(t), u(t), \omega, t) \tag{10}$$

with boundary conditions

$$p(t_1) = -\varphi_x(x(t_1), \omega) - q, \tag{11}$$

$$\langle \varphi_x(x(t_1), \omega) + q, y(t_1) - x(t_1)\rangle = \Delta_{y(t_1)} \varphi(x(t_1), \omega), \tag{12}$$

in which by definition we set $r(t) = 0$, $q = 0$ in the case of linearity of the functions f, F, φ with respect to x (problem (4), (5) linear by state), and also in the case of $y(t) = x(t)$ for the corresponding $t \in T$.

In the problem linear in the state (4), (5) the modified conjugate system (9)–(12) by definition coincides with the standard conjugate system (8).

In the non-linear problem (4), (5), the algebraic equations (10) and (12) can always be analytically resolved with respect to $r(t)$ and q in the form of explicit or conditional formulas (perhaps not in a unique way).

Thus, the differential-algebraic conjugate system (9)–(12) can always be reduced (possibly not the only way) to a differential conjugate system with uniquely determined values $r(t)$ and q.

For the available controls $\sigma \in \Omega$, $\sigma^I \in \Omega$, let $p(t, \sigma^I, \sigma)$, $t \in T$ be the solution of the modified conjugate system (9)–(12) for $x(t) = x(t, \sigma^I)$, $y(t) = x(t, \sigma)$, $u(t) = u^I(t)$, $\omega = \omega^I$. The definition implies the obvious equality $p(t, \sigma, \sigma) = \psi(t, \sigma)$, $t \in T$.

The projection conditions for improving the available control $\sigma^I \in \Omega$ with the specified projection parameter $\alpha > 0$ in accordance with [11] have the form:

$$u(t) = P_U(u^I(t) + \alpha(H_u(p(t, \sigma^I, \sigma), x(t, \sigma), u^I(t), \omega^I, t) + s_1(t))), t \in T, \tag{13}$$

$$\Delta_{u(t)} H(p(t, \sigma^I, \sigma), x(t, \sigma), u^I(t), \omega^I, t) = \left\langle H_u(p(t, \sigma^I, \sigma), x(t, \sigma), u^I(t), \omega^I, t) + s_1(t), u(t) - u^I(t)\right\rangle, \tag{14}$$

$$\omega = P_W(\omega^I + \alpha(-\varphi_\omega(x(t_1, \sigma), \omega^I) + \int_T H_\omega(p(t, \sigma^I, \sigma), x(t, \sigma), u(t), \omega^I, t)dt + s_2)), \tag{15}$$

$$\Delta_\omega\{-\varphi(x(t_1,\sigma),\omega^I) + \int_T H(p(t,\sigma^I,\sigma),x(t,\sigma),u(t),\omega^I,t)dt\}$$
$$= \langle -\varphi_\omega(x(t_1,\sigma),\omega^I)$$
$$+ \int_T H_\omega(p(t,\sigma^I,\sigma),x(t,\sigma),u(t),\omega^I,t)dt + s_2, \omega - \omega^I \rangle, \qquad (16)$$

in which in Eq. (14) by definition is assumed $s_1(t) = 0$ in the case of linearity of the function f, F in u (problem (4), (5) linear by the control u), or in the case $u(t) = u^I(t)$ for the corresponding $t \in T$. Similarly, in (16), by definition $s_2 = 0$, in the case of linearity of functions f, F, φ by ω (linear by the parameter ω problem (4), (5)), and also for $\omega = \omega^I$.

Equations (14) and (16) can always be uniquely resolved with respect to the quantities $s_1(t)$ and s_2 (perhaps not the only way). Thus, conditions (13)–(16) can be reduced to a system of equations in the form (13), (15) with respect to the vector $\sigma = (u, \omega)$ uniquely identifiable right side. The resulting system can be interpreted as a fixed point problem with respect to the control $\sigma = (u, \omega)$ for the control operator, uniquely defined by the right side of the system.

According to [10], the solution $\sigma = (u, \omega)$ of the system (13)–(16) provides an improvement in control $\sigma^I \in \Omega$ for any parameter $\alpha > 0$ with an estimate of the improvement of the functional:

$$\Delta_\sigma J(\sigma^I) \leq -\frac{1}{\alpha}\int_T \left\|u(t) - u^I(t)\right\|^2 dt - \frac{1}{\alpha}\left\|\omega - \omega^I\right\|^2. \qquad (17)$$

At the same time, control improvement is guaranteed not only in a sufficiently small neighborhood of the initial control $\sigma^I \in \Omega$, i.e. the improvement procedure under consideration has the property of nonlocality, in contrast to known gradient methods and other local methods for improving control.

We will complete the problem (13)–(16) with the condition for the exact fulfillment of constraint (3). As a result, we obtain the conditions for improving control with exact fulfillment of the constraint (13)–(16), (3) in the problem (1)–(3).

Conditions for improving control (13)–(16), (3), can be considered as a fixed point problem with an additional algebraic equation (3) with respect to control σ. This allows you to apply and modify known fixed point search algorithms to implement the conditions for improving control and constructing iterative methods for approximate solving the original problem (1)–(3).

The proposed fixed point approach for constructing a relaxation sequence of admissible controls consists of successively solving the problems of improving admissible control in the form of constructed fixed point problems of a uniquely defined control operator.

4 Iterative Algorithm

For the numerical solution of the fixed point problem (13)–(16), (3) various modifications of the known methods of successive approximations can be used

[14]. As an example, we consider an analog of the method of simple iteration [14] with $k \geq 0$ and given initial control available $\sigma^0 \in \Omega$ at $k = 0$:

$$u^{k+1}(t) = P_U(u^I(t) + \alpha(H_u(p(t,\sigma^I,\sigma^k), x(t,\sigma^k), u^I(t), \omega^I, t) + s_1^k(t))), t \in T, \tag{18}$$

$$\begin{aligned} &\Delta_{u^k(t)} H(p(t,\sigma^I,\sigma^k), x(t,\sigma^k), u^I(t), \omega^I, t) \\ &= \langle H_u(p(t,\sigma^I,\sigma^k), x(t,\sigma^k), u^I(t), \omega^I, t) + s_1^k(t), u^k(t) - u^I(t) \rangle, \end{aligned} \tag{19}$$

$$\omega^{k+1} = P_W(\omega^I + \alpha(-\varphi_\omega(x(t_1,\sigma^k), \omega^I) + \int_T H_\omega(p(t,\sigma^I,\sigma^k), x(t,\sigma^k), u^k(t), \omega^I, t)dt + s_2^k)), \tag{20}$$

$$\begin{aligned} &\Delta_{\omega^k}\{-\varphi(x(t_1,\sigma^k), \omega^I) + \textstyle\int_T H(p(t,\sigma^I,\sigma^k), x(t,\sigma^k), u^k(t), \omega^I, t)dt\} \\ &= \langle -\varphi_\omega(x(t_1,\sigma^k), \omega^I) \\ &+ \textstyle\int_T H_\omega(p(t,\sigma^I,\sigma^k), x(t,\sigma^k), u^k(t), \omega^I, t)dt + s_2^k, \omega^k - \omega^I \rangle, \end{aligned} \tag{21}$$

$$\Phi_1(\sigma^{k+1}) = \varphi_1(x(t_1,\sigma^{k+1})) = 0, \tag{22}$$

where the quantities $s_1^k(t)$ and s_2^k are determined according to a given method of a uniquely determined control operator for the fixed point problem (13)–(16).

For extension functionals (6) or (7) control σ^{k+1}, $k \geq 0$, calculated by rule (18)–(21), depends respectively on the penalty parameter $\mu > 0$ or on the Lagrange multiplier $\lambda \in R$. Thus, Eq. (22) at each iteration of the process reduces to a scalar equation for the penalty parameter or Lagrange multiplier, respectively. For the numerical solution of the specified equation with a given accuracy, you can use the known methods of scalar optimization (method of interval bisection, broken method, etc.) [4,5].

Iterations on the index $k \geq 0$ with the implementation of the constraints are carried out before the first condition:

$$J\left(\sigma^{k+1}\right) < J\left(\sigma^I\right).$$

In this case, a new problem (13)–(16), (3) is built to improve the obtained calculated control, which is taken as a new σ^I, and the iterative algorithm is repeated. At the same time as the initial approximation of the control $\sigma^0 \in \Omega$ at $k = 0$ to iteratively process (18)–(22) can choose the obtained calculated control.

Note that the calculated control obtained satisfies the constraint with a given accuracy. Thus, starting with the second computational improvement problem (13)–(16), (3), the sequence of computational controls forms a relaxation sequence of controls that are admissible with a given accuracy. At the same time, setting the accuracy of the constraint, you can achieve the desired relaxation accuracy for the target functional of the original problem:

$$\Phi_0(\sigma^{k+1}) \approx J\left(\sigma^{k+1}\right) < J\left(\sigma^I\right) \approx \Phi_0(\sigma^I).$$

If a strict improvement of control in the process of iterations is not achieved, then a numerical calculation of the fixed point problem (13)–(16), (3) can be carried out before the condition:

$$\max\{\, \|u^{k+1} - u^k\|_{C(T)}, |\omega^{k+1} - \omega^k| \,\} \leq \varepsilon,$$

where $\varepsilon > 0$ - given the accuracy of the calculation of the fixed point problem. This concludes the construction and calculation of successive tasks of control improvement.

As a result, we obtain a relaxation sequence of controls $\sigma^k \in \Omega$, satisfying the constraint (3) with a given accuracy.

The conditions of convergence of the iterative process (18)–(22) can be obtained similarly to the works [7,14] based on the requirements that provide the well-known "squeezing" property for the operator of the right-hand side of the fixed point problem.

The conditions of convergence of the relaxation sequence of controls to the optimal solution can be justified on the basis of sufficient conditions for the existence of a minimizing sequence of controls in problems with constraints, similarly to [1].

5 Conditions for Optimal Control

On the basis of the fixed point problem (13)–(16), it is possible to formulate the necessary optimality conditions for control in the auxiliary extension problem (4), (5), similarly to [7–13]. These necessary conditions for optimal control can be successfully used to test control for optimality in the extension problem (4), (5).

Sufficient conditions for optimal control in the original problem (1)–(3) can be formulated on the basis of the extension principle [1].

Theorem 1. *(Sufficient condition for optimal control based on a problem with extension functional).*

Let the admissible control $\sigma \in D$ be optimal in the problem (4), (5) and $\Phi_0(\sigma) = J(\sigma)$. Then $\sigma \in D$ is optimal in the problem (1)–(3).

Proof. Indeed, by virtue of optimality $\sigma \in D$ in the problem (4), (5) we have $\Phi_0(\sigma) = J(\sigma) \leq J(\tilde{\sigma})$, $\tilde{\sigma} \in \Omega$. Hence, by virtue of the extension property, we get $\Phi_0(\sigma) \leq J(\tilde{\sigma}) \leq \Phi_0(\tilde{\sigma})$, $\tilde{\sigma} \in D \subset \Omega$.

Corollary 1. *Let the admissible control $\sigma \in D$ be optimal in auxiliary problems without restrictions (4), (6) with some penalty parameter $\mu > 0$ or in problem (4), (7) with some Lagrange multiplier $\lambda \in R$. Then this control will be optimal in the initial problem with constraints (1)–(3).*

Proof. Indeed, for optimal $\sigma \in D$ by the definition of extension functionals in auxiliary problems, we obtain $\Phi_0(\sigma) = J(\sigma)$.

The optimal control in the problem (1)–(3) may not exist, but there is always exists a minimizing sequence of admissible controls, which is defined as a generalized solution of problem (1)–(3) according to [1]. A sufficient condition for the existence of a generalized solution in the problem (1)–(3) can also be formulated on the basis of the extension principle.

Theorem 2. *(a sufficient condition for the existence of a minimizing sequence of controls based on a problem with an extension functional).*

Let the sequence of admissible controls $\sigma^k \in D$, $k \geq 0$ *be minimizing in the problem (4), (5) and* $\Phi_0(\sigma^k) \to j = \inf_{\tilde{\sigma}\in\Omega} J(\tilde{\sigma})$. *Then* $\sigma^k \in D$ *is a minimizing sequence in the problem (1)–(3). Wherein* $j = i = \inf_{\tilde{\sigma}\in D} \Phi_0(\tilde{\sigma})$.

Proof. Indeed, for any $\varepsilon > 0$, by virtue of the convergence and extension property, we obtain

$$\Phi_0(\sigma^k) \leq j + \varepsilon \leq \inf_{\tilde{\sigma}\in D} J(\tilde{\sigma}) + \varepsilon \leq i + \varepsilon$$

for $k \to \infty$.

Corollary 2. *Let the relaxation sequence of admissible controls* $\sigma^k \in D$, $k \geq 0$ *be minimizing in auxiliary problem without restrictions (4), (6) with some penalty parameter* $\mu > 0$ *or in the auxiliary problem (4), (7) with some Lagrange multiplier* $\lambda \in R$. *Then the given sequence will be minimizing in the original problem with constraints (1)–(3).*

Proof. Indeed, for any $\sigma^k \in D$ by the definition of extension functionals in auxiliary problems, we obtain $\Phi_0(\sigma^k) = J(\sigma^k)$.

A generalization of Theorem 2 is the statement used to substantiate the proposed fixed point approach.

Theorem 3. *(a sufficient condition for the existence of a minimizing sequence of controls based on a sequence of problems with extension functionals).*

Let exist:

(1) a sequence of extension tasks without restrictions with extension functionals

$$J_s(\sigma) \leq \Phi_0(\sigma), \sigma \in D \subset \Omega_s, s \geq 0;$$

(2) a sequence of lower bounds j_s *for extension functionals*

$$j_s \leq J_s(\tilde{\sigma}), \tilde{\sigma} \in \Omega_s, s \geq 0;$$

(3) the sequence of admissible controls $\sigma^k \in D$, $k \geq 0$, *for which the limiting relation is satisfied*

$$\lim_{k\to\infty} \Phi_0(\sigma^k) = \lim_{s\to\infty} j_s.$$

Then $\sigma^k \in D$ *is a minimizing sequence in the problem (1)–(3). Wherein*

$$\lim_{s\to\infty} j_s = i = \inf_{\tilde{\sigma}\in D} \Phi_0(\tilde{\sigma}).$$

Proof. Indeed, for any $\varepsilon > 0$, by virtue of the conditions of the theorem, we obtain

$$\Phi_0(\sigma^k) \le \lim_{s\to\infty} j_s + \varepsilon \le \inf_{\tilde\sigma\in D} J_s(\tilde\sigma) + \varepsilon \le i + \varepsilon$$

at $k \to \infty$.

Corollary 3. *Let there be:*

(1) relaxation sequence of admissible controls $\sigma^k \in D$, $k \ge 0$, obtained by the fixed point method based on a sequence of auxiliary problems without restrictions (4), (6) with some penalty parameters $\mu_s > 0$ or problems (4), (7) with some Lagrange multipliers $\lambda_s \in R$ at $s \ge 0$;

(2) a sequence of lower bounds j_s, $s \ge 0$ for extension functionals in the indicated auxiliary problems without restrictions, such that $\lim_{k\to\infty} \Phi_0(\sigma^k) = \lim_{s\to\infty} j_s$. Then the sequence $\sigma^k \in D$, $k \ge 0$ will be minimizing in the original problem with constraints (1)–(3).

6 Example

The work on the proposed algorithm is demonstrated in the well-known model problem of Mehri-Davis [15,16], which "can serve as a good test for testing numerical methods" according to the recommendations of [15].

The initial problem with the control function and the phase inequality constraint is reduced by the technique [16] to an equivalent problem with terminal equality constraint:

$$\begin{cases} \dot x_1(t) = x_2(t), & x_1(0) = 0, \\ \dot x_2(t) = u(t) - x_2(t), & x_2(0) = -1, \\ \dot x_3(t) = x_1^2(t) + x_2^2(t) + 0.005u^2(t), & x_3(0) = 0, \\ \dot x_4(t) = Q(x_2(t), t), & x_4(0) = 0, \\ Q(x_2, t) = \begin{cases} 0, \quad if \quad \Gamma(x_2, t) \le 0, \\ \Gamma^2(x_2, t), \; if \quad \Gamma(x_2, t) > 0, \end{cases} \\ \Gamma(x_2, t) = x_2 - 8(t-0.5)^2 + 0.5, \quad t \in T = [0, 1], \end{cases}$$

$$\Phi(u) = x_3(1) \to \inf_{u\in V},$$

$$x_4(1) = 0, V = \{u \in PC(T) : u(t) \in R, t \in T\}.$$

For the obtained problem, we consider an auxiliary problem without restriction with the extension functional based on a regular Lagrange functional with a factor $\lambda \in R$

$$L^\lambda(u) = x_3(1) + \lambda x_4(1) \to \inf_{u\in V}.$$

Note that when $\lambda > 0$ extension functionality can be interpreted as a penalty functional with a linear penalty for violating the equality constraint since $x_4(1) \ge 0$.

The Pontryagin function and differential-algebraic conjugate system for the auxiliary problem take the following form:

$$H(p,x,u,t) = p_1 x_2 + p_2 (u - x_2) + p_3 \left(x_1^2 + x_2^2 + 0.005u^2\right) + p_4 Q(x_2,t),$$

$$\begin{cases} \dot{p}_1(t) = -2p_3(t)x_1(t) - r_1(t), \\ \dot{p}_2(t) = -p_1(t) + p_2(t) - 2p_3(t)x_2(t) - \\ -p_4(t)G(x_2(t),t) - r_2(t), \\ \dot{p}_3(t) = -r_3(t), \\ \dot{p}_4(t) = -r_4(t), \end{cases} \quad \begin{cases} p_1(1) = 0, \\ p_2(1) = 0, \\ \\ p_3(1) = -1, \\ p_4(1) = -\lambda, \end{cases}$$

where $G(x_2,t) = \begin{cases} 0, & if \quad \Gamma(x_2,t) \le 0, \\ 2\Gamma(x_2,t), & if \quad \Gamma(x_2,t) > 0. \end{cases}$

In this case, the variable $r(t) = (r_1(t), \ r_2(t), \ r_3(t), r_4(t))$ determined from an algebraic equation with an additional phase variable for which the notation is used $z(t) = (z_1(t), z_2(t), z_3(t), z_4(t))$:

$$\begin{gathered} p_1(t)(z_2(t) - x_2(t)) + p_2(t)(x_2(t) - z_2(t)) \\ +p_3(t)\left(z_1^2(t) + z_2^2(t) - x_1^2(t) - x_2^2(t)\right) + p_4(t)(Q(z_2(t),t) - Q(x_2(t),t)) \\ = (2p_3(t)x_1(t) + r_1(t))(z_1(t) - x_1(t)) \\ +(p_1(t) - p_2(t) + 2p_3(t)x_2(t) + p_4(t)G(x_2(t),t) + r_2(t))(z_2(t) - x_2(t)) \\ +r_3(t)(z_3(t) - x_3(t)) + r_4(t)(z_4(t) - x_4(t)). \end{gathered}$$

We fix the following method of uniquely resolving the value $r(t) = (r_1(t), r_2(t), \ r_3(t), \ r_4(t))$:

1. if $z_1(t) \ne x_1(t)$, then $r_2(t) = 0, r_3(t) = 0, r_4(t) = 0$

$$\begin{gathered} r_1(t) = \frac{p_1(t)(z_2(t) - x_2(t)) + p_2(t)(x_2(t) - z_2(t))}{z_1(t) - x_1(t)} \\ + \frac{p_3(t)\left(z_1^2(t) + z_2^2(t) - x_1^2(t) - x_2^2(t)\right)}{z_1(t) - x_1(t)} \\ + \frac{p_4(t)(Q(z_2(t),t) - Q(x_2(t),t))}{z_1(t) - x_1(t)} \\ - \frac{(p_1(t) - p_2(t) + 2p_3(t)x_2(t) + p_4(t)G(x_2(t),t))(z_2(t) - x_2(t))}{z_1(t) - x_1(t)} \\ -2p_3(t)x_1(t); \end{gathered}$$

2. if $z_1(t) = x_1(t)$, then

   2.1. if $z_2(t) \ne x_2(t)$, then $r_1(t) = 0, r_3(t) = 0, r_4(t) = 0$

$$r_2(t) = p_3(t)(z_2(t) - x_2(t))$$

$$+\frac{p_4(t)\left(Q(z_2(t),t)-Q(x_2(t),t)\right)}{z_2(t)-x_2(t)}-p_4(t)\,G(x_2(t),t);$$

2.2 if $z_2(t)=x_2(t)$, then $r_1(t)=0, r_2(t)=0, r_3(t)=0, r_4(t)=0$. This implies $p_3(t)=-1$, $p_4(t)=-\lambda$, $t\in[0,1]$.

Denote $p(t,u^I,u,\lambda)$- solution of the conjugate system with $x(t)=x(t,u^I)$, $z(t)=x(t,u)$, $u^I\in V$, $u\in V$.

Fixed point problem to improve a given $u^I\in V$ by functionality $L^\lambda(u)$ has the form:

$$u(t)=P_U\left(u^I(t)+\alpha\left(p_2\left(t,u^I,u,\lambda\right)-0.01u^I(t)+s_1(t)\right)\right),t\in[0,1],$$

where $s_1(t)$ determined from the equation:

$$p_2\left(t,u^I,u,\lambda\right)\left(u(t)-u^I(t)\right)-0.005\left(u^2(t)-(u^I)^2(t)\right)$$
$$=\left(p_2\left(t,u^I,u,\lambda\right)-0.01u^I(t)+s_1(t)\right)\left(u(t)-u^I(t)\right).$$

Choose the following method of unique resolution of the value $s_1(t)$ from an algebraic equation:

1. if $u(t)\neq u^I(t)$, then $s_1(t)=p_2\left(t,u^I,u,\lambda\right)-0.005\left(u(t)+u^I(t)\right)$
$$-p_2\left(t,u^I,u,\lambda\right)+0.01u^I(t)=0.005\left(u^I(t)-u(t)\right);$$
2. if $u(t)=u^I(t)$, then $s_1(t)=0$.

Thus, the problem is simplified and takes the form

$$u(t)=P_U\left(u^I(t)+\alpha\left(p_2\left(t,u^I,u,\lambda\right)-0.005(u^I(t)+u(t))\right)\right),t\in[0,1].$$

The parameter $\lambda\in R$ is determined by the condition of exact fulfillment of the terminal equality constraint: $x_4(1,u)=0$.

To solve the fixed-point problem with exact fulfillment of the constraint, an iteration process was considered at $k\geq 0$ with initial approximation $u^0\in V$ for $k=0$:

$$u^{k+1}(t)=P_U\left(u^I(t)+\alpha\left(p_2\left(t,u^I,u^k,\lambda\right)-0.005(u^I(t)+u^k(t))\right)\right),$$

at each iteration of which the equation was solved for $\lambda\in R$: $x_4\left(1,u^{k+1}\right)=0$.

Numerical solution of the Cauchy phase and conjugate problems was performed using the Runge-Kutta-Werner method of the variable (5)–(6) order accuracy using the DVIPRK library program IMSL Fortran PowerStation 4.0 [17]. The values of the controlled, phase and conjugate variables were memorized at the nodes of the fixed uniform grid T_h with the discretization step $h>0$ on the interval T. In the intervals between neighboring grid nodes T_h, the value of the control function was assumed to be constant and equal to the value at the left node.

The numerical solution of the equation for the parameter $\lambda\in R$ was carried out using the DUMPOL [17] program, realizes the deformable polyhedron

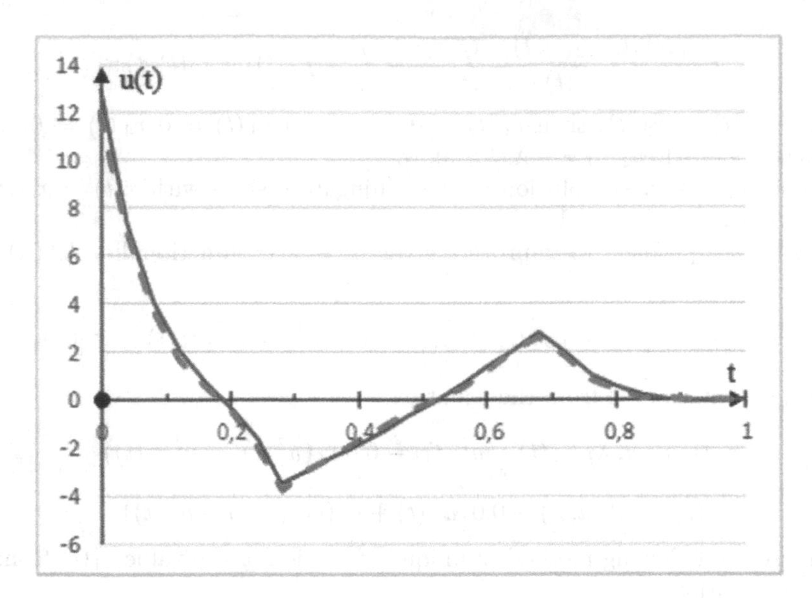

Fig. 1. Calculated control $u^* = u^*(t)$ is a dashed line. Control $\hat{u} = \hat{u}(t)$ is a solid line.

method. The accuracy of the solution of the equation was controlled by the criterion $\Gamma_1 = \max\{\Gamma(x_2(t, u^{k+1}), t),\ t \in T_h\} \leq \varepsilon_1$, where $\varepsilon_1 > 0$ - specified accuracy of phase limiting. The iterations of the calculation of the fixed-point problem for $k \geq 0$ continued until the first condition for improving the control $u^I \in V$ was met:

$$L^\lambda\left(u^{k+1}\right) < L^\lambda\left(u^I\right).$$

In this case, a new fixed point problem was constructed to improve the obtained computational control, and the iterative algorithm was repeated. At the same time, as the initial approximation of control $u^0 \in V$ with $k = 0$ for the iteration process, the calculated control obtained was chosen.

If the improvement of control in the indicated sense was not achieved, then the numerical calculation of the fixed point problem is carried out until the condition $\max\{|u^{k+1}(t) - u^k(t)|, t \in T_h\} \leq \varepsilon_2$, where $\varepsilon_2 > 0$ is given the accuracy of the calculation of the fixed point problem. The process of building and calculating successive control improvement problems ended there.

For a comparative analysis of the computational efficiency of the proposed fixed-point approach with known methods [15,16], the following calculation parameters were chosen: $h = 10^{-2}, \varepsilon_1 = 10^{-12}, \varepsilon_2 = 10^{-10}$. When calculating the initial problem of improving control, it was assumed that $u^0 = u^I$ with $k = 0$. The results of the calculation of successive improvement tasks by the proposed iterative algorithm with a fixed design parameter $\alpha = 0,1$ for various first initial (start-up) controls $u^0 \in V$ are presented in Table 1.

In the table Φ^0 and Φ^* is the initial and calculated values of the objective functional of the problem; Γ_1^0 and Γ_1^* is the initial and calculated values of the

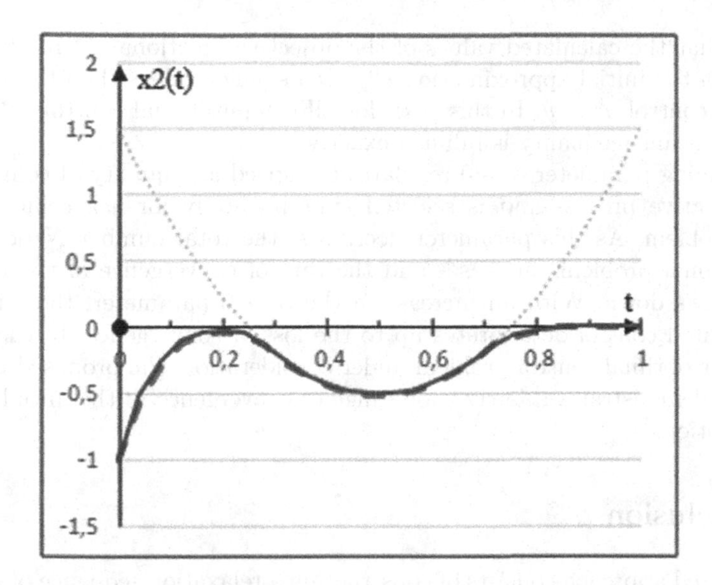

Fig. 2. The calculated trajectory $x_2(t, u^*)$ is the dashed line. The trajectory $x_2(t, \hat{u})$ is a solid line. Parabola $\Gamma(x_2, t)$ is a dotted line.

Table 1. Initial and calculated indicators of numerical experiments.

u^0	Φ^0	Φ^*	Γ_1^0	Γ_1^*	N
0	483,93	0,1684	6,5412	−0.0955	25108
1	49152,9957	0,1690	14.485	−0.0201	22536
5	52139,8664	0.1699	30,2017	−0.0524	24462
10	56033,8718	0,1702	21,324	−0.0840	34518
15	60106,1185	0,1702	24,547	−0,1247	43014
$\hat{u}$	0,1719	0,1683	−0,0025	−0.0021	2004

criterion for performing the phase constraint of the task; N- the total number of calculated phase and related Cauchy problems; $u^0 = u^0(t)$, $t \in T_h$ is starting control; $\hat{u} = \hat{u}(t)$, $t \in T_h$ is the approximate optimal control obtained in [15] by combined iterative algorithms with initial control approximation $u(t) \equiv 0$.

Figure 1 shows the calculated control $u^*(t)$, obtained for the start-up control $u^0 \equiv 1$, and the control $\hat{u}(t)$. Figure 2 shows the corresponding phase trajectories $x_2(t, u^*)$ and $x_2(t, \hat{u})$.

For the starting controls $u^0 \equiv 0$ and $u^0 = \hat{u}$ the calculated controls $u^*(t)$ and the trajectories $x_2(t, u^*)$ visually practically coincide in the figures with $\hat{u}(t)$ and $x_2(t, \hat{u})$. With increasing $u^0 \equiv const > 1$ the deviation of the calculated controls $u^*(t)$ and the trajectories $x_2(t, u^*)$ from $\hat{u}(t)$ and $x_2(t, \hat{u})$ visually quantitatively increases with a complete analogy of the qualitative dynamics.

Note that the calculated values of the objective functional Φ^* in all experiments with the initial approximation u^0 reach smaller values than the value Φ^0 under the control $u^0 = \hat{u}$. In this case, for all computational controls $u^*(t)$, the phase constraint-inequality is fulfilled exactly.

The tuning parameter $\alpha > 0$ regulates the speed and quality of convergence of the iterative process and is selected experimentally for a specific optimal control problem. As this parameter decreases, the total number N of Cauchy computational problems increases and the rate of convergence of the iterative process slows down. With an increase in the $\alpha > 0$ parameter, the quality of the calculated control deteriorates up to the loss of convergence. In the framework of the optimal control problem under consideration, the proposed iterative algorithm demonstrates a fairly wide range of convergence in the initial control approximation.

7 Conclusion

The proposed approach consists of constructing a relaxation sequence of controls with the exact fulfillment of the constraints of the problem based on solving a system of control improvement conditions for the auxiliary extension problem without constraints. The novelty of the proposed approach lies in presenting the system of control improvement conditions in the constructive form of the fixed-point problem of the control operator.

The developed form of the system for improving control in the form of a fixed point problem allows one to apply the theory and methods of fixed points to construct relaxation sequences for admissible improving controls in optimal control problems with constraints.

The constructed approach is characterized by the nonlocality of improvement of controls; the absence of a time-consuming procedure of needle-shaped or convex variation of control in a small neighborhood of improved control, characteristic of gradient methods; exact fulfillment of restrictions; the presence of one main tuning parameter $\alpha > 0$ regulating the speed, quality and region of convergence of the iterative process. These properties are essential factors for increasing the efficiency of solving nonlinear optimal control problems with constraints.

References

1. Gurman, V.I.: The Principle of Extension in Control Tasks, p. 288. Nauka, Moscow (1997). (in Russian)
2. Alekseev, V.M., Tikhomirov, V.M., Fomin, S.V.: Optimal Control, p. 428. Nauka, Moscow (1979). (in Russian)
3. Afanasyev, A.P., Dikusar, V.V., Milyutin, A.A., Chukanov, S.A.: Necessary Condition in Optimal Control, p. 320. Nauka, Moscow (1990). (in Russian)
4. Vasiliev, F.P.: Numerical Methods for Solving Extremal Problems, p. 518. Nauka, Moscow (1980). (in Russian)

5. Vasiliev, O.V.: Lectures on Optimization Methods, p. 340. Publishing House of ISU, Irkutsk (1994). (in Russian)
6. Srochko, V.A.: Iterative Methods for Solving Optimal Control Problems, p. 160. Fizmatlit, Moscow (2000). (in Russian)
7. Buldaev, A.S.: Perturbation Methods in Problems of Improvement and Optimization of Controlled Systems, p. 260. Publishing House of the Buryat State University, Ulan-Ude (2008). (in Russian)
8. Buldaev, A.S., Khishektueva, I.-K.: The fixed point method in parametric optimization problems for systems. Autom. Remote Control **74**(12), 1927–1934 (2013)
9. Buldaev, A.S., Daneev, A.V.: New approaches to optimization of parameters of dynamic systems on the basis of problems about fixed points. Far East J. Math. Sci. (FJMS) **99**(3), 439–454 (2016)
10. Buldaev, A.S.: Fixed point methods based on design operations in optimization problems of control functions and parameters of dynamic systems. Bull. Buryat State University. Math. Comput. Sci. **1**, 38–54 (2017). (in Russian). https://doi.org/10.18101/2304-5728-2017-1-38-54
11. Buldaev, A.S.: Fixed point method for searching of extremal controls. In: CEUR-WS Proceedings - 2017, vol. 1987: 8th International Conference on Optimization and Applications (OPTIMA-2017), Petrovac, Montenegro, 2–7 October, pp. 101–107 (2017)
12. Buldaev, A.S., Burlakov, I.D.: About one approach to numerical solution of nonlinear optimal speed problems. Bull. South Ural State Univ. Math. Model. Program. **11**(4), 55–66 (2018). https://doi.org/10.14529/mmp180404
13. Buldaev, A.S., Burlakov, I.D.: The method of non-local control improvement in optimal control problems with constraints. In: DEStech Transactions on Computer Science and Engineering. - IX International Conference on Optimization and Applications (OPTIMA-2018) (Supplementary volume), pp. 114–127 (2018)
14. Samarskiy, A.A., Gulin, A.V.: Numerical Methods, p. 432. Nauka, Moscow (1989). (in Russian)
15. Evtushenko, Yu.G.: Methods for Solving Extremal Problems and Their Application in Optimization Systems, p. 432. Nauka, Moscow (1982). (in Russian)
16. Tyatyushkin, A.I.: Numerical Methods and Software for Optimizing Managed Systems, p. 192. Nauka, Novosibirsk (1992). (in Russian)
17. Bartenev, O.V.: Fortran for Professionals. IMSL Math. Library. Part 2, p. 320. Dialog-MIFI, Moscow (2001). (in Russian)

Optimization "In Windows" for Routing Problems with Constraints

Alexander G. Chentsov[1,2(✉)], Alexey M. Grigoryev[1], and Alexey A. Chentsov[1]

[1] Krasovskii Institute of Mathematics and Mechanics, Ekaterinburg, Russia
chentsov@imm.uran.ru, ag@uran.ru, chentsov_a_a@mail.ru
[2] Ural Federal University, Ekaterinburg, Russia

Abstract. We investigate the problem of sequentially visiting a number of megalopoleis while satisfying precedence constraints where the travel cost functions depend on the set of pending tasks. It is supposed that the dimension of the investigated problem is sufficiently large, therefore, an exact solution is practically impossible; in these circumstances, heuristics are used very widely. We investigate some possibilities for local improvement of results achievable in a class of heuristics. For such improvement of a result, optimizing insertions and finite systems of optimizing insertions are used. We view these systems as multi-insertions. The given approach is combined with the employment of a parallel algorithm implemented for a multiprocessor computing system. The optimizing insertions are designed by means of a broadly understood dynamic programming.

Keywords: Insertion · Multi-insertion · Precedence constraints · Route

1 Introduction

In many applied problems, settings with routing elements arise. Very often, these settings are complicated by different constraints and task list dependence. In particular, such problems arise in nuclear power engineering and mechanical engineering. In the first case, the goal is to decrease the dose loads for nuclear power plant workers who perform complex operations in areas with increased radioactivity. The second case is connected with plate cutting. In both cases, different constraints arise; moreover, in the first case, travel cost functions depend on the pending task set (a worker is only affected by the radiation sources that are not dismantled yet). Some other complicated circumstances take place in the above-mentioned applied problems. So, these problems differ from their prototype, the Traveling Salesman Problem (TSP); see [1,2].

A specialized theory is necessary. We connect this theory with the Bellman approach [3], although, in DP solutions of TSP, the Held–Karp scheme [4] is prevalent. However, dynamic programming (DP) can be realized globally only for problems of small dimension; this singularity especially develops for our problems

I. Bykadorov et al. (Eds.): MOTOR 2019, CCIS 1090, pp. 470–485, 2019.
https://doi.org/10.1007/978-3-030-33394-2_36

with megalopoleis although precedence constraints help to significantly reduce computational complexity; see e.g. [5,6,8,24]. Nevertheless, one can realistically use DP for local improvement of heuristic solutions. This can be achieved by optimizing insertions and multi-insertions; see [7–14] and other. On this base, iterated procedures [13] and parallel algorithms [14] can be constructed. So, DP can be used in routing problems of large dimensions for improvement of heuristics.

Let us recall some studies connected with TSP and problems of the TSP type: monographs [1,2,15], review article [16], and original article [17]. But, in the considered problem, significant qualitative singularities arise, which are connected with the requirements of engineering applications. Therefore, the methods and algorithms used in TSP cannot be realized in our application-oriented task connected with the radiation safety problem, where the complicated travel cost functions is just one issue among many. Moreover, the construction of these functions is a part of our solution.

Namely, to determine these functions, one must integrate over fragments of worker's actual trajectory to find the specific radiation dose received by him, which results into a nonmetric routing problem. Another complication of our formulation is the fact that the radiation dose is the sum of radiation produced by the sources that are not dismantled at the time the worker moves, which is the reason why travel cost functions become dependent on the list of tasks.

In addition, we consider the routing problem with megalopoleis and precedence constraints (these conditions are natural for engineering problems). The above-mentioned singularities generate serious difficulties of mathematical character and require strict formalization. In the DP constructions, we follow [5,6,8,18,21]. We are not aware of the works of other authors that study the DP procedures for problems of such complexity.

Let us now describe the issues that complicate a direct application of insertions or multi-insertions. First, we have global precedence constraints for the initial big problem. For insertions and multi-insertions, fragments of general precedence constraints are required. These fragments must be coordinated with global conditions. Namely, a new solution with optimizing insertions must be admissible by precedence in the initial problem. Further, in our initial problem, we have very complicated travel cost functions, which take into account the effect of radiation. This effect is defined by sources not dismantled at the moment of travel, giving rise to a complex dependence on the pending tasks set. However, the insertion constructions act on fragments of the task sets. Here, it is required to exclude the crossed influence on results. In other words, we must devise a valid transformation of the initial route and trajectory by insertions. Therefore, these insertions must be specially organized. For this, precise definitions are required. We note that routing problems with cost functions admitting dependence on task lists are practically never considered (we note only a heuristic solution for TSP in [23]). Therefore, serious comparison with other investigations of our problem is impossible: the known works are connected with less complicated routing problems.

2 General Notations and Definitions

We will need a summary of the general notations, which is the essence of a valid mathematical formulation of the problem focused on engineering applications. We use standard set-theoretical symbolism and the notation of [5–7,12–14]; the equality by definition is denoted $\triangleq$. For every set T, we denote by $\mathcal{P}(T)$ and $\mathcal{P}'(T)$ the families of all and all nonempty subsets of T, respectively; in addition, $\mathrm{Fin}(T)$ denotes the family of all finite sets from $\mathcal{P}'(T)$. As usual, $\mathbb{R}$ is the real line, $\mathbb{R}_+ \triangleq \{\xi \in \mathbb{R} | 0 \leqslant \xi\}$, $\mathbb{N} \triangleq \{1; 2; \ldots\}$, $\mathbb{N}_0 \triangleq \{0; 1; 2; \ldots\} = \mathbb{N} \cup \{0\}$ (here and below we denote $\{h\}$ the singleton set containing the object h), and

$$\overline{p, q} \triangleq \{k \in \mathbb{N}_0 | (p \leqslant k) \& (k \leqslant q)\}$$

for any $p \in \mathbb{N}_0$ and $q \in \mathbb{N}_0$. For any two objects x and y, by (x, y) we denote the ordered pair (OP) with the first element x and the second element y; for every OP z, by $\mathrm{pr}_1(z)$ and $\mathrm{pr}_2(z)$, we denote the first and the second elements of z, respectively. If a, b, and c are objects, then $(a, b, c) \triangleq ((a, b), c)$; therefore, for sets A, B, and C, we have $A \times B \times C \triangleq (A \times B) \times C$ and, for $u \in A \times B$ and $v \in C$, $(u, v) \in A \times B \times C$. For any nonempty set S, we denote the set of all functions from S into $\mathbb{R}_+$ by $\mathcal{R}_+[S]$.

3 The Initial Problem

Fix a nonempty set X, $\mathbf{x}_0 \in X$, and $\mathbf{n} \in \mathbb{N}$ such that $\mathbf{n} \geqslant 3$; the sets

$$\mathbf{L}_1 \in \mathrm{Fin}(X), \ldots, \mathbf{L}_n \in \mathrm{Fin}(X)$$

are called megalopoleis. Assume that $\mathbf{x}_0 \notin \mathbf{L}_j \ \forall j \in \overline{1, \mathbf{n}}$. Moreover, let $\mathbf{L}_p \cap \mathbf{L}_q = \varnothing$ under $p \neq q$. We consider the processes of the type

$$\begin{aligned} \mathbf{x}_0 \to (x_{1,1} \in \mathbf{L}_{\alpha(1)} \rightsquigarrow x_{1,2} \in \mathbf{L}_{\alpha(1)}) \to \ldots \\ \to (x_{\mathbf{n},1} \in \mathbf{L}_{\alpha(\mathbf{n})} \rightsquigarrow x_{\mathbf{n},2} \in \mathbf{L}_{\alpha(\mathbf{n})}), \end{aligned} \tag{1}$$

where α is a permutation of the index set $\overline{1, \mathbf{n}}$. In addition, fix the relations

$$\mathbb{L}_1 \in \mathcal{P}'(\mathbf{L}_1 \times \mathbf{L}_1), \ldots, \mathbb{L}_\mathbf{n} \in \mathcal{P}'(\mathbf{L}_\mathbf{n} \times \mathbf{L}_\mathbf{n}).$$

It is required that, in (1),

$$(x_{1,1}, x_{1,2}) \in \mathbb{L}_{\alpha(1)}, \ldots, (x_{\mathbf{n},1}, x_{\mathbf{n},2}) \in \mathbb{L}_{\alpha(\mathbf{n})}.$$

We have to select

$$\alpha, (x_{1,1}, x_{1,2}), \ldots, (x_{\mathbf{n},1}, x_{\mathbf{n},2}).$$

Let $\mathbf{P}$ denote the set of all permutations of $\overline{1, \mathbf{n}}$ (in (1), $\alpha \in \mathbf{P}$). Assume that precedence constraints are given, specifically, let us fix the relation $\mathcal{K} \in \mathcal{P}(\overline{1, \mathbf{n}} \times \overline{1, \mathbf{n}})$ for which

$$\forall \mathcal{K}_0 \in \mathcal{P}'(\mathcal{K}) \ \exists z_0 \in \mathcal{K}_0 : \mathrm{pr}_1(z_0) \neq \mathrm{pr}_2(z) \ \forall z \in \mathcal{K}_0;$$

then,

$$\mathcal{A} \triangleq \{\alpha \in \mathbf{P} | \ \forall z \in \mathcal{K} \ \forall t_1 \in \overline{1,\mathbf{n}} \ \forall t_2 \in \overline{1,\mathbf{n}} \\ (z = (\alpha(t_1), \alpha(t_2))) \Rightarrow (t_1 < t_2)\} \neq \varnothing \tag{2}$$

is the set of all routes admissible by precedence (we define routes as index permutations). Our condition on $\mathcal{K}$ is fulfilled in typical situations of applied character; see [18, Chap. 2]. As seen from (1), a route (index permutation) is not enough to completely define our process. Therefore, we introduce trajectories, for which, beforehand, we introduce the set $\mathfrak{X}$ as the union of $\{\mathbf{x}_0\}$ and all sets $\mathbf{L}_t$, $t \in \overline{1,\mathbf{n}}$; then, $\mathfrak{X} \in \text{Fin}(X)$. By $\tilde{\mathfrak{Z}}$, we denote the set of all mappings from $\overline{0,\mathbf{n}}$ into $\mathfrak{X} \times \mathfrak{X}$. Then, for $\alpha \in \mathbf{P}$,

$$\mathfrak{Z}_\alpha \triangleq \{(z_i)_{i \in \overline{0,\mathbf{n}}} \in \tilde{\mathfrak{Z}} | (z_0 = (\mathbf{x}_0, \mathbf{x}_0)) \& (z_t \in \mathbb{L}_{\alpha(t)} \ \forall t \in \overline{1,\mathbf{n}})\} \in \text{Fin}(\tilde{\mathfrak{Z}}) \tag{3}$$

is the set of all α–trajectories (i.e. trajectories coordinated with α). Then,

$$\mathbf{D} \triangleq \{(\alpha, \mathbf{z}) \in \mathcal{A} \times \tilde{\mathfrak{Z}} | \mathbf{z} \in \mathfrak{Z}_\alpha\} \in \text{Fin}(\mathcal{A} \times \tilde{\mathfrak{Z}}) \tag{4}$$

is the set of all admissible solutions (see (1)). So, our solutions are OPs with components from sets (2) and (3).

4 Cost Functions

For $\mathbf{N} \triangleq \mathcal{P}'(\overline{1,\mathbf{n}})$, let

$$\mathbf{c}^\natural \in \mathcal{R}_+[\mathfrak{X} \times \mathfrak{X} \times \mathbf{N}], \ c_1^\natural \in \mathcal{R}_+[\mathfrak{X} \times \mathfrak{X} \times \mathbf{N}], \ldots, \\ c_N^\natural \in \mathcal{R}_+[\mathfrak{X} \times \mathfrak{X} \times \mathbf{N}], \ f^\natural \in \mathcal{R}_+[\mathfrak{X}]. \tag{5}$$

We view the sets from the family $\mathbf{N}$ as task lists. In terms of (5), we define an additive criterion: for $\alpha \in \mathbf{P}$ and $(\mathbf{z}_i)_{i \in \overline{0,\mathbf{n}}} \in \mathfrak{Z}_\alpha$,

$$\mathfrak{G}_\alpha[(\mathbf{z_t})_{\mathbf{t} \in \overline{0,\mathbf{n}}}] \triangleq \sum_{t=1}^{n} [\mathbf{c}^\natural(\text{pr}_2(\mathbf{z_{t-1}}), \text{pr}_1(\mathbf{z_t}), \{\alpha(k) : k \in \overline{t,\mathbf{n}}\}) \\ + c_{\alpha(t)}^\natural(\mathbf{z_t}, \{\alpha(k) : k \in \overline{t,\mathbf{n}}\})] + f^\natural(\text{pr}_2(z_\mathbf{n})). \tag{6}$$

Here, the values of $\mathbf{c}^\natural$ measure the exterior travel (i.e., the travel between megalopoleis and from x^0 to a megalopoleis), the values of $c_1^\natural, \ldots, c_\mathbf{n}^\natural$ measure the interior works done when visiting a megalopolis, and the values of $f^\natural$ measure the terminal state of process. As a corollary (see (2) and (6)), the following principal problem is defined:

$$\hat{\mathfrak{G}}_\alpha[(\mathbf{z_t})_{\mathbf{t} \in \overline{0,\mathbf{n}}}] \to \min, \ (\alpha, (\mathbf{z_t})_{\mathbf{t} \in \overline{0,\mathbf{n}}}) \in \mathbf{D}. \tag{7}$$

Clearly, for problem (3), the corresponding extremum V is well-defined: V is the smallest of the numbers

$$\hat{\mathfrak{G}}_\alpha[(\mathbf{z_t})_{\mathbf{t} \in \overline{0,\mathbf{n}}}], (\alpha, (\mathbf{z_t})_{\mathbf{t} \in \overline{0,\mathbf{n}}}) \in \mathbf{D};$$

in (7), an optimal solution exists. The structure of such solutions is defined by the DP procedure from [5,6,19], however, there are certain difficulties of computational character. Therefore, we use the insertions and multi-insertions methods to improve the results. Specifically, we optimize fragments of a global solution.

5 Optimizing Individual Insertions

In construction of the local improvement procedure in the class of individual optimizing insertions, we follow [7,12,13]. Let us sketch the scheme of this method; a more detailed presentation is provided in [7,12].

We follow the notation of [7,12]. Fix $N \in \overline{2, \mathbf{n}-1}$ as the "length" of the used insertion and $\nu \in \overline{0, \mathbf{n}-N}$ as the "beginning" of this insertion. For $s \in \overline{\nu+1, \nu+N}$, we transform some initial route and trajectory by the scheme of [7,12]. Now, we restrict ourselves to considering only the idea. First of all, we create local precedence constraints defining the set $\mathbf{K}_\nu[\alpha] \in \mathcal{P}(\overline{1,N} \times \overline{1,N})$, for which

$$\forall \mathbf{K}_0 \in \mathcal{P}'(\mathbf{K}_\nu[\alpha]) \; \exists z_0 \in \mathbf{K}_0 : \mathrm{pr}_1(z_0) \neq \mathrm{pr}_2(z) \; \forall z \in \mathbf{K}_0;$$

here, $\alpha \in \mathcal{A}$. In these terms, for $\alpha \in \mathcal{A}$ and $\nu \in \overline{0, \mathbf{n}-N}$, the nonempty set $\mathbf{A}_\nu[\alpha]$ of all locally admissible (by precedence) routes is defined; $\mathbf{A}_\nu[\alpha] \neq \varnothing$. In the form of $\mathbf{A}_\nu[\alpha]$, we have the set of admissible routes for the given insertion. If $\alpha \in \mathcal{A}$, $\nu \in \overline{0, \mathbf{n}-N}$, and $\beta \in \mathbf{A}_\nu[\alpha]$, then [12, (3.4)], the glued route

$$(\beta - \mathrm{sew})[\alpha; \nu] \in \mathcal{A} \tag{8}$$

is defined; so, we indicate the replacement $\alpha \to (\beta - \mathrm{sew})[\alpha;\nu]$ preserving admissibility. An analogous replacement is used for trajectories, thus, for $\alpha \in \mathcal{A}$, $(\mathbf{z_t})_{\mathbf{t} \in \overline{\mathbf{0},\mathbf{n}}} \in \mathfrak{Z}_\alpha$, and $\nu \in \overline{0, \mathbf{n}-N}$, we construct a set of local trajectories. In addition, for such trajectories, coordination with local routes is essential; we consider $\mathrm{pr}_2(\mathbf{z}_\nu)$ as the starting point for the local routing problem. If $\beta \in \mathbf{A}_\nu[\alpha]$ and h is a trajectory coordinated with β, then the glued trajectory (we consider the gluing for $(\mathbf{z_t})_{\mathbf{t} \in \overline{\mathbf{0},\mathbf{n}}}$ and h) is coordinated with $(\beta - \mathrm{sew})[\alpha;\nu]$ (8). So, we paste the local solution (β, h) into the initial solution $(\alpha, (\mathbf{z_t})_{\mathbf{t} \in \overline{\mathbf{0},\mathbf{n}}})$; this paste is realized within $\overline{\nu+1, \nu+N}$. In addition, as (β, h), we use an optimal solution in the local routing problem with additive criterion. For construction of this optimal solution, a variant of DP [5,6,20] is realized; naturally, we assume that the value of N is moderate. The concrete improvement of global result is defined by [12, (4.9),(4.10)].

Now, we consider the basic constructions for the above-mentioned steps of our investigation. So, for $\alpha \in \mathcal{A}$ and $\nu \in \overline{0, \mathbf{n}-N}$, we define $\Lambda_\nu[\alpha]$ as the mapping

$$s \longmapsto \alpha(\nu+s) : \overline{1,N} \to \overline{1,\mathbf{n}};$$

this mapping forms our insertion. Clearly, $\Lambda_\nu[\alpha]$ is a bijection from $\overline{1,N}$ onto

$$\Gamma_\nu[\alpha] \triangleq \{\Lambda_\nu[\alpha](s) : s \in \overline{1,N}\} = \{\alpha(\nu+s) : s \in \overline{1,N}\} \in \mathcal{P}'(\overline{1,\mathbf{n}}).$$

Here $\Gamma_\nu[\alpha]$ is the realization of the window of precedence constraints. To this end, we cut out the following part of the set $\mathcal{K}$:

$$Q_\nu[\alpha] \triangleq \{z \in \mathcal{K} | (\mathrm{pr}_1(z) \in \Gamma_\nu[\alpha]) \& (\mathrm{pr}_2(z) \in \Gamma_\nu[\alpha])\}.$$

Now, we form the above-mentioned set $\mathbf{K}_\nu[\alpha]$:

$$\mathbf{K}_\nu[\alpha] \triangleq \{(\Lambda_\nu[\alpha]^{-1}(\mathrm{pr}_1(z)), \Lambda_\nu[\alpha]^{-1}(\mathrm{pr}_2(z))) : z \in Q_\nu[\alpha]\}$$

(here $\Lambda_\nu[\alpha]^{-1}$ is the inverse of the bijection $\Lambda_\nu[\alpha]$; then,

$$\Lambda_\nu[\alpha]^{-1} : \Gamma_\nu[\alpha] \to \overline{1, N}$$

forms $\mathbf{K}_\nu[\alpha]$ in $\overline{1, N} \times \overline{1, N}$). Finally, we define the above-mentioned set of admissible local routes by the rule

$$\mathbf{A}_\nu[\alpha] \triangleq \{\beta \in \mathbb{P} | \beta^{-1}(\mathrm{pr}_1(z)) < \beta^{-1}(\mathrm{pr}_2(z)) \ \forall z \in \mathbf{K}_\nu[\alpha]\},$$

where $\mathbb{P}$ is the set of all permutations of the index set $\overline{1, N}$. Returning to (8), we note that, by [12, (3.3)], for $\beta \in \mathbf{A}_\nu[\alpha]$,

$$((\beta - \mathrm{sew})[\alpha; \nu](t) \triangleq \alpha(t) \ \ \forall t \in \overline{1, \mathbf{n}} \setminus \overline{\nu + 1, \nu + N}) \& ((\beta - \mathrm{sew})[\alpha; \nu](t)$$
$$\triangleq (\Lambda_\nu[\alpha] \circ \beta)(t - \nu) \ \ \forall t \in \overline{\nu + 1, \nu + N}).$$

So, we obtain the pasted route. Let us now define the pasted trajectory. To do this, we need to construct a localized phase space, a local system of megalopoleis. Indeed, (for $\alpha \in \mathcal{A}$ and $\nu \in \overline{0, \mathbf{n} - N}$) for $s \subset \overline{1, N}$, let

$$(M_s[\alpha; \nu] \triangleq \mathbf{L}_{\Lambda_\nu[\alpha](s)}) \& (\mathbb{M}_s[\alpha; \nu] \triangleq \mathbb{L}_{\Lambda_\nu[\alpha](s)}).$$

In the form of $M_1[\alpha; \nu], ..., M_N[\alpha; \nu]$, we obtain the local system of megalopoleis. Now, we suppose that, for $\mathbf{z} \in \mathfrak{Z}_\alpha$,

$$\mathbb{X}[\alpha; \mathbf{z}; \nu] \triangleq \{\mathrm{pr}_2(\mathbf{z}(\nu))\} \cup (\bigcup_{s=1}^{N} M_s[\alpha; \nu]);$$

$\mathbb{X}[\alpha; \mathbf{z}; \nu] \in \mathrm{Fin}(\mathfrak{X})$. Then, we use the scheme of [12, (3.7)]: for $\beta \in \mathbb{A}_\nu[\alpha]$, let $Z_\beta[\alpha; \mathbf{z}; \nu]$ denote the set of all collections

$$(h_i)_{i \in \overline{0, N}} : \overline{0, N} \to \mathbb{X}[\alpha; \mathbf{z}; \nu] \times \mathbb{X}[\alpha; \mathbf{z}; \nu]$$

for which $h_0 = (\mathrm{pr}_2(\mathbf{z}(\nu)), \mathrm{pr}_2(\mathbf{z}(\nu)))$ and $h_t \in \mathbb{M}_{\beta(t)}[\alpha; \nu] \ \ \forall t \in \overline{1, N}$. Clearly, $Z_\beta[\alpha; \mathbf{z}; \nu]$ is a nonempty finite set. Thus, in the form $Z_\beta[\alpha; \mathbf{z}; \nu]$, we obtain the set of all local trajectories coordinated with the (local) route β.

Now, for $\alpha \in \mathcal{A}$, $\mathbf{z} \in \mathfrak{Z}_\alpha$, and $\nu \in \overline{0, \mathbf{n} - N}$, we introduce the set $\widetilde{\mathbf{D}}[\alpha; \mathbf{z}; \nu]$ of all the pairs

$$(\beta, h), \ \ \beta \in \mathbf{A}_\nu[\alpha], \ \ h \in Z_\beta[\alpha; \mathbf{z}; \nu].$$

It is easy to see that $\widetilde{\mathbf{D}}[\alpha; \mathbf{z}; \nu]$ is a nonempty finite set. Elements of $\widetilde{\mathbf{D}}[\alpha; \mathbf{z}; \nu]$ are admissible local solutions (solutions of the insertion), and there are no other elements in this set.

6 Local Cost Functions

Let $\mathfrak{N} \triangleq \mathcal{P}'(\overline{1,N})$. For all $\alpha \in \mathcal{A}$, $\mathbf{z} \in \mathfrak{Z}_\alpha$, and $\nu \in \overline{0, \mathbf{n} - N}$, we form functions

$$\mathbf{c}[\alpha;\mathbf{z};\nu] \in \mathcal{R}_+[\mathbb{X}[\alpha;\mathbf{z};\nu] \times \mathbb{X}[\alpha;\mathbf{z};\nu] \times \mathfrak{N}],$$

$$c_1[\alpha;\mathbf{z};\nu] \in \mathcal{R}_+[\mathbb{X}[\alpha;\mathbf{z};\nu] \times \mathbb{X}[\alpha;\mathbf{z};\nu] \times \mathfrak{N}], \dots,$$

$$\dots, c_N[\alpha;\mathbf{z};\nu] \in \mathcal{R}_+[\mathbb{X}[\alpha;\mathbf{z};\nu] \times \mathbb{X}[\alpha;\mathbf{z};\nu] \times \mathfrak{N}],$$

$$f[\alpha;\mathbf{z};\nu] \in \mathcal{R}_+[\mathbb{X}[\alpha;\mathbf{z};\nu]].$$

In addition, $\mathbf{c}[\alpha;\mathbf{z};\nu]$ is used to measure the external movements;

$$c_1[\alpha;\mathbf{z};\nu], \dots, c_N[\alpha;\mathbf{z};\nu]$$

measure the internal works, and $f[\alpha;\mathbf{z};\nu]$ measures the terminal state of our local problem corresponding to the considered insertion.

Like in [12], we consider two different variants of definition for the above-mentioned functions in the cases $\nu < \mathbf{n} - N$ and $\nu = \mathbf{n} - N$. Namely, for $\nu < \mathbf{n} - N$, we use [14, (3.14),(3.15)], and, for $\nu = \mathbf{n} - N$, the representations [14, (3.16),(3.17)] are used. For both cases, we suppose that

$$\mathfrak{B}_\beta[h|\alpha;\mathbf{z};\nu] \in \mathbb{R}_+, \quad \beta \in \mathbf{A}_\nu[\alpha], \quad h \in Z_\beta[\alpha;\mathbf{z};\nu],$$

are defined by [14, (3.18)]. Namely, we use the additive criterion defined by summing the values for our cost functions corresponding to every local solution $(\beta, h) \in \widetilde{\mathbf{D}}[\alpha;\mathbf{z};\nu]$. As a result, we obtain the following local problem:

$$\mathfrak{B}_\beta[h|\alpha;\mathbf{z};\nu] \to \min, \quad (\beta, h) \in \widetilde{\mathbf{D}}[\alpha;\mathbf{z};\nu]$$

with the value $\mathbb{V}[\alpha;\mathbf{z};\nu] \in \mathbb{R}_+$ [14, (3.20)] and the solution set [14, (3.21)]. The obtained improvement of global criterion is the number $\varkappa[\alpha;\mathbf{z};\nu] \in \mathbb{R}_+$; this number is defined by [14, (3.30)]. As a result, we obtain the following estimate for the global extremum V:

$$V \leqslant \hat{\mathfrak{G}}_\alpha[\mathbf{z}] - \varkappa[\alpha;\mathbf{z};\nu],$$

where $\varkappa[\alpha;\mathbf{z};\nu]$ is defined in explicit form in terms of the initial solution $(\alpha, \mathbf{z})$ and index ν; see [12, (4.9)]. To determine $\mathbb{V}[\alpha;\mathbf{z};\nu]$ and the optimal solutions of insertion, we use an economical version of the DP procedure corresponding to [5–7,12], going back to [18, § 4.9] (in this connection, see also [19]). We note that the given constructions were derived in greater detail in [12], in a somewhat different notation. Thus, the constructions of [7,12] are a natural base of the above-mentioned scheme. For a more tangible improvement of the global result, we use the iterated procedure (see [13]). In addition, we vary the parameter ν: $\nu = \nu_1$, $\nu = \nu_2$, etc. Note that in [13], the iterations were constructed through different algorithms, and those algorithms were implemented on a personal computer.

7 Optimizing Multi-insertions

Let us return to problem (7) and consider a different approach to the question of how to optimize the fragments of the initial solution. Fix $(\lambda, \mathbf{h}) \in \mathbf{D}$; thus, $\lambda \in \mathcal{A}$ and $\mathbf{h} \in \mathfrak{Z}_\lambda$. In other words, we have an admissible global solution for the initial problem. Now, we form a "disjoint" system of insertions. Namely, we fix numbers $\nu_1 \in \overline{0, \mathbf{n} - N}, \ldots, \nu_m \in \overline{0, \mathbf{n} - N}$, $m \in \mathbb{N}$, for which

$$\nu_j + N < \nu_{j+1} \ \forall j \in \overline{1, m-1}.$$

Let us now consider insertions on the intervals $\overline{\nu_j + 1, \nu_j + N}, j \in \overline{1, m}$. For every such individual insertion, we use the DP procedure of [7, 12] to solve the corresponding local routing problem. As a result, we obtain a collection of local optimal solutions $(\rho_1, u_1), \ldots, (\rho_m, u_m)$; here, for $j \in \overline{1, m}$, ρ_j is the local route and u_j is the local trajectory coordinated with ρ_j. These constructions are realized by scheme of Sects. 5 and 6 for individual insertions. The procedures of local optimization are realized independently. A more detailed discussion is given in [14, Section 5]. We only note that the construction of local optimal solutions is implemented in parallel, specifically, each individual insertion is considered by its own processor, and all these processors work independently. In addition, the joint improvement of the additive criterion is realized as a sum of individual improvements for separate insertions. So, we obtain the following property: for solution (η, w) obtained after pasting the local solutions $(\rho_1, u_1), \ldots, (\rho_m, u_m)$ into the initial solution $(\lambda, \mathbf{h})$,

$$V \leqslant \hat{\mathfrak{G}}_\eta[w] = \hat{\mathfrak{G}}_\lambda[\mathbf{h}] - \sum_{i=1}^{m} \chi[\lambda; \mathbf{h}; \nu_i], \tag{9}$$

where $\chi[\lambda; \mathbf{h}; \nu_k], k \in \overline{1, m}$, are defined by [14, (3.4)] and characterize the local improvement of results when using the DP procedure.

8 The Problem of Dismantling the Radiating Elements

We used the above-mentioned construction with iterated procedures for optimizing insertions and multi-insertions in the problem of dismantling a finite system of radiating elements. Now, at a meaningful level, we consider the corresponding scheme of the insertions and multi–insertions application.

Let us first note that we used the iterated procedure with insertions for a metric routing problem with megalopoleis. We were able to achieve the improvement up to 6, 5–8, 5% over the original result after 5–9 iterations (we also used a PC). For a more complicated instance of the dismantlement problem, the analogous improvement was 3–4% after 7–8 iterations.

Returning to the description of the dismantlement problem, we note that, for every radiating element, the near zone stood out. For this near zone, the zone boundary was sampled. As a result, we obtained a megalopolis. The internal work for each megalopolis consisted of traveling from the entry point to the radiating element, dismantling it in a proper fashion, and traveling to the exit point. It was assumed that, during these operations, the radiation is higher than during external movements.

At each time, the worker is affected by radiating elements that are not dismantled at the time of his movement, hence the need for travel cost functions that depend on the list of tasks and express the radiation dose incurred by the worker during the mentioned movements. See the formal derivation of these functions in [20].

9 Computational Experiment

This section describes the practical implementation of a solver for the routing problem with precedence constraints and task list-dependent travel cost functions, specifically, a problem of dismantling a system of radiating elements (see [20]). To solve this problem, we use a scheme based on the above-mentioned construction with multi-insertions (see Sect. 7). This method allows one to solve routing problems of significant dimension.

The following is an algorithm for solving the problem (7) with the aid of multi-insertions.

1. At the initial stage, the source data is read, such as
   - coordinates of cities in the megalopoleis and the starting point,
   - coordinates of radiating elements,
   - power of radiating elements,
   - employee movement speed,
   - precedence constraints (address pairs).

   Then, the cost functions (5) are calculated using the formulas given in [20]. Since the process of calculating these functions is rather laborious for a large number of megalopoleis, a parallel algorithm based on the OpenMP library (shared memory) is used to speed up the calculations.
2. At this stage, the heuristic solution is constructed using the algorithm [21, § 6]. This algorithm takes into account precedence constraints and cost functions depending on the task list.
3. Next, we attempt to improve the route found at the previous stage with the help of a multi-insertion. To do this, we divide the obtained route into 20-megalopoleis fragments, while leaving between the fragments the "jumpers" consisting of one megalopolis. Each such fragment is transmitted by MPI protocol to a separate computational node. Then, each computational node separately builds the optimal route for the fragment allocated to it. In addition, DP is used as a tool for calculating the optimal route. To speed up the computations, we also use the OpenMP shared memory parallelization API. Since the originally transmitted fragments of the route were calculated heuristically, after the calculations, the results are improved.

4. After each computational node has processed its fragment of the original solution, it transfers the result to the main control machine, where the improved fragments are glued into one large solution. This solution is not guaranteed to be optimal, but it is "closer" to the optimal due to optimization of its fragments.

Consider a model example of solving the routing problem describing dismantlement of radioactive equipment on a plane (see [20]). The megalopoleis simulating the possible entrances/exits of rooms with radiation sources are obtained by discretizing the circles: on each circle, 40 points are located at an equal angular distance starting from the point with the angular coordinate 0. Each megalopolis corresponding to a point object then simulates a radiation source in the room. Let the starting point (it is the base of the dismantlement process) coincide with the origin, $\mathbf{x}^0 = (0, 0)$. We assume that, outside a megalopolis, the workers move 4 times faster than inside to simulate the complexity of moving within each megalopolis, due to the presence of certain structures and mechanisms that interfere with rapid movement inside the premises.

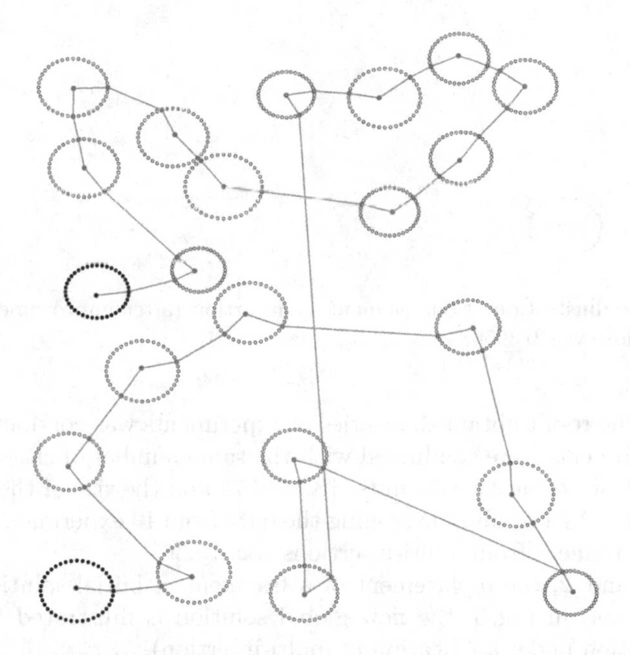

Fig. 1. Graphic illustration of one elementary insertion (before improvement), value of the radiation dose was 0.49771

The following is the result of single model experiment on the URAN supercomputer. For the model experiment, we used the variant with 255 megalopoleis and 45 address pairs corresponding to the precedence constraints, the size of the insertion fragment was 20 megalopoleis. The following results were obtained:

(1) the total dose rate obtained using the heuristic algorithm was 3.360531.
(2) the total value of the radiation dose obtained after the use of multi-insertion was 2.90207.
(3) the original result was improved by 15.8%.
(4) the computation time was 25 min 55 s.

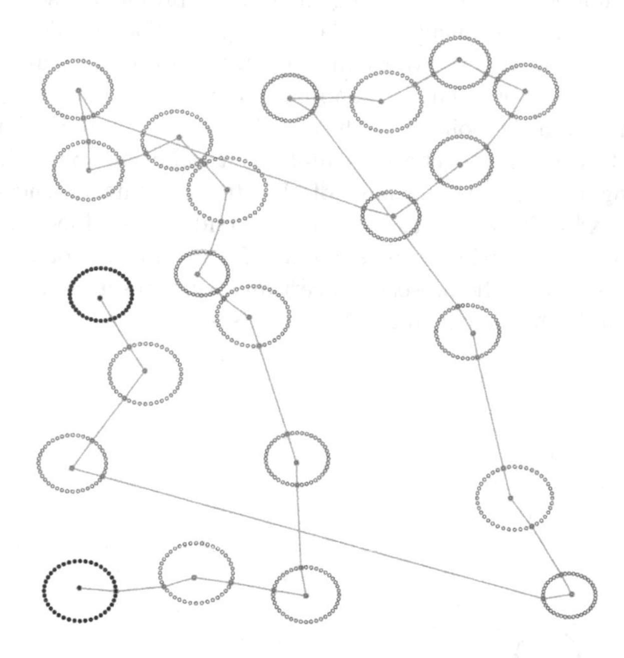

Fig. 2. Graphic illustration of one elementary insertion (after improvement), value of the radiation dose was 0.40568

To verify the result obtained, a series of experiments was conducted. Specifically, 10 experiments were conducted with the same number of cities ($\mathbf{n} = 255$), the number of precedence constraints ($|\mathbf{K}| = 45$), and the size of the optimizing insert ($N = 20$). As a result of averaging the data from 10 experiments, we found that the improvement from multi-insertions was 15.36%.

In Figs. 1 and 2, the replacement of a fragment of initial solution is represented. Moreover, in Fig. 3, the new global solution is illustrated (we keep in mind the solution under application of multi-insertion).

To study the dependence of the improvement on the size of multi-insertion fragments, we did another series of experiments. The fragment size ranged from 15 to 25 megalopoleis. For each fragment size of the multi-insert, 10 experiments were run, and the average value was computed. Figure 4 shows the dependence of improvement on the multi-insert fragment size. From this graph, it follows that the improvement percentage increases with fragment size.

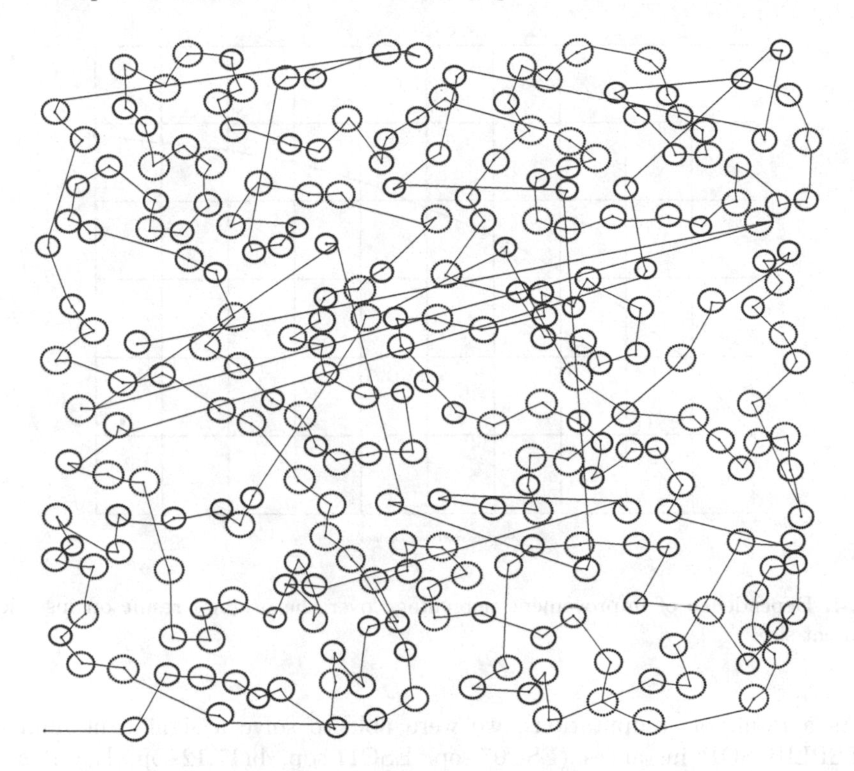

Fig. 3. The result of the computational experiment with 255 megalopoleis after the application of the improved multi-insertion ($f(x) \equiv 0$)

To test the parallel DP-based algorithm used to find the optimal route (see [14]), which we employed to improve the result over multi-inserts, we used the SOP instances from the TSPLIB library. Unfortunately, for our problem statement (GTSP with precedence constraints and cost functions depending on the set of pending tasks), the authors are not aware of an open library of problem instances. There are instance libraries for SOP (sequential ordering problem), which is the same as TSP-PC (TSP with precedence constraints), e.g. the SOP part of TSPLIB; and GTSP (generalized TSP), e.g. the library maintained by Karapetyan[1]. Both these libraries only partially correspond to our task. Since we are more interested in problem statements with precedence constraints, it was decided to test the algorithm on SOP instances from TSPLIB.

In order to adapt the library to our solver (and problem statement) we viewed the SOP cities as singleton megalopoleis and removed the dependence of travel costs function on the list of pending tasks. This approach significantly simplified our initial formulation but still allowed us to test the performance of the proposed algorithm for finding the optimal route.

[1] see http://www.cs.nott.ac.uk/~pszdk/gtsp.html.

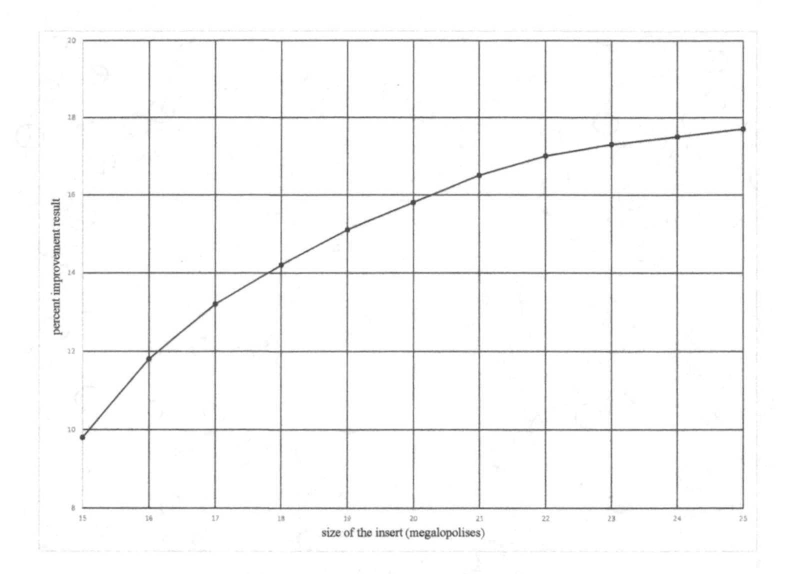

Fig. 4. Dependence of improvement percentage over the original result on insertion fragment size

As a result of computations, we were able to solve a significant number of TSPLIB SOP instances (ESC07.sop, ESC11.sop, br17.12.sop, ESC12.sop, br17.10.sop, ry48p.4.sop, rbg109a.sop, p43 .4.sop, ft53.4.sop, rbg150a.sop, ft70.4.sop, ESC25.sop, p43.3.sop, ft53.3.sop). For all these problems, an optimal solution was found, which was compared with the results available at the TSPLIB website. We managed to solve to optimality two instances that were listed as open on the site, and not solved in state-of-the-art paper on SOP [25], ry48p.3 (which was apparently first solved to optimality in [24]) and kro124p.4.

The following results were obtained for ry48p.3 (48 cities and 179 precedence constraints):

- The value of the optimal solution is $V = 19{,}894$.
- Found the best route 21,10,14,16,42,29,36,5,43,30,39,15,40,4,41,47,38,20,46,32, 45,35,26,18,27,22,2,13,24,12,31,23,9,44,34,3,25,1,28,33,11,19,6,17,37,8,7,48.
- Calculation time was 2,331 s.
- The amount of RAM used 12,003 MB.

For kro124p.4 (100 cities and 2404 precedence constraints), the following results were obtained:

- the value of the optimal solution is $V = 76{,}103$.
- Found the best route 5,89,87,9,62,27,57,39,1,81,32,4,84,56,76,7,88,96,65,25, 64,74,48,66,68,49,47,13,54,26,11,31,20,16,44,82,83,14,10,92,29,86,50,42,59,71, 35,98,52,15,21,69,93,17,79,30,55,91,78,61,34,6,8,60,95,77,12,75,94,51,40,99, 70,2,28,33,85,22,90,73,24,67,80,97,58,37,23,18,3,41,53,63,72,43,36,38,45,19, 46,100.

- Calculation time was 84,603 s.
- The amount of RAM used 128,248 MB.

The algorithm demonstrates good performance and efficiency, thereby proving its applicability for a certain range of tasks where other algorithms have no way of finding the optimal solution.

10 Conclusion

The considered routing problem contains constraints and complicated travel cost functions, which are connected with the requirements of real-life problems. Naturally, the mathematical setting is complicated. However, the structure of exact (optimal) solutions for this setting is known (see [5–8,12–14,18–21]). This structure is defined by the DP procedure, which is a variant and a development of the Bellman procedure [3]. It is useful to recall that the above-mentioned general procedure permits to optimize (see [11,22]) the initial state by means of a unique DP procedure.

But, for the routing problem of a large dimension, the computation cost of the above-mentioned global DP procedure becomes practically impossible. For such problems, heuristics are used very widely. However, DP can be used for local improvements. In particular, one could consider individual insertions and iterated procedures, where local optima are obtained by means of DP.

Another approach to local improvement of the result reached by heuristics is connected with the employment of multi-insertions and parallel algorithms. Both approaches are reflected in this article. Along with theoretical schemes, a software implementation is considered. It is significant that the above-mentioned implementation is connected with a very complicated real-life problem from nuclear power generation. As the experiment shows, for a sufficiently large dimension of the initial routing problem, our procedure with optimizing multi-insertions significantly improves the results.

Thus, this article deals with a very general approach to solving complicated routing problems arising from engineering applications. Specifically, we consider the process of sequentially dismantling a number of radiation sources, during which the worker who does gets a big dose of radiation. The dose depends on the sequence in which the dismantlement is conducted (i.e., on the worker's route) and the specific trajectory of the worker's movements. This gives rise to an actual extremal problem, which has difficulties, as discussed in the introduction. Let us now discuss the results.

Expression (9) is worth a separate mention. This relation means that our insertions are realized "optimally". It would certainly be nice to apply the global DP procedure and obtain the (global) extremum. But, we optimize within the limits of possible: we use exact algorithms for every insertion. Naturally, we use a more complicated DP procedure [5–8,12,19–21]; this procedure is different from the DP procedure for TSP and TSP-PC. It is clear that the global DP procedure is laborious. Therefore, we apply DP procedures for every insertion. Namely, we strive to "optimally" realize a variant of global DP. Under this realization, we are able to use parallel and, what is more important, independent calculation procedure (see [14]; other parallel algorithms were considered in [10,11]). As a result, we obtain (9). So, our approach is not only heuristic; in this approach, an essential regular component is present. This component is connected with the developed DP procedure and parallel algorithms. In particular, the global DP variant was realized for 50 megalopoleis with 30 cities in each.

References

1. Gutin, G., Punnen, A.P.: The Traveling Salesman Problem and its Variations. Springer, Berlin (2002)
2. William, J.C.: In Pursuit of the Traveling Salesman: Mathematics at the Limits of Computation, p. 248. Princeton University Press, Princeton (2012)
3. Bellman, R.: Dynamic programming treatment of the travelling salesman problem. J. Assoc. Comput. Mach. **9**, 61–63 (1962)
4. Held, M., Karp, R.M.: A dynamic programming approach to sequencing problems. J. Soc. Ind. Appl. Math. **10**(1), 196–210 (1962)
5. Chentsov, A.G.: Problem of successive megalopolis traversal with the precedence conditions. Autom. Remote Control **75**(4), 728–744 (2014). https://doi.org/10.1134/S0005117914040122
6. Chentsov, A.G., Chentsov, A.A.: Route problem with constraints depending on a list of tasks. Doklady Math. **92**(3), 685–688 (2015). https://doi.org/10.1134/S1064562415060083
7. Chentsov, A.G.: The Bellman insertions in the route problem with constraints and complicated cost functions. Vestnik Udmurtskogo Universiteta. Matematika. Mekhanika. Komp'yuternye Nauki **2014**(4), 122–141 (2014). https://doi.org/10.20537/vm140410. (in Russian)
8. Chentsov, A.G., Chentsov, P.A., Petunin, A.A., Sesekin, A.N.: Model of megalopolises in the tool path optimisation for CNC plate cutting machines. Int. J. Prod. Res. **56**(14), 4819–4830 (2018). https://doi.org/10.1080/00207543.2017.1421784
9. Petunin, A.A., Chentsov, A.G., Chentsov, P.A.: Optimizing inserts in a routing task with constraints and complicated cost functions. Izvestija RAN. Teorija i sistemy upravlenija **1**, 117–130 (2019). (in Russian)
10. Chentsov, A.G., Grigoryev, A.M.: Dynamic programming method in a routing problem: a scheme of independent computations. Mekhatronika, Avtomatizatsiya, Upravlenie **17**(12), 834–846 (2016). (in Russian)
11. Chentsov, A.G., Grigoryev, A.M., Chentsov, A.A.: Optimizing the starting point in a precedence constrained routing problem with complicated travel cost functions. Ural Math. J. **4**(2), 43–55 (2018). https://doi.org/10.15826/umj.2018.2.006
12. Chentsov, A.G.: The Bellmann insertions in route problems with constraints and complicated cost functions. II. Vestnik Udmurtskogo Universiteta. Matematika. Mekhanika. Komp'yuternye Nauki **26**(4), 565–578 (2016). https://doi.org/10.20537/vm160410. (in Russian)
13. Petunin, A.A., Chentsov, A.A., Chentsov, A.G., Chentsov, P.A.: Elements of dynamic programming in local improvement constructions for heuristic solutions of routing problems with constraints. Autom. Remote Control **78**(4), 666–681 (2017). https://doi.org/10.1134/S0005117917040087

14. Chentsov, A.G., Grigoryev, A.M.: Optimizing multi-inserts in routing problems with constraints. Vestnik Udmurtskogo Universiteta. Matematika. Mekhanika. Komp'yuternye Nauki **28**(4), 513–530 (2018). https://doi.org/10.20537/vm180406. (in Russian)
15. Gimadi, E.Kh., Khachai, M.Yu.: Extremal Problems on Sets of Permutations, p. 220. UMC UPI, Yekaterinburg (2016). (in Russian)
16. Melamed, I.I., Sergeev, S.I., Sigal, I.Kh.: The traveling salesman problem. I Issues in theory; II Exact methods; III Approximate algorithms. Autom. Remote Control **50**(9), 1147–1173; **50**(10), 1303–1324; **50**(11), 1459–1479 (1989)
17. Litl, Dzh., Murti, K., Suini, D., Kerel, K.: Algorithm for the traveling salesman problem. Ekon. Mat. Metod. **1**(1), 94–107 (1965)
18. Chentsov, A.G.: Extreme Problems of Routing and Tasks Distribution: Regular and Chaotic Dynamics, 240 p. Izhevsk Institute of Computer Research (2008). (in Russian)
19. Chentsov, A.G., Chentsov, P.A.: Routing under constraints: problem of visit to megalopolises. Autom. Remote Control **77**(11), 1957–1974 (2016). https://doi.org/10.1134/S0005117916110060
20. Chentsov, A.G., Chentsov, A.A.: A model variant of the problem about radiation sources utilization (iterations based on optimization insertions). Izvestiya Instituta Matematiki i Informatiki Udmurtskogo Gosudarstvennogo Universiteta **50**, 83–109 (2017). https://doi.org/10.20537/2226-3594-2017-50-08. (in Russian)
21. Chentsov, A.A., Chentsov, A.G., Chentsov, P.A.: Extremal routing problem with internal losses. Proc. Steklov Inst. Math. **264**(Suppl. 1), 87–106 (2009)
22. Chentsov, A.G., Chentsov, P.A.: Optimization of the start point in the GTSP with the precedence conditions. Bull. South Ural State Univ.: Series Math. Model., Prog. Comput. Soft. **11**(2), 83–95 (2018). https://doi.org/10.14529/mmp180207
23. Alkaya, A.F., Duman, E.: Combining and solving sequence dependent traveling salesman and quadratic assignment problems in PCB assembly. Discrete Appl. Math. **192**, 2–16 (2015)
24. Salii, Y.V.: Revisiting dynamic programming for precedence-constrained traveling salesman problem and its time-dependent generalization. Eur. J. Oper. Res. **272**(1), 32–42 (2019)
25. Gouveia, L., Ruthmair, M.: Load-dependent and precedence-based models for pickup and delivery problems. Comput. Oper. Res. **63**, 56–71 (2015)

The Stochastic Coverings Algorithm for Solving Applied Optimal Control Problems

Alexander Gornov, Tatiana Zarodnyuk, Anton Anikin, and Pavel Sorokovikov

Matrosov Institute for System Dynamics and Control Theory SB RAS, Irkutsk, Russia
tzarodnyuk@gmail.com
http://idstu.irk.ru/en

Abstract. The paper considers a heuristic method for a global extremum search in an optimal control problem based on the idea of covering a reachable set by n-dimensional balls, including the built-in mechanisms for Lipschitz constant estimating of the objective functional. A step-by-step description of the coverage algorithm and the proposed method for generating start and auxiliary controls are presented. The proposed technique was used for solving applied optimal control problems: the problem of investment programs in Buryatia Republic and the problem of restoring the Black Lands in Kalmykia.

Keywords: Optimal control · Numerical algorithms · Global optimization · Control applications

1 Introduction

A global extremum search for functionals defined on dynamic systems trajectories remains one of the most difficult extremal problems. In the theory of optimal control, it is not found approaches that guarantee to obtain a global optimal solution for nonlinear systems, based on which it is possible to build effective algorithms. Application direct methods of reduction to finite-dimensional problems lead to the appearance of approximative problems of mathematical programming, including hundreds and thousands of variables. Even the local extremum search for problems of such dimensionality in a number of cases is a serious problem. The study this problem for searching a global extremum may turn out to be unrealistic and require in practice the astronomical expenses of the processor time using most modern supercomputers. According to the classification proposed by [1], the problem of finding a global extremum of a multiextremal functions with dimensions of 200–300 variables should be considered super-hard.

Supported by Russian Foundation for Basic Research, grant No. 17-07-00627.

I. Bykadorov et al. (Eds.): MOTOR 2019, CCIS 1090, pp. 486–496, 2019.
https://doi.org/10.1007/978-3-030-33394-2_37

In the theory of global optimization, it is conventionally divided methods into "mathematical, rational", based on a specific model of the aim functions, and "heuristic"—all the rest (see, for example, [2]). The attitude of many specialists to heuristic methods have changed significantly: "heuristic methods: once scorned, are highly respectable" [3,4]. The "final judge" in this multi-year discussion is the practice of applying methods in solving complex applied problems.

One of the most reliable mathematical methods for finite-dimensional problems are the coverings methods, it is based on the hypothesis of the boundedness of the growth rate of the optimized function (see, for example, [1,3,5]). Most often this hypothesis is formalized in the form of the Lipschitz constants estimates for function or its derivatives. The algorithms constructed using this approach are among the most reliable, although require significant computational costs.

In the terminology discussed above, all known algorithms for a global extremum search in nonlinear optimal control problems: methods of genetic search (see, for example, [6,7]), random multistart methods [8,9], convexification methods [10], methods for stochastic approximations of the reachable set [11], curvilinear search methods [12] and others should be considered heuristic. In our opinion, by now no theoretical results have been obtained that can serve as the ideological basis for constructing guaranteed algorithms for global extremum search of functional in optimal control problems. Nevertheless, the problem of achieving guarantees, and more precisely, to increase the reliability of calculations (decreasing "degree heuristics") continues to be relevant.

The paper considers a heuristic method for a global extremum search in an optimal control problem based on the idea of covering a reachable set by n-dimensional balls, including the built-in mechanisms for Lipschitz constant estimating of the objective functional.

2 Statement of the Optimal Control Problem

The controlled process is described by a system of ordinary differential equations with initial conditions

$$\dot{x} = f(x(t), u(t), t), \quad x(t_0) = x^0, \tag{1}$$

where t is the time from the interval $[t_0, t_1]$, $x(t) = (x_1(t), x_2(t), \ldots, x_n(t))$ is vector of phase coordinates, $u(t) = (u_1(t), u_2(t), \ldots, u_r(t))$ is control vector. Vector function $f(x(t), u(t), t)$ is assumed to be continuously differentiable with respect to all arguments except t. We call piecewise-continuous control functions $u(t)$ belonging to set U for any time values t:

$$U = \{u(t) \in R^r : u_l \leq u(t) \leq u_g\}, \tag{2}$$

$u_l, u_g \in R^r$ are vectors of lower and upper constraints on control.

The optimal control problem with a free right end is the search for an admissible control $u^*(t)$ that delivers a minimum to the objective functional

$$I(u) = \varphi(x(t_1)). \tag{3}$$

Terminal function $\varphi(x)$ is assumed to be Lipschitz.

3 The Basic Optimization Algorithm

The creation of the stochastic coverings method in full accordance with tradition is reduced to generate a sequence of test points and the corresponding sets sequences whose union must cover the reachable set. However, unlike the unconditional optimization problem, the generated quasi-stochastic test points are constructed in space controls, and the covering sets are in the terminal phase space. The main task of the numerical algorithm is the complete covering of the system reachable set. The elementary covering sets are the balls $B(R,u)$ in n-dimensional Euclidean space, where R is the radius of the ball, u is the control, the end of the corresponding trajectory is the center of the ball. On iterations with a set of information about the problem, the estimation of Lipshitz constant is also corrected. To increase of the covering reliability is introduced, as in classical papers (see, for example, [13]), "safety factor" K_s, multiple increasing the value of the Lipshitz constant estimate. The sequence of records, in this case, it turns out to be monotonically decreasing. However, the total volume of covering can how to increase at the occurrence of new covering balls, and decrease on iterations as the Lipschitz constant is corrected. The proposed algorithm is divided into iterations, each of which is generated the specified number of test points. At the iteration, the number of auxiliary points that enter into the already constructed balls of coverage, as well as the number of new samples, in the balls of which there may be points less than the current record are calculated. These characteristics can serve as heuristic criteria for the end of the search. Traditionally, the "accuracy by functionals" ε_φ is specified. Thus, it is generated stochastic covering by different sizes balls, which depends on the value functional at the test point, the current estimate of the Lipschitz constant and the value ε_φ: the radius of the ball $R^j = (I^j - I_{REC} + \varepsilon_\varphi)/(K_s \cdot L)$, $j = \overline{1, M_X}$, where I_{REC} is the current record value, M_X is the number of balls in the covering.

3.1 Turn-Based Algorithm Scheme

Step 0. Algorithmic parameters are specified:

M is the test points number on the iteration,

M_{max} is the maximum number of samples,

L_0 is the initial value of the Lipschitz constant estimation,

ε_φ is the "accuracy by functionals",

K_s is the "safety factor".

Step 1. Set a record value $I_{REC} = \infty$, estimate the Lipschitz constant $L = L_0$, an empty initial covering is specified X_0, $M_X = 0$, the value k is equal 1.

Step 2. On the k th iteration, M test points are generated $\{u^1, \ldots, u^M\}$.

Step 3. For all u^j, the direct differential equations system is integrated, it is remembered $x^j(t_1)$, $j = \overline{1, M}$.

Step 4. For all $x^j(t_1)$, the values of the objective functional I^j, $j = \overline{1, M}$ are calculated.

Step 5. The record value is improving, if

$\exists j : I^j < I_{REC}, I_{REC} = I^j, \ j = \overline{1, M}$.

Step 6. Calculated the samples number in the iteration M_y that did not fall into any of the already available in the covering balls. Found the samples number M_q in which the ball can improve record value: $I^j < I_{REC} + \varepsilon_\varphi$, $j = \overline{1, M}$.

Step 7. For all $x^j(t_1)$, local estimates of the Lipschitz constant are calculated in comparison with the samples already available in the covering balls

$L^j = K_s \cdot \left|I^j - I^i\right| / \left\|x^j(t_1) - x^i(t_1)\right\|$, where $j = \overline{1, M}$, $i = \overline{1, M_X}$, the largest estimate is chosen.

Step 8. The estimate of the Lipschitz constant is refined if

$\exists j : L^j > L, \ L = L^j, \ j = \overline{1, M}$.

Step 9. It is set $X_{k+1} = X_k \cup \{B(R^j, u^1), \ldots, B(R^j, u^M)\}$, $M_X = M_X + M$. The iteration is complete.

The number of algorithm iterations depends on the set value of the algorithmic parameter M_{max}. The parameters M_y and M_q allows us evaluate the quality of the resulting coating. For this, when issuing results, their total values are presented.

3.2 The Stopping Criteria

The stopping criterion for the algorithm can be based on the obtained at the iterations numbers M_y. It is intuitively clear that the less frequently new test points fall into the still uncovered part of the attainable set, the less probability that a global extremum has not yet been found. An additional criterion for evaluating the quality of the obtained solution can be the number of samples per iteration, near which it is possible in principle to improve the values record of M_q. At zero values of these values for a given number of the last iterations can make a decision about the stop of the proposed algorithm.

3.3 The Algorithm Operation Modes

The full operation mode of the algorithm is described above. In many cases, it is advisable to save the computation time, which is performed with other algorithm operation modes. Currently, the following technologies options are implemented.

A Bounded Evaluation Mode for the Lipschitz Constant. In this mode, calculations of Lipschitz constants are estimated only on the indicated number of initial iterations (M_0 is an algorithmic parameter). In many cases, during this time the algorithm has time to get a good estimate, which in the future does not improve, and the calculation time increases substantially with an increase in the size of the sampling points basis.

Mode Without Evaluation of Lipschitz Constant. The application of this mode is reasonable if the estimate of the constant is already known, for example, with a linear terminal functional.

Uncoated Mode. In this mode, there is complete savings time at intermediate calculations, but the method under consideration turns into the usual grids method [8].

Different presented operation modes of the algorithm allow us to obtain the minimum value or the objective functional. It was confirmed by the results of numerical experiments. Let us give one of the test examples solved by using the stochastic covering algorithm.

3.4 Controls Generation

We proposed algorithms for generating relay, piecewise-linear, tabular and spline control functions to obtain quasi-random admissible controls that allow one to construct the inner approximation of the reachable set [11,14]. Taking into account the earlier computational experience a new algorithm for generating relay-type functions is proposed that allows randomly obtain a different number of switching points. Wherein the mathematical expectation of the number of control switching points, sequentially generated at the algorithm iteration, is a priori given algorithmic parameter.

We define a time-discretization grid from N nodes $T = \{t_0 = \tau_0 < \tau_1 < \ldots < \tau_N = t_1\}$. We denote by S the next quasi-random number uniformly distributed in the interval $[0, 1]$, generated by the standard URAND algorithm (see [15]).

Step 0. The recommended number of the control switching K_p is given.

Step 1. The value of the "switching probability level" is computed $P_p = K_p/(N-1)$.

Step 2. For i-th components of the control vector, $i = \overline{1,r}$ it is performed:

if $S < 1/2$ then $u_i(t_0) = (u_l)_i$, else $u_i(t_0) = (u_g)_i$.

Step 3. For j-th grid node T, $j = \overline{1,N}$ it is performed:

if $S < P_p$, then $u_i(\tau_j) = u_l(\tau_{j-1})$ (in interval $[\tau_{j-1}, \tau_j]$ there is no switching), else

if $u_i(\tau_{j-1}) = (u_l)_i$, then $u_i(\tau_j) = (u_g)_i$,

if $u_i(\tau_{j-1}) = (u_g)_i$, then $u_i(\tau_j) = (u_l)_i$ (in interval $[\tau_{j-1}, \tau_j]$ there is switching).

The algorithm is complete.

The success of the algorithm depends very much on the parameter K_p. To obtain an acceptable approximation of the reachable set, and, consequently, the successful operation of the algorithm, the number of recommended points switching K_p should not be either too small or too large. When a small number of K_p algorithm can not build a sufficiently representative set of trial controls, with a large value of K_p test points on the reachable set can be condensed in one local area. In both cases, the approximation of the reachable set will not be too good. Based on the computational experience by default, $K_p = 5$ is recommended. For increase the reliability of the algorithm, it is advisable to perform calculations with different $K_p \in [2, 10]$.

3.5 The Test Problem

We demonstrate the results of applying the proposed approach to a well-known test optimal control problem of nonlinear oscillation of pendulum (see, for example, [16–18]). The nonlinear system is described by following differential equations

$$\dot{x}_1 = x_2,\ \dot{x}_2 = u - \sin x_1 \tag{4}$$

The values of phase coordinates at the initial time are given $x_0 = (5, 0)$, $t \in [0, 5]$, control satisfies the parallelepiped constraints $u \in [-1, 1]$. It is necessary to minimize functional

$$I(u) = x_1^2(t_1) + x_2^2(t_1) \to \min. \tag{5}$$

The optimal value $I^* = 11.90876$ is achieved by optimal control and corresponding trajectories presented in Fig. 1

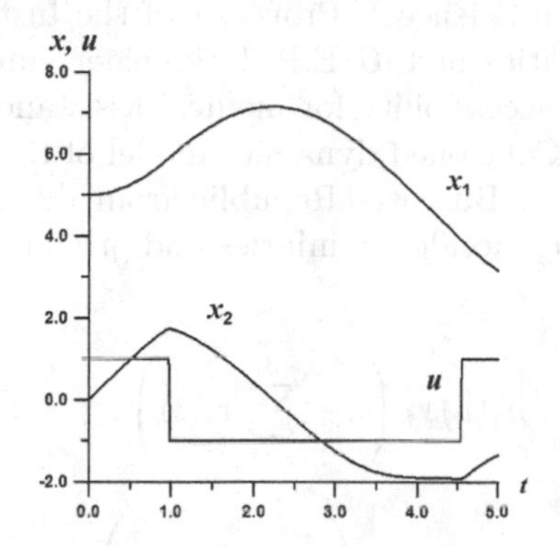

Fig. 1. Optimal control and corresponding trajectories in nonlinear pendulum oscillation problem

The results of the solving problem (4)–(5) by using the developed approach on different iterations of the algorithm are shown in Fig. 2.

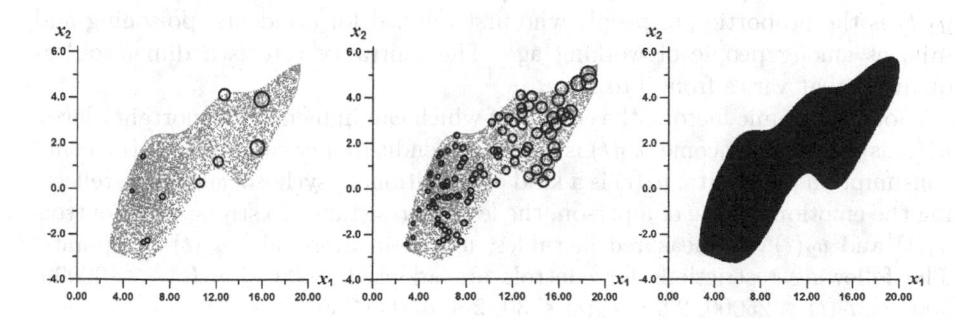

Fig. 2. The reachable set at 10, 100 and 10000 algorithm iterations

Optimal trajectories and control, reachable set and minimum functional value coincided with the result known from the original source.

4 Applied Optimal Control Problems

The developed approach is used to solve a number of applied nonlinear optimal control problems.

4.1 Optimization Model of Investment Programs in Buryatia Republic

The problem was posed by prof. M.P. Dyakovich (Research Institute of Occupational Health and Human Ecology, National Center of Scientific and Technical Information of the All-Union Scientific Center of the Siberian Branch of the Russian Academy of Medical Sciences). Professor of the Institute of Mathematics, Economics, and Informatics of ISU E.P. Bokmelder simulated of the population mortality rate and accessibility for medical assistance, taking into account socio-economic factors. Controlled dynamics model of the mortality rate for the able-bodied population of Buryatia Republic from diseases of the circulatory system, alcohol poisoning, accidents, injuries and primary care rates for medical care is as follows [19]:

$$\dot{x}_j = \mu_j(u)x_j\left(1 - \sum_{i=1}^{3} x_i(t)\right), \; j = \overline{1,3} \tag{6}$$

$$\dot{y}_k = \mu_{k+3}(u)y_k(t)\left(1 - \sum_{i=1}^{3} y_i(t)\right), \; k = 1, 2. \tag{7}$$

Here $x_1(t)$ is the proportion of patients of working age (from 20 to 60 years) who died from diseases of the circulatory system, $x_2(t)$ is the proportion of working age patients died from accidents, poisoning and injuries, $x_3(t)$ is number of deaths from alcohol poisoning, $y_1(t)$ is the proportion of people who for the first time this year applied for medical help for diseases of the circulatory system, $y_2(t)$ is the proportion of people who first applied for accidents, poisoning and injuries among people of working age. The mortality rate is a dimensionless quantity that varies from 0 to 1.

Socio-economic factors, the change of which can influence the mortality level: $u_1(t)$ is per capita income, $u_2(t)$ is health expenditure per capita, $u_3(t)$ is alcohol consumption per capita, $u_4(t)$ is a kind of conditional psychosocial factor, reflecting the emotional state of a person, the level of resistance to stress, etc. Controls $u_1(t)$ and $u_2(t)$ are measured in rubles, $u_3(t)$—in liters, and $u_4(t)$—in points. The following restrictions for controls are adopted: $3000 \leq u_1(t) \leq 100000$, $3000 \leq u_2(t) \leq 25000$, $2.5 \leq u_3(t) \leq 30$, $2 \leq u_4(t) \leq 30$

The initial values, calculated using the available statistical data, are as follows: $x_1(0) = 0.003112$, $x_2(0) = 0.00462$, $x_3(0) = 0.000772$, $y_1(0) = 0.34558$,

$y_2(0) = 0.091261$, $u_1(0) = u_1^0 = 5914$ (rbl), $u_2(0) = u_2^0 = 4398.9$ (rbl), $u_3(0) = u_3^0 = 7.4$ (litres), $u_4(0) = u_4^0 = 10$ (points).

The coefficient of proportionality for the mortality rate and the circulation of medical care $\mu_p(u)$ is found by the formula

$$\mu_p(u) = v_p + l_p e^{-\varepsilon_{p_1} u_1 - \varepsilon_{p_2} u_2} - q_p e^{-\varepsilon_{p_3} u_3 - \varepsilon_{p_4} u_4}, \tag{8}$$

here v_p, l_p, q_p, ε_{p_1}, ε_{p_2}, ε_{p_3}, ε_{p_4} are numeric parameters, $p = \overline{1,5}$.

It is necessary to minimize the total mortality and turnover rates for medical care at the lowest total costs, so the functional will take the form

$$I(u) = \int_0^T e^{-0.08t} \left(\sum_{i=1}^{4} b_i \right) dt \rightarrow min, \tag{9}$$

here $b_1 = p_1 (x_1(t) + x_2(t) + x_3(t))$, $b_2 = p_2 (y_1(t) + y_2(t))$, $b_3 = p_3 u_1(t)$, $b_4 = p_4 u_2(t)$, $e^{-0.08t}$ is discount multiplier, 0.08 is average bank interest rate in the republic.

Coefficients $p_1 = 1.2$, $p_2 = 0.05$, $p_3 = 5 * 10^{-7}$, $p_4 = 10^{-7}$ designed to balance the scale of damage from mortality and seeking medical attention for leading classes of diseases, were found as a result of a series of computational experiments.

Calculations showed that alcohol consumption u_3 and psycho-emotional stress u_4 so strongly affect mortality, that the optimal is always their lower level. Increased costs on health and a slight decrease in wages allow reducing the values of the studied indicators. At the same time, mortality from diseases of the circulatory system $x_1(t)$, accidents $x_2(t)$ and alcohol poisoning $x_3(t)$ will decrease by 5.24%, 5.08% and 4.55% respectively. While maintaining the same trends of socio-economic development $x_1(t)$, $x_2(t)$ and $x_3(t)$ will increase 10.3%, 2.6% and 6%

The obtained results testify to the necessity of social orientation of expenditures provided that the high rates of economic development are ensured, which will help reduce the mortality of the working-age population of Buryatia Republic.

4.2 The Problem of Restoring the Black Lands in Kalmykia

The model of desertification dynamics, created by Professor A.K. Cherkashin (IGSiDV, Siberian Branch of the Academy of Sciences of the USSR)—[20], describes the dynamic process of the transition of the area of lands from one state to another as a result of degradation of pastures and their natural restoration.

The system of differential equations has the form

$$\dot{s}_1 = -\alpha_1^0 s_1 + \alpha_2 s_2 + v_2 + v_3 + v_4 \tag{10}$$

$$\dot{s}_2 = \alpha_1^0 s_1 - \alpha_2^0 s_2 + \alpha_3 s_3 - \alpha_2 s_2 - v_2 \tag{11}$$

$$\dot{s}_3 = \alpha_2^0 s_2 - \alpha_3^0 s_3 + \alpha_4 s_4 - \alpha_3 s_3 - v_3 \tag{12}$$

$$\dot{s}_4 = \alpha_3^0 s_3 - \alpha_4 s_4 - v_4 \tag{13}$$

where $s_i(t)$ are the land areas that are at time t in the i-th degradation stage (s_1 is intact, s_2 is slightly broken, s_3 is weakly fixed, s_4 is loose land areas) in % of total area 216 thousand hectares. α_i^0, α_i are coefficients of intensity of pastures degradation and restoration, $i = \overline{1,4}$, v_2, v_3, v_4 are rate of phytomeliorative measures for the land restoration at the time t (% in a year) and are controls in the presented problem. The initial conditions for the system are the land structure at present $s(0) = (2.6, 8, 37.1, 52.3)$ in %.

According to aerospace survey data, areas s_i were determined in adjacent observation periods and the probability of transition from i-th to j-th state during the period between two observations. After identifying the model and scaling, the expressions for α_i^0, α_i have the following form

$$\alpha_i = P_i/\Delta\tau_i, \alpha_i^0 = (1 - P_i)/\Delta\tau_i, i = \overline{1,4}, \tag{14}$$

where $P_3 = \min\{1, \exp(-8.4s_1 + 5.76 - 5.76/C(\Sigma_{i=1}^4 \omega_i s_i))\}$, $P_1 = 0$, $P_4 = 1$, $P_2 = \min(1, \exp(8.4s_1 - 5.76))$, $\Delta\tau_4 = \exp(-8.3s_4 + 7.18)$, $\Delta\tau_1 = \Delta\tau_2 = \Delta\tau_3 = \exp(-4.4s_4 + 3.04)(1 - \exp(2.5N_b))$, here N_b is pasture load.

Natural variables were imposed on the variables $s_i \geq 0$, $i = \overline{1,4}$, $v_j \geq 0$, $j = \overline{2,4}$, $N_b \geq 0$.

The objective functional (profit) was described by the formula

$$I(u) = \int_0^{20} \left[aN_b - \sum_2^4 b_i v_i(t) \right] dt, \tag{15}$$

where a is average annual income from sheep maintenance, b_i is the hectare cost of land reclamation at i-th degradation stage, $a = 15.4$ RUR/pcs, $b_2 = 25$, $b_3 = 40$, $b_4 = 80$ RUR/hectare.

It was also assumed that all vegetation is eaten by sheep

$$\sum_{i=1}^4 \omega_i s_i = CN_b, \tag{16}$$

where ω_i is vegetation cover biomass at i-th stage of pasture degradation (centners/hectare), C is average annual need of one sheep in feed (centners). These variables take the following values $C = 5$, $\omega_1 = 8.5$, $\omega_2 = 4.5$, $\omega_3 = 2.0$, $\omega_4 = 0$.

A restriction $s_1 \geq 68\%$ are imposed on structure land, which was necessary to obtain. This condition corresponds to the area of critical values of undisturbed areas. Expressing N_b from the equality constraint and introducing penalty coordinates, we reduce the problem to the standard form. Parallelepipic constraints on three-dimensional control were given in the following form: $0 \leq v_i t \leq 1$, $i = \overline{2,4}$.

At the beginning of calculations, the quality of the system discretization was investigated and an acceptable integration grid was found. The first stage of the solution was to overcome the "nonphysicality of the model" area by artificially reducing the range of permissible control values and ended when direct control restrictions became inactive. At the second stage, the discretization was significantly improved and the main efforts of the algorithm were directed at satisfying the phase constraints.

The following record value of the functional is obtained $I_0(u) = -355.68$, with the maximum violation of the phase limitation $5.2 \cdot 10^{-3}$. The resulting solution means that in this way it is possible to restore the land structure within 10 years. The average income from sheep breeding, minus the cost of works on phytomelioration, will be 2.2 million rubles in year for 20 years.

5 Conclusion

The proposed algorithm makes it possible to improve the reliability of global extremum in the optimal control problems and get heuristic estimates of the probability of achieving the optimal result. The simplicity of the constructed technique allows us to hope for simple development of its modifications for parallel computing systems. The effectiveness of the presented approach is demonstrated in test problems and in the solution of applied optimal control problems.

References

1. Evtushenko, Y.G., Polovinkin, M.A.: Parallel methods for solving global optimization problems. In: Proceedings of IV International Conference Parallel Computing and control problems, pp. 18–39, Moscow (2008)
2. Zhigljavsky, A.A., Zhilinskas, A.G.: Searching Methods for a Global Extremum. Nauka, Moscow (1991)
3. Zhigljavsky, A.A., Zilinskas, A.G.: Stochastic Global Optimization. Springer, New York (2008). https://doi.org/10.1007/978-0-387-74740-8
4. Torn, A., Zhilinskas, A.: Global Optimization. Springer, Heidelberg (1989). https://doi.org/10.1007/3-540-50871-6
5. Strongin, R.G., Sergeyev, Y.D.: Global Optimization with Non-Convex Constraints: Sequential and Parallel Algorithms. Kluwer Academic Publishers, Dordrecht (2000)
6. Floudas, C.A., Gounaris, C.E.: A review of recent advanced in global optimization. J. Glob. Optim. **1**, 3–38 (2009)
7. Lopez Cruz, I.L., Van Willigenburg, L.G., Van Straten, G.: Efficient differential evolution algorithms for multimodal optimal control problems. Appl. Soft Comput. **3**, 97–122 (2003)
8. Gornov, A.Y.: Computational Technologies for Solving Optimal Control Problems. Nauka, Novosibirsk (2009)
9. Tyatyushkin, A., Zarodnyuk, T.: Numerical method for solving optimal control problems with phase constraints. Numer. Algebra Control Optim. **7**(4), 483–494 (2017)
10. Tolstonogov, A.A.: Differential Inclusions in a Banach Space. Nauka, Novosibirsk (1986). (in Russian)
11. Gornov, A.Y., Zarodnyuk, T.S., Finkelshtein, E.A., Anikin, A.S.: The method of uniform monotonous approximation of the reachable set border for a controllable system. J. Glob. Optim. **66**(1), 53–64 (2016)

12. Gornov, A.Y., Zarodnyuk, T.S.: The curvilinear search method for a global extremum search in the optimal control problem. modern technology. system analysis. Modeling **3**, 19–26 (2009). (in Russian)
13. Strongin, R.G.: Numerical Methods in Multi-Extremal Problems. Information-Statistical Approach. Nauka, Moscow (1978)
14. Zarodnyuk, S.: Algorithm for the numerical solution of multi-extremal optimal control problems with parallelepiped constraints. Comput. Technol. **18**(2), 46–54 (2013). (in Russian)
15. Forsythe, J., Malcolm, M., Moler, K.: Machine Methods Mathematical Calculation. Mir, Moscow (1980)
16. Afanasev, V.N., Kolmanovskii, V.B., Nosov, V.R.: Mathematical Theory of Control System Design. Visshaya shkola, Moscow (2003)
17. Tyatyushkin, A.I.: Numerical Methods and Software for Optimization of Controlled Systems. Nauka, Novosibirsk (1992)
18. Gornov, A.Y., Zarodnyuk, T.S., Madzhara, T.I., Daneyeva, A.V., Veyalko, I.A.: A collection of test multiextremal optimal control problems. In: Chinchuluun, A., Pardalos, P., Enkhbat, R., Pistikopoulos, E. (eds.) Optimization, Simulation and Control. Springer Optimization and Its Applications, vol. 76, pp. 257–274. Springer, New York (2013). https://doi.org/10.1007/978-1-4614-5131-0_16
19. Bokmelder, E.P., Dyakovich, M.P., Efimova, N.V., Gornov, A.Y., Zarodnyuk, T.S.: Experience of application of dynamic systems models for solving medico-social and medico-ecological problems. Inform. Control Syst. **2**(24), 161–164 (2010)
20. Vinogradov, B.V., Cherkashin, A.K., Gornov, A.Y., Kulik, K.N.: Dynamic monitoring of degradation and restoration of pastures in the Black Lands of Kalmykia. Probl. Desert Dev. **1**, 7–14 (1990)

Deterministic Approximation of Stochastic Programming Problems with Probabilistic Constraints

Yuri S. Kan and Sofia N. Vasil'eva(✉)

Moscow Aviation Institute (National Research University),
Volokolamskoe shosse, 4, Moscow 125993, Russia
yu_kan@mail.ru, sofia_mai@mail.ru
http://www.mai.ru

Abstract. The work is devoted to the development of a method for solving the stochastic programming problem with a deterministic objective function and individual probabilistic constraints. Each probabilistic constraint is a constraint on the probability of inequality for a certain loss function that is linear on random parameters. In this case, the loss function may be non-linear in strategies. It is proposed to replace each probabilistic constraint by an equivalent inequality for the quantile function. This inequality is approximated using the notion of the probability measure kernel. The kernel is defined as the intersection of all closed confidence half-spaces. It is known that if the kernel satisfies the regularity property and the loss function is linear in random parameters then the quantile function can be found as the maximum of the loss function in realizations of random parameters on the probability measure kernel. To evaluate quantiles an external polyhedral approximation [1] of the probability measure kernel is used. When replacing a kernel by its approximation the maximum mentioned above is an upper estimate of the exact value of the quantile function. As a result, each quantile constraint is replaced by several deterministic inequalities.

Keywords: The probability measure kernel · The stochastic programming · The probabilistic constraints · The quantile function

1 Introduction

First, we introduce the basic concepts and notation. Let η be a random variable with distribution function $F_\eta(y) = \mathbf{P}\{\eta \leq y\}$, where $\mathbf{P}$ denotes probability. *The p-quantile* for the distribution of the random variable η for a given level $p \in (0,1)$ is determined by the standard relation

$$[\eta]_p = \min\{y : \ F_\eta(y) \geq p\}.$$

Supported by the Russian Foundation for Basic Research (project No.18-08-00595).

I. Bykadorov et al. (Eds.): MOTOR 2019, CCIS 1090, pp. 497–507, 2019.
https://doi.org/10.1007/978-3-030-33394-2_38

Let $f(u,\xi)$ be a real *loss function* depending on the strategy vector u and the random vector ξ. If the loss function is Borel-measurable in ξ, then $\eta_u = f(u,\xi)$ is a random variable. Its distribution function is called *a probability function* for the loss function $f(u,\xi)$ and the p-quantile $[\eta_u]_p$ as a function of u, is called *a quantile function* for the same loss function. The role of the probability and quantile functions in stochastic programming are reflected in [2]. The state of the art in the theory of optimization problems with such functions is described sufficiently in [3]. The quantile optimization problem is close to the stochastic programming problems with probabilistic constraints. These constraints can be defined in two ways. The first one is associated with so-called *joint probabilistic constraints*. The strategy u is admissible for such a constraint iff

$$\mathbf{P}\{g(u,\xi) \le 0\} \ge p, \tag{1}$$

where $g(u,x)$ is a vector function, $p \in (0,1)$ is the given probability and the inequality $g(u,\xi) \le 0$ is understood componentwise. Therefore, joint probabilistic constraints restrict the probability of the system of inequalities depending on random parameters. The second way is associated with setting *individual probabilistic constraints* which form the next system of probabilistic inequalities:

$$\mathbf{P}\{g_i(u,\xi) \le 0\} \ge p_i, \qquad \forall i = \overline{1,k}, \tag{2}$$

where the real functions $g_i(u,\xi)$ can be interpreted as components of the vector functions $g(u,\xi)$. Note that the joint probabilistic constraints can be converted from the formal point of view into an individual probabilistic constraint

$$\mathbf{P}\left\{\max_{i=\overline{1,k}} g_i(u,\xi) \le 0\right\} \ge p. \tag{3}$$

However, in this way one can lose the "good" properties of the functions g_i, e. g. their linearity.

We should note that for the real-valued loss function $g(u,\xi)$ the quantile optimization problem

$$[g(u,\xi)]_p \to \min_{u\in U} \tag{4}$$

studied in [3] is a special case of the stochastic programming problem with a single individual probabilistic constraint

$$\varphi \to \min_{u\in U,\varphi\in\mathbf{R}^1} \tag{5}$$

subject to

$$\mathbf{P}\{g(u,\xi) \le \varphi\} \ge p. \tag{6}$$

For the first time, joint probabilistic constraints were introduced and studied in [4], where the function $g(u,\xi) = Tu - \xi$ has a linear structure and T is the technological matrix. The early study of these constraints was focused mainly on the deterministic equivalents. Their essence is the probability transformation in (1) into the deterministic function of the strategy vector [5]. Unfortunately, a class of problems in which such equivalents can be constructed, is narrow enough. The most difficult case arises if random parameters which are the components of the vector ξ are mutually dependent. This obstacle was overcome in [6,7] by using methods of the integer programming and the notion of p-efficient points of multivariate probability distribution in the case where ξ is discrete and $g(u,\xi) = Tu-\xi$ has a deterministic technological matrix T. This result was later generalized to the random technological matrix [8], see also [9].

The great breakthrough in this field is associated with the Hungarian mathematician Prékopa who obtained sufficient conditions for the convexity of the feasible set given by the individual probabilistic constraints. The conditions are based on the logarithmic concavity of many multivariate distributions. This fact allowed to use the methods of convex programming for the construction of numerical techniques for solving the stochastic programming problems with probabilistic constraints. The main results on this issue are collected in the book [10]. We would like also to note other results achieved at the end of the last century, namely efficient verification algorithms of probabilistic constraints fulfillment. A good survey of these algorithms can be found in [11].

Among the recent results that determine the current state of the art in the field of the stochastic programming problems with probabilistic constraints, first of all, it is worth noting the algorithms based on the Monte Carlo method (SAA - Sample Average Approximation), see for example [12–19], the method of stochastic approximation [20,21] and the mathematical tools based on the p-efficient points concept [22–24]. The latter turned out to be especially constructive for stochastic programming problems with probabilistic constraints, in which random parameters have a discrete distribution. Note that the concept of p-efficient points is actually an extension of the concept of the p-quantile in the multivariate case.

We should also mention the works [25–28], where a method was developed for solving the quantile optimization problem with a linear loss function. The method reduces the original problem to a problem of mixed linear programming of large dimension. In contrast to these papers, this article discusses a wider class of loss functions, namely, the class of bilinear functions.

The motivation of authors is connected with two reasons. First, most of the publications on probabilistic constraints consider the case where the functions $g(u,\xi)$ and $g_i(u,\xi)$ are linear in random parameters. Secondly, some recent results of the authors in the field of stochastic programming with probabilistic criteria were focused only on this class of problems. These results can be found in [1,29]. They offer algorithms for solving the minimization problem of the quantile function. In this paper, we consider the algorithms that are based on the concept of the p-kernel of the probability distribution. Its definition and properties are described below in Sect. 2. Section 3 shows that individual probabilistic constraints can be written as inequalities for the quantile function(s) and represented in the deterministic form using the p-kernel. In Sect. 4, a stochastic programming problem with deterministic linear loss function and several individual probabilistic constraints inequalities in which are linear in random parameters is considered. The probabilistic constraints are transformed first into the quantile inequalities. Further these inequalities are approximated on the basis of an external polyhedral approximation of the p-kernel [1].

2 The p-Kernel of n-Dimensional Random Vector

This section introduces the concept of p-kernel [2] for an n-dimensional random vector ξ. This concept plays a key role in the construction of deterministic equivalents or convex approximations of probabilistic constraints in which the functions $g_i(u, \xi), i = \overline{1, k}$, are linear in ξ. In the sequel, the probability measure $\mathbf{P}$ is associated with distribution of the vector ξ, i.e. it is defined on all measurable subsets of $\mathbb{R}^n$. We also assume that the vectors from $\mathbb{R}^n$ are columns.

The Borel measurable set S is *p-confidence* if $\mathbf{P}(S) \geq p$. The *p-kernel* $K(p)$ is defined as intersection of all closed convex p-confidence sets [3]. On the other hand, the following representation is valid [3]:

$$K(p) = \bigcap_{\|c\|=1} \{x \in \mathbb{R}^n : c^T x \leq b_p(c)\} \tag{7}$$

where $\|\cdot\|$ is the Euclidean norm of the vector and $b_p(c) = [c^T \xi]_p$. Thus the p-kernel coincides with the intersection of all closed p-confidence half-spaces corresponding to the unit vectors of the external normal c. As shown in [1], the set $K(p)$ is always (that is, for any $\mathbf{P}$ distribution) not empty if $p > n/(n+1)$. Obviously, a non-empty p-kernel is a convex compact set. Also it was proposed in [1] to approximate $K(p)$ by the p-kernel convex polyhedron

$$K_N(p) = \bigcap_{c \in C_N} \{x \in \mathbb{R}^n : c^T x \leq b_p(c)\}, \tag{8}$$

where C_N is a finite set of N unit vectors. In [1], there is proposed an algorithm for setting C_N. The algorithm generates a uniform dense net of N points at the surface of the unit cube with center in the origin. Then the points of the net are projected onto the unit sphere. The resulting points are the elements of C_N.

The algorithm [1] for constructing a dense set of vectors C_N is implemented in the program module ProKer (Probabilistic Kernel) [30] for the MATLAB package in the case of $n = 2$, where the components of the vector ξ are independent and common distributed. ProKer software module is designed for research goals. It allows us to get the visual presentations of p-kernels for various p.

In some cases the function $b_p(c)$ can be found in an analytical form. In general, this is problematic. In such situations, we can try to use its sample estimate $\widehat{b}_p(c)$ [3] instead of $b_p(c)$ and approximate the p-kernel by the set

$$\widehat{K}_N(p) = \bigcap_{c \in C_N} \{x \in \mathbb{R}^n : c^T x \leq \widehat{b}_p(c)\}. \tag{9}$$

A theoretical study of such a possibility is off the scope of this article.

A similar idea of approximating the convex p-confidence sets using polyhedra was previously used in [26]. Note that the p-kernel is not a p-confidence set.

It is obvious that $K(p) \subseteq K(q)$ for all $p < q$, since every p-confidential half-space with the normal vector c is a subset of q-confidence half-spaces with the same external normal vector.

The most important and fundamental property for deterministic approximation of the probability constraints by means of the p-kernel of continuous distributions is the regularity [3]. The p-kernel is *regular* iff every closed half-space containing it is p-confidence.

The next result is proved in [3, Corollary 3.13].

Theorem 1. *Let the random vector ξ have a regular p-kernel for some $p \in (0, 1)$. Then*

$$\left[a^T(u)\xi + b(u)\right]_p = b(u) + \max_{x \in K(p)} a^T(u)x.$$

This theorem is a base of deterministic equivalents or approximations discussed in the next section. The regularity property implies the triangle inequality for quantiles, i. e.

$$[\xi_1 + \xi_2]_p \leq [\xi_1]_p + [\xi_2]_p$$

if the distribution p-kernel of the random vector $\xi = (\xi_1, \xi_2)^T$ is regular. This result follows from the following chain of inequalities:

$$[\xi_1 + \xi_2]_p = \max_{(x_1, x_2) \in K(p)} (x_1 + x_2) \leq \max_{(x_1, x_2) \in K(p)} x_1 + \max_{(x_1, x_2) \in K(p)} x_2 = [\xi_1]_p + [\xi_2]_p.$$

The next two lemmas are proved in [31].

Lemma 1. *For any point x_0 belonging to the boundary of the p-kernel $K(p)$, there exists a p-confidence half-space for which this point is also a boundary one.*

Lemma 2. *Let the quantile function $b_p(c)$ be continuous. Then the set $K_N(p)$ constructed using the algorithm proposed in [1] converges in the Hausdorff metrics to the set $K(p)$ as $N \to \infty$.*

It follows from Lemma 2 that the polyhedral model constructed using the algorithm proposed in [1] arbitrarily accurately approximates the p-kernel in both the regular and irregular cases.

The continuity condition for the quantile function $b_p(c)$ can be verified using the well-known results from [3]. For example, if the random vector ξ has a bounded support, then the quantile function $b_p(c)$ is continuous according to [3, Theorem 2.5].

3 Stochastic Programming Problem with Individual Probabilistic Constraints

In this section we consider the stochastic programming problem

$$h(u) \to \max_{u \in U} \tag{10}$$

with individual probabilistic constraints (2), where $h(u)$ is a deterministic real objective function and U is a compact set. As noted in [32], individual probabilistic constraints (2) are easily reduced to a system of deterministic inequalities if the functions $g_i(u, \xi)$ are separable in u and ξ. We generalize this result for the case, where these functions are linear only in ξ.

Consider the individual probabilistic constraints of the form (2). They are related to inequalities for probability functions corresponding to the loss functions $g_i(u, \xi)$. Let η be a random variable with the distribution function $F_\eta(y)$. In [33], it is established that the inequality $F_\eta(y) \geq p$ can be replaced by $[\eta]_p \leq y$. Therefore, each probability constraint in (2) can be represented in the equivalent quantile form

$$[g_i(u, \xi)]_{p_i} \leq 0. \tag{11}$$

Consider the case where the functions $g_i(u, \xi)$ are linear in ξ, i.e.

$$g_i(u, \xi) = \alpha_i^T(u)\xi + \beta_i(u) \tag{12}$$

and the p_i-kernel of the ξ distribution are regular. Then, applying Theorem 1, we can represent expression (11) in the following form:

$$\beta_i(u) + \max_{x \in K(p_i)} \alpha_i^T(u)x \leq 0. \tag{13}$$

Note that if the function $\beta_i(u)$ is convex and the function $\alpha_i^T(u)x$ is convex in u for every $x \in K(p_i)$, then the left-hand of the inequality (13) is a convex function and, therefore, the admissible set of strategies u is convex.

Next, we replace the p_i-kernel in the expression (13) with its polyhedral approximation $K_N(p_i)$ where $v_i^j \in J_i$, $j = \overline{1, N}$, is the j-th vertex of the multi-faceted approximation of the p-kernel for the value $p = p_i$, J_i is the set of the vertices of the polyhedral approximation of the p-kernel for the value $p = p_i$. By the linearity of the maximized function in (13), we have

$$\max_{x \in K_N(p_i)} \alpha_i^T(u)x = \max_{j \in J_i} \alpha_i^T(u)v_i^j.$$

In consequence of this, it can be concluded that each individual probability constraint from (2) can be approximated by the system of inequalities

$$\beta_i(u) + \alpha_i^T(u)v_i^j \leq 0, \quad j \in J_i. \tag{14}$$

This system determines a convex admissible set if the left-hand of each inequality is convex in u. In the particular case where the functions $\beta_i(u)$ and $\alpha_i(u)$ are linear in u, this condition is satisfied. Consider this in more detail in the next section.

4 Approximation of the Stochastic Programming Problem with Loss Function Linear in Random Parameters

Consider a special case of stochastic programming problem

$$d^T u \to \max_{u \in U} \tag{15}$$

with individual probabilistic constraints

$$\mathbf{P}\{\alpha_i^{\mathrm{T}}(u)\xi + \beta_i(u) \leq 0\} \geq p_i, \quad i = \overline{1,k}, \tag{16}$$

where u is optimized strategy of $\mathbb{R}^m$, d is a deterministic vector of dimension m, $p_i \in (0,1)$ is a given confidence probability, ξ is a n-dimensional random vector, $\alpha_i(u), \beta_i(u)$ are deterministic functions on u. Taking into account the results of the previous section, we approximate the problem in question by the non-linear programming problem (NLP).

Taking into account the fact that restrictions (14) are the linear inequalities, we conclude that the initial problem of stochastic programming (15)–(16) is approximated by the (14)–(15).

Theorem 2. *Let U be a compact set, functions $\alpha_i(u)$ and $\beta_i(u)$ be continuous in u, $b_{p_i}(c)$ be continuous in c, p_i-kernels be regular, and $\alpha_i(u) \neq 0$, $\forall u \in U$. Then the solution of the problem (14)–(15) converges w. r. t. the criterion value to the solution of the problem (15) - (16) as $N \to \infty$.*

Proof. For the proof, it is necessary to determine the δ-neighbourhood of the set $K_N(p)$ of radius δ:

$$K^{\delta}(p_i) = \bigcup_{x \in K(p_i)} B_{\delta}(x), \tag{17}$$

where $B_{\delta}(x) = \{y : \|y - x\| \leq \delta\}$. Denote $g_i(u,\xi) = \alpha_i^T(u)\xi + \beta_i(u)$. As noted above, the constraint (16) is equivalent to the inequality

$$[g_i(u,\xi)]_{p_i} \leq 0. \tag{18}$$

Then, taking into account the regularity condition of the p_i-kernel $K(p_i)$ and the formula (13), we can represent the inequality (18) in the equivalent form

$$\max_{x \in K(p_i)} g_i(u,x) \leq 0. \tag{19}$$

Since the functions $b_{p_i}(c)$ are continuous, by the condition of the theorem, using Lemma 2, we conclude that $K_N(p_i) \xrightarrow[N\to\infty]{} K(p_i)$ in the Hausdorff metric and $\forall \delta > 0 \; \exists \, N_0 : \forall N \geq N_0$, $K(p_i) \subseteq K_N(p_i) \subseteq K^{\delta}(p_i), \forall i = \overline{1,k}$. Moreover, $K^{\delta}(p_i) \xrightarrow[\delta\to 0]{} K(p_i)$ in the Hausdorff metric. Denote $U_i(\delta) = \{u \in U : h_i(u,\delta) \leq 0\}$, where $h_i(u,\delta) = \max_{x \in K^{\delta}(p_i)} g_i(u,x)$. Since the function $h_i(u,\delta)$ is continuous in $u \in U$ and in δ at $\delta = 0$, then according to [34, Lemma 1.1 (II)], to complete the proof of the theorem, it is enough to show [34, Lemma 1.1 (II)] that the multi-valued mapping $U_i(\delta)$ is continuous in the Hausdorff metric δ at the point $\delta = 0$. For this, it suffices to check that the function $h_i(u,\delta)$ is strictly monotone in δ. To do this, we show that the function $g_i(u,x)$ reaches x on the set $K^{\delta}(p_i) \backslash K(p_i)$ which exceeds $\max_{x \in K(p_i)} g_i(u,x)$ for a fixed u. The gradient of the function $g_i(u,x)$ over x does not depend on x and is equal to

$$\nabla_x (g_i(u,x)) = \alpha_i(u).$$

By the condition of the theorem, the gradient is nonzero that ensures the fulfillment of the condition being proved. Thus, the solution of the non-linear programming problem (14)–(15) converges according to the criterion value to the solution of the problem (15)–(16).

The theorem is proved.

5 Example

Let us consider the following stochastic programming problem:

$$3u_1 - 2u_2 \to \max_{u_1, u_2} \tag{20}$$

subject to deterministic constraints

$$u_1 \geq 0, \quad u_2 \geq 0, \quad u_1 + u_2 \leq 1 \tag{21}$$

and probabilistic ones

$$\mathbf{P}(2u_1\xi_1 - u_2\xi_2 - u_1 \leq 0) \geq 0,8, \tag{22}$$

$$\mathbf{P}(3u_1\xi_1 - u_2\xi_2 + 0,5u_2 \leq 2) \geq 0,99, \tag{23}$$

where ξ_1, ξ_2 are random variables which are independent and uniformly distributed over the segment $[0, 1]$.

According to (11), the probabilistic constraints (22) and (23) can be represented in the quantile form

$$[2u_1\xi_1 - u_2\xi_2 - u_1]_{0,8} \leq 0, \qquad [3u_1\xi_1 - u_2\xi_2 + 0,5u_2]_{0,99} \leq 2. \tag{24}$$

Since the p-kernel for the uniform distribution over the square is regular [1], these inequalities can be approximately replaced by

$$\max_{(x_1,x_2)\in K_N(0,8)} (2u_1x_1 - u_2x_2 - u_1) \leq 0, \tag{25}$$

$$\max_{(x_1,x_2)\in K_N(0,99)} (3u_1x_1 - u_2x_2 + 0,5u_2) \leq 2, \tag{26}$$

As shown above, each of constraints (25) and (26) is equivalent to a system of linear inequalities. Thus the original problem (20)–(23) is approximated by the linear programming problem (20) with constraints (21) and

$$u_1v_1^j - u_2v_2^j \leq 1 \ \forall j \in J_1, \tag{27}$$

$$3u_1v_1^j - u_2v_2^j \leq 1 \ \forall j \in J_2, \tag{28}$$

where $K_N(p)$ is polyhedral approximation of the p-kernel $K(p)$, J_1, J_2 are sets of numbers of vertices of the polyhedrons $K_N(0,8)$ and $K_N(0,99)$.

The function $b_p(c)$ can be found in the analytical form, see e.g. [1]. For $N = 16$, the optimal value of objective function is equal to 0,2376 and optimal strategies are $u = (0,4475; 0,5525)$. For $N = 64$, the optimal value of the objective function is equal to 0,2224 and the optimal strategies are $u = (0,4445; 0,5555)$. For $N = 512$, the optimal value of objective function is equal to 0,2222 and the optimal strategies are $u = (0,4444; 0,5556)$.

This example illustrates the appropriate precision of the presented approach.

6 Conclusion

We have showed that the optimization problem with individual probabilistic constraints can be approximated by an NLP problem. The convergence theorem proves that the NLP solution converges to the solution of the original problem w. r. t. the criterion value. The numerical example given in the article illustrates the convergence of the proposed method for improving the approximation of the uncertainty set for a random vector.

References

1. Vasileva, S.N., Kan, Y.S.: A method for solving quantile optimization problems with a bilinear loss function. Autom. Remote Control **76**(9), 1582–1597 (2015). https://doi.org/10.1134/S0005117915090052
2. Kibzun, A., Kan, Y.: Stochastic Programming Problems with Probability and Quantile Functions. Wiley, Chichester (1996). https://doi.org/10.1057/palgrave.jors.2600833
3. Kibzun, A.I., Kan, Y.S.: Zadachi stokhasticheskogo programmirovaniya s veroyatnostnymi kriteriyami (Stochastic programming problems with probabilistic criteria). Fizmatlit, Moscow (2009). (in Russian)
4. Charnes, A., Cooper, W.W.: Chance-constrained programming. Manag. Sci. **6**, 73–79 (1959). https://doi.org/10.1287/mnsc.6.1.73
5. Miller, B.L., Wagner, H.M.: Chance constrained programming with joint constraints. Oper. Res. **13**, 930–945 (1965). https://doi.org/10.1287/opre.13.6.930
6. Lejeune, M.A.: Pattern-based modeling and solution of probabilistically constrained optimization problems. Oper. Res. **60**, 1356–1372 (2012). https://doi.org/10.1287/opre.1120.1120
7. Lejeune, M.A.: Pattern definition of the p-efficiency concept. Ann. Oper. Res. **200**, 23–36 (2012). https://doi.org/10.1007/s10479-010-0803-1
8. Kogan, A., Lejeune, M.A.: Threshold boolean form for joint probabilistic constraints with random technology matrix. Math. Program. **147**, 391–427 (2014). https://doi.org/10.1007/s10107-013-0728-y
9. Henrion, R.: Structural properties of linear probabilistic constraints. Optimization **56**(4), 425–440 (2007). https://doi.org/10.1080/02331930701421046
10. Prékopa, A.: Stochastic Programming. Kluwer, Dordrecht (1995). https://doi.org/10.1007/978-94-017-3087-7
11. Genz, A., Bretz, F.: Computation of Multivariate Normal and *t*-Probabilities. Springer, Heidelberg (2009). https://doi.org/10.1007/978-3-642-01689-9
12. Barrera, J., Homem-de-Mello, T., Moreno, E., Pagnoncelli, B.K., Canessa, G.: Chance-constrained problems and rare events: an importance sampling approach. Math. Program. Ser. B **157**, 153–189 (2016). https://doi.org/10.1007/s10107-015-0942-x
13. Guigues, V., Juditsky, A., Nemirovski, A.: Non-asymptotic confidence bounds for the optimal value of a stochastic program. Optim. Methods. Softw. **32**(5), 1033–1058 (2017). https://doi.org/10.1080/10556788.2017.1350177
14. Kleywegt, A.J., Shapiro, A., Mello-de-Homem, T.: The sample average approximation method for stochastic discrete optimization. SIAM J. Optim. **12**, 479–502 (2002). https://doi.org/10.1137/S1052623499363220

15. Linderoth, J., Shapiro, A., Wright, S.: The empirical behavior of sampling methods for stochastic programming. Ann. Oper. Res. **142**, 215–241 (2006). https://doi.org/10.1007/s10479-006-6169-8
16. Mak, W.-K., Morton, D.P., Wood, R.K.: Monte Carlo bounding techniques for determining solution quality in stochastic programs. Oper. Res. Lett. **24**, 47–56 (1999). https://doi.org/10.1016/S0167-6377(98)00054-6
17. Shapiro, A.: Monte Carlo sampling methods. In: Ruszczyński, A., Shapiro, A. (eds.) Handbooks in Operations Research and Management Science, vol. 10, pp. 353–425. Elsevier, Amsterdam (2003). https://doi.org/10.1016/S0927-0507(03)10006-0
18. Shapiro, A., Nemirovski, A.: On complexity of stochastic programming problems. In: Jeyakumar, V., Rubinov, A. (eds.) Continuous Optimization: Current Trends and Applications, pp. 111–146. Springer, Boston (2005). https://doi.org/10.1007/0-387-26771-9_4
19. Verweij, B., Ahmed, S., Kleywegt, A.J., Nemhauser, G., Shapiro, A.: The sample average approximation method applied to stochastic routing problems: a computational study. Comput. Optim. Appl. **24**, 289–333 (2003). https://doi.org/10.1023/A:1021814225969
20. Bottou, L.: Large-scale machine learning with stochastic gradient descent. In: Proceedings of COMPSTAT 2010. Springer, pp. 177–186 (2010). https://doi.org/10.1007/978-3-7908-2604-3_16
21. Nemirovski, A., Juditsky, A., Lan, G., Shapiro, A.: Robust stochastic approximation approach to stochastic programming. SIAM J. Optim. **19**, 1574–1609 (2009). https://doi.org/10.1137/070704277
22. Beraldi, P., Ruszczyński, A.: A branch and bound method for stochastic integer problems under probabilistic constraints. Optim. Methods Softw. **17**(3), 359–382 (2002). https://doi.org/10.1080/1055678021000033937
23. Prékopa, A., Vizvári, D., Badics, T.: Programming under probabilistic constraint with discrete random variable. In: Giannesi, F. (ed.) New Trends in Mathematical Programming, pp. 235–255. Kluwer Academic Publishers, Boston (1998). https://doi.org/10.1007/978-1-4757-2878-1_18
24. Dentcheva, D., Prékopa, A., Ruszczyński, A.: Concavity and efficient points of discrete distributions in probabilistic programming. Math. Program. **89**, 55–77 (2000). https://doi.org/10.1007/PL00011393
25. Ivanov, S.V., Kibzun, A.I.: On the convergence of sample approximations for stochastic programming problems with probabilistic criteria. Autom. Remote Control **79**(2), 216–228 (2018). https://doi.org/10.1134/S0005117918020029
26. Ivanov, S.V., Naumov, A.V.: Algorithm to optimize the quantile criterion for the polyhedral function and discrete distribution for random parameters. Autom. Remote Control **73**(1), 105–117 (2012). https://doi.org/10.1134/S0005117912010080
27. Naumov, A.V., Ivanov, S.V.: On stochastic linear programming problems with the quantile criterion. Autom. Remote Control **72**(2), 353–369 (2011). https://doi.org/10.1134/S0005117911020123
28. Kibzun, A.I., Naumov, A.V., Norkin, V.I.: On reducing a quantile optimization problem with discrete distribution to a mixed integer programming problem. Autom. Remote Control **74**(6), 951–967 (2013). https://doi.org/10.1134/S0005117913060064
29. Vasileva, S.N., Kan, Y.S.: Linearization method for solving quantile optimization problems with loss function depending on a vector of small random parameters. Autom. Remote Control **78**(7), 1251–1263 (2017). https://doi.org/10.1134/S0005117917070074

30. Vasileva, S.N., Kan, Y.S.: Algoritm vizualizacii ploskogo yadra veroyatnostnoj mery. Informatica i ee primeneniya. **12**(2), 60–68 (2018). (in Russian)
31. Vasileva, S.N., Kan, Yu.S.: Approksimaciya veroyatnostnyh ogranichenij v zadachah stohasticheskogo programmirovaniya s ispolzovaniem yadra veroyatnostnoj mery. Avtomatika i telemekhanika (2019, in print). (in Russian)
32. Guigues, V., Henrion, R.: Joint dynamic probabilistic constraints with projected linear decision rules. Optim. Methods Softw. **32**(5), 1006–1032 (2017). https://doi.org/10.1080/10556788.2016.1233972
33. Rosenblatt-Roth, M.: Quantiles and medians. Ann. Math. Stat. **36**, 921–925 (1965). https://doi.org/10.1214/aoms/1177700064
34. Fedorov, V.V.: Chislennye metody maksimina. Nauka, Moscow (1979). (in Russian)

On Estimates of the Solutions of Inverse Problems of Optimal Control

Evgenii A. Krupennikov[1,2(✉)]

[1] N.N. Krasovskii Institute of Mathematics and Mechanics of the Ural Branch of the Russian Academy of Sciences (IMM UB RAS), 16 S. Kovalevskaya Str., Ekaterinburg 620990, Russia
krupennikov@imm.uran.ru

[2] Ural Federal University named after the first President of Russia B.N.Yeltsin, 19 Mira Street, Ekaterinburg 620002, Russia

Abstract. This paper is devoted to the problem of reconstruction of the normal control generating a realized trajectory of a dynamic control system by using known inaccurate measurements of this trajectory. A class of dynamic control systems with dynamics linear in controls and non-linear in state coordinates is considered. A new method, suggested in earlier publications, for solving such problems is discussed. This approach relies on necessary optimality conditions in an auxiliary variational problem on extremum of an integral discrepancy functional. The distinguishing feature of the method is using a functional which is convex in control variables and concave in state variables discrepancy. This form of the functional allows to obtain oscillating solutions. In this paper the estimates of the error of the discussed method are exposed and validated.

Keywords: Normal control · Nonlinear systems · Normal control · Inverse problem · Calculus of variations · Hamiltonian systems

1 Introduction

The inverse problems of reconstruction of the normal control (that is the control with the least possible norm) generating a realized trajectory of a dynamic control system by using known inaccurate measurements of this trajectory are considered in this paper. Such reconstruction problems occur in many areas of mathematics such as optimal control theory, differential games, and others. They have applications in such areas as decision making, economics, medicine, robotics, and others.

This work was supported by the Russian Foundation for Basic Research (project no. 17-01-00074) and by the Ural Branch of the Russian Academy of Sciences (project no. 18-1-1-10).

I. Bykadorov et al. (Eds.): MOTOR 2019, CCIS 1090, pp. 508–523, 2019.
https://doi.org/10.1007/978-3-030-33394-2_39

The inverse problems have been considered by many authors. One of the most well-known approaches was developed by A. V. Kryazhimskii and Yu. S. Osipov [7,9]. The method suggested by them reconstructs the normal control by using a regularized (a variation of Tikhonov regularization, see [14]) procedure of extremal aiming on a stable motion. It is originated from the works of N. N. Krasovskii's school on the theory of optimal feedback control [2,3] and relies on the "control with a guide" procedure. This approach was afterwards developed also by V. I. Maximov, M. S. Blizorukova, et al. (see, in particular, [1,10]).

Another approach to construction of approximations of the normal control has been suggested by N. N. Subbotina, E. A. Krupennikov and T. B. Tokmantsev. It has been previously suggested and discussed in [4–6,11–13]. This method relies on auxiliary variational problems of optimization of a regularized integral discrepancy functional. This paper continues developing this approach. The innovation of the suggested method consists of using in the auxiliary problem a functional, which is convex in control variables and concave in state variables discrepancy, instead of a functional, which convex in all variables.

The new results presented in this paper include the explicit estimate of the method's error. This result is obtained for control systems that are linear in controls and non-linear in state coordinates with the dimension of the control parameter greater than or equal to the dimension of the state coordinates. Previously, the theorem about convergence of the suggested method has been formulated in [4] (where the scheme of the proof was presented). In this paper, a detailed proof of the theorem is presented.

1.1 Notation

We adopt the following notation:

$\|\cdot\|$	the Euclidean norm in the space $\mathbb{R}^n$, $n \in \mathbb{N}$;
$\langle\cdot,\cdot\rangle$	the scalar product in $\mathbb{R}^n$;
$\|\cdot\|_{C[a,b]}$	the norm in the Banach space $C[a,b]$ of continuous functions mapping $[a,b] \subseteq \mathbb{R}$ to $\mathbb{R}^n$;
$\|\cdot\|_{L^2[a,b]}$	the norm in the Hilbert space $L_2[a,b]$ of functions integrable with the square of norm mapping $[a,b]$ to $\mathbb{R}^n$;
$\|M\|_2$	the spectral norm of a matrix M;
M^{-1}	the inverse of a square non-degenerate matrix M;
M^T	the transpose of a matrix M;
$M^g = M^T(MM^T)^{-1}$	the generalized inverse of a matrix M with linearly independent rows;
$\omega_f(\cdot)$	the modulus of continuity of a function $f(\cdot) : [a,b] \to \mathbb{R}$. Moreover, $\omega_g(\cdot) \triangleq \omega_{\|g\|}(\cdot)$, $g : [a,b] \to \mathbb{R}^n$, $\omega_M(\cdot) \triangleq \omega_{\|M\|_2}(\cdot)$, $M(\cdot) : [a,b] \to \mathbb{R}^{n\times n}$.
I_n	an $n \times n$ identity matrix;
O_n	an $n \times n$ zero matrix;

2 Dynamics

We consider control systems with dynamics of the form

$$\dot{x}(t) = G(x(t),t)u(t) + f(x(t),t),$$
$$x(\cdot):[0,T]\to \mathbb{R}^n,\quad u(\cdot):[0,T]\to \mathbb{R}^m,\quad m\geq n,\quad t\in[0,T],\quad T<\infty. \tag{1}$$

The elements of the matrix function $G(x,t) : \mathbb{R}^n \times [0,T] \to \mathbb{R}^{n\times m}$ and the vector function $f(x,t) : \mathbb{R}^n \times [0,T] \to \mathbb{R}^n$ are continuously differentiable with respect to each variable.

The parameter $x(t)$ is the vector of the state coordinates and $u(t)$ is the vector of the control parameters. The set of admissible controls $U_{adm} \subseteq C[0,T]$ consists of continuous functions satisfying the restriction

$$u(t) \in \mathbf{U},\quad t\in[0,T], \tag{2}$$

where $\mathbf{U} \subset \mathbb{R}^m$ is a convex compact set.

3 Input Data

It is supposed that some trajectory $x^*(\cdot) : [0,T] \to \mathbb{R}^n$ of system (1) has been generated by an admissible control $v(\cdot)$.

We assume that continuous inaccurate measurements $y(\cdot,\delta) = y^\delta(\cdot) : [0,T] \to \mathbb{R}^n$ of the trajectory $x^*(\cdot)$ are known with the error δ:

$$\|y^\delta(\cdot) - x^*(\cdot)\|_{C[0,T]} \leq \delta. \tag{3}$$

Assume that the following assumptions are true:

Assumption 1. *There exists a constant $\delta_0 > 0$ such that for any $\delta \in (0,\delta_0]$, the functions $y^\delta(\cdot)$ are twice continuously differentiable and*

$$\lim_{\delta\to 0}\delta(\overline{Y}^{1,\delta}+\overline{Y}^{2,\delta})\to 0,\quad \overline{Y}^{1,\delta}=\max_{t\in[0,T]}\|\dot{y}^\delta(t)\|,\quad \overline{Y}^{2,\delta}=\max_{t\in[0,T]}\|\ddot{y}^\delta(t)\|. \tag{4}$$

Remark 1. In practice, the input data has usually the form of a set of discrete sample points. In this case, spline interpolations of the input data can be considered as the functions $y^\delta(\cdot)$. For a finite set of sample points, spline interpolation provides the fulfillment of Assumption 1.

Assumption 2. *The rows of the matrix $G(x,t)$ are linearly independent for $(x,t) \in \Psi \times [0,T]$, where*

$$\Psi = \bigcup_{t\in[0,T]} B_{2\delta_0}[x^*(t)], \tag{5}$$

$B_{2\delta_0}[x]$ being the closed ball of the radius $2\delta_0$ with the center in x.

4 Reconstruction Problem

One can prove (see [4]) that for input data (1)–(3) satisfying Assumptions 1 and 2, there exists a unique normal control $u^*(\cdot)$, which is the admissible control that generates the trajectory $x^*(\cdot)$ and has the least possible norm in the $L_2[0,T]$ space:

$$\|u^*(\cdot)\|_{L^2[0,T]} = \min_{u(\cdot)\in U_{adm}:\ \dot{x}^*(t)=G(x^*(t),t)u(t)+f(x^*(t),t)} \|u(\cdot)\|_{L^2[0,T]}. \quad (6)$$

Let us consider the following inverse reconstruction problem: for any given $\delta \in (0, \delta_0]$ and a given measurement function $y^\delta(\cdot)$ satisfying inequalities (3) and Assumption 1, find a function $u(\cdot, \delta) = u^\delta(\cdot) : [0,T] \to \mathbb{R}^m$ that satisfies the following conditions:

C1. $u^\delta(\cdot) \in U_{adm}$.

C2. The function $u^\delta(\cdot)$ generates a trajectory $x(\cdot,\delta) = x^\delta(\cdot) : [0,T] \to \mathbb{R}^n$ of system (1) for the boundary condition $x^\delta(0) = y^\delta(0)$ that satisfies

$$\lim_{\delta\to 0} \|x^\delta(\cdot) - x^*(\cdot)\|_{C[0,T]} = 0. \quad (7)$$

C3. The function $u^\delta(\cdot)$ satisfies the condition

$$\lim_{\delta\to 0} \|u^\delta(\cdot) - u^*(\cdot)\|_{L^2[0,T]} = 0. \quad (8)$$

Let us denote the stated inverse reconstruction problem as the problem **C1–C3**.

5 Constructing a Solution of the Reconstruction Problem

5.1 Auxilliary Problem

To solve the problem **C1–C3**, we introduce an auxiliary variational problem (AVP) for fixed parameters $\delta \in (0, \delta_0]$, $\alpha > 0$, dynamics (1) satisfying Assumption 2, and a given measurement function $y^\delta(\cdot)$ satisfying inequalities (3) and Assumption 1.

We consider the set of pairs of continuously differentiable functions $F_{xu} = \{\{x(\cdot), u(\cdot)\} : x(\cdot) : [0,T] \to \mathbb{R}^n,\ u(\cdot) : [0,T] \to \mathbb{R}^m\}$ that satisfy differential equations (1) and the following boundary conditions:

$$x(0) = y^\delta(0), \quad \dot{x}(0) = \dot{y}^\delta(0). \quad (9)$$

Remark 2. The conditions (9) are equivalent to the conditions

$$x(0) = y^\delta(0), \quad u(0) = G^g(y^\delta(0), 0)(\dot{y}^\delta(0) - f(y^\delta(0), 0)). \quad (10)$$

Due to Assumption 2, the generalised inverse $G^g(y^\delta(0), 0)$ exists.

The AVP is to find a pair of functions $\{x(\cdot), u(\cdot)\} \in F_{xu}$ that provide an extremum (minimum) for the integral functional

$$I(x(\cdot), u(\cdot)) = \int_0^T \left[-\frac{\|x(t) - y^\delta(t)\|^2}{2} + \frac{\alpha^2 \|u(t)\|^2}{2} \right] dt. \tag{11}$$

Here α is a small regularising (see [14]) parameter. The distinguishing feature of the suggested approach is the use of a functional which is convex with respect to the controls and concave with respect to the discrepancy of the state variables. The advantage of such approach is explained in the Sect. 6.

Remark 3. The suggested below algorithm for solving the problem **C1–C3** uses only some of the constructions from the AVP. Thus, we will only derive the necessary conditions of minimum of functional (11) without verifying if this minimum is actually reached in the AVP.

5.2 Necessary Optimality Conditions in the AVP

The necessary optimality conditions for AVP (1), (10), (11) can be written (see [6]) in the form of Hamiltonian equations, where the vector $s(t)$ plays the role of the adjoint variables vector

$$\begin{aligned} \dot{x}(t) &= -\alpha^{-2} G(x(t), t) G^T(x(t), t) s(t) + f(x(t), t), \\ \dot{s}_i(t) &= x_i(t) - y_i^\delta(t) + \alpha^{-2} \langle s(t), \tfrac{\partial G}{\partial x_i}(x(t), t) G^T(x(t), t) s(t) \rangle \\ &\quad + \langle s(t), \tfrac{\partial f}{\partial x_i}(x(t), t) \rangle, \quad i = 1, \dots, n \end{aligned} \tag{12}$$

with the boundary conditions

$$\begin{aligned} x(0) &= y^\delta(0), \\ s(0) &= -\alpha^2 \big(G(y^\delta(0), 0) G^T(y^\delta(0), 0)\big)^{-1} (\dot{y}^\delta(0) - f(y^\delta(0), 0)). \end{aligned} \tag{13}$$

5.3 Solution of the Inverse Problem

In order to construct the solution of the problem **C1–C3**, let us first consider a linearized version of system (12)

$$\begin{aligned} \dot{x}(t) &= -\alpha^{-2} \mathbf{G}^\delta(t) s(t) + f(y^\delta(t), t), \\ \dot{s}(t) &= x(t) - y^\delta(t), \quad t \in [0, T] \end{aligned} \tag{14}$$

with boundary conditions (13), where $\mathbf{G}^\delta(\cdot) \triangleq G(y^\delta(\cdot), \cdot) G^T(y^\delta(\cdot), \cdot)$. It is a heterogeneous linear system of ODEs with variable continuous coefficients. So, its solution $\{x(\cdot), s(\cdot)\} : [0, T] \to \mathbb{R}^{2n}$ exists and is unique on $[0, T]$. We will use this solution as the basis for constructing a solution of the problem **C1–C3**.

Let us introduce the function

$$u^{\delta,\alpha}(\cdot) = -\alpha^{-2} G^T(y^\delta(\cdot), \cdot) s(\cdot), \tag{15}$$

where $s(\cdot)$ is the solution of system (14) with boundary conditions (13). We also introduce the so-called cut-off vector function $\hat{u}^{\delta,\alpha}(\cdot) : [0,T] \to R^m$

$$\hat{u}^{\delta,\alpha}(t) = \begin{cases} u^{\delta,\alpha}(t), & u^{\delta,\alpha}(t) \in \mathbf{U}, \\ \hat{u} \in \mathbf{U} : \|u^{\delta,\alpha}(t) - \hat{u}\| = \min\limits_{v \in \mathbf{U}} \|u^{\delta,\alpha}(t) - v\|, & u^{\delta,\alpha}(t) \notin \mathbf{U}, \end{cases} \tag{16}$$

were $\mathbf{U}$ is the set from (2). Let us prove that the function $\hat{u}^{\delta,\alpha}(\cdot)$ is the solution to the problem **C1–C3**.

First, we introduce the following proposition and lemma.

Proposition 1. *Let matrix functions* $M_1(\cdot) : [0,T] \to \mathbb{R}^{n\times n}, \ldots, M_k(\cdot) : [0,T] \to \mathbb{R}^{n\times n}$, $k \in \mathbb{N}$ *have the moduli of continuity* $\omega_{M_1}(\cdot), \ldots, \omega_{M_k}(\cdot)$ *and a vector function* $h(\cdot) : [0,T] \to \mathbb{R}^n$ *have the modulus of continuity* $\omega_h(\cdot)$. *Then, their product* $\prod\limits_{i=1}^{k} \big[M_i(\cdot)\big] h(\cdot)$ *has the modulus of continuity*

$$\begin{aligned} \omega_{\prod}(\cdot) = \sum_{i=1}^{k} \Big[\omega_{M_i}(\cdot) \prod_{j=1,\ldots,k,\ j\neq i} \big[\max_{t\in[0,T]} \|M_j(t)\|_2\big]\Big] \max_{t\in[0,T]} \|h(t)\| \\ + \prod_{i=1}^{h} \big[\max_{t\in[0,T]} \|M_i(t)\|_2\big] \omega_h(\cdot). \end{aligned} \tag{17}$$

Lemma 1. *Let Assumptions 1 and 2 hold. Then, there exist parameters*

$$0 < \alpha(\delta) \xrightarrow{\delta\to 0} 0, \quad 0 < R^\delta = R^\delta(\delta) \xrightarrow{\delta\to 0} 0, \quad 0 < R_z^\delta = R_z^\delta(\delta) \xrightarrow{\delta\to 0} 0,$$

that depend on the constant δ_0 *from Assumption (1), the parameter* δ, *and the properties of the functions* $G(x,t)$, $f(x,t)$, *and the given measurements* $y^\delta(t)$, *such that the functions* $u^{\delta,\alpha}(\cdot)$, *defined in (15), and the solution* $x(\cdot)$ *of system (21) with boundary conditions (13) satisfy the conditions*

$$\begin{aligned} \|u^{\delta,\alpha}(\cdot) - G^g(y^\delta(t),t)(\dot{y}^\delta(\cdot) - f(y^\delta(\cdot),\cdot))\|_{C[0,T]} \le R^\delta, \\ \|x(\cdot) - y^\delta(\cdot)\|_{C[0,T]} \le 2\alpha R_z^\delta, \end{aligned} \tag{18}$$

provided that $\alpha = \alpha(\delta)$.

Proof. First, let us introduce some constants that will be used later in the proof.

$$\begin{aligned} \overline{G}^0 &= \max_{t\in[0,T],\ x\in\Psi} \|G(x,t)G^T(x,t)\|_2, \quad \tilde{G}^0 = \max_{t\in[0,T],\ x\in\Psi} \|\left(G(x,t)G^T(x,t)\right)^{-1}\|_2, \\ \overline{G}^1 &= \max_{i=1,\ldots,n,\ t\in[0,T],\ x\in\Psi} \left\{ \left\| \tfrac{\delta}{\delta x_i} \left[G(x,t)G^T(x,t)\right]\right\|_2, \left\| \tfrac{\delta}{\delta t} \left[G(x,t)G^T(x,t)\right]\right\|_2 \right\}, \\ \overline{F}^0 &= \max_{t\in[0,T],\ x\in\Psi} \|f(x,t)\|, \quad \overline{F}^1 = \max_{t\in[0,T],\ x\in\Psi} \|\dot{f}(x,t)\|. \end{aligned} \tag{19}$$

The matrix function $G(x,t)G^T(x,t)$ is positive definite for $t \in [0,T]$, $x \in \Psi$ by Assumption 2, so, $\tilde{G}^0 < \infty$.

We introduce new variables

$$\begin{aligned} z(\cdot) &= x(\cdot) - y^\delta(\cdot), \\ w(\cdot) &= s(\cdot) + \alpha^2\big(\mathbf{G}^\delta(\cdot)\big)^{-1}(\dot{y}^\delta(\cdot) - f(y^\delta(\cdot),\cdot)). \end{aligned} \tag{20}$$

System (14) with boundary conditions (13) has the following form in variables (20):

$$\begin{aligned} \dot{z}(t) &= -\alpha^{-2}\mathbf{G}^\delta(t)w(t), \\ \dot{w}(t) &= z(t) + \alpha^2 g(t), \\ z(0) &= w(0) = \mathbf{0}, \quad t \in [0,T], \end{aligned} \tag{21}$$

where

$$g(\cdot) = \frac{d}{dt}\left[\big(\mathbf{G}^\delta(\cdot)\big)^{-1}\right](\dot{y}^\delta(\cdot) - f(y^\delta(\cdot),\cdot)) + \big(\mathbf{G}^\delta(\cdot)\big)^{-1}(\ddot{y}^\delta(\cdot) - \dot{f}(y^\delta(\cdot),\cdot)). \tag{22}$$

System (21) is a heterogeneous linear system of ODEs with continuous coefficients. So, its solution $\{z(\cdot), w(\cdot)\} : [0,T] \to \mathbb{R}^{2n}$ exists and is unique.

We will now prove by contradiction that there exist parameters $R_w^\delta > 0$, $R_z^\delta > 0$ that depend on the constant δ_0 from Assumption 1, the parameter δ, and the properties of the functions $G(x,t)$, $f(x,t)$, and the given measurements $y^\delta(t)$ such that

$$\begin{aligned} &\|z(t)\| \leq \alpha 2R_z^\delta \leq \delta_0, \quad R_z^\delta \xrightarrow{\delta \to 0} 0 \\ &\|w(t)\| \leq \alpha^2 2R_w^\delta, \quad R_w^\delta \xrightarrow{\delta \to 0} 0, \quad \delta \in (0,\delta_0], \quad t \in [0,T]. \end{aligned} \tag{23}$$

Based on the continuity of the solution $\{z(\cdot), w(\cdot)\}$ and the boundary conditions $z(0) = w(0) = \mathbf{0}$, we can assume that, on the contrary to (23), for any R_w^δ and R_z^δ such that $R_w^\delta \xrightarrow{\delta \to 0} 0$ and $\delta_0 \geq R_z^\delta \xrightarrow{\delta \to 0} 0$ it holds

$$\exists T_1 \in (0,T) : \quad \left[\begin{array}{c} \|w(T_1)\| = \alpha^2 2R_w^\delta \\ \|z(T_1)\| = \alpha 2R_z^\delta \end{array}\right., \quad \left\{\begin{array}{c} \|w(t)\| \leq \alpha^2 2R_w^\delta \\ \|z(t)\| \leq \alpha 2R_z^\delta \end{array}\right., \quad t \in (0,T_1]. \tag{24}$$

Introduce the following partition of the interval $[0,T_1]$ with step α^2:

$$\{t_0, t_1, \ldots, t_K : t_0 = 0,\ t_1 = \alpha^2, \ldots, t_{h-1} = (h-1)\alpha^2,\ t_h = T_1, \quad h = \lceil T_1/\alpha^2 \rceil\}. \tag{25}$$

We will estimate the solution of system (21) on each interval $[t_i, t_{i+1}]$, $i = 0, \ldots, h-1$ subsequently beginning from $[t_0, t_1]$.

Let us, first, consider a so-called frozen system for the interval $[t_0, t_1]$:

$$\begin{aligned} \dot{\overline{z}}_0(t) &= -\alpha^{-2}\mathbf{G}_0^\delta \overline{w}_0(t), \\ \dot{\overline{w}}_0(t) &= \overline{z}_0(t) + \alpha^2 \overline{g}_0, \\ \overline{z}_0(0) &= z(0) = \mathbf{0}, \quad \overline{w}_0(T) = w(0) = \mathbf{0}, \quad t \in [t_0, T], \end{aligned} \tag{26}$$

where $\mathbf{G}_0^\delta = \mathbf{G}^\delta(y^\delta(0), 0)$, $\overline{g}_0 = g(0)$. The solution of this system will be used to estimate the solution of system (21).

Remark 4. Note that system (26) is considered on $[t_0, T]$, because the estimates of it's solutions will be used on the next partition steps.

System (26) is a heterogeneous linear system of ODEs with constant coefficients. So, its solution $\{\overline{z}_0(\cdot), \overline{w}_0(\cdot)\} : [t_0, t_1] \to \mathbb{R}^{2n}$ exists and is unique. This solution can be written in the following form with help of the Cauchy formula for solutions of heterogenous systems of linear ODEs with constant coefficients:

$$\begin{pmatrix} \overline{z}_0(t) \\ \overline{w}_0(t) \end{pmatrix} = \Phi_0(t) \begin{pmatrix} \overline{z}_0(0) \\ \overline{w}_0(0) \end{pmatrix} + \Phi_0(t) \int_{t_0}^{t} (\Phi_0(\tau))^{-1} \begin{pmatrix} \mathbf{0} \\ \alpha^2 \overline{g}_0 \end{pmatrix} d\tau, \quad t \in [t_0, T]. \tag{27}$$

Remark 5. Hereinafter $\begin{pmatrix} a \\ b \end{pmatrix} \triangleq (a_1, a_2, \ldots, a_n, b_1, b_2, \ldots, b_n)^T \in \mathbb{R}^{2n}, \ a, b \in \mathbb{R}^n$.

In (27) $\Phi_0(\cdot)$ is a $2n \times 2n$ normalized (that is $\Phi_0(0) = I_{2n}$) fundamental matrix of solutions of the homogenous part of system (26). It can be chosen as

$$\Phi_0(t) = \exp\left[(t - t_0)A_0\right] \stackrel{def}{=} \sum_{k=0}^{\infty} \frac{1}{k!} \left((t - t_0)A_0\right)^k, \tag{28}$$

where the $2n \times 2n$ matrix A_0 can be written in the block form

$$A_0 = \left(\begin{array}{c|c} O_n & -\alpha^{-2}\mathbf{G}_0^{\delta} \\ \hline I_n & O_n \end{array}\right). \tag{29}$$

One can substitute (29) into (28). Then, after collapsing the rows by the Maclaurin series for $\sin(\cdot)$ and $\cos(\cdot)$, one can obtain (see [4,8]) that

$$\Phi_0(t) = \left(\begin{array}{c|c} -Q_0\Lambda_0^{cos}(t)(Q_0)^T & \alpha^{-1}Q_0\Lambda_0^{sqrt}\Lambda_0^{sin}(t)(Q_0)^T \\ \hline -\alpha Q_0(\Lambda_0^{sqrt})^{-1}\Lambda_0^{sin}(t)(Q_0)^T & -Q_0\Lambda_0^{cos}(t)(Q_0)^T \end{array}\right) \tag{30}$$

(in the block form), were Q_0 is an orthogonal matrix, Λ_0^{sqrt} is a diagonal matrix with the square roots of the positive eigenvalues Λ_j, $j = 1, \ldots, n$ of the matrix $\mathbf{G}_0^{\delta}$ on the diagonal, $\Lambda_0^{cos}(t)$ and $\Lambda_0^{sin}(t)$ are diagonal matrices, where $\cos(\alpha^{-1}\sqrt{\Lambda_j}(t - t_0))$ and $\sin(\alpha^{-1}\sqrt{\Lambda_j}(t - t_0))$, $j = 1, \ldots, n$ are the diagonal elements, respectively. One can also check that

$$(\Phi_0(t))^{-1} = \left(\begin{array}{c|c} -Q_0\Lambda_0^{cos}(t)(Q_0)^T & -\alpha^{-1}Q_0\Lambda_0^{sqrt}\Lambda_0^{sin}(t)(Q_0)^T \\ \hline \alpha Q_0(\Lambda_0^{sqrt})^{-1}\Lambda_0^{sin}(t)(Q_0)^T & -Q_0\Lambda_0^{cos}(t)(Q_0)^T \end{array}\right). \tag{31}$$

So, by substituting (30) and (31) into (27), we get that

$$\begin{aligned} \overline{z}_0(t) &= \alpha^2 Q_0 \int_{t_0}^{t} \left[\alpha^{-1}\Lambda_0^{sqrt}\Lambda_0^{sin}(t - \tau)\overline{g}_0\right] d\tau Q_0^T \\ &= -\alpha^2 Q_0 \left(\Lambda_0^{cos}(\tau - t)\Big|_{t_0}^{t}\right)(Q_0)^T\overline{g}_0, \\ \overline{w}_0(t) &= -\int_{t_0}^{t} \left[\alpha^2 Q_0\Lambda_0^{cos}(t - \tau)Q_0^T\overline{g}_0\right] d\tau \\ &= \alpha^3 Q_0(\Lambda_0^{sqrt})^{-1}\left(\Lambda_0^{sin}(\tau - t)\Big|_{t_0}^{t}\right)(Q_0)^T\overline{g}_0, \quad t \in [t_0, T], \end{aligned} \tag{32}$$

since $\overline{z}_0(0) = \overline{w}_0(0) = z(0) = w(0) = \mathbf{0}$.

We will now estimate $\|\overline{z}_0(t)\|$ and $\|\overline{w}_0(t)\|$. The matrix Q_0 in (32) is orthogonal, therefore $\|Q_0\|_2 = 1$. Moreover, by definitions,

$$\|\Lambda_0^{sin}(t-\tau)\|_2 \leq 1, \quad \|\Lambda_0^{sqrt}\|_2 \leq \overline{G}^0, \quad \tau, t \in [t_0, T]. \tag{33}$$

It is known [8] that

$$\frac{d}{dt}\left[\left(\mathbf{G}^\delta(\cdot)\right)^{-1}\right] = \left(\mathbf{G}^\delta(\cdot)\right)^{-1}\frac{d}{dt}\left[\mathbf{G}^\delta(\cdot)\right]\left(\mathbf{G}^\delta(\cdot)\right)^{-1}. \tag{34}$$

Therefore, it follows from the definition (22) that

$$\|\overline{g}_0\| \leq (\tilde{G}^0)^2\overline{G}^1(\overline{Y}^1 + \overline{F}^0) + \tilde{G}^0(\overline{Y}^{2,\delta} + \overline{F}^1). \tag{35}$$

By substituting estimates (33) and (35) into (32), we get that

$$\begin{aligned} \|\overline{z}_0(t)\| &\leq \alpha^2 2\big((\tilde{G}^0)^2\overline{G}^1(\overline{Y}^1 + \overline{F}^0) + \tilde{G}^0(\overline{Y}^{2,\delta} + \overline{F}^1)\big) \triangleq \alpha^2 R_z, \\ \|\overline{w}_0(t)\| &\leq \alpha^3\tilde{G}^0\big((\tilde{G}^0)^2\overline{G}^1(\overline{Y}^1 + \overline{F}^0) + \tilde{G}^0(\overline{Y}^{2,\delta} + \overline{F}^1)\big) \triangleq \alpha^3 R_w, \quad t \in [t_0, T]. \end{aligned} \tag{36}$$

Let us now consider the discrepancies

$$\triangle z_0(t) = z(t) - \overline{z}_0(t), \quad \triangle w_0(t) = w(t) - \overline{w}_0(t), \quad t \in [t_0, t_1] \tag{37}$$

between the solutions of systems (21) and (26). The following equations hold:

$$\begin{aligned} \dot{\triangle z}_0(t) &= -\alpha^{-2}\mathbf{G}^\delta(t)w(t) + \alpha^{-2}\mathbf{G}_0^\delta\overline{w}_0(t) \\ &= -\alpha^{-2}\mathbf{G}_0^\delta\triangle w_0(t) + \alpha^{-2}(\mathbf{G}_0^\delta - \mathbf{G}^\delta(t))w(t), \\ \dot{\triangle w}_0(t) &= \triangle z_0(t) + \alpha^2(g(t) - \overline{g}_0), \\ \triangle z_0(0) &= \mathbf{0}, \quad \triangle w_0(0) = \mathbf{0}, \quad t \in [t_0, t_1]. \end{aligned} \tag{38}$$

A solution of system (38) can be written in the form

$$\begin{pmatrix} \triangle z_0(t) \\ \triangle w_0(t) \end{pmatrix} = \Phi_0(t) \cdot \int_{t_0}^{t} (\Phi_0)^{-1}(\tau) \begin{pmatrix} \alpha^{-2}(\mathbf{G}_0^\delta - \mathbf{G}^\delta(t))w(\tau) \\ \alpha^2(g(t) - \overline{g}_0) \end{pmatrix} d\tau, \quad t \in [t_0, t_1], \tag{39}$$

where the fundamental matrix $\Phi_0(\cdot)$ can be chosen as in (28) since the homogeneous parts (that is the part $\dot{x}(t) = Ax(t)$ of a system of the form $\dot{x}(t) = Ax(t) + f(t)$, $A = const$) of systems (26) and (38) coincide.

We get by substituting (30) and (31) into (39) that

$$\begin{aligned} \triangle z_0(t) &= \int_{t_0}^{t} \Big[\alpha^{-2}Q_0\Lambda_0^{cos}(t-\tau)Q_0^T(\mathbf{G}_0^\delta - \mathbf{G}^\delta(\tau))w(\tau) \\ &\quad -\alpha Q_0\Lambda_0^{sqrt}\Lambda_0^{sin}(t-\tau)Q_0^T(g(\tau) - \overline{g}_0)\Big]d\tau, \\ \triangle w_0(t) &= \int_{t_0}^{t} \Big[\alpha^{-1}Q_0(\Lambda_0^{sqrt})^{-1}\Lambda_0^{sin}(t-\tau)Q_0^T(\mathbf{G}_0^\delta - \mathbf{G}^\delta(\tau))w(\tau) \\ &\quad -\alpha^2 Q_0\Lambda_0^{cos}(t-\tau)Q_0^T(g(\tau) - \overline{g}_0)\Big]d\tau, \quad t \in [t_0, t_1]. \end{aligned} \tag{40}$$

Let us estimate $\|\triangle z_0(t)\|$ and $\|\triangle w_0(t)\|$. Since $\mathbf{G}_0^\delta = \mathbf{G}^\delta(0)$ and $t_1 - t_0 \le \alpha^2$,

$$\begin{aligned}
&\left\|\mathbf{G}_0^\delta - \mathbf{G}^\delta(\tau)\right\|_2 \le \alpha^2 \max_{t\in[t_0,t_1]} \left\|\dot{\mathbf{G}}^\delta(t)\right\|_2 \\
&= \alpha^2 \max_{t\in[t_0,t_1]} \left\| \tfrac{d}{dt}\left[G(y^\delta(t),t)G^T(y^\delta(t),t)\right] \right\|_2 \\
&= \alpha^2 \max_{t\in[t_0,t_1]} \left\| \Big(\sum_{i=1}^n \left[\tfrac{\delta}{\delta y_i}\left[G(y^\delta(t),t)\right] \dot{y}_i^\delta(t) \right] + \tfrac{\delta}{\delta t}\left[G(y^\delta(t),t)\right] \Big) G^T(y^\delta(t),t) \right. \\
&\quad \left. + G(y^\delta(t),t) \Big(\sum_{i=1}^n \left[\tfrac{\delta}{\delta y_i}\left[G(y^\delta(t),t)\right] \dot{y}_i^\delta(t) \right] + \tfrac{\delta}{\delta t}\left[G(y^\delta(t),t)\right] \Big)^T \right\|_2 \\
&\le \alpha^2 2\overline{G}^0 \overline{G}^1 (\overline{Y}^{1,\delta} + 1) \triangleq \alpha^2 R_G^\delta, \quad \tau \in [t_0, t_1].
\end{aligned} \tag{41}$$

Thus, by substituting (24) and (41) into (40), we get that

$$\begin{aligned}
&\|\triangle z_0(t)\ \| \le \alpha^3 \big(2\alpha R_G^\delta R_w^\delta + \omega_g(\alpha^2)\overline{G}^0\big) \triangleq \alpha^3 R_{\triangle z}^{\delta,\alpha}, \\
&\|\triangle w_0(t)\ \| \le \alpha^4 \big(2\alpha \tilde{G}^0 R_G^\delta R_w^\delta + \omega_g(\alpha^2)\big) \triangleq \alpha^4 R_{\triangle w}^{\delta,\alpha}, \\
&t \in [t_0, t_1],
\end{aligned} \tag{42}$$

where $g(\cdot)$ is defined in (22). Taking into account (34), for the continuously differentiable matrix function $\left(\mathbf{G}^\delta(\cdot)\right)^{-1}$, we obtain

$$\omega_{\mathbf{G}^{\delta-1}}(\triangle t) \le (\tilde{G}^0)^2 \overline{G}^1 (n\overline{Y}^{1,\delta} + 1)\triangle t. \tag{43}$$

Then, Lemma 1 provides that

$$\begin{aligned}
\omega_g(\triangle t) \le &\Big(2\omega_{\mathbf{G}^{\delta-1}}(\triangle t)\overline{G}^1 \tilde{G}^0 (\overline{Y}^{1,\delta} + \overline{F}^0) + \omega_{dG}(\triangle t)(\overline{Y}^{1,\delta} + 1)(\tilde{G}^0)^2(\overline{Y}^{1,\delta} + \overline{F}^0) \\
&+ \overline{G}^1 (\tilde{G}^0)^2 (\overline{Y}^{2,\delta} + \overline{F}^1)\Big) + \Big(\omega_{\mathbf{G}^{\delta-1}}(\triangle t)(\overline{Y}^{2,\delta} + \overline{F}^1) \\
&+ \tilde{G}^0 (\omega_{d^2y}(\triangle t) + \omega_{df}(\triangle t)(n\overline{Y}^{1,\delta} + 1))\Big).
\end{aligned} \tag{44}$$

The estimates (36), (37) and (42) provide that

$$\begin{aligned}
&\|z(t)\| \le \alpha^2 R_z + \alpha^3 R_{\triangle z}^{\delta,\alpha}, \\
&\|w(t)\| \le \alpha^3 R_w + \alpha^4 R_{\triangle w}^{\delta,\alpha}, \\
&t \in [t_0, t_1].
\end{aligned} \tag{45}$$

Let us now proceed to the next interval $[t_1, t_2]$. We introduce the frozen system for the interval $[t_1, t_2]$ as follows:

$$\begin{aligned}
&\dot{\overline{z}}_1(t) = -\alpha^{-2}\mathbf{G}_1^\delta \overline{w}_1(t), \\
&\dot{\overline{w}}_1(t) = \overline{z}_1(t) + \alpha^2 \overline{g}_1, \\
&\overline{z}_1(t_1) = z(t_1), \quad \overline{w}_1(t_1) = w(t_1), \quad t \in [t_1, T],
\end{aligned} \tag{46}$$

where $\mathbf{G}_1^\delta = \mathbf{G}^\delta(t_1)$, $\overline{g}_1 = g(t_1)$.

We can introduce auxiliary discrepancies between the solutions of frozen systems (26) and (46):

$$\triangle\overline{z}_1(\cdot) = \overline{z}_1(\cdot) - \overline{z}_0(\cdot), \quad \triangle\overline{w}_1(\cdot) = \overline{w}_1(\cdot) - \overline{w}_0(\cdot), \quad t \in [t_1, T]. \tag{47}$$

Consider a system of equations

$$\begin{aligned} \dot{\triangle\overline{z}}_1(t) &= -\alpha^{-2}\mathbf{G}_1^\delta \triangle\overline{w}_1(t) + \alpha^{-2}(\mathbf{G}_0^\delta - \mathbf{G}_1^\delta)\overline{w}_0(t), \\ \dot{\triangle\overline{w}}_1(t) &= \triangle\overline{z}_1(t) + \alpha^2(\overline{g}_1 - \overline{g}_0), \\ \triangle\overline{z}_1(t_1) &= \triangle z_0(t_1), \quad \triangle\overline{w}_1(t_1) = \triangle w_0(t_1), \quad t \in [t_1, T]. \end{aligned} \tag{48}$$

A solution of (48) is

$$\begin{aligned} \begin{pmatrix} \triangle\overline{z}_1(t) \\ \triangle\overline{w}_1(t) \end{pmatrix} &= \Phi_1(t) \begin{pmatrix} \triangle z_0(t_1) \\ \triangle w_0(t_1) \end{pmatrix} \\ &+ \Phi_1(t) \int_{t_1}^{t} (\Phi_1)^{-1}(\tau) \begin{pmatrix} \alpha^{-2}(\mathbf{G}_0^\delta - \mathbf{G}_1^\delta)\overline{w}_0(\tau) \\ \alpha^2(\overline{g}_1 - \overline{g}_0) \end{pmatrix} d\tau, \quad t \in [t_1, t_2], \end{aligned} \tag{49}$$

where the matrix $\Phi_1(\cdot)$ is defined as in (28) and calculations (29)–(31) can be applied to it, the only difference is that the second lower index changes from 0 for 1.

Since the constants (19), which define $R_{\triangle z}^{\delta,\alpha}$ and $R_{\triangle w}^{\delta,\alpha}$, are uniform for $t \in [0, T]$, one can obtain the following estimates for (49) using formulae (40)–(42) with (36), (42):

$$\begin{aligned} \|\triangle\overline{z}_1(t)\| &\leq \alpha^3 \big(R_{\triangle z}^{\delta,\alpha} + R_{\triangle w}^{\delta,\alpha}\overline{G}^0 + \alpha R_G^\delta R_w + \omega_g(\alpha^2)\overline{G}^0\big) \triangleq \alpha^3 R_{\triangle\overline{z}}, \\ \|\triangle\overline{w}_1(t)\| &\leq \alpha^4 \big(R_{\triangle z}^{\delta,\alpha}\tilde{G}^0 + R_{\triangle w}^{\delta,\alpha} + \alpha\tilde{G}^0 R_G^\delta R_w + \omega_g(\alpha^2)\big) \triangleq \alpha^4 R_{\triangle\overline{w}}, \\ & t \in [t_1, t_2]. \end{aligned} \tag{50}$$

Taking into account (36), (47) and (50), we get

$$\begin{aligned} \|\overline{z}_1(t)\| &\leq \alpha^2 R_z + \alpha^3 R_{\triangle\overline{z}}, \\ \|\overline{w}_1(t)\| &\leq \alpha^3 R_w + \alpha^4 R_{\triangle\overline{w}}, \quad t \in [t_1, t_2]. \end{aligned} \tag{51}$$

One can use the samc algorithm as defined by (40)–(42) to obtain

$$\begin{aligned} \|\triangle z_1(t)\| &\triangleq \|z(t) - \overline{z}_1(t)\| \leq \alpha^3 R_{\triangle z}^{\delta,\alpha}, \\ \|\triangle w_1(t)\| &\triangleq \|w(t) - \overline{w}_1(t)\| \leq \alpha^4 R_{\triangle w}^{\delta,\alpha}, \quad t \in [t_1, t_2]. \end{aligned} \tag{52}$$

Now, by gathering (51), (52), we get the estimates

$$\begin{aligned} \|z(t)\| &\leq \alpha^2 R_z + \alpha^3 R_{\triangle z}^{\delta,\alpha} + \alpha^3 R_{\triangle\overline{z}}, \\ \|w(t)\| &\leq \alpha^3 R_w + \alpha^4 R_{\triangle w}^{\delta,\alpha} + \alpha^4 R_{\triangle\overline{w}}, \quad t \in [t_1, t_2]. \end{aligned} \tag{53}$$

On the third step ($t \in [t_2, t_3]$) we obtain in the same way the estimates

$$\begin{aligned}
\|\triangle z_2(t)\| &\triangleq \|z(t) - \overline{z}_2(t)\| \le \alpha^3 R^{\delta,\alpha}_{\triangle z}, \\
\|\triangle w_2(t)\| &\triangleq \|w(t) - \overline{w}_2(t)\| \le \alpha^4 R^{\delta,\alpha}_{\triangle w}, \\
\|\triangle \overline{z}_2(t)\| &\triangleq \|\overline{z}_2(t) - \overline{z}_1(t)\| \le \alpha^3 R_{\triangle \overline{z}}, \\
\|\triangle \overline{w}_2(t)\| &\triangleq \|\overline{w}_2(t) - \overline{w}_1(t) \le \alpha^4 R_{\triangle \overline{w}}, \\
\|\overline{z}_2(t)\| &\le \alpha^2 R_z + \alpha^3 2 R_{\triangle \overline{z}}, \\
\|\overline{w}_2(t)\| &\le \alpha^3 R_w + \alpha^4 2 R_{\triangle \overline{w}}, \\
\|z(t)\| &\le \alpha^2 R_z + \alpha^3 R^{\delta,\alpha}_{\triangle z} + \alpha^3 2 R_{\triangle \overline{z}}, \\
\|w(t)\| &\le \alpha^3 R_w + \alpha^4 R^{\delta,\alpha}_{\triangle w} + \alpha^4 2 R_{\triangle \overline{w}}, \quad t \in [t_2, t_3].
\end{aligned} \tag{54}$$

By building estimates of the form (54) subsequently for the intervals $[t_3, t_4], \ldots, [t_{h-1}, t_h]$, we get that

$$\begin{aligned}
\|z(t)\| &\le \alpha^2 R_z + \alpha^3 R^{\delta,\alpha}_{\triangle z} + \alpha^3 i R_{\triangle \overline{z}}, \\
\|w(t)\| &\le \alpha^3 R_w + \alpha^4 R^{\delta,\alpha}_{\triangle w} + \alpha^4 i R_{\triangle \overline{w}}, \\
& t \in [t_i, t_{i+1}], \quad i = 0, \ldots, h-1.
\end{aligned} \tag{55}$$

By (25), we get $h \le T\alpha^{-2}$. Therefore,

$$\begin{aligned}
\|z(t)\| &\le \alpha^2 R_z + \alpha^3 R^{\delta,\alpha}_{\triangle z} + \alpha T R_{\triangle \overline{z}}, \\
\|w(t)\| &\le \alpha^3 R_w + \alpha^4 R^{\delta,\alpha}_{\triangle w} + \alpha^2 T R_{\triangle \overline{w}}, \\
& t \in [t_i, t_{i+1}], \quad i = 0, \ldots, h-1.
\end{aligned} \tag{56}$$

One can check that for

$$\begin{aligned}
R^\delta_w &= \frac{\alpha R_w + \alpha^2 \omega_g(\alpha^2) + \omega_g(\alpha^2) T(\tilde{G}^0 \overline{G}^0 + 2) + \alpha T \tilde{G}^0 R^\delta_G R_w}{1 - 2\alpha^3 \tilde{G}^0 R^\delta_G - 2\alpha T \tilde{G}^0 R^\delta_G}, \\
R^\delta_z &= \alpha R_z + 2\alpha^3 R^\delta_G R^\delta_w + \alpha \omega_g(\alpha^2) \overline{G}^0 + 2\alpha T R^\delta_G R^\delta_w (2 + \overline{G}^0 \tilde{G}^0) + 2\alpha T \omega_g(\alpha^2) \overline{G}^0
\end{aligned} \tag{57}$$

under condition

$$\alpha = \alpha(\delta) : \; 2\alpha^3 \tilde{G}^0 R^\delta_G + 2\alpha T \tilde{G}^0 R^\delta_G < 1, \quad \alpha(\delta) \xrightarrow{\delta \to 0} 0, \quad \omega_g(\alpha(\delta)^2) \xrightarrow{\delta \to 0} 0 \tag{58}$$

estimates (56) provide that

$$\|z(t)\| \le \alpha R^\delta_z < 2\alpha R^\delta_z, \quad \|w(t)\| \le \alpha^2 R^\delta_w < 2\alpha^2 R^\delta_w, \quad t \in [0, T_1]. \tag{59}$$

But this contradicts (for the moment $t = T_1$) initial hypothesis (24). Therefore, estimate (23) holds for R^δ_w, R^δ_z defined in (57) (provided condition (58) is true). Since (23) holds, calculations (25)–(56) are valid for $T_1 = T$ and, therefore,

$$\|z(t)\| \le 2\alpha R^\delta_z, \quad \|w(t)\| \le 2\alpha^2 R^\delta_w, \quad t \in [0, T]. \tag{60}$$

Let us now return to functions (15) and express them in variables (20). Taking into account (60) and under conditions (57), (58), we have

$$\begin{aligned}
\|u^{\delta,\alpha}(t) &- G^g(y^\delta(t), t)(\dot{y}^\delta(t) - f(y^\delta(t), t))\| \\
&= \|\alpha^{-2} G^T(y^\delta(t), t) w(t)\| \le \sqrt{\overline{G}^0}\, 2R^\delta_w \triangleq R^\delta \xrightarrow{\delta \to 0} 0, \quad t \in [0, T].
\end{aligned} \tag{61}$$

The lemma is proven.

Theorem 1. *Let Assumptions 1 and 2 hold. Then the function $\hat{u}^{\delta,\alpha}(\cdot)$, defined in (16), satisfy conditions* **C1–C3** *for $\alpha = \alpha(\delta)$ satisfying condition (58).*

Proof. The condition **C1** is fulfilled by definition (16).

One can use the scheme of the proof of Lemma 7.2 in [5] to prove that the control $u^{\delta,\alpha}(\cdot)$ generates a trajectory $\hat{x}(\cdot))$ such that

$$\lim_{\delta\to 0,\alpha\to 0} \|\hat{x}(\cdot) - x(\cdot)\|_{C[0,T]} = 0, \tag{62}$$

where $x(\cdot)$ is the solution of system (14) with boundary conditions (13). Therefore, Lemma 1 together with (3) provide that

$$\begin{aligned} &\|\hat{x}(\cdot) - x^*(\cdot)\|_{C[0,T]} \\ \le\ &\|\hat{x}(\cdot) - x(\cdot)\|_{C[0,T]} + \|x(\cdot) - y^\delta(\cdot)\|_{C[0,T]} + \|y^\delta(\cdot) - x^*(\cdot)\|_{C[0,T]} \xrightarrow{\delta\to 0} 0. \end{aligned} \tag{63}$$

Thus, the condition **C2** is fulfilled.

For the clearness of notations, the following abridgments will be used in calculations in a subsequent:

$$\begin{gathered} G^g(x^*(\cdot),\cdot) \triangleq G^{g*}, \quad G(x^*(\cdot),\cdot) \triangleq G^*, \quad G^g(y^\delta(\cdot),\cdot) \triangleq G^g_\delta, \quad f(y^\delta(\cdot),\cdot)) \triangleq f_\delta, \\ f(x^*(t),t)) \triangleq f^*, \quad u^*(\cdot) \triangleq u^*, \quad u^{\delta,\alpha}(\cdot) \triangleq u^{\delta,\alpha}, \\ v(\cdot) \triangleq v, \quad x^*(\cdot) \triangleq x^*, \quad y^\delta(\cdot) \triangleq y^\delta. \end{gathered} \tag{64}$$

Using the scheme of the proof in [4] for the case $f(x,t) \equiv 0$, one can prove that the normal control u^* for the trajectory x^*, generated by an admissible control v, can be found by the formula

$$u^* = G^{g*} G^* v. \tag{65}$$

Thus,

$$\begin{aligned} \|u^* - u^{\delta,\alpha}\|_{L_2[0,T]} \le\ &\|G^{g*} G^* v - G^g_\delta(\dot{y}^\delta - f_\delta)\|_{L_2[0,T]} \\ &+ \|u^{\delta,\alpha} - G^g_\delta(\dot{y}^\delta - f_\delta)\|_{L_2[0,T]}. \end{aligned} \tag{66}$$

The Weierstrass Approximation Theorem guarantees that for the continuous vector function $G^* v$ and any $\delta > 0$, there exists a vector function $\psi(\cdot)$ with polynomial elements such that

$$\begin{gathered} \|(G^* v + f^* - \psi(\cdot)\|_{C[0,T]} \le \delta, \\ \|\Psi(\cdot) - x^*\|_{C[0,T]} \le \delta T, \ \text{ where } \Psi(t) = x^*(0) + \int_0^t \psi(\tau) d\tau. \end{gathered} \tag{67}$$

By integration by parts, taking into account (4), we get

$$\begin{aligned}
&\| (G^*v+f^*)-\dot y^\delta\|_{L_2[0,T]}\\
&\le \|(G^*v+f^*)-\psi(\cdot)\|_{L_2[0,T]}+\|\psi(\cdot)-\dot y^\delta\|_{L_2[0,T]}\\
&\le \delta\sqrt{T}+\|\psi(\cdot)-\dot y^\delta\|_{L_2[0,T]} =\le \delta\sqrt{T}+\int_0^T\langle\underbrace{\psi(\tau)-\dot y^\delta(\tau)}_{\mathbf{U}},\underbrace{\psi(\tau)-\dot y^\delta(\tau)}_{\mathbf{dV}}\rangle d\tau\\
&=\delta\sqrt{T}+\langle\underbrace{\psi(\tau)-\dot y^\delta(\tau)}_{\mathbf{U}},\underbrace{\Psi(t)-y^\delta(\tau)}_{\mathbf{V}}\rangle\Big|_0^T+\int_0^T\langle\underbrace{\Psi(\tau)-y^\delta(\tau)}_{\mathbf{V}},\underbrace{\dot\psi(\tau)-\ddot y^\delta(\tau)}_{\mathbf{dU}}\rangle d\tau\\
&\le 2(\overline{G}^0\overline{U}+\delta+\overline{Y}^{1,\delta})\delta(T+1)+T\delta(T+1)(\max_{t\in[0,T]}\|\psi(t)\|+\overline{Y}^{2,\delta})\triangleq R_y^\delta \overset{\delta\to 0}{\longrightarrow} 0.
\end{aligned} \tag{68}$$

Therefore, since (3), (34) and (68), it holds

$$\begin{aligned}
\|G^{g*}G^*v-G^g_\delta(\dot y^\delta-f_\delta)\|_{L_2[0,T]}&\le \|\left(G^{g*}G^*-G^g_\delta G^\delta\right)(\dot y^\delta-f_\delta)\|_{L_2[0,T]}\\
+\|G^{g*}G^*\left(\dot y^\delta-(G^*v+f^*)\right)\|_{L_2[0,T]}&+\|G^{g*}G^*\left(f^*-f^\delta\right)\|_{L_2[0,T]}\\
\le n\delta\left(2\sqrt{\overline{G}^0\overline{G}^1}\tilde G^0+\overline{G}^1(\tilde G^0)^2\overline{G}^1\right)&\left(\max_{t\in[0,T]}\|x^*(t)\|+\delta+\overline{F}^0\right)\\
+\tilde G^0\overline{G}^0R_y^\delta+n\delta\tilde G^0\overline{G}^0\overline{F}^1&\triangleq R_u^\delta.
\end{aligned} \tag{69}$$

Applying formula (61) obtained in Lemma 1 and (69) to inequality (66), we conclude that

$$\|u^*(\cdot)-u^{\delta,\alpha}(\cdot)\|_{L_2[0,T]}\le R_u^\delta+R^\delta \overset{\delta\to 0}{\longrightarrow} 0 \tag{70}$$

under conditions (57), (58).

By definition (16), the inequality $\|u^*(t)-\hat u^{\delta,\alpha}(t)\|\le\|u^*(t)-u^{\delta,\alpha}(t)\|$, $t\in[0,T]$ is satisfied. Therefore, under conditions (57), (58) it holds

$$\|u^*(\cdot)-\hat u^{\delta,\alpha}(\cdot)\|_{L_2[0,T]} \overset{\delta\to 0}{\longrightarrow} 0. \tag{71}$$

Thus, the condition **C3** is also fulfilled and the theorem is proven.

Theorem 1 means that the functions $\hat u^{\delta,\alpha}(\cdot)$, defined in (16), form a solution of the problem **C1**–**C3** (under conditions (58), (57)).

6 Remarks on the Suggested Method

6.1 Comparison with Another Modern Approaches. The method considered in this paper is close to another approach, which is based on A. V. Kryazhimskii's and Yu. S. Osipov's works [7,9]. This approach has been developed by V. I. Maximov, M. S. Blizorukova, et al. as well (see, in particular, [1,10]). A survey of algorithms based on this approach is presented in [10]. The distinguishing feature of the new method consists in using functional (11), which is convex with respect to the control variable and concave with respect to the discrepancy of the state variables. Meanwhile, the approach discussed in [10] applies the minimizing of constructions that are convex in all variables. This difference allows to obtain the oscillating character of the solutions instead of the exponential one. A more accurate comparison of the effectiveness of this two approaches is the matter of the future research.

6.2 The Problem of Dynamic Reconstrcution. The suggested approach can be used to solve dynamic reconstruction problems, where discrete measurements of the trajectory arrive in real-time. Spline interpolation can be used to fulfill Assumption 1. This smaterial will be published in future papers.

6.3 Comparison with Inverse Matrices Approach. The result (68) means, in particular, that the desired normal control $u^*(\cdot)$ can be approximated by the cut-off functions

$$\hat{u}^\delta(t) = \begin{cases} u^\delta(t), & u^\delta(t) \in \mathbf{U}, \\ \hat{u} \in \mathbf{U} : \|u^\delta(t) - \hat{u}\| = \min\limits_{v \in \mathbf{U}} \|u^\delta(t) - v\|, & u^\delta(t) \notin \mathbf{U}, \end{cases} \quad u^\delta(t) = G^g(y^\delta(t), t)(\dot{y}^\delta(t) - f(t)). \tag{72}$$

Yet, this approach reduces the inverse problem to the problem of constructing the inverse matrices $G^g(y^\delta(t), t)$, while the approach, suggested in this paper, reduces it to the problem of integrating a system of linear heterogenous ODEs with variable coefficients (14). In some applications, the numerical integration can be more preferable than the numerical matrix inversion. It is illustrated by a numerical simulation in [4]. An example of a dynamic control system, for which the suggested approach is more effective than the matrix inversion, is suggested in [4].

6.4 Numerical Simulations. The examples, illustrating numerical simulations of the suggested method are exposed in [4–6]. In [5], it is shown that the suggested method can be applied to solving the inverse problems for macroeconomic models with real economic statistics.

7 Conclusion

This paper continues the study [4–6,11–13] of the new method for solving the inverse normal control reconstruction problems. Namely, the estimates for the method's error are obtained. The detailed proof of the convergence of the method is presented.

The perspective directions for the future research on the suggested method include expanding the class of the admissible dynamics, including the removal of the condition that the dimension of the controls is larger or equal than the dimension of the state variables. Accurate comparison with another known methods for solving the inverse problems is also the matter of the future publications.

References

1. Blizorukova, M.: On the reconstruction of the trajectory and control in a nonlinear second-order system. Proc. Steklov Inst. Math. **275**(1), 1–11 (2011). https://doi.org/10.1134/S008154381109001X
2. Krasovskii, N.: Teoriya upravleniya dvizheniem [Theory of motion control]. Nauka, Moscow (1968). (in Russian)
3. Krasovskii, N., Subbotin, A.: Positcionnie differentcialnie igri [Positional differential games]. Nauka, Moscow (1974). (in Russian)
4. Krupennikov, E.: A new approximate method for construction of the normal control. IFAC-PapersOnLine **51**(32), 343–348 (2018). https://doi.org/10.1016/j.ifacol.2018.11.407
5. Krupennikov, E.: On control reconstruction problems for dynamic systems linear in controls. In: Static & Dynamic Game Theory Foundations & Applications. Birkhauser, Cham (2018)
6. Krupennikov, E.: Solution of inverse problems for control systems with large control parameter dimension. IFAC-PapersOnLine **51**(32), 434–438 (2018). https://doi.org/10.1016/j.ifacol.2018.11.423
7. Kryazhimskij, A., Osipov, Y.: Modelling of a control in a dynamic system. Eng. Cybern. **21**(2), 38–47 (1983)
8. Magnus, J., Neudecker, H.: Matrix Differential Calculus with Applications in Statistics and Econometrics, 3rd edn. Wiley, Hoboken (2019). https://doi.org/10.1002/9781119541219
9. Osipov, Y., Kryazhimskii, A.: Inverse Problems for Ordinary Differential Equations: Dynamical Solutions. Gordon and Breach, London (1995)
10. Osipov, Y., Kryazhimskii, A., Maksimov, V.: Some algorithms for the dynamic reconstruction of inputs. Proc. Steklov Inst. Math. **275**(1), 86–120 (2011). https://doi.org/10.1134/S0081543811090082
11. Subbotina, N., Krupennikov, E.: Dynamic programming to identification problems. World J. Eng. Technol. **4**(3B), 228–234 (2016)
12. Subbotina, N., Tokmantsev, T., Krupennikov, E.: On the solution of inverse problems of dynamics of linearly controlled systems by the negative discrepancy method. Proc. Steklov Inst. Math. **291**, 253–262 (2015). https://doi.org/10.1134/S0081543815080209
13. Subbotina, N., Tokmantsev, T., Krupennikov, E.: Dynamic programming to reconstruction problems for a macroeconomic model. In: IFIP Advances in Information and Communication Technology. [S.l.], vol. 494, pp. 472–481 (2017)
14. Tikhonov, A.: Ob ustoichivosti obratnih zadach [on the stability of inverse problems]. Doklady Academii Nauk SSSR **39**, 195–198 (1943). (in Russian)

Optimization Problem in an Integral Model of the Developing System Without Prehistory

Evgeniia Markova(✉) and Inna Sidler

Melentiev Energy Systems Institute SB RAS,
Lermontov str. 130, 664033 Irkutsk, Russia
{markova,krlv}@isem.irk.ru

Abstract. This paper addresses an integral model of the developing system consisting of elements of n age groups. The model is described by means of the Volterra equation of the first kind with variable limits of integration. The case is considered when the moment of the system origin coincides with the beginning of the modeling, therefore there is no prehistory and for $t = 0$ all age groups of the elements are empty. Based on this model, we set the problem of optimizing the system age structure and the moment when the elements are decommissioned. In order to study a new integral model of developing systems as applied to the problem of forecasting the electric power system development, several model examples are considered. The results of numerical calculations are presented. They confirm the adequacy of the proposed model.

Keywords: Developing system · Age groups · Volterra equation of the first kind · Optimization problem

1 Introduction

In models of developing systems of the Glushkov type [1,2] the Volterra operator with variable limits of integration is used to describe the process of replacing obsolete elements with new ones. The use of integral equations is associated with economic models of technical progress that take into account the aging of the production system (vintage capital models) [3–6]. Since integral models allow us to take into account the heterogeneity of economic factors involved in the production, this leads to the formulation of problems of optimal development of the system [5,7–10]. The greatest interest (and the greatest complexity in mathematical terms) is provoked by special optimal control problems of dynamical systems with integral constraints and control functional parameters that fall within the lower limits of integration. The one-sector version of the Glushkov

The research was carried out under State Assignment, Project 17.3.1 (reg. no. AAAA-A17-117030310442-8) of the Fundamental Research of Siberian Branch of the Russian Academy of Sciences.

I. Bykadorov et al. (Eds.): MOTOR 2019, CCIS 1090, pp. 524–535, 2019.
https://doi.org/10.1007/978-3-030-33394-2_40

model became the basis for searching optimal strategies for commissioning generating capacities in the electric power system (EPS) of Russia with varying degrees of aggregation [11–14].

The problem of finding input capacities is reduced to solving a nonclassical Volterra integral equation of the first kind

$$\int_{a(t)}^{t} K(t,s)x(s)ds = y(t), \ \ t \in [0,T]. \tag{1}$$

Here $x(t)$ is the commissioning of electric capacities at time t; $K(t,s)$ is the efficiency coefficient of using at time t the capacity commissioned previously at time s; $t - a(t)$ is the lifetime of the oldest at time t capacity of the EPS; $a(t)$ is such non-decreasing function that $a(t) < t \ \forall t > 0$; $y(t)$ is total available capacity of the electric power system specified by the experts for the future. The Eq. (1) plays a key role in integral models of developing systems as a balance equation between the required level of system development $y(t)$ and the possibility of achieving it.

The specific features of (1) are largely determined by the value of $a(0)$. If $a(0) < 0$, then for the closure (1), we need to set the system prehistory:

$$x(t) \equiv x^0(t), \ \ t \in [a(0), 0), \tag{2}$$

so that the value of the right-hand side of (1) under $t = 0$ takes the form

$$y(0) = \int_{a(0)}^{0} K(0,s)x^0(s)ds. \tag{3}$$

Precisely this form is considered in the majority of works that address integral models of developing systems with prehistory.

A fundamentally different situation corresponds to the case $a(0) = 0$. It means that the beginning of modeling coincides with the moment of the system emergence. In this case, there is no prehistory, and $y(0) = 0$. The theory and numerical methods for solving (1) for different cases are quite diverse. A detailed analysis of the equations of the form (1), both for the case of $a(0) < 0$ and for $a(0) = 0$, was carried out in the monograph [15].

A new integral model for the developing system was proposed in [16,17]. It allows us to describe in greater detail the technical and economic parameters of the generating power plants equipment, taking into account its age structure. The equipment is divided into several age groups with different indicators of the effectiveness of their functioning.

In [16,17] an equation that generalizes (1) is considered for the case of n age groups

$$\sum_{i=1}^{n} \int_{a_i(t)}^{a_{i-1}(t)} K_i(t,s)x(s)ds = y(t), \ \ t \in [0,T], \tag{4}$$

$a_0(t) \equiv t > a_1(t) > \ldots > a_n(t) \geq 0$, $\forall t > 0$; $K_i(t,s)$ is the efficiency coefficient in age group i; $t - a_i(t)$ is the upper age limit of group i.

The values of $a_i(0)$ play an key role in (4), just like in the case of one age group (1). If $a_i(0) < 0$, $i = 1, \ldots, n$, then for the correct formulation of the problem it is necessary to set the desired function on the prehistory $x(t) = x^0(t)$, $t \in [a_n(0), 0)$. In particular, if $a_i(t) = t - T_i$, $0 < T_1 < \ldots < T_n$,

$$x(t) = x^0(t), \quad t \in [-T_n, 0). \tag{5}$$

This case for $n = 3$ is considered in [16–18] with the reference to the problem of finding long-term development strategies for the EPS of Russia. An equation of type (4) had the form

$$\int_{t-T_1}^{t} x(s)ds + 0.97 \int_{t-T_2}^{t-T_1} x(s)ds + 0.9 \int_{t-T_3}^{t-T_2} x(s)ds = y(t), \quad t \in [0, T]. \tag{6}$$

It seems promising to consider the case of $a_i(0) = 0$ for studying the problems of developing EPS, which corresponds to the coincidence of the beginning of modeling and the moment of the system emergence. At the same time, there is no history, $x(0) = 0$, so all age groups are empty. This is the case that the present research focuses on.

The goal of our research is to consider an integral model of the development of the electric power system without a prehistory [19–21] and the optimization problem of the age structure and decommissioning equipment based on it [22]. To solve this problem, we will apply the methods that were previously used to solve the optimization problem in the model of the developing system with a prehistory [11–14].

2 Integral Model of the EPS Development

For the purpose of modeling the EPS development, consider equations with constant Volterra kernels. The papers [19,21,23] are devoted to the theory of equations of the type (4) with the condition $a_i(0) = 0$. In particular, for the case of constant kernels $K_i(t,s) = \beta_i$, $i = 1, \ldots, n$, $\beta_1 = 1$, a detailed theoretical analysis was carried out in [19]. We will assume that $1 \equiv \beta_1 \geq \ldots \geq \beta_n \geq 0$. Also, suppose that $a_i(t) = \alpha_i t$, $1 = \alpha_0 > \alpha_1 > \ldots > \alpha_n \geq 0$, so (4) has the form

$$\sum_{i=1}^{n} \beta_i \int_{\alpha_i t}^{\alpha_{i-1} t} x(s)ds = y(t), \quad t \in [0, T]. \tag{7}$$

Using the classical results of functional analysis, it was shown in [19] that if the following inequality holds

$$\sum_{i=2}^{n} |\beta_{i-1} - \beta_i| \alpha_{i-1} < 1 \tag{8}$$

the Eq. (7) is well-posed by Hadamard on the pair $(C_{[0,T]}, \overset{\circ}{C}{}^{(1)}_{[0,T]})$ ($\overset{\circ}{C}{}^{(1)}_{[0,T]}$ is space of continuous differentiable functions on $[0, T]$ and $y(0) = 0$).

The transition to the equivalent functional equation

$$x(t) = \sum_{i=2}^{n}(\beta_{i-1} - \beta_i)\alpha_{i-1}x(\alpha_{i-1}t) + y'(t), \ \ t \in [0, T]. \tag{9}$$

shows that inequality (8) is the contraction condition of the operator in the right-hand side of (9), acting in the space $C_{[0,T]}$. It is essential that this condition is completely determined by the parameters α_i, β_i and does not depend on the time interval on which the Eq. (7) is considered.

Similarly (6), we set $\beta_1 = 1$, $\beta_2 = 0.97$, $\beta_3 = 0.9$ and consider Eq. (7) [22]

$$\int_{\alpha_1 t}^{t} x(s)ds + 0.97\int_{\alpha_2 t}^{\alpha_1 t} x(s)ds + 0.9\int_{\alpha_3 t}^{\alpha_2 t} x(s)ds = y(t), \ \ t \in [0, T]. \tag{10}$$

When applied to the EPS development problem, it can be considered as the equation of balance between the desired power consumption level $y(t)$ and the set of generating capacities $x(t)$ commissioned at time $t > 0$. All generating equipment is divided into three age groups. The monotonous decrease in the efficiency coefficients β_i, $i = 1, 2, 3$, reflects the natural process of aging of elements in the middle and older age groups. The values of $\alpha_{i-1} - \alpha_i$, $i = \overline{1, 3}$, determine the share of generating capacity included in the age group i (see Fig. 1).

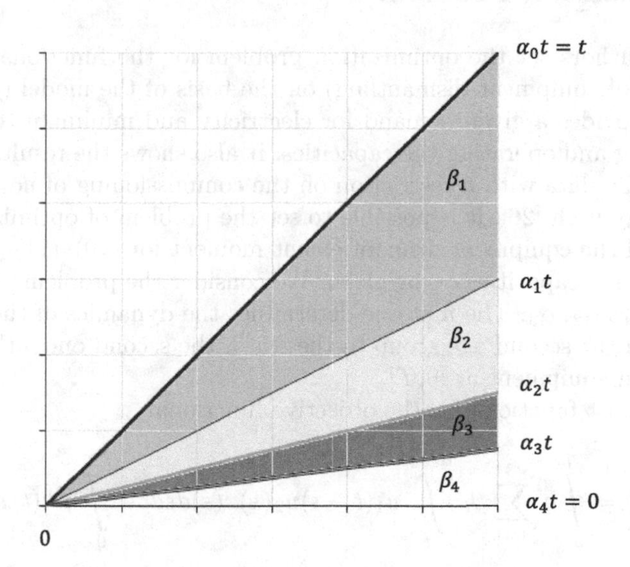

Fig. 1. Age groups of the developing system.

The model for the EPS development is represented by Eq. (10) with restrictions on the commissioning of capacities

$$x(t) \geq 0,\ t \in [t_0, T]. \tag{11}$$

Condition $\alpha_3 \geq 0$ implies the impossibility of restoring the retired capacities.

The analytical solution to Eq. (4) can be obtained only for some special cases.

For the numerical solution of (4) with the condition $a_i(0) = 0$, modified versions of the left and middle rectangles methods were developed [24,25].

The use of standard quadrature methods for the numerical solution of the non-classical integral equation (4) might result in an equation with n variables, which happens because the regular grid nodes do not match the values of the integration limits.

Modifications of the quadrature methods of the left and middle point rules are based on the transformation of the original equation (4) to the equivalent one with only upper variable limits. The constructed numerical schemes have the same order of convergence with respect to the grid step as the classical case. Their codes were used to obtain the following results.

On the basis of the described model, it is possible to formulate various economic problems. For example, if $x(t)$ is the decision variable, and all other functions are assumed to be known, we obtain the problem of forecasting the EPS development.

The model (10)–(11) can be used to perform various optimization problems.

3 Optimization Problem

In [26] the authors set the optimization problem for the functional parameter T_3 (the time of equipment dismantling) on the basis of the model (5)–(6). This parameter provides a given demand for electricity and minimum total costs of commissioning and operating the capacities. It also shows the results of its solution on real-life data with a restriction on the commissioning of new capacities.

By analogy with [26], it is possible to set the problem of optimizing the age structure and the equipment dismantlement moment for (10)–(11) [22]. Let the share of "young" capacities α_1 be given. We consider the problem of optimizing the parameters α_2, α_3. The first one determines the dynamics of the transfer of capacity from the second age group to the third, the second one – the dynamics of dismantling equipment at $[0, T]$.

Take the cost functional as the objective functional

$$I(x, \alpha_2, \alpha_3) = \int_0^T q^t \sum_{i=1}^{3} \beta_i \int_{\alpha_i t}^{\alpha_{i-1} t} u_1(t-s) u_2(s) x(s) ds dt + \int_0^T q^t k(t) x(t) dt. \tag{12}$$

Here the first summand corresponds to the operating costs; the second summand corresponds to the costs of putting capacities into service during the whole forecast period.

The following functions are known:

β_i is the efficiency coefficient in age group i;

$u_1(t-s)$ is the coefficient of increase in the costs of operating the capacities at time t that are commissioned at time s;

$u_2(t)$ is the specific annual costs of operating the capacity commissioned at time t;

$k(t)$ is the specific capital costs of commissioning a capacity unit at time t;

q^t is the costs discount coefficient, $0 < q < 1$.

The control parameters α_2 and α_3 belong to the feasible set

$$A = \{\alpha_2, \alpha_3 : 0 \leq \alpha_3 < \alpha_2 < \alpha_1\}. \tag{13}$$

It is required to find

$$(\alpha_2^*, \alpha_3^*) = arg \min_{\alpha_2, \alpha_3 \in A} I\big(x(\alpha_2, \alpha_3)\big) \tag{14}$$

under the conditions (10)–(13).

The problem (10)–(14) is nonlinear, the required parameters are within the integration limits in (10) and (12). In addition, there are restrictions on the phase variable (11). Therefore, we will seek a solution to the problem using numerical algorithms.

To solve the optimal control problem, we used a heuristic algorithm based on the discretization of all elements of the problem on a grid and replacement of the feasible set A by the set

$$A_h = \{\alpha_2, \alpha_3 : \alpha_3 = ih, i = \overline{0, N_1 - 1},\ \alpha_2 = \alpha_3 + jh, j = \overline{1, N_2 - 1}\}, \tag{15}$$

where $N_1 = \left[\frac{\alpha_2}{h}\right]$, $N_2 = \left[\frac{\alpha_1 - \alpha_3}{h}\right]$.

Firstly we choose the pair (α_2, α_3) from the feasible set A_h, substitute it into a discrete analog of the model (10) and find the numerical solution for $x(t)$. Then we substitute the obtained solution into a discrete analog of (12) and find the value of the objective functional for the given pair. Thus, we use the enumerative technique among all possible pairs (α_2, α_3) from feasible set (15) and find the optimal values (α_2^*, α_3^*) for this class. We used the modified version of left point rule for the approximation of integrals in (10), (12). We employ the forecast values of economic indices provided by experts.

4 Study of the Optimal Control Problem of Commissioning of Capacities on the Examples

To study a new integral model of developing systems with respect to the problem of forecasting the EPS development of EPS, we consider model examples.

Let in (10) $\alpha_1 = 1/2$, $\alpha_2 = 1/4$, $\alpha_3 = 1/8$. This means that at any time moment the part of capacity operating at 100% efficiency is 1/2 of the total capacity, the part of capacity operating at 97% efficiency is 1/4, and the part operating at 90% efficiency is 1/8. The capacity with efficiency lower than 90% is removed. We take these parameters as basic.

The optimization problem used the same data as the similar optimization problem for the model (6). The specific growth functions of the operating costs $u_1(t-s) \equiv u_1(\tau)$ are given:

$$u_1(\tau) = \begin{cases} 1, & \tau \leqslant u_c, \\ 1.03^{\tau - u_c}, & \tau > u_c, \end{cases} \tag{16}$$

(operating costs increase with the growth rate 3% per year after u_c years of service). The functions $k(t)$ and $u_2(t)$ were assumed to be constant: $k(t) = 1000$ (USD/MW), $u_2(t) = 189$ (USD/MW) $t \in [0, T]$.

To solve the optimal control problem, we used the heuristic algorithm described above.

Example 1. Let the right-hand side of the equation have the form $y(t) = t^2/2$ according to the optimistic electricity demand forecast, $T = 60$.

Figure 2 shows the dynamics of commissioning capacities $x(t)$ and the right-hand side $y(t)$, corresponding to the basic variant (without optimization by α_2, α_3).

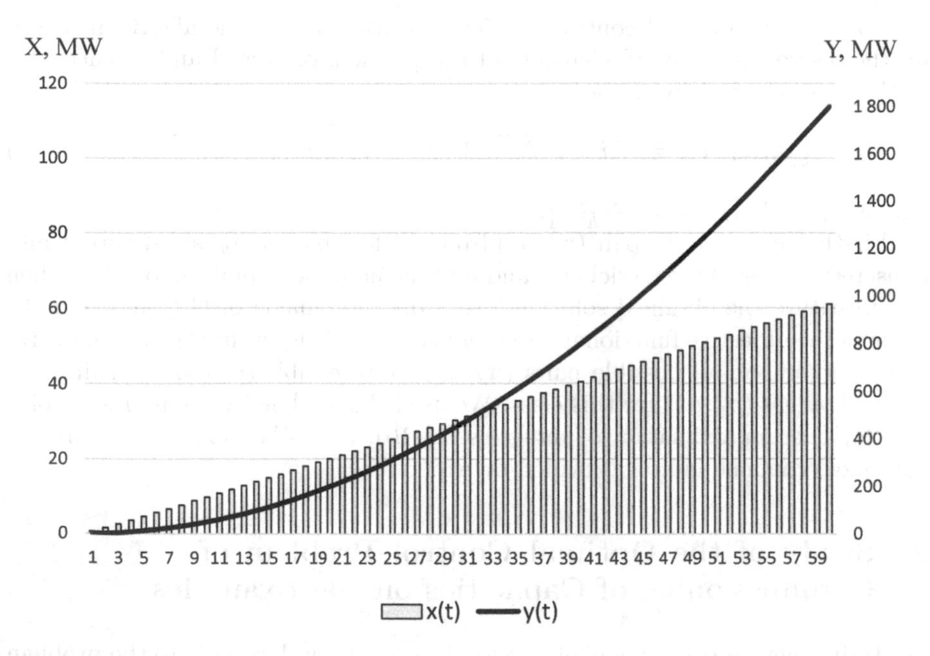

Fig. 2. Commissioning of capacities for $y(t) = t^2/2$ (base variant).

To begin, explore the solution to the problem, optimizing one of the parameters. Let α_3 be fixed.

Figure 3 shows the dynamics of the cost functional (12) for the case $u_c = 45$ with the changing parameter α_2. It can be seen that the optimal value of α_2^* tends to α_3, i.e. it is advantageous to reduce the share of capacities working with the efficiency of 0.9 due to an increase in the share of capacities working with the efficiency of 0.97.

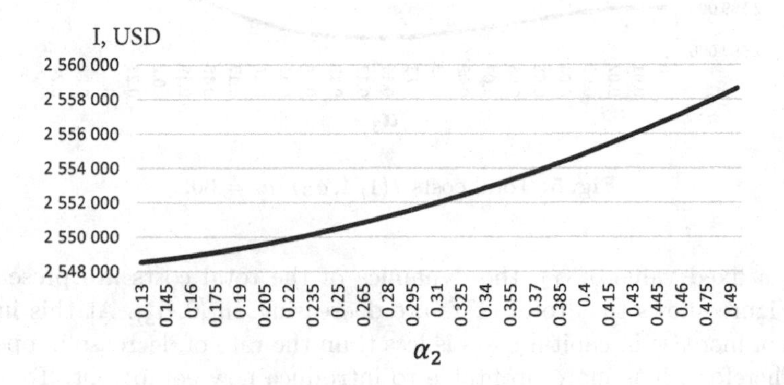

Fig. 3. Total costs $I(\alpha_2, 1/8)$, $u_c = 45$.

Now fix α_2. Figure 4 shows the dynamics of total costs for the case $u_c = 45$ with the changing parameter α_3. It can be seen that the optimal value of α_3^* tends to 0. This means that with a decrease of a lifetime (an increase of α_3), the capital costs of commissioning new capacities grow faster than the operating costs decrease. With these economic indicators, it is beneficial not to dismantle capacities.

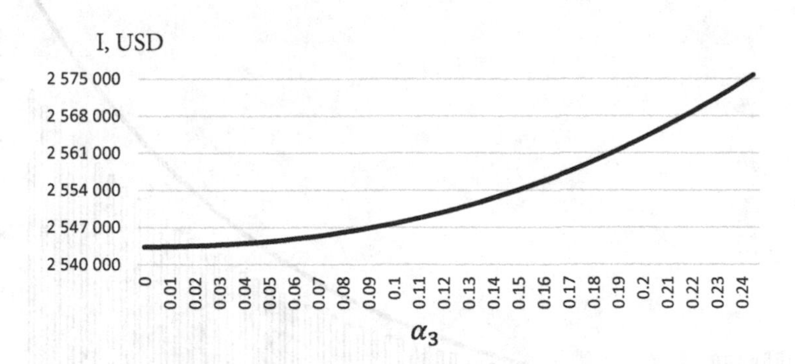

Fig. 4. Total costs $I(1/4, \alpha_3)$, $u_c = 45$.

Remark 1. In Example 1, set $u_c = 30$, i.e. we suppose that operating costs increase faster. For the optimization option α_2 with a fixed value of α_3, we get the dynamics of the total costs similar one to Fig. 3. For the optimization option

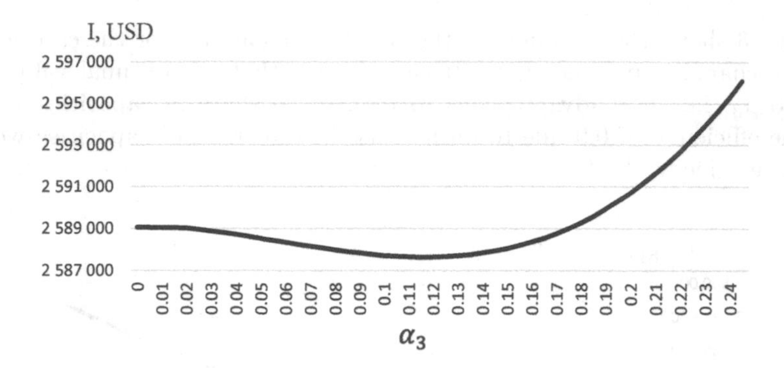

Fig. 5. Total costs $I(1/4, \alpha_3)$, $u_c = 30$.

α_3 with a fixed value of α_2, the dynamics of the total costs are presented in Fig. 5. Figure shows that total costs are decreasing on $[0, \alpha_3^*]$. At this interval, the rate of increase in capital costs is less than the rate of decrease in operating costs. Therefore, it is more profitable to introduce new equipment. Total costs increase on $[\alpha_3^*, \alpha_2]$.

Example 2. Now suppose that the right-hand side has the form $y(t) = 1.05^t y_1$, $T = 60$. It is similar to the right-hand side of the model (6) based on real-life data. The value of $y_1 = 100$ (MW), which is close to the real-life situation.

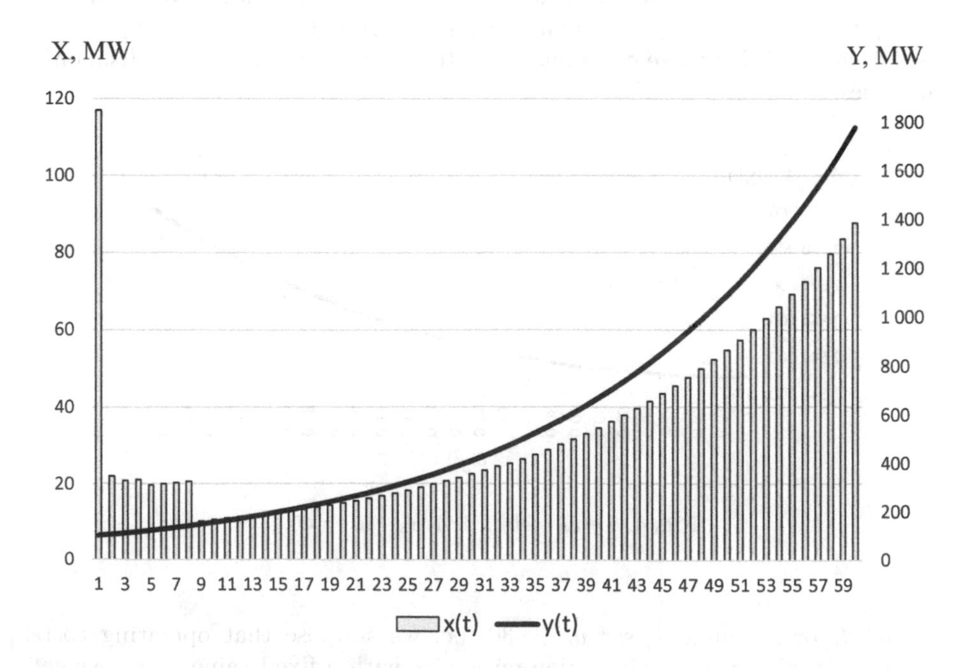

Fig. 6. Commissioning capacities (base variant).

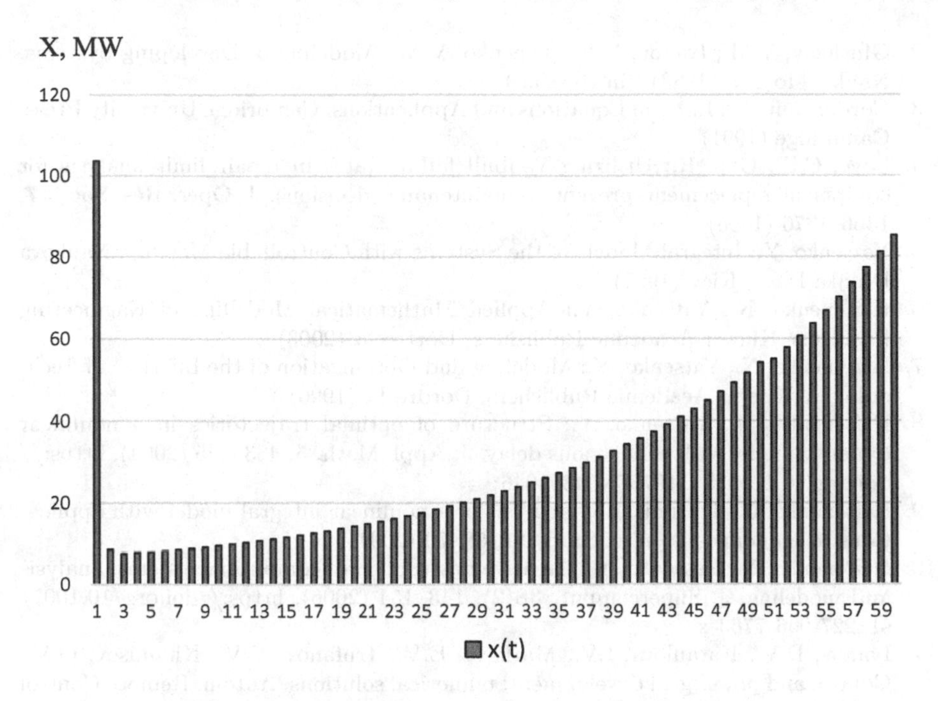

Fig. 7. Commissioning capacities for $(\alpha_2^*, \alpha_3^*) = (0.025, 0.02)$.

Figure 6 demonstrates the dynamics of commissioning capacities $x(t)$ and $y(t)$, corresponding to the basic case of Example 2. The jump in inputs in the first forecast year is necessary to meet the power requirement of y_1.

Figure 7 shows a graph of the solution $x^*(t)$ to a forecast problem, corresponding to the optimal $(\alpha_2^*, \alpha_3^*) = (0.025, 0.02)$. The economic benefit with respect to the basic variant will have been 3% by 60.

5 Conclusion

The problem of optimizing the lifetime of capacities in the integral model of the EPS development has been considered. The model describes a system consisting of n age groups. The moment of the beginning of modeling coincides with the moment of the system creation. To solve the optimization problem, a heuristic algorithm based on the model discretization is used. For the study of the problem, model examples typical for the real-life situation in the EPS of Russia are considered. The obtained numerical results show the adequacy of the model.

References

1. Glushkov, V.M.: On one class of dynamic macroeconomic models. Control Syst. Mach. **2**, 3–6 (1977). (in Russian)

2. Glushkov, V.M., Ivanov, V.V., Yanenko, V.M.: Modeling of Developing Systems. Nauka, Moscow (1983). (in Russian)
3. Corduneanu, C.: Integral Equations and Applications. Cambridge University Press, Cambridge (1991)
4. Love, C.E., Guo, R.: Utilizing Weibull failure rates in repair limit analysis for equipment replacement/preventive maintenance decisions. J. Oper. Res. Soc. **47**, 1366–1376 (1996)
5. Yatsenko, Y.: Integral Models of the Systems with Controllable Memory. Naukova Dumka Press, Kiev (1991)
6. Hritonenko, N., Yatsenko, Y.: Applied Mathematical Modelling of Engineering Problems. Kluwer Academic Publishers, Dortrecht (2003)
7. Hritonenko, N., Yatsenko, Y.: Modeling and Optimization of the Lifetime of Technologies. Kluwer Academic Publishers, Dordrecht (1996)
8. Hritonenko, N., Yatsenko, Y.: Structure of optimal trajectories in a nonlinear dynamic model with endogenous delay. J. Appl. Math. **5**, 433–445 (2004). https://doi.org/10.1155/S1110757X04311046
9. Hritonenko, N.: Optimization analysis of a nonlinear integral model with applications to economics. Nonlin. Stud. **11**, 59–70 (2004)
10. Hritonenko, N., Yatsenko, Y.: Creative destruction of computing systems: analysis and modeling. J. Supercomput. **38**(2), 143–154 (2006). https://doi.org/10.1007/s11227-006-7763-x
11. Ivanov, D.V., Karaulova, I.V., Markova, E.V., Trufanov, V.V., Khamisov, O.V.: Control and power grid development: numerical solutions. Autom. Remote Control **65**(3), 472–482 (2004)
12. Apartsyn, A.S., Karaulova, I.V., Markova, E.V., Trufanov, V.V.: Application of Volterra integral equations for modeling strategies of power industry technical re-equipment. Electrichestvo **10**, 69–75 (2005). (in Russian)
13. Karaulova, I.V., Markova, E.V.: Optimal control problem of development of an electric power system. Autom. Remote Control **69**(4), 637–644 (2008). https://doi.org/10.1134/S0005117908040103
14. Markova, E.V., Sidler, I.V., Trufanov, V.V.: On models of developing systems and their applications. Autom. Remote Control **72**(7), 1371–1379 (2011). https://doi.org/10.1134/S0005117911070046
15. Apartsyn, A.S.: Nonclassical Linear Volterra Equations of the First Kind. VSP, Utrecht, Boston (2003)
16. Apartsin, A.S., Sidler, I.V.: Using the nonclassical Volterra equations of the first kind to model the developing systems. Autom. Remote Control **74**(6), 899–910 (2013). https://doi.org/10.1134/S0005117913060015
17. Apartsin, A.S., Sidler, I.V.: Integral models development of electric power systems with allowance for ageing of equipments of electric power plants. Electron. Model. **4**, 81–88 (2014). (in Russian)
18. Trufanov, V.V., Apartsin, A.S., Markova, E.V., Sidler, I.V.: Integral models for the development of technical modernization of generating capacities strategy. Electrichestvo **3**, 4–11 (2017). (in Russian, Abstr. in Engl.)
19. Apartsin, A.S., Sidler, I.V.: On the test Volterra equations of the first kind in the integral models of developing systems. Autom. Remote Control **79**(4), 604–616 (2018). https://doi.org/10.1134/S0005117918040033
20. Apartsin, A.S., Sidler, I.V.: Study of test Volterra equations of the first kind in integral models of developing systems. Trudy Inst. Mat. i Mekh. UrO RAN **24**(2), 24–33 (2018). https://doi.org/10.21538/0134-4889-2018-24-2-24-33. (in Russian, Abstr. in Engl.)

21. Apartsyn, A.S., Sidler, I.V.: The test Volterra equation of the first kind in integral models of developing systems containing n age groups. Tambov Univ. Rep. Series: Nat. Tech. Sci. **23**(122), 168–179 (2018). https://doi.org/10.20310/1810-0198-2018-23-122-168-179. (in Russian, Abstr. in Engl.)
22. Apartsyn, A.S., Markova, E.V., Sidler, I.V.: Integral model of developing system without prehistory. Tambov Univ. Rep. Series: Nat. Tech. Sci. **23**(123), 361–367 (2018). https://doi.org/10.20310/1810-0198-2018-23-123-361-367. (in Russian, Abstr. in Engl.)
23. Apartsin, A.S.: To a study on stability of solutions to the test nonclassical Volterra equations of the first kind. Siberian Electron. Math. Rep. **12**(S), 15–20 (2015). (in Russian, Abstr. in Engl.)
24. Apartsyn, A.S., Sidler, I.V.: Numerical solution of the Volterra equations of the first kind in integral models of developing systems. In: Proceedings of the VII International Symposium on Generalized Statements and Solutions of Control Problems, GSSCP - 2014, pp. 21–25. ANO, Moscow (2014). (in Russian)
25. Apartsyn, A.S., Sidler, I.V.: On the numerical solution of the nonclassical Volterra equations of the first kind. In: Proceedings of the 9th International Conference on Analytical and Numerical Methods of Modeling of Natural Science and Social Problems, pp. 59–64. Penza State University, Penza (2014). (in Russian)
26. Apartsyn, A.S., Markova, E.V., Sidler, I.V., Trufanov, V.V.: On age structure control in integral model of EPS of Russia. Tambov Univ. Rep. Series: Natural Tech. Sci. **20**(5), 1006–1009 (2015). (in Russian)

A Modified Duality Scheme for Solving a 3D Elastic Problem with a Crack

Robert Namm and Georgiy Tsoy(✉)

Computing Center of Far Eastern Branch Russian Academy of Sciences, Kim Yu Chen 65, 680000 Khabarovsk, Russia
rnamm@yandex.ru, tsoy.dv@mail.ru
http://www.ccfebras.ru

Abstract. We consider an equilibrium problem for a 3D elastic body with a crack. Inequality-type boundary conditions are considered at the crack faces to prevent mutual penetration between them. This leads to the formulation of a problem with an unknown contact area, which admits a variational formulation in the form of a problem of minimization of energy functional in a set of feasible displacements. To solve the problem, we consider the Uzawa algorithm based on the modified Lagrange functional and compare it with the classical analog. Numerical results illustrating the efficiency of the proposed algorithm are presented.

Keywords: Elastic problem · Crack · Duality scheme · Modified Lagrange functional · Uzawa algorithm

1 Introduction

We consider a 3D elastic body with a planar crack. As it is known, classical linear models of the crack theory are characterized by linear boundary conditions on the crack faces. From the standpoint of mechanics, these models have the defect that the opposite faces of the crack can penetrate each other. In order to guarantee the mutual nonpenetration, the nonlinear boundary conditions of inequality type should be imposed on the crack faces, which leads to a nonlinear boundary value problems. The analysis of such models can be found in works [1–6].

The variational formulation of these models consists in minimizing the potential energy's functional on a closed convex subset of the original Hilbert space, which, in turn, is equivalent to a variational inequality. To solve variational problems we use the duality methods, in which, simultaneously with the solution of the original problem, the solution of the dual problem is found [4,5,7]. However, in an elastic problem with a crack, the regularity of the solution in the vicinity of the crack ends may be insufficient for the solvability of the dual problem. In this connection, the authors propose to use modified duality methods, the effectiveness of which is justified theoretically and demonstrated in numerical experiments.

This study was supported by the Russian Foundation for Basic Research (Project 17-01-00682 A).

I. Bykadorov et al. (Eds.): MOTOR 2019, CCIS 1090, pp. 536–547, 2019.
https://doi.org/10.1007/978-3-030-33394-2_41

2 Problem Statement

Let $\Omega \subset R^3$ be a bounded domain with a Lipschitz boundary Γ, and let $\gamma \subset R^3$ be a crack. We consider the following rectangular geometry of a solid in the unit cube $\{0 < x_i < 1,\ i = 1, 2, 3\}$ with a planar crack $\gamma = \{0.25 < x_1 < 1, 0 < x_2 < 1, x_3 = 0.5\}$. Denote by $\Gamma_D = \{x_1 = 0, 0 < x_i < 1,\ i = 2, 3\}$ the part of Γ where the body is clamped and by $\Gamma_N^{\pm}$ the parts of Γ where body is loaded by a surface force. In Fig. 1, the parts $\Gamma_N^{\pm}$ are marked and the arrows indicate the directions of the acting forces. Let $\nu = (0, 0, 1)$ be a unit normal vector on γ. According to the vector ν, denote the positive (upper) face of the crack γ by γ^+ and the negative (lower) face by γ^-. Suppose that $\Omega_\gamma = \Omega \setminus \overline{\gamma}$.

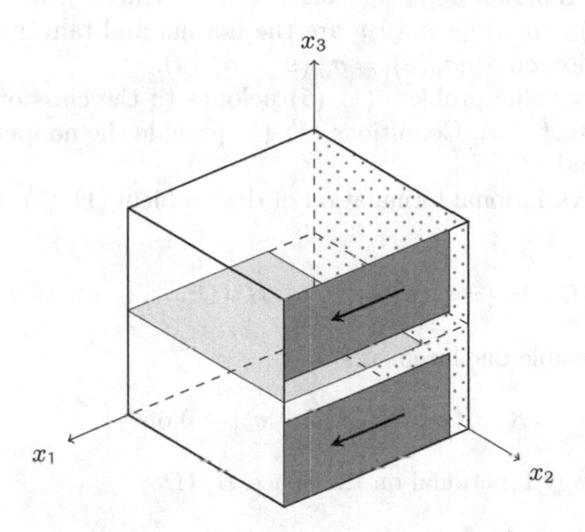

Fig. 1. Geometry and loading of an elastic body with a crack

For the displacement vector $u = (u_1, u_2, u_3)$, we introduce the stress tensor $\sigma = \{\sigma_{ij}\}$ and the strain tensor $\varepsilon = \{\varepsilon_{ij}\}$ which are related by the linear Hooke law:

$$\sigma_{ij}(u) = c_{ijkm}\varepsilon_{km}(u),\ \varepsilon_{ij}(u) = \frac{1}{2}\left(\frac{\partial u_i}{\partial x_j} + \frac{\partial u_j}{\partial x_i}\right),\ i, j, k, m = 1, 2, 3.$$

Here, c_{ijkm} are the components of the elasticity tensor with usual properties of symmetry and positive definiteness

$$c_{ijkm} = c_{jimk} = c_{kmij},\ c_{ijkm}\xi_{km}\xi_{ij} \geq c_0|\xi|^2 \quad \forall \xi_{ij} = \xi_{ji}, \quad c_0 = const > 0.$$

Summation over repeated indices is assumed.

Let us specify the vector-function of the body and surface force $f \in L_2(\Omega_\gamma)^3$ and $p \in L_2(\Gamma_N)^3$, respectively.

We consider the following boundary value problem describing the equilibrium of an elastic body Ω_γ with a crack [2,3].

$$-\operatorname{div}\sigma(u) = f \text{ in } \Omega_\gamma, \tag{1}$$

$$u = 0 \text{ on } \Gamma_D, \tag{2}$$

$$\sigma(u)n = p \text{ on } \Gamma_N, \tag{3}$$

$$[u_\nu] \geq 0, \; [\sigma_\nu(u)] = 0, \; \sigma_\nu(u)[u_\nu] = 0 \text{ on } \gamma, \tag{4}$$

$$\sigma_\nu(u) \leq 0, \; \sigma_\tau(u) = 0 \text{ on } \gamma^\pm. \tag{5}$$

Here $n = (n_1, n_2, n_3)$ is the unit outward normal vector to Γ, $[u_\nu] = u_\nu^+ - u_\nu^-$ is a jump of the function $u_\nu = u_i\nu_i$ on γ; $u_\nu^\pm$ are traces of u_ν on $\gamma^\pm$; $\sigma_\nu(u) = \sigma_{ij}(u)\nu_i\nu_j$, $\sigma_\tau(u) = \sigma(u)\nu - \sigma_\nu(u)\nu$ are the normal and tangent components of the surface traction on γ; $[\sigma_\nu(u)] = \sigma_\nu^+(u) - \sigma_\nu^-(u)$.

The boundary value problem (1)–(5) belongs to the class of problems with an unknown contact area. Conditions (4)–(5) provide the nonpenetration of the crack faces γ^+ and γ^-.

Let us give a variational formulation of the problem (1)–(5). Define the functional space

$$H_\Gamma^1(\Omega_\gamma) = \left\{v = (v_1, v_2, v_3) \in H^1(\Omega_\gamma)^3 \mid v = 0 \text{ on } \Gamma_D\right\}$$

and the set of feasible displacements

$$K = \left\{v \in H_\Gamma^1(\Omega_\gamma) \mid [v_\nu] \geq 0 \text{ on } \gamma\right\}.$$

Consider the energy functional on the space $H_\Gamma^1(\Omega_\gamma)$

$$J(v) = \frac{1}{2}\int\limits_{\Omega_\gamma} \sigma_{ij}\varepsilon_{ij}(v)\,d\Omega - \int\limits_{\Omega_\gamma} f_i v_i\,d\Omega - \int\limits_{\Gamma_N} p_i v_i\,d\Gamma.$$

The boundary value problem (1)–(5) corresponds to the following variational problem [2,5]:

$$\begin{cases} J(v) \to \min, \\ v \in K. \end{cases} \tag{6}$$

Thus, we need to find a function $u \in K$ such that

$$J(u) = \inf_{v \in K} J(v).$$

It is known that there exists a unique solution $u \in K$ of the problem (6), which satisfies Eq. (1) and boundary conditions (4)–(5) in a weak sense [2,3].

3 Classical and Modified Duality Schemes

To solve problem (6), we introduce the classical Lagrange functional on the space $H^1_\Gamma(\Omega_\gamma) \times L_2(\gamma)$

$$L(v,l) = J(v) + \int\limits_\gamma l(-[v_\nu])\, d\Gamma.$$

Denote by $(L_2(\gamma))^+$ the set of nonnegative on γ square integrable functions.

Definition 1. *A pair $(v^*, l^*) \in H^1_\Gamma(\Omega_\gamma) \times (L_2(\gamma))^+$ is called a saddle point of the Lagrange functional $L(v,l)$ if the following two-sided inequality takes place [5, 7, 12]*

$$L(v^*,l) \le L(v^*,l^*) \le L(v,l^*) \quad \forall (v,l) \in H^1_\Gamma(\Omega_\gamma) \times (L_2(\gamma))^+.$$

If (v^*, l^*) is the saddle point of $L(v,l)$, then v^* is a solution of the problem (6) and l^* is a solution of the corresponding dual problem

$$\begin{cases} \underline{L}(l) \to \sup, \\ l \in (L_2(\gamma))^+, \end{cases} \tag{7}$$

where

$$\underline{L}(l) = \inf_{v \in H^1_\Gamma(\Omega_\gamma)} L(v,l).$$

Note that the dual problem (7) is solvable if the solution u of the initial problem belongs to the space $[H^2(\Omega_\gamma)]^3$.

Since measure $\mu(\Gamma_D) > 0$, then the Lagrange functional $L(v,l)$ is strongly convex with respect to $v \in H^1_\Gamma(\Omega_\gamma)$ for fixed $l \in L_2(\gamma)$. Therefore,

$$\inf_{v \in H^1_\Gamma(\Omega_\gamma)} L(v,l) = \min_{v \in H^1_\Gamma(\Omega_\gamma)} L(v,l).$$

Also, we define the modified Lagrange functional on the space $H^1_\Gamma(\Omega_\gamma) \times L_2(\gamma)$ [4]

$$M(v,l) = J(v) + \frac{1}{2r} \int\limits_\gamma \left(\left[(l - r[v_\nu])^+ \right]^2 - l^2 \right) d\Gamma,$$

where $r > 0$ is an arbitrary positive constant.

Definition 2. *A pair $(v^*, l^*) \in H^1_\Gamma(\Omega_\gamma) \times L_2(\gamma)$ is called a saddle point of the Lagrange functional $M(v,l)$ if the following two-sided inequality takes place:*

$$M(v^*,l) \le M(v^*,l^*) \le M(v,l^*) \quad \forall (v,l) \in H^1_\Gamma(\Omega_\gamma) \times L_2(\gamma).$$

It can be shown that sets of saddle points of the classical Lagrange functional and the modified one coincide [5, 7]. So the second component of the saddle point l^* is the solution to the corresponding dual problem:

$$\begin{cases} \underline{M}(l) \to \sup, \\ l \in L_2(\gamma), \end{cases} \tag{8}$$

where

$$\underline{M}(l) = \inf_{v \in H^1_\Gamma(\Omega_\gamma)} M(v,l).$$

We consider the following Uzawa algorithm (9)–(10) to determine the saddle point of the classical Lagrange functional [8–11,14]. At the initial step, $k=0$, specify an arbitrary function $l^0 \in (L_2(\gamma))^+$, then:

$$\text{find } u^k \in H^1_\Gamma(\Omega_\gamma) : L(u^k, l^k) \le L(v, l^k) \ \forall v \in H^1_\Gamma(\Omega_\gamma), \tag{9}$$

$$l^{k+1} = \mathrm{P}_{(L_2(\gamma))^+}(l^k - \rho\kappa[u^k_\nu]) = (l^k - \rho\kappa[u^k_\nu])^+, \tag{10}$$

where $\mathrm{P}_{(L_2(\gamma))^+}$ is the projection operator of $L_2(\gamma)$ onto $(L_2(\gamma))^+$ with respect to the norm in $L_2(\gamma)$, $\rho > 0$, κ is the trace operator from $H^1_\Gamma(\Omega_\gamma)$ to $L_2(\gamma)$, $(l^k - \rho\kappa[u^k_\nu])^+ = \max\{0, l^k - \rho\kappa[u^k_\nu]\}$.

Theorem 1. *The dual functional $\underline{L}(l)$ is Gateaux differentiable in $L_2(\gamma)$ and its derivative is given by*

$$\nabla \underline{L}(l) = -\kappa[v_\nu(l)], \ \textit{where } v(l) = \arg\min_{v \in H^1_\Gamma(\Omega_\gamma)} L(v,l),$$

herewith

$$\|\nabla \underline{L}(l^{'}) - \nabla \underline{L}(l^{''})\|_{L_2(\gamma)} \le \frac{1}{\alpha}\|l^{'} - l^{''}\|_{L_2(\gamma)} \ \ \forall l^{'}, l^{''} \in L_2(\gamma),$$

where α is the constant of strong convexity of $J(v)$.

Proof. For simplicity of notation, we omit the trace operator κ below. It is easy to see that for $\forall v \in H^1_\Gamma(\Omega_\gamma)$ it holds

$$J(v(l)) - \int_\gamma l[v_\nu(l)]\,d\Gamma + \frac{\alpha}{2}\|v - v(l)\|^2_{H^1_\Gamma(\Omega_\gamma)} \le J(v) - \int_\gamma l[v_\nu]\,d\Gamma \ \ \forall v \in H^1_\Gamma(\Omega_\gamma).$$

Denote by $\hat{v} = v(\hat{l}), \check{v} = v(\check{l})$. Then

$$J(\hat{v}) - \int_\gamma \hat{l}[\hat{v_\nu}]\,d\Gamma + \frac{\alpha}{2}\|\check{v} - \hat{v}\|^2_{H^1_\Gamma(\Omega_\gamma)} \le J(\check{v}) - \int_\gamma \hat{l}[\check{v_\nu}]\,d\Gamma,$$

$$J(\check{v}) - \int_\gamma \check{l}[\check{v_\nu}]\,d\Gamma + \frac{\alpha}{2}\|\check{v} - \hat{v}\|^2_{H^1_\Gamma(\Omega_\gamma)} \le J(\hat{v}) - \int_\gamma \check{l}[\hat{v_\nu}]\,d\Gamma. \tag{11}$$

Summing the above inequalities, we obtain

$$\alpha\|\check{v} - \hat{v}\|^2_{H^1_\Gamma(\Omega_\gamma)} \le \int_\gamma (\hat{l} - \check{l})([\check{v_\nu}] - [\hat{v_\nu}])\,d\Gamma,$$

$$\alpha\|[\check{v_\nu}] - [\hat{v_\nu}]\|^2_{L_2(\gamma)} \le \int_\gamma (\hat{l} - \check{l})([\check{v_\nu}] - [\hat{v_\nu}])\,d\Gamma,$$

$$\|[\check{v_\nu}] - [\hat{v_\nu}]\|^2_{L_2(\gamma)} \le \frac{1}{\alpha}\|\check{l} - \hat{l}\|^2_{L_2(\gamma)}. \tag{12}$$

It follows from the embedding $H^1_\Gamma(\Omega_\gamma) \subset L_2(\gamma)$ and (11) that the following two-sided inequality holds

$$\int_\gamma \hat{l}([\check{v_\nu}] - [\hat{v_\nu}])\, d\Gamma + \frac{\alpha}{2}\|[\check{v_\nu}] - [\hat{v_\nu}]\|^2_{L_2(\gamma)} \le J(\check{v}) - J(\hat{v})$$

$$\le \int_\gamma \check{l}([\check{v_\nu}] - [\hat{v_\nu}])\, d\Gamma - \frac{\alpha}{2}\|[\check{v_\nu}] - [\hat{v_\nu}]\|^2_{L_2(\gamma)}.$$

Together with (12) it provides

$$\lim_{\check{l}\to\hat{l}} J(\check{v}) = J(\hat{v}).$$

This means that the dual concave functional $\underline{L}(l)$ is continuous in $L_2(\gamma)$. Therefore, the subdifferential $\partial(-\underline{L}(l))$ of the convex functional $(-\underline{L}(l))$ is a non-empty set for any $l \in L_2(\gamma)$. To prove that $\underline{L}(l)$ is differentiable, it suffices to verify that $\partial(-\underline{L}(l))$ consists of a single element. This element will be the derivative of the functional $\underline{L}(l)$ [9].

Let $l \in L_2(\gamma)$ be a fixed element and $t \in \partial(-\underline{L}(l))$. Then, for any $m \in L_2(\gamma)$, it holds

$$\underline{L}(m) \le \underline{L}(l) - \int_\gamma t(m - l)\, d\Gamma,$$

i.e.

$$J(v(m)) - \int_\gamma m[v_\nu(m)]\, d\Gamma \le J(v(l)) - \int_\gamma l[v_\nu(l)]\, d\Gamma - \int_\gamma t(m - l)\, d\Gamma$$

$$\le J(v(m)) - \int_\gamma l[v_\nu(m)]\, d\Gamma - \int_\gamma t(m - l)\, d\Gamma,$$

$$\int_\gamma [v_\nu(m)](l - m)\, d\Gamma + \int_\gamma t(m - l)\, d\Gamma \le 0,$$

$$\int_\gamma ([v_\nu(m)] - t)(l - m)\, d\Gamma \le 0.$$

We set $m = l - \beta p$, $\beta > 0$, where $p \in L_2(\gamma)$ is an arbitrary function. Then

$$\int_\gamma ([v_\nu(l - \beta p)] - t)\beta p\, d\Gamma \le 0 \quad \forall p \in L_2(\gamma),$$

$$\beta \int_\gamma ([v_\nu(l - \beta p)] - t)p\, d\Gamma \le 0,$$

$$\int_\gamma ([v_\nu(l - \beta p)] - t)p\, d\Gamma \le 0.$$

Passing to the limit as $\beta \to 0$, we deduce from (12) that $t = [v_\nu(l)]$.

□

Thus, at step (10) we obtain the gradient projection method.

It should be noted that the Uzawa algorithm with the classical Lagrange functional converges with respect to the primal variable v and its convergence is ensured only by a sufficiently small step size ρ for the dual variable l, namely if the step size is chosen according to the rule

$$0 < \rho < \frac{2}{\|\kappa\|^2},$$

where $\|\kappa\|$ is the norm of the trace operator. In general, the value of this norm is rather difficult to determine.

Using the property of weak lower semicontinuity of the sensitivity functional [4], we can show that functional $\underline{M}(l)$ is Gateaux differentiable in $L_2(\gamma)$, and its derivative $\nabla \underline{M}(l)$ satisfies the Lipschitz condition with constant $\frac{1}{r}$, i.e., for any $l', l'' \in L_2(\gamma)$, it holds that

$$\|\nabla \underline{M}(l') - \nabla \underline{M}(l'')\|_{L_2(\gamma)} \le \frac{1}{r}\|l' - l''\|_{L_2(\gamma)} \quad \forall\, l', l'' \in L_2(\gamma).$$

and subdifferential $\underline{M}(l)$ consists of the single element $\partial \underline{M}(l) = m(l)$

$$m(l) = \max\{-[u_\nu], -\frac{l}{r}\} \;\forall l \in L_2(\gamma).$$

Then the Uzawa algorithm with the modified Lagrange functional can be described as follows:

$$\text{find } u^k \in H^1_\Gamma(\Omega_\gamma) : M(u^k, l^k) \le M(v, l^k) \;\forall v \in H^1_\Gamma(\Omega_\gamma), \tag{13}$$

$$l^{k+1} = l^k + r m(l^k) = (l^k - r[u^k_\nu])^+. \tag{14}$$

So we obtain the gradient method at the step (14).

Thus, we have shown that the modified Lagrange functionals allow us to solve a constrained minimization problems of type (6) more efficiently than their classical analogs. First, they allow proving the theoretical convergence of the Uzawa algorithm both with respect to the primal and dual variables. Secondly, to find the second component of the saddle point of the Lagrange functional, (pure) gradient method is used, which has a more faster convergence rate than the gradient projection method.

4 Numerical Experiment

For the numerical solution of the problem, we use the finite element method. We discretize the domain Ω_γ with a crack by a nearly uniform triangulation and apply standard piecewise affine basis functions. In zones of the maximum stress (near the crack), the mesh was thickened to improve the accuracy of the numerical solution. The domain Ω_γ was subdivided into 1749882 tetrahedrons with $N = 310636$ nodes. By $h = 0.01$ we denote the corresponding mesh size on γ.

As in [1], we assume that a volume load $f = 0$, the solid occupying the domain Ω_γ is clamped at Γ_D and it is loaded by a traction force at $\Gamma_N^{\pm}$. The remaining part of the boundary of Ω is assumed to be stress-free. The boundary loading is taken as in Example 2 in [1]

$$-\sigma_{12}(u) = -g, \quad \sigma_{22}(u) = \sigma_{32}(u) = 0 \text{ on } \Gamma_N^{\pm},$$
$$\Gamma_N^{+} = \{0.1 < x_1 < 1, x_2 = 1, 0.6 < x_3 < 1\},$$
$$\Gamma_N^{-} = \{0.1 < x_1 < 1, x_2 = 1, 0 < x_3 < 0.4\},$$

with constant $g = 27$ MPa, the Young elasticity modulus $E = 73000$ MPa and the Poisson coefficient $\mu = 0.34$.

After a suitable discretization of (6) we have the stiffness matrix $A = (a_{ij}) \in \mathbb{R}^{3N\times 3N}$, the column vector of right side $F = (f_i)^T \in \mathbb{R}^{3N}$, vector $t = (t_1, \ldots, t_{3N})^T \in \mathbb{R}^{3N}$ of unknowns, an assembling in an appropriate way of the components of the displacement vector $(u_1(x^m), u_2(x^m), u_3(x^m))^T$ at the grid points x^m with $m = \overline{1, N}$, $\alpha^k = (\alpha_1^k, \cdots, \alpha_{N_\gamma}^k)$, the approximate value of the dual variable l^k. Then the finite-dimensional minimization problem is reduced to finding the optimal values of t_i. For this, we use the generalized Newton method [13].

Let us introduce the gradient $g(t)$ of the corresponding finite-dimensional functional

$$g(t) = At - F + \beta(t), \tag{15}$$

where $\beta(t) = (b_i)^T \in \mathbb{R}^{3N}$. For convenience, we denote by $\{i_j^+\}, \{i_j^-\}_{j=1,N_\gamma}$ the numbers of the nodes lying respectively on the upper and lower faces of the crack. Then

$$b_i = 0, \text{ for all } i \neq \{i_j^+ + 2N, i_j^- + 2N\}.$$

Otherwise, if

$$\lambda_j(t) = \alpha_j^k - r(t_{i_j^+ + 2N} - t_{i_j^- + 2N}),$$

then

$$b_{i_j^+ + 2N} = -\delta_j(\lambda_j(t))^+,$$
$$b_{i_j^- + 2N} = \delta_j(\lambda_j(t))^+.$$

Here $\delta = (\delta_j)^T \in \mathbb{R}^{N_\gamma}$ is the vector of coefficients obtained after discretization of the boundary integral over γ.

The generalized Newton method is applied in the following way:

(1) At the initial step, set t^0.
(2) For every $m = 0, 1, 2, \cdots$ calculate

$$t^{m+1} = t^m - (\partial g(t^m))^{-1} g(t^m). \tag{16}$$

(3) Check

$$\|t^{m+1} - t^m\|_\infty < \varepsilon_t, \varepsilon_t = 10^{-12}.$$

Here $\partial g(t)$ is the generalized Jacobian of $g(t)$:

$$\partial g(t) = A + D(t) \ ,$$

where $D(t) = (d_{ij}) \in \mathbb{R}^{3N\times 3N}$ is a symmetric sparse matrix. If $\lambda_j(t) > 0$, then matrix D will have nonzero elements:

$$d_{i_j^+ + 2N, i_j^+ + 2N} = r\delta_j, d_{i_j^+ + 2N, i_j^- + 2N} = -r\delta_j,$$
$$d_{i_j^- + 2N, i_j^- + 2N} = r\delta_j, d_{i_j^- + 2N, i_j^+ + 2N} = -r\delta_j.$$

The components of the dual variable are computed using the formula:

$$\alpha_j^{k+1} = (\alpha_j^k - r(t^k_{i_j^+ + 2N} - t^k_{i_j^- + 2N}))^+, \ j = \overline{1, N_\gamma}. \tag{17}$$

The stop criterion for the Uzawa algorithm has the following form:

$$\|\alpha^{k+1} - \alpha^k\|_\infty < \varepsilon_\alpha, \varepsilon_\alpha = 10^{-8}.$$

Numerical experiments were conducted on a hybrid computing cluster based on the OpenPOWER architecture. It should be noted that the generalized Newton method is easy and well parallelized and its main computational complexity consists in finding the inverse matrix. But for large sparse matrices, the matrix inversion is not efficient, therefore, we can replace (16) with

$$\partial g(t^m) z^m = -g(t^m),$$

where $z^m = t^{m+1} - t^m$, and solve it using the conjugate gradient method.

The calculations were performed on NVIDIA Tesla P100 GPU using the cuSPARSE, cuBLAS libraries. This allows to significantly accelerate the speed of execution compared to the CPU version.

Figure 2 shows the dependence of the number of iterations (17) on the parameter r for the classical and modified Lagrange functionals. The experiments show that with an increase of the parameter r, the number of iterations by dual variable decreases. The number of iterations of both algorithms is almost the same for relatively small r. But the Uzawa algorithm with the classical Lagrange functional ceases to converge for the values of $r > 13000$, in contrast to the modified analog.

Table 1 presents the results of the numerical solution using the modified Lagrange functional for different values of the parameter r and shows the average number of iterations (16) for the primal variable and the number of iterations (17) for the dual variable.

Table 1. Number of iterations of the Uzawa algorithm with the modified functional.

r	iter. by t	iter. by α
10^6	3	223
10^7	4	29
10^8	4	8
10^9	4	4
10^{10}	5	3

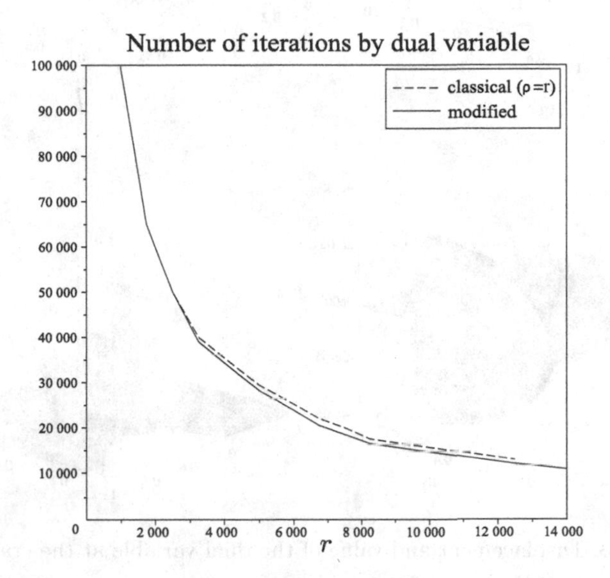

Fig. 2. Number of iterations depending on the parameter r.

It can be seen that the generalized Newton method converges in a small number of iterations and as the parameter r increases, the number of iterations for the dual variable decreases significantly.

The results of the numerical experiments are presented graphically in Fig. 3. The graphs show the values of the components of the displacement vector u and the dual variable l. It can be seen that there is no mutual penetration of the crack faces and the value of the dual variable is greater than zero at the points where the crack faces are stuck together. This indicates the presence of the normal stress in these nodes.

Thus, the numerical calculations confirm that the modified Lagrange functionals make it possible to efficiently solve equilibrium problems for a 3D elastic body with a planar crack.

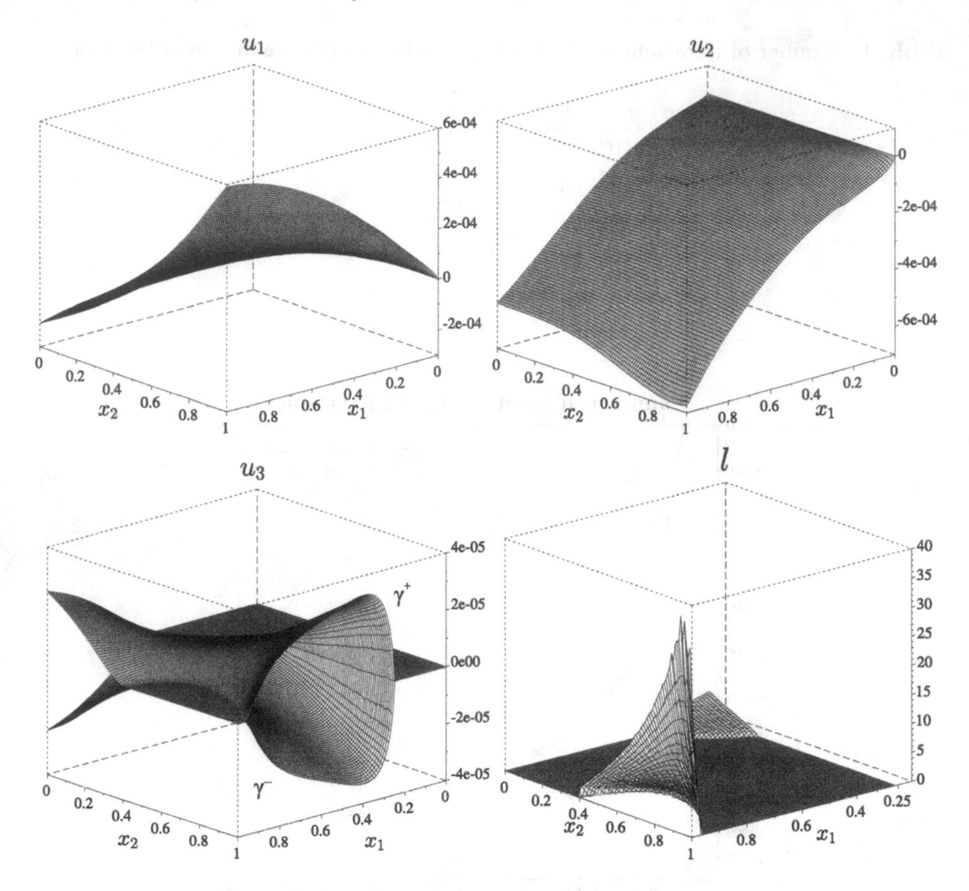

Fig. 3. Displacement and value of the dual variable at the crack.

Acknowledgements. This research was supported through computational resources provided by the Shared Facility Center "Data Center of FEB RAS" (Khabarovsk) [15].

References

1. Hintermüller, M., Kovtunenko, V.A., Kunisch, K.: A Papkovich-Neuber-based numerical approach to cracks with contact in 3D. IMA J. Appl. Math. **74**(3), 325–343 (2009)
2. Khludnev, A.M.: Problems of Elasticity Theory in Nonsmooth Domains. Fizmatlit, Moscow (2010)
3. Khludnev, A.M., Kovtunenko, V.A.: Analysis of Crack in Solids. WIT-Press, Southampton (2000)
4. Namm, R., Tsoy, G., Vikhtenko, E.: A modified duality method for solving an elasticity problem with a crack extending to the outer boundary. Commun. Comput. Inf. Sci. **974**, 35–48 (2019)
5. Namm, R.V., Tsoy, G.I.: A modified dual scheme for solving an elastic crack problem. Num. Anal. Appl. **10**(1), 37–46 (2017)

6. Rudoy, E.M.: Domain decomposition method for a model crack problem with a possible contact of crack edges. Comput. Math. Math. Phys. **55**(2), 305–316 (2015)
7. Vikhtenko, E.M., Maksimova, N.N., Namm, R.V.: Modified Lagrange functionals to solve the variational and quasivariational inequalities of mechanics. Autom. Remote Control **73**(4), 605–615 (2012)
8. Bertsekas, D.: Constrained Optimization and Lagrange Multiplier Methods. Academic Press, New York (1982)
9. Ekeland, I., Temam, R.: Convex Analysis and Variational Problems. North-Holland, Amsterdam (1976)
10. Glowinski, R., Lions, J.L., Tremoliers, R.: Numerical Analysis of Variational Inequalities. North-Holland, Amsterdam (1981)
11. Gol'shtein, E.G., Tret'yakov, N.V.: Modified Lagrangian Functions: Theory and Optimization Methods. Nauka, Moscow (1989)
12. Hlavačhek, I., Haslinger, Y., Nechas, I., Lovišhek, Y.: Numerical Solution of Variational Inequalities. Springer, New York (1988). https://doi.org/10.1007/978-3-8348-9546-2
13. Mangasarian, O.L.: A generalized Newton method for absolute value equations. Optim. Lett. **3**(1), 101–108 (2009)
14. Polyak, B.T.: Introduction to Optimization. Optimization Software, Publications Division, New York (1987)
15. Sorokin, A.A., Makogonov, S.V., Korolev, S.P.: The information infrastructure for collective scientific work in the far east of Russia. Sci. Tech. Inform. Process. **44**(4), 302–304 (2017)

Control of the Oscillations Through Nonlinear Interactions

Lev F. Petrov(✉)

Plekhanov Russian University of Economics, Stremianniy per., 36, 115998 Moscow, Russia
lfp@mail.ru

Abstract. The use of nonlinear effects in the control of stationary oscillations in nonlinear dynamic systems is considered. To find periodic solutions of corresponding ordinary differential equation systems, an interactive algorithm is used, based on minimizing the solution's deviation from the periodic form. The possibility of the system behavior controlling due to the mutual nonlinear influence of various types of oscillations is considered. For nonlinear dynamical systems with one and more degrees of freedom, examples of various types of oscillations control are given.

Keywords: Oscillations control · Nonlinear dynamical systems · Evolution of solutions

1 Introduction

The mathematical apparatus of nonlinear dynamical systems is a common tool in many branches of science: physics, mechanics, oscillation theory, biology, economics, and others [1,2].

In constructing of mathematical models that correspond to the initial problem, various systems of ordinary differential equations are used: conservative and dissipative, autonomous and non-autonomous, systems with a different number of degrees of freedom. In such systems, oscillations of various kinds can arise: forced [3], parametric [4], self-oscillations [5], resonant [6] and non-resonant [7], chaos [8], nonlinear interaction of various oscillation modes [9].

Mathematical models of dynamic systems have historically gone through several stages of development. The first results were obtained using models based on linear differential equations. Further development of dynamical systems studies is associated with the analysis of quasi-linear differential equations. Modern approaches to the analysis of dynamic systems are based on the study of essential nonlinear systems of differential equations. It should be noted that in the English-language scientific literature, different variants of terms are used: essentially, strongly, highly nonlinear dynamic system.

From the whole variety of approaches and models, we will consider the deterministic dynamical systems based on ordinary strongly nonlinear differential equations. We will study the control of various steady-state oscillations in such

I. Bykadorov et al. (Eds.): MOTOR 2019, CCIS 1090, pp. 548–561, 2019.
https://doi.org/10.1007/978-3-030-33394-2_42

systems, taking into account the nonlinear interaction of various types of oscillations and the possibility of chaotic solutions in deterministic systems.

As the control object, we will consider not only the amplitude and frequency of a periodic solution but also its other characteristics, such as components of Fourier series, stability, solutions evolution under the influence of control actions.

The system parameters and (or) external influence are considered as the solution control tools. Also, as the solution type control tools, we will consider the nonlinear effects and nonlinear interaction between oscillations of the different type in systems with one or more degrees of freedom.

2 Variety of Periodic Solutions in Strongly Nonlinear Dynamical Systems

We shall consider both stable and unstable periodic solutions $x(t)$ of strongly nonlinear systems of ordinary differential equations. The external influences, if they exist, are considered to be periodic with a known period T.

The periodic solutions in the systems under consideration are generated by various factors. First of all, let us single out autonomous and non-autonomous systems in which there are periodic solutions that are different in their nature.

2.1 Autonomous Systems

For autonomous systems without external influences, the oscillation period is determined by the system properties, this period is not known in advance. This type of periodic solution is called self-oscillation (auto-oscillation, self-excited vibration).

For a nonlinear system of ordinary differential equations in the form

$$\frac{dx_i}{dt} = X_i(x_1(t), x_2(t), ..., x_n(t)), i = 1, 2, 3, ..., n, \tag{1}$$

where $X_i(x_1(t), x_2(t), ..., x_n(t)), i = 1, 2, 3, ..., n$ are known functions of their arguments, explicitly time-independent, we are looking for periodic solution $x_i(t) = x_i(t + T)(i = 1, 2, 3, ..., n)$, T being an unknown period.

2.2 Non-autonomous Systems

For a nonlinear system of ordinary differential equations in the form

$$\frac{dx_i(t)}{dt} = X_i(x_1(t), x_2(t), ..., x_n(t), t), i = 1, 2, 3, ..., n, \tag{2}$$

where $X_i(x_1(t), x_2(t), ..., x_n(t), t)$ are known functions of their arguments and

$$X_i(x_1(t), x_2(t), ..., x_n(t), t) = X_i(x_1(t), x_2(t), ..., x_n(t), t + T),$$

T is a known period, we are looking for one or more periodic solutions $x_i(t)$ with period kT:

$$x_i(t) = x_i(t + kT)(\text{k - given number, k} = 1,\ 2,\ ...,\ k < \infty).$$

In this variant, the rational choice of the solution's period multiplicity k is of great importance. In case of an unsuccessful choice k, the user has the option to change its value interactively. After determining the appropriate initial conditions $(Y_1, Y_2, ..., Y_n)$, the periodic solution is computed numerically for one period kT.

For non-autonomous systems, the solution period is related to the period of the external influence. The simplest form of the periodic solution, which is present in linear, quasilinear, and highly nonlinear models, has a period coinciding with the external influence one.

But in nonlinear systems, the occurrence of ultraharmonic oscillations is possible, when the solution period T_s is less than the period of the external influence T and is determined by the relation $T_s = T/m, m = 1, 2, 3, ..., m < \infty$.

There is also a possibility of the subharmonic oscillations occurrence, when the solution period T_s is longer than the period of the external influence T : $T_s = T * k, k = 1, 2, 3, ..., k < \infty$.

And, finally, sub-ultraharmonic oscillations may arise when the solution's period T_s is related to the period of the external influence T by the relation $T_s = k * T/m, k = 1, 2, 3, ..., k < \infty, m = 1, 2, 3, ..., m < \infty$.

Let's consider an example. We fix $k = 12, m = 1$ (see Fig. 1). In this case, the existence of solutions with the periods $T, 2T, 3T, 4T, 6T, 12T$ is possible. However, the solutions of other periods ($5T, 7T, 8T, 9T, 10T, 11T$ and all solutions for $k > 12$) which may exist at such fixed values of $k = 12$ and $m = 1$ cannot be found. To find them, the appropriate values k and m should be selected. This problem is caused by the nonlinearity of the system. Note that all ultraharmonic periodic solutions, if exist, naturally correspond to the periods $T, 2T, 3T, ...$, and to find these ultraharmonics solutions there is no need in choosing of the appropriate solution's period.

Note also that different types of external influences in non-autonomous systems lead to the emergence of the steady oscillations of different nature, such as forced, parametric, and, possibly, to their combinations.

Finally, there may be an external periodic influence on the system, where self-oscillation are possible. In this case, a combination of forced, parametric and self-oscillations can be considered. The solution can have both a period as close to the period of self-oscillations as well as a period associated with the external influence. When changing the system's parameters and (or) the external periodic influence, the evolution of solutions may include bifurcations, the transition from stable states to unstable regimes and vice versa, the appearance of solutions with different periods. In the absence of stable periodic solutions, a strange attractor is realized in a strongly nonlinear deterministic dynamical system.

For nonlinear dynamical systems with one degree of freedom, the interaction of various types of oscillations - forced, parametric, self-oscillation, is possible.

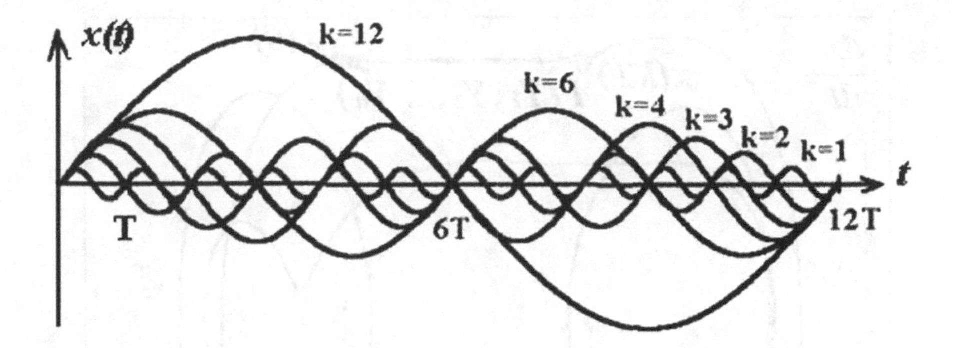

Fig. 1. Subharmonic solutions of different periods

For nonlinear dynamic systems with several degrees of freedom, the possibility of mutual influence of oscillations corresponding to different degrees of freedom is added.

3 Search for Periodic Solutions of Strongly Nonlinear Dynamical Systems

3.1 Finding Periodic Solutions by Minimizing the Discrepancy

The main idea [10] of the method for constructing periodic solutions is to find the initial conditions $Y_1 = x_1(0), Y_2 = x_2(0), ..., Y_n = x_n(0)$ of the Cauchy problem corresponding to the periodic solution. This corresponds to minimizing the distance between the beginning and the end of the trajectory on one solution period $F(Y_1, Y_2, ..., Y_n)$ (see Fig. 2), where

$$F(Y_1, Y_2, ..., Y_n) = \sqrt{\sum_{i=1}^{n} (Y_i - x_i(kT))^2}. \tag{3}$$

The Cauchy problem for one solution period is solved numerically using standard programs with a controlled error. Given a limited time interval (one period), the solution's computation in one period is realized with high accuracy. It is obvious that for a periodic solution $F(Y_1, Y_2, ..., Y_n) = 0$.

To determine the initial conditions corresponding to the periodic solution $Y_1, Y_2, ..., Y_n$, one can use optimization algorithms [11] to solve the problem:

$$F(Y_1, Y_2, ..., Y_n) \to \min \tag{4}$$

3.2 Finding Periodic Solutions by Solving a System of Nonlinear Algebraic Equations

The algorithm of constructing periodic solutions used in this work has two varieties: one for autonomous and another for non-autonomous systems of ordinary differential equations.

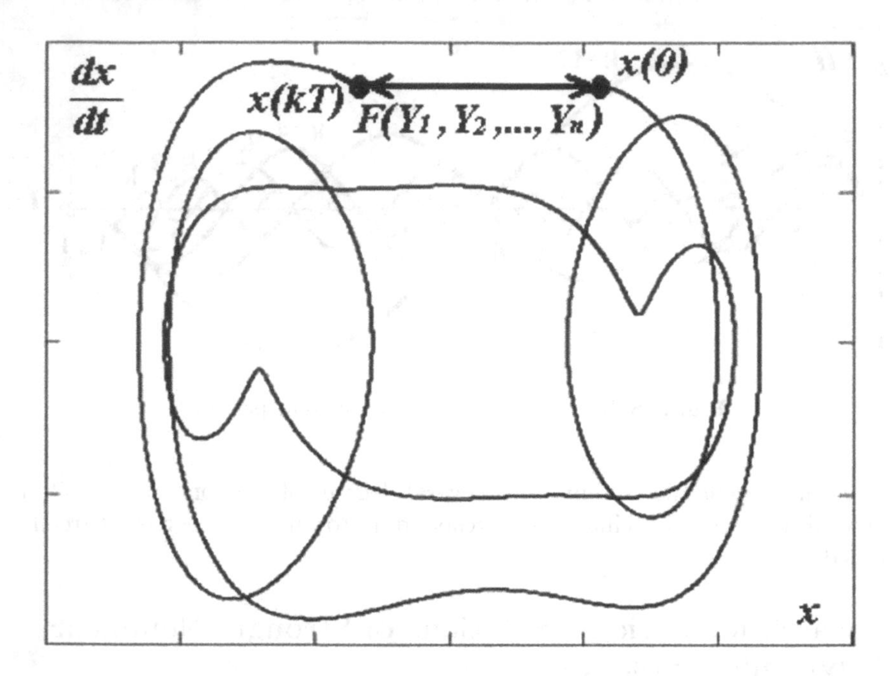

Fig. 2. Finding a periodic solution

For the autonomous system (1), in which the right side does not depend explicitly on time t, we can use the substitution $t = t_1 + c$ to choose the point $t = 0$ in such a way that the condition $Y_n = 0$ is satisfied. In this case, we seek the solution period T and the $(n-1)$ initial condition $Y_1, Y_2, ..., Y_{n-1}$, corresponding to the $T-$ periodic solution. These initial conditions $Y_1, Y_2, ..., Y_{n-1}$ and solution period T can be found from a nonlinear algebraic equation system in the form

$$\begin{cases} Y_i = x_i(T), \\ x_n(T) = 0 \end{cases} \quad i = 1, 2, ..., n-1. \tag{5}$$

To solve the system of nonlinear algebraic equations (5), one can use Newton's method, realized in an interactive form [10,12]. Partial derivatives in the software implementation of the Newton method are determined numerically.

For non-autonomous systems of ordinary differential equations, we solve a system of nonlinear algebraic equations of the form [10]

$$Y_i = x_i(kT), i = 1, 2, ..., n. \tag{6}$$

Note that the functions $x_i(kT), i = 1, 2, ..., n$ are not written analytically, but are computed at each step using the Cauchy problem's numerical solution on one period of the desired solution. Also note that the systems (5) and (6) does not decompose into separate equations.

To calculate stable and unstable periodic solutions of various types and periods, an interactive algorithm [10,12] based on finding the initial conditions corresponding to different periodic solutions is used. Thus, a unified approach has been implemented to construct periodic solutions of fundamentally different (autonomous and non-autonomous) dynamical systems. The interactive form of organizing the computational procedure allows the user to track the evolution of solutions that can not be traced automatically, for example, in the bifurcation point area. Also, the user can interactively modify the algorithm to find a solution in the case when the convergence is not guaranteed. To find the initial conditions corresponding to the periodic solution, we use Newton's method for solving a nonlinear system of algebraic equations (5) or (6) (or an optimization procedure [11] for (4)). After determining the appropriate initial conditions, the periodic solution is computed numerically in one period.

3.3 Fourier Analysis of Periodic Solutions

Given that we investigate periodic solutions, it is natural to consider their Fourier series decomposition:

$$x_i(t) = \frac{a_0}{2} + \sum_{j=1}^{N} [a_j \cos(jt) + b_j \sin(jt)]. \tag{7}$$

Initially, periodic solutions for essentially nonlinear systems of ordinary differential equations were sought in the form of the Fourier series long segments (7) [3]. Using the Galerkin procedure, the problem was reduced to a system of nonlinear algebraic equations. Even in a dynamical system with one degree of freedom, the number of the Fourier harmonics N in the solution decomposition (7) to achieve the required accuracy [13] was about hundreds. The required number of harmonics N was determined by the nonlinearity level of the system. After applying the Galerkin procedure, the system of nonlinear algebraic equations for determining the coefficients of the Fourier series also had the corresponding dimension of hundreds of equations. Even for solving one particular problem for a dynamic system with one degree of freedom, it was necessary to overcome large computational problems [3].

The approach used in this paper to find periodic solutions for strongly nonlinear systems of ordinary differential equations is fundamentally different. Thus, for example, the dimension of nonlinear algebraic equations system in the form (6) for a non-autonomous system or (5) for an autonomous system, coincides with the dimension of the original differential equation system. For a dynamical system with one degree of freedom, this dimension is two. At the same time, it is possible to calculate a sufficiently long segment of the Fourier series for the found periodic solution. The periodic solution is calculated numerically for several hundred or thousand points in one period. Using the standard Fourier expansion programs for a periodic function given numerically, we can obtain the number of Fourier coefficients which is equal to the number of points in the function approximation. Thus, using standard computational procedures, we can carry out a full Fourier analysis of periodic solutions for strongly nonlinear differential equation systems. This allows us to consider dynamical systems of higher dimension.

3.4 Stability of Periodic Solutions

In order to analyze the stability of the found kT-periodic solution [10] of the nonautonomous system (2), we consider the perturbed motion of the system

$$x_i(t) = x_i^p(t) + \delta x_i(t), \tag{8}$$

where $x_i(t)$ are the perturbed solutions, $x_i^p(t)$ are kT-periodic solutions, $\delta x_i(t)$ are small variations. Given (8) and (2), we obtain the system in variations:

$$\frac{d\delta x_i(t)}{dt} = \sum_{j=1}^{n} \delta x_j(t) \frac{\partial X_i(x_1^p(t), x_2^p(t), ..., x_n^p(t), t)}{\partial x_j}, i = 1, 2, ..., n. \tag{9}$$

We note that system (9) can be constructed not numerically, but analytically, if the analytical expressions for $X_i(x_1(t), x_2(t), ..., x_n(t), t)$ from the initial system (2) are known. The system (9) is a linear system of ordinary differential equations with kT-periodic coefficients. According to the Lyapunov theorem, the periodic solution of the system (2) is asymptotically stable if all the characteristic exponents λ_j of the variational system (9) corresponding to this periodic solution have negative real parts:

$$Re(\lambda_j) < 0, j = 1, 2, ..., n. \tag{10}$$

In the framework of the used algorithm, it is more convenient to calculate not Lyapunov's characteristic exponents λ_j, but the multipliers ρ_j.

Given the relationship between λ_j and ρ_j

$$\lambda_j = \frac{1}{kT} Ln(\rho_j) = \frac{1}{kT}[\ln(|\rho_j| + i(arg(\rho_j) + 2m\pi)], j = 1, 2, ..., n. \tag{11}$$

Here i is imaginary unit. Thus

$$Re(\lambda_j) < 0 \Leftrightarrow |\rho_j| < 1, j = 1, 2, ..., n. \tag{12}$$

The multipliers ρ_j are the eigenvalues of the monodromy matrix $M(kT)$, where $M(t)$ is the fundamental solution's matrix of the system in variations (9). For the linear system in variations with kT-periodic coefficients (9), the expression for the fundamental matrix of solutions $M(t)$ is known:

$$M(t) = \Phi(t) \exp(\Lambda t), \tag{13}$$

where $\Phi(t)$ is a piecewise smooth kT - periodic nonsingular matrix, $\Phi(t)(0) = E$ (E is a unit matrix, Λ is a constant matrix), all these matrices have the dimension n*n.

Hence, we see a constructive way for calculating multipliers: fix n initial conditions in the form of the unit matrix columns, numerically solve the Cauchy problem for the system (9) on the interval [0, kT] n times, and fix the columns of the solution at the terminal point $t = kT$. This will form the monodromy matrix $M(kT)$. And, finally, using standard programs, the eigenvalues of the monodromy matrix $M(kT)$ (the multipliers $\rho_j (j = 1, 2, ..., n)$) are determined.

In order to analyze the stability of the found T - periodic solution of the autonomous system (1), we also consider the perturbed motion of the system (1) in the form (8) and in this case $x_i^p(t)$ are T-periodic solutions. In this variant, the variational system has the form:

$$\frac{d\delta x_i(t)}{dt} = \sum_{j=1}^{n} \delta x_j(t) \frac{\partial X_i(x_1^p(t), x_2^p(t), ..., x_n^p(t))}{\partial x_j}, i = 1, 2, ..., n. \quad (14)$$

If the initial autonomous system (1) has a nontrivial T-periodic solution $x_i(t)$, then the corresponding system in variations (14) is a linear periodic system and at least one of its characteristic exponents is $\lambda_m = 0$ (corresponding multipliers $\rho_m = 1$.) According to the Andronov-Witt theorem, the periodic solution of the autonomous system (1) is asymptotically stable by Lyapunov if other characteristic exponents have negative real parts $Re(\lambda_j) < 0, j = 1, 2, ..., n, (j \neq m)$ (or multipliers modules $|\rho_j| < 1$, $j = 1, 2, ..., n, (j \neq m)$). The multipliers for the system of variations (14) corresponding to the autonomous system (1) are calculated in a similar way to the nonautonomous variant (2), (9). Thus, despite the use of various theorems for analyzing the stability of autonomous and non-autonomous systems, the algorithms for these variants are similar.

4 Strongly Nonlinear Dynamical Systems. The Use of Nonlinear Effects to Control Oscillations

Various nonlinear effects associated with the oscillations of strongly nonlinear dynamical systems are known [14]. In this section, we give some examples of oscillation control using nonlinear effects.

4.1 Oscillations of a Strongly Nonlinear System with Two Degrees of Freedom

Consider a two-dimensional nonlinear vibration isolation system (see Fig. 3). We will consider the motion of a protected object with two degrees of freedom - along the x axis and its rotation around the center of mass C. In general, the vibration isolation system can be asymmetric, that is, the characteristics of the nonlinear springs $F_1(x, \frac{dx}{dt})$ and $F_2(x, \frac{dx}{dt})$ may be different, the distances from the center of mass C to the spring attachment points d_1 and d_2 may also be different.

The protected object is subject to the influence of external periodic force $W(t) = W(t + T)$, T - known period. The dynamic equations of this system have the form:

$$\begin{cases} m\frac{d^2x(t)}{dt^2} + F_1(x(t) - d_1\varphi(t)) + F_2(x(t) + d_2\varphi(t)) = W(t), \\ J\frac{d^2\varphi(t)}{dt^2} - d_1F_1(x(t) - d_1\varphi(t)) + d_2F_2(x(t) + d_2\varphi(t)) = 0. \end{cases} \quad (15)$$

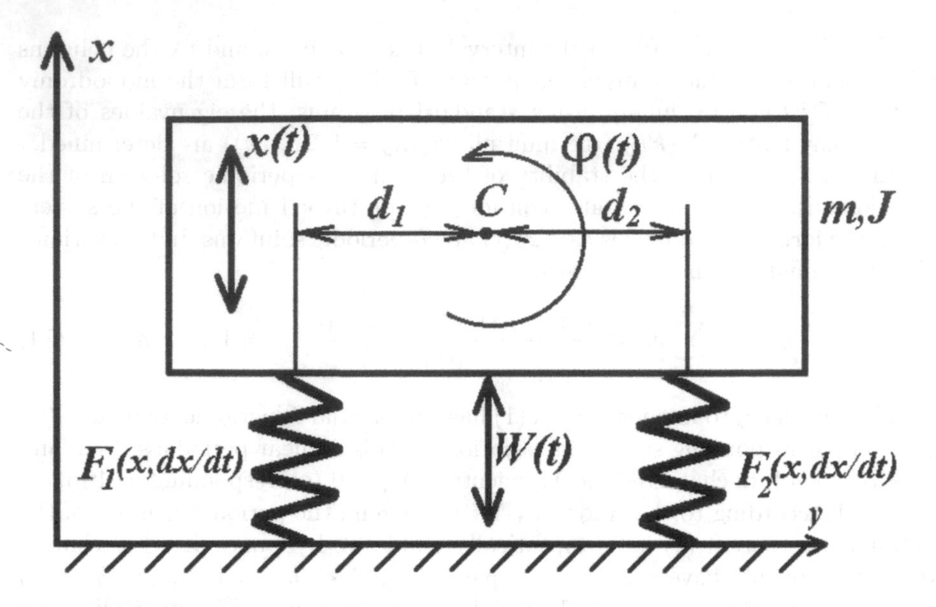

Fig. 3. Two-dimensional asymmetric nonlinear vibration isolation system

Here m is the mass of the protected object, J is its moment of inertia about the center of mass C, $x(t)$ is the center of mass's deviation from the equilibrium position, $\varphi(t)$ is rotation's angle around the center of mass C. The motions along the y axis are not considered. Without the loss of generality, in the computations we set

$$\begin{cases} d_1 = d_2 = d, \\ F_1(x, \frac{dx}{dt}) = F_2(x) = cx + \gamma x^3 + b\frac{dx}{dt}, \\ W(t) = W_0 \sin(\omega t). \end{cases} \tag{16}$$

We note that these assumptions do not follow from the method used to find the periodic solutions, but they are adopted for a clearer identification of nonlinear effects. We also note that if conditions (16) are satisfied, the system (15) becomes symmetric and it can be represented in the following form:

$$\begin{cases} m\frac{d^2x(t)}{dt^2} + 2cx(t) + 2\gamma x^3(t) + 6\gamma d^2 x(t)\varphi^2(t) + 2b\frac{dx(t)}{dt} = W_0\sin(\omega t), \\ J\frac{d^2\varphi(t)}{dt^2} + 2d^2c\varphi(t) + 2d^4\gamma\varphi^3(t) + 6d^2\gamma x^2(t)\varphi(t) + 2d^2b\frac{d\varphi(t)}{dt} = 0. \end{cases} \tag{17}$$

The corresponding system in variations has the form:

$$\begin{cases} m\frac{d^2\delta x(t)}{dt^2} + 2c\delta x(t) + 6\gamma x^2(t)\delta x(t) + \\ 6\gamma d^2\varphi^2(t)\delta x(t) + 12\gamma d^2 x(t)\varphi(t)\delta\varphi(t) + 2b(\frac{d\delta x(t)}{dt}) = 0, \\ J\frac{d^2\delta\varphi(t)}{dt^2} + 2d^2c\delta\varphi(t) + 6d^4\gamma\varphi^2(t)\delta\varphi(t) + \\ 12d^2\gamma x(t)\varphi(t)\delta x(t) + 6d^2\gamma x^2(t)\delta\varphi(t) + 2d^2b\frac{d\delta\varphi(t)}{dt} = 0, \end{cases} \tag{18}$$

where $\delta x(t), \delta\varphi(t)$ are small variations.

We note that the oscillations of rotation $\varphi(t)$ around the center of mass C are parametric, and the parametric influence is the function $x(t)$, which is related to the forced oscillations of the object along the x axis.

Using the above algorithm for constructing and analyzing the stability of periodic solutions for a nonlinear model of vibration isolation (17), (18), amplitude-frequency characteristics were calculated. Here are the results (see Fig. 4) obtained with $m = 1$; $c = 1$; $d = 1$; $\gamma = 1$; $W_0 = 1$; $J = 3$; $b = 0,2$; $T = 2\pi/\omega$.

A non-trivial stable periodic solution for the function $\varphi(t)$ is found in the frequency range $0,94 < \omega < 1,15$. In this frequency range, the oscillation amplitude in the direction of the x axis decreases, that is, the energy is redistributed between different modes of oscillation. In the same frequency range $0,94 < \omega < 1,15$., an another unstable trivial solution $\varphi(t) \equiv 0$ is found, and the amplitude of the unstable solution $x(t)$ does not decrease (it is marked by the dashed line in Fig. 4). In the range of higher frequencies ($2 \lessapprox \omega \lessapprox 3$), a stable solution with a smaller amplitude for oscillations along the x-axis without a rotation around the center of mass was detected. Such solutions are characteristic for nonlinear systems, and in the simplest form are found in the classical Duffing equation with one degree of freedom.

For physical reasons, it can be expected that stable parametric oscillations of a nonzero amplitude $\varphi(t) \not\equiv 0$ may not occur with a smaller value of the moment of inertia J (or at a greater dissipation b). The numerical experiment confirmed these assumptions. For $J = 1$, the trivial solution $\varphi(t) \equiv 0$, associated with rotation around the center of mass C is stable over the entire frequency range. Rotation does not arise, the system with two degrees of freedom behaves like a one-dimensional system.

Note that the discussed results are obtained from the symmetric system. It can be expected that for the asymmetric system ($d_1 \neq d_2, F_1(x) \neq F_2(x)$) the mutual influence of oscillation different forms will be stronger.

A similar effect was observed in the study of a more complex model of forced strongly nonlinear oscillations with the dynamic jumping of a beam that lost its static stability [9]. In this case, the interaction of several forms and modes of oscillations was also observed, leading to a decrease in the amplitude of the basic form's oscillations. Thus, we can use the observed effect of nonlinear interaction of various forms and modes of oscillations in a multidimensional nonlinear dynamical system for controlling its oscillations. This feature complements the traditional methods of controlling oscillations in multidimensional nonlinear dynamic systems.

We note the resulting reduction of the amplitude oscillations. This effect due to the mutual influence of different types of oscillations. A similar effect may appear in a multidimensional linear dynamic system (for example, a linear dynamic vibration absorber).

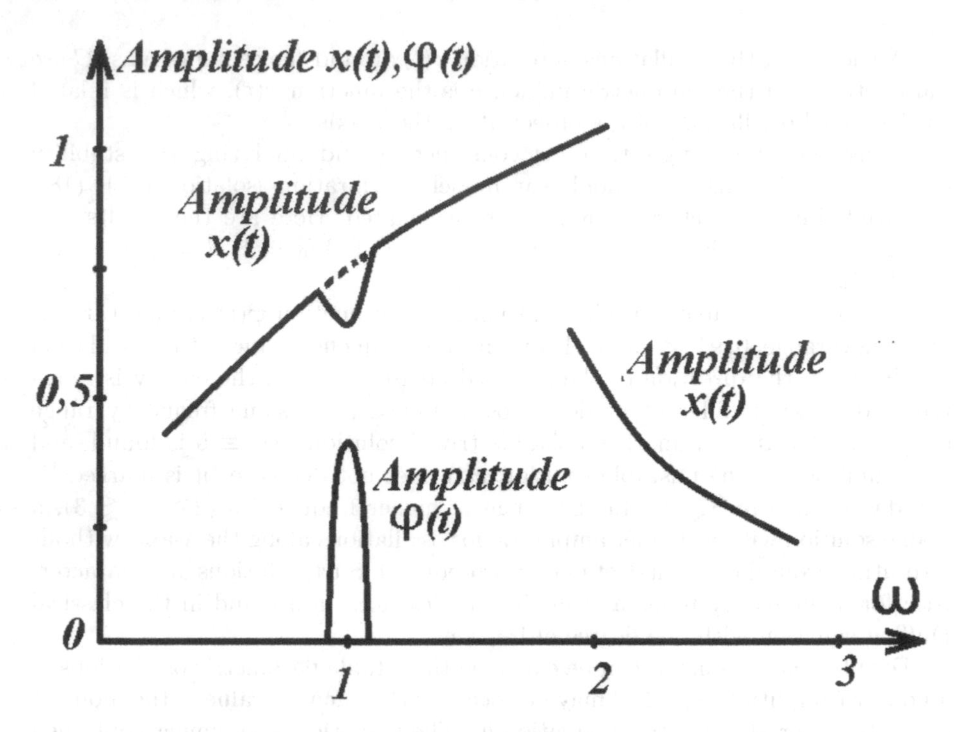

Fig. 4. Amplitude-frequency characteristic of two-dimensional nonlinear vibration isolation system

4.2 Control of Oscillations in a One-Dimensional Dynamic System Due to the Nonlinear Interaction of Different Types of Oscillations

Earlier [10] it was found that in a one-dimensional dynamic system of the form

$$\frac{d^2x(t)}{dt^2} + [\omega_0^2 + Psin(\omega t)]x(t) + b\frac{dx(t)}{dt} + \gamma x^3(t) = W_0 sin(\omega t) \tag{19}$$

it is possible to reduce the total amplitude of the oscillations due to the nonlinear interaction of forced and parametric oscillations. In this system, $x(t)$ is the required periodic function of time t, $Psin(\omega t)$ corresponds to a parametric influence with amplitude P and frequency ω, $W_0 sin(\omega t)$ corresponds to a simultaneous external influence with the same frequency ω and amplitude W_0, ω_0 is the natural frequency of the corresponding linear system, γ is the coefficient of nonlinearity, b is dissipation coefficient. The solution's period T_s may correspond to forced or parametric oscillations.

We also note that the interaction phenomenon of forced and parametric oscillations in a nonlinear system can be used to control the oscillations of the systems under consideration. The control tools are the amplitudes of parametric (P) and (or) external (W_0) influences.

Note that this effect is fundamentally impossible in a linear one-dimensional dynamic system, in which solutions are formed in accordance with the superposition principle.

We consider a generalization of this problem when the interactions of forced, parametric and self-oscillations are possible in a nonlinear dynamical system. Consider a nonlinear dynamical system with one degree of freedom in the form

$$\begin{aligned} \frac{d^2x(t)}{dt^2} + [\omega_0^2 + Psin(\omega t)]x(t) + b\frac{dx(t)}{dt} + \gamma x^3(t) \\ = W_0 sin(\omega t) + s_1\frac{dx(t)}{dt} - s_3(\frac{dx(t)}{dt})^3. \end{aligned} \tag{20}$$

Simultaneous existence of the forced oscillations (excited by external influence $W_0 sin(\omega t)$, parametric oscillations (excited by periodic influence $Psin(\omega t)$, and self-oscillations (determined by components $s_1\frac{dx(t)}{dt}$ *and* $s_3(\frac{dx(t)}{dt})^3$) is possible in this system. Such equations arise in the electromechanical systems modeling. Various combinations of stable and unstable forced and parametric oscillations, as well as self-oscillations, may occur depending on the system parameters values. For example, in Fig. 5 the zones corresponding mainly to forced and to parametric oscillations are given.

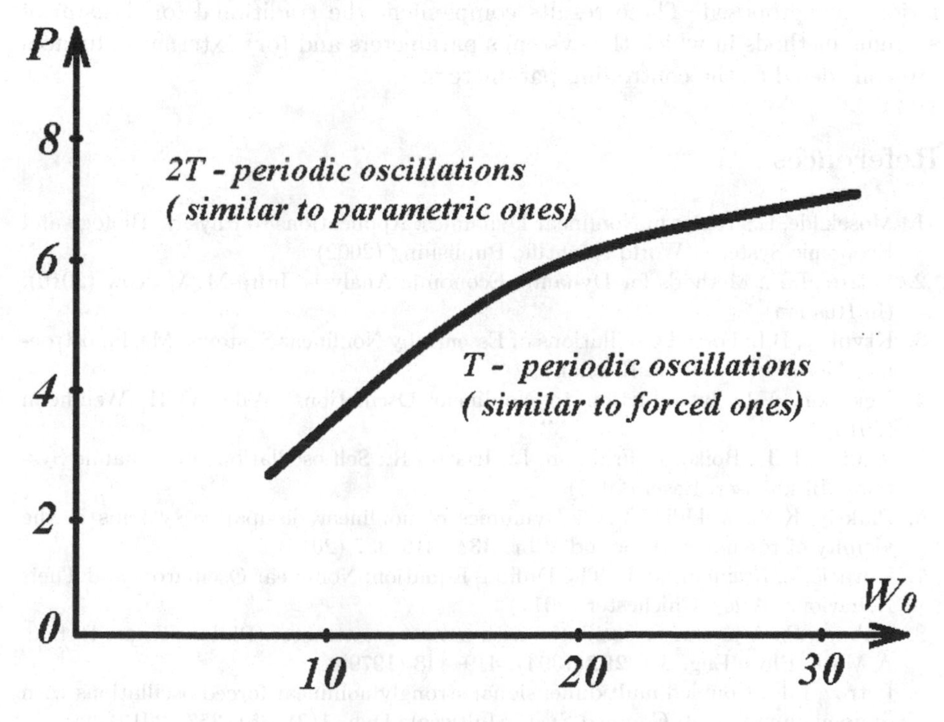

Fig. 5. Zones of mostly forced and predominantly parametric stable oscillations in the self-oscillating system with forced and parametric external influences

In this one-dimensional nonlinear system, the same paradoxical effect of resulting oscillation amplitude reducing with the external influence amplitude W_0 increase (with other unchanged parameters of the system) is noticed. This phenomenon can also be used to control the oscillations. In this case the control tools are the amplitudes of parametric (P) and (or) external (W_0) influences and (or) self-oscillations parameters (s_1 and s_3). Thus, we can use the effects of interaction between different types of oscillations in a nonlinear dynamic system with one degree of freedom to control the oscillations.

5 Conclusion

For the analysis and control of oscillations in nonlinear dynamical systems, the numerical-analytical method for the search and the stability analysis of periodic solutions for autonomous and non-autonomous strongly nonlinear systems of ordinary differential equations is presented. This method implements a unified approach to the study of various oscillations (forced, parametric, self-oscillation, resonant and non-resonant, chaos, nonlinear interaction of different oscillation modes and types) for strongly nonlinear multidimensional dynamical systems.

On the basis of this approach, new methods for control of oscillations in such systems based on the nonlinear interaction between different types of oscillations are proposed. These results complement the traditional for dynamical systems methods in which the system's parameters and (or) external influences are considered as the controlling parameters.

References

1. Mosekilde, E.: Topics in Nonlinear Dynamics; Applications to Physics, Biology and Economic Systems. World Scientific Publishing (2002)
2. Petrov, L.F.: Methods for Dynamic Economic Analysis. Infra-M, Moscow (2010). (in Russian)
3. Kryukov, B.I.: Forced Oscillations of Essentially Nonlinear Systems. Mashinostroenie, Moscow (1984). (in Russian)
4. Nekorkin, V.I.: Introduction to Nonlinear Oscillations. Wiley-VCH, Weinheim (2015)
5. Aguilar, L.T., Boiko, I., Fridman, L., Iriarte, R.: Self-oscillations in Dynamic Systems. Birkhäuser, Basel (2015)
6. Plaksiy, K.Y., Mikhlin, Y.V.: Dynamics of nonlinear dissipative systems in the vicinity of resonance. J. Sound Vibr. **334**, 319–337 (2015)
7. Kovacic, I., Brennan, M.J.: The Duffing Equation: Nonlinear Oscillators and Their Behaviour. Wiley, Chichester (2011)
8. Holmes, P.: A nonlinear oscillator with a strange attractor. Philos. Trans. R. Soc. A Math. Phys. Eng. Sci. **292**(1394), 419–448 (1979)
9. Petrov, L.F.: Coupled multidimensional strongly nonlinear forced oscillations with dynamic jumping. J. Coupled Syst. Multiscale Dyn. **1**(3), 351–357 (2013)
10. Petrov, L.F.: Search for periodic solutions of highly nonlinear dynamical systems. Comput. Math. Math. Phys. **58**(384), 403–413 (2018)

11. Gornov, AYu.: Computational Techniques for Solving Optimal Control Problems. Nauka, Novosibirsk (2009). (in Russian)
12. Petrov, L.F.: Interactive computational search strategy of periodic solutions in essentially nonlinear dynamics. In: Cojocaru, M., Kotsireas, I., Makarov, R., Melnik, R., Shodiev, H. (eds.) Interdisciplinary Topics in Applied Mathematics, Modeling and Computational Science, Springer Proceedings in Mathematics and Statistics, vol. 117, pp. 355–360. Springer, Cham (2015). https://doi.org/10.1007/978-3-319-12307-3_51
13. Urabe, M.: Galerkin's procedure for nonlinear periodic systems. Arch. Ration. Mech. Anal. **20**, 120–152 (1965)
14. Petrov, L.F.: Nonlinear effects in economic dynamic models. Nonlinear Anal. **71**, 2366–2371 (2009)

Risk Management in Gaussian Stochastic Systems as an Optimization Problem

Al'fiya A. Surina[1(✉)] and Alexander N. Tyrsin[2,3]

[1] South Ural State University (National Research University), Lenina Ave. 76, 454080 Chelyabinsk, Russia
dallila87@mail.ru

[2] Ural Federal University named after the first President of Russia B.N.Yeltsin, Mira St. 19, 620002 Ekaterinburg, Russia
at2001@yandex.ru

[3] Moscow Institute of Physics and Technology (State University), Institutskiy per. 9, 141701 Dolgoprudny, Moscow Region, Russia

Abstract. In article the risk management algorithm in Gaussian stochastic system is describes. The model risk management represents an optimization problem. The risk management algorithm is realized on the basis of the barrier functions method. The features of this nonlinear programming problem are not the convexity of the accessible solution region and the presence of stochastic restrictions on the required risk. The software implementation of the algorithm in the form of a separate module is performed. Using the Monte Carlo statistical test method, the algorithm was investigated. The algorithm showed stable control. Its efficiency is proved. Results of a research are presented in article. Recommendations on practical application of the algorithm are given.

Keywords: Model · Risk · Control · Stochastic system · Random vector · Normal distribution · Monitoring · Optimization

1 Introduction

The development of adequate techniques of monitoring and risk management for complex systems is immediate problem. Usually, risk modeling comes down to selection of dangerous outcomes, the quantitative assignment of consequences from their occurrence and estimation of the probabilities of these outcomes [1]. The contribution of the components of a multidimensional system is combined and the one-dimensional system is considered as a random variable [2–5]. For relatively simple objects for which it is possible to specify a priori all dangerous outcomes in the presence of statistical information or expert estimates on chances of their emergence in general, this approach yields the results acceptable

The study was supported by the Russian Foundation for Basic Research (project no. 17-01-00315).

I. Bykadorov et al. (Eds.): MOTOR 2019, CCIS 1090, pp. 562–577, 2019.
https://doi.org/10.1007/978-3-030-33394-2_43

in practice. Usually it is possible to accumulate sufficient statistics to assess the probability of occurrence of dangerous outcomes, and the form of the relationship between the elements of the system is quite simple and can be described, for example, with the help of logical-probabilistic risk models [6] in the framework of the theory of structurally complex systems [7]. However, complex systems, for example in the economy, society, health care, etc., as a rule, are multidimensional. Its functioning is largely stochastic in nature, and at them it is often possible to allocate tens of various risk factors [8]. It is usually not possible to single out all these dangerous outcomes. In [9] the approach to modeling of risk management procedures of multidimensional stochastic systems based on simultaneous use of two submodels – system effectiveness assessment model and the risk assessment model of its functioning is described. Risk management is reduced to solving the problem of nonlinear programming, which allows finding the optimal management strategy in terms of efficiency and sustainability. However, this approach is difficult to implement in practice, since the interpretation of the concept "system efficiency" is not disclosed and is, as the authors of the article indicate, ambiguous. Secondly, it is not clear how to solve the multicriteria problem associated with the simultaneous use of two criteria - efficiency and risk. The risk model of multidimensional stochastic systems according to which the system is presented in the form of a random vector with mutually correlative components is offered in [10]. Instead of the conventional selection of concrete dangerous situations we will set geometrical areas of failures. Its can look arbitrarily depending on a specific objective, and are determined on the basis of the available a priori information. In many cases larger and improbable deviations of selective values of any of the risk factors relative to the best safety values are dangerous situations. This model was tested in relation to the tasks of risk analysis in the field of energy, regional economic development and population health [11,15]. The purpose of the article is to describe and study the risk management algorithm based on this approach.

2 Model Risk Management in Gaussian Stochastic Systems

Let's present m-dimensional Gaussian stochastic system S in the form of a random vector of $\mathbf{X}$ with a probability density $p_{\mathbf{x}}(\mathbf{x})$.For the Gaussian random vector $\mathbf{X} = (X_1, X_2, ..., X_m)$, the numerical characteristics are the covariance matrix $\mathbf{\Sigma} = \{\sigma_{ij}\}_{m \times m}$ and the expectations vector $a = (a_1, a_2, .., a_m)^T$. According to offered model we will exercise risk management of a Gaussian system by means of the matrix of $\mathbf{\Sigma}$ and the vector of $\mathbf{a}$. At the risk management it is possible to solve two problems [12]:

1. Risk minimization when performing restrictions for $\mathbf{\Sigma}$ and $\mathbf{a}$.
2. Reaching an acceptable risk level at minimum(minimal) change of $\mathbf{\Sigma}$ and $\mathbf{a}$.

The first task is to reach a minimum risk level. However for its effective use it is required to set rather "hard" restrictions for $\mathbf{\Sigma}$ and $\mathbf{a}$, that is not always possible. Too general restrictions lead to the decisions which are a little realized in practice. Therefore we will consider the second task – reaching acceptable risk of r^* at minimum(minimal) changes of numerical characteristics of Gaussian system X_0:

$$\begin{cases} f(\mathbf{a},\mathbf{\Sigma}) = \sum_{j=1}^{m}\sum_{k=j}^{m}(\sigma_{jk}-\sigma_{jk}^0)^2 + \sum_{i=1}^{m}(a_i-a_i^0)^2 \to \min\limits_{\mathbf{a},\mathbf{\Sigma}}, \\ r(X) = r^*,\ \mathbf{\Sigma}\in G(\mathbf{\Sigma}),\ \mathbf{a}\in H(\mathbf{a}). \end{cases} \quad (1)$$

The risk in (1) is described as [13]:

$$r(X) = \int\int_{R^m}\ldots\int g(\mathbf{x})p_{\mathbf{x}}(\mathbf{x})d\mathbf{x}, \quad (2)$$

where $g(\mathbf{x})$ is a function of consequences from dangerous situations (risk function). Varying $g(\mathbf{x})$ in (2) it is possible to receive various estimates of risk [12].

As it was stated above, under dangerous situations it is considered larger and improbable deviations of selective values of any of risk factors concerning the values, best in sense of safety. For definiteness, in (2) we will accept $g(\mathbf{x}) = 1\ \forall\mathbf{x}\in D$ and $g(\mathbf{x}) = 0\ \forall\mathbf{x}\notin D$, then $r(\mathbf{X}) = P(\mathbf{X}\in D)$, i.e. the risk level will be equal to failure probability. Here D is the area of dangerous situations (failures)

$$D = \left\{\mathbf{x} = (x_1,x_2,\ldots,x_m) : \sum_{j=1}^{m}\frac{(x_j-a_j')^2}{d_j^2} \geq 1\right\}, a_j' = \frac{a_j^- + a_j^+}{2}.$$

Then the problem of reaching acceptable risk of r^* (1) will be rewritten taking into account restrictions:

$$\sum_{j=1}^{m}\sum_{k=j}^{m}\left(\sigma_{jk}-\sigma_{jk}^0\right)^2 + \sum_{i=1}^{m}\left(a_i-a_i^*\right)^2 \to \min_{\mathbf{a},\,\mathbf{\Sigma}} \quad (3)$$

at restrictions

$$\begin{cases} \sigma_{jj}\sigma_{kk} > \sigma_{jk}^2,\ \ \sigma_{jk} = \sigma_{kj},\ \ \sigma_{jj} > 0, \quad 1\leq j,k\leq m, \\ \sigma_{jk}^- < \sigma_{jk} < \sigma_{jk}^+, \quad 1\leq j,k\leq m, \\ a_i^- < a_i < a_i^+, \quad 1\leq i\leq m, \\ p_{\text{targ}} = p^*(\mathbf{\Sigma},\mathbf{a}) + \varepsilon. \end{cases}$$

Let's note features of a task (3).

First, there is a stochastic restriction for the required risk $p_{\text{targ}} = p^*(\mathbf{\Sigma},\mathbf{a})+\varepsilon$. It is bound to the fact that probability $p^*(\mathbf{\Sigma},\mathbf{a})$ – isn't calculated analytically. Therefore this condition is checked separately.

Secondly, generally the area of admissible decisions isn't convex. It results in dependence of efficiency of the solution algorithm of the task (3) on the choice of the initial point.

Thirdly, the number of managing variables in (3) grows at a quadratic rate: $m(m+3)/2$.

3 Description of the Risk Management Algorithm

The solution algorithm of the task (3) is based on a barrier functions method [13] and consists of 4 steps:

Step 1. Let's set initial values of elements of the covariance matrix $\mathbf{\Sigma}_0$ and the expectations vector of $\mathbf{a}_0$, $b_k \geq 0$, $C > 1$, $\varepsilon > 0$, p_{targ}.

Step 2. Let's make support function

$$F(\mathbf{\Sigma}, \mathbf{a}, b_k) = f(\mathbf{\Sigma},\ \mathbf{a}) + P(\mathbf{\Sigma}, \mathbf{a}, b_k),$$

where $P(\mathbf{\Sigma},\ \mathbf{a}, b_k)$ – penalty function, $F(\mathbf{\Sigma},\ \mathbf{a}, b_k)$ – support function.

The inverse function is selected as a penalty function: $P(\mathbf{\Sigma},\ \mathbf{a}, b_k) = b_k \sum_{j=1}^{m} \frac{1}{t_j(\mathbf{\Sigma},\ \mathbf{a})}$, where b_k – the positive value chosen randomly.

Then taking into account all restrictions the support function will take the form:

$$\begin{aligned} F(\mathbf{\Sigma},\ \mathbf{a}, b_k) &= \sum_{j=1}^{m}\sum_{k=j}^{m}\left(\sigma_{jk} - \sigma_{jk}^{0}\right)^2 + \sum_{i=1}^{m}\left(a_i - a_i^0\right)^2 \\ &+ b^k \cdot \left(\sum_{j=1}^{m}\frac{1}{\sigma_{jj}} + \sum_{j=1}^{m}\sum_{k=j}^{m}\frac{1}{\sigma_{jj}\sigma_{kk} - \sigma_{jk}^2} + \sum_{j=1}^{m}\sum_{k=j}^{m}\frac{1}{\sigma_{jk}^{-} - \sigma_{jk}}\right. \\ &\left. + \sum_{j=1}^{m}\sum_{k=j}^{m}\frac{1}{\sigma_{jk} - \sigma_{jk}^{+}} + \sum_{i=1}^{m}\frac{1}{a_i^{-} - a_i} + \sum_{i=1}^{m}\frac{1}{a_i - a_j^{+}}\right). \end{aligned}$$

Step 3. Let's find minimum point $\mathbf{x}k$ of the function $F(\mathbf{\Sigma},\ \mathbf{a}, b_k)$ by means of the unconstrained optimization method based on the Nelder–Mead method [13] (see Fig. 1).

Step 4. Let's check the fulfillment of the condition of the end:

(a) if $\left|P(\mathbf{\Sigma}^k,\ \mathbf{a}^k,\ b_k)\right| \leq \varepsilon$, process of searching is completed: $\mathbf{\Sigma}^k = \mathbf{\Sigma}^*$, $\mathbf{a}^k = \mathbf{a}^*$, $P(\mathbf{\Sigma}^k, \mathbf{a}^k) = p^*(\mathbf{\Sigma}^*,\ \mathbf{a}^*)$;
(b) if $\left|P(\mathbf{\Sigma}^k,\ \mathbf{a}^k,\ b_k)\right| > \varepsilon$, put $b_{k+1} = \frac{b_k}{C}$, $\mathbf{\Sigma}^{k+1} = \mathbf{\Sigma}^*$, $\mathbf{a}^{k+1} = \mathbf{a}^*$, $k = k+1$, and proceed to step 2.

The *Init2* function searches for the ideal point $\mathbf{x}_{ideal}$ and vertices of a polyhedron. The point at which the values of the elements of the covariance matrix $\mathbf{\Sigma}$ are close to or equal to zero, and the values of the expectation vector $\mathbf{a}$ – the best values specified earlier is chosen as the ideal point of the ideal. Changing values of the elements of $\mathbf{\Sigma}$, $\mathbf{a}$ and, calculating value p^*, at some moment the value of the probability of p^* becomes equal p_{targ}. Infinitude number of such sets of values of the elements $\mathbf{\Sigma}$ and $\mathbf{a}$ form m-dimensional convex figure on which surface probability is equal p_{targ}. In the center of a figure the point $\mathbf{x}ideal$ with probability $p_{ideal} = 0$.

The *BWPoint* function carries out calculation of weight coefficients of points to find minimum and maximum values. Then the "best" point of $\mathbf{x}^l$ and "worst"

$\mathbf{x}^h$ points where $F(x^l) = \min_{k=1,\dots,m+1} F(x^k)$ and $F(x^h) = \max_{k=1,\dots,m+1} F(x^k)$, and also the point of $\mathbf{x}^s$ in which the second largest is reached after maximal value of function is chosen.

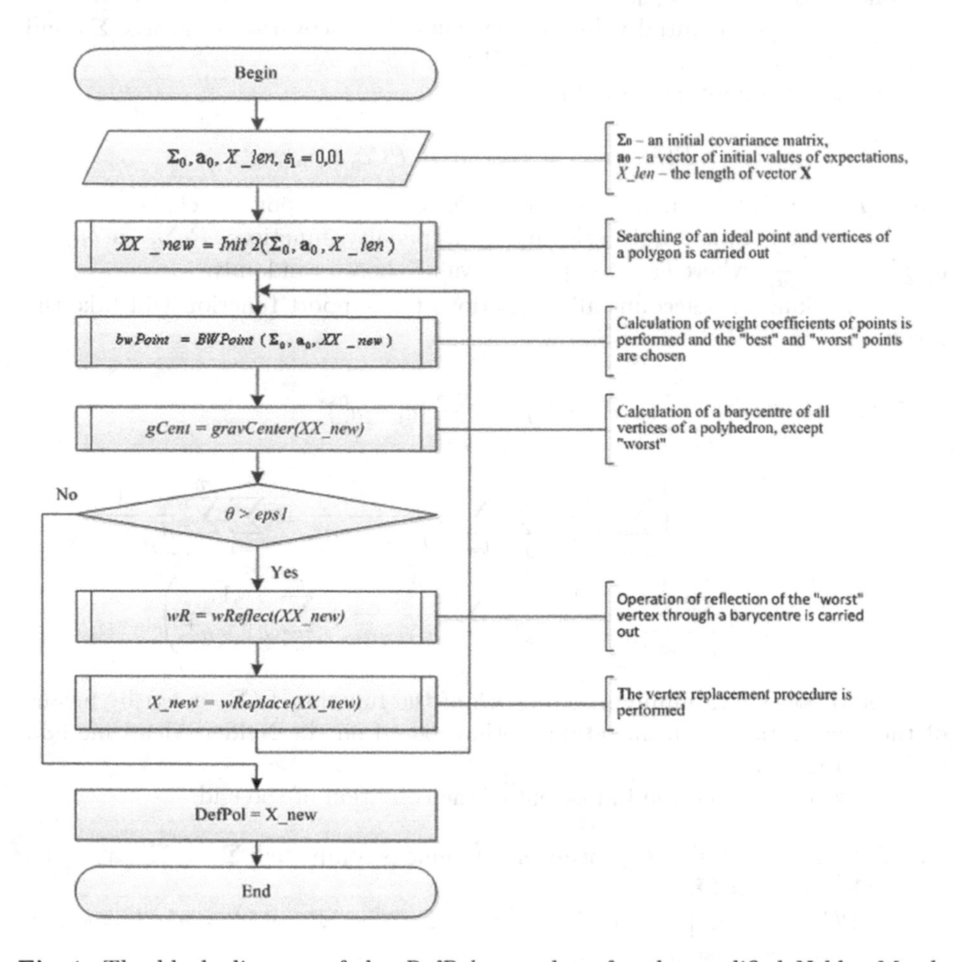

Fig. 1. The block diagram of the *DefPol* procedure for the modified Nelder–Mead method.

The *gravCenter* function finds "barycenter" of all vertices of a polyhedron except for the "worst" $\mathbf{x}^h$:

$$x^{m+2} = \frac{1}{m+1}\left(\sum_{k=1}^{m+2} x^k - x^h\right) = \frac{1}{m+1}\sum_{\substack{k=1\\k\neq h}}^{m+2} x^k.$$

The *wReflect* function carries out operation of reflection of the "worst" vertex $\mathbf{x}^h$ through the barycenter x^{m+2}: $x^{m+3} = x^{m+2} - x^h$ in case

$$\theta = \left\{ \frac{1}{m+1} \sum_{k=1}^{m+2} \left[F(x^k) - F(x^{m+2}) \right]^2 \right\}^{\frac{1}{2}} > \varepsilon.$$

Let's receive the vector $x^k = x^{m+3} - x^{m+2}$. Otherwise the algorithm leaves the *DefPol* procedure, having kept the found values.

The *wReplace* function carries out operation of replacement of vertices after the reflection operation.

Further carries out transition to the *BWPoint* function described earlier.

After the values satisfying (3) have been found, the stochastic restriction on the required risk is checked. If the found values meet a condition, then the algorithm finishes work and removes results. Otherwise, the barrier functions method starts the work anew.

Let's review the following example for check of work of the algorithm.

Example 1. Let's set initial values of elements of the covariance matrix $\mathbf{\Sigma}$ and expectations vector of $\mathbf{a}$ (for computing experiments of values of elements of the $\mathbf{\Sigma}$, $\mathbf{a}$ and area of unfavorable outcomes of D are taken from [15], the calculations are executed by means of the program risk_model_calc.xlsm [16]):

$$\mathbf{\Sigma} = \begin{pmatrix} 397.053 & 27.243 & 3.094 \\ 27.243 & 18.357 & 0.400 \\ 3.094 & 0.400 & 1.353 \end{pmatrix}, \mathbf{a} = \begin{pmatrix} 133.850 \\ 25.889 \\ 5.162 \end{pmatrix}, \quad P_0 = 0.99, b^k = 1,$$
$$C > 1, \varepsilon_1 = 0.05.$$

The initial point was selected at the boundary of area D. The decision is considered found if the target risk value p_{targ} and for the found risk value of P^*: $|P^* - p_{targ}| \leq \varepsilon_1$.

Let's check work of the algorithm offered earlier for amount of risk factors of $m = 3$ on the basis of the given parameters. For this purpose let's carry out about 100 tests with initial value of the probability of an unfavorable outcome of $P_0 = 0,99$. Let's show that at the obviously poor value $P_0 = 0.99$ algorithm is efficient, and allows to achieve an acceptable risk p_{targ} at minimal changes of numerical characteristics of Gaussian system $\mathbf{X}_0$.

Results are given below. The value p_{targ} changed on 0,1, i.e. $p_{targ} = 0,9; 0,8; \ldots; 0,1$ and the last the value $p_{targ} = 0,01$ was chosen:

$$p_{targ} = 0.9; \; P^* = 0.867; \; \mathbf{a} = \begin{pmatrix} 134.613 \\ 23.843 \\ 3.830 \end{pmatrix}; \mathbf{\Sigma} = \begin{pmatrix} 291.892 & 23.525 & 5.491 \\ 23.525 & 20.029 & 0.650 \\ 5.491 & 0.650 & 1.042 \end{pmatrix};$$

$$f^*(\mathbf{a}, \mathbf{\Sigma}) = 0.241,$$

$$p_{targ} = 0.8;\ P^* = 0.803;\ \mathbf{a} = \begin{pmatrix} 135.798 \\ 23.719 \\ 3.435 \end{pmatrix};\ \mathbf{\Sigma} = \begin{pmatrix} 326.300 & 17.878 & 5.347 \\ 17.878 & 14.070 & 1.215 \\ 5.347 & 1.215 & 0.882 \end{pmatrix};$$

$f^*(\mathbf{a},\ \mathbf{\Sigma}) = 0.472,$

$$p_{targ} = 0.7;\ P^* = 0.707;\ \mathbf{a} = \begin{pmatrix} 134.066 \\ 22.024 \\ 3.468 \end{pmatrix};\ \mathbf{\Sigma} = \begin{pmatrix} 314.673 & 25.512 & 6.570 \\ 25.512 & 15.491 & 1.207 \\ 6.570 & 1.207 & 0.882 \end{pmatrix};$$

$f^*(\mathbf{a},\ \mathbf{\Sigma}) = 0.453,$

$$p_{targ} = 0.6;\ P^* = 0.606;\ \mathbf{a} = \begin{pmatrix} 124.552 \\ 23.263 \\ 3.311 \end{pmatrix};\ \mathbf{\Sigma} = \begin{pmatrix} 297.345 & 6.971 & 6.304 \\ 6.971 & 12.173 & 0.822 \\ 6.304 & 10.822 & 0.705 \end{pmatrix};$$

$f^*(\mathbf{a}, \mathbf{\Sigma}) = 0.722,$

$$p_{targ} = 0.5;\ P^* = 0.544;\ \mathbf{a} = \begin{pmatrix} 127.645 \\ 21,910 \\ 3,087 \end{pmatrix};\ \mathbf{\Sigma} = \begin{pmatrix} 241.987 & 12.697 & 1.759 \\ 12.697 & 12.439 & 1.292 \\ 1.759 & 1.292 & 0.992 \end{pmatrix};$$

$f^*(\mathbf{a},\ \mathbf{\Sigma}) = 0.605,$

$$p_{targ} = 0.4;\ P^* = 0.400;\ \mathbf{a} = \begin{pmatrix} 121.706 \\ 21.965 \\ 3.180 \end{pmatrix};\ \mathbf{\Sigma} = \begin{pmatrix} 151.527 & 10.956 & 1.346 \\ 10.956 & 10.318 & 0.631 \\ 1.346 & 0.631 & 0.753 \end{pmatrix};$$

$f^*(\mathbf{a},\ \mathbf{\Sigma}) = 0.973,$

$$p_{targ} = 0.3;\ P^* = 0.302;\ \mathbf{a} = \begin{pmatrix} 121.540 \\ 20.675 \\ 3.065 \end{pmatrix};\ \mathbf{\Sigma} = \begin{pmatrix} 318.736 & 10.113 & 4.318 \\ 10.113 & 7.607 & 0.631 \\ 4.318 & 0.631 & 0.844 \end{pmatrix};$$

$f^*(\mathbf{a},\ \mathbf{\Sigma}) = 0.986,$

$$p_{targ} = 0.2;\ P^* = 0.207;\ \mathbf{a} = \begin{pmatrix} 112.325 \\ 20.960 \\ 2.990 \end{pmatrix};\ \mathbf{\Sigma} = \begin{pmatrix} 194.454 & 17.664 & 3.590 \\ 17.664 & 6.898 & 0.674 \\ 3.590 & 0.674 & 0.528 \end{pmatrix};$$

$f^*(\mathbf{a},\ \mathbf{\Sigma}) = 1.208,$

$$p_{targ} = 0.1;\ P^* = 0.100;\ \mathbf{a} = \begin{pmatrix} 111.179 \\ 19.902 \\ 2.896 \end{pmatrix};\ \mathbf{\Sigma} = \begin{pmatrix} 117.745 & 9.342 & 0.798 \\ 9.342 & 6.898 & 0.674 \\ 0.978 & 0.674 & 0.528 \end{pmatrix};$$

$f^*(\mathbf{a},\ \mathbf{\Sigma}) = 1.607,$

$$p_{targ} = 0.01;\ P^* = 0.011;\ \mathbf{a} = \begin{pmatrix} 109.351 \\ 19.483 \\ 2.734 \end{pmatrix};\ \mathbf{\Sigma} = \begin{pmatrix} 91.773 & 1.058 & 0.674 \\ 1.058 & 2.022 & 0.167 \\ 0.674 & 0.167 & 0.294 \end{pmatrix};$$

$$f^*(\mathbf{a},\ \mathbf{\Sigma}) = 2.389$$

We see that with decrease of risk, the minimum of target function $f^*(\mathbf{a},\ \mathbf{\Sigma})$ grows.

As it became clear during the test, the algorithm at rather great value of P_0 can reach $p_{targ} \ll P_0$, however, significant temporary and computing resources for this purpose are required. In addition it is necessary to investigate an algorithm and to reveal what sample size it is necessary to set and how many tests to carry out to receive necessary value p_{targ} with the given accuracy of calculations ε_1.

Let's note that too sharp decrease in the value ΔP of P_0 isn't enough realizable in practice. it is expedient to set it no more $\Delta P \leq 0.3$.

4 Research of Accuracy of the Risk Management Algorithm

The greatest difficulties at realization of the risk management algorithm are caused by existence of stochastic restriction. For estimation of the improper integral (2) we use the Monte Carlo statistical test method [14]. Let's conduct a research of dependence of calculation accuracy on a sample size of M at a numerical integration and from the number of tests of *col.* Under accuracy of calculations we will consider the value of a 95% confidence interval Δp_{targ} for the calculated values of the required risk value (probability of a unfavorable outcome) p_{targ}.

Let's set two risk factors ($m = 2$). We will take initial values of elements of the covariance matrix $\mathbf{\Sigma}$ and expectations vector of $\mathbf{a}$ from the results received in [15] for two risk factors:

$$\mathbf{\Sigma} = \begin{pmatrix} 397.053 & 27.2427 \\ 27.2427 & 18.357 \end{pmatrix}, \mathbf{a} = \begin{pmatrix} 133.850 \\ 25.889 \end{pmatrix}, P_0 = 0.732.$$

We model samples with the number of points of M = 100. For $p_{targ} = 0.7$ we touch values of number of tests col: 10, 20, 50 and 100. For each set M and *col* we repeat calculations of 200 times and form a 95% confidence interval, discarding 5 greatest and 5 least values. Further we repeat for $p_{targ} = 0.6; 0.5; ...; 0.1$.

Similarly we find accuracy with a 95% confidence interval for $M = 100$, 1000, 10000, 100000 points. The received results are given in Tables 1 and 2.

Similar tests are carried out for three ($m = 3$) and four ($m = 4$) risk factors. The initial values of elements of the covariance matrix $\mathbf{\Sigma}$ and expectations vector of $\mathbf{a}$ from the results received in [15] for three and four risk factors respectively:

$$m = 3 : \mathbf{\Sigma} = \begin{pmatrix} 397.053 & 27.243 & 3.094 \\ 27.243 & 18.357 & 0.400 \\ 3.094 & 0.400 & 1.353 \end{pmatrix}, \mathbf{a} = \begin{pmatrix} 133.85 \\ 25.889 \\ 5.162 \end{pmatrix}, P_0 = 0.929;$$

$$m = 4 : \mathbf{\Sigma} = \begin{pmatrix} 397.053 & 27.243 & 3.094 & 2.777 \\ 27.243 & 18.357 & 0.400 & 0.291 \\ 3.094 & 0.400 & 1.353 & 0.046 \\ 2.777 & 0.291 & 0.046 & 0.766 \end{pmatrix}, \mathbf{a} = \begin{pmatrix} 133.850 \\ 25.889 \\ 5.162 \\ 4.919 \end{pmatrix}, P_0 = 0.978.$$

Calculation results are given in Tables 3, 4, 5 and 6.

From Tables 1, 2, 3, 4, 5 and 6 we see that:

- for each M with increasing *col* confidence intervals are decreased,
- for each *col* the confidence intervals decrease with increasing M.

Thus, accuracy of calculations increases with increase in number of tests of *col* and sample size of M for different p_{targ}. However with increase in number of risk factors fast increase of computing cost is observed. It does necessary to set for the given accuracy acceptable *col* and M values for the purpose of restriction of computing cost. To set the dependence of the calculation accuracy on a sample size and the number of tests an additional research is needed at the fixed value p_{targ}.

Let's fix the value $p_{targ} = 0.5$. Let's change a sample size of M. We find values of calculation accuracy for risk factors of $m = 2$. We repeat for $col = 20$, 50 and 100. Results are given in Table 7 and in Fig. 2.

Let's record value $p_{targ} = 0,7$ for $m = 3$ and $m = 4$ risk factors. That is change will be that is more rational at the solution of real problems. Calculation results of accuracy for $m = 3$ and $m = 4$ respectively are given in Tables 8 and 9 and Figs. 3 and 4.

Curves in the drawing show how the calculation accuracy changed with change of a sample size. Values of number of tests in the drawing are composed in different colors.

On the basis of the data obtained during the researches the regression equation dependence of calculations accuracy (size of a 95% confidence interval) of Δp_{targ} on a sample size of M, numbers of tests of *col* and amount of risk factors of m were worked out:

$$\ln \Delta p_{targ} = -3.132 - 2.13 \cdot 10^{-5} M - 0.439 \ln col + 0.273m. \quad (4)$$

The coefficients of the regression Eq. (4) were statistically significant: p-level of the constant term is equal 3.7 • 10-6, coefficient p-level at variable M is equal 1.4 • 10-10, coefficient p-level at the ln *col* variable is equal 8.3 • 10-4, coefficient p-level at variable m is equal 0.0436, the coefficient of determination is equal to 0.663.

Table 1. Values of 95% confidence intervals for calculated values p_{targ} for $m = 2$; $P_0 = 0.732$ for various values M at variation of *col*.

p_{targ}	*col*			
	10	20	50	100
	$M = 100$			
0,1	0,0255	0,0240	0,0237	0,0168
0,2	0,0270	0,0400	0,0251	0,0245
0,3	0,0360	0,0388	0,0405	0,0312
0,4	0,0325	0,0405	0,0350	0,0289
0,5	0,0325	0,0393	0,0308	0,0249
0,6	0,0325	0,0223	0,0273	0,0219
0,7	0,0315	0,0418	0,0320	0,0247
	$M = 1000$			
0,1	0,0233	0,0165	0,0070	0,0058
0,2	0,0244	0,0165	0,0088	0,0062
0,3	0,0297	0,0217	0,0150	0,0136
0,4	0,0269	0,0241	0,0192	0,0180
0,5	0,0257	0,0214	0,0188	0,0155
0,6	0,0219	0,0164	0,0105	0,0083
0,7	0,0304	0,0207	0,0131	0,0076
	$M = 10000$			
0,1	0,0075	0,0046	0,0040	0,0028
0,2	0,0063	0,0063	0,0042	0,0033
0,3	0,0129	0,0129	0,0087	0,0069
0,4	0,0180	0,0142	0,0124	0,0059
0,5	0,0164	0,0143	0,0095	0,0043
0,6	0,0066	0,0043	0,0028	0,0023
0,7	0,0131	0,0072	0,0047	0,0035
	$M = 100000$			
0,1	0,0030	0,0016	0,0013	0,0011
0,2	0,0033	0,0038	0,0012	0,0006
0,3	0,0076	0,0034	0,0027	0,0012
0,4	0,0069	0,0049	0,0030	0,0018
0,5	0,0052	0,0036	0,0023	0,0013
0,6	0,0027	0,0019	0,0011	0,0008
0,7	0,0032	0,0019	0,0014	0,0011

Table 2. Values of 95% confidence intervals for calculated values p_{targ} for $m = 2$; $P_0 = 0.732$ for various values *col* at variation of M.

p_{targ}	M			
	100	1000	10000	100000
	$col = 10$			
0,1	0,0255	0,0233	0,0075	0,0030
0,2	0,0270	0,0244	0,0063	0,0033
0,3	0,0360	0,0297	0,0129	0,0076
0,4	0,0325	0,0269	0,0180	0,0069
0,5	0,0325	0,0257	0,0164	0,0052
0,6	0,0325	0,0219	0,0066	0,0027
0,7	0,0315	0,0304	0,0131	0,0032
	$col = 20$			
0,1	0,0240	0,0165	0,0046	0,0016
0,2	0,0400	0,0165	0,0063	0,0038
0,3	0,0388	0,0217	0,0129	0,0034
0,4	0,0405	0,0241	0,0142	0,0049
0,5	0,0393	0,0214	0,0143	0,0036
0,6	0,0223	0,0164	0,0043	0,0019
0,7	0,0418	0,0207	0,0072	0,0019
	$col = 50$			
0,1	0,0237	0,0070	0,0040	0,0013
0,2	0,0251	0,0088	0,0042	0,0012
0,3	0,0405	0,0150	0,0087	0,0027
0,4	0,0350	0,0192	0,0124	0,0030
0,5	0,0308	0,0188	0,0095	0,0023
0,6	0,0273	0,0105	0,0028	0,0011
0,7	0,0320	0,0131	0,0047	0,0014
	$col = 100$			
0,1	0,0168	0,0058	0,0028	0,0011
0,2	0,0245	0,0062	0,0033	0,0006
0,3	0,0312	0,0136	0,0069	0,0012
0,4	0,0289	0,0180	0,0059	0,0018
0,5	0,0249	0,0155	0,0043	0,0013
0,6	0,0219	0,0083	0,0023	0,0008
0,7	0,0247	0,0076	0,0035	0,0011

Table 3. Values of 95% confidence intervals for calculated values p_{targ} for $m = 3$; $P_0 = 0.929$ for various values M at variation of *col*.

p_{targ}	*col*			
	10	20	50	100
	$M = 100$			
0,1	0,0210	0,0203	0,0203	0,0179
0,2	0,0345	0,0300	0,0294	0,0259
0,3	0,0465	0,0428	0,0425	0,0321
0,4	0,0500	0,0378	0,0462	0,0282
0,5	0,0505	0,0447	0,0308	0,0349
0,6	0,0555	0,0358	0,0368	0,0307
0,7	0,0315	0,0320	0,0297	0,0240
0,8	0,0200	0,0218	0,0172	0,0112
0,9	0,0240	0,0160	0,0172	0,0130
	$M = 1000$			
0,1	0,0184	0,0110	0,0076	0,0055
0,2	0,0257	0,0163	0,0059	0,0045
0,3	0,0304	0,0232	0,0091	0,0062
0,4	0,0297	0,0211	0,0090	0,0070
0,5	0,0343	0,0207	0,0153	0,0090
0,6	0,0294	0,0187	0,0128	0,0090
0,7	0,0230	0,0177	0,0096	0,0055
0,8	0,0119	0,0108	0,0060	0,0062
0,9	0,0146	0,0086	0,0052	0,0056
	$M = 10000$			
0,1	0,0073	0,0047	0,0035	0,0017
0,2	0,0039	0,0035	0,0019	0,0015
0,3	0,0079	0,0070	0,0040	0,0030
0,4	0,0125	0,0062	0,0049	0,0049
0,5	0,0107	0,0074	0,0064	0,0033
0,6	0,0112	0,0080	0,0052	0,0036
0,7	0,0096	0,0068	0,0061	0,0021
0,8	0,0052	0,0037	0,0028	0,0027
0,9	0,0048	0,0047	0,0042	0,0029
	$M = 100000$			
0,1	0,0036	0,0014	0,0009	0,0005
0,2	0,0017	0,0011	0,0008	0,0005
0,3	0,0035	0,0021	0,0013	0,0007
0,4	0,0044	0,0032	0,0019	0,0015
0,5	0,0042	0,0027	0,0018	0,0013
0,6	0,0039	0,0023	0,0013	0,0009
0,7	0,0057	0,0026	0,0010	0,0007
0,8	0,0022	0,0021	0,0012	0,0005
0,9	0,0026	0,0029	0,0017	0,0010

Table 4. Values of 95% confidence intervals for calculated values p_{targ} for $m = 3$; $P_0 = 0.929$ for various values *col* at variation of *M*.

p_{targ}	*M*			
	100	1000	10000	100000
	$col = 10$			
0,1	0,0210	0,0184	0,0073	0,0036
0,2	0,0345	0,0257	0,0039	0,0017
0,3	0,0465	0,0304	0,0079	0,0035
0,4	0,0500	0,0297	0,0125	0,0044
0,5	0,0505	0,0343	0,0107	0,0042
0,6	0,0555	0,0294	0,0112	0,0039
0,7	0,0315	0,0230	0,0096	0,0057
0,8	0,0200	0,0119	0,0052	0,0022
0,9	0,0240	0,0146	0,0048	0,0026
	$col = 20$			
0,1	0,0203	0,0110	0,0047	0,0014
0,2	0,0300	0,0163	0,0035	0,0011
0,3	0,0428	0,0232	0,0070	0,0021
0,4	0,0378	0,0211	0,0062	0,0032
0,5	0,0447	0,0207	0,0074	0,0027
0,6	0,0358	0,0187	0,0080	0,0023
0,7	0,0320	0,0177	0,0068	0,0026
0,8	0,0218	0,0108	0,0037	0,0021
0,9	0,0160	0,0086	0,0047	0,0029
	$col = 50$			
0,1	0,0203	0,0076	0,0035	0,0009
0,2	0,0294	0,0059	0,0019	0,0008
0,3	0,0425	0,0091	0,0040	0,0013
0,4	0,0462	0,0090	0,0049	0,0019
0,5	0,0308	0,0153	0,0064	0,0018
0,6	0,0368	0,0128	0,0052	0,0013
0,7	0,0297	0,0096	0,0061	0,0010
0,8	0,0172	0,0060	0,0028	0,0012
0,9	0,0172	0,0052	0,0042	0,0017
	$col = 100$			
0,1	0,0179	0,0055	0,0017	0,0005
0,2	0,0259	0,0045	0,0015	0,0005
0,3	0,0321	0,0062	0,0030	0,0007
0,4	0,0282	0,0070	0,0049	0,0015
0,5	0,0349	0,0090	0,0033	0,0013
0,6	0,0307	0,0090	0,0036	0,0009
0,7	0,0240	0,0055	0,0021	0,0007
0,8	0,0112	0,0062	0,0027	0,0005
0,9	0,0130	0,0056	0,0029	0,0010

Table 5. Values of 95% confidence intervals for calculated values p_{targ} for $m = 4$; $P_0 = 0.978$ for various values M at variation of *col*.

p_{targ}	*col*			
	10	20	50	100
	$M = 100$			
0,1	0,0780	0,0920	0,0546	0,0319
0,2	0,1265	0,1093	0,0941	0,0527
0,3	0,1505	0,1178	0,0775	0,0625
0,4	0,1875	0,1448	0,1329	0,1130
0,5	0,1705	0,1605	0,1428	0,1113
0,6	0,1485	0,1448	0,1304	0,0807
0,7	0,1685	0,1503	0,0762	0,0797
0,8	0,1415	0,1405	0,1121	0,0319
0,9	0,0865	0,0787	0,0605	0,0534
	$M = 1000$			
0,1	0,0298	0,0179	0,0092	0,0071
0,2	0,0464	0,0254	0,0126	0,0246
0,3	0,0788	0,0435	0,0168	0,0105
0,4	0,0553	0,0342	0,0210	0,0228
0,5	0,0565	0,0312	0,0181	0,0087
0,6	0,0480	0,0417	0,0264	0,0131
0,7	0,0843	0,0439	0,0150	0,0127
0,8	0,0459	0,0259	0,0152	0,0100
0,9	0,0269	0,0290	0,0186	0,0101
	$M = 10000$			
0,1	0,0086	0,0057	0,0044	0,0071
0,2	0,0108	0,0110	0,0058	0,0139
0,3	0,0452	0,0415	0,0052	0,0038
0,4	0,0580	0,0429	0,0104	0,0061
0,5	0,0194	0,0100	0,0069	0,0087
0,6	0,0203	0,0174	0,0223	0,0082
0,7	0,0167	0,0122	0,0079	0,0093
0,8	0,0322	0,0077	0,0063	0,0040
0,9	0,0209	0,0141	0,0082	0,0032
	$M = 100000$			
0,1	0,0036	0,0022	0,0026	0,0028
0,2	0,0043	0,0024	0,0022	0,0027
0,3	0,0142	0,0089	0,0030	0,0019
0,4	0,0112	0,0095	0,0085	0,0027
0,5	0,0058	0,0061	0,0037	0,0042
0,6	0,0051	0,0269	0,0076	0,0038
0,7	0,0056	0,0030	0,0040	0,0051
0,8	0,0034	0,0028	0,0014	0,0019
0,9	0,0047	0,0035	0,0013	0,0019

Table 6. Values of 95% confidence intervals for calculated values p_{targ} for $m = 4$; $P_0 = 0.978$ for various values *col* at variation of M.

p_{targ}	M			
	100	1000	10000	100000
	$col = 10$			
0,1	0,0780	0,0298	0,0086	0,0036
0,2	0,1265	0,0464	0,0108	0,0043
0,3	0,1505	0,0788	0,0452	0,0142
0,4	0,1875	0,0553	0,0580	0,0112
0,5	0,1705	0,0565	0,0194	0,0058
0,6	0,1485	0,0480	0,0203	0,0051
0,7	0,1685	0,0843	0,0167	0,0056
0,8	0,1415	0,0459	0,0322	0,0034
0,9	0,0865	0,0269	0,0209	0,0047
	$col = 20$			
0,1	0,0920	0,0179	0,0057	0,0022
0,2	0,1093	0,0254	0,0110	0,0024
0,3	0,1178	0,0435	0,0415	0,0089
0,4	0,1448	0,0342	0,0429	0,0095
0,5	0,1605	0,0312	0,0100	0,0061
0,6	0,1448	0,0417	0,0174	0,0269
0,7	0,1503	0,0439	0,0122	0,0030
0,8	0,1405	0,0259	0,0077	0,0028
0,9	0,0788	0,0290	0,0141	0,0035
	$col = 50$			
0,1	0,0546	0,0092	0,0044	0,0026
0,2	0,0941	0,0126	0,0058	0,0022
0,3	0,0775	0,0168	0,0052	0,0030
0,4	0,1329	0,0210	0,0104	0,0085
0,5	0,1428	0,0181	0,0069	0,0037
0,6	0,1304	0,0264	0,0223	0,0076
0,7	0,0762	0,0150	0,0079	0,0040
0,8	0,1121	0,0152	0,0063	0,0014
0,9	0,0605	0,0186	0,0082	0,0013
	$col = 100$			
0,1	0,0319	0,0071	0,0071	0,0028
0,2	0,0527	0,0246	0,0139	0,0027
0,3	0,0625	0,0105	0,0038	0,0019
0,4	0,1130	0,0228	0,0061	0,0027
0,5	0,1113	0,0087	0,0087	0,0042
0,6	0,0807	0,0131	0,0082	0,0038
0,7	0,0797	0,0127	0,0093	0,0051
0,8	0,0319	0,0100	0,0040	0,0019
0,9	0,0534	0,0101	0,0032	0,0019

Table 7. Dependence of the calculation accuracy on a sample size and number of tests for $m = 2$.

lg$M$	Number of tests for $p_{targ} = 0.5$			
	$col = 10$	$col = 20$	$col = 50$	$col = 100$
2	0,033	0,039	0,031	0,025
3	0,026	0,021	0,019	0,015
4	0,016	0,014	0,010	0,004
5	0,005	0,004	0,002	0,001

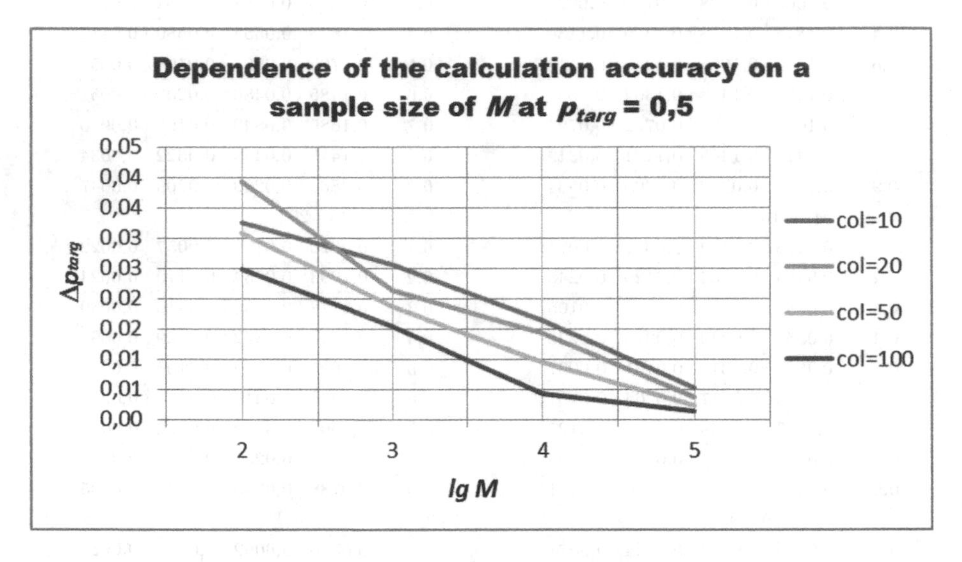

Fig. 2. Dependence of the calculation accuracy on a sample size of M for $m = 2$ risk factors at $p_{targ} = 0.5$.

Table 8. Dependence of the calculation accuracy on a sample size and number of tests for $m = 3$.

lg$M$	Number of tests for $p_{targ} = 0.7$			
	$col = 10$	$col = 20$	$col = 50$	$col = 100$
2	0,032	0,032	0,030	0,024
3	0,023	0,018	0,010	0,005
4	0,010	0,007	0,006	0,002
5	0,006	0,003	0,001	0,001

Thus, the statistical significance of the Eq. (4) allows us to conclude that depending on amount of risk factors it is possible to choose the number of tests and a sample size according to the required accuracy.

Table 9. Dependence of the calculation accuracy on a sample size and number of tests for $m = 4$.

lg$M$	Number of tests for $p_{targ} = 0.7$			
	$col = 10$	$col = 20$	$col = 50$	$col = 100$
2	0,169	0,150	0,076	0,080
3	0,084	0,044	0,015	0,013
4	0,017	0,012	0,008	0,009
5	0,006	0,003	0,004	0,005

Follows from the Eq. (4) that at constancy of other factors:

- increase in M on 100 reduces Δp_{targ} on average by 0.213%;
- increase in col by 1% reduces Δp_{targ} on average by 0.439%;
- increase in m on 1 increases Δp_{targ} on average by 31.36%.

The execution time of the algorithm directly depends on the technical characteristics of the personal computer. For example, on a PC with a performance of 20 GFLOPS in single-threaded mode, execution time of the algorithm was $t =$ 21.67 s with initial characteristics $m = 2$, $M = 10000$, $col = 10$, $p_{targ} = 0.5$. By increasing the sample size M in 10 times, the execution time of the algorithm is increasing in 10 times. Similarly for col. However, when using parallel computing in the implementation of the algorithm, the execution speed can be increased in times, up to the number of threads involved. Therefore, for a larger number of

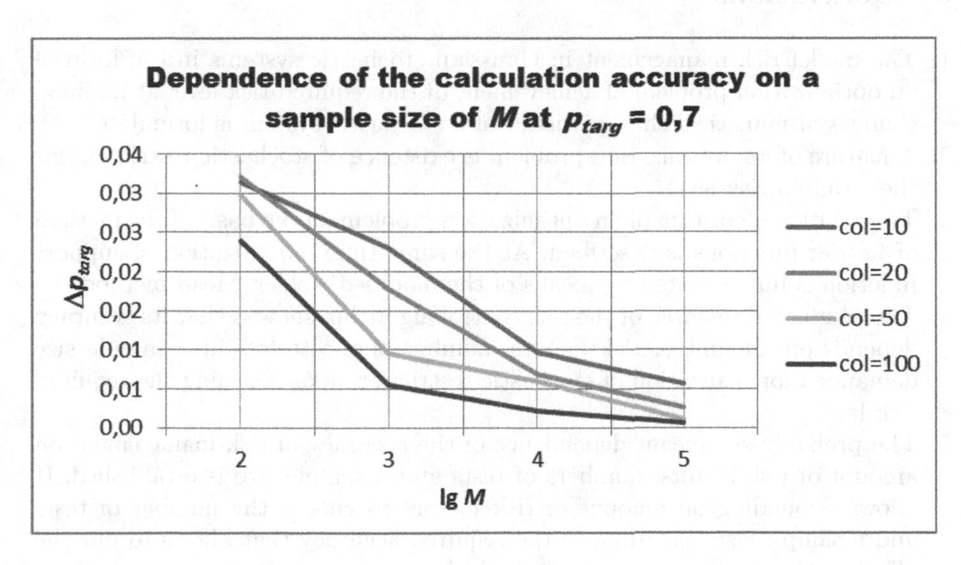

Fig. 3. Dependence of the calculation accuracy on a sample size of M for $m = 3$ risk factors at $p_{targ} = 0.7$.

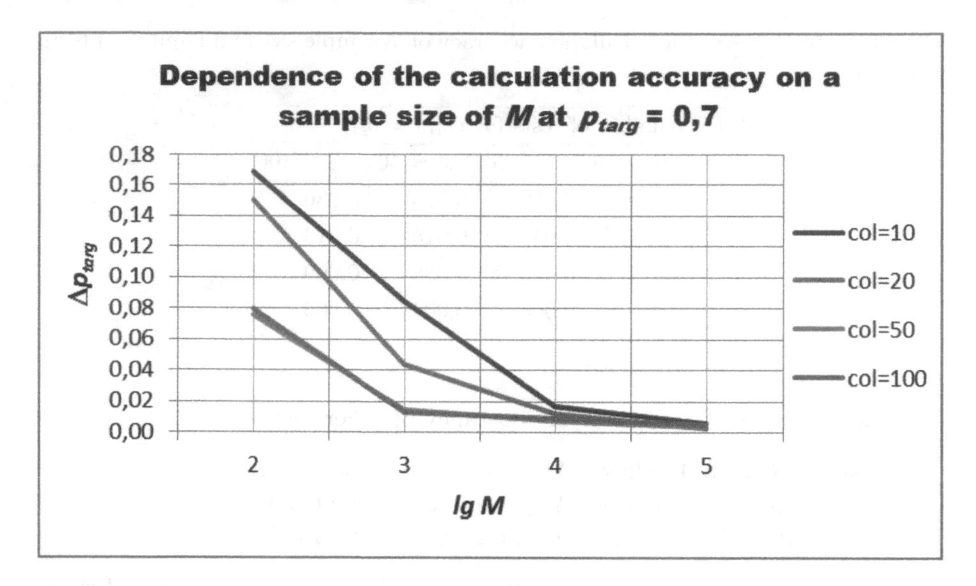

Fig. 4. Dependence of the calculation accuracy on a sample size of M for $m = 4$ risk factors at $p_{targ} = 0.7$.

the risk factors m, a reasonable result can be obtained by using large computing resources in the form of parallel computing on GPU or cluster computing.

5 Conclusions

1. The model risk management in Gaussian stochastic systems in the form of an optimization problem of achievement of the required risk level at minimal changes of numerical characteristics of a stochastic system is formulated.
2. A feature of an optimization problem is existence of stochastic restriction for the required risk level.
3. The solution algorithm of an optimization problem on the basis of the method of barrier functions is described. At the same time minimization of support function is implemented by means of the modified Nelder–Mead method.
4. The conducted research of the risk control algorithm showed that its accuracy depends on amount of risk factors, number of the tests and a sample size demanded for realization of stochastic restriction upon reaching the required risk level.
5. The probably significant dependence of the accuracy of risk management on amount of risk factors, numbers of tests and a sample size is established. It allows depending on amount of risk factors to choose the number of tests and a sample size according to the required accuracy that allows to put the offered risk management into practice.

References

1. Vishnyakov, Ya.D., Radayev, N.N.: Common Theory of Risks, 2nd edn. Academy, Moscow (2008)
2. Akimov, V.A., Lesnykh, V.V., Radaev, N.N.: Riski v prirode, tehnosfere, obshhestve i ekonomike [Risks in nature, technosphere, society, and the economy]. Business Express, Moscow (2004)
3. Mun, J.: Modeling Risk, 2nd edn. Wiley, Hoboken (2010)
4. Scheule, H., Rosch, D. (eds.): Model Risk: Identification, Measurement, and Management. Risk Books, London (2010)
5. Madera, A.G.: Risks and Chances. Uncertainty, Forecasting, and Assessment. URSS, Moscow (2014)
6. Solozhentsev, E.D.: Stsenarnoe logiko-veroyatnostnoe upravlenie riskom v biznesei tekhnike [Scenario logic and probabilistic management of risk in business and engineering], 2nd edn. Business Press, Saint Petersburg (2006)
7. Ryabinin, I.A.: Nadezhnost' i bezopasnost' strukturno-slozhnykh sistem [Reliability and safety of the structural and composite systems]. Polytechnique, Saint Petersburg (2000)
8. Vorobyov, Yu.L., Malinetsky, G.G., Makhutov, N.A.: Management of risk and sustainable development: human measurement. News of higher education institutions. Appl. Nonlinear Dyn. **8**(6), 12–26 (2000)
9. Gorelik, V.A., Zolotova, T.V.: General approach to modeling of risk management procedures and its application to stochastic and hierarchic systems. Manage. Large Syst. **37**, 5–24 (2012). (in Russian)
10. Tyrsin, A.N.: About model operation of risk in the systems of critical infrastructures. In: Economic and technical aspects of safety of structural critical infrastructures: Materials of the international conference, pp. 205–208. URFU, Yekaterinburg (2015)
11. Tyrsin, A.N., Surina, A.A.: Monitoring of risk of multidimensional stochastic system as tools for research of sustainable development of regions. IOP Conf. Series Earth Environ. Sci. **1**(124), 776–783 (2018)
12. Tyrsin, A.N., Surina, A.A.: Models of monitoring and management of risk in Gaussian stochastic systems. Series Nat. Tech. Sci. **23**(124), 776–783 (2018). The Bulletin of the Tambov University
13. Panteleev, A.V., Letova, T.A.: Optimization methods in examples and tasks. 3rd prod. The Higher School, Moscow (2008)
14. Mikhaylov, G.A., Voytishek, A.V.: Numerical Statistical Model Operation. Monte-Carlo Methods. Academy, Moscow (2006)
15. Tyrsin, A.N., Kalev, O.F., Yashin, D.A., Surina, A.A.: Model of risk of a multidimensional stochastic system as tools of a research of the state of health of population. Syst. Anal. Manage. Biomed. Syst. **17**(4), 948–957 (2018)
16. Surina, A.A.: The program of risk management of a multidimensional Gaussian stochastic system: certificate on the state filing of the computer program No. 2018661134; stat. 21.08.2018; publ. 03.09.2018

The Accuracy of Approximate Solutions for a Boundary Value Inverse Problem with Final Overdetermination

Elena Tabarintseva(✉)

South Ural State University, Chelyabinsk, Russia
eltab@rambler.ru

Abstract. The paper aims to investigate the accuracy of the methods for approximate solving a boundary value inverse problem with final overdetermination for a parabolic equation. We use the technique of the continuation to the complex domain and the expansion of the unknown function into a Dirichlet series (exponential series) to formulate the inverse problem as a linear operator equation of the first kind in the appropriate linear normed spaces. This allows us to estimate the continuity module for the inverse problem through classical spectral technique and investigate the order-optimal approximate methods for the boundary value inverse problem under study.

Keywords: Parabolic equation · Boundary value inverse problem · Module of continuity of the inverse operator · Exponential series

1 Introduction

We study a boundary value inverse problem with final overdetermination (the problem of the most accurate heating of a rod). Namely, we should recover the boundary condition in a mixed boundary value problem for the heat transfer equation from the knowledge of the solution at the final time moment.

Originally, the problem was formulated in [8] in the form of an optimization problem. In applications, a large number of optimization problems associated with parabolic equations arise. One of such problems in thermophysical terms can be formulated as follows. Consider a homogeneous rod with a thermally insulated lateral surface, the left end of which is thermally insulated, and the given temperature $h(t)$ s maintained at the right end. We need, by controlling the temperature at the right end of the rod, make the temperature distribution in the rod as close as possible to the specified distribution $g(x)$ by a given point in time. Namely, let $u(x,t) = u(x,t,h)$ define the distribution of the temperature

The work was supported by Act 211 Government of the Russian Federation, contract No 02.A03.21.0011.

I. Bykadorov et al. (Eds.): MOTOR 2019, CCIS 1090, pp. 578–589, 2019.
https://doi.org/10.1007/978-3-030-33394-2_44

in the rod at the time moment t. It is required, by controlling the function $h \in L_2[0.T]$, to minimize the function

$$J(h) = \|u(x,T) - g(x)\|^2$$

under the condition that $u(x,t) = u(x,t,h)$ solves the boundary value problem

$$\frac{\partial u}{\partial t} = \frac{\partial^2 u}{\partial x^2}; \quad t \in (0;T), \quad x \in (0;l)$$

$$u(x,0) = 0 \quad (0 < x < l), \; u(0,t) = 0, \quad u(l,t) = h(t) \quad (0 < t < T)$$

$h(t)$ satisfies the conditions $h(0) = 0$; $\|h'(t)\|_{L_2[0,T]} \leq r$.

Aviation, rocket and space technology, energy and metallurgy use experimental studies, bench and field studies of thermal conditions, the creation of effective diagnostic methods and the results of heat exchange processes are based on the results of experiments and tests. These methods are based on the problems associated with inverse problems of different types, the boundary value inverse problems being among the most important classes o the inverse problems of thermal conductivity [10,11,16]. Various statements of the boundary value inverse problems were studied also in [12].

For linear ill-posed problems, the classical spectral technique is widely used to obtain estimates of the error for the approximate solutions on compact sets (correctness classes) and the error of the optimal approximate methods.

For such problems, the technique of calculating the error of the optimal method is based on the relation between the error of the optimal method and the module of continuity of the inverse operator, which can be calculated for specific operators and correctness classes. The continuity module for some classes of nonlinear inverse problems was estimated, for example, in [1–3,7].

In the classical spectral technique, the commuting of the operator of the problem with the operator defining the correctness class (reflecting a priori information about the exact solution of the inverse problem) plays the main role but for some important inverse problems in the classical formulation, these operators do not commute.

We use the technique of extending of the domain to the complex domain and expansion of the unknown function into a Dirichlet series (exponential series) to formulate the inverse problem as an operator equation of the first kind in the space that is isometric to the space of the initial data and the space of the solutions. This allows us to calculate the module of continuity and investigate the accuracy of optimal and order-optimal approximate methods for the inverse problem under study. The obtained estimate for the modulus of continuity makes it possible to investigate the optimal methods for the approximate solution of the inverse problem and construct the order-optimal methods.

The paper is organized as follows.

In Sect. 2 we formulate the boundary value inverse problem and the corresponding direct problem and introduce the functional spaces which will be used in the study of the inverse problem.

In Sect. 3 we obtain the linear operator equation which gives the equivalent formulation of the inverse problem and the corresponding direct problem in l_2 space.

Section 4 is aimed at the estimation of the function which plays the basic role for the investigation of the accuracy of the approximate solution to the inverse problem - its continuity module.

2 The Inverse Problem with Final Overdetermination

Direct Problem. Consider the following boundary value problem for a heat conductivity equation. We have to determine a function $u = u(x,t)$,

$$u \in C([0,l]; W_2^1[0;T]) \cap C^2((0,l); L_2[0;T])$$

which meets the conditions

$$\frac{\partial u}{\partial t} = \frac{\partial^2 u}{\partial x^2}; \quad t \in (0;T), \quad x \in (0;l), \tag{1}$$

$$u(x,0) = 0 \quad (0 < x < l),$$

$$u(0,t) = 0, \quad u(l,t) = h(t) \quad (0 < t < T).$$

(we will consider the case $T = 2\pi$).

Inverse Problem. We study the following inverse problem for a parabolic equation. Suppose that $u(x,t)$ satisfies the conditions (1) and the additional condition

$$u(x,T) = g(x). \tag{2}$$

We have to recover the function $h \in L_2[0;T]$ (the boundary condition), $g \in C[0;l]$ is the given function.

The problem was formulated and studied in [8,15] as an optimization problem.

Recall the well-known theorem on approximation of continuous functions by polynomials [5].

Theorem 1. *Let* $0 < \lambda_1 < \lambda_2 < ..., \lambda_i \to \infty$, $0 < a < b$. *For any function* $f \in C[a,b]$ *and for any* $\varepsilon > 0$ *there exists a linear combination*

$$P_n(x) = \sum_{i=1}^{n} c_i x^{\lambda_i}$$

such that

$$\|f - P_n\|_{C[a,b]} < \varepsilon$$

if and only if the condition

$$\sum_{i=1}^{\infty} \frac{1}{\lambda_i} = \infty$$

is satisfied.

The next lemma follows directly from the theorem and gives the possibility to approximate continuous functions by exponential polynomials [6].

Lemma 1. *Suppose that the real valued functions* $\{e_i(x)\}$; $e_0(x) = 1$, $e_i(x) = e^{-\lambda_i x}$ *are defined with the help of the numerical series* $0 < \lambda_1 < \lambda_2 < ..., \lambda_i \to \infty$ *that satisfies the condition*

$$\sum_{i=1}^{\infty} \frac{1}{\lambda_i} = \infty.$$

Then the system of functions $\{e_i(x)\}_{n=0}^{\infty}$ *is closed in the space* $C[0, l]$.

Denote by $g(x)$ a continuous function on $[0, l]$ which can be expanded into the uniformly converging exponential series

$$G(x) = \sum_{n=1}^{\infty} b_n e^{-\sqrt{n}x}. \tag{3}$$

If the series (3) converges for $x \geq -l$, then the odd extension of $g(x)$ defined on the segment $[-l, l]$ can be expanded into the uniformly converging series

$$g(x) = \sum_{n=1}^{\infty} b_n \operatorname{sh} \sqrt{n}x. \tag{4}$$

We conclude from the elementary properties of the exponential series that the function

$$G_0(z) = \sum_{n=1}^{\infty} b_n \operatorname{sh} \sqrt{n}z \tag{5}$$

is analytical and bounded in the domain $-l < \mathrm{Re}z < l$.

Let a function $G(z)$ have a Dirichlet series expansion

$$G_0(z) = \sum_{n=1}^{\infty} b_n \; sh\sqrt{n}z,$$

convergent uniformly in the strip $-l < Rez < l$, i.e. it is an analytic almost periodic function in the strip. Then the function $p(y) = G(x + iy)$ for any $-l < y < l$ is a uniform almost periodic function, whose norm can be defined by the formula [14]

$$\|p\|_0^2 = \lim_{Y \to \infty} \frac{1}{2Y} \int_{-Y}^{Y} \|p(iy)\|^2 dy.$$

Let X denote the linear space of functions $G(z)$ which are analytical and bounded in the domain $0 < \mathrm{Re}z < l$, equipped with the norm

$$\|G\|^2 = \lim_{Y \to \infty} \frac{1}{2Y} \int_{-Y}^{Y} \|G(iy)\|^2 dy.$$

Suppose that for the given continuous function $g(x)$ there exists an exact solution $h(t)$ to the inverse problem (1) which belongs to the set

$$M = \{h(t) : \|h''(t)\|_{L_2[0,\infty)} \leq r\}.$$

Suppose also that the values of $g(x)$ are unknown but instead we know its approximate values. That means that we know the function $g_\delta \in C[0;l]$ such that for the analytical extensions of this function and the exact function $g(x)$ in the domain $0 < \text{Re}z < l$, the inequality $\|g - g_\delta\|_X < \delta$ holds. We have to determine an approximate solution h_δ to the boundary value inverse problem and estimate its deviation from the exact solution.

3 Reducing the Boundary Value Inverse Problem to the Operator Equation

We will find the solution to the direct problem in the form of the complex Fourier series

$$u(x,t) = \sum_{n=-\infty}^{\infty} c_n(x)e^{int} = 2\text{Re}U(x,t),$$

where

$$U(x,t) = \frac{c_0(x)}{2} + \sum_{n=1}^{\infty} c_n(x)e^{int}.$$

Consider the expansion into the Fourier series

$$h(t) = \sum_{n=-\infty}^{\infty} h_n e^{int} = 2\text{Re}H(t).$$

Here

$$H(t) = \frac{h_0}{2} + \sum_{n=1}^{\infty} h_n e^{int}.$$

Consider the particular solution $u_n(x,t)$ to the mixed boundary value problem which corresponds to the boundary condition $u_n(l,t) = e^{int}$. We find the particular solution in the form $u_n(x,t) = w_n(x,t) + v_n(x,t)$ (see, for example, [13]),

$$w_n(x,t) = \frac{\text{sh}\,\mu_0\sqrt{n}x}{\text{sh}\,\mu_0\sqrt{n}l}e^{int},$$

where $\mu_0 = \frac{1}{\sqrt{2}}(1+i)$ and the function $v_n(x,t)$ satisfies the conditions

$$\frac{\partial v_n}{\partial t} = \frac{\partial^2 v_n}{\partial x^2}; \quad t \in (0;T), \quad x \in (0;l), \tag{6}$$

$$v_n(x,0) = -\frac{\text{sh}\,\mu_0\sqrt{n}x}{\text{sh}\,\mu_0\sqrt{n}l} \quad (0 < x < l),$$

$$v_n(0,t) = 0, \quad v_n(l,t) = 0 \quad (0 < t < T).$$

Make sure that the function

$$Y(x,t) = \sum_{n=1}^{\infty} h_n \frac{\operatorname{sh} \mu_0 \sqrt{n} x}{\operatorname{sh} \mu_0 \sqrt{n} l} e^{int} \tag{7}$$

satisfies the conditions

$$\frac{\partial Y}{\partial t} = \frac{\partial^2 Y}{\partial x^2}; \quad t \in (0;T), \quad x \in (0;l), \tag{8}$$

$$Y(x,0) = \Phi(x),$$

$$Y(0,t) = 0, \quad Y(l,t) = H(t) \quad (0 < t < T),$$

where

$$\Phi(x) = \sum_{n=1}^{\infty} h_n \frac{\operatorname{sh} \mu_0 \sqrt{n} x}{\operatorname{sh} \mu_0 \sqrt{n} l}.$$

Actually, for $0 < x < l - \delta$ it holds

$$\left| \frac{\operatorname{sh} \mu_0 \sqrt{n} x}{\operatorname{sh} \mu_0 \sqrt{n} l} \right| \leq e^{-\sqrt{\frac{n}{2}(l-x)}},$$

so for the series (7) the majorizing series is

$$\sum_{n=1}^{\infty} |h_n| e^{-\sqrt{\frac{n}{2}(l-x)}} \tag{9}$$

and for the series which we obtain by termwise differentiation of (7), the majorizing series is

$$\sum_{n=1}^{\infty} n |h_n| e^{-\sqrt{\frac{n}{2}(l-x)}}. \tag{10}$$

The series (9) and (10) are converging, therefore the series (7) is uniformly converging on any interval $0 < x < l - \delta$, and the function $Y(x,t)$ satisfies the conditions (8).

Hence, we can represent the solution of the boundary value problem in the form

$$U(x,t) = \sum_{n=1}^{\infty} h_n \frac{\operatorname{sh} \mu_0 \sqrt{n} x}{\operatorname{sh} \mu_0 \sqrt{n} l} e^{int} - Z(x,t),$$

$Z(x,t)$ satisfies the conditions

$$\frac{\partial Z}{\partial t} = \frac{\partial^2 Z}{\partial x^2}; \quad t \in (0;T), \quad x \in (0;l), \tag{11}$$

$$Z(x,0) = -\Phi(x) \quad (0 < x < l),$$

$$Z(0,t)=0,\quad Z(l,t)=0\quad (0<t<T).$$

Here

$$\Phi(x)=\sum_{n=1}^{\infty}h_n\frac{\operatorname{sh}\mu_0\sqrt{n}x}{\operatorname{sh}\mu_0\sqrt{n}l}.$$

We have to recover the function $G(x)=\Phi(x)+Z(x,T)$.

Denote by

$$a_n=\frac{h_n}{\operatorname{sh}\mu_0\sqrt{n}l}$$

the coefficients of the expansion of $\Phi(x)$ in the Dirichlet series. Considering the expansion of a hyperbolic sine in the Fourier series, we find the Fourier coefficients of the function $\Phi(x)$:

$$\Phi(x)=\sum_{k=1}^{\infty}\varphi_k\sin\frac{\pi kx}{l},$$

where

$$\varphi_k=2(-1)^{k+1}\sum_{n=1}^{\infty}a_n\frac{\pi k\operatorname{sh}\mu_0\sqrt{n}l}{(\pi k)^2+inl^2}. \tag{12}$$

Consider the function

$$F(\lambda)=2\sum_{n=1}^{\infty}a_n\frac{\lambda\operatorname{sh}\mu_0\sqrt{n}l}{\lambda^2+inl^2}.$$

It is evident that

$$F(\pi k)=(-1)^{k+1}\varphi_k.$$

We show that the function $F(\lambda)$ is analytic in the domain

$$D_k=\left\{0<|\lambda-\mu_1\sqrt{k}l|<\frac{1}{k}\right\},$$

$\mu_1=\frac{1}{\sqrt{2}}(1-i)$.

Consider the series

$$\sum_{n=1}^{\infty}a_n\frac{\lambda\operatorname{sh}\mu_0\sqrt{n}l}{\lambda^2+inl^2}=\sum_{n=1}^{\infty}h_n\frac{\lambda}{\lambda^2+inl^2}..$$

For $n\neq k$ the inequalities hold

$$|\lambda|\leq\sqrt{k}+\frac{1}{k}\leq\sqrt{k}l+1;$$

$$|\lambda-\mu_1\sqrt{n}l|\geq|\mu_1\sqrt{n}l-\mu_1\sqrt{k}l|-|\lambda-\mu_1\sqrt{k}l|$$
$$\geq l|\sqrt{n}-\sqrt{k}|-\frac{1}{k}.$$

Similarly,

$$|\lambda - \mu_1\sqrt{n}l| \geq l|\sqrt{n} - \sqrt{k}| - \frac{1}{k}.$$

Then,

$$|\lambda^2 + inl^2| = |\lambda - \mu_1\sqrt{n}l||\lambda + \mu_1\sqrt{n}l|$$
$$\geq \left(|\sqrt{n} - \sqrt{k}| - \frac{1}{k}\right)^2$$

and

$$|\frac{h_n\lambda}{\lambda^2 + inl^2}| \leq \frac{|h_n|(\sqrt{k}l + 1)}{(|\sqrt{n} - \sqrt{k}| - \frac{1}{k})^2}.$$

Note that

$$\frac{|h_n|(\sqrt{k}l + 1)}{(\sqrt{n} - \sqrt{k}| - \frac{1}{k})^2}. \simeq \frac{|h_n|(\sqrt{k}l + 1)}{l^2 n}$$

for $n \to \infty$. Hence the majorizing series

$$\sum_{n=1}^{\infty} \frac{|h_n|(\sqrt{k}l + 1)}{(\sqrt{n} - \sqrt{k}| - \frac{1}{k})^2}$$

uniformly converges in every domain D_k ($k = 1, 2, ...$). Hence, the function $F(\lambda)$ is analytical in the domain

$$D = \bigcup_{n=1}^{\infty} D_k.$$

The function $F(\lambda)$ has simple poles at the points $\lambda_n = \mu_1\sqrt{n}l$. Hence, the equality (12) implies

$$\lim_{\lambda \to \mu_1\sqrt{n}l} F(\lambda)(\lambda - \mu_1\sqrt{n}l) = a_n \operatorname{sh} \mu_0\sqrt{n}l. \tag{13}$$

Consider the expansion in Dirichlet series of the function $G(x)$:

$$G(x) = \sum_{n=1}^{\infty} b_n \operatorname{sh} \mu_0\sqrt{n}x.$$

Taking into account the expansion of $G(x)$ in the Fourier series we can conclude that

$$G(x) = \sum_{k=1}^{\infty} g_k \sin\frac{\pi k x}{l},$$

where

$$g_k = \varphi_k(1 + e^{-(\frac{\pi k}{l})^2 T}).$$

Consider the function

$$P(\lambda) = F(\lambda)(1 + e^{-\lambda^2 T}).$$

The function $G(\lambda)$ has simple poles at the points $\lambda_n = \mu_1\sqrt{n}l$ if the condition $nl^2T \neq \pi + 2\pi m$ is met. Similarly to the equality (13), we obtain

$$\lim_{\lambda \to \mu_1\sqrt{n}l} P(\lambda)(\lambda - \mu_1\sqrt{n}l) = b_n \operatorname{sh} \mu_0\sqrt{n}. \tag{14}$$

The equalities (13) and (14) imply

$$\frac{b_n}{a_n} = \lim_{\lambda \to \mu_1\sqrt{n}l} (1 + e^{-\lambda^2 T}) = 1 + e^{-inT}.$$

Let us establish a correspondence between the function $G(z) \in X$ and the sequence of coefficients of its Dirichlet series $\hat{G} = \{b_n\} \in l_1 \cap l_2$. Set also a correspondence between the function $h(t) \in L_2[0, T]$ and the sequence of coefficients of its Fourier series $\hat{H} = \{h_n\} \in l_2$.

The mean value theorem [4] states that

$$\|G\|^2 = \lim_{Y \to \infty} \frac{1}{2Y} \int_{-Y}^{Y} \|G(iy)\|^2 dy = \sum_{n=1}^{\infty} b_n^2.$$

That is, the operator $E : X \to l_2$ which acts according to the rule

$$EG = \{b_n\}$$

is an isometry. Denote $A : l_2 \to l_2$ a linear operator acting according to the rule

$$A\hat{H} = \left\{ \frac{h_n}{2 \operatorname{sh} \mu_0\sqrt{n}l} \right\}_{n=1}^{\infty}.$$

Thus, the inverse problem with final overdetermination can be formulated as an operator equation

$$A\hat{H} = \hat{G}.$$

4 The Continuity Module for the Inverse Problem

Denote by

$$\hat{\omega}(M, \delta) = \sup\{\|h_1 - h_2\| : h_1, h_2 \in M, \|\hat{g}_1 - \hat{g}_2\| \leq \delta\}$$

the continuity module for the boundary value inverse problem with final overdetermination.

We use the scheme proposed in [9] to estimate the continuity module. The following theorem holds.

Theorem 2. *There exists $\delta_0 > 0$, such that for all $0 < \delta < \delta_0$ the inequalities*

$$C_1 \frac{rl^2}{(\ln \delta)^2} \leq \omega(M, \delta) \leq C_2 \frac{rl^2}{(\ln \delta)^2}$$

are true.

Proof. Let $H = \{h_n\}_{n=1}^{\infty} \in l_2$. Consider the linear bounded operator A acting in the space l_2 according to the rule

$$A\hat{H} = \left\{ \frac{h_n}{2\,\mathrm{sh}\,\mu_0\sqrt{n}l} \right\}_{n=1}^{\infty},$$

B is a linear bounded operator acting in the space l_2 as follows

$$B\hat{H} = \left\{ \frac{h_n}{n^2 l} \right\}_{n=1}^{\infty}.$$

Denote $B_1 = B^*B$. Then the operator B_1 acts in l_2 according to the rule

$$B_1\hat{H} = \left\{ \frac{h_n}{n^4 l} \right\}_{n=1}^{\infty}.$$

Denote $C = AB$, $C_1 = C^*C$. Then C_1 is the operator acting in l_2 according to the rule

$$C_1\hat{H} = \left\{ \frac{h_n}{4n^2|\,\mathrm{sh}\,\mu_0\sqrt{n}l|^2} \right\}_{n=1}^{\infty}.$$

It follows from the definitions of the operators that C_1 is a function of the operator B_1. Namely,

$$C_1 = \lambda(B_1),$$

where

$$\lambda(\sigma) = \frac{\sigma^2}{4\left|\mathrm{sh}\,\frac{\mu_0 l}{\sigma^{1/4}}\right|^2}. \tag{15}$$

Calculating the module of a hyperbolic sine and denoting

$$s = \frac{1}{\sigma^{1/4}},$$

we write the equality (15) as

$$\lambda(s) = \frac{1}{s^8(\mathrm{sh}^2\,\frac{l}{2}s + \sin^2\frac{l}{2}s)}. \tag{16}$$

Further,

$$\lim_{s\to 0}\lambda(s) = +\infty, \quad \lim_{s\to\infty}\lambda(s) = 0,$$

the function $\lambda(s)$ is continuous and monotone on $(0,\infty)$. Thus, the Eq. (16) has a unique solution $s = s(\lambda)$ for every $\lambda > 0$.

Consider the elementary inequalities

$$\mathrm{sh}^2\,\frac{l}{2}s + \sin^2\frac{l}{2}s \le e^{\sqrt{2}ls}; \tag{17}$$

$$\mathrm{sh}^2\,\frac{l}{2}s + \sin^2\frac{l}{2}s \ge \frac{e^{\sqrt{2}ls}}{4} - \frac{1}{2}. \tag{18}$$

For $s > \frac{2\sqrt{2}}{l}\ln 2$, we obtain

$$e^{\sqrt{2}ls} > 4$$

and inequality (18) implies

$$\operatorname{sh}^2 \frac{l}{2}s + \sin^2 \frac{l}{2}s \geq \frac{e^{\sqrt{2}ls}}{8}. \tag{19}$$

Applying the inequalities (17) and (19) for $s > \frac{2\sqrt{2}}{l}\ln 2$ we obtain

$$\frac{e^{\sqrt{2}ls}}{8} \leq \lambda(s) \leq e^{\sqrt{2}ls}. \tag{20}$$

Taking the logarithm of both sides in the inequality (20) and considering that $s > \frac{2\sqrt{2}}{l}\ln 2$, we get the inequality

$$\frac{\sqrt{2}}{2}l - 8\varepsilon \leq \frac{\ln\lambda}{s} \leq \sqrt{l} + 8\varepsilon, \tag{21}$$

where

$$\varepsilon(s) = \frac{\ln s}{s}, \quad \varepsilon.(s(\lambda)) \to 0$$

as $\lambda \to 0$. Consequently,

$$s \simeq \frac{\ln\lambda}{\sqrt{2}l}$$

for $\lambda \to 0$. Finally,

$$\sigma = p(\lambda) \simeq \frac{4l^4}{(\ln\lambda)^4}. \tag{22}$$

as $\lambda \to 0$. The relation (20) by the Theorem proved in [9], implies

$$s \simeq \frac{\ln\lambda}{\sqrt{2}l}$$

as $\lambda \to 0$. Hence,

$$\omega(r,\delta) \simeq r\sqrt{p\left(\frac{r^2}{\delta^2}\right)} \simeq \frac{rl^2}{(\ln\delta)^2} \tag{23}$$

as $\delta \to 0$. That is, the statement of the theorem holds.

References

1. Tabarintseva, E.V.: On an estimate for the modulus of continuity of a nonlinear inverse problem. Ural Math. J. **1**(1), 87–92 (2015)
2. Tabarintseva, E.V.: On methods to solve an inverse problems for a nonlinear differential equation. Siberian Electron. Math. Rep. **1417**, 199–209 (2017)

3. Tabarintseva, E.V.: Estimating the accuracy of a method of auxiliary boundary conditions in solving an inverse boundary value problem for a nonlinear equation. Numer. Anal. Appl. **11**(3), 236–255 (2018). https://doi.org/10.1134/S1995423918030059
4. Leontev, A.F.: Tselye funktsii: Ryady eksponent. Nauka, Moscow (1983)
5. Müntz, C.H.: Über den approximationssatz von weierstraß. In: Carathéodory, C., Hessenberg, G., Landau, E., Lichtenstein, L. (eds.) Mathematische Abhandlungen Hermann Amandus Schwarz, pp. 303–312. Springer, Heidelberg (1914). https://doi.org/10.1007/978-3-642-50735-9_22
6. Vasin, V.V., Skorik, G.G.: Solution of the deconvolution problem in the general statement. Trudy Inst. Mat. i Mekh. UrO RAN **22**(2), 79–90 (2016)
7. Tanana, V.P.: Methods for Solving Operator Equations. VSP, Utrecht (2002)
8. Ivanov, V.K., Vasin, V.V., Tanana, V.P.: Theory of Linear Ill-Posed Problems and its Applications. VSP, Utrecht (2002)
9. Ivanov, V.K., Korolyuk, T.I.: Error estimates for solutions of incorrectly posed linear problems. USSR Comput. Math. Math. Phys. **9**(1), 35–49 (1969)
10. Solodusha, S.V., Yaparova, N.M.: A numerical solution of an inverse boundary value problem of heat conduction using the Volterra equations of the first kind. Numer. Anal. Appl. **8**(3), 267–274 (2015). https://doi.org/10.1134/S1995423915030076
11. Alifanov, O.M.: Inverse Heat Transfer Problems. Springer, Heidelberg (1994). https://doi.org/10.1007/978-3-642-76436-3
12. Denisov, A.M.: Elements of the Theory of Inverse Problems. VSP, Utrecht (1999)
13. Il'in, A.M.: Uravnenija matematicheskoj fiziki (The Equations of Mathematical Physics). Izdatel'skij centr ChelGU, Chelyabinsk (2005)
14. Levitan, B.M.: Almost-Periodic Functions. GITTL, Moscow (1953)
15. Vasil'ev, F.P.: Optimization Methods. Factorial Press, Moscow (2002)
16. Ukhobotov, V.I., Izmest'ev, I.V.: A control problem for a rod heating process with unknown temperature at the right end and unknown density of the heat source. Trudy Inst. Mat. i Mekh. UrO RAN **25**(1), 297–305 (2019)

On the Issue of Comparison of Fuzzy Numbers

Viktor Ukhobotov[1], Irina Stabulit[2], and Konstantin Kudryavtsev[3](✉)

[1] Chelyabinsk State University, Bratiev Kashirinykh st. 129, Chelyabinsk 454001, Russia
ukh@csu.ru

[2] South Ural State Agrarian University, Gagarin st. 13, Troitsk 457100, Russia
irisku76@mail.ru

[3] South Ural State University, Lenin prospekt 76, Chelyabinsk 454080, Russia
kudrkn@gmail.com

Abstract. In the class of decision-making problems with fuzzy information concerning criterion values, the problem of comparing fuzzy numbers is relevant. There are various approaches to solving it. They are determined by the specific character of the problem under consideration. This paper proposes one approach to comparing fuzzy numbers. The proposed approach is as follows. At first, a rule is constructed for comparing a real number with a level set of a fuzzy number. Then, with the help of a procedure for constructing the exact lower approximation for the collection of sets, a fuzzy set is constructed. This fuzzy set determine the rule for comparing a real number with a fuzzy number. Using this rule and the approach based on separating two fuzzy numbers with a real number, the procedure is chosen for comparing two fuzzy numbers. As an example, fuzzy numbers with trapezoidal membership functions are considered, and the geometric interpretation of the results being given.

Keywords: Fuzzy numbers · Fuzzy payoff · Fuzzy numbers comparison

1 Introduction

For a number of economic and social problems, the information about the study objects has fuzzy forms. For investigating these problems, the theory of fuzzy sets is applied. Since Zadeh has published his work about fuzzy sets [1], a large number of papers have appeared. In papers [2–4], some sections of this theory are developed. Its applications to research problems from other fields of knowledge are considered in [5–8].

In decision-making and in game problems [9–12], the payoff can be given by a fuzzy number which is a special case of a fuzzy set. It raises the problem of defining operations with fuzzy numbers [13–15] and their comparison. A large number of different methods for comparison of fuzzy numbers have been

I. Bykadorov et al. (Eds.): MOTOR 2019, CCIS 1090, pp. 590–603, 2019.
https://doi.org/10.1007/978-3-030-33394-2_45

proposed in [16–21], no one of them being universal. There is a problem of integration of the proposed methods. When we choose a method for comparison of fuzzy numbers, we have to bear in mind the specific character of the considered problem.

In this paper, the research started in [10,11,19–23] is continued.

2 Construction of a Fuzzy Set

Let a set X and a function $\mu_A : \mathrm{X} \to [0;1]$ be given. A fuzzy set A is [3] the collection of pairs $(x, \mu_A(x))$, $x \in \mathrm{X}$.

The set X is called the universal set of the fuzzy set A, and the function $\mu_A : \mathrm{X} \to [0;1]$ is called the membership function of the fuzzy set A.

For each number $\alpha \in [0;1]$ the set

$$A(\alpha) = \{x \in \mathrm{X} \; : \; \mu_A(x) \geq \alpha\} \tag{1}$$

is called the α-cut of the fuzzy set A.

The α-cuts have the following properties:

$$\begin{gathered} A(0) = \mathrm{X}, \\ 0 \leq \alpha_1 \leq \alpha_2 \leq 1 \Rightarrow A(\alpha_2) \subset A(\alpha_1), \\ 0 < \alpha \leq 1 \Rightarrow \bigcap_{0<t\leq 1} A(t) = A(\alpha). \end{gathered} \tag{2}$$

Let for every $\alpha \in [0;1]$ the set $B(\alpha) \subset \mathrm{X}$ be defined.

Lemma 1. *If the totality of sets $B(\alpha)$ has the properties (2), then $\forall \alpha \in [0;1]$ the set $B(\alpha)$ is the level set of the fuzzy set B, the membership function of which is*

$$\mu_B(x) = \sup\{\alpha \in [0;1] \; : \; x \in B(\alpha)\}. \tag{3}$$

The proof of this fact is simple and can be found in [24].

Further we use the definition from [25, p. 45].

Definition 1. *The fuzzy set A is called the lower approximation of the collection of sets $A^*(\alpha) \subset \mathrm{X}$, $0 \leq \alpha \leq 1$ if the α-cuts $A(\alpha)$ of the fuzzy set A satisfy the inclusion*

$$A(\alpha) \subset A^*(\alpha), \quad \forall\, 0 \leq \alpha \leq 1. \tag{4}$$

Lemma 2. *If the fuzzy set A is the lower approximation of the collection of sets $A^*(\alpha) \subset \mathrm{X}$, then*

$$A(\alpha) \subset A_*(\alpha),$$

where

$$A_*(\alpha) = \bigcap_{0<t\leq \alpha} A^*(t) \quad \forall\, 0 < \alpha \leq 1. \tag{5}$$

The collection of sets $A_*(\alpha)$ with $0 < \alpha \leq 1$ has the properties (2).

The proof is due to the inclusion (4) and from formula (5).

According to Lemmas 1 and 2, the sets $A_*(\alpha)$ are the α-cuts of a fuzzy set A_* when $0 < \alpha \leq 1$. According to (3) and (5), its membership function is

$$\mu_{A_*}(x) = \sup \{\alpha \in [0;1] \, : \, x \in A^*(t) \ \forall \, 0 \leq t \leq \alpha\}. \tag{6}$$

The fuzzy set A_* is called [25, p. 47] the exact lower approximation of the collection of sets $A^*(\alpha)$.

Further, we will consider fuzzy numbers, for which, when any $0 < \alpha \leq 1$, the α-cuts (1) have the form of segments

$$A(\alpha) = [g_A(\alpha), G_A(\alpha)] \subset \mathbf{R}. \tag{7}$$

We assume that the functions $g_A : (0,1] \to \mathbf{R}$ and $G_A : (0,1] \to \mathbf{R}$ have the following properties:

$$\begin{gathered} g_A(\alpha) \leq g_A(\beta) < G_A(\beta) \leq G_A(\alpha) \quad \forall \, 0 < \alpha \leq \beta < 1, \\ \lim_{t \to \alpha - 0} g_A(t) = g_A(\alpha), \quad \lim_{t \to \alpha - 0} G_A(t) = G_A(\alpha) \quad \forall \, 0 < \alpha \leq 1. \end{gathered} \tag{8}$$

If these conditions are implemented, then the line segments (7) have properties (2). Note that conditions (8) imply that $g_A(1) \leq G_A(1)$.

3 Description of the Rules for Comparison Fuzzy Numbers

For a fixed fuzzy number A, we construct a fuzzy number $D(A)$ whose linguistic description is $D(A)$ = "the number $y \in \mathbf{R}$ is less than or equal to the fuzzy number A".

For a fixed $y \in \mathbf{R}$ and $\alpha \in (0,1]$, we introduce the measure $\gamma(\alpha, y) \in [0,1]$ of the fact that y is less than or equal to the line segment $[g_A(\alpha), G_A(\alpha)]$, which is the α-cut of the fuzzy number A, assuming that

$$\gamma(\alpha, y) = \begin{cases} 1, & if \ \ y \leq g_A(\alpha); \\ 0, & if \ \ G_A(\alpha) \leq y; \\ \frac{G_A(\alpha) - y}{G_A(\alpha) - g_A(\alpha)}, & if \ \ g_A(\alpha) < y < G_A(\alpha). \end{cases} \tag{9}$$

Note that for $g_A(\alpha) < y < G_A(\alpha)$ the number $\gamma(\alpha, y)$ determines the proportion of those numbers from the line segment $[g_A(\alpha), G_A(\alpha)]$ that are greater than or equal to y. If $y = g_A(1) = G_A(1)$, then we assume that $\gamma(\alpha, y) = 1$.

Let us assume that $\mathrm{Y}(0) = \mathbf{R}$ and for $0 < \alpha \leq 1$

$$\mathrm{Y}(\alpha) = \{y \in \mathbf{R} \, : \, \gamma(\alpha, y) \geq \alpha\}. \tag{10}$$

As a fuzzy number $D(A)$ we take the exact lower approximation of the collection of sets $\mathrm{Y}(\alpha)$. From formulas (6), (9) and (10), we obtain that its membership function is equal to

$$\mu_{D(A)}(y) = \sup \{\alpha \in [0,1] \, : \, \gamma(t, y) \geq t, \ \forall \, 0 \leq t \leq \alpha\}. \tag{11}$$

According to formula (9), $\gamma(\alpha, y_2) \leq \gamma(\alpha, y_1)$ for all $y_1 < y_2$ and $\alpha \in (0, 1]$. Then, taking into account (11), we get the inequality

$$\mu_{D(A)}(y_1) \geq \mu_{D(A)}(y_2), \quad if \ y_1 < y_2. \tag{12}$$

Let us assume that

$$\begin{aligned} &\rho(t, y) = 1 - t, && if \ \ y \leq g_A(t); \\ &\rho(t, y) = -t, && if \ \ G_A(t) \leq y; \\ &\rho(t, y) = \psi(t) - y, && if \ \ g_A(t) < y < G_A(t). \end{aligned} \tag{13}$$

If denote

$$\psi(t) = tg_A(t) + (1 - t)G_A(t), \tag{14}$$

then the formula (11) has the form

$$\mu_{D(A)}(y) = \sup \{\alpha \in [0, 1] : \rho(t, y) \geq 0, \ \forall 0 \leq t \leq \alpha\}. \tag{15}$$

Similarly, we can construct a fuzzy number $F(A)$ the linguistic description of which has the form $F(A)$ = "the number $y \in \mathbf{R}$ is greater than or equal to the fuzzy number A". The measure $\beta(\alpha, y) \in [0, 1]$ with $y \in \mathbf{R}$ and $\alpha \in (0, 1]$ that a number y is greater than or equal to the line segment $[g_A(\alpha), G_A(\alpha)]$ is given by the following formula:

$$\beta(\alpha, y) = 1 - \gamma(\alpha, y). \tag{16}$$

Similarly to (11), we define the membership function of the fuzzy number $F(A)$:

$$\mu_{F(A)}(y) = \sup \{\alpha \in [0, 1] \ : \ \beta(t, y) \geq t, \ \forall 0 \leq t \leq \alpha\}. \tag{17}$$

According to (16), $\beta(\alpha, y_2) \geq \beta(\alpha, y_1)$ when $y_1 < y_2$ and $\alpha \in (0, 1]$. Therefore, similarly to (12), we obtain from (17) the following inequality:

$$\mu_{F(A)}(y_1) \leq \mu_{F(A)}(y_2), \quad if \ \ y_1 < y_2. \tag{18}$$

Assume that

$$\begin{aligned} &\tau(t, y) = -t, && if \ \ y \leq g_A(t); \\ &\tau(t, y) = 1 - t, && if \ \ G_A(t) \leq y; \\ &\tau(t, y) = \varphi(t) + y, && if \ \ g_A(t) < y < G_A(t). \end{aligned} \tag{19}$$

We denote here

$$\varphi(t) = -(1 - t)g_A(t) - tG_A(t). \tag{20}$$

Then formula (17) has the form

$$\mu_{F(A)}(y) = \sup \{\alpha \in [0, 1] \ : \ \tau(t, y) \geq 0, \ \forall 0 \leq t \leq \alpha\}. \tag{21}$$

Let us consider a fuzzy number $L(A)$ whose linguistic description has the form $L(A)$ = "the number $y \in \mathbf{R}$ is less than or equal to a fuzzy number A and the number $y \in \mathbf{R}$ is greater than or equal to the fuzzy number A". This fuzzy number $L(A)$ is the intersection of the fuzzy numbers $D(A)$ and $F(A)$. Therefore, [1] its membership function is equal to

$$\mu_{L(A)}(y) = \min\left\{\mu_{D(A)}(y),\ \mu_{F(A)}(y)\right\}.$$

According to the properties of monotonicity (12) and (18), it can be obtained that if at some point $y_A \in \mathbf{R}$,

$$\mu_{D(A)}(y_A) = \mu_{F(A)}(y_A), \tag{22}$$

then

$$\mu_{L(A)}(y_A) \geq \mu_{L(A)}(y) \quad \forall\, y \in \mathbf{R}.$$

Therefore, applying the maximum method in the defuzzification procedure [25, p. 87] of the fuzzy number A, we get the real number y_A.

So, two fuzzy numbers A and B can be compared, for example, using real numbers y_A and y_B assuming that *the fuzzy number A is less than or equal to the fuzzy number B then and only then when $y_A \leq y_B$*.

Now we describe another approach to comparing fuzzy numbers A and B. We consider a fuzzy set $G = G(A, B)$, which linguistic description has the form G = "the number $y \in \mathbf{R}$ is less than or equal to the fuzzy number A and the number $y \in \mathbf{R}$ is greater than or equal to the fuzzy number B". This fuzzy set G is the intersection of fuzzy numbers $D(A)$ and $F(B)$. Therefore, its membership function is

$$\mu_{G(A,B)}(y) = \min\left\{\mu_{D(A)}(y), \mu_{F(B)}(y)\right\}.$$

As the measure $d(A \geq B)$ of the fact that the fuzzy number A is greater than or equal to the fuzzy number B, we can take

$$d(A \geq B) = \sup_{y \in \mathbf{R}} \min\left\{\mu_{D(A)}(y), \mu_{F(B)}(y)\right\}. \tag{23}$$

If $d(A \geq B) \geq d(B \geq A)$, then we assume that the fuzzy number A is greater than or equal to the fuzzy number B. We denote this $A \succeq B$.

4 Comparison of the Trapezoidal Fuzzy Numbers

Let us consider the case of a trapezoidal fuzzy number A, for which functions (8) have the form [25, p. 79]

$$g_A(\alpha) = a + (b - a)\alpha, \quad G_A(\alpha) = f - (f - c)\alpha, \quad 0 < \alpha \leq 1. \tag{24}$$

Here $a < b \leq c < f$ are given real numbers. From formulas (24) and (14) we get

$$\psi(t) = (f - c + b - a)t^2 - (2f - a - c)t + f. \tag{25}$$

Lemma 3. *If* $y \le a$, *then* $\rho(t, y) = 1 - t$ *for all* $0 \le t \le 1$.
If $a < y \le b$, *then*

$$\begin{aligned} &\rho(t, y) = \psi(t) - y, \quad \text{if } 0 \le t < \tfrac{y-a}{b-a}, \\ &\rho(t, y) = 1 - t, \quad \text{if } \tfrac{y-a}{b-a} \le t \le 1. \end{aligned} \tag{26}$$

If $c < y \le f$, *then*

$$\begin{aligned} &\rho(t, y) = \psi(t) - y, \quad \text{if } 0 \le t < \tfrac{f-y}{f-c}, \\ &\rho(t, y) = -t, \quad \text{if } \tfrac{f-y}{f-c} \le t \le 1. \end{aligned} \tag{27}$$

If $f \le y$, *then* $\rho(t, y) = -t$ *for all* $0 \le t \le 1$.

The proof follows from formulas (13) and (24).

According to (25), the condition of non-negativity of the discriminant of the quadratic equation

$$\psi(t) - y = 0 \tag{28}$$

is the inequality

$$y \ge \frac{4fb - (a + c)^2}{4(f - c + b - a)} = f - \frac{(2f - a - c)^2}{4(f - c + b - a)}. \tag{29}$$

If $y \le f$ and the inequality (29) is true, then the minimal positive root of the Eq. (28) is

$$t_{D(A)}(y) = \frac{2f - a - c - \sqrt{(2f - a - c)^2 - 4(f - y)(f - c + b - a)}}{2(f - c + b - a)}. \tag{30}$$

First, we consider the case when

$$2b > a + c. \tag{31}$$

It can be shown that if the inequality (31) is true, then

$$a < \varepsilon < b, \quad \text{where } \varepsilon = \frac{4fb - (a + c)^2}{4(f - c + b - a)}; \tag{32}$$

$$0 < t_* < 1, \quad \text{where } t_* = \frac{2f - a - c}{2(f - c + b - a)}. \tag{33}$$

We note that at the point t_*, the derivative of function (25) vanishes. At the point t_*, function (25) reaches its minimum value.

$$\min_{t \in \mathbf{R}} \psi(t) = \psi(t_*) = \varepsilon. \tag{34}$$

Theorem 1. *Let the inequality (31) be satisfied. Then*

$$\begin{aligned} \mu_{D(A)}(y) = 1, \quad & \text{if } y \le \varepsilon, \\ \mu_{D(A)}(y) = 0, \quad & \text{if } f \le y, \\ \mu_{D(A)}(y) = t_{D(A)}(y), \quad & \text{if } \varepsilon < y < f. \end{aligned} \tag{35}$$

Proof. Let $y \le a$. Then, according to Lemma 3, $\rho(t, y) = 1 - t$ for all $0 \le t \le 1$. From the last equality and from the formula (15) we obtain that $\mu_{D(A)}(y) = 1$.

Let $a < y \le b$. The function $\rho(t, y)$ is defined by the formula (26). According to the inequality (32), the cases $a < y \le \varepsilon$ and the case $\varepsilon < y \le b$ are possible.

If $a < y \le \varepsilon$, then from (34) we get that $\psi(t) - y \ge 0$ for all $t \in \mathbf{R}$. According to (26), the inequality $\rho(t, y) \ge 0$ is true for all $0 \le t \le 1$. This inequality together with (15) give $\mu_{D(A)}(y) = 1$.

Let $\varepsilon < y < f$. Then from (29) and (32), it follows that the quadratic equation (28) has a root (30). From (25) and (34), we obtain $\psi(0) - y > 0$ and $\psi(t_*) - y < 0$. From the last inequalities and from the fact that the number $t_{D(A)}(y)$ is the minimal positive root of the Eq. (28), it follows that

$$\psi(t) - y > 0, \text{ if } 0 \le t < t_{D(A)}(y) \quad \text{and} \quad \psi(t) - y < 0, \text{ if } t_{D(A)}(y) < t \le t_*. \tag{36}$$

If $\varepsilon < y \le b$, then the function $\rho(t, y)$ is given by formula (26). Using the inequality (31) and formulas (32), (33), it can be shown that $t_* < \frac{y-a}{b-a}$. Therefore, substituting the function (26) into formula (21), we obtain the required equality in (35).

For $b < y \le c$, the function $\rho(t, y) = \psi(t) - y$ for all $0 \le t \le 1$. Substituting this function into the formula (21) and taking into account the inequalities (36), we obtain the equality from (35).

If $c < y < f$, then the function $\rho(t, y)$ is given by the formula (27). Let us show that

$$\psi\left(\tfrac{f-y}{f-c}\right) - y < 0. \tag{37}$$

Then from (36) it will follow that $t_{D(A)}(y) < \frac{f-y}{f-c}$.

Substituting the function (27) into the formula (21), we obtain the equality from (35). From (25), it follows

$$\begin{gathered} \psi\left(\tfrac{f-y}{f-c}\right) - y = \tfrac{f-y}{(f-c)^2} g(y), \\ g(y) = (f - c + b - a)(f - y) - (2f - a - c)(f - c) + (f - c)^2. \end{gathered}$$

We have $g(f) = -(f - a)(f - c) < 0$, $g(c) = -(f - c)(c - b) \le 0$. Since the function $g(y)$ is linear, then $g(y) < 0$ for all $c < y < f$. Therefore, inequality (37) is true.

For $y \ge f$ the function $\rho(t, y) = -t$ for all $0 \le t \le 1$. Then, according to (21) we obtain that $\mu_{D(A)}(y) = 0$.

Let us analyze the function $\mu_{D(A)}(y)$. From (30), it follows $t_{D(A)}(f) = 0$ and $t_{D(A)}(\varepsilon) = t_* < 1$. Further, for $\varepsilon < y < f$ the function $t_{D(A)}(y)$ strictly decreases and is convex.

Note also that for $2b = a + c$ from formulas (32) and (33), the equalities $\varepsilon = b$ and $t_* = 1$ appear.

Theorem 2. *Let the inequality be true*

$$2b \le a + c.$$

Then

$$\mu_{D(A)}(y) = 1, \quad \text{if } y \le b,$$
$$\mu_{D(A)}(y) = 0, \quad \text{if } f \le y,$$
$$\mu_{D(A)}(y) = t_{D(A)}(y), \quad \text{if } b < y < f.$$

The proof is similar to the one of Theorem 1.

Let us construct of a fuzzy number $F(A)$. Let us substitute functions (24) into formulas (19) and (20). We get the function in the form

$$\varphi(t) = (f - c + b - a)t^2 - (f + b - 2a)t - a \tag{38}$$

and the following lemma.

Lemma 4. *If $y \le a$, then $\tau(t, y) = -t$ for all $0 \le t \le 1$. If $a < y \le b$, then $\tau(t, y) = \varphi(t) + y$ for $0 \le t < \frac{y-a}{b-a}$ and $\tau(t, y) = -t$ for $\frac{y-a}{b-a} < t \le 1$.*

If $b < y < c$, then $\tau(t, y) = \varphi(t) + y$ for all $0 \le t \le 1$.

If $c \le y < f$, then $\tau(t, y) = \varphi(t) + y$ for $0 \le t < \frac{f-y}{f-c}$ and $\tau(t, y) = 1 - t$ for $\frac{f-y}{f-c} \le t \le 1$.

If $f \le y$, then $\tau(t, y) = 1 - t$ for all $0 \le t \le 1$.

According to formula (38), it follows that the nonnegativity condition of the discriminant of the quadratic equation

$$\varphi(t) + y = 0 \tag{39}$$

takes the form

$$y \le \frac{(f + b)^2 - 4ac}{4(f - c + b - a)} = a + \frac{(f + b - 2a)^2}{4(f - c + b - a)}. \tag{40}$$

If $y \ge a$ and the inequality (40) holds, then the minimal positive root of the Eq. (39) is

$$t_{F(A)}(y) = \frac{f + b - 2a - \sqrt{(f + b - 2a)^2 + 4(a - y)(f - c + b - a)}}{2(f - c + b - a)}. \tag{41}$$

First, consider the case when

$$f + b > 2c. \tag{42}$$

It can be shown that if the inequality (42) is true, then

$$c < \delta < f, \quad \text{where } \delta = \frac{(f + b)^2 - 4ac}{4(f - c + b - a)}; \tag{43}$$

$$0 < t^* < 1, \quad \text{where } t^* = \frac{f + b - 2a}{2(f - c + b - a)}. \tag{44}$$

Note that at the point t^*, the derivative of the function (38) vanishes. At the point t_*, the function (38) reaches its minimum value.

Theorem 3. *Let inequality (42) be true. Then*

$$\mu_{F(A)}(y) = 0, \quad \text{if } y \leq a,$$
$$\mu_{F(A)}(y) = 1, \quad \text{if } \delta \leq y,$$
$$\mu_{F(A)}(y) = t_{F(A)}(y), \quad \text{if } a < y < \delta.$$

The proof of this theorem is carried out by analogy with the proof of Theorem 1.

Let us analyze the function $\mu_{F(A)}(y)$. According to (41), $t_{F(A)}(a) = 0$ and $t_{F(A)}(\delta) = t^* < 1$. Further, for $a < y < b$, the function $t_{F(A)}(y)$ is strictly increasing and convex.

Note that for $f + b = 2c$ from formulas (43) and (44), the equalities $\delta = c$ and $t^* = 1$ follow.

Theorem 4. *Let the following inequality be true:*

$$f + b \leq 2c.$$

Then

$$\mu_{F(A)}(y) = 0, \quad \text{if } y \leq a,$$
$$\mu_{F(A)}(y) = 1, \quad \text{if } y \geq c,$$
$$\mu_{F(A)}(y) = t_{F(A)}(y), \quad \text{if } a < y < c.$$

The proof is similar to that of Theorem 1.

Remark 1. For the triangular fuzzy number A [25, p. 16], the equality is $b = c$. In this case, inequalities (31) and (42) hold.

The graphs of the functions $\mu_{D(A)}(y)$ and $\mu_{F(A)}(y)$ for the case when the inequalities (31) and (42) are true are shown in Fig. 1.

From (33) and (44), it follows that $t^* < t_*$ if and only if $b - a < f - c$.

Let us proceed to comparing the trapezoidal fuzzy numbers. Equality (22) takes the form

$$t_{D(A)}(y) = t_{F(A)}(y). \tag{45}$$

Since function (30) is strictly decreasing, and function (41) is strictly increasing, then Eq. (45) can have only one solution.

From (30) and (41) it follows that

$$t_{D(A)}(y_A) = t_{F(A)}(y_A) = \frac{1}{2}, \quad \text{if} \quad y_A = \frac{a + b + c + f}{4}. \tag{46}$$

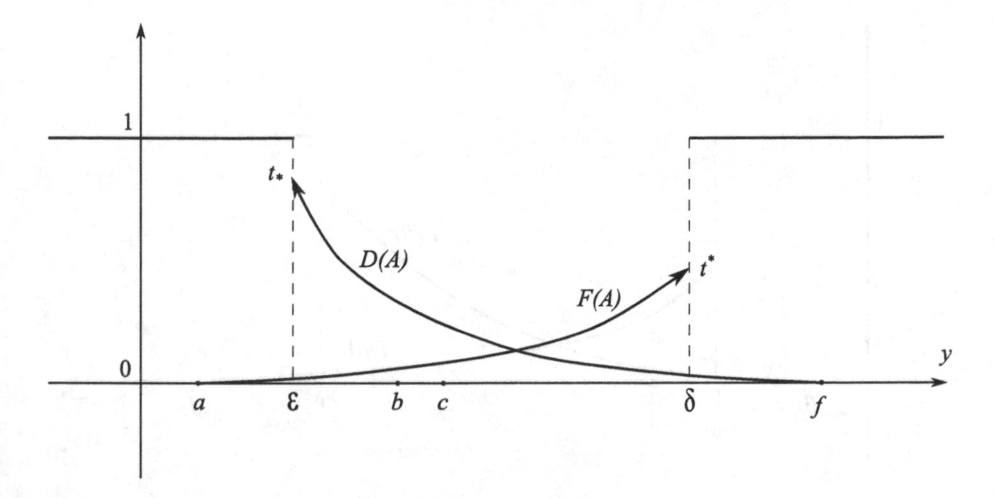

Fig. 1. $\mu_{D(A)}(y)$ and $\mu_{F(A)}(y)$ for the case $2b > a + c$, $f + b > 2c$

Further, from (32) and (43) we obtain $\varepsilon < y_A < \delta$.

It should be noted that the number y_A obtained in (46) has the following property: a straight line passing through the point y_A and perpendicular to the y axis divides the figure formed by the y axis and the graph of the membership function $\mu_A(y)$ into two equal parts [25, p. 92].

Let us consider two fuzzy trapezoidal numbers $A = (a_1, b_1, c_1, f_1)$, $B = (a_2, b_2, c_2, d_2)$ and calculate the volume $d(A \geq B) \in [0, 1]$, which is defined by the formula (23). Let us denote

$$u(y) = \min\left\{\mu_{D(A)}(y), \mu_{F(B)}(y)\right\}. \tag{47}$$

Case 1. Let $\delta_2 \leq \varepsilon_1$. Then (see Fig. 1) at any point $y \in [\delta_2, \varepsilon_1]$, the equality $u(y) = 1$ is true. According to (23) and (47), it will be $d(A \geq B) = 1$.
Case 2. Let $f_1 \leq a_2$. Then (see Fig. 1) $u(y) = 0$ for all $y \in \mathbf{R}$. According to (23) and (47), it will be $d(A \geq B) = 0$.
Case 3. Let $\delta_2 > \varepsilon_1$, $f_1 > a_2$ and $t_{F(B)}(\varepsilon_1) \geq t_{D(A)}(\varepsilon_1)$ (see Fig. 2). Then $u(y) = t_{F(B)}(y)$ for $y \leq \varepsilon_1$ and $u(y) = t_{D(A)}(y)$ for $\varepsilon_1 < y$. Therefore, $d(A \geq B) = t_{F(B)}(\varepsilon_1)$.
Case 4. Let $\delta_2 > \varepsilon_1$, $f_1 > a_2$, $t_{F(B)}(\varepsilon_1) < t_{D(A)}(\varepsilon_1)$. In this case, the equation

$$t_{D(A)}(y) = t_{F(B)}(y) \tag{48}$$

has the only solution $y_0 \in (\varepsilon_1, f_1) \cap (a_2, \delta_2)$. Then $d(A \geq B) = z$, where $z = t_{D(A)}(y_0)$.

From formulas (30), (41) and (48), it can be obtained that this number z satisfies the quadratic equation

$$\begin{aligned}((f_1 - c_1 + b_1 - a_1) + (f_2 - c_2 + b_2 - a_2))\, z^2 \\ - ((2f_1 - a_1 - c_1) + (f_2 + b_2 - 2a_2))\, z + f_1 - a_2 = 0.\end{aligned} \tag{49}$$

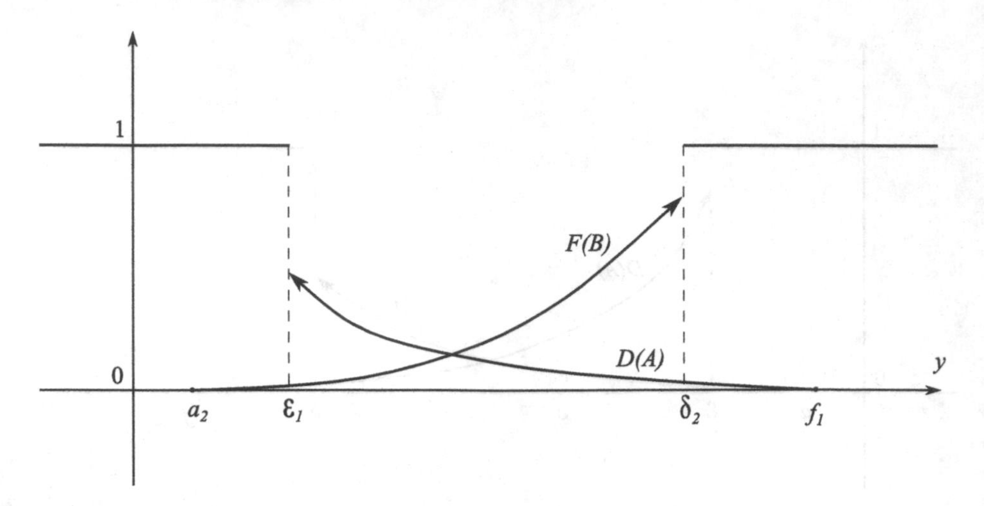

Fig. 2. $\mu_{D(A)}(y)$ and $\mu_{F(B)}(y)$ for the case $\delta_2 > \varepsilon_1$, $f_1 > a_2$, $t_{F(B)}(\varepsilon_1) \geq t_{D(A)}(\varepsilon_1)$

Example 1. Let two trapezoidal fuzzy numbers $A = (20, 30, 60, 100)$, $B = (10, 40, 70, 80)$ be given. Their membership functions are shown in Fig. 3.

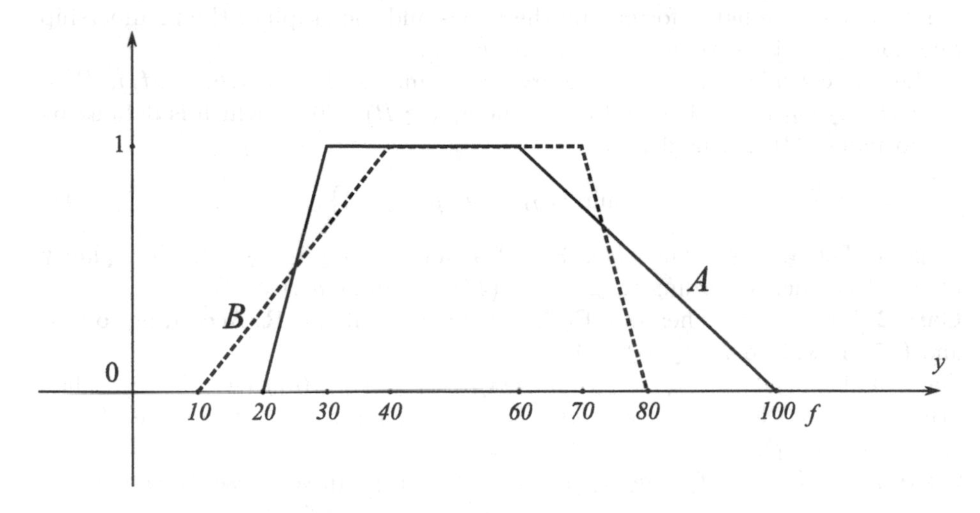

Fig. 3. Comparison of two trapezoidal fuzzy numbers

According to (24)

$$g_A(\alpha) = 20 + 10\alpha, \quad G_A(\alpha) = 100 - 40\alpha, \quad 0 < \alpha \leq 1,$$

and

$$g_B(\alpha) = 10 + 30\alpha, \quad G_B(\alpha) = 80 - 10\alpha, \quad 0 < \alpha \leq 1.$$

Since for the trapezoidal fuzzy number A the condition $2b_1 \leq a_1 + c_1$ is true, then, according to Theorem 2 and (30),

$$\mu_{D(A)}(y) = \begin{cases} 1, & if \;\; y \leq 30, \\ t_{D(A)}(y), & if \;\; 30 < y < 100, \\ 0, & if \;\; y \geq 100, \end{cases}$$

where

$$t_{D(A)}(y) = \frac{12 - \sqrt{2y - 56}}{10}. \tag{50}$$

Consequently, $\varepsilon_1 = 30$.

Since the condition $f_2 + b_2 \leq 2c_2$ is true for the trapezoidal fuzzy number B, then according to Theorem 4 and (41),

$$\mu_{F(B)}(y) = \begin{cases} 0, & if \;\; y \leq 10, \\ t_{F(B)}(y), & if \;\; 10 < y < 70, \\ 1, & if \;\; y \geq 70, \end{cases}$$

where

$$t_{F(B)}(y) = \frac{25 - \sqrt{725 - 10y}}{20}. \tag{51}$$

Hence, $\delta_2 = 70$.

According to (50), (51)

$$t_{D(A)}(\varepsilon_1) = t_{D(A)}(30) = 1,$$
$$t_{F(B)}(\varepsilon_1) = t_{F(B)}(30) = \tfrac{5-\sqrt{17}}{4}.$$

Since $\delta_2 > \varepsilon_1$, $f_1 > a_2$ and $t_{F(B)}(\varepsilon_1) < t_{D(A)}(\varepsilon_1)$, then (see case 4) $d(A \geq B) = z$, where z is the root of Eq. (49)

$$90z^2 - 220z + 90 = 0.$$

According to $z \in [0, 1]$, we obtain

$$d(A \geq B) = \frac{11 - 2\sqrt{10}}{9} \approx 0,51949.$$

Now we find $d(B \geq A)$. Since the condition $2b_2 = a_2 + c_2$ is satisfied for the trapezoidal fuzzy number B, then according to Theorem 2 and (30), we get

$$\mu_{D(B)}(y) = \begin{cases} 1, & if \;\; y \leq 40, \\ t_{D(B)}(y), & if \;\; 40 < y < 80, \\ 0, & if \;\; y \geq 80, \end{cases}$$

where

$$t_{D(B)}(y) = 1 - \frac{\sqrt{10y - 400}}{20}. \tag{52}$$

Hence, $\varepsilon_2 = 40$.

The condition (42), $f_1 + b_1 > 2c_1$, is true for the trapezoidal fuzzy number A. Therefore, from (43) it follows that $\delta_1 = 60\frac{1}{2}$. According to Theorem 3 and (41),

$$\mu_{F(A)}(y) = \begin{cases} 0, & if \ \ y \leq 20, \\ t_{F(A)}(y), & if \ \ 20 < y < 60\frac{1}{2}, \\ 1, & if \ \ y \geq 60\frac{1}{2}, \end{cases}$$

where

$$t_{F(A)}(y) = \frac{9 - \sqrt{121 - 2y}}{10}. \tag{53}$$

According to (52), (53) we obtain

$$t_{D(B)}(\varepsilon_2) = t_{D(B)}(40) = 1,$$
$$t_{F(A)}(\varepsilon_2) = t_{F(A)}(40) = \tfrac{9-\sqrt{41}}{10}.$$

Since $\delta_1 > \varepsilon_2$, $f_2 > a_1$ and $t_{F(A)}(\varepsilon_2) < t_{D(B)}(\varepsilon_2)$, then (see case 4) $d(B \geq A) = z$, where z is the root of Eq. (49)

$$90z^2 - 170z + 60 = 0.$$

According to $z \in [0, 1]$, we obtain

$$d(B \geq A) = \frac{17 - \sqrt{73}}{18} \approx 0,46978.$$

Finally, since $d(A \geq B) \geq d(B \geq A)$, then $A \succeq B$.

5 Conclusion

In this paper, we present a new approach to comparison of fuzzy numbers. We plan to apply this approach to fuzzy games [10,11] and decision-making problem with fuzzy uncertainty [23].

References

1. Zadeh, L.A.: Fuzzy sets. Inf. Control **8**, 338–353 (1963)
2. Zadeh, L.A.: Fuzzy logic. Computer **21**(4), 83–93 (1988)
3. Zimmermann, H.J.: Fuzzy Set Theory - and its Applications. Kluwer Academic Publishers, New York (1996)
4. Kosko, B.: Fuzzy systems as universal approximators. IEEE Trans. Comput. **43**(11), 1329–1333 (1994)
5. Li, L., Lai, K.K.: A fuzzy approach to the multiobjective transportation problem. Comput. Oper. Res. **27**(1), 43–57 (2000)
6. Chen, C.T., Lin, C.T., Huang, S.F.: A fuzzy approach for supplier evaluation and selection in supply chain management. Int. J. Prod. Econ. **102**(2), 289–301 (2006)

7. Bahri, O., Talbi, E.G., Amor, N.B.: A generic fuzzy approach for multi-objective optimization under uncertainty. Swarm Evol. Comput. **40**, 166–183 (2018)
8. Korzhov, A.V., Korzhova, M.E.: A method of accounting for fuzzy operational factors influencing 6 (10) kV power cable insulation longevity. In: 2016 2nd International Conference on Industrial Engineering, Applications and Manufacturing (ICIEAM), pp. 1–4. IEEE, Chelyabinsk (2016). https://doi.org/10.1109/ICIEAM.2016.7911429
9. Larbani, M.: Non cooperative fuzzy games in normal form: a survey. Fuzzy Sets Syst. **160**(22), 3184–3210 (2009)
10. Kudryavtsev, K.N., Stabulit, I.S., Ukhobotov, V.I.: A bimatrix game with fuzzy payoffs and crisp game. In: CEUR Workshop Proceedings 1987, pp. 343–349 (2017)
11. Kudryavtsev, K.N., Stabulit, I.S., Ukhobotov, V.I.: One approach to fuzzy matrix games. In: CEUR Workshop Proceedings 2098, pp. 228–238 (2018)
12. Verma, T., Kumar, A.: Ambika methods for solving matrix games with Atanassov's intuitionistic fuzzy payoffs. IEEE Trans. Fuzzy Syst. **26**(1), 270–283 (2018)
13. Dutta, P., Boruah, H., Ali, T.: Fuzzy arithmetic with and without using α-cut method: a comparative study. Int. J. Latest Trends Comput. **2**(1), 99–107 (2011)
14. Bansal, A.: Trapezoidal fuzzy numbers (a, b, c, d): arithmetic behavior. Int. J. Phys. Math. Sci. **2**(1), 39–44 (2011)
15. Gallyamov, E.R., Ukhobotov, V.I.: Computer implementation of operations with fuzzy numbers. Bull. South Ural State Universit. **3**(3), 97–108 (2014). Series "Vychislitelnaya Matematika i Informatika". (in Russian)
16. Yager, R.R.: A procedure for ordering fuzzy subsets of the unit interval. Inf. Sci. **24**(2), 143–161 (1981)
17. Ibanez, L.M.C., Munoz, A.G.: A subjective approach for ranking fuzzy numbers. Fuzzy Sets Syst. **29**(2), 145–153 (1989)
18. Chen, S.J., Hwang, C.L.: Fuzzy multiple attribute decision making methods. In: Fuzzy Multiple Attribute Decision Making, pp. 289–486. Springer, Heidelberg (1992). https://doi.org/10.1007/978-3-642-46768-4_5
19. Ukhobotov, V.I., Shchichko, P.V.: An approach to ranking fuzzy numbers. Bull. South Ural State Universit. **10**, 54–62 (2011). Series "Matematicheskoe modelirovanie I programmirovanie". (in Russian)
20. Ukhobotov, V.I., Mikhailova, E.S.: An approach to the comparison of fuzzy numbers in decision-making problems. Bull. South Ural State University **71**, 32–37 (2015). Series "Mathematics. Mechanics. Physics". (in Russian)
21. Ukhobotov, V.I., Mikhailova, E.S.: Comparison of fuzzy numbers in decision-making problems. Vestnik Udmurtskogo Universiteta. Matematika. Mekhanika. Komp'yuternye Nauki **26**(1), 87–94 (2016). https://doi.org/10.20537/vm160108. (in Russian)
22. Ukhobotov, V.I., Stabulit, I.S., Kudryavtsev, K.N.: Comparison of triangular fuzzy numbers. Vestnik Udmurtskogo Universiteta. Matematika. Mekhanika. Komp'yuternye Nauki **29**(2), 197–210 (2019). https://doi.org/10.20537/vm190205. (in Russian)
23. Ukhobotov, V.I., Stabulit, I.S., Kudryavtsev, K.N.: On decision making under fuzzy information about an uncontrolled factor. Procedia Comput. Sci. **150**, 524–531 (2019). https://doi.org/10.1016/j.procs.2019.02.088
24. Ramik, J., Vlach, M.: Generalized Concavity in Fuzzy Optimization and Decision Analysis. Kluwer Academic Publishers, Boston (2001)
25. Ukhobotov, V.I.: The Selected Chapters of Fuzzy Set Theory: Study Guide. Chelyabinsk State University, Chelyabinsk (2011). (in Russian)

Author Index

ıted in the United States
3ookmasters